LIST OF SYMBOLS

DISCRETE
MATHEMATICS

DISCRETE MATHEMATICS

Fourth Edition

Richard Johnsonbaugh

DePaul University, Chicago

Prentice Hall, Upper Saddle River, New Jersey 07458

Library of Congress Cataloging-in-Publication Data

Johnsonbaugh, Richard, date
    Discrete mathematics / Richard Johnsonbaugh — 4th ed.
      p.   cm.
    Includes bibliographical references and index.
    ISBN 0–13–518242–5
      1. Mathematics. 2. Computer Science—Mathematics.   I. Title.
QA39.2.J65 1997   96–23749
510—dc20   CIP

Editorial Director: Tim Bozik
Editor-in-Chief: Jerome Grant
Acquisition Editor: George Lobell
Director, Production and Manufacturing: David Riccardi
Executive Managing Editor: Kathleen Schiaparelli
Managing Editor: Linda Mihatov Behrens
Editorial/Production Supervision: Nicholas Romanelli
Manufacturing Manager: Trudy Pisciotti
Manufacturing Buyer: Alan Fischer
Creative Director: Paula Maylahn
Art Director: Amy Rosen
Assistant to Art Director: Rod Hernandez
Interior Designer: Donna Wickes
Cover Designer: Bruce Kenselaar
Art Manager: Gus Vibal
Editorial Assistants: Gale Epps/Nancy Bauer
Director of Marketing: John Tweeddale
Marketing Assistant: Diana Penha

ISBN 0-13-518242-5

Prentice-Hall International (UK) Limited, *London*
Prentice-Hall of Australia Pty. Limited, *Sydney*
Prentice-Hall Canada, Inc., *Toronto*
Prentice-Hall Hispanoamericana, S.A., *Mexico*
Prentice-Hall of India Private Limited, *New Delhi*
Prentice-Hall of Japan, Inc., *Tokyo*
Simon & Schuster Asia Pte. Ltd., *Singapore*
Editoria Prentice-Hall do Brasil, Ltda., *Rio de Janeiro*

CONTENTS

† Sections preceded by † sign can be omitted without loss of continuity

6 GRAPH THEORY   304

7 TREES   376

8 NETWORK MODELS AND PETRI NETS   455

PREFACE

This book is intended for a one- or two-term introductory course in discrete mathematics. Formal mathematics prerequisites are minimal; calculus is *not* required. There are no computer science prerequisites. The book includes examples, exercises, figures, tables, sections on problem solving, notes, chapter reviews, and self-tests to help the reader master introductory discrete mathematics. In addition, an *Instructor's Guide* is available from the publisher.

Overview

In the early 1980s there were almost no books appropriate for an introductory course in discrete mathematics. At the same time, there was a need for a course that extended students' mathematical maturity and ability to deal with abstraction and included useful topics such as combinatorics, algorithms, and graphs. The original edition of this book (1984) addressed this need. Subsequently, discrete mathematics courses were endorsed by many groups for several different audiences, including mathematics and computer science majors. A panel of the MAA (Mathematical Association of America) endorsed a year-long course in discrete mathematics. The Educational Activities Board of IEEE (Institute of Electrical and Electronics Engineers) recommended a freshman discrete mathematics course. ACM (Association for Computing Machinery) and IEEE accreditation guidelines mandated a discrete mathematics course. This edition, like its predecessors, includes topics such as algorithms, combinatorics, sets, functions, and mathematical induction endorsed by these groups. It also addresses understanding and doing proofs and, generally, expanding mathematical maturity.

About This Book

This book includes:

- Logic (including quantifiers), proofs, proofs by resolution, and mathematical induction (Chapter 1).

- Sets, sequences, strings, sum and product notations, number systems, relations, and functions, including motivating examples such as an application of partial orders to task scheduling (Section 2.4), relational databases (Section 2.7), and an introduction to hash functions (Section 2.8).

- A thorough discussion of algorithms, recursive algorithms, and the analysis of algorithms (Chapter 3). In addition, an algorithmic approach is taken throughout this book. The algorithms are written in a flexible form of pseudocode. (The book does not assume any computer science prerequisites; the description of the pseudocode used is self-contained.) Among the algorithms presented are the Euclidean algorithm for finding the greatest common divisor (Section 3.3), tiling (Section 3.4), the RSA public-key encryption algorithm (Section 3.7), generating combinations and permutations (Section 4.3), merge sort (Section 5.3), Dijkstra's shortest-path algorithm (Section 6.4), backtracking algorithms (Section 7.3), breadth-first and depth-first search (Section 7.3), tree traversals (Section 7.6), evaluating a game tree (Section 7.9), finding a maximal flow in a network (Section 8.2), finding a closest pair of points (Section 11.1), and computing the convex hull (Section 11.3).

- A full discussion of the "big oh," omega, and theta notations for the growth of functions (Section 3.5). Having all of these notations available makes it possible to give precise statements about the growth of functions and the complexity of algorithms.

- Combinations, permutations, and the pigeonhole principle (Chapter 4).

- Recurrence relations and their use in the analysis of algorithms (Chapter 5).

- Graphs, including coverage of graph models of parallel computers, the knight's tour, Hamiltonian cycles, graph isomorphisms, and planar graphs (Chapter 6). Theorem 6.4.3 gives a simple, short, elegant proof of the correctness of Dijkstra's algorithm.

- Trees, including binary trees, tree traversals, minimal spanning trees, decision trees, the minimum time for sorting, and tree isomorphisms (Chapter 7).

- The maximal flow algorithm, matching, and Petri nets (Chapter 8).

- A treatment of Boolean algebras that emphasizes the relation of Boolean algebras to combinatorial circuits (Chapter 9).

- An approach to automata emphasizing modeling and applications (Chapter 10). The *SR* flip-flop circuit is discussed in Example 10.1.11. Fractals, including the von Koch snowflake, are described by special kinds of grammars (Example 10.3.19).

- An introduction to computational geometry (Chapter 11).

- An appendix on matrices.

- A strong emphasis on the interplay among the various topics. As examples, mathematical induction is closely tied to recursive algorithms (Section 3.4); the Fibonacci sequence is used in the analysis of the Euclidean algorithm (Section 3.6); many exercises throughout the book require mathematical induction; we show how to characterize the components of a graph by defining an equivalence relation on the set of vertices (see the discussion following Example 6.2.13); and we count the number of n-vertex binary trees (Theorem 7.8.12).

- A strong emphasis on reading and doing proofs. Most proofs of theorems are illustrated with annotated figures. Separate sections, called *Problem-Solving Corner*, show students how to solve problems and how to do proofs.

- Numerous worked examples throughout the book. (There are more than 430 worked examples.)

- A large number of applications, especially applications to computer science.

- Nearly 2400 exercises, with answers to about one-third of them in the back of the book. (Exercises with numbers in **color** have an answer in the back of the book.)

- More than 650 figures and tables to illustrate the concepts, to show how algorithms work, to elucidate proofs, and to motivate the material.

- *Notes* sections with suggestions for further reading.

- *Chapter Review* sections.

- *Chapter Self-Test* sections.

- A reference section containing over 100 references.

- Front and back endpapers that summarize the mathematical and algorithm notation used in the book.

Changes from the Third Edition

- Eleven *Problem-Solving Corner* sections have been added. These sections show students how to attack and solve problems and how to do proofs.

- Proof by resolution is the topic of new Section 1.5. This proof technique, which can be automated and is, therefore, important in artificial intelligence, helps students gain additional insight into logic, in general, and reading and constructing proofs, in particular.

- A new section on binary and hexadecimal number systems has been added (Section 2.3). Binary and hexadecimal number systems are introduced, and conversions between the various systems are discussed. Arithmetic in the various systems is also covered.

- The new Section 3.7 is devoted to the RSA public-key cryptosystem, named after its inventors, Ronald L. Rivest, Adi Shamir, and Leonard M. Adleman. In the RSA system, each participant makes public an encryption key and hides a decryption key. To send a message, one looks up the recipient's encryption key in the publicly distributed table. The recipient then decrypts the message using the hidden decryption key.

- Several figures have been added to illustrate proofs of theorems. All figures now have captions, and the captions of the figures that illustrate proofs provide additional explanation and insight into the proofs.
- A number of recent books and articles have been added to the list of references.
- The number of worked examples has been increased to over 430.
- The number of exercises has been increased to nearly 2400.
- A World Wide Web site has been established to provide up-to-date support for the book.

Chapter Structure

Each chapter is organized as follows:

- Overview
- Section
- Section Exercises
- Section
- Section Exercises
  . . .
- Notes
- Chapter Review
- Chapter Self-Test

The *Notes* sections contain suggestions for further reading. The *Chapter Review* sections provide reference lists of the key concepts of the chapters. The *Chapter Self-Test* sections contain four exercises per section, with answers in the back of the book. In addition, most chapters have *Problem-Solving Corner* sections.

Exercises

The book contains nearly 2400 exercises. Exercises felt to be more challenging than average are indicated with a star, ★. Exercise numbers in **color** (approximately a third of the exercises) indicate that the exercise has a hint or solution in the back of the book. The solutions to the remaining exercises may be found in the *Instructor's Guide*. A handful of exercises are clearly identified as requiring calculus. No calculus concepts are used in the main body of the book and, except for these marked exercises, no calculus is needed to solve the exercises. Ends of proofs are marked with the symbol ■.

Examples

The book contains over 430 worked examples. These examples show students how to tackle problems in discrete mathematics, demonstrate applications of the theory, clarify proofs, and help motivate the material. Ends of examples are marked with the symbol □.

Problem-Solving Corners

The new *Problem-Solving Corner* sections help students attack and solve problems and show them how to do proofs. Written in an informal style, each is a self-contained section following the discussion of the subject of the problem. Rather than simply presenting a proof or a solution to a problem, in these sections the intent is to show different ways of attacking a problem, to discuss what to look for in trying to obtain a solution to a problem, and to present problem-solving and proof techniques.

Each *Problem-Solving Corner* begins with a statement of a problem. After stating the problem, ways to attack the problem are discussed. This discussion is followed by techniques for finding a solution. After a solution is found, a formal solution is given to show how to write up a formal solution correctly. Finally, the problem-solving techniques used in the section are summarized. In addition, some of these sections end with a *Comments* subsection, which discusses connections with other topics in mathematics and computer science, provides motivation for the problem, and lists references for further reading about the problem.

Instructor Supplement

An *Instructor's Guide* is available from the publisher, at no cost to adopters of this book. The *Instructor's Guide* contains solutions to the exercises not included in the book, computer exercises, and transparency masters. More than 100 computer exercises are included in the *Instructor's Guide*.

World Wide Web Site

The World Wide Web site http://condor.depaul.edu/~rjohnson contains information about the book, including computer programs, transparencies, computer exercises, a program to generate random graphs of various types, and an errata list.

Acknowledgments

I received helpful comments from many persons, including Gary Andrus, Robert Busby, David G. Cantor, Tim Carroll, Joseph P. Chan, Hon-Wing Cheng, Robert Crawford, Henry D'Angelo, Jerry Delazzer, Br. Michael Driscoll, Carl E. Eckberg, Susanna Epp, Gerald Gordon, Jerrold Grossman, Mark Herbster, Steve Jost, Nicholas Krier, Warren Krueger, Glenn Lancaster, Donald E. G. Malm, Kevin Phelps, James H. Stoddard, Michael Sullivan, Edward J. Williams, and Hanyi Zhang.

Special thanks for this edition go to Martin Kalin for his comments on the new *Problem-Solving Corners* and for his advice on the new section on resolution proofs; to Gregory Brewster and I-Ping Chu for discussions about flows in computer networks; to Gregory Bachelis for reviewing the new section on the RSA encryption system; and to Sam Stueckle, Northeastern University; Towanna Roller, Asbury College; Feng-Eng Lin, George Mason University; Gordon D. Prichett, Babson College; and Donald Bein, Fairleigh Dickinson University for reviewing the manuscript for this edition.

I am indebted to Helmut Epp, Dean of the School of Computer Science, Telecommunications and Information Systems at DePaul University, for providing time and encouragement for the development of this edition and its predecessors.

I have received consistent support from the staff at Prentice Hall. Special thanks for their help go to George Lobell, executive editor, and Nicholas Romanelli, production supervisor.

R.J.

1

LOGIC AND PROOFS

My dear Mr. Marlowe, I notice in you an unpleasant tendency toward abrupt transitions. A characteristic of your generation, but in this case, I must ask you to follow some sort of logical progression.

—from Murder, My Sweet

Logic is the study of reasoning; it is specifically concerned with whether reasoning is correct. Logic focuses on the relationship among statements as opposed to the content of any particular statement. Consider, for example, the argument:

> All mathematicians wear sandals.
>
> Anyone who wears sandals is an algebraist.
>
> Therefore, all mathematicians are algebraists.

Technically, logic is of no help in determining whether any of these statements is true; however, if the first two statements are true, logic assures us that the statement

> All mathematicians are algebraists.

is also true.

Logical methods are used in mathematics to prove theorems and in computer science to prove that programs do what they are alleged to do.

---

† This section can be omitted without loss of continuity.

In the latter part of the chapter, we discuss some general methods of proof, one of which, mathematical induction, is used throughout mathematics and computer science. Mathematical induction is especially useful in discrete mathematics.

1.1 PROPOSITIONS

Which of the sentences are either true or false (but not both)?

(a) The only positive integers that divide 7 are 1 and 7 itself.

(b) Alfred Hitchcock won an Academy Award in 1940 for directing *Rebecca*.

(c) For every positive integer n, there is a prime number larger than n.

(d) Earth is the only planet in the universe that has life.

(e) Buy two tickets to the Unhinged Universe rock concert for Friday.

Sentence (a) is true. We call an integer n *prime* if $n > 1$ and the only positive integers that divide n are 1 and n itself. Sentence (a) is another way to say that 7 is prime.

Sentence (b) is false. Although *Rebecca* won the Academy Award for best picture in 1940, John Ford won the directing award for *The Grapes of Wrath*. It is a surprising fact that Alfred Hitchcock never won an Academy Award for directing.

Sentence (c), which is another way to say that there are an infinite number of primes, is true.

Sentence (d) is either true or false (but not both), but no one knows which at this time.

Sentence (e) is neither true nor false [sentence (e) is a command].

A sentence that is either true or false, but not both, is called a **proposition**. Sentences (a)–(d) are propositions, whereas sentence (e) is not a proposition. A proposition is typically expressed as a declarative sentence (as opposed to a question, command, etc.). Propositions are the basic building blocks of any theory of logic.

We will use lowercase letters, such as p, q, and r, to represent propositions. We will also use the notation

$$p\colon \ 1 + 1 = 3$$

to define p to be the proposition $1 + 1 = 3$.

In ordinary speech and writing, we combine propositions using connectives such as *and* and *or*. For example, the propositions "It is raining" and "I will take my umbrella" can be combined to form the single proposition "It is raining and I will take my umbrella." The formal definitions of *and* and *or* follow.

DEFINITION 1.1.1

Let p and q be propositions.

The *conjunction* of p and q, denoted $p \wedge q$, is the proposition

$$p \quad \text{and} \quad q.$$

The *disjunction* of p and q, denoted $p \vee q$, is the proposition

$$p \quad \text{or} \quad q.$$

Propositions such as $p \wedge q$ and $p \vee q$ that result from combining propositions are called **compound propositions**.

EXAMPLE 1.1.2

If

$$p: 1 + 1 = 3,$$
$$q: \text{A decade is 10 years,}$$

then the conjunction of p and q is

$$p \wedge q: 1 + 1 = 3 \text{ and a decade is 10 years.}$$

The disjunction of p and q is

$$p \vee q: 1 + 1 = 3 \text{ or a decade is 10 years.} \qquad \square$$

The truth values of propositions such as conjunctions and disjunctions can be described by **truth tables**. The truth table of a proposition P made up of the individual propositions $p_1, \ldots, p_n$ lists all possible combinations of truth values for $p_1, \ldots, p_n$, T denoting true and F denoting false, and for each such combination lists the truth value of P.

DEFINITION 1.1.3

The truth value of the compound proposition $p \wedge q$ is defined by the truth table

p	q	$p \wedge q$
T	T	T
T	F	F
F	T	F
F	F	F

Notice that in the truth table in Definition 1.1.3 all four possible combinations of truth assignments for p and q are given.

Definition 1.1.3 states that the conjunction $p \wedge q$ is true provided that p and q are both true; $p \wedge q$ is false otherwise.

EXAMPLE 1.1.4

If
$$p:\ 1+1=3,$$
$$q:\ \text{A decade is 10 years,}$$

then p is false, q is true, and the conjunction

$$p \wedge q:\ 1+1=3 \text{ and a decade is 10 years}$$

is false.   □

EXAMPLE 1.1.5

If
$$p:\ \text{Benny Goodman recorded classical music,}$$
$$q:\ \text{The Baltimore Orioles used to be the St. Louis Browns,}$$

then p and q are both true. Although Benny Goodman is best known for his jazz recordings, he recorded much classical music (e.g., the Weber Clarinet Concertos, numbers 1 and 2, with the Chicago Symphony Orchestra). The St. Louis Browns moved to Baltimore in 1954 and were renamed the Orioles. The conjunction

$$p \wedge q:\ \text{Benny Goodman recorded classical music and the}$$
$$\text{Baltimore Orioles used to be the St. Louis Browns}$$

is true.   □

EXAMPLE 1.1.6

If
$$p:\ 1+1=3,$$
$$q:\ \text{Minneapolis is the capital of Minnesota,}$$

then p and q are both false and the conjunction

$$p \wedge q:\ 1+1=3 \text{ and Minneapolis is the capital of Minnesota}$$

is false.   □

DEFINITION 1.1.7

The truth value of the compound proposition $p \vee q$ is defined by the truth table

p	q	$p \vee q$
T	T	T
T	F	T
F	T	T
F	F	F

The *or* in the disjunction $p \vee q$ is used in the *inclusive* sense; that is, $p \vee q$ is considered to be true if either p or q or *both* are true and $p \vee q$ is false only if both p and q are false. There is also an **exclusive-or** (see Exercise 31) in which $p\ exor\ q$ is true if either p or q but *not* both is true.

EXAMPLE 1.1.8

If
$$p: 1 + 1 = 3,$$
$$q: \text{A decade is 10 years,}$$

then p is false, q is true, and the disjunction

$$p \vee q: \ 1 + 1 = 3 \text{ or a decade is 10 years}$$

is true.   □

EXAMPLE 1.1.9

If
$$p: \text{Benny Goodman recorded classical music,}$$
$$q: \text{The Baltimore Orioles used to be the St. Louis Browns,}$$

then p and q are both true and the disjunction

$$p \vee q: \text{Benny Goodman recorded classical music or the}$$
$$\text{Baltimore Orioles used to be the St. Louis Browns}$$

is also true.   □

EXAMPLE 1.1.10

If
$$p: 1 + 1 = 3,$$
$$q: \text{Minneapolis is the capital of Minnesota,}$$

then p and q are both false and the disjunction

$$p \vee q: \ 1 + 1 = 3 \text{ or Minneapolis is the capital of Minnesota}$$

is false.   □

The final operation on a proposition p that we discuss in this section is the **negation** of p.

> **DEFINITION 1.1.11**
>
> The *negation* of p, denoted $\overline{p}$, is the proposition
>
> $$\text{not } p.$$
>
> The truth value of the proposition $\overline{p}$ is defined by the truth table

p	$\overline{p}$
T	F
F	T

> **EXAMPLE 1.1.12**

If

$$p: \quad \text{Cary Grant starred in } Rear\ Window,$$

the negation of p is the proposition

$\overline{p}$:   It is not the case that Cary Grant starred in *Rear Window*.

Since p is false, $\overline{p}$ is true. (James Stewart played the leading male role in *Rear Window*.) The negation would normally be written:

$$\text{Cary Grant did not star in } Rear\ Window. \qquad \square$$

> **EXAMPLE 1.1.13**

Let

p: Blaise Pascal invented several calculating machines,

q : The first all-electronic digital computer was constructed in the twentieth century,

r : π was calculated to $1,000,000$ decimal digits in 1954.

Represent the proposition

Either Blaise Pascal invented several calculating machines and it is not the case that the first all-electronic digital computer was constructed in the twentieth century; or π was calculated to $1,000,000$ decimal digits in 1954,

symbolically and determine whether it is true or false.

The proposition may be written symbolically as

$$(p \wedge \overline{q}) \vee r.$$

We first note that p and q are true and r is false. (It was not until 1973 that 1,000,000 decimal digits of π were computed. Since then over 2,000,000,000 decimal digits of π have been computed.) If we replace each symbol by its truth value, we find that

$$(p \wedge \overline{q}) \vee r = (\text{T} \wedge \overline{\text{T}}) \vee \text{F}$$
$$= (\text{T} \wedge \text{F}) \vee \text{F}$$
$$= \text{F} \vee \text{F}$$
$$= \text{F}.$$

Therefore, the given proposition is false.   □

∽ ∽ ∽

Exercises

Determine whether each sentence in Exercises 1–8 is a proposition. If the sentence is a proposition, write its negation. (You are not being asked for the truth values of the sentences that are propositions.)

†**1.** $2 + 5 = 19$

2. Waiter, will you serve the nuts—I mean, would you serve the guests the nuts?

3. For some positive integer n, $19340 = n \cdot 17$.

4. Audrey Meadows was the original "Alice" in "The Honeymooners."

5. Peel me a grape.

6. The line "Play it again, Sam" occurs in the movie *Casablanca*.

7. Every even integer greater than 4 is the sum of two primes.

8. The difference of two primes.

Evaluate each proposition in Exercises 9–14 for the truth values

$$p = \text{F}, \qquad q = \text{T}, \qquad r = \text{F}.$$

9. $p \vee q$

10. $\overline{p} \vee \overline{q}$

11. $\overline{p} \vee q$

12. $\overline{p} \vee (q \wedge r)$

13. $\overline{(p \vee q)} \wedge (\overline{p} \vee r)$

14. $(p \vee \overline{r}) \wedge (q \vee t) \vee \overline{(r \vee p)}$

Write the truth table of each proposition in Exercises 15–22.

15. $p \wedge \overline{q}$

16. $(\overline{p} \vee \overline{q}) \vee p$

17. $(p \vee q) \wedge \overline{p}$

18. $(p \wedge q) \wedge \overline{p}$

19. $(p \wedge q) \vee (\overline{p} \vee q)$

20. $\overline{(p \wedge q)} \vee (r \wedge \overline{p})$

21. $(p \vee q) \wedge (p \vee q) \wedge (p \vee \overline{q}) \wedge (\overline{p} \vee \overline{q})$

22. $\overline{(p \wedge q)} \vee (\overline{q} \vee r)$

† Exercise numbers in color indicate that a hint or solution appears at the back of the book in the section following the References.

In Exercises 23–25, represent the given statement symbolically by letting

$$p: \ 5 < 9, \qquad q: \ 9 < 7, \qquad r: \ 5 < 7.$$

Determine whether each statement is true or false.

23. $5 < 9$ and $9 < 7$.

24. It is not the case that ($5 < 9$ and $9 < 7$).

25. $5 < 9$ or it is not the case that ($9 < 7$ and $5 < 7$).

In Exercises 26–30, formulate the symbolic expression in words using

$$p: \ \text{Today is Monday.}$$
$$q: \ \text{It is raining.}$$
$$r: \ \text{It is hot.}$$

26. $p \vee q$ **27.** $\overline{p} \wedge (q \vee r)$ **28.** $\overline{p \vee q} \wedge r$

29. $(p \wedge q) \wedge \overline{(r \vee p)}$

30. $(p \wedge (q \vee r)) \wedge (r \vee (q \vee p))$

31. Give the truth table for the exclusive-or of p and q in which p *exor* q is true if either p or q but not both is true.

32. At one time, the following ordinance was in effect in Naperville, Illinois: "It shall be unlawful for any person to keep more than three [3] dogs and three [3] cats upon his property within the city." Was Charles Marko, who owned five dogs and no cats, in violation of the ordinance? Explain.

1.2 CONDITIONAL PROPOSITIONS AND LOGICAL EQUIVALENCE

The dean has announced that

> If the Mathematics Department gets an additional \$20,000, then it will hire one new faculty member.   (1.2.1)

Statement (1.2.1) states that on the condition that the Mathematics Department gets an additional \$20,000, then the Mathematics Department will hire one new faculty member. A proposition such as (1.2.1) is called a **conditional proposition**.

DEFINITION 1.2.1

If p and q are propositions, the compound proposition

$$\text{if } p \text{ then } q \tag{1.2.2}$$

is called a *conditional proposition* and is denoted

$$p \rightarrow q.$$

The proposition p is called the *hypothesis* (or *antecedent*) and the proposition q is called the *conclusion* (or *consequent*).

EXAMPLE 1.2.2

If we define

   p: The Mathematics Department gets an additional $20,000,
   q: The Mathematics Department hires one new faculty member,

then statement (1.2.1) assumes the form (1.2.2). The hypothesis is the statement "The Mathematics Department gets an additional $20,000," and the conclusion is the statement "The Mathematics Department hires one new faculty member."   □

Some statements not of the form (1.2.2) may be rephrased as conditional propositions, as the next example illustrates.

EXAMPLE 1.2.3

Restate each proposition in the form (1.2.2) of a conditional proposition.
   (a) Mary will be a good student if she studies hard.
   (b) John may take calculus only if he has sophomore, junior, or senior standing.
   (c) When you sing, my ears hurt.
   (d) A necessary condition for the Cubs to win the World Series is that they sign a right-handed relief pitcher.
   (e) A sufficient condition for Ralph to visit California is that he goes to Disneyland.

   (a) The hypothesis is the clause following *if*; thus an equivalent formulation is

        If Mary studies hard, then she will be a good student.

   (b) The *only if* clause is the conclusion; that is,

                        if p then q

   is considered logically the same as

                        p only if q.

   An equivalent formulation is

        If John takes calculus, then he has sophomore, junior, or
           senior   standing.

        The "if p then q" formulation emphasizes the hypothesis, whereas the "p only if q" formulation emphasizes the conclusion; the difference is only stylistic.
   (c) *When* means the same as *if*; thus an equivalent formulation is

                 If you sing, then my ears hurt.

(d) The conclusion expresses a **necessary condition**; thus an equivalent formulation is

> If the Cubs win the World Series, then they sign a right-handed   relief pitcher.

(e) The hypothesis expresses a **sufficient condition**; thus an equivalent formulation is

> If Ralph goes to Disneyland, then he visits California.   □

Consider the problem of assigning a truth value to the dean's proposition

> If the Mathematics Department gets an additional $20,000, then it will hire one new faculty member.

The statement is really of interest only when the hypothesis "the Mathematics Department gets an additional $20,000" is true. If it is true that the Mathematics Department gets an additional $20,000 and it is also true that the Mathematics Department hires one new faculty member, we would regard the dean's statement as true. On the other hand, if it is true that the Mathematics Department gets an additional $20,000, but it is false that the Mathematics Department hires one new faculty member, we would regard the dean's statement as false. When the hypothesis is true, the conditional proposition's truth value, as a whole, depends on the conclusion's truth value. In general, when the hypothesis p is true, the conditional proposition $p \to q$ is true if q is true and false if q is false. If the hypothesis p is false, the only intuitively clear point is that the truth value of the conditional statement $p \to q$ should not hinge on the conclusion's truth value. We would not consider the dean's statement to be false simply because the Mathematics Department did not get an additional $20,000. Nonetheless, the conditional proposition, like any other proposition, must have a truth value, even if the hypothesis is false. The standard definition declares $p \to q$ to be true if p is false. The preceding discussion is summarized in the next definition.

DEFINITION 1.2.4

The truth value of the conditional proposition $p \to q$ is defined by the following truth table:

p	q	$p \to q$
T	T	T
T	F	F
F	T	T
F	F	T

EXAMPLE 1.2.5

Let

$$p: \ 1 > 2, \quad q: \ 4 < 8.$$

Then p is false and q is true. Therefore,

$$p \rightarrow q \text{ is true}, \quad q \rightarrow p \text{ is false}. \qquad \Box$$

EXAMPLE 1.2.6

Assuming that p is true, q is false, and r is true, find the truth value of each proposition.

(a) $(p \wedge q) \rightarrow r$

(b) $(p \vee q) \rightarrow \bar{r}$

(c) $p \wedge (q \rightarrow r)$

(d) $p \rightarrow (q \rightarrow r)$

We replace each symbol $p, q,$ and r by its truth value to obtain the truth value of the proposition:

(a) $(T \wedge F) \rightarrow T = F \rightarrow T = \text{true}$

(b) $(T \vee F) \rightarrow \bar{T} = T \rightarrow F = \text{false}$

(c) $T \wedge (F \rightarrow T) = T \wedge T = \text{true}$

(d) $T \rightarrow (F \rightarrow T) = T \rightarrow T = \text{true} \qquad \Box$

In ordinary language, the hypothesis and conclusion in a conditional proposition are normally related, but in logic, the hypothesis and conclusion in a conditional proposition are not required to refer to the same subject matter. For example, in logic, we permit propositions such as:

If $5 < 3$, then Nelson Rockefeller was president of the United States.

Logic is concerned with the form of propositions and the relation of propositions to each other and not with the subject matter itself. (In fact, since the hypothesis is false, this proposition is true. Notice that a true conditional proposition is different from a conditional proposition with a true conclusion.)

Example 1.2.5 shows that the proposition $p \rightarrow q$ can be true while the proposition $q \rightarrow p$ is false. We call the proposition $q \rightarrow p$ the **converse** of the proposition $p \rightarrow q$. Thus a conditional proposition can be true while its converse is false.

EXAMPLE 1.2.7

Write each conditional proposition symbolically. Write the converse of each statement symbolically and in words. Also, find the truth value of each conditional proposition and its converse.

(a) If $1 < 2$, then $3 < 6$.
(b) If $1 > 2$, then $3 < 6$.

(a) Let
$$p:\ 1 < 2, \quad q:\ 3 < 6.$$

The given statement may be written symbolically as

$$p \rightarrow q.$$

Since p and q are both true, this statement is true. The converse may be written symbolically as

$$q \rightarrow p$$

and in words as

$$\text{If } 3 < 6, \text{ then } 1 < 2.$$

Since p and q are both true, the converse $q \rightarrow p$ is true.

(b) Let
$$p:\ 1 > 2, \quad q:\ 3 < 6.$$

The given statement may be written symbolically as

$$p \rightarrow q.$$

Since p is false and q is true, this statement is true. The converse may be written symbolically as

$$q \rightarrow p$$

and in words as

$$\text{If } 3 < 6, \text{ then } 1 > 2.$$

Since q is true and p is false, the converse $q \rightarrow p$ is false.   □

Another useful compound proposition is

$$p \text{ if and only if } q. \tag{1.2.3}$$

Such a statement is considered to be true precisely when p and q have the same truth values (i.e., p and q are both true or p and q are both false).

DEFINITION 1.2.8

If p and q are propositions, the compound proposition

$$p \text{ if and only if } q$$

is called a *biconditional proposition* and is denoted

$$p \leftrightarrow q.$$

The truth value of the proposition $p \leftrightarrow q$ is defined by the following truth table:

p	q	$p \leftrightarrow q$
T	T	T
T	F	F
F	T	F
F	F	T

An alternative way to state "p if and only if q" is "p is a necessary and sufficient condition for q." "p if and only if q" is sometimes written "p iff q."

EXAMPLE 1.2.9

The statement

$$1 < 5 \text{ if and only if } 2 < 8 \tag{1.2.4}$$

can be written symbolically as

$$p \leftrightarrow q$$

if we define

$$p: \ 1 < 5, \quad q: \ 2 < 8.$$

Since both p and q are true, the statement $p \leftrightarrow q$ is true.   □

An alternative way to state (1.2.4) is: A necessary and sufficient condition for $1 < 5$ is that $2 < 8$.

In some cases, two different compound propositions have the same truth values no matter what truth values their constituent propositions have. Such propositions are said to be **logically equivalent**.

DEFINITION 1.2.10

Suppose that the compound propositions P and Q are made up of the propositions $p_1, \ldots, p_n$. We say that P and Q are *logically equivalent* and write

$$P \equiv Q,$$

provided that given any truth values of $p_1, \ldots, p_n$, either P and Q are both true or P and Q are both false.

| EXAMPLE 1.2.11 | *De Morgan's Laws for Logic* |

We will verify the first of **De Morgan's laws**

$$\overline{p \vee q} \equiv \overline{p} \wedge \overline{q}, \quad \overline{p \wedge q} \equiv \overline{p} \vee \overline{q}$$

leaving the second as an exercise (see Exercise 44).

By writing the truth tables for $P = \overline{p \vee q}$ and $Q = \overline{p} \wedge \overline{q}$, we can verify that given any truth values of p and q, either P and Q are both true or P and Q are both false:

p	q	$\overline{p \vee q}$	$\overline{p} \wedge \overline{q}$
T	T	F	F
T	F	F	F
F	T	F	F
F	F	T	T

Thus P and Q are logically equivalent.   □

Our next example gives a logically equivalent form of the negation of $p \rightarrow q$.

| EXAMPLE 1.2.12 |

Show that the negation of $p \rightarrow q$ is logically equivalent to $p \wedge \overline{q}$.

We must show that

$$\overline{p \rightarrow q} \equiv p \wedge \overline{q}.$$

By writing the truth tables for $P = \overline{p \rightarrow q}$ and $Q = p \wedge \overline{q}$, we can verify that given any truth values of p and q, either P and Q are both true or P and Q are both false:

p	q	$\overline{p \rightarrow q}$	$p \wedge \overline{q}$
T	T	F	F
T	F	T	T
F	T	F	F
F	F	F	F

Thus P and Q are logically equivalent.   □

We now show that according to our definitions, $p \leftrightarrow q$ is logically equivalent to $p \rightarrow q$ *and* $q \rightarrow p$.

EXAMPLE 1.2.13

The truth table shows that

$$p \leftrightarrow q \equiv (p \rightarrow q) \wedge (q \rightarrow p).$$

p	q	$p \leftrightarrow q$	$p \rightarrow q$	$q \rightarrow p$	$(p \rightarrow q) \wedge (q \rightarrow p)$
T	T	T	T	T	T
T	F	F	F	T	F
F	T	F	T	F	F
F	F	T	T	T	T

☐

We conclude this section by defining the **contrapositive** of a conditional proposition. We will see (Theorem 1.2.16) that the contrapositive is an alternative, logically equivalent form of the conditional proposition. Exercise 45 gives another logically equivalent form of the conditional proposition.

DEFINITION 1.2.14

The *contrapositive* (or *transposition*) of the conditional proposition $p \rightarrow q$ is the proposition $\overline{q} \rightarrow \overline{p}$.

Notice the difference between the contrapositive and the converse. The converse of a conditional proposition merely reverses the roles of p and q, whereas the contrapositive reverses the roles of p and q *and* negates each of them.

EXAMPLE 1.2.15

Write the proposition

$$\text{If } 1 < 4, \text{ then } 5 > 8$$

symbolically. Write the converse and the contrapositive both symbolically and in words. Find the truth value of each proposition.

If we define

$$p: \ 1 < 4, \quad q: \ 5 > 8,$$

then the given proposition may be written symbolically as

$$p \rightarrow q.$$

The converse is

$$q \rightarrow p,$$

or, in words,

$$\text{If } 5 > 8, \text{ then } 1 < 4.$$

The contrapositive is

$$\overline{q} \rightarrow \overline{p},$$

or, in words,

If 5 is not greater than 8, then 1 is not less than 4.

We see that $p \rightarrow q$ is false, $q \rightarrow p$ is true, and $\overline{q} \rightarrow \overline{p}$ is false.   □

An important fact is that a conditional proposition and its contrapositive are logically equivalent.

THEOREM 1.2.16

The conditional proposition $p \rightarrow q$ and its contrapositive $\overline{q} \rightarrow \overline{p}$ are logically equivalent.

Proof. The truth table

p	q	$p \rightarrow q$	$\overline{q} \rightarrow \overline{p}$
T	T	T	T
T	F	F	F
F	T	T	T
F	F	T	T

shows that $p \rightarrow q$ and $\overline{q} \rightarrow \overline{p}$ are logically equivalent.   ■

Exercises

In Exercises 1–7, restate each proposition in the form (1.2.2) of a conditional proposition.

1. Joey will pass the discrete mathematics exam if he studies hard.
2. Rosa may graduate if she has 160 quarter-hours of credits.
3. A necessary condition for Fernando to buy a computer is that he obtain $2000.
4. A sufficient condition for Katrina to take the algorithms course is that she pass discrete mathematics.
5. When better cars are built, Buick will build them.
6. The audience will go to sleep if the chairperson gives the lecture.
7. The program is readable only if it is well structured.
8. Write the converse of each proposition in Exercises 1–7.
9. Write the contrapositive of each proposition in Exercises 1–7.

Assuming that p and r are false and that q and s are true, find the truth value of each proposition in Exercises 10–17.

10. $p \rightarrow q$
11. $\overline{p} \rightarrow \overline{q}$

12. $\overline{p \to q}$

13. $(p \to q) \land (q \to r)$

14. $(p \to q) \to r$

15. $p \to (q \to r)$

16. $(s \to (p \land \overline{r})) \land ((p \to (r \lor q)) \land s)$

17. $((p \land \overline{q}) \to (q \land r)) \to (s \lor \overline{q})$

In Exercises 18–21, represent the given statement symbolically by letting

$$p:\ 4 < 2, \quad q:\ 7 < 10, \quad r:\ 6 < 6.$$

18. If $4 < 2$, then $7 < 10$.

19. If ($4 < 2$ and $6 < 6$), then $7 < 10$.

20. If it is not the case that ($6 < 6$ and 7 is not less than 10), then $6 < 6$.

21. $7 < 10$ if and only if ($4 < 2$ and 6 is not less than 6).

In Exercises 22–27, formulate the symbolic expression in words using

$p:$ Today is Monday,
$q:$ It is raining,
$r:$ It is hot.

22. $p \to q$

23. $\overline{q} \to (r \land p)$

24. $\overline{p} \to (q \lor r)$

25. $\overline{p \lor q} \leftrightarrow r$

26. $(p \land (q \lor r)) \to (r \lor (q \lor p))$

27. $\left(p \lor (\overline{p} \land \overline{(q \lor r)})\right) \to \left(p \lor \overline{(r \lor q)}\right)$

In Exercises 28–31, write each conditional proposition symbolically. Write the converse and contrapositive of each statement symbolically and in words. Also, find the truth value of each conditional proposition, its converse, and its contrapositive.

28. If $4 < 6$, then $9 > 12$.

29. If $4 > 6$, then $9 > 12$.

30. $|1| < 3$ if $-3 < 1 < 3$.

31. $|4| < 3$ if $-3 < 4 < 3$.

For each pair of propositions P and Q in Exercises 32–41, state whether or not $P \equiv Q$.

32. $P = p$, $Q = p \lor q$

33. $P = p \land q$, $Q = \overline{p} \lor \overline{q}$

34. $P = p \to q$, $Q = \overline{p} \lor q$

35. $P = p \land (\overline{q} \lor r)$, $Q = p \lor (q \land \overline{r})$

36. $P = p \land (q \lor r)$, $Q = (p \lor q) \land (p \lor r)$

37. $P = p \to q$, $Q = \overline{q} \to \overline{p}$

38. $P = p \rightarrow q$, $Q = p \leftrightarrow q$

39. $P = (p \rightarrow q) \wedge (q \rightarrow r)$, $Q = p \rightarrow r$

40. $P = (p \rightarrow q) \rightarrow r$, $Q = p \rightarrow (q \rightarrow r)$

41. $P = (s \rightarrow (p \wedge \overline{r})) \wedge ((p \rightarrow (r \vee q)) \wedge s)$, $Q = p \vee t$

42. Define the truth table for *imp1* by

p	q	$p\ imp1\ q$
T	T	T
T	F	F
F	T	F
F	F	T

Show that
$$p\ imp1\ q \equiv q\ imp1\ p.$$

43. Define the truth table for *imp2* by

p	q	$p\ imp2\ q$
T	T	T
T	F	F
F	T	T
F	F	F

(a) Show that
$$(p\ imp2\ q) \wedge (q\ imp2\ p) \not\equiv p \leftrightarrow q. \tag{1.2.5}$$

(b) Show that (1.2.5) remains true if we alter *imp2* so that if p is false and q is true, then $p\ imp2\ q$ is false.

44. Verify the second De Morgan law, $\overline{p \wedge q} \equiv \overline{p} \vee \overline{q}$.

45. Show that $(p \rightarrow q) \equiv (\overline{p} \vee q)$.

1.3 *QUANTIFIERS*

The logic in Sections 1.1 and 1.2 that deals with propositions is incapable of describing most of the statements in mathematics and computer science.

Consider, for example, the statement:

$$p:\quad n \text{ is an odd integer.} \tag{1.3.1}$$

A proposition is a statement that is either true or false. The statement p is not a proposition because whether p is true or false depends on the value of n. For example, p is true if $n = 103$ and false if $n = 8$. Since most of the statements in mathematics and computer science use variables, we must extend the system of logic to include such statements.

DEFINITION 1.3.1

Let $P(x)$ be a statement involving the variable x and let D be a set. We call P a *propositional function* (with respect to D) if for each x in D, $P(x)$ is a proposition. We call D the *domain of discourse* of P.

EXAMPLE 1.3.2

Let $P(n)$ be the statement

$$n \text{ is an odd integer}$$

and let D be the set of positive integers. Then P is a propositional function with domain of discourse D since for each n in D, $P(n)$ is a proposition (i.e., for each n in D, $P(n)$ is true or false but not both). For example, if $n = 1$, we obtain the proposition

$$1 \text{ is an odd integer}$$

(which is true). If $n = 2$, we obtain the proposition

$$2 \text{ is an odd integer}$$

(which is false).   □

A propositional function P, by itself, is neither true nor false. However, for each x in its domain of discourse, $P(x)$ is a proposition and is, therefore, either true or false. We can think of a propositional function as defining a class of propositions, one for each element of its domain of discourse. For example, if P is a propositional function with domain of discourse equal to the set of positive integers, we obtain the class of propositions

$$P(1), P(2), \ldots .$$

Each of $P(1), P(2), \ldots$ is either true or false.

EXAMPLE 1.3.3

The following are propositional functions.
(a) $n^2 + 2n$ is an odd integer (domain of discourse = set of positive integers).
(b) $x^2 - x - 6 = 0$ (domain of discourse = set of real numbers).
(c) The baseball player hit over .300 in 1974 (domain of discourse = set of baseball players).
(d) The restaurant rated over two stars in *Chicago* magazine (domain of discourse = restaurants rated in *Chicago* magazine).

In statement (a), for each positive integer n, we obtain a proposition; therefore, statement (a) is a propositional function.

Similarly, in statement (b), for each real number x, we obtain a proposition; therefore, statement (b) is a propositional function.

We can regard the variable in statement (c) as "baseball player." Whenever we substitute a particular baseball player for the variable "baseball player," the statement is a proposition. For example, if we substitute "Willie Stargell" for "baseball player," statement (c) is

Willie Stargell hit over .300 in 1974

which is true. If we substitute "Carlton Fisk" for "baseball player," statement (c) is

Carlton Fisk hit over .300 in 1974

which is false. Thus statement (c) is a propositional function.

Statement (d) is similar in form to statement (c): here the variable is "restaurant." Whenever we substitute a restaurant rated in *Chicago* magazine for the variable "restaurant," the statement is a proposition. For example, if we substitute "Yugo Inn" for "restaurant," statement (d) is

Yugo Inn rated over two stars in *Chicago* magazine

which is false. If we substitute "Le Français" for "restaurant," statement (d) is

Le Français rated over two stars in *Chicago* magazine

which is true. Thus statement (d) is a propositional function.   □

Most of the statements in mathematics and computer science use terms such as "for every" and "for some." For example, in mathematics we have the theorem:

For every triangle T, the sum of the angles of T is equal to $180°$.

In computer science, we have the theorem:

For some program P, the output of P is P itself.

We now extend the logical system of Sections 1.1 and 1.2 so that we can handle statements that include "for every" and "for some."

DEFINITION 1.3.4

Let P be a propositional function with domain of discourse D. The statement

for every x, $P(x)$

is said to be a *universally quantified statement*. The symbol ∀ means "for every." Thus the statement

for every x, $P(x)$

may be written

$\forall x, P(x)$.

The symbol ∀ is called a *universal quantifier*.

The statement
$$\text{for every } x, P(x)$$

is true if $P(x)$ is true for every x in D. The statement
$$\text{for every } x, P(x)$$

is false if $P(x)$ is false for at least one x in D.
   The statement
$$\text{for some } x, P(x)$$

is said to be an *existentially quantified statement*. The symbol $\exists$ means "for some." Thus the statement
$$\text{for some } x, P(x)$$

may be written
$$\exists x, P(x).$$

The symbol $\exists$ is called an *existential quantifier*.
   The statement
$$\text{for some } x, P(x)$$

is true if $P(x)$ is true for at least one x in D. The statement
$$\text{for some } x, P(x)$$

is false if $P(x)$ is false for every x in D.
   We call the variable x in the propositional function $P(x)$ a *free variable*. (The idea is that x is "free" to roam over the domain of discourse.) We call the variable x in the universally quantified statement

$$\forall x, P(x) \tag{1.3.2}$$

or in the existentially quantified statement

$$\exists x, P(x) \tag{1.3.3}$$

a *bound variable*. (The idea is that x is "bound" by the quantifier $\forall$ or $\exists$.) We previously pointed out that a propositional function does not have a truth value. On the other hand, Definition 1.3.4 assigns a truth value to the quantified statements (1.3.2) and (1.3.3). In sum, a statement with free (unquantified) variables is not a proposition and a statement with no free variables (no unquantified variables) is a proposition.
   Alternative ways to write

$$\text{for every } x, P(x)$$

are
$$\text{for all } x, P(x)$$

and
$$\text{for any } x, P(x).$$

The symbol $\forall$ may be read "for every," "for all," or "for any."

Alternative ways to write

$$\text{for some } x,\, P(x)$$

are

$$\text{for at least one } x,\, P(x)$$

and

$$\text{there exists } x \text{ such that, } P(x).$$

The symbol $\exists$ may be read "for some," "for at least one," or "there exists."

Sometimes, to specify the domain of discourse D, we write a universally quantified statement as

$$\text{for every } x \text{ in } D,\, P(x)$$

and we write an existentially quantified statement as

$$\text{for some } x \text{ in } D,\, P(x).$$

EXAMPLE 1.3.5

The statement

$$\text{for every real number } x,\, x^2 \geq 0$$

is a universally quantified statement. The domain of discourse is the set of real numbers. The statement is true because, *for every* real number x, it is true that the square of x is positive or zero.   □

EXAMPLE 1.3.6

The universally quantified statement

$$\text{for every real number } x,\ \text{if } x > 1,\ \text{then } x + 1 > 1$$

is true. This time we must verify that the statement

$$\text{if } x > 1,\ \text{then } x + 1 > 1$$

is true *for every* real number x.

Let x be any real number whatsoever. It is true that for any real number x, either $x \leq 1$ or $x > 1$. In case $x \leq 1$, the conditional proposition

$$\text{if } x > 1,\ \text{then } x + 1 > 1$$

is true because the hypothesis $x > 1$ is false. (Recall that when the hypothesis is false, the conditional proposition is true regardless of whether the conclusion is true or false.)

Now suppose that $x > 1$. Regardless of the specific value of x, $x + 1 > x$. Since

$$x + 1 > x \quad \text{and} \quad x > 1,$$

we conclude that $x + 1 > 1$, so the conclusion is true. If $x > 1$, the hypothesis and conclusion are both true; hence the conditional proposition

$$\text{if } x > 1, \text{ then } x + 1 > 1$$

is true.

We have shown that for every real number x, the proposition

$$\text{if } x > 1, \text{ then } x + 1 > 1$$

is true. Therefore, the universally quantified statement

$$\text{for every real number } x, \text{ if } x > 1, \text{ then } x + 1 > 1$$

is true. ◻

Example 1.3.6 provides further motivation for defining a conditional proposition $p \rightarrow q$ to be true when p is false. For the universally quantified statement

$$\text{for every real number } x, \text{ if } x > 1, \text{ then } x + 1 > 1$$

to be true, the conditional proposition

$$\text{if } x > 1, \text{ then } x + 1 > 1$$

must be true *no matter what value x has*. In particular, the proposition

$$\text{if } x > 1, \text{ then } x + 1 > 1$$

must be true if $x > 1$ is false.

According to Definition 1.3.4, the universally quantified statement

$$\text{for every } x, \ P(x)$$

is false if *for at least one x* in the domain of discourse, the proposition $P(x)$ is false. A value x in the domain of discourse that makes $P(x)$ false is called a **counterexample** to the statement

$$\text{for every } x, \ P(x).$$

EXAMPLE 1.3.7

The universally quantified statement

$$\text{for every real number } x, \ x^2 - 1 > 0$$

is false since, if $x = 1$, the proposition

$$1^2 - 1 > 0$$

is false. The value 1 is a counterexample to the statement

$$\text{for every real number } x, \ x^2 - 1 > 0.$$
◻

To show that the universally quantified statement

$$\text{for every } x, \, P(x)$$

is *false*, it is sufficient to find *one* value x in the domain of discourse for which the proposition $P(x)$ is false. The method of disproving the statement

$$\text{for every } x, \, P(x)$$

is quite different from the method used to prove that the statement is true. To prove that

$$\text{for every } x, \, P(x)$$

is *true*, we must, in effect, examine *every* value of x in the domain of discourse and show that for every x, $P(x)$ is true.

EXAMPLE 1.3.8

The universally quantified statement

$$\text{for every positive integer } n, \text{ if } n \text{ is even, then } n^2 + n + 19 \text{ is prime}$$

is false; a counterexample is obtained by taking $n = 38$. The conditional proposition

$$\text{if 38 is even, then } 38^2 + 38 + 19 \text{ is prime}$$

is false because the hypothesis

$$38 \text{ is even}$$

is true, but the conclusion

$$38^2 + 38 + 19 \text{ is prime}$$

is false. $38^2 + 38 + 19$ is not prime since it can be factored:

$$38^2 + 38 + 19 = 38 \cdot 38 + 38 + 19 = 19(2 \cdot 38 + 2 + 1) = 19 \cdot 79. \qquad \square$$

We turn next to existentially quantified statements. According to Definition 1.3.4, the existentially quantified statement

$$\text{for some } x \text{ in } D, \, P(x)$$

is true if $P(x)$ is true for *at least one* x in D. If $P(x)$ is true for some values of x, it is certainly possible that $P(x)$ is false for other values of x.

EXAMPLE 1.3.9

The existentially quantified statement

$$\text{for some real number } x, \quad \frac{x}{x^2 + 1} = \frac{2}{5}$$

is true because it is possible to find *at least one* real number x for which the proposition

$$\frac{x}{x^2 + 1} = \frac{2}{5}$$

is true. For example, if $x = 2$, we obtain the true proposition

$$\frac{2}{2^2 + 1} = \frac{2}{5}.$$

It is not the case that *every* value of x results in a true proposition. For example, the proposition

$$\frac{1}{1^2 + 1} = \frac{2}{5}$$

is false.   □

EXAMPLE 1.3.10

The existentially quantified statement

for some positive integer n, if n is prime, then $n + 1, n + 2, n + 3$,
    and $n + 4$ are not prime

is true because we can find *at least one* integer n that makes the conditional proposition

if n is prime, then $n + 1, n + 2, n + 3$, and $n + 4$ are not prime

true. For example, if $n = 23$, we obtain the true proposition

if 23 is prime, then 24, 25, 26, and 27 are not prime.

(This conditional proposition is true because both the hypothesis "23 is prime" and the conclusion "24, 25, 26, and 27 are not prime" are true.) Some values of n make the conditional proposition true (e.g., $n = 23, n = 4, n = 47$), while others make it false (e.g., $n = 2, n = 101$). The point is that we found *one* value that makes the conditional proposition

if n is prime, then $n + 1, n + 2, n + 3$, and $n + 4$ are not prime

true. For this reason the existentially quantified statement

for some positive integer n, if n is prime, then $n + 1, n + 2, n + 3$,
    and $n + 4$ are not prime

is true.   □

According to Definition 1.3.4, the existentially quantified statement

$$\text{for some } x, \ P(x)$$

is false if for every x in the domain of discourse, the proposition $P(x)$ is false.

EXAMPLE 1.3.11

To verify that the existentially quantified statement

$$\text{for some real number } x, \quad \frac{1}{x^2 + 1} > 1$$

is false, we must show that

$$\frac{1}{x^2 + 1} > 1$$

is false for every real number x. Now

$$\frac{1}{x^2 + 1} > 1$$

is false precisely when

$$\frac{1}{x^2 + 1} \leq 1$$

is true. Thus, we must show that

$$\frac{1}{x^2 + 1} \leq 1$$

is true for every real number x. To this end, let x be any real number whatsoever. Since $0 \leq x^2$, we may add 1 to both sides of this inequality to obtain $1 \leq x^2 + 1$. If we divide both sides of this last inequality by $x^2 + 1$, we obtain

$$\frac{1}{x^2 + 1} \leq 1.$$

Therefore the statement

$$\frac{1}{x^2 + 1} \leq 1$$

is true for every real number x. Thus the statement

$$\frac{1}{x^2 + 1} > 1$$

is false for every real number x. We have shown that the existentially quantified statement

$$\text{for some } x, \quad \frac{1}{x^2 + 1} > 1$$

is false.   □

In Example 1.3.11, we showed that an existentially quantified statement was false by proving that a related universally quantified statement was true. The following theorem makes this relationship precise. The theorem generalizes De Morgan's laws of logic (Example 1.2.11).

| THEOREM 1.3.12 | *Generalized De Morgan Laws for Logic* |

If P is a propositional function, each pair of propositions in (a) and (b) has the same truth values (i.e., either both are true or both are false).

(a) $\overline{\forall x,\ P(x)}$; $\exists x,\ \overline{P(x)}$

(b) $\overline{\exists x,\ P(x)}$; $\forall x,\ \overline{P(x)}$

Proof. We prove only part (a) and leave the proof of part (b) to the reader (Exercise 50).

Suppose that the proposition $\overline{\forall x,\ P(x)}$ is true. Then the proposition $\forall x,\ P(x)$ is false. By Definition 1.3.4, the proposition $\forall x,\ P(x)$ is false precisely when $P(x)$ is false for at least one x in the domain of discourse. But if $P(x)$ is false for at least one x in the domain of discourse, $\overline{P(x)}$ is true for at least one x in the domain of discourse. Again by Definition 1.3.4, when $\overline{P(x)}$ is true for at least one x in the domain of discourse, the proposition $\exists x,\ \overline{P(x)}$ is true. Thus if the proposition $\overline{\forall x,\ P(x)}$ is true, the proposition $\exists x,\ \overline{P(x)}$ is true. Similarly, if the proposition $\overline{\forall x,\ P(x)}$ is false, the proposition $\exists x,\ \overline{P(x)}$ is false.

Therefore the pair of propositions in part (a) always has the same truth values. ∎

| EXAMPLE 1.3.13 |

Let $P(x)$ be the statement

$$\frac{1}{x^2+1} > 1.$$

In Example 1.3.11 we showed that

for some real number x, $P(x)$

is false by verifying that

for every real number x, $\overline{P(x)}$ (1.3.4)

is true.

The technique can be justified by appealing to Theorem 1.3.12. After we prove that proposition (1.3.4) is true, we may negate (1.3.4) and conclude that

$$\overline{\text{for every real number } x,\ \overline{P(x)}}$$

is false. By Theorem 1.3.12, part (a),

for some real number x, $\overline{\overline{P(x)}}$

or equivalently,

for some real number x, $P(x)$

is also false. □

A universally quantified proposition generalizes the compound proposition

$$P_1 \wedge P_2 \wedge \cdots \wedge P_n \qquad (1.3.5)$$

in the sense that the individual propositions $P_1, P_2, \ldots, P_n$ are replaced by an arbitrary family $P(x)$, where x is a member of the domain of discourse, and (1.3.5) is replaced by

$$\text{for every } x, P(x). \qquad (1.3.6)$$

The proposition (1.3.5) is true if and only if P_i is true for every $i = 1, \ldots, n$. The truth value of proposition (1.3.6) is defined similarly: (1.3.6) is true if and only if $P(x)$ is true for every x in the domain of discourse.

Similarly, an existentially quantified proposition generalizes the compound proposition

$$P_1 \vee P_2 \vee \cdots \vee P_n \qquad (1.3.7)$$

in the sense that the individual propositions $P_1, P_2, \ldots, P_n$ are replaced by an arbitrary family $P(x)$, where x is a member of the domain of discourse, and (1.3.7) is replaced by

$$\text{for some } x, P(x).$$

The preceding observations explain how Theorem 1.3.12 generalizes De Morgan's laws for logic (Example 1.2.11). Recall that the first De Morgan law for logic states that the propositions

$$\overline{P_1 \vee P_2 \vee \cdots \vee P_n} \quad \text{and} \quad \overline{P_1} \wedge \overline{P_2} \wedge \cdots \wedge \overline{P_n}$$

have the same truth values. In Theorem 1.3.12, part (b),

$$\overline{P_1} \wedge \overline{P_2} \wedge \cdots \wedge \overline{P_n}$$

is replaced by

$$\forall x, \overline{P(x)}$$

and

$$\overline{P_1 \vee P_2 \vee \cdots \vee P_n}$$

is replaced by

$$\overline{\exists x, P(x)}.$$

EXAMPLE 1.3.14

Statements in words often have more than one possible interpretation. Consider the well-known quotation from Shakespeare

All that glitters is not gold.

One possible interpretation of this quotation is: Nothing that glitters is gold (i.e., a gold object never glitters). However, this is surely not what Shakespeare intended. The correct interpretation is: Something that glitters is not gold.

If we let $P(x)$ be the propositional function "x glitters" and $Q(x)$ be the propositional function "x is gold," the first interpretation becomes

$$\text{for all } x, \ P(x) \to \overline{Q(x)}, \qquad (1.3.8)$$

and the second interpretation becomes

$$\text{for some } x, \ P(x) \wedge \overline{Q(x)}.$$

Using the result of Example 1.2.12, we see that the truth values of

$$\text{for some } x, \ P(x) \wedge \overline{Q(x)}$$

and

$$\text{for some } x, \ \overline{P(x) \to Q(x)}$$

are the same. By Theorem 1.3.12, the truth values of

$$\text{for some } x, \ \overline{P(x) \to Q(x)}$$

and

$$\overline{\text{for all } x, \ P(x) \to Q(x)}$$

are the same. Thus an equivalent way to represent the second interpretation is

$$\overline{\text{for all } x, \ P(x) \to Q(x)}. \qquad (1.3.9)$$

Comparing (1.3.8) and (1.3.9), we see that the ambiguity results from whether the negation applies to $Q(x)$ (the first interpretation) or to the entire statement

$$\text{for all } x, \ P(x) \to Q(x)$$

(the second interpretation). The correct interpretation of the statement

$$\text{All that glitters is not gold}$$

results from negating the entire statement.

In positive statements, "any," "all," "each," and "every" have the same meaning. In negative statements, the situation changes:

$$\text{Not all } C_1 \text{ is } C_2$$

$$\text{Not each } C_1 \text{ is } C_2$$

$$\text{Not every } C_1 \text{ is } C_2$$

are considered to have the same meaning as

$$\text{Some } C_1 \text{ is not } C_2,$$

whereas

$$\text{Not any } C_1 \text{ is } C_2$$

means

$$\text{No } C_1 \text{ is } C_2.$$

See Exercises 47 and 48 for other examples.   □

Our next example shows how it is possible to mix universal and existential quantifiers within a single statement and also to quantify over more than one variable.

EXAMPLE 1.3.15

Suppose that the domain of discourse is the set of real numbers. Consider the statement

for every x, for some y, $x + y = 0$.

The meaning of this statement is that for any x whatsoever, there is at least one y, which may depend on the choice of x, such that $x + y = 0$. We can show that the statement

for every x, for some y, $x + y = 0$

is true. For any x, we can find at least one y, namely, $y = -x$, such that $x + y = 0$ is true.   □

Suppose that we revise the statement of Example 1.3.15 to read

for some y, for every x, $x + y = 0$.

If this statement is true, then it is possible to select some value of y such that the statement

for every x, $x + y = 0$

is true. However, we can disprove this last statement with a counterexample. For example, we might take $x = 1 - y$. We then obtain the false statement

$$1 - y + y = 0.$$

Therefore, the statement

for some y, for every x, $x + y = 0$

is false.

EXAMPLE 1.3.16

Let $P(x, y)$ be the statement

if $x^2 < y^2$, then $x < y$.

The domain of discourse is the set of real numbers.

The statement

for every x, for every y, $P(x, y)$

is false. A counterexample is $x = 1$, $y = -2$. In this case, we obtain the false proposition

if $1^2 < (-2)^2$, then $1 < -2$.

The statement

for every x, for some y, $P(x, y)$

is true. We show that for every x, the proposition

for some y, if $x^2 < y^2$, then $x < y$

is true by exhibiting a value of y for which

if $x^2 < y^2$, then $x < y$

is true. Indeed, if we set $y = 0$, we obtain the true proposition

$$\text{if } x^2 < 0, \text{ then } x < 0.$$

(The conditional proposition is true, because the hypothesis $x^2 < 0$ is false.)

The statement

$$\text{for every } y, \text{ for some } x, \ P(x, y)$$

is true. We show that for every y, the proposition

$$\text{for some } x, \text{ if } x^2 < y^2, \text{ then } x < y$$

is true by exhibiting a value of x for which

$$\text{if } x^2 < y^2, \text{ then } x < y$$

is true. Indeed, if we set $x = |y| + 1$, we obtain the true proposition

$$\text{if } (|y| + 1)^2 < y^2, \text{ then } |y| + 1 < y.$$

(The conditional proposition is true, because the hypothesis is false.)   □

We summarize the rules for proving or disproving universally or existentially quantified statements:

- To prove that the universally quantified statement

$$\text{for every } x, \ P(x)$$

is true, show that for *every* x in the domain of discourse, the proposition $P(x)$ is true. Showing that $P(x)$ is true for a *particular* value x does *not* prove that

$$\text{for every } x, \ P(x)$$

is true.

- To prove that the existentially quantified statement

$$\text{for some } x, \ P(x)$$

is true, find *one* value of x in the domain of discourse for which $P(x)$ is true. *One* value suffices.

- To prove that the universally quantified statement

$$\text{for every } x, \ P(x)$$

is false, find *one* value of x (a counterexample) in the domain of discourse for which $P(x)$ is false.

- To prove that the existentially quantified statement

$$\text{for some } x, \ P(x)$$

is false, show that for *every* x in the domain of discourse, the proposition $P(x)$ is false. Showing that $P(x)$ is false for a *particular* value x does *not* prove that

$$\text{for some } x, \ P(x)$$

is false.

❧ ❧ ❧

Exercises

In Exercises 1–6, tell whether the statement is a propositional function. For each statement that is a propositional function, give a domain of discourse.

1. $(2n + 1)^2$ is an odd integer.
2. Choose an integer between 1 and 10.
3. Let x be a real number.
4. The movie won the Academy Award as the best picture of 1955.
5. $1 + 3 = 4$
6. There exists x such that $x < y$ (x, y real numbers).

Let $P(n)$ be the propositional function "n divides 77." Write each proposition in Exercises 7–11 in words and tell whether it is true or false. The domain of discourse is the set of positive integers.

7. $P(11)$ 8. $P(1)$ 9. $P(3)$ 10. For every n, $P(n)$.
11. For some n, $P(n)$.

Let $T(x, y)$ be the propositional function "x is taller than y." The domain of discourse consists of three students: Garth, who is 5 feet 11 inches tall; Erin, who is 5 feet 6 inches tall; and Marty who is 6 feet tall. Write each proposition in Exercises 12–15 in words and tell whether it is true or false.

12. $\forall x \, \forall y \, T(x, y)$ 13. $\forall x \, \exists y \, T(x, y)$
14. $\exists x \, \forall y \, T(x, y)$ 15. $\exists x \, \exists y \, T(x, y)$
16. Write the negation of each proposition in Exercises 12–15 in words and symbolically.

Let $L(x, y)$ be the propositional function "x loves y." The domain of discourse is the set of all living people. Write each proposition in Exercises 17–20 symbolically. Which do you think are true?

17. Someone loves everybody.
18. Everybody loves everybody.
19. Somebody loves somebody.
20. Everybody loves somebody.
21. Write the negation of each proposition in Exercises 17–20 in words and symbolically.

Determine the truth value of each statement in Exercises 22–45. The domain of discourse is the set of real numbers. Justify your answers.

22. For every x, $x^2 > x$.
23. For some x, $x^2 > x$.
24. For every x, if $x > 1$, then $x^2 > x$.
25. For some x, if $x > 1$, then $x^2 > x$.
26. For every x, if $x > 1$, then $x/(x^2 + 1) < \frac{1}{3}$.
27. For some x, if $x > 1$, then $x/(x^2 + 1) < \frac{1}{3}$.
28. For every x, for every y, $x^2 < y + 1$.

29. For every x, for some y, $x^2 < y + 1$.

30. For some x, for every y, $x^2 < y + 1$.

31. For some x, for some y, $x^2 < y + 1$.

32. For some y, for every x, $x^2 < y + 1$.

33. For every y, for some x, $x^2 < y + 1$.

34. For every x, for every y, $x^2 + y^2 = 9$.

35. For every x, for some y, $x^2 + y^2 = 9$.

36. For some x, for every y, $x^2 + y^2 = 9$.

37. For some x, for some y, $x^2 + y^2 = 9$.

38. For every x, for every y, $x^2 + y^2 \geq 0$.

39. For every x, for some y, $x^2 + y^2 \geq 0$.

40. For some x, for every y, $x^2 + y^2 \geq 0$.

41. For some x, for some y, $x^2 + y^2 \geq 0$.

42. For every x, for every y, if $x < y$, then $x^2 < y^2$.

43. For every x, for some y, if $x < y$, then $x^2 < y^2$.

44. For some x, for every y, if $x < y$, then $x^2 < y^2$.

45. For some x, for some y, if $x < y$, then $x^2 < y^2$.

46. Write the negation of each proposition in Exercises 22–45.

47. The following proposition appeared in the *Dear Abby* column: All men do not cheat on their wives. What exactly is the meaning of this statement? Do think that the proposition is true or false?

48. Economist Robert J. Samuelson was quoted as saying that "Every environmental problem is not a tragedy." What exactly is the meaning of this statement? Clarify the statement by rephrasing it.

49. (a) Use a truth table to prove that if p and q are propositions, one of $p \rightarrow q$ or $q \rightarrow p$ is true.

   (b) Let $P(x)$ be the propositional function "x is a rational number" and let $Q(x)$ be the propositional function "x is a positive number." The domain of discourse is the set of all real numbers. Comment on the following argument, which allegedly proves that all rational numbers are positive or all positive real numbers are rational.

   By part (a),

$$\forall x \big((P(x) \rightarrow Q(x)) \vee (Q(x) \rightarrow P(x)) \big)$$

   is true. In words: For all x, if x is rational, then x is positive, or if x is positive, then x is rational. Therefore all rational numbers are positive or all positive real numbers are rational.

50. Prove Theorem 1.3.12, part (b).

1.4 PROOFS

A **mathematical system** consists of **axioms**, **definitions**, and **undefined terms**. The axioms are assumed true. Definitions are used to create new concepts in terms of existing ones. Some terms are not explicitly defined but rather are implicitly defined by the axioms. Within a mathematical system we can derive theorems. A **theorem** is a proposition that has been proved to be true. Special kinds of theorems are referred to as lemmas and corollaries. A **lemma** is a theorem that is usually not too interesting in its own right but is useful in proving another theorem. A **corollary** is a theorem that follows quickly from another theorem.

An argument that establishes the truth of a theorem is called a **proof**. Logic is a tool for the analysis of proofs. In this section, we describe some general methods of proof, and we use logic to analyze valid and invalid arguments. In Sections 1.5 and 1.6, we discuss resolution and mathematical induction, which are special proof techniques. We begin by giving some examples of mathematical systems.

> **EXAMPLE 1.4.1**

Euclidean geometry furnishes an example of a mathematical system. Among the axioms are

- Given two distinct points, there is exactly one line that contains them.
- Given a line and a point not on the line, there is exactly one line parallel to the line through the point.

The terms *point* and *line* are undefined terms that are implicitly defined by the axioms that describe their properties.

Among the definitions are

- Two triangles are *congruent* if their vertices can be paired so that the corresponding sides and corresponding angles are equal.
- Two angles are *supplementary* if the sum of their measures is $180°$.   ◻

> **EXAMPLE 1.4.2**

The real numbers furnish another example of a mathematical system. Among the axioms are

- For all real numbers x and y, $xy = yx$.
- There is a subset **P** of real numbers satisfying

  (a) If x and y are in **P**, then $x + y$ and xy are in **P**.
  (b) If x is a real number, then exactly one of the following statements is true:

$$x \text{ is in } \mathbf{P} \qquad x = 0 \qquad -x \text{ is in } \mathbf{P}.$$

Multiplication is implicitly defined by the first axiom and others that describe the properties multiplication is assumed to have.

Among the definitions are

- The elements in **P** (of the preceding axiom) are called *positive real numbers.*
- The *absolute value* $|x|$ of a real number x is defined to be x if x is positive or 0 and $-x$ otherwise. □

We give several examples of theorems, corollaries, and lemmas in Euclidean geometry and in the system of real numbers.

EXAMPLE 1.4.3

Examples of theorems in Euclidean geometry are

- If two sides of a triangle are equal, then the angles opposite them are equal.
- If the diagonals of a quadrilateral bisect each other, then the quadrilateral is a parallelogram. □

EXAMPLE 1.4.4

An example of a corollary in Euclidean geometry is

- If a triangle is equilateral, then it is equiangular.

This corollary follows immediately from the first theorem of Example 1.4.3.
□

EXAMPLE 1.4.5

Examples of theorems about real numbers are

- $x \cdot 0 = 0$ for every real number x.
- For all real numbers x, y, and z, if $x \leq y$ and $y \leq z$, then $x \leq z$. □

EXAMPLE 1.4.6

An example of a lemma about real numbers is

- If n is a positive integer, then either $n - 1$ is a positive integer or $n - 1 = 0$.

Surely this result is not that interesting in its own right, but it can be used to prove other results. □

Theorems are often of the form

For all $x_1, x_2, \ldots, x_n$, if $p(x_1, x_2, \ldots, x_n)$, then
$q(x_1, x_2, \ldots, x_n)$.

This universally quantified statement is true provided that the conditional proposition

$$\text{if } p(x_1, x_2, \ldots, x_n), \text{ then } q(x_1, x_2, \ldots, x_n) \qquad (1.4.1)$$

is true for all $x_1, x_2, \ldots, x_n$ in the domain of discourse. To prove (1.4.1), we assume that $x_1, x_2, \ldots, x_n$ are arbitrary members of the domain of discourse. If $p(x_1, x_2, \ldots, x_n)$ is false, by Definition 1.2.4, (1.4.1) is true; thus, we need only consider the case that $p(x_1, x_2, \ldots, x_n)$ is true. A **direct proof** assumes that $p(x_1, x_2, \ldots, x_n)$ is true and then, using $p(x_1, x_2, \ldots, x_n)$ as well as other axioms, definitions, and previously derived theorems, shows directly that $q(x_1, x_2, \ldots, x_n)$ is true.

EXAMPLE 1.4.7

We will give a direct proof of the following statement. For all real numbers d, d_1, d_2, and x,

$$\text{If } d = \min \{d_1, d_2\} \text{ and } x \le d, \text{ then } x \le d_1 \text{ and } x \le d_2.$$

Proof.   We assume that d, d_1, d_2, and x are arbitrary real numbers. The preceding discussion shows that it suffices to assume that

$$d = \min \{d_1, d_2\} \text{ and } x \le d$$

is true and then prove that

$$x \le d_1 \text{ and } x \le d_2$$

is true.

From the definition of min, it follows that $d \le d_1$ and $d \le d_2$. From $x \le d$ and $d \le d_1$, we may derive $x \le d_1$ from a previous theorem (the second theorem of Example 1.4.5). From $x \le d$ and $d \le d_2$, we may derive $x \le d_2$ from the same previous theorem. Therefore, $x \le d_1$ and $x \le d_2$.   ■ □

A second technique of proof is **proof by contradiction**. A proof by contradiction establishes (1.4.1) by assuming that the hypothesis p is true and that the conclusion q is false and then, using p and $\overline{q}$ as well as other axioms, definitions, and previously derived theorems, derives a **contradiction**. A contradiction is a proposition of the form $r \wedge \overline{r}$ (r may be any proposition whatever). A proof by contradiction is sometimes called an **indirect proof** since to establish (1.4.1) using proof by contradiction, one follows an indirect route: derive $r \wedge \overline{r}$, then conclude that (1.4.1) is true.

The only difference between the assumptions in a direct proof and a proof by contradiction is the negated conclusion. In a direct proof the negated conclusion is not assumed, whereas in a proof by contradiction the negated conclusion is assumed.

Proof by contradiction may be justified by noting that the propositions

$$p \to q \qquad \text{and} \qquad p \wedge \overline{q} \to r \wedge \overline{r}$$

are equivalent. The equivalence is immediate from a truth table:

p	q	r	$p \to q$	$p \wedge \overline{q}$	$r \wedge \overline{r}$	$p \wedge \overline{q} \to r \wedge \overline{r}$
T	T	T	T	F	F	T
T	T	F	T	F	F	T
T	F	T	F	T	F	F
T	F	F	F	T	F	F
F	T	T	T	F	F	T
F	T	F	T	F	F	T
F	F	T	T	F	F	T
F	F	F	T	F	F	T

EXAMPLE 1.4.8

We will give a proof by contradiction of the following statement:

For all real numbers x and y, if $x + y \geq 2$, then either $x \geq 1$ or $y \geq 1$.

Proof.   Suppose that the conclusion is false. Then, $x < 1$ and $y < 1$. (Remember that negating *or*'s results in *and*'s; see Example 1.2.11, De Morgan's laws for logic.) Using a previous theorem, we may add these inequalities to obtain

$$x + y < 1 + 1 = 2.$$

At this point, we have derived the contradiction $p \wedge \overline{p}$, where

$$p: \quad x + y \geq 2.$$

Thus we conclude that the statement is true.   ■ □

Suppose that we give a proof by contradiction of (1.4.1) in which, as in Example 1.4.8, we deduce $\overline{p}$. In effect, we have proved

$$\overline{q} \to \overline{p}. \tag{1.4.2}$$

This special case of proof by contradiction is called **proof by contrapositive**.

In constructing a proof, we must be sure that the arguments used are **valid**. In the remainder of this section we make precise the concept of a valid argument and explore this concept in some detail.

Consider the following sequence of propositions.

The bug is either in module 17 or in module 81.
The bug is a numerical error.
Module 81 has no numerical error.   (1.4.3)

Assuming that these statements are true, it is reasonable to conclude:

The bug is in module 17.   (1.4.4)

This process of drawing a conclusion from a sequence of propositions is called **deductive reasoning**. The given propositions, such as (1.4.3), are called **hypotheses** or **premises** and the proposition that follows from the hypotheses, such as (1.4.4), is called the **conclusion**. A **(deductive) argument** consists of hypotheses together with a conclusion. Many proofs in mathematics and computer science are deductive arguments.

Any argument has the form

$$\text{If } p_1 \text{ and } p_2 \text{ and } \cdots \text{ and } p_n, \text{ then } q.$$ (1.4.5)

Argument (1.4.5) is said to be valid if the conclusion follows from the hypotheses; that is, if p_1 and p_2 and $\cdots$ and p_n are true, then q must also be true. This discussion motivates the following definition.

DEFINITION 1.4.9

An *argument* is a sequence of propositions written

$$
\begin{array}{c}
p_1 \\
p_2 \\
\vdots \\
\underline{p_n} \\
\therefore p
\end{array}
$$

or

$$p_1, p_2, \ldots, p_n / \therefore q.$$

The propositions $p_1, p_2, \ldots, p_n$ are called the *hypotheses* (or *premises*) and the proposition q is called the *conclusion*. The argument is *valid* provided that if p_1 and p_2 and $\cdots$ and p_n are all true, then q must also be true; otherwise, the argument is *invalid* (or a *fallacy*).

In a valid argument, we sometimes say that the conclusion follows from the hypotheses. Notice that we are not saying that the conclusion is true; we are only saying that if you grant the hypotheses, you must also grant the conclusion. An argument is valid because of its form, not because of its content.

EXAMPLE 1.4.10

Determine whether the argument
$$p \to q$$
$$p$$
$$\therefore q$$
is valid.

[First solution.] We construct a truth table for all the propositions involved:

p	q	$p \to q$	p	q
T	T	T	T	T
T	F	F	T	F
F	T	T	F	T
F	F	T	F	F

We observe that whenever the hypotheses $p \to q$ and p are true, the conclusion q is also true; therefore, the argument is valid.

[Second solution.] We can avoid writing the truth table by directly verifying that whenever the hypotheses are true, the conclusion is also true.

Suppose that $p \to q$ and p are true. Then q must be true, for otherwise $p \to q$ would be false. Therefore, the argument is valid.   □

EXAMPLE 1.4.11

Represent the argument
$$\text{If } 2 = 3, \text{ then I ate my hat.}$$
$$\text{I ate my hat.}$$
$$\therefore 2 = 3$$
symbolically and determine whether the argument is valid.

If we let

$$p: \ 2 = 3, \qquad q: \ \text{I ate my hat.}$$

the argument may be written
$$p \to q$$
$$q$$
$$\therefore p$$

If the argument is valid, then whenever $p \to q$ and q are both true, p must also be true. Suppose that $p \to q$ and q are true. This is possible if p is false and q is true. In this case, p is not true; thus the argument is invalid.   □

We can also determine the validity of the argument in Example 1.4.11 by examining the truth table of Example 1.4.10. In the third row of the table, the hypotheses are true and the conclusion is false; thus the argument is invalid.

ॐ ॐ ॐ

Exercises

1. Give an example (different from those of Example 1.4.1) of an axiom in Euclidean geometry.

2. Give an example (different from those of Example 1.4.2) of an axiom in the system of real numbers.

3. Give an example (different from those of Example 1.4.1) of a definition in Euclidean geometry.

4. Give an example (different from those of Example 1.4.2) of a definition in the system of real numbers.

5. Give an example (different from those of Example 1.4.3) of a theorem in Euclidean geometry.

6. Give an example (different from those of Example 1.4.5) of a theorem in the system of real numbers.

7. Justify each step of the following direct proof, which shows that if x is a real number, then $x \cdot 0 = 0$. Assume that the following are previous theorems: If a, b, and c are real numbers, then $b + 0 = b$ and $a(b + c) = ab + ac$. If $a + b = a + c$, then $b = c$.
   Proof. $x \cdot 0 + 0 = x \cdot 0 = x \cdot (0 + 0) = x \cdot 0 + x \cdot 0$; therefore, $x \cdot 0 = 0$. ∎

8. Justify each step of the following proof by contradiction, which shows that if $xy = 0$, then either $x = 0$ or $y = 0$. Assume that if a, b, and c are real numbers with $ab = ac$ and $a \neq 0$, then $b = c$.
   Proof. Suppose that $xy = 0$ and $x \neq 0$ and $y \neq 0$. Since $xy = 0 = x \cdot 0$ and $x \neq 0$; $y = 0$, which is a contradiction. ∎

9. Show, by giving a proof by contradiction, that if 100 balls are placed in nine boxes, some box contains 12 or more balls.

Formulate the arguments of Exercises 10–14 symbolically and determine whether each is valid. Let

$$p: \text{I study hard.} \quad q: \text{I get A's.} \quad r: \text{I get rich.}$$

10. If I study hard, then I get A's.
    I study hard.
    _____
    ∴ I get A's.

11. If I study hard, then I get A's.
    If I don't get rich, then I don't get A's.
    _____
    ∴ I get rich.

12. I study hard if and only if I get rich.
    I get rich.
    _____
    ∴ I study hard.

13. If I study hard or I get rich, then I get A's.
    I get A's.
    _____
    ∴ If I don't study hard, then I get rich.

14. If I study hard, then I get A's or I get rich.
I don't get A's and I don't get rich.
_____

∴ I don't study hard.

In Exercises 15–19, write the given argument in words and determine whether each argument is valid. Let

p: 64K is better than no memory at all.
q: We will buy more memory.
r: We will buy a new computer.

15. $p \to r$
$p \to q$
_____
∴ $p \to (r \wedge q)$

16. $p \to (r \vee q)$
$r \to \overline{q}$
_____
∴ $p \to r$

17. $p \to r$
$r \to q$
_____
∴ q

18. $\overline{r} \to \overline{p}$
r
_____
∴ p

19. $p \to r$
$r \to q$
p
_____
∴ q

Determine whether each argument in Exercises 20–24 is valid.

20. $p \to q$
$\overline{p}$
_____
∴ $\overline{q}$

21. $p \to q$
q
_____
∴ $\overline{p}$

22. $p \wedge \overline{p}$
_____
∴ q

23. $p \to (q \to r)$
$q \to (p \to r)$
_____
∴ $(p \vee q) \to r$

24. $(p \to q) \wedge (r \to s)$
$p \vee r$
_____
∴ $q \vee s$

25. Show that if

$$p_1, p_2 / \therefore p \quad \text{and} \quad p, p_3, \ldots, p_n / \therefore c$$

are valid arguments, the argument

$$p_1, p_2, \ldots, p_n / \therefore c$$

is also valid.

26. Comment on the following argument.

Floppy disk storage is better than nothing.
Nothing is better than a hard disk drive.
_____

∴ Floppy disk storage is better than a hard disk drive.

†1.5 RESOLUTION PROOFS

(Skip)

In this section, we will write $a \wedge b$ as ab.

Resolution is a proof technique proposed by J. A. Robinson in 1965 (see [Robinson]) that depends on a single rule

$$\text{If } p \vee q \text{ and } \overline{p} \vee r \text{ are both true, then } q \vee r \text{ is true.} \qquad (1.5.1)$$

(1.5.1) can be verified by writing the truth table (see Exercise 1). Because resolution depends on this single, simple rule, it is the basis of many computer programs that reason and prove theorems.

In a proof by resolution, the hypotheses and the conclusion are written as **clauses**. A clause consists of terms separated by *or*'s, where each term is a variable or the negation of a variable.

EXAMPLE 1.5.1

The expression

$$a \vee b \vee \overline{c} \vee d$$

is a clause since the terms $a, b, \overline{c}$, and d are separated by *or*'s, and each term is a variable or the negation of a variable. □

EXAMPLE 1.5.2

The expression

$$xy \vee w \vee \overline{z}$$

is *not* a clause even though the terms are separated by *or*'s since the term xy consists of two variables—not a single variable. □

EXAMPLE 1.5.3

The expression

$$p \rightarrow q$$

is *not* a clause since the terms are separated by $\rightarrow$. Each term is, however, a variable. □

A direct proof by resolution proceeds by repeatedly applying (1.5.1) to pairs of statements to derive new statements until the conclusion is derived. In applying (1.5.1), p must be a single variable, but q and r can be expressions. Notice that when (1.5.1) is applied to clauses, the result $q \vee r$ is a clause. (Since q and r each consist of terms separated by *or*'s, where each term is a variable or the negation of a variable, $q \vee r$ also consists of terms separated by *or*'s, where each term is a variable or the negation of a variable.)

---

† This section can be omitted without loss of continuity.

EXAMPLE 1.5.4

We prove the following using resolution

$$
\begin{array}{ll}
1. & a \vee b \\
2. & \overline{a} \vee c \\
3. & \overline{c} \vee d \\
\hline
& \therefore b \vee d
\end{array}
$$

Applying (1.5.1) to expressions 1 and 2, we derive

$$4. \quad b \vee c$$

Applying (1.5.1) to expressions 3 and 4, we derive

$$5. \quad b \vee d$$

the desired conclusion. Given the hypotheses 1, 2, and 3, we have proved the conclusion $b \vee d$.   □

Special cases of (1.5.1) are

$$
\begin{array}{l}
\text{If } p \vee q \text{ and } \overline{p} \text{ are true, then } q \text{ is true.} \\
\text{If } p \text{ and } \overline{p} \vee r \text{ are true, then } r \text{ is true.}
\end{array}
\tag{1.5.2}
$$

EXAMPLE 1.5.5

We prove the following using resolution

$$
\begin{array}{ll}
1. & a \\
2. & \overline{a} \vee c \\
3. & \overline{c} \vee d \\
\hline
& \therefore d
\end{array}
$$

Applying (1.5.2) to expressions 1 and 2, we derive

$$4. \quad c$$

Applying (1.5.2) to expressions 3 and 4, we derive

$$5. \quad d$$

the desired conclusion. Given the hypotheses 1, 2, and 3, we have proved the conclusion d.   □

If a hypothesis is not a clause, it must be replaced by an equivalent expression that is either a clause or the *and* of clauses. For example, suppose that one of the hypotheses is $\overline{a \vee b}$. Since the bar is over more than one variable, we use the first of De Morgan's laws (see Example 1.2.11)

$$\overline{a \vee b} \equiv \overline{a}\,\overline{b}, \qquad \overline{ab} \equiv \overline{a} \vee \overline{b} \tag{1.5.3}$$

to obtain an equivalent expression with the bar over single variables

$$\overline{a \vee b} \equiv \overline{a}\,\overline{b}.$$

We then replace the original hypothesis $\overline{a \vee b}$ by the two hypotheses $\overline{a}$ and $\overline{b}$. This replacement is justified by recalling that individual hypotheses h_1 and h_2 are equivalent to $h_1 h_2$ (see Definition 1.4.9 and the discussion that precedes it). Repeated use of De Morgan's laws will result in each bar applied to only one variable.

An expression that consists of terms separated by *or*'s, where each term consists of the *and* of *several* variables may be replaced by an equivalent expression that consists of the *and* of clauses by using the equivalence

$$a \vee bc \equiv (a \vee b)(a \vee c). \tag{1.5.4}$$

In this case, we may replace the single hypothesis $a \vee bc$ by the two hypotheses $a \vee b$ and $a \vee c$. By using first De Morgan's laws (1.5.3) and then (1.5.4), we can obtain equivalent hypotheses each of which is a clause.

> ### EXAMPLE 1.5.6

We prove the following using resolution

$$
\begin{array}{ll}
1. & a \vee \overline{b}c \\
2. & \overline{a \vee d} \\
\hline
& \therefore \overline{b}
\end{array}
$$

We use (1.5.4) to replace hypothesis 1 with the two hypotheses

$$a \vee \overline{b}$$

$$a \vee c$$

We use the first of De Morgan's laws (1.5.3) to replace hypothesis 2 with the two hypotheses

$$\overline{a}$$

$$\overline{d}$$

The argument becomes

$$
\begin{array}{ll}
1. & a \vee \overline{b} \\
2. & a \vee c \\
3. & \overline{a} \\
4. & \overline{d} \\
\hline
& \therefore \overline{b}
\end{array}
$$

Applying (1.5.1) to expressions 1 and 3, we immediately derive the conclusion

$$\overline{b} \qquad\qquad \square$$

In automated reasoning systems, proof by resolution is combined with proof by contradiction. We write the negated conclusion as clauses and add the clauses to the hypothesis. We then repeatedly apply (1.5.1) until we derive a contradiction.

EXAMPLE 1.5.7

We reprove Example 1.5.4 by combining resolution with proof by contradiction.

We first negate the conclusion and use the first of De Morgan's laws (1.5.3) to obtain

$$\overline{b \vee d} \equiv \overline{b}\,\overline{d}.$$

We then add the clauses $\overline{b}$ and $\overline{d}$ to the hypotheses to obtain

1. $a \vee b$
2. $\overline{a} \vee c$
3. $\overline{c} \vee d$
4. $\overline{b}$
5. $\overline{d}$

Applying (1.5.1) to expressions 1 and 2, we derive

6. $b \vee c$

Applying (1.5.1) to expressions 3 and 6, we derive

7. $b \vee d$

Applying (1.5.1) to expressions 4 and 7, we derive

8. d

Now 5 and 8 combine to give a contradiction, and the proof is complete.   □

It can be shown that resolution is *correct* and *refutation complete*. Resolution is correct means that it can only derive a contradiction from a set of inconsistent clauses (i.e., a set of clauses that cannot all be true). Resolution is refutation complete means that if a set of clauses is inconsistent, resolution will be able to derive a contradiction. Thus, if a conclusion follows from a set of hypotheses, resolution will be able to derive a contradiction from the hypotheses and the negation of the conclusion. Unfortunately, resolution does not tell us which clauses to combine in order to deduce the contradiction. A key challenge in automating a reasoning system is to help guide the search for clauses to combine. References on resolution and automated reasoning are [Gallier; Genesereth; and Wos].

≈ ≈ ≈

Exercises

1. Write a truth table that proves (1.5.1).

Use resolution to derive each conclusion in Exercises 2–6. *Hint*: In Exercises 5 and 6, replace $\rightarrow$ and $\leftrightarrow$ with logically equivalent expressions that use *or* and *and*.

2. $\overline{p} \vee q \vee r$
$\overline{q}$
$\overline{r}$ ————
$\therefore \overline{p}$

3. $\overline{p} \vee r$
$\overline{r} \vee q$
p ————
$\therefore q$

4. $\overline{p} \vee t$
$\overline{q} \vee s$
$\overline{r} \vee st$
$p \vee q \vee r \vee u$ ————
$\therefore s \vee t \vee u$

5. $p \rightarrow q$
$p \vee q$ ————
$\therefore q$

6. $p \leftrightarrow r$
r ————
$\therefore p$

7. Use resolution and proof by contradiction to reprove Exercises 2–6.

8. Use resolution and proof by contradiction to reprove Example 1.5.6.

1.6 MATHEMATICAL INDUCTION

Suppose that a sequence of blocks numbered 1, 2, . . . sits on an (infinitely) long table (see Figure 1.6.1) and that some blocks are marked with an "X." (All of the blocks visible in Figure 1.6.1 are marked.) Suppose that

> The first block is marked.   (1.6.1)

> If all the blocks preceding the $(n + 1)$st block
> are marked, then the $(n + 1)$st block is also marked.   (1.6.2)

We will show that (1.6.1) and (1.6.2) imply that every block is marked by examining the blocks one by one.

Statement (1.6.1) explicitly states that block 1 is marked. Consider block 2. All of the blocks preceding block 2, namely block 1, are marked; thus, according to (1.6.2), block 2 is also marked. Consider block 3. All of the blocks preceding block 3, namely blocks 1 and 2, are marked; thus, according to (1.6.2), block 3 is also marked. In this way we can show that every block is marked. For example, suppose that we have verified that blocks 1–5 are marked, as shown in Figure 1.6.1. To show that block 6, which is not shown in Figure 1.6.1, is marked, we note that all the blocks that precede block 6 are marked, so by (1.6.2), block 6 is also marked.

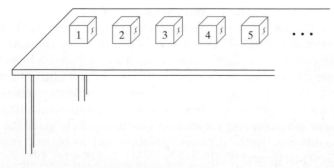

FIGURE 1.6.1   Numbered blocks on a table.

The preceding example illustrates the **Principle of Mathematical Induction**. To show how mathematical induction can be used in a more profound way, let S_n denote the sum of the first n positive integers

$$S_n = 1 + 2 + 3 + \cdots + n. \tag{1.6.3}$$

Suppose that someone claims that

$$S_n = \frac{n(n+1)}{2} \qquad \text{for } n = 1, 2, \ldots \tag{1.6.4}$$

A sequence of statements is really being made, namely,

$$S_1 = \frac{1(2)}{2} = 1$$

$$S_2 = \frac{2(3)}{2} = 3$$

$$S_3 = \frac{3(4)}{2} = 6$$

$$\vdots$$

Suppose that each true equation has an "$\times$" placed beside it (see Figure 1.6.2). Since the first equation is true, it is marked. Now suppose we can show that if all the equations preceding a particular equation, say the $(n+1)$st equation, are marked, then the $(n+1)$st equation is also marked. Then, as in the example involving the blocks, all of the equations are marked; that is, all the equations are true and the formula (1.6.4) is verified.

We must show that if all of the equations preceding the $(n+1)$st equation are true, then the $(n+1)$st equation is also true. Assuming that all of the equations preceding the $(n+1)$st equation are true, then, in particular, the nth equation is true:

$$S_n = \frac{n(n+1)}{2}. \tag{1.6.5}$$

We must show that the $(n+1)$st equation

$$S_{n+1} = \frac{(n+1)(n+2)}{2}$$

is true. According to definition (1.6.3),

$$S_{n+1} = 1 + 2 + \cdots + n + (n+1).$$

We note that S_n is contained within S_{n+1}, in the sense that

$$S_{n+1} = 1 + 2 + \cdots + n + (n+1)$$
$$= S_n + (n+1). \tag{1.6.6}$$

Because of (1.6.5) and (1.6.6), we have

$$S_{n+1} = S_n + n + 1 = \frac{n(n+1)}{2} + n + 1 = \frac{(n+1)(n+2)}{2}.$$

$$\begin{array}{lll}
S_1 = & \dfrac{1(2)}{2} & \times \\[2mm]
S_2 = & \dfrac{2(3)}{2} & \times \\[1mm]
\vdots & & \\[1mm]
S_{n-1} = & \dfrac{(n-1)n}{2} & \times \\[2mm]
S_n = & \dfrac{n(n+1)}{2} & \times \\[2mm]
S_{n+1} = & \dfrac{(n+1)(n+2)}{2} & ? \\[1mm]
\vdots & &
\end{array}$$

FIGURE 1.6.2
A sequence of statements. True statements are marked with $\times$.

Our proof using mathematical induction consisted of two steps. First, we verified that the statement corresponding to $n = 1$ was true. Second, we *assumed* that statements $1, 2, \ldots, n$ were true and then *proved* that the $(n+1)$st statement was also true. In proving the $(n+1)$st statement, we were permitted to make use of statements $1, 2, \ldots, n$; indeed, the trick in constructing a proof using mathematical induction is to relate statements $1, 2, \ldots, n$ to the $(n+1)$st statement.

We next formally state the Principle of Mathematical Induction.

Principle of Mathematical Induction

Suppose that for each positive integer n we have a statement $S(n)$ that is either true or false. Suppose that

$$S(1) \text{ is true;} \tag{1.6.7}$$

$$\text{if } S(i) \text{ is true, for all } i < n + 1, \text{ then } S(n + 1) \text{ is true.} \tag{1.6.8}$$

Then $S(n)$ is true for every positive integer n.

Condition (1.6.7) is sometimes called the **Basis Step** and condition (1.6.8) is sometimes called the **Inductive Step**. Hereafter, "induction" will mean "mathematical induction."

At this point, we illustrate the Principle of Mathematical Induction with another example.

EXAMPLE 1.6.1

Use induction to show that

$$n! \geq 2^{n-1} \qquad \text{for } n = 1, 2, \ldots. \tag{1.6.9}$$

BASIS STEP.   [Condition (1.6.7)] We must show that (1.6.9) is true if $n = 1$. This is easily accomplished, since $1! = 1 \geq 1 = 2^{1-1}$.

INDUCTIVE STEP.   [Condition (1.6.8)] We must show that if $i! \geq 2^{i-1}$ for $i = 1, \ldots, n$, then

$$(n + 1)! \geq 2^n. \tag{1.6.10}$$

Assume that $i! \geq 2^{i-1}$ for $i = 1, \ldots, n$. Then, in particular, for $i = n$, we have

$$n! \geq 2^{n-1}. \tag{1.6.11}$$

We can relate (1.6.10) and (1.6.11) by observing that

$$(n + 1)! = (n + 1)(n!).$$

Now

$$
\begin{aligned}
(n + 1)! &= (n + 1)(n!) \\
&\geq (n + 1)2^{n-1} &&\text{by (1.6.11)} \\
&\geq 2 \cdot 2^{n-1} &&\text{since } n + 1 \geq 2 \\
&= 2^n.
\end{aligned}
$$

Therefore, (1.6.10) is true. We have completed the Inductive Step.

Since the Basis Step and the Inductive Step have been verified, the Principle of Mathematical Induction tells us that (1.6.9) is true for every positive integer n. □

To verify the Inductive Step (1.6.8), we assume that $S(i)$ is true for all $i < n+1$ and then prove that $S(n+1)$ is true. This formulation of mathematical induction is called the **strong form of mathematical induction.** Often, as was the case in the preceding examples, we can deduce $S(n + 1)$ assuming only $S(n)$. Indeed, the Inductive Step is often stated:

If $S(n)$ is true, then $S(n + 1)$ is true.

In these two formulations, the Basis Step is unchanged. It can be shown (see Exercise 45) that the two forms of mathematical induction are logically equivalent.

If we want to verify that the statements

$$S(n_0), S(n_0 + 1), \ldots,$$

where $n_0 \neq 1$, are true, we must change the Basis Step to

$$S(n_0) \text{ is true.}$$

The Inductive Step is unchanged.

EXAMPLE 1.6.2   *Geometric Sum*

Use induction to show that if $r \neq 1$,

$$a + ar^1 + ar^2 + \cdots + ar^n = \frac{a(r^{n+1} - 1)}{r - 1} \qquad (1.6.12)$$

for $n = 0, 1, \ldots$.

The sum on the left is called a **geometric sum**. In a geometric sum, the ratio of consecutive terms $(ar^{i+1}/ar^i = r)$ is constant.

BASIS STEP.   The Basis Step, which in this case is obtained by setting $n = 0$, is

$$a = \frac{a(r^1 - 1)}{r - 1},$$

which is true.

INDUCTIVE STEP.   Assume that statement (1.6.12) is true for n. Now

$$
\begin{aligned}
a + ar^1 + ar^2 + \cdots + ar^n + ar^{n+1} &= \frac{a(r^{n+1} - 1)}{r - 1} + ar^{n+1} \\
&= \frac{a(r^{n+1} - 1)}{r - 1} + \frac{ar^{n+1}(r - 1)}{r - 1} \\
&= \frac{a(r^{n+2} - 1)}{r - 1}.
\end{aligned}
$$

Since the modified Basis Step and the Inductive Step have been verified, the Principle of Mathematical Induction tells us that (1.6.12) is true for $n = 0, 1, \ldots$.   □

As an example of the use of the geometric sum, if we take $a = 1$ and $r = 2$ in (1.6.12), we obtain the formula

$$1 + 2 + 2^2 + 2^3 + \cdots + 2^n = \frac{2^{n+1} - 1}{2 - 1} = 2^{n+1} - 1.$$

The reader has surely noticed that in order to prove the previous formulas, one has to be given the correct formulas in advance. A reasonable question is: How does one come up with the formulas? There are many answers to this question. One technique to derive a formula is to experiment with small values and try to discover a pattern. For example, consider the sum $1 + 3 + \cdots + (2n - 1)$. The following table gives the values of this sum for $n = 1, 2, 3, 4$.

n	$1 + 3 + \cdots + (2n - 1)$
1	1
2	4
3	9
4	16

Since the second column consists of squares, we conjecture that

$$1 + 3 + \cdots + (2n - 1) = n^2 - \quad \text{for every positive integer } n.$$

The conjecture is correct and the formula can be proved by mathematical induction (see Exercise 1).

Our final two examples show that induction is not limited to proving formulas for sums and verifying inequalities.

EXAMPLE 1.6.3

Use induction to show that $5^n - 1$ is divisible by 4 for $n = 1, 2, \ldots$.

Basis Step.   If $n = 1$, $5^n - 1 = 5^1 - 1 = 4$, which is divisible by 4.

Inductive Step.   Assume that $5^n - 1$ is divisible by 4. We must show that $5^{n+1} - 1$ is divisible by 4. To relate the $(n + 1)$st case to the nth case, we write

$$5^{n+1} - 1 = 5 \cdot 5^n - 1 = (5^n - 1) + 4 \cdot 5^n.$$

By assumption, $5^n - 1$ is divisible by 4 and, since $4 \cdot 5^n$ is divisible by 4, the sum

$$(5^n - 1) + 4 \cdot 5^n = 5^{n+1} - 1$$

is divisible by 4. Since the Basis Step and the Inductive Step have been verified, the Principle of Mathematical Induction tells us that $5^n - 1$ is divisible by 4 for $n = 1, 2, \ldots$. $\square$

EXAMPLE 1.6.4   *A Tiling Problem*

FIGURE 1.6.3
A tromino.

A *right tromino*, hereafter called simply a *tromino*, is an object made up of three squares, as shown in Figure 1.6.3. A tromino is a type of polyomino. Since polyominoes were introduced by Solomon W. Golomb in 1954 (see [Golomb, 1954]), they have been a favorite topic in recreational mathematics. A *polyomino of order s* consists of s squares joined at the edges. A tromino is a polyomino of order 3. Three squares in a row form the only other type of polyomino of order 3. (No one has yet found a simple formula for the number of polyominoes of order s.) Numerous problems using polyominoes have been devised (see [Martin]).

We give Golomb's inductive proof (see [Golomb, 1954]) that if we remove one square from an $n \times n$ board, where n is a power of 2, we can tile the remaining squares with right trominoes (see Figure 1.6.4). By a *tiling* of a figure by trominoes, we mean an exact covering of the figure by trominoes without having any of the trominoes overlap each other or extend outside the figure. We call a board with one square missing a *deficient board*.

We now use induction on k to prove that we can tile a $2^k \times 2^k$ deficient board with trominoes.

BASIS STEP.   If $k = 1$, the 2×2 deficient board is itself a tromino and can therefore be tiled with one tromino.

INDUCTIVE STEP.   Assume that we can tile a deficient $2^k \times 2^k$ board. We show that we can tile a deficient $2^{k+1} \times 2^{k+1}$ board.

Consider a deficient $2^{k+1} \times 2^{k+1}$ board. Divide the board into four $2^k \times 2^k$ boards, as shown in Figure 1.6.5. Rotate the board so that the missing square is in the upper left quadrant. By the inductive assumption, the upper left $2^k \times 2^k$ board can be tiled. Place one tromino T in the center, as shown in Figure 1.6.5, so that each square of T is in each of the other quadrants. If we consider the squares covered by T as missing, each of these quadrants is a deficient $2^k \times 2^k$ board. Again, by the inductive assumption, these boards can be tiled. We now have a tiling of the $2^{k+1} \times 2^{k+1}$ board. By the Principle of Mathematical Induction, it follows that any deficient $2^k \times 2^k$ board can be tiled with trominoes, $k = 1, 2, \ldots$.

If we can tile a deficient $n \times n$ board, where n is not necessarily a power of 2, then the number of squares, $n^2 - 1$, must be divisible by 3. [Chu] showed that the converse is true, except when n is 5. More precisely, if $n \neq 5$, any deficient $n \times n$ board can be tiled with trominoes if and only if 3 divides $n^2 - 1$. [Some deficient 5×5 boards can be tiled and some cannot (see Exercises 32 and 33).]   ☐

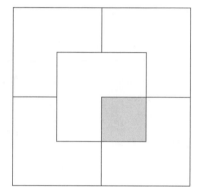

FIGURE 1.6.4
Tiling a deficient 4×4 board with trominoes.

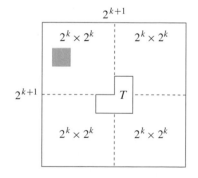

FIGURE 1.6.5
Using mathematical induction to tile a deficient $2^{k+1} \times 2^{k+1}$ board with trominoes.

Exercises

In Exercises 1–11, using induction, verify that each equation is true for every positive integer n.

1. $1 + 3 + 5 + \cdots + (2n - 1) = n^2$

2. $1 \cdot 2 + 2 \cdot 3 + 3 \cdot 4 + \cdots + n(n + 1) = \dfrac{n(n + 1)(n + 2)}{3}$

3. $1(1!) + 2(2!) + \cdots + n(n!) = (n+1)! - 1$

4. $1^2 + 2^2 + 3^2 + \cdots + n^2 = \dfrac{n(n+1)(2n+1)}{6}$

5. $1^2 - 2^2 + 3^2 - \cdots + (-1)^{n+1}n^2 = \dfrac{(-1)^{n+1}n(n+1)}{2}$

6. $1^3 + 2^3 + 3^3 + \cdots + n^3 = \left[\dfrac{n(n+1)}{2}\right]^2$

7. $\dfrac{1}{1\cdot 3} + \dfrac{1}{3\cdot 5} + \dfrac{1}{5\cdot 7} + \cdots + \dfrac{1}{(2n-1)(2n+1)} = \dfrac{n}{2n+1}$

8. $\dfrac{1}{2\cdot 4} + \dfrac{1\cdot 3}{2\cdot 4\cdot 6} + \dfrac{1\cdot 3\cdot 5}{2\cdot 4\cdot 6\cdot 8} + \cdots + \dfrac{1\cdot 3\cdot 5\cdots(2n-1)}{2\cdot 4\cdot 6\cdots(2n+2)} = \dfrac{1}{2} - \dfrac{1\cdot 3\cdot 5\cdots(2n+1)}{2\cdot 4\cdot 6\cdots(2n+2)}$

9. $\dfrac{1}{2^2-1} + \dfrac{1}{3^2-1} + \cdots + \dfrac{1}{(n+1)^2-1} = \dfrac{3}{4} - \dfrac{1}{2(n+1)} - \dfrac{1}{2(n+2)}$

†★ **10.** $\cos x + \cos 2x + \cdots + \cos nx = \dfrac{\cos[(x/2)(n+1)]\sin(nx/2)}{\sin(x/2)}$ provided that $\sin(x/2) \neq 0$.

★ **11.** $1\sin x + 2\sin 2x + \cdots + n\sin nx =$

$$\dfrac{\sin[(n+1)x]}{4\sin^2(x/2)} - \dfrac{(n+1)\cos\left(\frac{2n+1}{2}x\right)}{2\sin(x/2)}$$

provided that $\sin(x/2) \neq 0$.

In Exercises 12–17, using induction, verify the inequality.

12. $\dfrac{1}{2n} \leq \dfrac{1\cdot 3\cdot 5\cdots(2n-1)}{2\cdot 4\cdot 6\cdots(2n)}, n = 1, 2, \ldots$

★ **13.** $\dfrac{1\cdot 3\cdot 5\cdots(2n-1)}{2\cdot 4\cdot 6\cdots(2n)} \leq \dfrac{1}{\sqrt{n+1}}, n = 1, 2, \ldots$

14. $2n + 1 \leq 2^n, n = 3, 4, \ldots$

★ **15.** $2^n \geq n^2, n = 4, 5, \ldots$

★ **16.** $(a_1 a_2 \cdots a_{2^n})^{1/2^n} \leq \dfrac{a_1 + a_2 + \cdots + a_{2^n}}{2^n}, n = 1, 2, \ldots$ and the a_i are positive numbers.

17. $(1 + x)^n \geq 1 + nx$, for $x \geq -1$ and $n = 1, 2, \ldots$

In Exercises 18–21, use induction to prove the statement.

18. $7^n - 1$ is divisible by 6, for $n = 1, 2, \ldots$.

19. $11^n - 6$ is divisible by 5, for $n = 1, 2, \ldots$.

20. $6 \cdot 7^n - 2 \cdot 3^n$ is divisible by 4, for $n = 1, 2, \ldots$.

★ **21.** $3^n + 7^n - 2$ is divisible by 8, for $n = 1, 2, \ldots$.

22. By experimenting with small values of n, guess a formula for the given sum,

$$\dfrac{1}{1\cdot 2} + \dfrac{1}{2\cdot 3} + \cdots + \dfrac{1}{n(n+1)},$$

then use induction to verify your formula.

† A starred exercise indicates a problem of above-average difficulty.

23. Use induction to show that n straight lines in the plane divide the plane into $(n^2 + n + 2)/2$ regions. Assume that no two lines are parallel and that no three lines have a common point.

24. Show that postage of 5 cents or more can be achieved by using only 2-cent and 5-cent stamps.

25. Show that postage of 24 cents or more can be achieved by using only 5-cent and 7-cent stamps.

The Egyptians of antiquity expressed a fraction as a sum of fractions whose numerators were 1. For example, $\frac{5}{6}$ might be expressed as

$$\frac{5}{6} = \frac{1}{2} + \frac{1}{3}.$$

We say that a fraction p/q, where p and q are positive integers, is in *Egyptian form* if

$$\frac{p}{q} = \frac{1}{n_1} + \frac{1}{n_2} + \cdots + \frac{1}{n_k}, \qquad (1.6.13)$$

where $n_1, n_2, \ldots, n_k$ are positive integers satisfying $n_1 < n_2 < \cdots < n_k$.

26. Show that the representation (1.6.13) need not be unique by representing $\frac{5}{6}$ in two different ways.

★ 27. Show that the representation (1.6.13) is never unique.

28. By completing the following steps, give a proof by induction on p to show that every fraction p/q with $0 < p/q < 1$ may be expressed in Egyptian form.

   (a) Verify the Basis Step ($p = 1$).

   (b) Suppose that $0 < p/q < 1$ and that all fractions i/q', with $1 \le i < p$ and q' arbitrary, can be expressed in Egyptian form. Choose the smallest positive integer n with $1/n \le p/q$. Show that

$$n > 1 \qquad \text{and} \qquad \frac{p}{q} < \frac{1}{n-1}.$$

   (c) Show that if $p/q = 1/n$, the proof is complete.

   (d) Assume that $1/n < p/q$. Let

$$p_1 = np - q \qquad \text{and} \qquad q_1 = nq.$$

   Show that

$$\frac{p_1}{q_1} = \frac{p}{q} - \frac{1}{n}, \qquad 0 < \frac{p_1}{q_1} < 1, \qquad \text{and} \qquad p_1 < p.$$

   Conclude that
$$\frac{p_1}{q_1} = \frac{1}{n_1} + \frac{1}{n_2} + \cdots + \frac{1}{n_k}$$

   with $n_1, n_2, \ldots, n_k$ distinct.

   (e) Show that $p_1/q_1 < 1/n$.

   (f) Show that
$$\frac{p}{q} = \frac{1}{n} + \frac{1}{n_1} + \cdots + \frac{1}{n_k}$$

   and $n, n_1, \ldots, n_k$ are distinct.

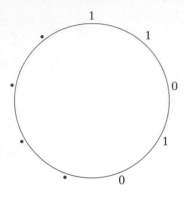

29. Use the method of Exercise 28 to find Egyptian forms of $\frac{3}{8}, \frac{5}{7}$ and $\frac{13}{19}$.

★ **30.** Show that any fraction p/q, where p and q are positive integers, can be written in Egyptian form. (We are not assuming that $p/q < 1$.)

31. Given an equal number of 0's and 1's distributed around a circle (see adjacent figure), show that it is possible to start at some number and proceed around the circle to the original starting position in such a way that, at any point during the cycle, one has seen at least as many 0's as 1's.

32. Give a tiling of a 5×5 board with trominoes in which the upper left square is missing.

33. Show a deficient 5×5 board that it is impossible to tile with trominoes. Explain why your board cannot be tiled with trominoes.

34. Show that any $(2i) \times (3j)$ board, where i and j are positive integers, with no square missing, can be tiled with trominoes.

★ **35.** Show that any 7×7 deficient board can be tiled with trominoes.

★ **36.** Show that any $n \times n$ deficient board can be tiled with trominoes if n is odd, $n > 5$, and 3 divides $n^2 - 1$.

A *3D-septomino* is a three-dimensional $2 \times 2 \times 2$ cube with one $1 \times 1 \times 1$ corner cube removed. A *deficient cube* is a $k \times k \times k$ cube with one $1 \times 1 \times 1$ cube removed.

37. Prove that a deficient $2^n \times 2^n \times 2^n$ cube can be tiled by 3D-septominoes.

38. Prove that if a $k \times k \times k$ deficient cube can be tiled by 3D-septominoes, then 7 divides one of $k - 1, k - 2, k - 4$.

39. Suppose that $S_n = (n + 2)(n - 1)$ is (incorrectly) proposed as a formula for

$$2 + 4 + \cdots + 2n.$$

(a) Show that the Inductive Step is satisfied but that the Basis Step fails.

★ (b) If S_n' is an arbitrary expression that satisfies the Inductive Step, what form must S_n' assume?

★ **40.** What is wrong with the following argument, which allegedly shows that any two positive integers are equal?

> We use induction on n to "prove" that if a and b are positive integers and $n = \max\{a, b\}$, then $a = b$.
>
> **BASIS STEP** $(n = 1)$. If a and b are positive integers and $1 = \max\{a, b\}$, we must have $a = b = 1$.
>
> **INDUCTIVE STEP.** Assume that if a' and b' are positive integers and $n = \max\{a', b'\}$, then $a' = b'$. Suppose that a and b are positive integers and that $n + 1 = \max\{a, b\}$. Now $n = \max\{a - 1, b - 1\}$. By the inductive hypothesis, $a - 1 = b - 1$. Therefore $a = b$.
>
> Since we have verified the Basis Step and the Inductive Step, by the Principle of Mathematical Induction, any two positive integers are equal!

41. What is wrong with the following "proof" that

$$\frac{1}{2} + \frac{2}{3} + \cdots + \frac{n}{n + 1} \neq \frac{n^2}{n + 1}$$

for all $n \geq 2$?

Suppose by way of contradiction that

$$\frac{1}{2} + \frac{2}{3} + \cdots + \frac{n}{n + 1} = \frac{n^2}{n + 1}. \tag{1.6.14}$$

Then also

$$\frac{1}{2} + \frac{2}{3} + \cdots + \frac{n}{n+1} + \frac{n+1}{n+2} = \frac{(n+1)^2}{n+2}.$$

We could prove statement (1.6.14) by induction. In particular, the Inductive Step would give

$$\left(\frac{1}{2} + \frac{2}{3} + \cdots + \frac{n}{n+1}\right) + \frac{n+1}{n+2} = \frac{n^2}{n+1} + \frac{n+1}{n+2}.$$

Therefore

$$\frac{n^2}{n+1} + \frac{n+1}{n+2} = \frac{(n+1)^2}{n+2}.$$

Multiplying each side of this last equation by $(n+1)(n+2)$ gives

$$n^2(n+2) + (n+1)^2 = (n+1)^3.$$

This last equation can be rewritten as

$$n^3 + 2n^2 + n^2 + 2n + 1 = n^3 + 3n^2 + 3n + 1$$

or

$$n^3 + 3n^2 + 2n + 1 = n^3 + 3n^2 + 3n + 1,$$

which is a contradiction. Therefore,

$$\frac{1}{2} + \frac{2}{3} + \cdots + \frac{n}{n+1} \neq \frac{n^2}{n+1}$$

as claimed.

42. Use mathematical induction to prove that

$$\frac{1}{2} + \frac{2}{3} + \cdots + \frac{n}{n+1} < \frac{n^2}{n+1}$$

for all $n \geq 2$. This inequality gives a correct proof of the statement of Exercise 41.

★ 43. Assume the form of mathematical induction where the Inductive Step is: If $S(n)$ is true, then $S(n+1)$ is true. Prove the Well-Ordering Theorem for Positive Integers.

Well-Ordering Theorem for Positive Integers

If X is a nonvoid set of positive integers, X contains a least element.

Assume that there is no positive integer less than 1 and that if n is a positive integer, there is no positive integer between n and $n+1$.

★ 44. Assume the Well-Ordering Theorem for Positive Integers (see Exercise 43). Prove the strong form of the Principle of Mathematical Induction.

★ 45. Show that the strong form of the Principle of Mathematical Induction and the form of mathematical induction where the Inductive Step is "if $S(n)$ is true, then $S(n+1)$ is true" are equivalent; that is, assume the strong form and prove the alternative form, then assume the alternative form and prove the strong form.

46. Show that if 40 coins are distributed among nine bags, at least two bags contain the same number of coins.

★ 47. [Carmony] Suppose that $n > 1$ people are positioned so that each has a unique nearest neighbor. Suppose further that each person has a pie that is hurled at the nearest neighbor. A *survivor* is a person that is not hit by a pie.
   (a) Give an example to show that if n is even, there may be no survivor.
   (b) Use induction on n to show that if n is odd, there is always at least one survivor.

Problem

Define

$$H_k = 1 + \frac{1}{2} + \frac{1}{3} + \cdots + \frac{1}{k} \tag{1}$$

for $k \geq 1$. Prove that

$$H_{2^n} \geq 1 + \frac{n}{2} \tag{2}$$

for all $n \geq 0$.

Attacking the Problem

It's often a good idea to begin by looking at some concrete examples of the expressions under consideration. Let's look at H_k for some small values of k. The smallest value of k for which H_k is defined is $k = 1$. In this case, the last term $1/k$ in the definition of H_k equals $1/1 = 1$. Since the first and last terms coincide,

$$H_1 = 1.$$

For $k = 2$, the last term $1/k$ in the definition of H_k equals $\frac{1}{2}$, so

$$H_2 = 1 + \frac{1}{2}.$$

Similarly, we find that

$$H_3 = 1 + \frac{1}{2} + \frac{1}{3},$$

$$H_4 = 1 + \frac{1}{2} + \frac{1}{3} + \frac{1}{4}.$$

We observe that H_1 appears as the first term of H_2, H_3, and H_4, that H_2 appears as the first two terms of H_3 and H_4, and that H_3 appears as the first three terms of H_4. In general, H_m appears as the first m terms of H_k, if $m \leq k$. This observation will help us later because the Inductive Step in a proof by induction must relate smaller instances of a problem to larger instances of the problem.

In general, it's a good strategy to delay combining terms and simplifying until as late as possible, which is why, for example, we left H_4 as the sum of four terms rather than writing $H_4 = 25/12$. Since we left H_4 as the sum of four terms, we were able to see that each of H_1, H_2, and H_3 appears in the expression for H_4.

Finding a Solution

The Basis Step is to prove the given statement for the smallest value of n, which here is $n = 0$. For $n = 0$, inequality (2) that we must prove becomes

$$H_{2^0} \geq 1 + \frac{0}{2} = 1. \tag{3}$$

We have already observed that $H_1 = 1$. Thus inequality (3) is true when $n = 0$; in fact, the inequality is an equality. (Recall that by definition, if $x = y$ is true, then $x \geq y$ is also true.)

Let's move to the Inductive Step. It's a good idea to write down what is assumed (here the case n)

$$H_{2^n} \geq 1 + \frac{n}{2} \tag{4}$$

and what needs to be proved (here the case $n + 1$)

$$H_{2^{n+1}} \geq 1 + \frac{n+1}{2}. \tag{5}$$

It's also a good idea to write the formulas for any expressions that occur. Using equation (1), we may write

$$H_{2^n} = 1 + \frac{1}{2} + \cdots + \frac{1}{2^n} \tag{6}$$

and

$$H_{2^{n+1}} = 1 + \frac{1}{2} + \cdots + \frac{1}{2^{n+1}}.$$

It's not so evident from the last equation that H_{2^n} appears as the first 2^n terms of $H_{2^{n+1}}$. Let's rewrite the last equation as

$$H_{2^{n+1}} = 1 + \frac{1}{2} + \cdots + \frac{1}{2^n} + \frac{1}{2^n + 1} + \cdots + \frac{1}{2^{n+1}} \tag{7}$$

to make it clear that H_{2^n} appears as the first 2^n terms of $H_{2^{n+1}}$.

For clarity, we have written the term that follows $1/2^n$. Notice that the denominators increase by one, so that the term that follows $1/2^n$ is $1/(2^n+1)$. Also notice that there's a big difference between $1/(2^n + 1)$, the term that follows $1/2^n$, and $1/2^{n+1}$, the last term in equation (7).

Using equations (6) and (7), we may relate H_{2^n} to $H_{2^{n+1}}$ explicitly by writing

$$H_{2^{n+1}} = H_{2^n} + \frac{1}{2^n + 1} + \cdots + \frac{1}{2^{n+1}}. \tag{8}$$

Combining (4) and (8), we obtain

$$H_{2^{n+1}} \geq 1 + \frac{n}{2} + \frac{1}{2^n + 1} + \cdots + \frac{1}{2^{n+1}}. \tag{9}$$

This inequality shows that $H_{2^{n+1}}$ is greater than or equal to

$$1 + \frac{n}{2} + \frac{1}{2^n + 1} + \cdots + \frac{1}{2^{n+1}},$$

but our goal (5) is to show that $H_{2^{n+1}}$ is greater than or equal to $1+(n+1)/2$. We will achieve our goal if we show that

$$1 + \frac{n}{2} + \frac{1}{2^n + 1} + \cdots + \frac{1}{2^{n+1}} \geq 1 + \frac{n+1}{2}.$$

In general, to prove an inequality, we replace terms in the larger expression with smaller terms so that the resulting expression equals the smaller expression, or we replace terms in the smaller expression with larger terms so that the resulting expression equals the larger expression. Here let's replace each of the terms in the sum

$$\frac{1}{2^n + 1} + \cdots + \frac{1}{2^{n+1}}$$

by the smallest term $1/2^{n+1}$ in the sum. We obtain

$$\frac{1}{2^n + 1} + \cdots + \frac{1}{2^{n+1}} \geq \frac{1}{2^{n+1}} + \cdots + \frac{1}{2^{n+1}}.$$

Since there are 2^n terms in the latter sum each equal to $1/2^{n+1}$, we may rewrite the preceding inequality as

$$\frac{1}{2^n + 1} + \cdots + \frac{1}{2^{n+1}} \geq \frac{1}{2^{n+1}} + \cdots + \frac{1}{2^{n+1}} = 2^n \frac{1}{2^{n+1}} = \frac{1}{2}. \tag{10}$$

Combining (9) and (10),

$$H_{2^{n+1}} \geq 1 + \frac{n}{2} + \frac{1}{2} = 1 + \frac{n+1}{2}. \tag{11}$$

We have the desired result, and the Inductive Step is complete.

Formal Solution

The formal solution could be written as follows.

Basis Step $(n = 0)$

$$H_{2^0} = 1 \geq 1 = 1 + \frac{0}{2}$$

Inductive Step.   We assume (2). Now

$$H_{2^{n+1}} = 1 + \frac{1}{2} + \cdots + \frac{1}{2^n} + \frac{1}{2^n + 1} + \cdots + \frac{1}{2^{n+1}}$$

$$= H_{2^n} + \frac{1}{2^n + 1} + \cdots + \frac{1}{2^{n+1}} \geq 1 + \frac{n}{2} + \frac{1}{2^{n+1}} + \cdots + \frac{1}{2^{n+1}}$$

$$= 1 + \frac{n}{2} + 2^n \frac{1}{2^{n+1}} = 1 + \frac{n}{2} + \frac{1}{2} = 1 + \frac{n+1}{2}.$$

Summary of Problem-Solving Techniques

- Look at concrete examples of the expressions under consideration, typically for small values of the variables.

- Look for expressions for small values of n to appear within expressions for larger values of n. In particular, the Inductive Step depends on relating case n to case $n + 1$.

- Delay combining terms and simplifying until as late as possible to help discover relationships among the expressions.

- Write out in full the specific cases to prove, specifically, the smallest value of n for the Basis Step, the case n that is assumed in the Inductive Step, and the case $n + 1$ to prove in the Inductive Step. Write out the formulas for the various expressions that appear.

- To prove an inequality, replace terms in the larger expression with smaller terms so that the resulting expression equals the smaller expression, or replace terms in the smaller expression with larger terms so that the resulting expression equals the larger expression.

Comments

The numbers H_k are called the *harmonic numbers*, and the series

$$1 + \frac{1}{2} + \frac{1}{3} + \cdots$$

which surfaces in calculus is called the *harmonic series*. Inequality (2) shows that the harmonic numbers increase without bound. In calculus terminology, the harmonic series *diverges*.

𝒩 NOTES

General references on discrete mathematics are [Dossey; Graham, 1988; Liu, 1985; Ross; Tucker]. [Knuth, 1973, Vols. 1 and 3; 1981] is the classic reference for much of this material.

[Barker; Copi; Edgar] are introductory logic textbooks. A more advanced treatment is found in [Davis]. The first chapter of the geometry book by [Jacobs] is devoted to basic logic. [Solow] addresses the problem of how to construct proofs. For a history of logic, see [Kline]. The role of logic in reasoning about computer programs is discussed by [Gries].

Tiling with polyominoes is the subject of the book by [Martin].

𝒩 CHAPTER REVIEW

Section 1.1

Logic

Proposition

Conjunction: p and q, $p \wedge q$

Disjunction: p or q, $p \vee q$

Negation: not p, $\overline{p}$

Compound proposition

Truth table

Exclusive-or of propositions p, q:
  p or q, but not both

Section 1.2

Conditional proposition: if p, then q; $p \rightarrow q$

Hypothesis

Conclusion

Necessary condition

Sufficient condition

Converse of $p \rightarrow q$: $q \rightarrow p$

Biconditional proposition:
  p if and only if q, $p \leftrightarrow q$

Logical equivalence: $P \equiv Q$

De Morgan's laws for logic: $\overline{p \vee q} \equiv \overline{p} \wedge \overline{q}$, $\overline{p \wedge q} \equiv \overline{p} \vee \overline{q}$

Contrapositive of $p \rightarrow q$: $\overline{q} \rightarrow \overline{p}$

Section 1.3

Propositional function

Domain of discourse

Universal quantifier

Universally quantified statement

Counterexample

Existential quantifier

Existentially quantified statement

Generalized De Morgan laws for logic:

$$\overline{\forall x,\, P(x)} \text{ and } \exists x,\, \overline{P(x)}$$
have same truth values.

$$\overline{\exists x,\, P(x)} \text{ and } \forall x,\, \overline{P(x)}$$
have same truth values.

To prove that the universally quantified statement

$$\text{for every } x,\, P(x)$$

is true, show that for every x in the domain of discourse, the proposition $P(x)$ is true.

To prove that the existentially quantified statement

$$\text{for some } x,\, P(x)$$

is true, find one value of x in the domain of discourse for which $P(x)$ is true.

To prove that the universally quantified statement

$$\text{for every } x,\, P(x)$$

is false, find one value of x (a counterexample) in the domain of discourse for which $P(x)$ is false.

To prove that the existentially quantified statement

$$\text{for some } x,\, P(x)$$

is false, show that for every x in the domain of discourse, the proposition $P(x)$ is false.

Section 1.4

Mathematical system

Axiom

Definition

Undefined term

Theorem

Proof

Lemma

Direct proof

Proof by contradiction

Indirect proof

Proof by contrapositive

Deductive reasoning

Hypothesis

Premises

Conclusion

Argument

Valid argument

Invalid argument

Section 1.5

Resolution proof; uses: If $p \vee q$ and $\overline{p} \vee r$ are both true, then $q \vee r$ is true.

Clause: consists of terms separated by *or*'s, where each term is a variable or a negation of a variable.

Section 1.6

Principle of Mathematical Induction

Basis Step: prove true for the first instance

Inductive Step: assume true for all instances less than n, then prove true for n

Formula for the sum of the first n positive integers:

$$1 + 2 + \cdots + n = \frac{n(n+1)}{2}$$

Geometric sum:

$$a + ar + ar^2 + \cdots + ar^n$$
$$= \frac{a(r^{n+1} - 1)}{r - 1}$$

CHAPTER SELF-TEST

Section 1.1

1. If p, q, and r are true, find the truth value of the proposition $(p \vee q) \wedge \overline{((\overline{p} \wedge r) \vee q)}$.

2. Write the truth table of the proposition $\overline{(p \wedge q)} \vee (p \vee \overline{r})$.

3. Formulate the proposition $p \wedge (\overline{q} \vee r)$ in words using

p: I take hotel management.
q: I take recreation supervision.
r: I take popular culture.

4. Assume that a, b, and c are real numbers. Represent the statement

$$a < b \text{ or } (b < c \text{ and } a \geq c)$$

symbolically, letting

$$p: \ a < b, \qquad q: \ b < c, \qquad r: \ a < c.$$

Section 1.2

5. Restate the proposition "A necessary condition for Leah to get an A in discrete mathematics is to study hard" in the form of a conditional proposition.

6. Write the converse and contrapositive of the proposition of Exercise 5.

7. If p is true and q and r are false, find the truth value of the proposition

$$(p \vee q) \to \overline{r}.$$

8. Represent the statement

$$\text{If } (a \geq c \text{ or } b < c), \text{ then } b \geq c$$

symbolically using the definitions of Exercise 4.

Section 1.3

9. Is the statement

The team won the 1996 National Basketball Association championship

a proposition? Explain.

10. Is the statement of Exercise 9 a propositional function? Explain.

Let $P(n)$ be the statement
$$n \text{ and } n + 2 \text{ are prime.}$$

In Exercises 11 and 12, write the statement in words and tell whether it is true or false.

11. For all positive integers n, $P(n)$.

12. For some positive integer n, $P(n)$.

Section 1.4

13. Show, by giving a proof by contradiction, that if four teams play seven games, some pair of teams plays at least two times.

14. Distinguish between the terms *axiom* and *definition*.

15. What is the difference between a direct proof and a proof by contradiction?

16. Determine whether the following argument is valid.

$$p \to q \vee r$$
$$p \vee \overline{q}$$
$$\underline{r \vee q}$$
$$\therefore q$$

Section 1.5

17. Find an expression, which is the *and* of clauses, equivalent to $(p \vee q) \to r$.

18. Find an expression, which is the *and* of clauses, equivalent to $(p \vee \overline{q}) \to \overline{r}s$.

19. Use resolution to prove

$$\overline{p} \vee q$$
$$\overline{q} \vee \overline{r}$$
$$\underline{p \vee \overline{r}}$$
$$\therefore \overline{r}$$

20. Reprove Exercise 19 using resolution and proof by contradiction.

Section 1.6

Use mathematical induction to prove that the statements in Exercises 21–24 are true for every positive integer n.

21. $2 + 4 + \cdots + 2n = n(n + 1)$

22. $2^2 + 4^2 + \cdots + (2n)^2 = \dfrac{2n(n + 1)(2n + 1)}{3}$

23. $\dfrac{1}{2!} + \dfrac{2}{3!} + \cdots + \dfrac{n}{(n + 1)!} = 1 - \dfrac{1}{(n + 1)!}$

24. $2^{n+1} < 1 + (n + 1)2^n$

2

THE LANGUAGE OF MATHEMATICS

When I use a word,
it means just what I choose it
to mean—neither more nor less.

—from Alice in Wonderland

This chapter deals with the language of mathematics. The topics, some of which may be familiar, are sets, sequences, number systems, relations, and functions. All of mathematics, as well as subjects that rely on mathematics, such as computer science and engineering, make use of these fundamental concepts.

A set is a collection of objects. Discrete mathematics is concerned with structures such as graphs (sets of vertices and edges) and Boolean algebras (sets with certain operations defined on them).

Unlike a set, a sequence takes order into account. A list of the letters as they appear in a word is an example of a sequence. (Order is obviously important since, for example, *form* and *from* are different words.)

Number systems include the familiar decimal (base 10) system, as well as the binary (base 2) and hexadecimal (base 16) systems.

† This section can be omitted without loss of continuity.

A relation is a set of ordered pairs. The presence of the ordered pair (a, b) in a relation is interpreted as indicating a relationship from a to b. The relational database model that helps users access information in a database (a collection of records manipulated by a computer) is based on the concept of relation.

A function, which is a special kind of relation, assigns to each member of a set X exactly one member of a set Y. Functions are used extensively in discrete mathematics; for example, functions are used to analyze the time needed to execute algorithms.

2.1 SETS

The concept of set is basic to all of mathematics and mathematical applications. A **set** is simply any collection of objects. If a set is finite and not too large, we can describe it by listing the elements in it. For example, the equation

$$A = \{1, 2, 3, 4\} \tag{2.1.1}$$

describes a set A made up of the four elements 1, 2, 3, and 4. A set is determined by its elements and not by any particular order in which the elements might be listed. Thus A might just as well be specified as

$$A = \{1, 3, 4, 2\} \, .$$

The elements making up a set are assumed to be distinct, and although for some reason we may have duplicates in our list, only one occurrence of each element is in the set. For this reason we may also describe the set A defined in (2.1.1) as

$$A = \{1, 2, 2, 3, 4\} \, .$$

If a set is a large finite set or an infinite set, we can describe it by listing a property necessary for membership. For example, the equation

$$B = \{x \mid x \text{ is a positive, even integer }\} \tag{2.1.2}$$

describes the set B made up of all positive, even integers; that is, B consists of the integers 2, 4, 6, and so on. The vertical bar "$\mid$" is read "such that." Equation (2.1.2) would be read "B equals the set of all x such that x is a positive, even integer." Here the property necessary for membership is "is a positive, even integer." Note that the property appears after the vertical bar.

If X is a finite set, we let

$$|X| = \text{ number of elements in } X.$$

Given a description of a set X such as (2.1.1) or (2.1.2) and an element x, we can determine whether or not x belongs to X. If the members of X are listed as in (2.1.1), we simply look to see whether or not x appears in the listing. In a description such as (2.1.2), we check to see whether the element x has the property listed. If x is in the set X, we write $x \in X$, and if x is not in X, we write $x \notin X$. For example, if $x = 1$, then $x \in A$, but $x \notin B$, where A and B are given by equations (2.1.1) and (2.1.2).

The set with no elements is called the **empty** (or **null** or **void**) **set** and is denoted $\emptyset$. Thus $\emptyset = \{ \}$.

Two sets X and Y are **equal** and we write $X = Y$ if X and Y have the same elements. To put it another way, $X = Y$ if whenever $x \in X$, then $x \in Y$ and whenever $x \in Y$, then $x \in X$.

EXAMPLE 2.1.1

If

$$A = \left\{ x \mid x^2 + x - 6 = 0 \right\}, \qquad B = \{2, -3\},$$

then $A = B$. □

Suppose that X and Y are sets. If every element of X is an element of Y, we say that X is a **subset** of Y and write $X \subseteq Y$.

EXAMPLE 2.1.2

If

$$C = \{1, 3\} \qquad \text{and} \qquad A = \{1, 2, 3, 4\},$$

then C is a subset of A. □

Any set X is a subset of itself, since any element in X is in X. If X is a subset of Y and X does not equal Y, we say that X is a **proper subset** of Y. The empty set is a subset of every set (see Exercise 56). The set of all subsets (proper or not) of a set X, denoted $\mathcal{P}(X)$, is called the **power set** of X.

EXAMPLE 2.1.3

If $A = \{a, b, c\}$, the members of $\mathcal{P}(A)$ are

$$\emptyset, \{a\}, \{b\}, \{c\}, \{a, b\}, \{a, c\}, \{b, c\}, \{a, b, c\}.$$

All but $\{a, b, c\}$ are proper subsets of A. For this example,

$$|A| = 3, \quad |\mathcal{P}(A)| = 2^3 = 8.$$ □

We give a proof using mathematical induction that the result of Example 2.1.3 holds in general; that is, the power set of a set with n elements has 2^n elements.

THEOREM 2.1.4

If $|X| = n$, then

$$|\mathcal{P}(X)| = 2^n. \tag{2.1.3}$$

Proof. The proof is by induction on n.

BASIS STEP. If $n = 0$, X is the empty set. The only subset of the empty set is the empty set itself; thus

$$|\mathcal{P}(X)| = 1 = 2^0 = 2^n.$$

Thus (2.1.3) is true for $n = 0$.

INDUCTIVE STEP.   Assume that (2.1.3) holds for n. Let X be a set with $n + 1$ elements. Choose $x \in X$. We claim that exactly half of the subsets of X contain x, and exactly half of the subsets of X do not contain x. To see this, notice that each subset S of X that contains x can be paired uniquely with the subset obtained by removing x from S (see Figure 2.1.1). Thus exactly half of the subsets of X contain x, and exactly half of the subsets of X do not contain x.

If we let Y be the set obtained from X by removing x, Y has n elements. By the inductive assumption, $|\mathcal{P}(Y)| = 2^n$. But the subsets of Y are precisely the subsets of X that do not contain x. From the argument in the preceding paragraph, we conclude that

$$|\mathcal{P}(Y)| = \frac{|\mathcal{P}(X)|}{2}.$$

Therefore,

$$|\mathcal{P}(X)| = 2\,|\mathcal{P}(Y)| = 2 \cdot 2^n = 2^{n+1}.$$

Thus (2.1.3) holds for $n + 1$ and the inductive step is complete. By the Principle of Mathematical Induction, (2.1.3) holds for all $n \geq 0$. ∎

In Section 4.1 (see Example 4.1.4) we will give another proof of Theorem 2.1.4.

Given two sets X and Y, there are various ways to combine X and Y to form a new set. The set

$$X \cup Y = \{x \mid x \in X \text{ or } x \in Y\}$$

is called the **union** of X and Y. The union consists of all elements belonging to either X or Y (or both).

The set

$$X \cap Y = \{x \mid x \in X \text{ and } x \in Y\}$$

is called the **intersection** of X and Y. The intersection consists of all elements belonging to both X and Y.

Sets X and Y are **disjoint** if $X \cap Y = \emptyset$. A collection of sets $\mathcal{S}$ is said to be **pairwise disjoint** if whenever X and Y are distinct sets in $\mathcal{S}$, X and Y are disjoint.

The set

$$X - Y = \{x \mid x \in X \text{ and } x \notin Y\}$$

is called the **difference** (or **relative complement**). The difference $X - Y$ consists of all elements in X that are not in Y.

Subsets of X that contain a	Subsets of X that do not contain a
$\{a\}$	$\emptyset$
$\{a, b\}$	$\{b\}$
$\{a, c\}$	$\{c\}$
$\{a, b, c\}$	$\{b, c\}$

FIGURE 2.1.1
Subsets of $X = \{a, b, c\}$ divided into two classes: those that contain a and those that do not contain a. Each subset in the right column is obtained from the corresponding subset in the left column by deleting the element a from it.

EXAMPLE 2.1.5

If $A = \{1, 3, 5\}$ and $B = \{4, 5, 6\}$, then

$$A \cup B = \{1, 3, 4, 5, 6\}$$

$$A \cap B = \{5\}$$

$$A - B = \{1, 3\}$$

$$B - A = \{4, 6\}.$$ □

EXAMPLE 2.1.6

The sets

$$\{1, 4, 5\} \quad \text{and} \quad \{2, 6\}$$

are disjoint. The collection of sets

$$\mathcal{S} = \{\{1, 4, 5\}, \{2, 6\}, \{3\}, \{7, 8\}\}$$

is pairwise disjoint. □

Sometimes we are dealing with sets all of which are subsets of a set U. This set U is called a **universal set** or a **universe**. The set U must be explicitly given or inferred from the context. Given a universal set U and a subset X of U, the set $U - X$ is called the **complement** of X and is written $\overline{X}$.

EXAMPLE 2.1.7

Let $A = \{1, 3, 5\}$. If U, a universal set, is specified as $U = \{1, 2, 3, 4, 5\}$, then $\overline{A} = \{2, 4\}$. If, on the other hand, a universal set is specified as $U = \{1, 3, 5, 7, 9\}$, then $\overline{A} = \{7, 9\}$. The complement obviously depends on the universe in which we are working. □

Our next theorem summarizes some useful properties of sets. The proof is left to the reader (see Exercise 70).

THEOREM 2.1.8

Let U be a universal set and let A, B, and C be subsets of U. The following properties hold.

(a) *Associative laws:*

$$(A \cup B) \cup C = A \cup (B \cup C), \quad (A \cap B) \cap C = A \cap (B \cap C)$$

(b) *Commutative laws:*

$$A \cup B = B \cup A, \quad A \cap B = B \cap A$$

(c) *Distributive laws:*

$$A \cap (B \cup C) = (A \cap B) \cup (A \cap C)$$
$$A \cup (B \cap C) = (A \cup B) \cap (A \cup C)$$

(d) *Identity laws:*

$$A \cup \emptyset = A, \quad A \cap U = A$$

(e) Complement laws:

$$A \cup \overline{A} = U, \quad A \cap \overline{A} = \emptyset$$

(f) Idempotent laws:

$$A \cup A = A, \quad A \cap A = A$$

(g) Bound laws:

$$A \cup U = U, \quad A \cap \emptyset = \emptyset$$

(h) Absorption laws:

$$A \cup (A \cap B) = A, \quad A \cap (A \cup B) = A$$

(i) Involution law:

$$\overline{\overline{A}} = A$$

(h) 0/1 laws:

$$\overline{\emptyset} = U, \quad \overline{U} = \emptyset$$

(i) De Morgan's laws for sets:

$$\overline{(A \cup B)} = \overline{A} \cap \overline{B}, \quad \overline{(A \cap B)} = \overline{A} \cup \overline{B}$$

Proof. See Exercise 70.   ∎

We define the union of an arbitrary family $\mathcal{S}$ of sets to be those elements x belonging to at least one set X in $\mathcal{S}$. Formally,

$$\cup \mathcal{S} = \{x \mid x \in X \text{ for some } X \in \mathcal{S}\}.$$

Similarly, we define the intersection of an arbitrary family $\mathcal{S}$ of sets to be those elements x belonging to every set X in $\mathcal{S}$. Formally,

$$\cap \mathcal{S} = \{x \mid x \in X \text{ for all } X \in \mathcal{S}\}.$$

If

$$\mathcal{S} = \{A_1, A_2, \ldots, A_n\},$$

we write

$$\bigcup \mathcal{S} = \bigcup_{i=1}^{n} A_i, \quad \bigcap \mathcal{S} = \bigcap_{i=1}^{n} A_i$$

and if

$$\mathcal{S} = \{A_1, A_2, \ldots\},$$

we write

$$\bigcup \mathcal{S} = \bigcup_{i=1}^{\infty} A_i, \quad \bigcap \mathcal{S} = \bigcap_{i=1}^{\infty} A_i.$$

EXAMPLE 2.1.9

If

$$A_n = \{n, n+1, \ldots\} \qquad \text{and} \qquad S = \{A_1, A_2, \ldots\},$$

then

$$\bigcup_{i=1}^{\infty} A_i = \bigcup S = \{1, 2, \ldots\}, \qquad \bigcap_{i=1}^{\infty} A_i = \bigcap S = \emptyset. \qquad \square$$

A partition of a set X divides X into nonoverlapping subsets. More formally, a collection S of nonempty subsets of X is said to be a **partition** of the set X if every element in X belongs to exactly one member of S. Notice that if S is a partition of X, S is pairwise disjoint and $\cup S = X$.

EXAMPLE 2.1.10

Since each element of

$$X = \{1, 2, 3, 4, 5, 6, 7, 8\}$$

is in exactly one member of

$$S = \{\{1, 4, 5\}, \{2, 6\}, \{3\}, \{7, 8\}\},$$

S is a partition of X. $\square$

At the beginning of this section we pointed out that a set is an unordered collection of elements, that is, a set is determined by its elements and not by any particular order in which the elements are listed. Sometimes, however, we do want to take order into account. An **ordered pair** of elements, written (a, b), is considered distinct from the ordered pair (b, a), unless, of course, $a = b$. To put it another way, $(a, b) = (c, d)$ if and only if $a = c$ and $b = d$. If X and Y are sets, we let $X \times Y$ denote the set of all ordered pairs (x, y) where $x \in X$ and $y \in Y$. We call $X \times Y$ the **Cartesian product** of X and Y.

EXAMPLE 2.1.11

If $X = \{1, 2, 3\}$ and $Y = \{a, b\}$, then

$$X \times Y = \{(1, a), (1, b), (2, a), (2, b), (3, a), (3, b)\}$$

$$Y \times X = \{(a, 1), (b, 1), (a, 2), (b, 2), (a, 3), (b, 3)\}$$

$$X \times X = \{(1, 1), (1, 2), (1, 3), (2, 1), (2, 2), (2, 3), (3, 1), (3, 2), (3, 3)\}$$

$$Y \times Y = \{(a, a), (a, b), (b, a), (b, b)\}. \qquad \square$$

Example 2.1.11 shows that, in general, $X \times Y \neq Y \times X$. Notice that $|X \times Y| = |X| \cdot |Y|$.

EXAMPLE 2.1.12

A restaurant serves four appetizers

$$r = \text{ribs}, \quad n = \text{nachos}, \quad s = \text{shrimp}, \quad f = \text{fried cheese}$$

and three main courses

$$c = \text{chicken}, \quad b = \text{beef}, \quad t = \text{trout}.$$

If we let $A = \{r, n, s, f\}$ and $M = \{c, b, t\}$, the Cartesian product $A \times M$ lists the 12 possible dinners consisting of one appetizer and one main course.   □

Ordered lists need not be restricted to two elements. An ***n*-tuple**, written $(a_1, a_2, \ldots, a_n)$, takes order into account:

$$(a_1, a_2, \ldots, a_n) = (b_1, b_2, \ldots, b_n)$$

if and only if

$$a_1 = b_1, a_2 = b_2, \ldots, a_n = b_n.$$

The Cartesian product of sets $X_1, X_2, \ldots, X_n$ is defined to be the set of all n-tuples $(x_1, x_2, \ldots, x_n)$ where $x_i \in X_i$ for $i = 1, \ldots, n$.

EXAMPLE 2.1.13

If

$$X = \{1, 2\}, \qquad Y = \{a, b\}, \qquad Z = \{\alpha, \beta\},$$

then

$$X \times Y \times Z = \{(1, a, \alpha), (1, a, \beta), (1, b, \alpha), (1, b, \beta), (2, a, \alpha),$$
$$(2, a, \beta), (2, b, \alpha), (2, b, \beta)\}.$$

□

Notice that in Example 2.1.13, $|X \times Y \times Z| = |X| \cdot |Y| \cdot |Z|$. In general, we have

$$|X_1 \times X_2 \times \cdots \times X_n| = |X_1| \cdot |X_2| \cdots |X_n|. \tag{2.1.4}$$

This last statement can be proved by induction on the number n of sets (see Exercise 71).

EXAMPLE 2.1.14

If A is a set of appetizers, M is a set of main courses, and D is a set of desserts, the Cartesian product $A \times M \times D$ lists all possible dinners consisting of one appetizer, one main course, and one dessert.   □

Exercises

In Exercises 1–16, let the universe be the set $U = \{1, 2, 3, \ldots, 10\}$. Let $A = \{1, 4, 7, 10\}$, $B = \{1, 2, 3, 4, 5\}$, and $C = \{2, 4, 6, 8\}$. List the elements of each set.

1. $A \cup B$ **2.** $B \cap C$

3. $A - B$ **4.** $B - A$

5. $\overline{A}$ **6.** $U - C$

7. $\overline{U}$ **8.** $A \cup \emptyset$

9. $B \cap \emptyset$ **10.** $A \cup U$

11. $B \cap U$ **12.** $A \cap (B \cup C)$

13. $\overline{B} \cap (C - A)$ **14.** $(A \cap B) - C$

15. $\overline{A \cap B} \cup C$ **16.** $(A \cup B) - (C - B)$

In Exercises 17–20, let $X = \{1, 2\}$ and $Y = \{a, b, c\}$. List the elements in each set.

17. $X \times Y$ **18.** $Y \times X$

19. $X \times X$ **20.** $Y \times Y$

In Exercises 21–24, let $X = \{1, 2\}$, $Y = \{a\}$, and $Z = \{\alpha, \beta\}$. List the elements of each set.

21. $X \times Y \times Z$ **22.** $X \times Y \times Y$

23. $X \times X \times X$ **24.** $Y \times X \times Y \times Z$

In Exercises 25–28, list all partitions of the set.

25. $\{1\}$ **26.** $\{1, 2\}$

27. $\{a, b, c\}$ **28.** $\{a, b, c, d\}$

In Exercises 29–32, answer true or false.

29. $\{x\} \subseteq \{x\}$ **30.** $\{x\} \in \{x\}$

31. $\{x\} \in \{x, \{x\}\}$ **32.** $\{x\} \subseteq \{x, \{x\}\}$

In Exercises 33–37, determine whether each pair of sets is equal.

33. $\{1, 2, 3\}$, $\{1, 3, 2\}$

34. $\{1, 2, 2, 3\}$, $\{1, 2, 3\}$

35. $\{1, 1, 3\}$, $\{3, 3, 1\}$

36. $\{x \mid x^2 + x = 2\}$, $\{1, -2\}$

37. $\{x \mid x$ is a real number and $0 < x \leq 2\}$, $\{1, 2\}$

38. List the members of $\mathcal{P}(\{a, b\})$. Which are proper subsets of $\{a, b\}$?

39. List the members of $\mathcal{P}(\{a, b, c, d\})$. Which are proper subsets of $\{a, b, c, d\}$?

40. If X has 10 members, how many members does $\mathcal{P}(X)$ have? How many proper subsets does X have?

41. If X has n members, how many proper subsets does X have?

42. If X and Y are nonempty sets and $X \times Y = Y \times X$, what can we conclude about X and Y?

In each of Exercises 43–55, write "true" if the statement is true; otherwise, give a counterexample. The sets X, Y, and Z are subsets of a universal set U. Assume that the universe for Cartesian products is $U \times U$.

43. For any sets X and Y, either X is a subset of Y or Y is a subset of X.

44. $X \cap (Y - Z) = (X \cap Y) - (X \cap Z)$ for all sets X, Y, and Z

45. $(X - Y) \cap (Y - X) = \emptyset$ for all sets X and Y

46. $X - (Y \cup Z) = (X - Y) \cup Z$ for all sets X, Y, and Z

47. $\overline{X - Y} = \overline{Y - X}$ for all sets X and Y

48. $\overline{X \cap Y} \subseteq X$ for all sets X and Y

49. $(X \cap Y) \cup (Y - X) = X$ for all sets X and Y

50. $X \times (Y \cup Z) = (X \times Y) \cup (X \times Z)$ for all sets X, Y, and Z

51. $\overline{X \times Y} = \overline{X} \times \overline{Y}$ for all sets X and Y

52. $X \times (Y - Z) = (X \times Y) - (X \times Z)$ for all sets X, Y, and Z

53. $X - (Y \times Z) = (X - Y) \times (X - Z)$ for all sets X, Y, and Z

54. $X \cap (Y \times Z) = (X \cap Y) \times (X \cap Z)$ for all sets X, Y, and Z

55. $X \times \emptyset = \emptyset$ for any set X

56. Show that for any set X, $\emptyset \subseteq X$.

For each condition in Exercises 57–60, what relation must hold between sets A and B?

57. $A \cap B = A$ **58.** $A \cup B = A$

59. $\overline{A} \cap U = \emptyset$ **60.** $\overline{A \cap B} = \overline{B}$

The *symmetric difference* of two sets A and B is the set

$$A \triangle B = (A \cup B) - (A \cap B).$$

61. If $A = \{1, 2, 3\}$ and $B = \{2, 3, 4, 5\}$, find $A \triangle B$.

62. Describe the symmetric difference of sets A and B in words.

63. Given a universe U, describe $A \triangle A$, $A \triangle \overline{A}$, $U \triangle A$, and $\emptyset \triangle A$.

64. Prove or disprove: If A, B, and C are sets satisfying $A \triangle C = B \triangle C$, then $A = B$.

65. Show that

$$|A \cup B| = |A| + |B| - |A \cap B|.$$

66. Find a formula for $|A \cup B \cup C|$ similar to the formula of Exercise 65. Show that your formula holds for all sets A, B, and C.

67. Let C be a circle and let $\mathcal{D}$ be the set of all diameters of C. What is $\cap \mathcal{D}$?

68. Let P denote the set of integers greater than 1. For $i \geq 2$, define

$$X_i = \{ik \mid k \geq 2, k \in P\}.$$

Describe $P - \displaystyle\bigcup_{i=2}^{\infty} X_i$.

69. Use induction to show that if $X_1, \ldots, X_n$ and X are sets, then

(a) $X \cap (X_1 \cup X_2 \cup \cdots \cup X_n) = (X \cap X_1) \cup (X \cap X_2) \cup \cdots \cup (X \cap X_n)$.

(b) $\overline{X_1 \cap X_2 \cap \cdots \cap X_n} = \overline{X_1} \cup \overline{X_2} \cup \cdots \cup \overline{X_n}$.

70. Prove Theorem 2.1.8.

71. Use induction to prove statement (2.1.4).

2.2 SEQUENCES AND STRINGS

Blue Taxi Inc. charges \$1 for the first mile and 50 cents for each additional mile. The adjacent table shows the cost of traveling from 1 to 10 miles. In general, the cost C_n of traveling n miles is 1.00 (the cost of traveling the first mile) plus 0.50 times the number $(n - 1)$ of additional miles. That is,

$$C_n = 1 + 0.5(n - 1).$$

As examples,

$$C_1 = 1 + 0.5(1 - 1) = 1 + 0.5 \cdot 0 = 1,$$
$$C_5 = 1 + 0.5(5 - 1) = 1 + 0.5 \cdot 4 = 1 + 2 = 3.$$

Mileage	Cost
1	\$1.00
2	1.50
3	2.00
4	2.50
5	3.00
6	3.50
7	4.00
8	4.50
9	5.00
10	5.50

   A **sequence** is a list in which order is taken into account. In the previous example, the list of fares

$$1.00, \quad 1.50, \quad 2.00, \quad 2.50, \quad 3.00, \ldots$$

is a sequence. Notice that order is indeed important. For example, if the first and fifth numbers were interchanged, a one-mile fare would then cost \$3.00— quite different from a \$1.00 one-mile fare.

   If s is a sequence, we frequently denote the first element of the sequence as s_1, the second element of the sequence as s_2, and so on. In general, s_n denotes the nth element of the sequence. We call n the **index** of the sequence.

EXAMPLE 2.2.1

The ordered list

$$2, \quad 4, \quad 6, \ldots, \quad 2n, \ldots$$

is a sequence. The first element of the sequence is 2, the second element of the sequence is 4, and so on. The nth element of the sequence is $2n$. If we let s denote this sequence, we have

$$s_1 = 2, \quad s_2 = 4, \quad s_3 = 6, \ldots, \quad s_n = 2n, \ldots. \qquad \square$$

EXAMPLE 2.2.2

The ordered list

$$a, \quad a, \quad b, \quad a, \quad b$$

is a sequence. The first element of the sequence is a, the second element of the sequence is a, and so on. If we let t denote this sequence, we have

$$t_1 = a, \quad t_2 = a, \quad t_3 = b, \quad t_4 = a, \quad t_5 = b. \qquad \square$$

Example 2.2.2 shows that a sequence (unlike a set) can have repetitions. A sequence may have an infinite number of elements (such as the sequence of Example 2.2.1) or a finite number of elements (such as the sequence of Example 2.2.2).

An alternative notation for the sequence s is $\{s_n\}$. Here s or $\{s_n\}$ denotes the entire sequence

$$s_1, \quad s_2, \quad s_3, \quad \ldots.$$

We use the notation s_n to denote the single, nth element of the sequence s.

EXAMPLE 2.2.3

Define a sequence $\{t_n\}$ by the rule

$$t_n = n^2 - 1, \qquad n \geq 1.$$

The first five terms of this sequence are

$$0, \quad 3, \quad 8, \quad 15, \quad 24.$$

The 55th term is

$$t_{55} = 55^2 - 1 = 3024. \qquad \square$$

EXAMPLE 2.2.4

Define a sequence u by the rule u_n, is the nth letter in the word *digital*. Then $u_1 = d$, $u_2 = u_4 = i$, and $u_7 = l$. This sequence is a finite sequence. $\square$

Although in this book we often denote the first element of a sequence s as s_1, in general, the first element may be indexed by any integer. For example, if v is a sequence whose first element is v_0, the elements of v would be

$$v_0, \quad v_1, \quad v_2, \quad \ldots$$

When we want to explicitly mention the initial index of an infinite sequence s, we write $\{s_n\}_{n=1}^{\infty}$. An infinite sequence v whose initial index is 0 is denoted $\{v_n\}_{n=0}^{\infty}$. A finite sequence x indexed from -1 to 4 is denoted $\{x_n\}_{n=-1}^{4}$.

EXAMPLE 2.2.5

If x is the sequence defined by

$$x_n = 1/2^n, \qquad -1 \leq n \leq 4,$$

the elements of x are

$$2, \quad 1, \quad 1/2, \quad 1/4, \quad 1/8, \quad 1/16. \qquad \square$$

Two important types of sequences are increasing sequences and decreasing sequences. [†] A sequence s is **increasing** if $s_n \leq s_{n+1}$ for all n. A sequence s is **decreasing** if $s_n \geq s_{n+1}$ for all n. Notice that in both definitions, equality between successive terms of the sequence is allowed.

EXAMPLE 2.2.6

The sequence, $2, 4, 6, \ldots$, of Example 2.2.1 is increasing since $s_n = 2n \leq 2(n+1) = s_{n+1}$ for all n. □

EXAMPLE 2.2.7

The sequence, $2, 1, 1/2, \ldots$, of Example 2.2.5 is decreasing since $x_n = 1/2^n \geq 1/2^{n+1} = x_{n+1}$ for all n. □

EXAMPLE 2.2.8

The sequence s

$$3, \quad 5, \quad 5, \quad 7, \quad 8, \quad 8, \quad 13$$

is increasing since $s_n \leq s_{n+1}$ for all n. Notice that $s_2 = s_3$ and $s_5 = s_6$ (we assume that the index of the first term in the sequence is 1). Equality is allowed in the definition of *increasing*. □

One way to form a new sequence from a given sequence is to retain only certain terms of the original sequence, maintaining the order of terms in the given sequence. The resulting sequence is called a **subsequence** of the original sequence.

DEFINITION 2.2.9

Let $\{s_n\}$ be a sequence defined for $n = m, m+1, \ldots$, and let $n_1, n_2, \ldots$ be an increasing sequence satisfying $n_k < n_{k+1}$, for all k, whose values are in the set $\{m, m+1, \ldots\}$. We call the sequence $\{s_{n_k}\}$ a *subsequence* of $\{s_n\}$.

EXAMPLE 2.2.10

The sequence

$$b, c \tag{2.2.1}$$

is a subsequence of the sequence

$$t_1 = a, \quad t_2 = a, \quad t_3 = b, \quad t_4 = c, \quad t_5 = q. \tag{2.2.2}$$

[†] In some books, what we call *increasing* is called *nondecreasing*, and what we call *decreasing* is called *nonincreasing*.

Subsequence (2.2.1) is obtained from sequence (2.2.2) by choosing the third and fourth terms. The expression n_k of Definition 2.2.9 tells us which terms of (2.2.2) to choose to obtain subsequence (2.2.1); thus, $n_1 = 3$, $n_2 = 4$. The subsequence (2.2.1) is

$$t_3, t_4 \qquad \text{or} \qquad t_{n_1}, t_{n_2}.$$

Notice that the sequence

$$c, b$$

is *not* a subsequence of sequence (2.2.2) since the order of terms in the sequence (2.2.2) is not maintained.   □

EXAMPLE 2.2.11

The sequence

$$2, \quad 4, \quad 8, \quad 16, \quad \ldots, \quad 2^k, \quad \ldots \tag{2.2.3}$$

is a subsequence of the sequence

$$2, \quad 4, \quad 6, \quad 8, \quad 10, \quad 12, \quad 14, \quad 16, \ldots, \quad 2n, \ldots. \tag{2.2.4}$$

Subsequence (2.2.3) is obtained from sequence (2.2.4) by choosing the first, second, fourth, eighth, and so on, terms; thus the value of n_k of Definition 2.2.9 is $n_k = 2^{k-1}$. If we define sequence (2.2.4) by $s_n = 2n$, the subsequence (2.2.3) is defined by

$$s_{n_k} = s_{2^{k-1}} = 2 \cdot 2^{k-1} = 2^k.$$ □

Two important ways to create new sequences from numerical sequences are by adding and by multiplying the terms together.

DEFINITION 2.2.12

If $\{a_i\}_{i=m}^n$ is a sequence, we define

$$\sum_{i=m}^n a_i = a_m + a_{m+1} + \cdots + a_n, \qquad \prod_{i=m}^n a_i = a_m \cdot a_{m+1} \cdots a_n.$$

The formalism

$$\sum_{i=m}^n a_i \tag{2.2.5}$$

is called the *sum* (or *sigma*) *notation* and

$$\prod_{i=m}^n a_i \tag{2.2.6}$$

is called the *product notation*.

In (2.2.5) or (2.2.6), i is called the *index*, m is called the *lower limit*, and n is called the *upper limit*.

EXAMPLE 2.2.13

Let a be the sequence defined by $a_n = 2n$, $n \geq 1$. Then

$$\sum_{i=1}^{3} a_i = a_1 + a_2 + a_3 = 2 + 4 + 6 = 12,$$

$$\prod_{i=1}^{3} a_i = a_1 \cdot a_2 \cdot a_3 = 2 \cdot 4 \cdot 6 = 48.$$ $\square$

EXAMPLE 2.2.14

The geometric sum (see Example 1.6.2)

$$a + ar + ar^2 + \cdots + ar^n$$

can be rewritten compactly using the sum notation as

$$\sum_{i=0}^{n} ar^i.$$ $\square$

The name of the index in (2.2.5) or (2.2.6) is irrelevant. For example,

$$\sum_{i=1}^{n} a_i = \sum_{j=1}^{n} a_j \quad \text{and} \quad \prod_{i=1}^{n} a_i = \prod_{x=1}^{n} a_x.$$

It is sometimes useful not only to change the name of the index, but to change its limits as well. (The process is analogous to changing the variable in an integral.)

EXAMPLE 2.2.15   *Changing the Index and Limits in a Sum*

Rewrite the sum

$$\sum_{i=0}^{n} i r^{n-i},$$

replacing the index i by j, where $i = j - 1$.

Since $i = j - 1$, the term $i r^{n-i}$ becomes

$$(j - 1) r^{n-(j-1)} = (j - 1) r^{n-j+1}.$$

Since $j = i + 1$, when $i = 0$, $j = 1$. Thus the lower limit for j is 1. Similarly, when $i = n$, $j = n + 1$, and the upper limit for j is $n + 1$. Therefore,

$$\sum_{i=0}^{n} i r^{n-i} = \sum_{j=1}^{n+1} (j - 1) r^{n-j+1}.$$ $\square$

EXAMPLE 2.2.16

Let a be the sequence defined by the rule $a_n = 2(-1)^n$, $n \geq 0$. Find a formula for the sequence s defined by

$$s_n = \sum_{i=0}^{n} a_i.$$

We find that

$$s_n = 2(-1)^0 + 2(-1)^1 + 2(-1)^2 + \cdots + 2(-1)^n$$
$$= 2 - 2 + 2 - \cdots \pm 2 = \begin{cases} 2, & \text{if } n \text{ is even} \\ 0, & \text{if } n \text{ is odd.} \end{cases} \qquad \square$$

Sometimes the sum and product notations are modified to denote sums and products indexed over arbitrary sets of integers. Formally, if S is a set of integers and a is a sequence,

$$\sum_{i \in S} a_i$$

denotes the sum of the elements $\{a_i \mid i \in S\}$. Similarly,

$$\prod_{i \in S} a_i$$

denotes the product of the elements $\{a_i \mid i \in S\}$.

EXAMPLE 2.2.17

If S denotes the set of prime numbers less than 20,

$$\sum_{i \in S} \frac{1}{i} = \frac{1}{2} + \frac{1}{3} + \frac{1}{5} + \frac{1}{7} + \frac{1}{11} + \frac{1}{13} + \frac{1}{17} + \frac{1}{19} = 1.455. \qquad \square$$

In certain contexts, a finite sequence is called a **string**.

DEFINITION 2.2.18

A *string over* X is a finite sequence of elements from X.

EXAMPLE 2.2.19

Let $X = \{a, b, c\}$. If we let

$$\beta_1 = b, \quad \beta_2 = a, \quad \beta_3 = a, \quad \beta_4 = c,$$

we obtain a string over X. This string is written *baac*. $\qquad \square$

Since a string is a sequence, order is taken into account. For example, the string *baac* is different from the string *acab*.

Repetitions in a string can be specified by superscripts. For example, the string *bbaaac* may be written b^2a^3c.

The string with no elements is called the **null string** and is denoted λ. We let X^* denote the set of all strings over X, including the null string, and we let X^+ denote the set of all nonnull strings over X.

EXAMPLE 2.2.20

Let $X = \{a, b\}$. Some elements in X^* are

$$\lambda, \quad a, \quad b, \quad abab, \quad b^{20}a^5ba. \qquad \square$$

The **length** of a string α is the number of elements in α. The length of α is denoted $|\alpha|$.

EXAMPLE 2.2.21

If $\alpha = aabab$ and $\beta = a^3b^4a^{32}$, then

$$|\alpha| = 5 \quad \text{and} \quad |\beta| = 39. \qquad \square$$

If α and β are two strings, the string consisting of α followed by β, written $\alpha\beta$, is called the **concatenation** of α and β.

EXAMPLE 2.2.22

If $\gamma = aab$ and $\theta = cabd$, then

$$\gamma\theta = aabcabd, \quad \theta\gamma = cabdaab, \quad \gamma\lambda = \gamma = aab, \quad \lambda\gamma = \gamma = aab. \qquad \square$$

Exercises

1. Answer (a)–(c) for the sequence s defined by

$$c, d, d, c, d, c.$$

(a) Find s_1. (b) Find s_4.

(c) Write s as a string.

2. Answer (a)–(k) for the sequence t defined by

$$t_n = 2n - 1, \qquad n \geq 1.$$

(a) Find t_3.
(b) Find t_7.
(c) Find t_{100}.
(d) Find t_{2077}.
(e) Find $\sum_{i=1}^{3} t_i$.
(f) Find $\sum_{i=3}^{7} t_i$.
(g) Find $\prod_{i=1}^{3} t_i$.
(h) Find $\prod_{i=3}^{6} t_i$.
(i) Find a formula that represents this sequence as a sequence whose lower index is 0.
(j) Is t increasing?
(k) Is t decreasing?

3. Answer (a)–(f) for the sequence v defined by

$$v_n = (n-1) \cdots 2 \cdot 1 + 2, \qquad n \geq 1.$$

(a) Find v_3.
(b) Find v_4.
(c) Find $\sum_{i=1}^{4} v_i$.
(d) Find $\sum_{i=3}^{3} v_i$.
(e) Is v increasing?
(f) Is v decreasing?

4. Compute the given quantity using the sequence a defined by

$$a_n = n^2 - 3n + 3, \qquad n \geq 1.$$

(a) $\sum_{i=1}^{4} a_i$
(b) $\sum_{j=3}^{5} a_j$
(c) $\sum_{i=4}^{4} a_i$
(d) $\sum_{k=1}^{6} a_k$
(e) $\prod_{i=1}^{2} a_i$
(f) $\prod_{i=1}^{3} a_i$
(g) $\prod_{n=2}^{3} a_n$
(h) $\prod_{x=3}^{4} a_x$

5. Answer (a) and (b) for the sequence a of Exercise 4.

(a) Is a increasing?
(b) Is a decreasing?

6. Answer (a)–(f) for the sequence b defined by $b_n = n(-1)^n$.

(a) Find $\sum_{i=1}^{4} b_i$.
(b) Find $\sum_{i=1}^{10} b_i$
(c) Find a formula for the sequence c defined by

$$c_n = \sum_{i=1}^{n} b_i.$$

(d) Find a formula for the sequence d defined by

$$d_n = \prod_{i=1}^{n} b_i.$$

(e) Is b increasing?
(f) Is b decreasing?

7. Answer (a)–(f) for the sequence Ω defined by $\Omega_n = 3$ for all n.

(a) Find $\displaystyle\sum_{i=1}^{3} \Omega_i$.

(b) Find $\displaystyle\sum_{i=1}^{10} \Omega_i$.

(c) Find a formula for the sequence c defined by

$$c_n = \sum_{i=1}^{n} \Omega_i.$$

(d) Find a formula for the sequence d defined by

$$d_n = \prod_{i=1}^{n} \Omega_i.$$

(e) Is Ω increasing?

(f) Is Ω decreasing?

8. Answer (a)–(e) for the sequence x defined by

$$x_1 = 2 \qquad x_n = 3 + x_{n-1}, \qquad n \geq 2.$$

(a) Find $\displaystyle\sum_{i=1}^{3} x_i$.

(b) Find $\displaystyle\sum_{i=1}^{10} x_i$.

(c) Find a formula for the sequence c defined by

$$c_n = \sum_{i=1}^{n} x_i.$$

(d) Is x increasing?

(e) Is x decreasing?

9. Answer (a)–(f) for the sequence w defined by

$$w_n = \frac{1}{n} - \frac{1}{n+1}, \qquad n \geq 1.$$

(a) Find $\displaystyle\sum_{i=1}^{3} w_i$.

(b) Find $\displaystyle\sum_{i=1}^{10} w_i$.

(c) Find a formula for the sequence c defined by

$$c_n = \sum_{i=1}^{n} w_i.$$

(d) Find a formula for the sequence d defined by

$$d_n = \prod_{i=1}^{n} w_i.$$

(e) Is w increasing?

(f) Is w decreasing?

10. Let u be the sequence defined by

$$u_1 = 3 \qquad u_n = 3 + u_{n-1}, \qquad n \geq 2.$$

Find a formula for the sequence d defined by

$$d_n = \prod_{i=1}^{n} u_i.$$

11. Define $\{s_n\}$ by the rule

$$s_n = 2n - 1, \qquad n \geq 1.$$

Consider the subsequence of s obtained by taking the first, third, fifth, ... terms.

(a) List the first seven terms of s.

(b) List the first seven terms of the subsequence.

(c) Find a formula for the expression n_k of Definition 2.2.9.

(d) Find a formula for the kth term of the subsequence.

12. Define $\{t_n\}$ by the rule

$$t_n = 2^n, \qquad n \geq 1.$$

Consider the subsequence of t obtained by taking the first, second, fourth, seventh, eleventh, ... terms.

(a) List the first seven terms of t.

(b) List the first seven terms of the subsequence.

(c) Find a formula for the expression n_k of Definition 2.2.9.

(d) Find a formula for the kth term of the subsequence.

13. Answer (a)–(d) using the sequences y and z defined by

$$y_n = 2^n - 1, \qquad z_n = n(n - 1).$$

(a) Find $\left(\displaystyle\sum_{i=1}^{3} y_i \right) \left(\displaystyle\sum_{i=1}^{3} z_i \right)$.

(b) Find $\left(\displaystyle\sum_{i=1}^{5} y_i \right) \left(\displaystyle\sum_{i=1}^{4} z_i \right)$.

(c) Find $\displaystyle\sum_{i=1}^{3} y_i z_i$.

(d) Find $\left(\displaystyle\sum_{i=3}^{4} y_i \right) \left(\displaystyle\prod_{i=2}^{4} z_i \right)$.

14. Answer (a)–(h) for the sequence r defined by

$$r_n = 3 \cdot 2^n - 4 \cdot 5^n, \qquad n \geq 0.$$

(a) Find r_0.

(b) Find r_1.

(c) Find r_2.

(d) Find r_3.

(e) Find a formula for r_p.

(f) Find a formula for r_{n-1}.

(g) Find a formula for r_{n-2}.

(h) Show that $\{r_n\}$ satisfies

$$r_n = 7r_{n-1} - 10r_{n-2}, \quad n \geq 2.$$

15. Answer (a)–(h) for the sequence z defined by,

$$z_n = (2 + n)3^n, \qquad n \geq 0.$$

(a) Find z_0.

(b) Find z_1.

(c) Find z_2.

(d) Find z_3.

(e) Find a formula for z_i.

(f) Find a formula for z_{n-1}.

(g) Find a formula for z_{n-2}.

(h) Show that $\{z_n\}$ satisfies

$$z_n = 6z_{n-1} - 9z_{n-2}, \qquad n \geq 2.$$

16. Find b_i, $i = 1, \ldots, 6$, where

$$b_n = 2[1 + (n-1)(n-2)(n-3)(n-4)(n-5)] + \frac{(n-1)n}{2}.$$

17. Rewrite the sum

$$\sum_{i=1}^{n} i^2 r^{n-i}$$

replacing the index i by k, where $i = k + 1$.

18. Rewrite the sum

$$\sum_{k=1}^{n} C_{k-1} C_{n-k}$$

replacing the index k by i, where $k = i + 1$.

19. Let a and b be sequences, and let

$$s_k = \sum_{i=1}^{k} a_i.$$

Show that

$$\sum_{k=1}^{n} a_k b_k = \sum_{k=1}^{n} s_k(b_k - b_{k+1}) + s_n b_{n+1}.$$

This equation, known as the *summation-by-parts formula*, is the discrete analog of the integration-by-parts formula in calculus.

20. Sometimes we generalize the notion of sequence as defined in this section by allowing more general indexing. Suppose that $\{a_{ij}\}$ is a sequence indexed over pairs of positive integers. Show that

$$\sum_{i=1}^{n} \left(\sum_{j=i}^{n} a_{ij} \right) = \sum_{j=1}^{n} \left(\sum_{i=1}^{j} a_{ij} \right).$$

21. Compute the given quantity using the strings

$$\alpha = baab, \qquad \beta = caaba, \qquad \gamma = bbab.$$

(a) $\alpha\beta$ (b) $\beta\alpha$ (c) $\alpha\alpha$ (d) $\beta\beta$

(e) $|\alpha\beta|$ (f) $|\beta\alpha|$ (g) $|\alpha\alpha|$ (h) $|\beta\beta|$

(i) $\alpha\lambda$ (j) $\lambda\beta$ (k) $\alpha\beta\gamma$ (l) $\beta\beta\gamma\alpha$

22. List all strings over $X = \{0, 1\}$ of length 2.

23. List all strings over $X = \{0, 1\}$ of length 2 or less.

24. List all strings over $X = \{0, 1\}$ of length 3.

25. List all strings over $X = \{0, 1\}$ of length 3 or less.

26. A string s is a **substring** of t if there are strings u and v with $t = usv$. Find all substrings of the string $babc$.

27. Find all substrings of the string $aabaabb$.

28. Use induction to show that

$$\sum \frac{1}{n_1 \cdot n_2 \cdots n_k} = n,$$

where the sum is taken over all nonempty subsets $\{n_1, n_2, \ldots, n_k\}$ of $\{1, 2, \ldots, n\}$.

2.3 NUMBER SYSTEMS

A **bit** is a *b*inary dig*it*, that is a 0 or a 1. In a digital computer, data and instructions are encoded as bits. (The term *digital* refers to the use of the digits 0 and 1.) Technology determines how the bits are physically represented within a computer system. Today's hardware relies on the state of an electronic circuit to represent a bit. The circuit must be capable of being in two states—one representing 1, the other 0. In this section we discuss the **binary number system**, which represents integers using bits, and the **hexadecimal number system**, which represents integers using 16 symbols. The **octal number system**, which represents integers using eight symbols, is discussed before Exercise 35.

In the decimal number system, to represent integers we use the ten symbols 0, 1, 2, 3, 4, 5, 6, 7, 8, and 9. In representing an integer, the position of the symbols is significant; reading from the right, the first symbol represents the number of 1's, the next symbol the number of 10's, the next symbol the number of 100's, and so on (see Figure 2.3.1). In general, the symbol in position n (with the rightmost symbol being in position 0) represents the number of 10^n's. Since $10^0 = 1$, the symbol in position 0 represents the number of 10^0's or 1's; since $10^1 = 10$, the symbol in position 1 represents the number of 10^1's or 10's; since $10^2 = 100$, the symbol in position 2 represents the number of 10^2's or 100's; and so on. We call the value on which the system is based (10 in the case of the decimal system) the **base** of the number system.

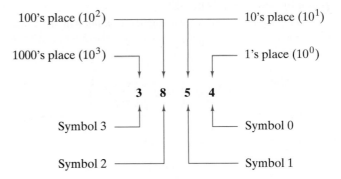

100's place (10^2) — 10's place (10^1)

1000's place (10^3) — 1's place (10^0)

3   8   5   4

Symbol 3 — Symbol 0

Symbol 2 — Symbol 1

FIGURE 2.3.1   The decimal number system.

In the binary (base 2) number system. to represent integers we need only two symbols, 0 and 1. In representing an integer, reading from the right, the first symbol represents the number of 1's, the next symbol the number of 2's, the next symbol the number of 4's, the next symbol the number of 8's, and so on (see Figure 2.3.2). In general, the symbol in position n (with the rightmost symbol being in position 0) represents the number of 2^n's. Since $2^0 = 1$, the symbol in position 0 represents the number of 2^0's, or 1's; since $2^1 = 2$, the symbol in position 1 represents the number of 2^1's or 2's; since $2^2 = 4$, the symbol in position 2 represents the number of 2^2's or 4's; and so on.

Without knowing which number system is being used, a representation is ambiguous; for example, 101101 represents one number in decimal and quite a different number in binary. Often the context will make clear which number system is in effect, but when we want to be absolutely clear, we subscript the number to specify the base—the subscript 10 denotes the decimal system and the subscript 2 denotes the binary system. For example, the binary number 101101 can be written 101101_2.

EXAMPLE 2.3.1   *Binary to Decimal*

The binary number 101101_2 represents the number consisting of one 1, no 2's, one 4, one 8, no 16's, and one 32 (see Figure 2.3.2). This representation may be expressed

$$101101_2 = 1 \cdot 2^5 + 0 \cdot 2^4 + 1 \cdot 2^3 + 1 \cdot 2^2 + 0 \cdot 2^1 + 1 \cdot 2^0.$$

Computing the right-hand side in decimal, we find that

$$101101_2 = 1 \cdot 32 + 0 \cdot 16 + 1 \cdot 8 + 1 \cdot 4 + 0 \cdot 2 + 1 \cdot 1$$
$$= 32 + 8 + 4 + 1 = 45_{10}.$$ □

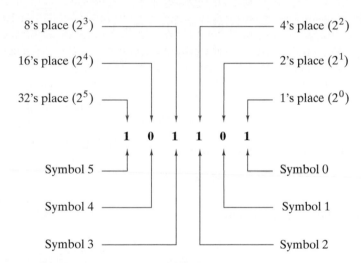

FIGURE 2.3.2   The binary number system.

Example 2.3.1 shows how to convert a binary number to decimal. Consider the reverse problem—converting a decimal number to binary. Suppose, for example, that we want to convert the decimal number 91 to binary. If we divide 91 by 2, we obtain

$$\begin{array}{r} 45 \\ 2\overline{)91} \\ \underline{8} \\ 11 \\ \underline{10} \\ 1 \end{array}$$

This computation shows that

$$91 = 2 \cdot 45 + 1. \qquad (2.3.1)$$

We are beginning to express 91 in powers of 2. If we next divide 45 by 2, we find

$$45 = 2 \cdot 22 + 1. \tag{2.3.2}$$

Substituting this expression for 45 into (2.3.1), we obtain

$$\begin{aligned}
91 &= 2 \cdot 45 + 1 \\
&= 2 \cdot (2 \cdot 22 + 1) + 1 \\
&= 2^2 \cdot 22 + 2 + 1.
\end{aligned} \tag{2.3.3}$$

If we next divide 22 by 2, we find

$$22 = 2 \cdot 11.$$

Substituting this expression for 22 into (2.3.3), we obtain

$$\begin{aligned}
91 &= 2^2 \cdot 22 + 2 + 1 \\
&= 2^2 \cdot (2 \cdot 11) + 2 + 1 \\
&= 2^3 \cdot 11 + 2 + 1.
\end{aligned} \tag{2.3.4}$$

If we next divide 11 by 2, we find

$$11 = 2 \cdot 5 + 1.$$

Substituting this expression for 11 into (2.3.4), we obtain

$$91 = 2^4 \cdot 5 + 2^3 + 2 + 1. \tag{2.3.5}$$

If we next divide 5 by 2, we find

$$5 = 2 \cdot 2 + 1.$$

Substituting this expression for 5 into (2.3.5), we obtain

$$\begin{aligned}
91 &= 2^5 \cdot 2 + 2^4 + 2^3 + 2 + 1 \\
&= 2^6 + 2^4 + 2^3 + 2 + 1 \\
&= 1011011_2.
\end{aligned}$$

The preceding computation shows that the *remainders*, as N is successively divided by 2, give the bits in the binary representation of N. The first division by 2 in (2.3.1) gives the 1's bit; the second division by 2 in (2.3.2) gives the 2's bit; and so on. We illustrate with another example.

EXAMPLE 2.3.2   *Decimal to Binary*

Write the decimal number 130 in binary.

The computation shows the successive divisions by 2 with the remainders recorded at the right.

$$
\begin{array}{llll}
2\,)\,\underline{130} & \text{remainder} = 0 & \text{1's bit} \\
2\,)\,\underline{65} & \text{remainder} = 1 & \text{2's bit} \\
2\,)\,\underline{32} & \text{remainder} = 0 & \text{4's bit} \\
2\,)\,\underline{16} & \text{remainder} = 0 & \text{8's bit} \\
2\,)\,\underline{8} & \text{remainder} = 0 & \text{16's bit} \\
2\,)\,\underline{4} & \text{remainder} = 0 & \text{32's bit} \\
2\,)\,\underline{2} & \text{remainder} = 0 & \text{64's bit} \\
2\,)\,\underline{1} & \text{remainder} = 1 & \text{128's bit} \\
0
\end{array}
$$

We may stop when the dividend is 0. Remembering that the first remainder gives the number of 1's, the second remainder gives the number of 2's, and so on, we obtain

$$130_{10} = 10000010_2. \qquad \qquad \square$$

Next we turn our attention to addition of numbers in arbitrary bases. The same method that we use to add decimal numbers can be used to add binary numbers; however, we must replace the decimal addition table with the binary addition table

+	0	1
0	0	1
1	1	10

(In decimal, $1 + 1 = 2$, and $2_{10} = 10_2$; thus, in binary, $1 + 1 = 10$.)

EXAMPLE 2.3.3   *Binary Addition*

Add the binary numbers 10011011 and 1011011.

We write the problem as

$$
\begin{array}{r}
10011011 \\
+\quad 1011011 \\
\hline
\end{array}
$$

As in decimal addition, we begin from the right adding 1 and 1. This sum is 10_2; thus we write 0 and carry 1. At this point the computation is

$$
\begin{array}{r}
1 \\
10011011 \\
+\quad 1011011 \\
\hline
0
\end{array}
$$

Next, we add 1 and 1 and 1, which is 11_2. We write 1 and carry 1. At this point, the computation is

$$
\begin{array}{r}
1 \\
10011011 \\
+\quad 1011011 \\
\hline
10
\end{array}
$$

Continuing in this way, we obtain

$$
\begin{array}{r}
10011011 \\
+\quad 1011011 \\
\hline
11110110
\end{array}
$$

$\square$

EXAMPLE 2.3.4

The addition problem of Example 2.3.3, in decimal, is

$$
\begin{array}{r}
155 \\
+\quad 91 \\
\hline
246
\end{array}
$$

$\square$

Other important bases for number systems in computer science are base 8 or **octal** and base 16 or **hexadecimal** (sometimes shortened to **hex**). We will discuss the hexadecimal system and leave the octal system to the exercises (see Exercises 35–40).

In the hexadecimal number system, to represent integers we use the symbols 0, 1, 2, 3, 4, 5, 6, 7, 8, 9, A, B, C, D, E, and F. The symbols A–F are interpreted as decimal 10–15. (In general, in the base N number system, N distinct symbols, representing $0, 1, 2, \ldots, N - 1$ will be required.) In representing an integer, reading from the right, the first symbol represents the number of 1's, the next symbol the number of 16's, the next symbol the number of 16^2's, and so on (see Figure 2.3.3). In general, the symbol in position n (with the rightmost symbol being in position 0) represents the number of 16^n's.

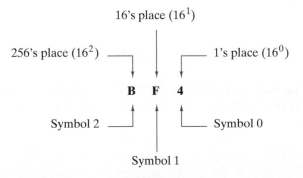

FIGURE 2.3.3   The hexadecimal number system.

EXAMPLE 2.3.5   *Hexadecimal to Decimal*

Convert the hexadecimal number B4F to decimal.
  We obtain

$$B4F_{16} = 11 \cdot 16^2 + 4 \cdot 16^1 + 15 \cdot 16^0$$
$$= 11 \cdot 256 + 4 \cdot 16 + 15 = 2816 + 64 + 15 = 2895_{10}. \qquad \square$$

  To convert a decimal number to hexadecimal, we divide successively by 16. The remainders give the hexadecimal symbols.

EXAMPLE 2.3.6   *Decimal to Hexadecimal*

Convert the decimal number 20385 to hexadecimal.
  The computation shows the successive divisions by 16 with the remainders recorded at the right.

$$
\begin{array}{rll}
16)\,\underline{20385} & \text{remainder} = 1 & \text{1's place} \\
16)\,\underline{1274} & \text{remainder} = 10 & \text{16's place} \\
16)\,\underline{79} & \text{remainder} = 15 & 16^2\text{'s place} \\
16)\,\underline{4} & \text{remainder} = 4 & 16^3\text{'s place} \\
0 & &
\end{array}
$$

We may stop when the dividend is 0. The first remainder gives the number of 1's, the second remainder gives the number of 16's, and so on; thus we obtain

$$20385_{10} = 4FA1_{16}. \qquad \square$$

  Our next example shows that we can add hexadecimal numbers in the same way that we add decimal or binary numbers.

EXAMPLE 2.3.7   *Hexadecimal Addition*

Add the hexadecimal numbers 84F and 42EA.
  The problem may be written

$$
\begin{array}{r}
84F \\
+ \ 42EA \\
\end{array}
$$

We begin in the rightmost column by adding F and A. Since F is 15_{10} and A is 10_{10}, $F + A = 15_{10} + 10_{10} = 25_{10} = 19_{16}$. We write 9 and carry 1:

$$
\begin{array}{r}
1 \\
84F \\
+ \ 42EA \\
\hline
9
\end{array}
$$

Next, we add 1, 4, and E, obtaining 13_{16}. We write 3 and carry 1:

$$
\begin{array}{r}
1 \\
84F \\
+\ 42EA \\
\hline
39
\end{array}
$$

Continuing in this way, we obtain

$$
\begin{array}{r}
84F \\
+\ 42EA \\
\hline
4B39
\end{array}
$$
□

EXAMPLE 2.3.8

The addition problem of Example 2.3.7, in decimal, is

$$
\begin{array}{r}
2127 \\
+\ 17130 \\
\hline
19257
\end{array}
$$
□

✧ ✧ ✧

Exercises

In Exercises 1–6, express each binary number in decimal.

1. 1001 **2.** 11011 **3.** 11011011

4. 100000 **5.** 11111111 **6.** 110111011011

In Exercises 7–12, express each decimal number in binary.

7. 34 **8.** 61 **9.** 223

10. 400 **11.** 1024 **12.** 12,340

In Exercises 13–18, add the binary numbers.

13. 1001 + 1111 **14.** 11011 + 1101

15. 110110 + 101101

16. 101101 + 11011

17. 110110101 + 1101101

18. 1101 + 101100 + 11011011

In Exercises 19–24, express each hexadecimal number in decimal.

19. 3A **20.** 1E9 **21.** 3E7C

22. A03 **23.** 209D **24.** 4B07A

25. Express each decimal number in Exercises 7–12 in hexadecimal.

26. Express each binary number in Exercises 1–6 in hexadecimal.

27. Express each hexadecimal number in Exercises 19, 20, and 22 in binary.

In Exercises 28–32, add the hexadecimal numbers.

28. 4A + B4   **29.** 195 + 76E   **30.** 49F7 + C66

31. 349CC + 922D   **32.** 82054 + AEFA3

33. Does 2010 represent a number in binary? in decimal? in hexadecimal?

34. Does 1101010 represent a number in binary? in decimal? in hexadecimal?

In the octal (base 8) number system, to represent integers we use the symbols 0, 1, 2, 3, 4, 5, 6, and 7. In representing an integer, reading from the right, the first symbol represents the number of 1's, the next symbol the number of 8's, the next symbol the number of 8^2's, and so on. In general, the symbol in position n (with the rightmost symbol being in position 0) represents the number of 8^n's. In Exercises 35–40, express each octal number in decimal.

35. 63   **36.** 7643   **37.** 7711

38. 10732   **39.** 1007   **40.** 537261

41. Express each decimal number in Exercises 7–12 in octal.

42. Express each binary number in Exercises 1–6 in octal.

43. Express each hexadecimal number in Exercises 19–24 in octal.

44. Express each octal number in Exercises 35–40 in hexadecimal.

45. Does 1101010 represent a number in octal?

46. Does 30470 represent a number in binary? in octal? in decimal? in hexadecimal?

47. Does 9450 represent a number in binary? in octal? in decimal? in hexadecimal?

48. Let T_n denote the highest power of 2 that divides n. Show that $T_{mn} = T_m + T_n$ for all $m, n \geq 1$.

49. Let S_n denote the number of 1's in the binary representation of n. Use induction to prove that $T_{n!} = n - S_n$ for all $n \geq 1$. (T_n is defined in Exercise 48.)

2.4 RELATIONS

A **relation** can be thought of as a table that lists the relationship of elements to other elements (see Table 2.4.1). Table 2.4.1 shows which students are taking which courses. For example, Bill is taking Computer Science and Art, and Mary is taking Mathematics. In the terminology of relations, we would say that Bill is related to Computer Science and Art, and that Mary is related to Mathematics.

Of course, Table 2.4.1 is really just a set of ordered pairs. Abstractly, we *define* a relation to be a set of ordered pairs. In this setting, we consider the first element of the ordered pair to be related to the second element of the ordered pair.

TABLE 2.4.1
Relation of Students
to Courses

Student	Course
Bill	CompSci
Mary	Math
Bill	Art
Beth	History
Beth	CompSci
Dave	Math

DEFINITION 2.4.1

A *(binary) relation R from a set X to a set Y* is a subset of the Cartesian product $X \times Y$. If $(x, y) \in R$, we write $x \, R \, y$ and say that *x is related to y*. In case $X = Y$, we call R a *(binary) relation on X*.

The set

$$\{x \in X \mid (x, y) \in R \text{ for some } y \in Y\}$$

is called the *domain* of R. The set

$$\{y \in Y \mid (x, y) \in R \text{ for some } x \in X\}$$

is called the *range* of R.

If a relation is given as a table, the domain consists of the members of the first column and the range consists of the members of the second column.

EXAMPLE 2.4.2

If we let

$$X = \{\text{Bill, Mary, Beth, Dave}\}$$

and

$$Y = \{\text{CompSci, Math, Art, History}\},$$

our relation R of Table 2.4.1 can be written

$$R = \{(\text{Bill, CompSci}), \quad (\text{Mary, Math}), \quad (\text{Bill, Art}),$$
$$(\text{Beth, History}), \quad (\text{Beth, CompSci}), \quad (\text{Dave, Math})\}.$$

Since (Beth, History) $\in R$, we may write Beth R History. The domain (first column) of R is the set X and the range (second column) of R is the set Y.   □

Example 2.4.2 shows that a relation can be given by simply specifying which ordered pairs belong to the relation. Our next example shows that sometimes it is possible to define a relation by giving a rule for membership in the relation.

EXAMPLE 2.4.3

Let

$$X = \{2, 3, 4\} \quad \text{and} \quad Y = \{3, 4, 5, 6, 7\}.$$

If we define a relation R from X to Y by

$$(x, y) \in R \text{ if } x \text{ divides } y \text{ (with zero remainder)},$$

we obtain

$$R = \{(2, 4), (2, 6), (3, 3), (3, 6), (4, 4)\}.$$

If we rewrite R as a table, we obtain

X	Y
2	4
2	6
3	3
3	6
4	4

The domain of R is the set $\{2, 3, 4\}$ and the range of R is the set $\{3, 4, 6\}$.   □

EXAMPLE 2.4.4

Let R be the relation on $X = \{1, 2, 3, 4\}$ defined by $(x, y) \in R$ if $x \leq y$, $x, y \in X$. Then

$$R = \{(1, 1), (1, 2), (1, 3), (1, 4), (2, 2), (2, 3), (2, 4), (3, 3), (3, 4), (4, 4)\}.$$

The domain and range of R are both equal to X.   □

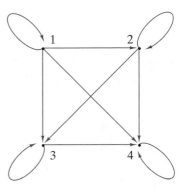

FIGURE 2.4.1
The digraph of the relation of Example 2.4.4.

An informative way to picture a relation on a set is to draw its **digraph**. (Digraphs are discussed in more detail in Chapter 6. For now, we mention digraphs only in connection with relations.) To draw the digraph of a relation on a set X, we first draw dots or **vertices** to represent the elements of X. In Figure 2.4.1, we have drawn four vertices to represent the elements of the set X of Example 2.4.4. Next, if the element (x, y) is in the relation, we draw an arrow (called a **directed edge**) from x to y. In Figure 2.4.1, we have drawn directed edges to represent the members of the relation R of Example 2.4.4. Notice that an element of the form (x, x) in a relation corresponds to a directed edge from x to x. Such an edge is called a **loop**. There is a loop at every vertex in Figure 2.4.1.

EXAMPLE 2.4.5

The relation R on $X = \{a, b, c, d\}$ given by the digraph of Figure 2.4.2 is

$$R = \{(a, a), (b, c), (c, b), (d, d)\}.$$ □

We next define several properties that some relations have.

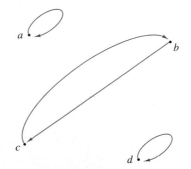

FIGURE 2.4.2
The digraph of the relation of Example 2.4.5.

DEFINITION 2.4.6

A relation R on a set X is called *reflexive* if $(x, x) \in R$ for every $x \in X$.

EXAMPLE 2.4.7

The relation R on $X = \{1, 2, 3, 4\}$ of Example 2.4.4 is reflexive because for each element $x \in X$, $(x, x) \in R$; specifically, $(1, 1)$, $(2, 2)$, $(3, 3)$, and $(4, 4)$ are each in R. The digraph of a reflexive relation has a loop at every vertex. Notice that the digraph of the reflexive relation of Example 2.4.4 (see Figure 2.4.1) has a loop at every vertex. □

EXAMPLE 2.4.8

The relation R on $X = \{a, b, c, d\}$ of Example 2.4.5 is not reflexive. For example, $b \in X$, but $(b, b) \notin R$. That this relation is not reflexive can also be seen by looking at its digraph (see Figure 2.4.2); vertex b does not have a loop. □

DEFINITION 2.4.9

A relation R on a set X is called *symmetric* if for all $x, y \in X$, if $(x, y) \in R$, then $(y, x) \in R$.

EXAMPLE 2.4.10

The relation of Example 2.4.5 is symmetric because for all x, y, if $(x, y) \in R$, then $(y, x) \in R$. For example, (b, c) is in R and (c, b) is also in R. The digraph of a symmetric relation has the property that whenever there is a directed edge from v to w, there is also a directed edge from w to v. Notice that the digraph of the relation of Example 2.4.5 (see Figure 2.4.2) has the property that for every directed edge from v to w, there is also a directed edge from w to v. □

EXAMPLE 2.4.11

The relation of Example 2.4.4 is not symmetric. For example, $(2, 3) \in R$, but $(3, 2) \notin R$. The digraph of this relation (see Figure 2.4.1) has a directed edge from 2 to 3, but there is no directed edge from 3 to 2. □

DEFINITION 2.4.12

A relation R on a set X is called *antisymmetric* if for all $x, y \in X$, if $(x, y) \in R$ and $x \neq y$, then $(y, x) \notin R$.

EXAMPLE 2.4.13

The relation of Example 2.4.4 is antisymmetric because for all x, y, if $(x, y) \in R$ and $x \neq y$, then $(y, x) \notin R$. For example, $(1, 2) \in R$, but $(2, 1) \notin R$. The digraph of an antisymmetric relation has the property that between any two vertices there is at most one directed edge. Notice that the digraph of the relation of Example 2.4.4 (see Figure 2.4.1) has at most one directed edge between each pair of vertices.   □

EXAMPLE 2.4.14

The relation of Example 2.4.5 is not antisymmetric because both (b, c) and (c, b) are in R. Notice that in the digraph of the relation of Example 2.4.5 (see Figure 2.4.2) there are two directed edges between b and c.   □

EXAMPLE 2.4.15

If a relation R on X has no members of the form (x, y) with $x \neq y$, then R is antisymmetric. In this case, if x and y are any elements in X, the proposition

$$\text{if } (x, y) \in R \text{ and } x \neq y, \text{ then } (y, x) \notin R$$

is true because the hypothesis is false. For example,

$$R = \{(a, a), (b, b), (c, c)\}$$

on $X = \{a, b, c\}$ is antisymmetric. The digraph of R shown in Figure 2.4.3 has at most one directed edge between each pair of vertices. Notice that R is also reflexive and symmetric. This example shows that "antisymmetric" is not the same as "not symmetric."   □

FIGURE 2.4.3   The digraph of the relation of Example 2.4.15.

DEFINITION 2.4.16

A relation R on a set X is called *transitive* if for all $x, y, z \in X$, if (x, y) and $(y, z) \in R$, then $(x, z) \in R$.

EXAMPLE 2.4.17

The relation R of Example 2.4.4 is transitive because for all x, y, z, if (x, y) and $(y, z) \in R$, then $(x, z) \in R$. To formally verify that this relation satisfies Definition 2.4.16, we would have to list all pairs of pairs of the form (x, y) and (y, z) in R and then verify that in every case, $(x, z) \in R$:

Pairs of Form			*Pairs of Form*		
(x, y), (y, z)		*(x, z)*	*(x, y), (y, z)*		*(x, z)*
(1, 1)	(1, 1)	(1, 1)	(2, 2)	(2, 2)	(2, 2)
(1, 1)	(1, 2)	(1, 2)	(2, 2)	(2, 3)	(2, 3)
(1, 1)	(1, 3)	(1, 3)	(2, 2)	(2, 4)	(2, 4)
(1, 1)	(1, 4)	(1, 4)	(2, 3)	(3, 3)	(2, 3)
(1, 2)	(2, 2)	(1, 2)	(2, 3)	(3, 4)	(2, 4)
(1, 2)	(2, 3)	(1, 3)	(2, 4)	(4, 4)	(2, 4)
(1, 2)	(2, 4)	(1, 4)	(3, 3)	(3, 3)	(3, 3)
(1, 3)	(3, 3)	(1, 3)	(3, 3)	(3, 4)	(3, 4)
(1, 3)	(3, 4)	(1, 4)	(3, 4)	(4, 4)	(3, 4)
(1, 4)	(4, 4)	(1, 4)	(4, 4)	(4, 4)	(4, 4)

In determining whether a relation R is transitive directly from Definition 2.4.16, in case $x = y$ or $y = z$ we need not explicitly verify that the condition

$$\text{if } (x, y) \text{ and } (y, z) \in R, \text{ then } (x, z) \in R$$

is satisfied since it will automatically be true. Suppose, for example, that $x = y$ and (x, y) and (y, z) are in R. Since $x = y$, $(x, z) = (y, z)$ is in R and the condition is satisfied. Eliminating the cases $x = y$ and $y = z$ leaves only the following to be explicitly checked to verify that the relation of Example 2.4.4 is transitive:

Pairs of Form (x, y), (y, z)		*(x, z)*
(1, 2)	(2, 3)	(1, 3)
(1, 2)	(2, 4)	(1, 4)
(1, 3)	(3, 4)	(1, 4)
(2, 3)	(3, 4)	(2, 4)

The digraph of a transitive relation has the property that whenever there are directed edges from x to y and from y to z, there is also a directed edge from x to z. Notice that the digraph of the relation of Example 2.4.4 (see Figure 2.4.1) has this property.   □

EXAMPLE 2.4.18

The relation of Example 2.4.5 is not transitive. For example, (b, c) and (c, b) are in R, but (b, b) is not in R. Notice that in the digraph (see Figure 2.4.2) of the relation of Example 2.4.5 there are directed edges from b to c and from c to b, but there is no directed edge from b to b.   □

Relations can be used to order elements of a set. For example, the relation R defined on the set of integers by

$$(x, y) \in R \qquad \text{if } x \leq y$$

orders the integers. Notice that the relation R is reflexive, antisymmetric, and transitive. Such relations are called **partial orders**.

DEFINITION 2.4.19

A relation R on a set X is called a *partial order* if R is reflexive, antisymmetric, and transitive.

EXAMPLE 2.4.20

Since the relation R defined on the positive integers by

$$(x, y) \in R \qquad \text{if } x \text{ divides } y \text{ (evenly)}$$

is reflexive, antisymmetric, and transitive, R is a partial order.   □

If R is a partial order on a set X, the notation $x \preceq y$ is sometimes used to indicate that $(x, y) \in R$. This notation suggests that we are interpreting the relation as an ordering of the elements in X.

Suppose that R is a partial order on a set X. If $x, y \in X$ and either $x \preceq y$ or $y \preceq x$, we say that x and y are **comparable**. If $x, y \in X$ and $x \npreceq y$ and $y \npreceq x$, we say that x and y are **incomparable**. If every pair of elements in X is comparable, we call R a **total order**. The less than or equals relation on the positive integers is a total order since, if x and y are integers, either $x \leq y$ or $y \leq x$. The reason for the term "partial order" is that in general some elements in X may be incomparable. The "divides" relation on the positive integers (see Example 2.4.20) has both comparable and incomparable elements. For example, 2 and 3 are incomparable (since 2 does not divide 3 and 3 does not divide 2), but 3 and 6 are comparable (since 3 divides 6).

One application of partial orders is to task scheduling.

EXAMPLE 2.4.21   *Task Scheduling*

Consider the set T of tasks that must be completed in order to take an indoor flash picture.

1. Remove lens cap.
2. Focus camera.
3. Turn off safety lock.
4. Turn on flash unit.
5. Push photo button.

Some of these tasks must be done before others. For example, task 1 must be done before task 2. On the other hand, other tasks can be done in either order. For example, tasks 2 and 3 can be done in either order.

The relation R' defined on T by

$$i R' j \text{ if and only if task } i \text{ must be done before task } j$$

orders the tasks. Although R' is antisymmetric and transitive, it is not reflexive, so R' is not a partial order. We can obtain a partial order by "throwing in" all pairs (i, i) for $i = 1, \ldots, 5$. Formally, the relation

$$R' \cup \{(1, 1), (2, 2), (3, 3), (4, 4), (5, 5)\}$$

is a partial order on T. A solution to the problem of scheduling the tasks so that we can take a picture is a total ordering of the tasks consistent with the partial order. More precisely, we require a total ordering of the tasks

$$t_1, \quad t_2, \quad t_3, \quad t_4, \quad t_5$$

such that if $t_i R' t_j$, then t_i precedes t_j in the list. □

Given a relation R from X to Y, we may define a relation from Y to X by reversing the order of each ordered pair in R. The formal definition follows.

DEFINITION 2.4.22

Let R be a relation from X to Y. The *inverse* of R, denoted R^{-1}, is the relation from Y to X defined by

$$R^{-1} = \{(y, x) \mid (x, y) \in R\}.$$

EXAMPLE 2.4.23

The inverse of the relation R of Example 2.4.3 is

$$R^{-1} = \{(4, 2), (6, 2), (3, 3), (6, 3), (4, 4)\}.$$

In words, we might describe this relation as "is divisible by." □

If we have a relation R_1 from X to Y and a relation R_2 from Y to Z, we can form a relation from X to Z by applying first relation R_1 and then relation R_2. The resulting relation is denoted $R_2 \circ R_1$. Notice the order in which the relations are written. The formal definition follows.

DEFINITION 2.4.24

Let R_1 be a relation from X to Y and R_2 be a relation from Y to Z. The *composition* of R_1 and R_2, denoted $R_2 \circ R_1$, is the relation from X to Z defined by

$$R_2 \circ R_1 = \{(x, z) \mid (x, y) \in R_1 \text{ and } (y, z) \in R_2 \text{ for some } y \in Y\}.$$

EXAMPLE 2.4.25

The composition of the relations

$$R_1 = \{(1, 2), (1, 6), (2, 4), (3, 4), (3, 6), (3, 8)\}$$

and

$$R_2 = \{(2, u), (4, s), (4, t), (6, t), (8, u)\}$$

is

$$R_2 \circ R_1 = \{(1, u), (1, t), (2, s), (2, t), (3, s), (3, t), (3, u)\}. \qquad \square$$

Exercises

In Exercises 1–4, write the relation as a set of ordered pairs.

1.

8840	Hammer
9921	Pliers
452	Paint
2207	Carpet

2.

a	3
b	1
b	4
c	1

3.

Sally	Math
Ruth	Physics
Sam	Econ

4.

a	a
b	b

In Exercises 5–8, write the relation as a table.

5. $R = \{(a, 6), (b, 2), (a, 1), (c, 1)\}$
6. $R = \{(\text{Roger, Music}), (\text{Pat, History}), (\text{Ben, Math}), (\text{Pat, PolySci})\}$
7. The relation R on $\{1, 2, 3, 4\}$ defined by $(x, y) \in R$ if $x^2 \geq y$.
8. The relation R from the set X of states whose names begin with the letter "M" to the set Y of cities defined by $(S, C) \in X \times Y$ if C is the capital of S.

In Exercises 9–12, draw the digraph of the relation.

9. The relation of Exercise 4 on $\{a, b, c\}$.
10. The relation $R = \{(1, 2), (2, 1), (3, 3), (1, 1), (2, 2)\}$ on $X = \{1, 2, 3\}$.

11. The relation $R = \{(1, 2), (2, 3), (3, 4), (4, 1)\}$ on $\{1, 2, 3, 4\}$.

12. The relation of Exercise 7.

In Exercises 13–16, write the relation as a set of ordered pairs.

13.

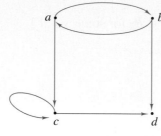

14.

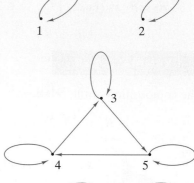

15. $1 \cdot$ $2 \cdot$ **16.** $a \cdot$ b ⟷ c d

17. Find the domain and range of each relation in Exercises 1–16.

18. Find the inverse (as a set of ordered pairs) of each relation in Exercises 1–16.

Exercises 19–24 refer to the relation R on the set $\{1, 2, 3, 4, 5\}$ defined by the rule $(x, y) \in R$ if 3 divides $x - y$.

19. List the elements of R.

20. List the elements of R^{-1}.   **21.** Find the domain of R.

22. Find the range of R.   **23.** Find the domain of R^{-1}.

24. Find the range of R^{-1}.

25. Repeat Exercises 19–24 for the relation R on the set $\{1, 2, 3, 4, 5\}$ defined by the rule $(x, y) \in R$ if $x + y \le 6$.

26. Repeat Exercises 19–24 for the relation R on the set $\{1, 2, 3, 4, 5\}$ defined by the rule $(x, y) \in R$ if $x = y - 1$.

27. Is the relation of Exercise 25 reflexive, symmetric, antisymmetric, transitive, and/or a partial order?

28. Is the relation of Exercise 26 reflexive, symmetric, antisymmetric, transitive, and/or a partial order?

In Exercises 29–33, determine whether each relation defined on the set of positive integers is reflexive, symmetric, antisymmetric, transitive, and/or a partial order.

29. $(x, y) \in R$ if $x = y^2$.

30. $(x, y) \in R$ if $x > y$.

31. $(x, y) \in R$ if $x \ge y$.

32. $(x, y) \in R$ if $x = y$.

33. $(x, y) \in R$ if 3 divides $x - y$.

34. Let X be a nonempty set. Define a relation on $\mathcal{P}(X)$, the power set of X, as $(A, B) \in R$ if $A \subseteq B$. Is this relation reflexive, symmetric, antisymmetric, transitive, and/or a partial order?

35. Let X be the set of all four-bit strings (e.g., 0011, 0101, 1000). Define a relation R on X as $s_1 R s_2$ if some substring of s_1 of length 2 is equal to some substring of s_2 of length 2. Examples: $0111 R 1010$ (because both 0111 and 1010 contain 01). $1110 \not{R} 0001$ (because 1110 and 0001 do not share a common substring of length 2). Is this relation reflexive, symmetric, antisymmetric, transitive, and/or a partial order?

36. Suppose that R_i is a partial order on X_i, $i = 1, 2$. Show that R is a partial order on $X_1 \times X_2$ if we define

$$(x_1, x_2) R(x_1', x_2') \qquad \text{if } x_1 R_1 x_1' \text{ and } x_2 R_2 x_2'.$$

37. Let R_1 and R_2 be the relations on $\{1, 2, 3, 4\}$ given by

$$R_1 = \{(1, 1), (1, 2), (3, 4), (4, 2)\}$$

$$R_2 = \{(1, 1), (2, 1), (3, 1), (4, 4), (2, 2)\}.$$

List the elements of $R_1 \circ R_2$ and $R_2 \circ R_1$.

Give examples of relations on $\{1, 2, 3, 4\}$ having the properties specified in Exercises 38–42.

38. Reflexive, symmetric, and not transitive
39. Reflexive, not symmetric, and not transitive
40. Reflexive, antisymmetric, and not transitive
41. Not reflexive, symmetric, not antisymmetric, and transitive
42. Not reflexive, not symmetric, and transitive

Let R and S be relations on X. Determine whether each statement in Exercises 43–58 is true or false. If the statement is false, give a counterexample.

43. If R and S are transitive, then $R \cup S$ is transitive.
44. If R and S are transitive, then $R \cap S$ is transitive.
45. If R and S are transitive, then $R \circ S$ is transitive.
46. If R is transitive, then R^{-1} is transitive.
47. If R and S are reflexive, then $R \cup S$ is reflexive.
48. If R and S are reflexive, then $R \cap S$ is reflexive.
49. If R and S are reflexive, then $R \circ S$ is reflexive.
50. If R is reflexive, then R^{-1} is reflexive.
51. If R and S are symmetric, then $R \cup S$ is symmetric.
52. If R and S are symmetric, then $R \cap S$ is symmetric.
53. If R and S are symmetric, then $R \circ S$ is symmetric.
54. If R is symmetric, then R^{-1} is symmetric.
55. If R and S are antisymmetric, then $R \cup S$ is antisymmetric.
56. If R and S are antisymmetric, then $R \cap S$ is antisymmetric.
57. If R and S are antisymmetric, then $R \circ S$ is antisymmetric.
58. If R is antisymmetric, then R^{-1} is antisymmetric.
59. What is wrong with the following argument, which supposedly shows that any relation R on X that is symmetric and transitive is reflexive?

Let $x \in X$. Using symmetry, we have (x, y) and (y, x) both in R. Since $(x, y), (y, x) \in R$, by transitivity we have $(x, x) \in R$. Therefore, R is reflexive.

Problem

Define the relation R on the set of positive integers by

$(x, y) \in R$ if the greatest common divisor of x and y is 1.

Determine whether R is reflexive, symmetric, antisymmetric, transitive, and/or a partial order.

Attacking the Problem

The relation R consists of ordered pairs of positive integers. Let's classify some ordered pairs as either in or not in R. Let's further systematically list all pairs in increasing order, in the sense that we'll first list all pairs (x, y) where $x + y = 2$, then all pairs (x, y) where $x + y = 3$, and so on:

Ordered Pair (x, y)	$x + y$	Greatest Common Divisor	In R?
$(1, 1)$	2	1	Yes
$(1, 2)$	3	1	Yes
$(2, 1)$	3	1	Yes
$(1, 3)$	4	1	Yes
$(2, 2)$	4	2	No
$(3, 1)$	4	1	Yes
$(4, 1)$	5	1	Yes
$(3, 2)$	5	1	Yes
$(2, 3)$	5	1	Yes
$(1, 4)$	5	1	Yes

You should add the entries for $x + y = 6$ before reading on.

In constructing the table, several patterns emerge. First, we see that the greatest common divisors of (x, y) and (y, x) are equal. Thus if (x, y) is in R (i.e., the greatest common divisor of x and y is 1), then (y, x) is also in R (since the greatest common divisor of y and x is also 1). Also, (x, x) is in R only if $x = 1$.

Finding a Solution

The first question is whether R is reflexive. Recall (see Definition 2.4.6) that R is reflexive on the set of positive integers if (x, x) is in R for all x in the set of positive integers. Our table shows that $(2, 2)$ is not in R. Thus $(2, 2)$ is a counterexample that shows that R is not reflexive.

The next question is whether R is symmetric. Recall (see Definition 2.4.9) that R is symmetric if for all x and y, if (x, y) is in R, then (y, x) is also in R. We observed in the previous subsection that this condition holds for R. Thus R is symmetric.

The next question is whether R is antisymmetric. Recall (see Definition 2.4.12) that R is antisymmetric if for all x and y, if (x, y) is in R and $x \neq y$, then (y, x) is not in R. Our table shows that $(2, 1)$ is in R and surely $2 \neq 1$, but $(1, 2)$ is in R. Thus $(2, 1)$ is a counterexample that shows that R is not antisymmetric.

The next question is whether R is transitive. Recall (see Definition 2.4.16) that R is transitive if for all x, y, and z, if (x, y) and (y, z) are in R, then (x, z) is in R. Our table shows that $(2, 1)$ and $(1, 2)$ are in R, but $(2, 2)$ is not in R. Thus if we take $x = 2$, $y = 1$, and $z = 2$, we have a counterexample that shows that R is not transitive.

The last question is whether R is a partial order. Recall (see Definition 2.4.19) that R is a partial order if R is reflexive, antisymmetric, and transitive. We have already shown that R is not reflexive, R is not antisymmetric, and R is not transitive. Therefore, R is not a partial order. If R were to fail only one or two of the three criteria, reflexive, antisymmetric, or transitive, R would not be a partial order. Here R happens to fail all three criteria.

Formal Solution

We have found that R is not reflexive, symmetric, not antisymmetric, not transitive, and not a partial order.

Summary of Problem-Solving Techniques

- In a problem involving a concrete relation, begin by listing several ordered pairs, and classify each as belonging or not belonging to the relation.

- When listing ordered pairs, do so in a systematic way. Your method of listing ordered pairs should generate all ordered pairs if it were continued indefinitely.

- When listing ordered pairs, watch for patterns. For example, in this problem we saw early on that (x, y) and (y, x) were either both in R or both not in R.

- Don't forget that a counterexample proves that a property fails to hold for a relation.

- To prove that a property holds for a relation, it is necessary to take arbitrary elements of the relation and prove the property for them. For example, to prove that a relation is symmetric, you must let (x, y) be an *arbitrary* ordered pair in R and give an argument that shows that (y, x) is in R.

FIGURE 2.5.1
A set of colored balls.

2.5 EQUIVALENCE RELATIONS

Suppose that we have a set X of 10 balls, each of which is either red, blue, or green (see Figure 2.5.1). If we divide the balls into sets R, B, and G according to color, the family $\{R, B, G\}$ is a partition of X. (Recall that in Section 2.1, we defined a partition of a set X to be a collection $\mathcal{S}$ of nonempty subsets of X such that every element in X belongs to exactly one member of $\mathcal{S}$.)

A partition can be used to define a relation. If $\mathcal{S}$ is a partition of X, we may define $x R y$ to mean that for some set $S \in \mathcal{S}$, both x and y belong to S. For the example of Figure 2.5.1, the relation obtained could be described as "is the same color as." The next theorem shows that such a relation is always reflexive, symmetric, and transitive.

THEOREM 2.5.1

Let $\mathcal{S}$ be a partition of a set X. Define $x R y$ to mean that for some set S in $\mathcal{S}$, both x and y belong to S. Then R is reflexive, symmetric, and transitive.

Proof. Let $x \in X$. By the definition of partition, x belongs to some member S of $\mathcal{S}$. Thus $x R x$ and R is reflexive.

Suppose that $x R y$. Then both x and y belong to some set $S \in \mathcal{S}$. Since both y and x belong to S, $y R x$ and R is symmetric.

Finally, suppose that $x R y$ and $y R z$. Then both x and y belong to some set $S \in \mathcal{S}$ and both y and z belong to some set $T \in \mathcal{S}$. Since y belongs to exactly one member of $\mathcal{S}$, we must have $S = T$. Therefore, both x and z belong to S and $x R z$. We have shown that R is transitive. ∎

EXAMPLE 2.5.2

Consider the partition

$$\mathcal{S} = \{\{1, 3, 5\}, \{2, 6\}, \{4\}\}$$

of $X = \{1, 2, 3, 4, 5, 6\}$. The relation R on X given by Theorem 2.5.1 contains the ordered pairs $(1, 1)$, $(1, 3)$, and $(1, 5)$ because $\{1, 3, 5\}$ is in $\mathcal{S}$. The complete relation is

$$R = \{(1, 1), (1, 3), (1, 5), (3, 1), (3, 3), (3, 5), (5, 1), (5, 3), (5, 5),$$

$$(2, 2), (2, 6), (6, 2), (6, 6), (4, 4)\}. \qquad \square$$

Let $\mathcal{S}$ and R be as in Theorem 2.5.1. If $S \in \mathcal{S}$, we can regard the members of S as equivalent in the sense of the relation R. For this reason, relations that are reflexive, symmetric, and transitive are called **equivalence relations**. In the example of Figure 2.5.1, the relation is "is the same color as"; hence *equivalent* means "is the same color as." Each set in the partition consists of all the balls of a particular color.

DEFINITION 2.5.3

A relation that is reflexive, symmetric, and transitive on a set X is called an *equivalence relation on X*.

EXAMPLE 2.5.4

The relation R of Example 2.5.2 is an equivalence relation on $\{1, 2, 3, 4, 5, 6\}$ because of Theorem 2.5.1. We can also verify directly that R is reflexive, symmetric, and transitive.

The digraph of the relation R of Example 2.5.2 is shown in Figure 2.5.2. Again, we see that R is reflexive (there is a loop at every vertex), symmetric (for every directed edge from v to w, there is also a directed edge from w to v), and transitive (if there is a directed edge from x to y and a directed edge from y to z, there is a directed edge from x to z). □

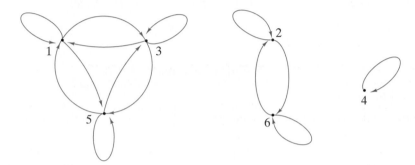

FIGURE 2.5.2   The digraph of the relation of Example 2.5.2.

EXAMPLE 2.5.5

Consider the relation

$$R = \{(1, 1), (1, 3), (1, 5), (2, 2), (2, 4), (3, 1), (3, 3), (3, 5), (4, 2),$$

$$(4, 4), (5, 1), (5, 3), (5, 5)\}$$

on $\{1, 2, 3, 4, 5\}$. R is reflexive because $(1, 1), (2, 2), (3, 3), (4, 4), (5, 5) \in R$. R is symmetric because whenever (x, y) is in R, (y, x) is also in R. Finally, R is transitive because whenever (x, y) and (y, z) are in R, (x, z) is also in R. Since R is reflexive, symmetric, and transitive, R is an equivalence relation on $\{1, 2, 3, 4, 5\}$. □

EXAMPLE 2.5.6

The relation R of Example 2.4.4 is not an equivalence relation because R is not symmetric. □

EXAMPLE 2.5.7

The relation R of Example 2.4.5 is not an equivalence relation because R is neither reflexive nor transitive. □

EXAMPLE 2.5.8

The relation R of Example 2.4.15 is an equivalence relation because R is reflexive, symmetric, and transitive.   □

Given an equivalence relation on a set X, we can partition X by grouping related members of X together. Elements related to one another may be thought of as equivalent. The next theorem gives the details.

THEOREM 2.5.9

Let R be an equivalence relation on a set X. For each $a \in X$, let

$$[a] = \{x \in X \mid x R a\}.$$

Then

$$\mathcal{S} = \{[a] \mid a \in X\}$$

is a partition of X.

Proof. We must show that every element in X belongs to exactly one member of $\mathcal{S}$.

Let $a \in X$. Since $a R a$, $a \in [a]$. Thus every element in X belongs to *at least one* member of $\mathcal{S}$. It remains to show that every element in X belongs to *exactly* one member of $\mathcal{S}$; that is,

$$\text{if } x \in X \text{ and } x \in [a] \cap [b], \text{ then } [a] = [b]. \tag{2.5.1}$$

We first show that if $a R b$, then $[a] = [b]$. Suppose that $a R b$. Let $x \in [a]$. Then $x R a$. Since $a R b$ and R is transitive, $x R b$. Therefore, $x \in [b]$ and $[a] \subseteq [b]$. The argument that $[b] \subseteq [a]$ is the same as that just given, but with the roles of a and b interchanged. Thus $[a] = [b]$.

We now prove (2.5.1). Assume that $x \in X$ and $x \in [a] \cap [b]$. Then $x R a$ and $x R b$. Our preceding result shows that $[x] = [a]$ and $[x] = [b]$. Thus $[a] = [b]$.   ■

DEFINITION 2.5.10

Let R be an equivalence relation on a set X. The sets $[a]$ defined in Theorem 2.5.9 are called the *equivalence classes of X given by the relation R*.

EXAMPLE 2.5.11

Consider the equivalence relation R of Example 2.5.2. The equivalence class [1] containing 1 consists of all x such that $(x, 1) \in R$. Therefore,

$$[1] = \{1, 3, 5\}.$$

The remaining equivalence classes are found similarly:

$$[3] = [5] = \{1, 3, 5\}, \qquad [2] = [6] = \{2, 6\}, \qquad [4] = \{4\}.   □$$

EXAMPLE 2.5.12

The equivalence classes appear quite clearly in the digraph of an equivalence relation. The three equivalence classes of the relation R of Example 2.5.2 appear in the digraph of R (shown in Figure 2.5.2) as the three subgraphs whose vertices are $\{1, 3, 5\}$, $\{2, 6\}$, and $\{4\}$. A subgraph G that represents an equivalence class is a largest subgraph of the original digraph having the property that for any vertices v and w in G, there is a directed edge from v to w. For example, if $v, w \in \{1, 3, 5\}$, there is a directed edge from v to w. Moreover, no additional vertices can be added to 1, 3, 5 so that the resulting vertex set has a directed edge between each pair of vertices. $\square$

EXAMPLE 2.5.13

There are two equivalence classes for the equivalence relation of Example 2.5.5, namely,
$$[1] = [3] = [5] = \{1, 3, 5\}, \qquad [2] = [4] = \{2, 4\}. \qquad \square$$

EXAMPLE 2.5.14

The equivalence classes for the equivalence relation of Example 2.4.15 are
$$[a] = \{a\}, \qquad [b] = \{b\}, \qquad [c] = \{c\}. \qquad \square$$

EXAMPLE 2.5.15

Let $X = \{1, 2, \ldots, 10\}$. Define $x R y$ to mean that 3 divides $x - y$. We can readily verify that the relation R is reflexive, symmetric, and transitive. Thus R is an equivalence relation on X.

Let us determine the members of the equivalence classes. The equivalence class $[1]$ consists of all x with $x R 1$. Thus
$$[1] = \{x \in X \mid 3 \text{ divides } x - 1\} = \{1, 4, 7, 10\}.$$

Similarly,
$$[2] = \{2, 5, 8\}$$
$$[3] = \{3, 6, 9\}.$$

These three sets partition X. Note that
$$[1] = [4] = [7] = [10], \qquad [2] = [5] = [8], \qquad [3] = [6] = [9].$$

For this relation, *equivalence* is "has the same remainder when divided by 3."
$$\square$$

We close this section by proving a special result that we will need later (see Sections 4.2 and 4.4). The proof is illustrated in Figure 2.5.3.

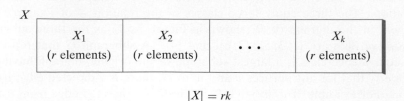

$$|X| = rk$$

FIGURE 2.5.3   The proof of Theorem 2.5.16.

THEOREM 2.5.16

Let R be an equivalence relation on a finite set X. If each equivalence class has r elements, there are $|X|/r$ equivalence classes.

Proof. Let $X_1, X_2, \ldots, X_k$ denote the distinct equivalence classes. Since these sets partition X,

$$|X| = |X_1| + |X_2| + \cdots + |X_k| = r + r + \cdots + r = kr$$

and the conclusion follows.   ■

Exercises

In Exercises 1–8, determine whether the given relation is an equivalence relation on $\{1, 2, 3, 4, 5\}$. If the relation is an equivalence relation, list the equivalence classes. (In Exercises 5–8, $x, y \in \{1, 2, 3, 4, 5\}$.)

1. $\{(1, 1)(2, 2), (3, 3), (4, 4), (5, 5), (1, 3), (3, 1)\}$

2. $\{(1, 1), (2, 2), (3, 3), (4, 4), (5, 5), (1, 3), (3, 1), (3, 4), (4, 3)\}$

3. $\{(1, 1), (2, 2), (3, 3), (4, 4)\}$

4. $\{(1, 1), (2, 2), (3, 3), (4, 4), (5, 5), (1, 5), (5, 1), (3, 5), (5, 3), (1, 3), (3, 1)\}$

5. $\{(x, y) \mid 1 \leq x \leq 5, 1 \leq y \leq 5\}$

6. $\{(x, y) \mid 4 \text{ divides } x - y\}$

7. $\{(x, y) \mid 3 \text{ divides } x + y\}$

8. $\{(x, y) \mid x \text{ divides } 2 - y\}$

In Exercises 9–14, list the members of the equivalence relation on $\{1, 2, 3, 4\}$ defined (as in Theorem 2.5.1) by the given partition. Also, find the equivalence classes [1], [2], [3], and [4].

9. $\{\{1, 2\}, \{3, 4\}\}$

10. $\{\{1\}, \{2\}, \{3, 4\}\}$

11. $\{\{1\}, \{2\}, \{3\}, \{4\}\}$

12. $\{\{1, 2, 3\}, \{4\}\}$

13. $\{\{1, 2, 3, 4\}\}$

14. $\{\{1\}, \{2, 4\}, \{3\}\}$

In Exercises 15–17, let $X = \{1, 2, 3, 4, 5\}$, $Y = \{3, 4\}$, and $C = \{1, 3\}$. Define the relation R on $\mathcal{P}(X)$, the set of all subsets of X, as

$$ARB \text{ if and only if } A \cup Y = B \cup Y.$$

15. Show that R is an equivalence relation.

16. List the elements of $[C]$, the equivalence class containing C.

17. How many distinct equivalence classes are there?

18. Let

$$X = \{\text{San Francisco, Pittsburgh, Chicago, San Diego,}$$
$$\text{Philadelphia, Los Angeles}\}.$$

Define a relation R on X as $x R y$ if x and y are in the same state.

(a) Show that R is an equivalence relation.

(b) List the equivalence classes of X.

19. Show that if R is an equivalence relation on X, then

$$\text{domain } R = \text{ range } R = X.$$

20. If an equivalence relation has only one equivalence class, what must the relation look like?

21. If R is an equivalence relation on a set X and $|X| = |R|$, what must the relation look like?

22. By listing ordered pairs, give an example of an equivalence relation on $\{1, 2, 3, 4, 5, 6\}$ having exactly four equivalence classes.

23. How many equivalence relations are there on the set $\{1, 2, 3\}$?

24. Let $X = \{1, 2, \ldots, 10\}$. Define a relation R on $X \times X$ by $(a, b) R (c, d)$ if $a + d = b + c$.

(a) Show that R is an equivalence relation on $X \times X$.

(b) List one member of each equivalence class of $X \times X$.

25. Let $X = \{1, 2, \ldots, 10\}$. Define a relation R on $X \times X$ by $(a, b) R (c, d)$ if $ad = bc$.

(a) Show that R is an equivalence relation on $X \times X$.

(b) List one member of each equivalence class of $X \times X$.

(c) Describe the relation R in familiar terms.

26. Let R be a reflexive and transitive relation on X. Show that $R \cap R^{-1}$ is an equivalence relation on X.

27. Let R_1 and R_2 be equivalence relations on X.

(a) Show that $R_1 \cap R_2$ is an equivalence relation on X.

(b) Describe the equivalence classes of $R_1 \cap R_2$ in terms of the equivalence classes of R_1 and the equivalence classes of R_2.

28. Suppose that S is a collection of subsets of a set X and $X = \cup S$. (It is not assumed that the family S is pairwise disjoint.) Define $x R y$ to mean that for some set $S \in S$, both x and y are in S. Is R necessarily reflexive, symmetric, or transitive?

29. Let S be a unit square including the interior, as shown in the accompanying figure. Define a relation R on S by $(x, y)R(x', y')$ if $(x = x'$ and $y = y')$ or $(y = y'$ and $x = 0$ and $x' = 1)$ or $(y = y'$ and $x = 1$ and $x' = 0)$.

(a) Show that R is an equivalence relation on S.

(b) If points in the same equivalence class are glued together, how would you describe the figure formed?

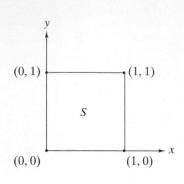

30. Let S be a unit square including the interior (as in Exercise 29). Define a relation R' on S by $(x, y)R'(x', y')$ if $(x = x'$ and $y = y')$ or $(y = y'$ and $x = 0$ and $x' = 1)$ or $(y = y'$ and $x = 1$ and $x' = 0)$ or $(x = x'$ and $y = 0$ and $y' = 1)$ or $(x = x'$ and $y = 1$ and $y' = 0)$. Let

$$R = R' \cup \{((0, 0), (1, 1)), ((0, 1), (1, 0)), ((1, 0), (0, 1)), ((1, 1), (0, 0))\}.$$

(a) Show that R is an equivalence relation on S.

(b) If points in the same equivalence class are glued together, how would you describe the figure formed?

Let R be a relation on a set X. Define

$$\rho(R) = R \cup \{(x, x) \mid x \in X\}$$
$$\sigma(R) = R \cup R^{-1}$$
$$R^n = R \circ R \circ R \circ \cdots \circ R \quad (n\ R\text{'s})$$
$$\tau(R) = \cup \{R^n \mid n = 1, 2, \ldots\}.$$

The relation $\tau(R)$ is called the *transitive closure* of R.

31. For the relations R_1 and R_2 of Exercise 37, Section 2.4, find $\rho(R_i)$, $\sigma(R_i)$, $\tau(R_i)$, and $\tau(\sigma(\rho(R_i)))$ for $i = 1, 2$.

32. Show that $\rho(R)$ is reflexive.   **33.** Show that $\sigma(R)$ is symmetric.

34. Show that $\tau(R)$ is transitive.

★ **35.** Show that $\tau(\sigma(\rho(R)))$ is an equivalence relation containing R.

★ **36.** Show that $\tau(\sigma(\rho(R)))$ is the smallest equivalence relation on X containing R; that is, show that if R' is an equivalence relation on X and $R' \supseteq R$, then $R' \supseteq \tau(\sigma(\rho(R)))$.

★ **37.** Show that R is transitive if and only if $\tau(R) = R$.

In Exercises 38–44, write "true" if the statement is true for all relations R_1 and R_2 on an arbitrary set X; otherwise, give a counterexample.

38. $\rho(R_1 \cup R_2) = \rho(R_1) \cup \rho(R_2)$ **39.** $\sigma(R_1 \cap R_2) = \sigma(R_1) \cap \sigma(R_2)$

40. $\tau(R_1 \cup R_2) = \tau(R_1) \cup \tau(R_2)$ **41.** $\tau(R_1 \cap R_2) = \tau(R_1) \cap \tau(R_2)$

42. $\sigma(\tau(R_1)) = \tau(\sigma(R_1))$ **43.** $\sigma(\rho(R_1)) = \rho(\sigma(R_1))$

44. $\rho(\tau(R_1)) = \tau(\rho(R_1))$

PROBLEM-SOLVING CORNER:

EQUIVALENCE RELATIONS

Problem

Answer the following questions for the relation R defined on the set of eight-bit strings by $s_1 R s_2$, provided that the first four bits of s_1 and s_2 coincide.

(a) Show that R is an equivalence relation.

(b) List one member of each equivalence class.

(c) How many equivalence classes are there?

Attacking the Problem

Let's begin by looking at some specific eight-bit strings that are related according to the relation R. Let's take an arbitrary string 01111010 and find strings related to it. A string s is related to 01111010 if the first four bits of 01111010 and s coincide. This means that s must begin 0111 and the last four bits can be anything. An example is $s = 01111000$.

Let's list all of the strings related to 01111010. In doing so, we must be careful to follow 0111 with every possible four-bit string:

01110000, 01110001, 01110010, 01110011,
01110100, 01110101, 01110110, 01110111,
01111000, 01111001, 01111010, 01111011,
01111100, 01111101, 01111110, 01111111.

Assuming for the moment that R is an equivalence relation, the equivalence class containing 01111010, denoted [01111010], consists of all strings related to 01111010. Therefore, what we have just computed are the members of [01111010].

Notice that if we take any string in [01111010], say 01111100, and compute its equivalence class [01111100], we will obtain exactly the same set of strings—namely, the set of eight-bit strings that begin 0111.

To obtain a different example, we would have to start with a string whose first four bits are different from 0111, say 1011. As an example, the strings related to 10110100 are

10110000, 10110001, 10110010, 10110011,
10110100, 10110101, 10110110, 10110111,
10111000, 10111001, 10111010, 10111011,
10111100, 10111101, 10111110, 10111111.

What we have just computed are the members of [10110100].

Before reading on, compute the members of some other equivalence class.

We see that [01111010] and [10110100] have no members in common. It is always the case that two equivalence classes are identical or have no members in common (see Theorem 2.5.9).

Finding a Solution

To show that R is an equivalence relation, we must show that R is reflexive, symmetric, and transitive (see Definition 2.5.3). For each property, we will go directly to the definition and check that the conditions specified in the definition hold.

For R to be reflexive, we must have sRs for every eight-bit string s. For sRs to be true, the first four bits of s and s must coincide. This is certainly the case!

For R to be symmetric, for all eight-bit strings s_1 and s_2, if $s_1 R s_2$, then $s_2 R s_1$. Using the definition of R, we may translate this condition to: If the first four bits of s_1 and s_2 coincide, then the first four bits of s_2 and s_1 coincide. This is also certainly the case!

For R to be transitive, for all eight-bit strings s_1, s_2, and s_3, if $s_1 R s_2$ and $s_2 R s_3$, then $s_1 R s_3$. Again using the definition of R, we may translate this condition to: If the first four bits of s_1 and s_2 coincide and the first four bits of s_2 and s_3 coincide, then the first four bits of s_1 and s_3 coincide. This too is certainly the case! We have proved that R is an equivalence relation.

In our earlier discussion, we found that each distinct four-bit string determines an equivalence class. For example, the string 0111 determines the equivalence class consisting of all eight-bit strings that begin 0111. Therefore, the number of equivalence classes is equal to the number of four-bit strings. We can simply list them all

$$0000, \quad 0001, \quad 0010, \quad 0011,$$
$$0100, \quad 0101, \quad 0110, \quad 0111,$$
$$1000, \quad 1001, \quad 1010, \quad 1011,$$
$$1100, \quad 1101, \quad 1110, \quad 1111$$

and then count them. There are 16 equivalence classes.

Consider the problem of listing one member of each equivalence class. The 16 four-bit strings listed previously determine the 16 equivalence classes. The first string 0000 determines the equivalence class consisting of all eight-bit strings that begin 0000; the second string 0001 determines the equivalence class consisting of all eight-bit strings that begin 0001; and so on. Thus to list one member of each equivalence class, we simply need to append some four-bit string to each of the strings in the previous list:

$$00000000, \quad 00010000, \quad 00100000, \quad 00110000,$$
$$01000000, \quad 01010000, \quad 01100000, \quad 01110000,$$
$$10000000, \quad 10010000, \quad 10100000, \quad 10110000,$$
$$11000000, \quad 11010000, \quad 11100000, \quad 11110000.$$

Formal Solution

(a) We have already presented a formal proof that R is an equivalence relation.

(b)

$$00000000, \quad 00010000, \quad 00100000, \quad 00110000,$$
$$01000000, \quad 01010000, \quad 01100000, \quad 01110000,$$
$$10000000, \quad 10010000, \quad 10100000, \quad 10110000,$$
$$11000000, \quad 11010000, \quad 11100000, \quad 11110000$$

lists one member of each equivalence class.

(c) There are 16 equivalence classes.

Summary of Problem-Solving Techniques

- List elements that are related.

- Compute some equivalence classes; that is, list *all* elements related to a particular element.

- It may help to solve the parts of a problem in a different order than that given in the problem statement. In our example, it was helpful in looking at some concrete cases to *assume* that the relation was an equivalence relation before actually proving that it was an equivalence relation.

- To show that a particular relation R is an equivalence relation, go directly to the definitions. Show that R is reflexive, symmetric, and transitive by directly verifying that R satisfies the definitions of reflexive, symmetric, and transitive.

- If the problem is to count the number of items satisfying some property (e.g., in our problem we were asked to count the number of equivalence classes) and the number is sufficiently small, just list all the items and count them directly.

Comments

In programming languages, usually only some specified number of characters of the names of variables and special terms (technically these are called *identifiers*) are significant. For example, in the C programming language, only the first 31 characters of identifiers are significant. This means that if two identifiers begin with the same 31 characters, the system is allowed to consider them identical.

If we define a relation R on the set of C identifiers by $s_1 R s_2$, provided that the first 31 characters of s_1 and s_2 coincide, then R is an equivalence relation. An equivalence class consists of identifiers that the system is allowed to consider identical.

2.6 *MATRICES OF RELATIONS*

A matrix is a convenient way to represent a relation R from X to Y. Such a representation can be used by a computer to analyze a relation. We label the rows with the elements of X (in some arbitrary order) and we label the columns with the elements of Y (again, in some arbitrary order). We then set the entry in row x and column y to 1 if $x R y$ and to 0 otherwise. This matrix is called the **matrix of the relation R** (relative to the orderings of X and Y).

EXAMPLE 2.6.1

The matrix of the relation

$$R = \{(1, b), (1, d), (2, c), (3, c), (3, b), (4, a)\}$$

from $X = \{1, 2, 3, 4\}$ to $Y = \{a, b, c, d\}$ relative to the orderings $1, 2, 3, 4$ and a, b, c, d is

$$
\begin{array}{c@{}c}
 & \begin{array}{cccc} a & b & c & d \end{array} \\
\begin{array}{c} 1 \\ 2 \\ 3 \\ 4 \end{array} &
\left(\begin{array}{cccc}
0 & 1 & 0 & 1 \\
0 & 0 & 1 & 0 \\
0 & 1 & 1 & 0 \\
1 & 0 & 0 & 0
\end{array}\right).
\end{array}
$$

□

EXAMPLE 2.6.2

The matrix of the relation R of Example 2.6.1 relative to the orderings $2, 3, 4, 1$ and d, b, a, c is

$$
\begin{array}{c@{}c}
 & \begin{array}{cccc} d & b & a & c \end{array} \\
\begin{array}{c} 2 \\ 3 \\ 4 \\ 1 \end{array} &
\left(\begin{array}{cccc}
0 & 0 & 0 & 1 \\
0 & 1 & 0 & 1 \\
0 & 0 & 1 & 0 \\
1 & 1 & 0 & 0
\end{array}\right).
\end{array}
$$

Obviously, the matrix of a relation from X to Y is dependent on the orderings of X and Y.   □

EXAMPLE 2.6.3

The matrix of the relation R from $\{2, 3, 4\}$ to $\{5, 6, 7, 8\}$, relative to the orderings $2, 3, 4$ and $5, 6, 7, 8$, defined by

$$x R y \qquad \text{if } x \text{ divides } y$$

is

$$
\begin{array}{c@{}c}
 & \begin{array}{cccc} 5 & 6 & 7 & 8 \end{array} \\
\begin{array}{c} 2 \\ 3 \\ 4 \end{array} &
\left(\begin{array}{cccc}
0 & 1 & 0 & 1 \\
0 & 1 & 0 & 0 \\
0 & 0 & 0 & 1
\end{array}\right).
\end{array}
$$

□

When we write the matrix of a relation R on a set X (i.e., from X to X), we use the same ordering for the rows as we do for the columns.

EXAMPLE 2.6.4

The matrix of the relation

$$R = \{(a, a), (b, b), (c, c), (d, d), (b, c), (c, b)\}$$

on $\{a, b, c, d\}$, relative to the ordering a, b, c, d, is

$$
\begin{array}{c@{}c}
 & \begin{array}{cccc} a & b & c & d \end{array} \\
\begin{array}{c} a \\ b \\ c \\ d \end{array} &
\left(\begin{array}{cccc}
1 & 0 & 0 & 0 \\
0 & 1 & 1 & 0 \\
0 & 1 & 1 & 0 \\
0 & 0 & 0 & 1
\end{array} \right).
\end{array}
$$
□

Notice that the matrix of a relation on a set X is always a square matrix.

We can quickly determine whether a relation R on a set X is reflexive by examining the matrix A of R (relative to some ordering). The relation R is reflexive if and only if A has 1's on the main diagonal. (The main diagonal of a square matrix consists of the entries on a line from the upper left to the lower right.) The relation R is reflexive if and only if $(x, x) \in R$ for all $x \in X$. But this last condition holds precisely when the main diagonal consists of 1's. Notice that the relation R of Example 2.6.4 is reflexive and that the main diagonal of the matrix of R consists of 1's.

We can also quickly determine whether a relation R on a set X is symmetric by examining the matrix A of R (relative to some ordering). The relation R is symmetric if and only if for all i and j, the ijth entry of A is equal to the jith entry of A. (Less formally, R is symmetric if and only if A is symmetric about the main diagonal.) The reason is that R is symmetric if and only if whenever (x, y) is in R, (y, x) is also in R. But this last condition holds precisely when A is symmetric about the main diagonal. Notice that the relation R of Example 2.6.4 is symmetric and that the matrix of R is symmetric about the main diagonal.

We can also quickly determine whether a relation R is antisymmetric by examining the matrix of R (relative to some ordering) (see Exercise 10). Unfortunately, there is no simple way to test whether a relation R on X is transitive by examining the matrix of R.

We conclude by showing how matrix multiplication relates to composition of relations.

EXAMPLE 2.6.5

Let R_1 be the relation from $X = \{1, 2, 3\}$ to $Y = \{a, b\}$ defined by

$$R_1 = \{(1, a), (2, b), (3, a), (3, b)\}$$

and let R_2 be the relation from Y to $Z = \{x, y, z\}$ defined by

$$R_2 = \{(a, x), (a, y), (b, y), (b, z)\}.$$

The matrix of R_1 relative to the orderings $1, 2, 3$ and a, b is

$$
A_1 = \begin{array}{c} \\ 1 \\ 2 \\ 3 \end{array}\begin{array}{c} a \quad b \\ \begin{pmatrix} 1 & 0 \\ 0 & 1 \\ 1 & 1 \end{pmatrix} \end{array}
$$

and the matrix of R_2 relative to the orderings a, b and x, y, z is

$$
A_2 = \begin{array}{c} \\ a \\ b \end{array}\begin{array}{c} x \quad y \quad z \\ \begin{pmatrix} 1 & 1 & 0 \\ 0 & 1 & 1 \end{pmatrix} \end{array}.
$$

The product of these matrices is

$$
A_1 A_2 = \begin{pmatrix} 1 & 1 & 0 \\ 0 & 1 & 1 \\ 1 & 2 & 1 \end{pmatrix}.
$$

Let us interpret this product.

The ikth entry in $A_1 A_2$ is computed as

$$
\begin{array}{c} \\ i \end{array}\begin{array}{c} a \quad b \\ \begin{pmatrix} s & t \end{pmatrix} \end{array}\begin{array}{c} k \\ \begin{pmatrix} u \\ v \end{pmatrix} \end{array} = su + tv.
$$

If this value is nonzero, then either su or tv is nonzero. Suppose that $su \neq 0$. (The argument is similar if $tv \neq 0$.) Then $s \neq 0$ and $u \neq 0$. This means that $(i, a) \in R_1$ and $(a, k) \in R_2$. This implies that $(i, k) \in R_2 \circ R_1$. We have shown that if the ikth entry in $A_1 A_2$ is nonzero, then $(i, k) \in R_2 \circ R_1$. The converse is also true, as we now show.

Assume that $(i, k) \in R_2 \circ R_1$. Then, either

1. $(i, a) \in R_1$ and $(a, k) \in R_2$

or

2. $(i, b) \in R_1$ and $(b, k) \in R_2$.

If 1 holds, then $s = 1$ and $u = 1$, so $su = 1$ and $su + tv$ is nonzero. Similarly, if 2 holds, $tv = 1$ and again we have $su + tv$ nonzero. We have shown that if $(i, k) \in R_2 \circ R_1$, then the ikth entry in $A_1 A_2$ is nonzero.

We have shown that $(i, k) \in R_2 \circ R_1$ if and only if the ikth entry in $A_1 A_2$ is nonzero; thus $A_1 A_2$ is "almost" the matrix of the relation $R_2 \circ R_1$. To obtain the matrix of the relation $R_2 \circ R_1$, we need only change all nonzero entries in $A_1 A_2$ to 1. Thus the matrix of the relation $R_2 \circ R_1$, relative to the previously chosen orderings $1, 2, 3$ and x, y, z, is

$$
\begin{array}{c} \\ 1 \\ 2 \\ 3 \end{array}\begin{array}{c} x \quad y \quad z \\ \begin{pmatrix} 1 & 1 & 0 \\ 0 & 1 & 1 \\ 1 & 1 & 1 \end{pmatrix} \end{array}. \qquad \square
$$

The argument given in Example 2.6.5 holds for any relations. We summarize this result as Theorem 2.6.6.

THEOREM 2.6.6

Let R_1 be a relation from X to Y and let R_2 be a relation from Y to Z. Choose orderings of X, Y, and Z. All matrices of relations are with respect to these orderings. Let A_1 be the matrix of R_1 and let A_2 be the matrix of R_2. The matrix of the relation $R_2 \circ R_1$ is obtained by replacing each nonzero term in the matrix product $A_1 A_2$ by 1.

Proof. The proof precedes the statement of the theorem.   ∎

Exercises

In Exercises 1–3, find the matrix of the relation R from X to Y relative to the orderings given.

1. $R = \{(1, \delta), (2, \alpha), (2, \Sigma), (3, \beta), (3, \Sigma)\}$
   Ordering of X: 1, 2, 3
   Ordering of Y: $\alpha, \beta, \Sigma, \delta$

2. R as in Exercise 1
   Ordering of X: 3, 2, 1
   Ordering of Y: $\Sigma, \beta, \alpha, \delta$

3. $R = \{(x, a), (x, c), (y, a), (y, b), (z, d)\}$
   Ordering of X: x, y, z
   Ordering of Y: a, b, c, d

In Exercises 4–6, find the matrix of the relation R on X relative to the ordering given.

4. $R = \{(1, 2), (2, 3), (3, 4), (4, 5)\}$
   Ordering of X: 1, 2, 3, 4, 5

5. R as in Exercise 1
   Ordering of X: 5, 3, 1, 2, 4

6. $R = \{(x, y) \mid x < y\}$
   Ordering of X: 1, 2, 3, 4

In Exercises 7–9, write the relation R, given by the matrix, as a set of ordered pairs.

7.
$$
\begin{array}{c c c c c}
  & w & x & y & z \\
a & 1 & 0 & 1 & 0 \\
b & 0 & 0 & 0 & 0 \\
c & 0 & 0 & 1 & 0 \\
d & 1 & 1 & 1 & 1
\end{array}
$$

8.
$$
\begin{array}{c c c c c}
  & 1 & 2 & 3 & 4 \\
1 & 1 & 0 & 1 & 0 \\
2 & 0 & 1 & 1 & 1
\end{array}
$$

9.
$$
\begin{array}{c c c c c}
  & w & x & y & z \\
w & 1 & 0 & 1 & 0 \\
x & 0 & 0 & 0 & 0 \\
y & 1 & 0 & 1 & 0 \\
z & 0 & 0 & 0 & 1
\end{array}
$$

10. How can we quickly determine whether a relation R is antisymmetric by examining the matrix of R (relative to some ordering)?

11. Tell whether the relation of Exercise 9 is reflexive, symmetric, transitive, antisymmetric, a partial order, and/or an equivalence relation.

12. Given the matrix of a relation R from X to Y, how can we find the matrix of the inverse relation R^{-1}?

13. Find the matrix of the inverse of each of the relations of Exercises 7 and 8.

In Exercises 14–16, find

(a) The matrix A_1 of the relation R_1 (relative to the given orderings).

(b) The matrix A_2 of the relation R_2 (relative to the given orderings).

(c) The matrix product $A_1 A_2$.

(d) Use the result of part (c) to find the matrix of the relation $R_2 \circ R_1$.

(e) Use the result of part (d) to find the relation $R_2 \circ R_1$ (as a set of ordered pairs).

14. $R_1 = \{(1, x), (1, y), (2, x), (3, x)\}$
$R_2 = \{(x, b), (y, b), (y, a), (y, c)\}$
Orderings: $1, 2, 3; x, y; a, b, c$

15. $R_1 = \{(x, y) \mid x \text{ divides } y\}$; R_1 is from X to Y
$R_2 = \{(y, z) \mid y > z\}$; R_2 is from Y to Z
Ordering of X and Y: $2, 3, 4, 5$
Ordering of Z: $1, 2, 3, 4$

16. $R_1 = \{(x, y) \mid x + y \leq 6\}$; R_1 is from X to Y
$R_2 = \{(y, z) \mid y = z + 1\}$; R_2 is from Y to Z
Ordering of X, Y, and Z: $1, 2, 3, 4, 5$

17. Given the matrix of an equivalence relation R on X, how can we easily find the equivalence class containing the element $x \in X$?

★ 18. Let R_1 be a relation from X to Y and let R_2 be a relation from Y to Z. Choose orderings of X, Y, and Z. All matrices of relations are with respect to these orderings. Let A_1 be the matrix of R_1 and let A_2 be the matrix of R_2. Show that the ikth entry in the matrix product $A_1 A_2$ is equal to the number of elements in the set

$$\{m \mid (i, m) \in R_1 \text{ and } (m, k) \in R_2\}.$$

†2.7 RELATIONAL DATABASES

The "bi" in a binary relation R refers to the fact that R has two columns when we write R as a table. It is often useful to allow a table to have an arbitrary number of columns. If a table has n columns, the corresponding relation is called an **n-ary relation**.

EXAMPLE 2.7.1

Table 2.7.1 represents a 4-ary relation. This table expresses the relationship among identification numbers, names, positions, and ages.   □

We can also express an n-ary relation as a collection of n-tuples.

This section can be omitted without loss of continuity.

TABLE 2.7.1
PLAYER

ID Number	Name	Position	Age
22012	Johnsonbaugh	c	22
93831	Glover	of	24
58199	Battey	p	18
84341	Cage	c	30
01180	Homer	1b	37
26710	Score	p	22
61049	Johnsonbaugh	of	30
39826	Singleton	2b	31

EXAMPLE 2.7.2

Table 2.7.1 can be expressed as the set

$$\{(22012, \text{Johnsonbaugh}, c, 22), \quad (93831, \text{Glover}, \text{of}, 24),$$
$$(58199, \text{Battey}, p, 18), \quad (84341, \text{Cage}, c, 30),$$
$$(01180, \text{Homer}, 1b, 37), \quad (26710, \text{Score}, p, 22),$$
$$(61049, \text{Johnsonbaugh}, \text{of}, 30), \quad (39826, \text{Singleton}, 2b, 31)\}$$

of 4-tuples.   □

A **database** is a collection of records that are manipulated by a computer. For example, an airline database might contain records of passengers' reservations, flight schedules, equipment, and so on. Computer systems are capable of storing large amounts of information in databases. The data are available to various applications. **Database management systems** are programs that help users access the information in databases. The **relational database model**, invented by E. F. Codd in 1970, is based on the concept of an n-ary relation. We will briefly introduce some of the fundamental ideas in the theory of relational databases. For more details on relational databases, the reader is referred to [Codd; Date; and Kroenke]. We begin with some of the terminology.

The columns of an n-ary relation are called **attributes**. The domain of an attribute is a set to which all the elements in that attribute belong. For example, in Table 2.7.1 the attribute Age might be taken to be the set of all positive integers less than 100. The attribute Name might be taken to be all strings over the alphabet having length 30 or less.

A single attribute or a combination of attributes for a relation is a **key** if the values of the attributes uniquely define an n-tuple. For example, in Table 2.7.1, we can take the attribute ID Number as a key. (It is assumed that each person has a unique identification number.) The attribute Name is not a key because different persons can have the same name. For the same reason, we cannot take the attribute Position or Age as a key. Name and Position, in combination, could be used as a key for Table 2.7.1, since in our example a player is uniquely defined by a name and a position.

A database management system responds to **queries**. A query is a request for information from the database. For example, "Find all persons who play outfield" is a meaningful query for the relation given by Table 2.7.1. We will discuss several operations on relations that are used to answer queries in the relational database model.

> **EXAMPLE 2.7.3**   *Select*

The selection operator chooses certain n-tuples from a relation. The choices are made by giving conditions on the attributes. For example, for the relation PLAYER given in Table 2.7.1,

$$\text{PLAYER [Position = c]}$$

will select the tuples

$$(22012, \text{Johnsonbaugh, c, } 22), \quad (84341, \text{Cage, c, } 30). \qquad \square$$

> **EXAMPLE 2.7.4**   *Project*

Whereas the selection operator chooses rows of a relation, the projection operator chooses columns. In addition, duplicates are eliminated. For example, for the relation PLAYER given by Table 2.7.1,

$$\text{PLAYER [Name, Position]}$$

will select the tuples

$$(\text{Johnsonbaugh, c}), \quad (\text{Glover, of}), \quad (\text{Battey, p}), \quad (\text{Cage, c}),$$
$$(\text{Homer, 1b}), \quad (\text{Score, p}), \quad (\text{Johnsonbaugh, of}), \quad (\text{Singleton, 2b}). \qquad \square$$

TABLE 2.7.2
ASSIGNMENT

PID	Team
39826	Blue Sox
26710	Mutts
58199	Jackalopes
01180	Mutts

> **EXAMPLE 2.7.5**   *Join*

The selection and projection operators manipulate a single relation; join manipulates two relations. The join operation on relations R_1 and R_2 begins by examining all pairs of tuples, one from R_1 and one from R_2. If the join condition is satisfied, the tuples are combined to form a new tuple. The join condition specifies a relationship between an attribute in R_1 and an attribute in R_2. For example, let us perform a join operation on Tables 2.7.1 and 2.7.2. As the condition we take

$$\text{ID Number} = \text{PID}.$$

We take a row from Table 2.7.1 and a row from Table 2.7.2 and if ID Number = PID, we combine the rows. For example, the ID Number 01180 in the fifth row (01180, Homer, 1b, 37) of Table 2.7.1 matches the PID in the fourth row (01180, Mutts) of Table 2.7.2. These tuples are combined by first writing the tuple from Table 2.7.1, following it by the tuple from Table 2.7.2, and eliminating the equal entries in the specified attributes to give

(01180, Homer, 1b, 37, Mutts).

This operation is expressed as

PLAYER [ID Number = PID] ASSIGNMENT.

The relation obtained by executing this join is shown in Table 2.7.3.   □

TABLE 2.7.3
PLAYER [ID Number = PID] ASSIGNMENT

ID Number	Name	Position	Age	Team
58199	Battey	p	18	Jackalopes
01180	Homer	1b	37	Mutts
26710	Score	p	22	Mutts
39826	Singleton	2b	31	Blue Sox

Most queries to a relational database require several operations to provide the answer.

EXAMPLE 2.7.6

Describe operations that provide the answer to the query "Find the names of all persons who play for some team."

If we first join the relations given by Tables 2.7.1 and 2.7.2 subject to the condition ID Number = PID, we will obtain Table 2.7.3, which lists all persons who play for some team as well as other information. To obtain the names, we need only project on the attribute Name. We obtain the relation

Name
Battey
Homer
Score
Singleton

Formally, these operations would be specified as

> TEMP := PLAYER [ID Number = PID] ASSIGNMENT
> TEMP [Name]   □

EXAMPLE 2.7.7

Describe operations that provide the answer to the query "Find the names of all persons who play for the Mutts."

If we first use the selection operator to pick the rows of Table 2.7.2 that reference Mutts' players, we obtain the relation

TEMP1

PID	Team
26710	Mutts
01180	Mutts

If we now join Table 2.7.1 and the relation TEMP1 subject to ID Number = PID, we obtain the relation

TEMP2

ID Number	Name	Position	Age	Team
01180	Homer	1b	37	Mutts
26710	Score	p	22	Mutts

If we project the relation TEMP2 on the attribute Name, we obtain the relation

Name
Homer
Score

We would formally specify these operations as follows:

> TEMP1 := ASSIGNMENT [Team = Mutts]
> TEMP2 := PLAYER [ID Number = PID] TEMP1
> TEMP2 [Name]   □

Notice that the operations

> TEMP1 := PLAYER [ID Number = PID] ASSIGNMENT
> TEMP2 := TEMP1 [Team = Mutts]
> TEMP2 [Name]

would also answer the query of Example 2.7.7.

❧ ❧ ❧
Exercises

1. Express the relation given by Table 2.7.4 as a set of *n*-tuples.

TABLE **2.7.4**
EMPLOYEE

ID	Name	Manager
1089	Suzuki	Zamora
5620	Kaminski	Jones
9354	Jones	Yu
9551	Ryan	Washington
3600	Beaulieu	Yu
0285	Schmidt	Jones
6684	Manacotti	Jones

2. Express the relation given by Table 2.7.5 as a set of *n*-tuples.

TABLE **2.7.5**
DEPARTMENT

Dept	Manager
23	Jones
04	Yu
96	Zamora
66	Washington

3. Express the relation given by Table 2.7.6 as a set of *n*-tuples.

TABLE **2.7.6**
SUPPLIER

Dept	Part No	Amount
04	335B2	220
23	2A	14
04	8C200	302
66	42C	3
04	900	7720
96	20A8	200
96	1199C	296
23	772	39

4. Express the relation given by Table 2.7.7 as a set of n-tuples.

TABLE 2.7.7
BUYER

Name	Part No
United Supplies	2A
ABC Unlimited	8C200
United Supplies	1199C
JCN Electronics	2A
United Supplies	335B2
ABC Unlimited	772
Danny's	900
United Supplies	772
Underhanded Sales	20A8
Danny's	20A8
DePaul University	42C
ABC Unlimited	20A8

In Exercises 5–20, write a sequence of operations to answer the query. Also, provide an answer to the query. Use Tables 2.7.4 to 2.7.7.

5. Find the names of all employees. (Do not include any managers.)

6. Find the names of all managers.

7. Find all part numbers.

8. Find the names of all buyers.

9. Find the names of all employees who are managed by Jones.

10. Find all part numbers supplied by department 96.

11. Find all buyers of part 20A8.

12. Find all employees in department 04.

13. Find the part numbers of parts of which there are at least 100 items on hand.

14. Find all department numbers of departments that supply parts to Danny's.

15. Find the part numbers and amounts of parts bought by United Supplies.

16. Find all managers of departments that produce parts for ABC Unlimited.

17. Find the names of all employees who work in departments that supply parts for JCN Electronics.

18. Find all buyers who buy parts in the department managed by Jones.

19. Find all buyers who buy parts that are produced by the department for which Suzuki works.

20. Find all part numbers and amounts for Zamora's department.

21. Make up at least three *n*-ary relations with artificial data that might be used in a medical database. Illustrate how your database would be used by posing and answering two queries. Also, write a sequence of operations that could be used to answer the queries.

22. Describe a *union* operation on a relational database. Illustrate how your operator works by answering the following query, using the relations of Tables 2.7.4 to 2.7.7: Find the names of all employees who work either in department 23 or department 96. Also, write a sequence of operations that could be used to answer the query.

23. Describe an *intersection* operation on a relational database. Illustrate how your operator works by answering the following query, using the relations of Tables 2.7.4 to 2.7.7: Find the names of all buyers who buy both parts 2A and 1199C. Also write a sequence of operations that could be used to answer the query.

24. Describe a *difference* operation on a relational database. Illustrate how your operator works by answering the following query, using the relations of Tables 2.7.4 to 2.7.7: Find the names of all employees who do not work in department 04. Also, write a sequence of operations that could be used to answer the query.

2.8 FUNCTIONS

A **function** is a special kind of relation. Recall (see Definition 2.4.1) that a relation R from X to Y is a subset of the Cartesian product $X \times Y$ and that

$$\text{domain } R = \{x \in X \mid (x, y) \in R \text{ for some } y \in Y\}.$$

If f is a relation from X to Y, in order for f also to be a function, the domain of f must equal X and if (x, y) and (x, y') are in f, we must have $y = y'$.

DEFINITION 2.8.1

A *function* f from X to Y is a relation from X to Y having the properties:

1. The domain of f is X.
2. If $(x, y), (x, y') \in f$, then $y = y'$.

A function from X to Y is sometimes denoted $f : X \rightarrow Y$.

EXAMPLE 2.8.2

The relation

$$f = \{(1, a), (2, b), (3, a)\}$$

from $X = \{1, 2, 3\}$ to $Y = \{a, b, c\}$ is a function from X to Y. The domain of f is X and the range of f is $\{a, b\}$. [Recall (see Definition 2.4.1) that the range of a relation R is the set

$$\{y \in Y \mid (x, y) \in R \text{ for some } x \in X\}.]$$ □

A sequence (see Section 2.2) is a special type of function. A sequence whose smallest index is 1 is a function whose domain is either the set of all positive integers or a set of the form $\{1, \ldots, n\}$. The sequence of Example 2.2.2 has domain $\{1, 2, 3, 4, 5\}$. The sequence of Example 2.2.1 has domain equal to the set of all positive integers.

EXAMPLE 2.8.3

The relation

$$R = \{(1, a), (2, a), (3, b)\} \tag{2.8.1}$$

from $X = \{1, 2, 3, 4\}$ to $Y = \{a, b, c\}$ is not a function from X to Y. Property 1 of Definition 2.8.1 is violated. The domain of R, $\{1, 2, 3\}$, is not equal to X. If (2.8.1) were regarded as a relation from $X' = \{1, 2, 3\}$ to $Y = \{a, b, c\}$, it would be a function from X' to Y.   □

EXAMPLE 2.8.4

The relation

$$R = \{(1, a), (2, b), (3, c), (1, b)\}$$

from $X = \{1, 2, 3\}$ to $Y = \{a, b, c\}$ is not a function from X to Y. Property 2 of Definition 2.8.1 is violated. We have $(1, a)$ and $(1, b)$ in R but $a \neq b$.   □

Given a function f from X to Y, according to Definition 2.8.1, for each element x of the domain X, there is exactly one $y \in Y$ with $(x, y) \in f$. This unique value y is denoted $f(x)$. In other words, $y = f(x)$ is another way to write $(x, y) \in f$.

EXAMPLE 2.8.5

For the function f of Example 2.8.2, we may write

$$f(1) = a, \qquad f(2) = b, \qquad f(3) = a.   □$$

Functions involving the **modulus operator** play an important role in mathematics and computer science.

DEFINITION 2.8.6

If x is a nonnegative integer and y is a positive integer, we define x mod y to be the remainder when x is divided by y.

EXAMPLE 2.8.7

$$6 \bmod 2 = 0, \quad 5 \bmod 1 = 0, \quad 8 \bmod 12 = 8, \quad 199673 \bmod 2 = 1   □$$

EXAMPLE 2.8.8

What day of the week will it be 365 days from Wednesday?

Seven days after Wednesday it is Wednesday again; 14 days after Wednesday it is Wednesday again; and in general, if n is a positive integer, after $7n$ days it is Wednesday again. Thus we need to remove as many 7's as possible from 365 and see how many days are left. But this is simply $365 \mod 7 = 1$. Thus 365 days from Wednesday, it will be one day later, namely Thursday. This explains why, except for leap year when an extra day is added to February, the identical month and date in consecutive years moves forward one day of the week. □

EXAMPLE 2.8.9   *International Standard Book Numbers*

An International Standard Book Number (ISBN) is a code of 10 characters separated by dashes such as 0–8065–0959–7. An ISBN consists of four parts: a group code, a publisher code, a code that uniquely identifies the book among those published by the particular publisher, and a check character. The check character is used to validate an ISBN.

For the ISBN 0–8065–0959–7, the group code is 0, which identifies the book as one from an English-speaking country. The publisher code 8065 identifies the book as one published by Citadel Press. The code 0959 uniquely identifies the book among those published by Citadel Press (Brode: *Woody Allen: His Films and Career*, in this case). The check character is $s \mod 11$, where s is the sum of the first digit plus two times the second digit plus three times the third digit, ..., plus nine times the ninth digit. If this value is 10, the check character is X. For example, the sum s for the ISBN 0–8065–0959–7 is

$$s = 0 + 2 \cdot 8 + 3 \cdot 0 + 4 \cdot 6 + 5 \cdot 5 + 6 \cdot 0 + 7 \cdot 9 + 8 \cdot 5 + 9 \cdot 9 = 249.$$

Thus the check character is $249 \mod 11 = 7$. □

EXAMPLE 2.8.10   *Hash Functions*

Suppose that we have cells in a computer memory indexed from 0 to 10 (see Figure 2.8.1). We wish to store and retrieve arbitrary nonnegative integers in these cells. One approach is to use a **hash function**. A hash function takes a data item to be stored or retrieved and computes the first choice for a location for the item. For example, for our problem, to store or retrieve the number n, we might take as the first choice for a location, $n \mod 11$. Our hash function becomes

$$h(n) = n \mod 11.$$

Figure 2.8.1 shows the result of storing 15, 558, 32, 132, 102, and 5, in this order, in initially empty cells.

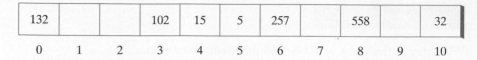

132			102	15	5	257		558		32
0	1	2	3	4	5	6	7	8	9	10

FIGURE 2.8.1   Cells in a computer memory.

Now suppose that we want to store 257. Since $h(257) = 4$, 257 should be stored at location 4; however, this position is already occupied. In this case we say that a **collision** has occurred. More precisely, a collision occurs for a hash function H if $H(x) = H(y)$, but $x \neq y$. To handle collisions, a **collision resolution policy** is required. One simple collision resolution policy is to find the next highest (with 0 assumed to follow 10) unoccupied cell. If we use this collision resolution policy, we would store 257 at location 6 (see Figure 2.8.1).

If we want to locate a stored value n, we compute $m = h(n)$ and begin looking at location m. If n is not at this position, we look in the next highest position (again, 0 is assumed to follow 10); if n is not in this position, we proceed to the next highest position, and so on. If we reach an empty cell or return to our original position, we conclude that n is not present; otherwise, we obtain the position of n.

If collisions occur infrequently, and if when one does occur it is resolved quickly, then hashing provides a very fast method of storing and retrieving data. As an example, personnel data are frequently stored and retrieved by hashing on employee identification numbers.   □

We next define the **floor** and **ceiling** of a real number.

DEFINITION 2.8.11

The *floor* of x, denoted $\lfloor x \rfloor$, is the greatest integer less than or equal to x. The *ceiling* of x, denoted $\lceil x \rceil$, is the least integer greater than or equal to x.

EXAMPLE 2.8.12

$$\lfloor 8.3 \rfloor = 8, \qquad \lceil 9.1 \rceil = 10,$$
$$\lfloor -8.7 \rfloor = -9, \qquad \lceil -11.3 \rceil = -11,$$
$$\lceil 6 \rceil = 6, \qquad \lceil -8 \rceil = -8 \qquad \square$$

The floor of x "rounds x down" while the ceiling of x "rounds x up." We will use the floor and ceiling functions throughout the book.

EXAMPLE 2.8.13

In 1996, the U.S. first-class postage rate for up to 11 ounces was 32 cents for the first ounce or fraction thereof and 23 cents for each additional ounce or fraction thereof. The postage $P(w)$ as a function of weight w is given by the equation

$$P(w) = 32 + 23\lceil w - 1 \rceil, \qquad 11 \geq w > 0.$$

The expression $\lceil w - 1 \rceil$ counts the number of additional ounces beyond 1 with a fraction counting as one additional ounce. As examples,

$$P(3.7) = 32 + 23\lceil 3.7 - 1 \rceil = 32 + 23\lceil 2.7 \rceil = 32 + 23 \cdot 3 = 101,$$
$$P(2) = 32 + 23\lceil 2 - 1 \rceil = 32 + 23\lceil 1 \rceil = 32 + 23 \cdot 1 = 55. \qquad \square$$

DEFINITION 2.8.14

A function f from X to Y is said to be *one-to-one* (or *injective*) if for each $y \in Y$, there is at most one $x \in X$ with $f(x) = y$.

The condition given in Definition 2.8.14 for a function to be one-to-one is equivalent to: If $x, x' \in X$ and $f(x) = f(x')$, then $x = x'$.

Because the amount of potential data is usually so much larger than the available memory, hash functions are usually not one-to-one. In other words, most hash functions produce collisions.

EXAMPLE 2.8.15

The function

$$f = \{(1, b), (3, a), (2, c)\}$$

from $X = \{1, 2, 3\}$ to $Y = \{a, b, c, d\}$ is one-to-one. $\qquad \square$

EXAMPLE 2.8.16

The function of Example 2.8.2 is not one-to-one since $f(1) = a = f(3)$. $\quad \square$

If the range of a function f is Y, the function is said to be **onto** Y.

DEFINITION 2.8.17

If f is a function from X to Y and the range of f is Y, f is said to be *onto Y* (or an *onto function* or a *surjective function*).

EXAMPLE 2.8.18

The function

$$f = \{(1, a), (2, c), (3, b)\}$$

from $X = \{1, 2, 3\}$ to $Y = \{a, b, c\}$ is one-to-one and onto Y. $\qquad \square$

EXAMPLE 2.8.19

The function f of Example 2.8.15 is *not* onto $Y = \{a, b, c, d\}$. It is onto $\{a, b, c\}$.   □

DEFINITION 2.8.20

A function that is both one-to-one and onto is called a *bijection*.

EXAMPLE 2.8.21

The function f of Example 2.8.18 is a bijection.   □

Suppose that f is a one-to-one, onto function from X to Y. It can be shown (see Exercise 60) that the inverse relation

$$\{(y, x) \mid (x, y) \in f\}$$

is a function from Y to X. This new function, denoted f^{-1}, is called f **inverse**.

EXAMPLE 2.8.22

For the function f of Example 2.8.18, we have

$$f^{-1} = \{(a, 1), (c, 2), (b, 3)\}.$$ □

Since functions are special kinds of relations, we can form the **composition** of two functions. Specifically, suppose that g is a function from X to Y and f is a function from Y to Z. Given $x \in X$, we may apply g to determine a unique element $y = g(x) \in Y$. We may then apply f to determine a unique element $z = f(y) = f(g(x)) \in Z$. The resulting function from X to Z is called the *composition of f with g* and is denoted $f \circ g$.

EXAMPLE 2.8.23

Given

$$g = \{(1, a), (2, a), (3, c)\},$$

a function from $X = \{1, 2, 3\}$ to $Y = \{a, b, c\}$, and

$$f = \{(a, y), (b, x), (c, z)\},$$

a function from Y to $Z = \{x, y, z\}$, the composition function from X to Z is the function

$$f \circ g = \{(1, y), (2, y), (3, z)\}.$$ □

A **binary operator** on a set X associates with each ordered pair of elements in X one element in X.

DEFINITION 2.8.24

A function from $X \times X$ into X is called a *binary operator* on X.

EXAMPLE 2.8.25

Let $X = \{1, 2, \ldots\}$. If we define

$$f(x, y) = x + y,$$

then f is a binary operator on X.   □

EXAMPLE 2.8.26

Let $X = \{a, b, c\}$. If we define

$$f(s, t) = st,$$

where s and t are strings over X and st is the concatenation of s and t, then f is a binary operator on X^*.   □

A **unary operator** on a set X associates with each single element of X one element in X.

DEFINITION 2.8.27

A function from X into X is called a *unary operator* on X.

EXAMPLE 2.8.28

Let U be a universal set. If we define

$$f(X) = \overline{X}, \qquad X \subseteq U,$$

then f is a unary operator on $\mathcal{P}(U)$.   □

ॐ ॐ ॐ

Exercises

Determine whether each relation in Exercises 1–5 is a function from $X = \{1, 2, 3, 4\}$ to $Y = \{a, b, c, d\}$. If it is a function, find its domain and range and determine if it is one-to-one or onto. If it is both one-to-one and onto, give the description of the inverse function as a set of ordered pairs and give the domain and range of the inverse function.

1. $\{(1, a), (2, a), (3, c), (4, b)\}$

2. $\{(1, c), (2, a), (3, b), (4, c), (2, d)\}$

3. $\{(1, c), (2, d), (3, a), (4, b)\}$

4. $\{(1, d), (2, d), (4, a)\}$

5. $\{(1, b), (2, b), (3, b), (4, b)\}$

6. Give an example of a function that is one-to-one but not onto.

7. Give an example of a function that is onto but not one-to-one.

8. Give an example of a function that is neither one-to-one nor onto.

9. Given

$$g = \{(1, b), (2, c), (3, a)\},$$

a function from $X = \{1, 2, 3\}$ to $Y = \{a, b, c, d\}$, and

$$f = \{(a, x), (b, x), (c, z), (d, w)\},$$

a function from Y to $Z = \{w, x, y, z\}$, write $f \circ g$ as a set of ordered pairs.

10. Given

$$f = \{(x, x^2) \mid x \in X\},$$

a function from $X = \{-5, -4, \ldots, 4, 5\}$ to the set of integers, write f as a set of ordered pairs. Is f one-to-one or onto?

11. How many functions are there from $\{1, 2\}$ into $\{a, b\}$? Which are one-to-one? Which are onto?

12. Given

$$f = \{(a, b), (b, a), (c, b)\},$$

a function from $X = \{a, b, c\}$ to X:

(a) Write $f \circ f$ and $f \circ f \circ f$ as sets of ordered pairs.

(b) Define

$$f^n = f \circ f \circ \cdots \circ f$$

to be the n-fold composition of f with itself. Find f^9 and f^{623}.

13. Let f be the function from $X = \{0, 1, 2, 3, 4\}$ to X defined by

$$f(x) = 4x \bmod 5.$$

Write f as a set of ordered pairs. Is f one-to-one or onto?

14. Let f be the function from $X = \{0, 1, 2, 3, 4, 5\}$ to X defined by

$$f(x) = 4x \bmod 6.$$

Write f as a set of ordered pairs. Is f one-to-one or onto?

★ 15. Let m and n be positive integers. Let f be the function from

$$X = \{0, 1, \ldots, m - 1\}$$

to X defined by

$$f(x) = nx \bmod m.$$

Find conditions on m and n that ensure that f is one-to-one and onto.

16. Verify the ISBN check character for this book.

17. Universal product codes (UPC) are the familiar bar codes that identify products so that they can be automatically priced at the checkout counter. A UPC is a 12-digit code in which the first digit characterizes the type of product (0 identifies an ordinary grocery item, 2 is an item sold by weight, 3 is a medical item, 4 is a special item, 5 is a coupon, and 6 and 7 are items not sold in retail stores). The next five digits identify the manufacturer, the next five digits identify the product, and the last digit is a check digit. (All UPC codes have a check digit. It is always present on the bar code, but it may not appear in the printed version.) For example, the UPC for a package of 10 Ortega taco shells is 0–54400–00800–5. The first zero identifies this as an ordinary grocery item, the next five digits 54400 identify the manufacturer Nabisco Foods, and the next five digits 00800 identify the product as a package of 10 Ortega taco shells.

The check digit is computed as follows. First compute s, where s is 3 times the sum of every other number starting with the first plus the sum of every other number, except the check digit, starting with the second. The check digit is the number c, between 0 and 9 satisfying $(c + s) \bmod 10 = 0$. For the code on the package of taco shells, we would have

$$s = 3(0 + 4 + 0 + 0 + 8 + 0) + 5 + 4 + 0 + 0 + 0 = 45.$$

Since $(5 + 45) \bmod 10 = 0$, the check digit is 5.

Find the check digit for the UPC whose first 11 digits are 3–41280–21414.

For each hash function in Exercises 18–21, show how the data would be inserted in the order given in initially empty cells. Use the collision resolution policy of Example 2.8.10.

18. $h(x) = x \bmod 11$; cells indexed 0 to 10; data: 53, 13, 281, 743, 377, 20, 10, 796

19. $h(x) = x \bmod 17$; cells indexed 0 to 16; data: 714, 631, 26, 373, 775, 906, 509, 2032, 42, 4, 136, 1028

20. $h(x) = x^2 \bmod 11$; cells and data as in Exercise 18

21. $h(x) = (x^2 + x) \bmod 17$; cells and data as in Exercise 19

22. Suppose that we store and retrieve data as described in Example 2.8.10. Will any problem arise if we delete data? Explain.

23. Suppose that we store data as described in Example 2.8.10 and that we never store more than 10 items. Will any problem arise when retrieving data if we stop searching when we encounter an empty cell? Explain.

24. Suppose that we store data as described in Example 2.8.10 and retrieve data as described in Exercise 23. Will any problem arise if we delete data? Explain.

Let g be a function from X to Y and let f be a function from Y to Z. For each statement in Exercises 25–30, write "true" if the statement is true. If the statement is false, give a counterexample.

25. If f is one-to-one, then $f \circ g$ is one-to-one.

26. If f and g are onto, then $f \circ g$ is onto.

27. If f and g are one-to-one and onto, then $f \circ g$ is one-to-one and onto.

28. If $f \circ g$ is one-to-one, then f is one-to-one.

29. If $f \circ g$ is one-to-one, then g is one-to-one.

30. If $f \circ g$ is onto, then f is onto.

If f is a function from X to Y and $A \subseteq X$ and $B \subseteq Y$, we define

$$f(A) = \{f(x) \mid x \in A\}, \qquad f^{-1}(B) = \{x \in X \mid f(x) \in B\}.$$

We call $f^{-1}(B)$ the *inverse image* of B under f.

31. Let

$$g = \{(1, a), (2, c), (3, c)\}$$

be a function from $X = \{1, 2, 3\}$ to $Y = \{a, b, c, d\}$. Let $S = \{1\}$, $T = \{1, 3\}$, $U = \{a\}$, and $V = \{a, c\}$. Find $g(S)$, $g(T)$, $g^{-1}(U)$, and $g^{-1}(V)$.

★ **32.** Let f be a function from X to Y. Prove that f is one-to-one if and only if

$$f(A \cap B) = f(A) \cap f(B)$$

for all subsets A and B of X.

33. Let f be a function from X to Y. Define a relation R on X by

$$x R y \qquad \text{if} \quad f(x) = f(y).$$

Show that R is an equivalence relation on X.

34. Let f be a function from X onto Y. Let

$$S = \{f^{-1}(\{y\}) \mid y \in Y\}.$$

[The definition of $f^{-1}(B)$, where B is a set, precedes Exercise 31.] Show that S is a partition of X. Describe an equivalence relation that gives rise to this partition.

35. Let R be an equivalence relation on a set A. Define a function f from A to the set of equivalence classes of A by the rule

$$f(x) = [x].$$

When do we have $f(x) = f(y)$?

36. Let R be an equivalence relation on a set A. Suppose that g is a function from A into a set X having the property that if $x R y$, then $g(x) = g(y)$. Show that

$$h([x]) = g(x)$$

defines a function from the set of equivalence classes of A into X. [What needs to be shown is that h *uniquely* assigns a value to $[x]$; that is, if $[x] = [y]$, then $g(x) = g(y)$.]

★ **37.** Let f be a function from X to Y. Show that f is one-to-one if and only if whenever g is a one-to-one function from any set A to X, $f \circ g$ is one-to-one.

★ **38.** Let f be a function from X to Y. Show that f is onto Y if and only if whenever g is a function from Y onto any set Z, $g \circ f$ is onto Z.

Let U be a universal set and let $X \subseteq U$. Define

$$C_X(x) = \begin{cases} 1 & \text{if } x \in X \\ 0 & \text{if } x \notin X. \end{cases}$$

We call C_X the *characteristic function* of X (in U).

39. Show that $C_{X \cap Y}(x) = C_X(x) C_Y(x)$ for all $x \in U$.

40. Show that $C_{X \cup Y}(x) = C_X(x) + C_Y(x) - C_X(x) C_Y(x)$ for all $x \in U$.

41. Show that $C_{\overline{X}}(x) = 1 - C_X(x)$ for all $x \in U$.

42. Show that $C_{X-Y}(x) = C_X(x)[1 - C_Y(x)]$ for all $x \in U$.

43. Show that if $X \subseteq Y$, then $C_X(x) \leq C_Y(x)$ for all $x \in U$.

44. Show that $C_{X \cup Y}(x) = C_X(x) + C_Y(x)$ for all $x \in U$, if and only if $X \cap Y = \emptyset$.

45. Find a formula for $C_{X \triangle Y}$. ($X \triangle Y$ is the symmetric difference of X and Y. The definition is given before Exercise 61, Section 2.1.)

46. Show that the function f from $\mathcal{P}(U)$ to the set of characteristic functions in U defined by

$$f(X) = C_X$$

is one-to-one and onto.

47. Let f be a characteristic function in X. Define a relation R on X by $x R y$ if $f(x) = f(y)$. According to Exercise 33, R is an equivalence relation. What are the equivalence classes?

If X and Y are sets, we define X to be *equivalent* to Y if there is a one-to-one, onto function from X to Y.

48. Show that set equivalence is an equivalence relation.

49. If X and Y are finite sets and X is equivalent to Y, what does this say about X and Y?

50. Show that the sets $\{1, 2, \ldots\}$ and $\{2, 4, \ldots\}$ are equivalent.

★ **51.** Show that for any set X, X is not equivalent to $\mathcal{P}(X)$, the power set of X.

52. Let X and Y be sets. Show that there is a one-to-one function from X to Y if and only if there is a function from Y onto X.

A binary operator f on a set X is *commutative* if $f(x, y) = f(y, x)$ for all $x, y \in X$. In Exercises 53–57, state whether the given function f is a binary operator on the set X. If f is not a binary operator, state why. State whether each binary operator is commutative or not.

53. $f(x, y) = x + y$, $X = \{1, 2, \ldots\}$

54. $f(x, y) = x - y$, $X = \{1, 2, \ldots\}$

55. $f(s, t) = st$, X is the set of strings over $\{a, b\}$

56. $f(x, y) = x/y$, $X = \{0, 1, 2, \ldots\}$

57. $f(x, y) = x^2 + y^2 - xy$, $X = \{1, 2, \ldots\}$

In Exercises 58 and 59, give an example of a unary operator (different from $f(x) = x$, for all x) on the given set.

58. $\{\ldots, -2, -1, 0, 1, 2, \ldots\}$

59. The set of strings over $\{a, b\}$

60. Show that if f is a one-to-one, onto function from X to Y, then

$$\{(y, x) \mid (x, y) \in f\}$$

is a one-to-one, onto function from Y to X.

61. How can we quickly determine whether a relation R is a function by examining the matrix of R (relative to some ordering)?

62. Let A be the matrix of a function f from X to Y (relative to some orderings of X and Y). What conditions must A satisfy for f to be onto Y?

63. Let A be the matrix of a function f from X to Y (relative to some orderings of X and Y). What conditions must A satisfy for f to be one-to-one?

In Exercises 64–66, write "true" if the statement is true for all real numbers. If the statement is false, give a counterexample.

64. $\lceil x + 3 \rceil = \lceil x \rceil + 3$

65. $\lceil x + y \rceil = \lceil x \rceil + \lceil y \rceil$

66. $\lfloor x + y \rfloor = \lfloor x \rfloor + \lceil y \rceil$

67. Show that if n is an odd integer,

$$\left\lfloor \frac{n^2}{4} \right\rfloor = \left(\frac{n-1}{2} \right) \left(\frac{n+1}{2} \right).$$

68. Show that if n is an odd integer,

$$\left\lceil \frac{n^2}{4} \right\rceil = \frac{n^2 + 3}{4}.$$

January 1 in year x occurs on the day of the week shown in the second column of row

$$y = \left(x + \left\lfloor \frac{x-1}{4} \right\rfloor - \left\lfloor \frac{x-1}{100} \right\rfloor + \left\lfloor \frac{x-1}{400} \right\rfloor \right) \bmod 7$$

in the following table (see [Ritter]).

y	January 1	Non-Leap Year	Leap Year
0	Sunday	January, October	January, April, July
1	Monday	April, July	September, December
2	Tuesday	September, December	June
3	Wednesday	June	March, November
4	Thursday	February, March, November	February, August
5	Friday	August	May
6	Saturday	May	October

The months with Friday the 13th in year x are found in row y in the appropriate column.

69. Find the months with Friday the 13th in 1945.

70. Find the months with Friday the 13th in the present year.

71. Find the months with Friday the 13th in 2000.

72. Let X denote the set of sequences with finite domain. Define a relation R on X as sRt if $|\text{domain } s| = |\text{domain } t|$ and, if the domain of s is $\{m, m+1, \ldots, m+k\}$ and the domain of t is $\{n, n+1, \ldots, n+k\}$, $s_{m+i} = t_{n+i}$ for $i = 0, \ldots, k$.

(a) Show that R is an equivalence relation.

(b) Explain in words what it means for two sequences in X to be equivalent under the relation R.

(c) A sequence is a function and so is a set of ordered pairs. Two sequences are equal if the two sets of ordered pairs are equal. Contrast the difference between two equivalent sequences in X and two equal sequences in X.

◈ NOTES

Most general references on discrete mathematics address the topics of this chapter. [Halmos; Lipschutz; and Stoll] are recommended to the reader wanting to study set theory, relations, and functions in more detail. [Codd; Date; Kroenke; and Ullman] are recommended references on databases in general and the relational model in particular.

෴ CHAPTER REVIEW

Section 2.1

Set: any collection of objects

Notation for sets:

$\{x \mid x \text{ has property } P\}$

$|X|$: the number of elements in the set X

$x \in X$: x is an element of the set X

$x \notin X$: x is not an element of the set X

Empty set: $\emptyset$ or $\{\}$

$X = Y$, where X and Y are sets: X and Y have the same elements

$X \subseteq Y$, X is a subset of Y: every element in X is also in Y

X is a proper subset of Y: $X \subseteq Y$ and $X \neq Y$

$\mathcal{P}(X)$, the power set of X: set of all subsets of X

$|\mathcal{P}(X)| = 2^{|X|}$

$X \cup Y$, X union Y: set of elements in X *or* Y

Union of a family $\mathcal{S}$ of sets: $\cup \mathcal{S} = \{x \mid x \in X$ for some $X \in \mathcal{S}\}$

$X \cap Y$, X intersect Y: set of elements in X *and* Y

Intersection of a family $\mathcal{S}$ of sets: $\cap \mathcal{S} = \{x \mid x \in X$ for all $X \in \mathcal{S}\}$

Disjoint sets X and Y: $X \cap Y = \emptyset$

Pairwise disjoint family of sets

$X - Y$, difference of X and Y, relative complement: set of elements in X but not in Y

Universal set, universe

$\overline{X}$, complement of X: $U - X$, where U is a universal set

Properties of sets (see Theorem 2.1.8)

De Morgan's laws for sets:
$$\overline{(A \cup B)} = \overline{A} \cap \overline{B},$$
$$\overline{(A \cap B)} = \overline{A} \cup \overline{B}$$

Partition of X: a collection $\mathcal{S}$ of nonempty subsets of X such that every element in X belongs to exactly one member of $\mathcal{S}$

Ordered pair: (x, y)

Cartesian product of X and Y:
$$X \times Y = \{(x, y) \mid x \in X, y \in Y\}$$

Cartesian product of $X_1, X_2, \ldots, X_n$:
$$X_1 \times X_2 \times \cdots \times X_n$$
$$= \{(a_1, a_2, \ldots, a_n) \mid a_i \in X_i\}$$

Section 2.2

Sequence: list in which order is taken into account

Index: in the sequence $\{s_n\}$, n is the index

Increasing sequence: $s_n \leq s_{n+1}$ for all n

Decreasing sequence: $s_n \geq s_{n+1}$ for all n

Subsequence s_{n_k} of the sequence $\{s_n\}$

Sum or sigma notation:
$$\sum_{i=m}^{n} a_i = a_m + a_{m+1} + \cdots + a_n$$

Product notation:
$$\prod_{i=m}^{n} a_i = a_m \cdot a_{m+1} \cdots a_n$$

String: finite sequence

Null string, λ: string with no elements

X^*: set of all strings over X, including the null string

X^+: set of all nonnull strings over X

Length of string α, $|\alpha|$: number of elements in α

Concatenation of strings α and β, $\alpha\beta$: α followed by β

Section 2.3

Decimal number system

Binary number system

Hexadecimal number system

Base of a number system

Convert binary to decimal

Convert decimal to binary

Convert hexadecimal to decimal

Convert decimal to hexadecimal

Add binary numbers

Add hexadecimal numbers

Section 2.4

Binary relation from X to Y:
  set of ordered pairs (x, y), $x \in X$,
  $y \in Y$

Domain of a binary relation R:
  $\{x \mid (x, y) \in R\}$

Range of a binary relation R:
  $\{y \mid (x, y) \in R\}$

Digraph of a binary relation

Reflexive relation R on X:
  $(x, x) \in R$ for all $x \in X$

Symmetric relation R on X:
  for all $x, y \in X$, if $(x, y) \in R$,
  then $(y, x) \in R$

Antisymmetric relation R on X:
  for all $x, y \in X$, if $(x, y) \in R$
  and $x \neq y$, then $(y, x) \notin R$

Transitive relation R on X:
  for all $x, y, z \in X$, if (x, y)
  and (y, z) are in R,
  then $(x, z) \in R$

Partial order: relation that is reflexive, antisymmetric, and transitive

Inverse relation R^{-1}:
  $\{(y, x) \mid (x, y) \in R\}$

Composition of relations $R_2 \circ R_1$:
  $\{(x, z) \mid (x, y) \in R_1$,
  $(y, z) \in R_2\}$

Section 2.5

Equivalence relation: relation that
  is reflexive, symmetric, and transitive

Equivalence class containing a
  given by equivalence relation R:
  $[a] = \{x \mid x R a\}$

Equivalence classes partition the set
  (Theorem 2.5.9)

Section 2.6

Matrix of a relation

R is a reflexive relation if and only if
  the main diagonal of the matrix of
  R consists of 1's.

R is a symmetric relation if and only
  if the matrix of R is symmetric
  about the main diagonal.

If A_1 is the matrix of the relation R_1
  and A_2 is the matrix of the relation R_2, the matrix of the relation
  $R_2 \circ R_1$ is obtained by replacing
  each nonzero term in the matrix
  product $A_1 A_2$ by 1.

Section 2.7

n-ary relation: Set of n-tuples
Database management system
Relational database
Key
Query
Select
Project
Join

Section 2.8

Function from X to Y, $f : X \rightarrow Y$:
  relation from X to Y satisfying
  domain of $f = X$ and
  if $(x, y), (x, y') \in f$, then $y = y'$

$x \bmod y$: remainder when x is divided by y

Hash function

Collision for a hash function H:
  $H(x) = H(y)$

Collision resolution policy

Floor of x, $\lfloor x \rfloor$: greatest integer
  less than or equal to x

Ceiling of x, $\lceil x \rceil$: least integer
  greater than or equal to x

One-to-one function f:
  if $f(x) = f(x')$, then $x = x'$

Onto function f from X to Y:
  range of $f = Y$

Bijection: one-to-one and
  onto function

Inverse f^{-1} of a one-to-one, onto
  function f: $\{(y, x) \mid (x, y) \in f\}$

Composition of functions:
  $f \circ g = \{(x, z) \mid (x, y) \in g$
  and $(y, z) \in f\}$

Binary operator on X:
  function from $X \times X$ into X

Unary operator on X: function from
  X into X

ᔥ CHAPTER SELF-TEST

Section 2.1

1. If $A = \{1, 3, 4, 5, 6, 7\}$, $B = \{x \mid x$ is an even integer $\}$, $C = \{2, 3, 4, 5, 6\}$, find $(A \cap B) - C$.

2. If X is a set and $|X| = 8$, how many members does $\mathcal{P}(X)$ have? How many proper subsets does X have?

3. If $A \cup B = B$, what relation must hold between A and B?

4. Are the sets

$$\{3, 2, 2\}, \qquad \{x \mid x \text{ is an integer and } 1 < x \le 3\}$$

equal? Explain.

Section 2.2

5. For the sequence a defined by $a_n = 2n + 2$, find

   (a) a_6 (b) $\displaystyle\sum_{i=1}^{3} a_i$ (c) $\displaystyle\prod_{i=1}^{3} a_i$

   (d) a formula for the subsequence of a obtained by selecting every other term of a starting with the first.

6. Rewrite the sum

$$\sum_{i=1}^{n}(n-i)r^i$$

   replacing the index i by k, where $i = k + 2$.

7. Let

$$b_n = \sum_{i=1}^{n}(i+1)^2 - i^2.$$

   (a) Find b_5 and b_{10}.   (b) Find a formula for b_n.

   (c) Is b increasing?   (d) Is b decreasing?

8. Let $\alpha = ccddc$ and $\beta = c^3d^2$. Find

   (a) $\alpha\beta$ (b) $\beta\alpha$ (c) $|\alpha|$ (d) $|\alpha\alpha\beta\alpha|$

Section 2.3

9. Write the binary number 10010110 in decimal.

10. Write the decimal number 430 in binary and hexadecimal.

11. Add the binary numbers 11001 and 101001.

12. Write the hexadecimal number C39 in decimal.

Section 2.4

In Exercises 13 and 14, determine whether the relation defined on the set of positive integers is reflexive, symmetric, antisymmetric, transitive, and/or a partial order.

13. $(x, y) \in R$ if 2 divides $x + y$ 14. $(x, y) \in R$ if 3 divides $x + y$

15. Give an example of a relation on $\{1, 2, 3, 4\}$ that is reflexive, not antisymmetric, and not transitive.

16. Suppose that R is a relation on X that is symmetric and transitive but not reflexive. Suppose also that $|X| \geq 2$. Define the relation $\overline{R}$ on X by

$$\overline{R} = X \times X - R.$$

Which of the following must be true? For each false statement, provide a counterexample.

(a) $\overline{R}$ is reflexive.   (b) $\overline{R}$ is symmetric.

(c) $\overline{R}$ is not antisymmetric.   (d) $\overline{R}$ is transitive.

Section 2.5

17. Is the relation
$$\{(1, 1), (1, 2), (2, 2), (4, 4), (2, 1), (3, 3)\}$$

an equivalence relation on $\{1, 2, 3, 4\}$? Explain.

18. Given that the relation

$$\{(1, 1), (2, 2), (3, 3), (4, 4), (1, 2), (2, 1), (3, 4), (4, 3)\}$$

is an equivalence relation on $\{1, 2, 3, 4\}$, find [3], the equivalence class containing 3. How many (distinct) equivalence classes are there?

19. Find the equivalence relation (as a set of ordered pairs) on $\{a, b, c, d, e\}$ whose equivalence classes are
$$\{a\}, \quad \{b, d, e\}, \quad \{c\}.$$

20. Let R be the relation defined on the set of eight-bit strings by $s_1 R s_2$ provided that s_1 and s_2 have the same number of zeros.

    (a) Show that R is an equivalence relation.

    (b) How many equivalence classes are there?

    (c) List one member of each equivalence class.

Section 2.6

Exercises 21–24 refer to the relations

$$R_1 = \{(1, x), (2, x), (2, y), (3, y)\}, \qquad R_2 = \{(x, a), (x, b), (y, a), (y, c)\}.$$

21. Find the matrix A_1 of the relation R_1 relative to the orderings

$$1, 2, 3; \quad x, y.$$

22. Find the matrix A_2 of the relation R_2 relative to the orderings

$$x, y; \quad a, b, c.$$

23. Find the matrix product $A_1 A_2$.

24. Use the result of Exercise 23 to find the matrix of the relation $R_2 \circ R_1$.

Section 2.7

In Exercises 25–28, write a sequence of operations to answer the query. Also, provide an answer to the query. Use Tables 2.7.1 and 2.7.2.

25. Find all teams.

26. Find all players' names and ages.

27. Find the names of all teams that have a pitcher.

28. Find the names of all teams that have players aged 30 years or older.

Section 2.8

29. Let X be the set of strings over $\{a, b\}$ of length 4 and let Y be the set of strings over $\{a, b\}$ of length 3. Define a function f from X to Y by the rule

$$f(\alpha) = \text{string consisting of the first three characters of } \alpha.$$

Is f one-to-one? Is f onto?

30. Find real numbers satisfying $\lfloor x \rfloor \lfloor y \rfloor = \lfloor xy \rfloor - 1$.

31. Give examples of functions f and g such that $f \circ g$ is onto, but g is not onto.

32. For the hash function

$$h(x) = x \bmod 13,$$

show how the data

$$784, \quad 281, \quad 1141, \quad 18, \quad 1, \quad 329, \quad 620, \quad 43, \quad 31, \quad 684$$

would be inserted in the order given in initially empty cells indexed 0 to 12.

3

ALGORITHMS

It's so simple.

Step 1: We find the worst play in the world—a sure fire flop.

Step 2: I raise a million bucks—there are a lot of little old ladies in the world.

Step 3: You go back to work on the books. Phony lists of backers—one for the government, one for us. You can do it, Bloom, you're a wizard.

Step 4: We open on Broadway and before you can say

Step 5: We close on Broadway.

Step 6: We take our million bucks and we fly to Rio de Janeiro.

—from The Producers

An algorithm is a step-by-step method of solving some problem. Adlai Stevenson's recipe for carp furnishes us an example of an algorithm:

1. Take a 1- to 2-pound carp and allow it to swim in clear water for 24 hours.

2. Scale and fillet the carp.

3. Rub fillets with butter and season with salt and pepper.

4. Place on board and bake in moderate oven for 20 minutes.

5. Throw away carp and eat board.

Examples of algorithms can be found throughout history, going back at least as far as ancient Babylonia. Indeed, the word "algorithm" derives from the name of the ninth-century Arabic mathematician al-Khowārizmī. Algorithms based on sound mathematical principles play

† This section can be omitted without loss of continuity.

a central role in mathematics and computer science. In order for a solution to a problem to be executed by a computer, the solution must be described as a sequence of precise steps.

After introducing algorithms and our notation for them, we discuss the greatest common divisor algorithm, an ancient Greek algorithm that is still much used. We then turn to complexity of algorithms, which refers to the time and space required to execute algorithms, and the analysis of the resources required for particular algorithms. We conclude by discussing the RSA public-key cryptosystem—a method of encoding and decoding messages whose security relies primarily on the lack of an efficient algorithm for finding the prime divisors of an arbitrary integer.

3.1 INTRODUCTION

An **algorithm** is a finite set of instructions having the following characteristics:

- *Precision*. The steps are precisely stated.
- *Uniqueness*. The intermediate results of each step of execution are uniquely defined and depend only on the inputs and the results of the preceding steps.
- *Finiteness*. The algorithm stops after finitely many instructions have been executed.
- *Input*. The algorithm receives input.
- *Output*. The algorithm produces output.
- *Generality*. The algorithm applies to a set of inputs.

As an example, consider the following algorithm:

1. $x := a$
2. If $b > x$, then $x := b$.
3. If $c > x$, then $x := c$.

which finds the maximum of three numbers a, b, and c. The idea of the algorithm is to inspect the numbers one by one and copy the largest value seen into a variable x. At the conclusion of the algorithm, x will then be equal to the largest of the three numbers.

The notation $y := z$ means "copy the value of z into y" or, equivalently, "replace the current value of y by the value of z." When $y := z$ is executed, the value of z is unchanged. We call $:=$ the **assignment operator**.

We show how the preceding algorithm executes for some specific values of a, b, and c. Such a simulation is called a **trace**. First suppose that

$$a = 1, \qquad b = 5, \qquad c = 3.$$

At line 1, we set x to a (1). At line 2, $b > x$ ($5 > 1$) is true, so we set x to b (5). At line 3, $c > x$ ($3 > 5$) is false, so we do nothing. At this point x is 5, the largest of a, b, and c.

Suppose that
$$a = 6, \qquad b = 1, \qquad c = 9.$$

At line 1, we set x to a (6). At line 2, $b > x$ (1 > 6) is false, so we do nothing. At line 3, $c > x$ (9 > 6) is true, so we set x to 9. At this point x is 9, the largest of a, b, and c.

We note that our example algorithm has the properties set forth at the beginning of this section.

The steps of an algorithm must be stated precisely. The steps of the example algorithm are stated sufficiently precisely so that the algorithm could be written in a programming language and executed by a computer.

Given values of the input, each intermediate step of an algorithm produces a unique result. For example, given the values

$$a = 1, \qquad b = 5, \qquad c = 3,$$

at line 2 of the example algorithm, x will be set to 5 regardless of what person or machine executes the algorithm.

An algorithm stops after finitely many steps answering the given question. For example, the example algorithm stops after three steps and produces the largest of the three given values.

An algorithm receives input and produces output. The example algorithm receives, as input, the values a, b, and c, and produces, as output, the value x.

An algorithm must be general. The example algorithm can find the largest value of *any* three numbers.

Our description of what an algorithm is will suffice for our needs in this book. However, it should be noted that it is possible to give a precise, mathematical definition of "algorithm" (see the Notes for Chapter 10).

In Section 3.2 we introduce a more formal way to specify algorithms and we give several additional examples of algorithms.

⤳ ⤳ ⤳

Exercises

1. Write an algorithm that finds the smallest element among a, b, and c.

2. Write an algorithm that finds the second smallest element among a, b, and c. Assume that the values of a, b, and c are distinct.

3. Write the standard method of adding two positive decimal integers, taught in elementary schools, as an algorithm.

4. Consult the telephone book for the instructions for making a long-distance call. Which of the properties of an algorithm—precision, uniqueness, finiteness, input, output, generality—are present? Which properties are lacking?

3.2 NOTATION FOR ALGORITHMS

Although ordinary language is sometimes adequate to specify an algorithm, many mathematicians and computer scientists prefer **pseudocode** because of its precision, structure, and universality. Pseudocode is so named because it resembles the actual code (programs) of languages such as Pascal and C. There are many versions of pseudocode. Unlike actual computer languages, which fuss over semicolons, uppercase and lowercase letters, special words, and so on, any version of pseudocode is acceptable as long as its instructions are unambiguous and it resembles in form, if not in exact syntax, the pseudocode described in this section.

As our first example of pseudocode, we rewrite the algorithm of Section 3.1, which finds the maximum of three numbers.

ALGORITHM 3.2.1   *Finding the Maximum of Three Numbers*

This algorithm finds the largest of the numbers a, b, and c.

Input: Three numbers a, b, and c

Output: x, the largest of a, b, and c

```
1.   procedure max(a, b, c)
2.      x := a
3.      if b > x then    // if b is larger than x, update x
4.         x := b
5.      if c > x then    // if c is larger than x, update x
6.         x := c
7.      return(x)
8.   end max
```

Our algorithms consist of a title, a brief description of the algorithm, the input to and output of the algorithm, and the procedures containing the instructions of the algorithm. Algorithm 3.2.1 consists of a single procedure. To make it convenient to refer to individual lines within a procedure, we will sometimes number some of the lines. The procedure in Algorithm 3.2.1 has eight numbered lines. The first line of a procedure will consist of the word **procedure**, then the name of the procedure, and then, in parentheses, the parameters supplied to the procedure. The parameters describe the data, variables, arrays, and so on, that are available to the procedure. In Algorithm 3.2.1, the parameters supplied to the procedure are the three numbers a, b, and c. The last line of a procedure consists of the word **end** followed by the name of the procedure. Between the **procedure** and **end** lines are the executable lines of the procedure. Lines 2–7 are the executable lines of the procedure in Algorithm 3.2.1.

When the procedure in Algorithm 3.2.1 executes, at line 2 we set x to a. At line 3, b and x are compared. If b is greater than x, we execute line 4

$$x := b$$

but if b is not greater than x, we skip to line 5. At line 5, c and x are compared. If c is greater than x, we execute line 6

$$x := c$$

but if c is not greater than x, we skip to line 7. Thus when we arrive at line 7, x will correctly hold the largest of a, b, and c.

At line 7 we return the value of x, which is equal to the largest of the numbers a, b, and c, to the invoker of the procedure and terminate the procedure. Algorithm 3.2.1 has correctly found the largest of three numbers.

In general, in the **if-then** structure

> **if** p **then**
>   *action*

if condition p is true, *action* is executed and control passes to the statement following *action*. If condition p is false, control immediately passes to the statement following *action*.

An alternative form is the **if-then-else** structure. In the if-then-else structure

> **if** p **then**
>   *action* 1
> **else**
>   *action* 2

if condition p is true, *action* 1 (but not *action* 2) is executed and control passes to the statement following *action* 2. If condition p is false, *action* 2 (but not *action* 1) is executed and control passes to the statement following *action* 2.

As shown, we use indentation to identify the statements that make up *action*. In addition, if *action* consists of multiple statements, we delimit those statements with the words **begin** and **end**. An example of a multiple-statement *action* in an if statement is

> **if** $x \geq 0$ **then**
>   **begin**
> $x := x - 1$
> $a := b + c$
>   **end**

Two slash marks // signal the beginning of a **comment**, which then extends to the end of the line. An example of a comment in Algorithm 3.2.1 is

> // if b is larger than x, update x

Comments help the reader understand the algorithm but are not executed.

The **return**(x) statement terminates a procedure and returns the value of x to the invoker of the procedure. The statement **return** [without (x)] simply terminates a procedure. If there is no return statement, the procedure terminates just before the **end** line.

A procedure that contains a **return**(x) statement is a function. The domain consists of all valid values for the parameters and the range is the set of all values that may be returned by the procedure.

When using pseudocode, we will use the usual arithmetic operators $+$, $-$, $*$ (for multiplication), and $/$ as well as the relational operators $=$, $\neq$, $<$, $>$, $\leq$, and $\geq$ and the logical operators **and**, **or**, and **not**. We will use $=$ to denote the equality operator and $:=$ to denote the assignment operator. We will sometimes use less formal statements (*Example*: Choose an element x in S.) when to do otherwise would obscure the meaning. In general, solutions to exercises that request algorithms should be written in the form illustrated by Algorithm 3.2.1.

The lines of a procedure, which are executed sequentially, are typically assignment statements, conditional statements (if statements), loops, return statements, and combinations of these statements. One useful loop structure is the **while loop**

> **while** p **do**
>   *action*

in which *action* is repeatedly executed as long as p is true. We call *action* the **body** of the loop. As in the if statement, if *action* consists of multiple statements, we delimit those statements with the words **begin** and **end**. We illustrate the while loop in Algorithm 3.2.2 that finds the largest value in a sequence. As in Algorithm 3.2.1, we step through the numbers one by one and update the variable that holds the largest. We use a while loop to step through the numbers.

ALGORITHM 3.2.2   *Finding the Largest Element in a Finite Sequence*

This algorithm finds the largest number in the sequence $s_1, s_2, \ldots, s_n$. This version uses a while loop.

 Input: The sequence $s_1, s_2, \ldots, s_n$ and the length n of the sequence

Output: *large*, the largest element in this sequence

```
 1.  procedure find_large(s, n)
 2.      large := s₁
 3.      i := 2
 4.      while i ≤ n do
 5.        begin
 6.          if sᵢ > large then    // a larger value was found
 7.              large := sᵢ
 8.          i := i + 1
 9.        end
10.      return(large)
11.  end find_large
```

We trace Algorithm 3.2.2 when $n = 4$ and s is the sequence

$$s_1 = -2, \qquad s_2 = 6, \qquad s_3 = 5, \qquad s_4 = 6.$$

At line 2 we set *large* to s_1; in this case we set *large* to -2. Next, at line 3, i is set to 2. At line 4 we test whether $i \leq n$; in this case we test whether $2 \leq 4$. Since this condition is true, we execute the body of the while loop (lines 5–9). At line 6 we test whether $s_i > large$; in this case we test whether $s_2 > large$ ($6 > -2$). Since the condition is true, we execute line 7; *large* is set to 6. At line 8, i is set to 3. We then return to line 4.

We again test whether $i \leq n$; in this case we test whether $3 \leq 4$. Since this condition is true, we execute the body of the while loop. At line 6 we test whether $s_i > large$; in this case we test whether $s_3 > large$ ($5 > 6$). Since the condition is false, we skip line 7. At line 8, i is set to 4. We then return to line 4.

We again test whether $i \leq n$; in this case we test whether $4 \leq 4$. Since this condition is true, we execute the body of the while loop. At line 6 we test whether $s_i > large$; in this case we test whether $s_4 > large$ ($6 > 6$). Since the condition is false, we skip line 7. At line 8, i is set to 5. We then return to line 4.

We again test whether $i \leq n$; in this case we test whether $5 \leq 4$. Since the condition is false, we terminate the while loop and arrive at line 10, where we return *large* (6). We have found the largest element in the sequence.

In Algorithm 3.2.2 we stepped through a sequence by using the variable i that took on the integer values 1 through n. This kind of loop is so common that a special loop, called the **for loop**, is often used instead of the while loop. The form of the for loop is

>**for** *var* := *init* **to** *limit* **do**
>   *action*

As in the previous if statement and while loop, if *action* consists of multiple statements, we delimit the statements with the words **begin** and **end**. When the for loop is executed, *action* is executed for values of *var* from *init* to *limit*. More precisely, *init* and *limit* are expressions that have integer values. The variable *var* is first set to the value *init*. If $var \leq limit$, we execute *action* and then add 1 to *var*. The process is then repeated. Repetition continues until $var > limit$. Notice that if $init > limit$, *action* will not be executed at all.

Algorithm 3.2.2 may be rewritten in the following way using a for loop.

ALGORITHM 3.2.3   *Finding the Largest Element in a Finite Sequence*

This algorithm finds the largest number in the sequence $s_1, s_2, \ldots, s_n$. This version uses a for loop.

  Input: The sequence $s_1, s_2, \ldots, s_n$ and the length n of the sequence

Output: *large*, th largest element in this sequence

1.   **procedure** *find_large*(s, n)
2.   *large* := s_1
3.   **for** $i := 2$ **to** n **do**
4.   **if** $s_i > large$ **then**   // a larger value was found
5.   *large* := s_i
6.   **return**(*large*)
7.   **end** *find_large*

In developing an algorithm, it is often a good idea to break the original problem into two or more subproblems. A procedure can be developed to solve each subproblem, after which these procedures can be combined to provide a solution to the original problem. Our final algorithms illustrate these ideas.

Suppose that we want an algorithm to find the least prime number that exceeds a given positive integer. More precisely, the problem is: Given a positive integer n, find the least prime p satisfying $p > n$. We might break this problem up into two subproblems. We could first develop an algorithm to determine whether a positive integer is prime. We could then use this algorithm to find the least prime greater than a given positive integer.

Algorithm 3.2.4 tests whether a positive integer m is prime. We simply test whether any integer between 2 and $m - 1$ divides m. If we find an integer between 2 and $m - 1$ that divides m, m is not prime. If we fail to find an integer between 2 and $m - 1$ that divides m, m is prime. (Exercise 17 shows that it suffices to check integers between 2 and $\sqrt{m}$ as possible divisors.) Algorithm 3.2.4 shows that we allow procedures to return **true** or **false**.

ALGORITHM 3.2.4   *Testing Whether a Positive Integer Is Prime*

This algorithm tests whether the positive integer m is prime. The output is **true** if m is prime and **false** if m is not prime.

Input: m, a positive integer

Output: **true**, if m is prime; **false**, if m is not prime

```
procedure is_prime(m)
   for i := 2 to m − 1 do
      if m mod i = 0 then    // i divides m
         return(false)
   return(true)
end is_prime
```

Algorithm 3.2.5, which finds the least prime exceeding the positive integer n, uses Algorithm 3.2.4. To invoke a procedure that returns a value as in Algorithm 3.2.4, we name it. To invoke a procedure named say, *proc*, that does not return a value, we write

$$\textbf{call } proc(p_1, p_2, \ldots, p_k),$$

where $p_1, p_2, \ldots, p_k$ are the arguments passed to *proc*.

ALGORITHM 3.2.5   *Finding a Prime Larger Than a Given Integer*

This algorithm finds the smallest prime that exceeds the positive integer n.

Input: n, a positive integer

Output: m, the smallest prime greater than n

```
procedure large_prime(n)
   m := n + 1
   while not is_prime(m) do
      m := m + 1
   return(m)
end large_prime
```

(handwritten note: Start loop at here go to $\sqrt{m}$)

Since the number of primes is infinite (see Exercise 18), the procedure in Algorithm 3.2.5 will eventually terminate.

෴ ෴ ෴

Exercises

Write all algorithms in the style of Algorithms 3.2.1–3.2.5.

1. Write an algorithm that outputs the smallest element in the sequence

$$s_1, \quad \ldots, \quad s_n.$$

2. Write an algorithm that outputs the largest and second largest elements in the sequence

$$s_1, \quad \ldots, \quad s_n.$$

3. Write an algorithm that outputs the smallest and second smallest elements in the sequence

$$s_1, \quad \ldots, \quad s_n.$$

4. Write an algorithm that outputs the largest and smallest elements in the sequence

$$s_1, \quad \ldots, \quad s_n.$$

5. Write an algorithm that outputs the index of the first occurrence of the largest element in the sequence

$$s_1, \quad \ldots, \quad s_n.$$

Example: If the sequence were

$$6.2 \quad 8.9 \quad 4.2 \quad 8.9,$$

the algorithm would output the value 2.

6. Write an algorithm that outputs the index of the last occurrence of the largest element in the sequence

$$s_1, \quad \ldots, \quad s_n.$$

Example: If the sequence were

$$6.2 \quad 8.9 \quad 4.2 \quad 8.9,$$

the algorithm would output the value 4.

7. Write an algorithm that outputs the index of the first occurrence of the value *key* in the sequence

$$s_1, \quad \ldots, \quad s_n.$$

If *key* is not in the sequence, the algorithm outputs the value 0.

Example: If the sequence were

‘MARY’   ‘JOE’   ‘MARK’   ‘RUDY’,

and *key* were ‘MARK’, the algorithm would output the value 3.

8. Write an algorithm that outputs the index of the last occurrence of the value *key* in the sequence

$$s_1, \quad \ldots, \quad s_n.$$

If *key* is not in the sequence, the algorithm outputs the value 0.

9. Write an algorithm that outputs the index of the first item that is less than its predecessor in the sequence

$$s_1, \quad \ldots, \quad s_n.$$

If the items are in increasing order, the algorithm outputs the value 0.

Example: If the sequence were

‘AMY’   ‘BRUNO’   ‘ELIE’   ‘DAN’   ‘ZEKE’,

the algorithm would output the value 4.

10. Write an algorithm that outputs the index of the first item that is greater than its predecessor in the sequence

$$s_1, \quad \ldots, \quad s_n.$$

If the items are in decreasing order, the algorithm outputs the value 0.

11. Write an algorithm that reverses the sequence

$$s_1, \quad \ldots, \quad s_n.$$

Example: If the sequence were

'AMY' 'BRUNO' 'ELIE',

the reversed sequence would be

'ELIE' 'BRUNO' 'AMY'.

12. Write the standard method of multiplying two positive decimal integers, taught in elementary schools, as an algorithm.

13. Write an algorithm that receives as input the matrix of a relation R and tests whether R is reflexive.

14. Write an algorithm that receives as input the matrix of a relation R and tests whether R is antisymmetric.

15. Write an algorithm that receives as input the matrix of a relation R and tests whether R is a function.

16. Write an algorithm that receives as input the matrix of a relation R and produces as output the matrix of the inverse relation R^{-1}.

17. Show that the positive integer $m \geq 2$ is prime if and only if no integer between 2 and $\sqrt{m}$ divides m.

18. Show that the number of primes is infinite by completing the following argument.
    It suffices to show that if p is prime, there is a prime larger than p. Let $p_1 < p_2 < \cdots < p_k = p$ denote the primes less than or equal to p, and let $n = p_1 p_2 \cdots p_k + 1$. Show that any prime that divides n is larger than p.

3.3 *THE EUCLIDEAN ALGORITHM*

An old and famous algorithm is the **Euclidean algorithm** for finding the greatest common divisor of two integers. The greatest common divisor of two integers m and n (not both zero) is the largest positive integer that divides both m and n. For example, the greatest common divisor of 4 and 6 is 2, and the greatest common divisor of 3 and 8 is 1. We use the notion of greatest common divisor when we check to see if a fraction m/n, where m and n are integers, is in lowest terms. If the greatest common divisor of m and n is 1, m/n is in lowest terms; otherwise, we can reduce m/n. For example, 4/6 is not in lowest terms because the greatest common divisor of 4 and 6 is 2, not 1. (We can divide both 4 and 6 by 2.) The fraction 3/8 is in lowest terms because the greatest common divisor of 3 and 8 is 1. After discussing the divisibility of integers, we examine the greatest common divisor in detail and present the Euclidean algorithm.

If a, b, and q are integers, $b \neq 0$, satisfying $a = bq$, we say that b **divides** a and write $b \mid a$. In this case, we call q the **quotient** and call b a **divisor** of a. If b does not divide a, we write $b \nmid a$.

EXAMPLE 3.3.1

Since $21 = 3 \cdot 7$, 3 divides 21 and we write $3 \mid 21$. The quotient is 7.   □

Let m and n be integers that are not both zero. Among all the integers that divide both m and n, there is a largest divisor known as the **greatest common divisor** of m and n.

DEFINITION 3.3.2

Let m and n be integers with not both m and n zero. A *common divisor* of m and n is an integer that divides both m and n. The *greatest common divisor*, written

$$\gcd(m, n),$$

is the largest common divisor of m and n.

EXAMPLE 3.3.3

The positive divisors of 30 are

$$1, \quad 2, \quad 3, \quad 5, \quad 6, \quad 10, \quad 15, \quad 30$$

and the positive divisors of 105 are

$$1, \quad 3, \quad 5, \quad 7, \quad 15, \quad 21, \quad 35, \quad 105;$$

thus the positive common divisors of 30 and 105 are

$$1, \quad 3, \quad 5, \quad 15.$$

It follows that the greatest common divisor of 30 and 105, $\gcd(30, 105)$, is 15.
□

The properties of common divisors given in the following theorem will be useful in our subsequent work in this section.

THEOREM 3.3.4

Let m, n, and c be integers.

(a) If c is a common divisor of m and n, then

$$c \mid (m + n).$$

(b) If c is a common divisor of m and n, then

$$c \mid (m - n).$$

(c) If $c \mid m$, then $c \mid mn$.

Proof. (a) Let c be a common divisor of m and n. Since $c \mid m$,

$$m = cq_1 \tag{3.3.1}$$

for some integer q_1. Similarly, since $c \mid n$,

$$n = cq_2 \tag{3.3.2}$$

for some integer q_2. If we add equations (3.3.1) and (3.3.2), we obtain

$$m + n = cq_1 + cq_2 = c(q_1 + q_2).$$

Therefore, c divides $m + n$ (with quotient $q_1 + q_2$). We have proved part (a).

The proofs of parts (b) and (c) are left to the reader (see Exercises 17 and 18). ∎

If we divide the nonnegative integer a by the positive integer b, we obtain a quotient q and a remainder r satisfying

$$a = bq + r, \qquad 0 \le r < b, \quad q \ge 0. \tag{3.3.3}$$

EXAMPLE 3.3.5

The quotient q and the remainder r in (3.3.3) are illustrated for various values of a and b:

$$
\begin{array}{llll}
a = 22, & b = 7, & q = 3, & r = 1; \quad 22 = 7 \cdot 3 + 1 \\
a = 24, & b = 8, & q = 3, & r = 0; \quad 24 = 8 \cdot 3 + 0 \tag{3.3.4} \\
a = 103, & b = 21, & q = 4, & r = 19; \quad 103 = 21 \cdot 4 + 19 \\
a = 4895, & b = 87, & q = 56, & r = 23; \quad 4895 = 87 \cdot 56 + 23 \\
a = 0, & b = 47, & q = 0, & r = 0; \quad 0 = 47 \cdot 0 + 0. \tag{3.3.5}
\end{array}
$$

□

In (3.3.4) and (3.3.5), the remainder r is zero and $b \mid a$. In all other cases, $b \nmid a$.

Now suppose that a is a nonnegative integer and b is a positive integer. We may divide a by b to obtain

$$a = bq + r, \qquad 0 \le r < b.$$

We show that the set of common divisors of a and b is equal to the set of common divisors of b and r.

Let c be a common divisor of a and b. By Theorem 3.3.4c, $c \mid bq$. Since $c \mid a$ and $c \mid bq$, by Theorem 3.3.4b, $c \mid a - bq \ (= r)$. Thus c is a common divisor of b and r. Conversely, if c is a common divisor of b and r, then $c \mid bq$ and $c \mid bq + r \ (= a)$ and c is a common divisor of a and b. Thus the set of common divisors of a and b is equal to the set of common divisors of b and r. It follows that

$$\gcd(a, b) = \gcd(b, r).$$

We summarize this result as a theorem.

THEOREM 3.3.6

If a is a nonnegative integer, b is a positive integer, and

$$a = bq + r, \qquad 0 \le r < b,$$

then

$$\gcd(a, b) = \gcd(b, r).$$

Proof. The proof precedes the statement of the theorem. ∎

EXAMPLE 3.3.7

If we divide 105 by 30, we obtain

$$105 = 30 \cdot 3 + 15.$$

The remainder is 15. By Theorem 3.3.6,

$$\gcd(105, 30) = \gcd(30, 15).$$

If we divide 30 by 15, we obtain

$$30 = 15 \cdot 2 + 0.$$

The remainder is 0. By Theorem 3.3.6,

$$\gcd(30, 15) = \gcd(15, 0).$$

By inspection, $\gcd(15, 0) = 15$. Therefore,

$$\gcd(105, 30) = \gcd(30, 15) = \gcd(15, 0) = 15. \qquad \square$$

In Example 3.3.3, we obtained the greatest common divisor of 105 and 30 by listing all of the divisors of 105 and 30. By using Theorem 3.3.6, two simple divisions produce the greatest common divisor. This computation illustrates the Euclidean algorithm.

In general, the Euclidean algorithm finds the greatest common divisor of a and b by repeatedly using Theorem 3.3.6 to replace the original problem of finding the greatest common divisor of a and b by the problem of finding the greatest common divisor of smaller numbers. We ultimately reduce the original problem to that of finding the greatest common divisor of two numbers, one of which is 0. Since $\gcd(m, 0) = m$, we have solved the original problem. We now delineate this technique precisely.

Let r_0 and r_1 be nonnegative integers with r_1 nonzero. If we divide r_0 by r_1, we obtain

$$r_0 = r_1 q_2 + r_2, \qquad 0 \le r_2 < r_1.$$

By Theorem 3.3.6,

$$\gcd(r_0, r_1) = \gcd(r_1, r_2).$$

If $r_2 \neq 0$, we can divide r_1 by r_2 to obtain

$$r_1 = r_2 q_3 + r_3, \qquad 0 \leq r_3 < r_2.$$

By Theorem 3.3.6,

$$\gcd(r_1, r_2) = \gcd(r_2, r_3).$$

We continue dividing r_i by r_{i+1}, provided that $r_{i+1} \neq 0$. Since $r_1, r_2, \ldots$ are nonnegative integers and

$$r_1 > r_2 > r_3 > \cdots,$$

eventually some r_i will be zero. Let r_n be the first zero remainder. Now

$$\gcd(r_0, r_1) = \gcd(r_1, r_2) = \gcd(r_2, r_3) = \cdots$$
$$= \gcd(r_{n-1}, r_n) = \gcd(r_{n-1}, 0).$$

The greatest common divisor of r_{n-1} and 0 is r_{n-1}; hence

$$\gcd(r_0, r_1) = \gcd(r_{n-1}, 0) = r_{n-1}.$$

Thus the greatest common divisor of r_0 and r_1 will be the last nonzero remainder.

We state the Euclidean algorithm as Algorithm 3.3.8.

ALGORITHM 3.3.8   *Euclidean Algorithm*

This algorithm finds the greatest common divisor of the nonnegative integers a and b, where not both a and b are zero.

  Input: a and b (nonnegative integers, not both zero)

Output: Greatest common divisor of a and b

```
 1.    procedure gcd(a, b)
 2.        // make a largest
 3.        if a < b then
 4.           swap(a, b)
              // that is, execute
              // temp := a
              // a := b
              // b := temp
 5.        while b ≠ 0 do
 6.           begin
 7.           divide a by b to obtain a = bq + r, 0 ≤ r < b
 8.           a := b
 9.           b := r
10.           end
11.        return(a)
12.    end gcd
```

> ### EXAMPLE 3.3.9

We show how Algorithm 3.3.8 finds gcd(504, 396).

Let $a = 504$ and $b = 396$. Since $a > b$, we move to line 5. Since $b \neq 0$, we proceed to line 7, where we divide a (504) by b (396) to obtain

$$504 = 396 \cdot 1 + 108.$$

We then move to line 8. We set a to 396 and b to 108 and return to line 5.

Since $b \neq 0$, we proceed to line 7, where we divide a (396) by b (108) to obtain

$$396 = 108 \cdot 3 + 72.$$

We then move to line 8. We set a to 108 and b to 72 and return to line 5.

Since $b \neq 0$, we proceed to line 7, where we divide a (108) by b (72) to obtain

$$108 = 72 \cdot 1 + 36.$$

We then move to line 8. We set a to 72 and b to 36 and return to line 5.

Since $b \neq 0$, we proceed to line 7, where we divide a (72) by b (36) to obtain

$$72 = 36 \cdot 2 + 0.$$

We then move to line 8. We set a to 36 and b to 0 and return to line 5.

This time $b = 0$, so we skip to line 11, where we return a (36), the greatest common divisor of 396 and 504.   □

Exercises

In Exercises 1–6, find integers q and r so that $a = bq + r$, with $0 \leq r < b$.

1. $a = 45$, $b = 6$ **2.** $a = 106$, $b = 12$

3. $a = 66$, $b = 11$ **4.** $a = 221$, $b = 17$

5. $a = 0$, $b = 31$ **6.** $a = 0$, $b = 47$

Use the Euclidean algorithm to find the greatest common divisor of each pair of integers in Exercises 7–16.

7. 60, 90   **8.** 110, 273

9. 220, 1400   **10.** 315, 825

11. 20, 40   **12.** 331, 993

13. 2091, 4807   **14.** 2475; 32,670

15. 67,942; 4209   **16.** 490,256; 337

17. Let m, n, and c be integers. Show that if c is a common divisor of m and n, then $c \mid (m - n)$.

18. Let m, n, and c be integers. Show that if $c \mid m$, then $c \mid mn$.

19. Suppose that a, b, and c are positive integers. Show that if $a \mid b$ and $b \mid c$, then $a \mid c$.

20. If a and b are positive integers, show that $\gcd(a, b) = \gcd(a, a + b)$.

★ **21.** Using the notation in the text following Example 3.3.7, show that we may successively write

$$r_{n-1} = s_{n-3}r_{n-2} + t_{n-3}r_{n-3}$$

$$r_{n-1} = s_{n-4}r_{n-3} + t_{n-4}r_{n-4}$$

$$\vdots$$

$$r_{n-1} = s_0r_1 + t_0r_0$$

$$\gcd(r_0, r_1) = s_0r_1 + t_0r_0,$$

where the s's and t's are integers.

★ **22.** Use the method of Exercise 21 to write the greatest common divisor of each pair of integers a and b in Exercises 7–16 in the form $ta + sb$.

★ **23.** Show that if p is a prime number and a and b are positive integers and $p \mid ab$, then $p \mid a$ or $p \mid b$.

24. Give an example of positive integers a, b, and c where $a \mid bc$, $a \nmid b$, and $a \nmid c$.

25. Show that if $a > b \geq 0$, then

$$\gcd(a, b) = \gcd(a - b, b).$$

26. Using Exercise 25, write an algorithm to compute the greatest common divisor of two nonnegative integers a and b, not both zero, that uses subtraction but not division.

3.4 RECURSIVE ALGORITHMS

A **recursive procedure** is a procedure that invokes itself. A **recursive algorithm** is an algorithm that contains a recursive procedure. Recursion is a powerful, elegant, and natural way to solve a large class of problems. A problem in this class can be solved using a *divide-and-conquer* technique in which the problem is decomposed into problems of the same type as the original problem. Each subproblem, in turn, is decomposed further until the process yields subproblems that can be solved in a straightforward way. Finally, solutions to the subproblems are combined to obtain a solution to the original problem.

EXAMPLE 3.4.1

n **factorial** is defined as

$$n! = \begin{cases} 1 & \text{if } n = 0 \\ n(n-1)(n-2)\cdots 2 \cdot 1 & \text{if } n \geq 1. \end{cases}$$

That is, if $n \geq 1$, $n!$ is equal to the product of all the integers between 1 and n inclusive. 0! is defined to be 1. As examples,

$$3! = 3 \cdot 2 \cdot 1 = 6, \qquad 6! = 6 \cdot 5 \cdot 4 \cdot 3 \cdot 2 \cdot 1 = 720.$$

Notice that n factorial can be written "in terms of itself" since, if we "peel off" n, the remaining product is simply $(n-1)!$; that is,

$$n! = n(n-1)(n-2)\cdots 2 \cdot 1 = n \cdot (n-1)!$$

For example,

$$5! = 5 \cdot 4 \cdot 3 \cdot 2 \cdot 1 = 5 \cdot 4!$$

The equation

$$n! = n \cdot (n-1)!$$

which happens to be true even when $n = 1$, shows how to decompose the original problem (compute $n!$) into increasingly simpler subproblems [compute $(n-1)!$, compute $(n-2)!$, . . .] until the process reaches the straightforward problem of computing 0!. The solutions to these subproblems can then be combined, by multiplying, to solve the original problem.

For example, the problem of computing 5! is reduced to computing 4!; the problem of computing 4! is reduced to computing 3!; and so on. Table 3.4.1 summarizes this process.

TABLE 3.4.1
Decomposing the factorial
problem

Problem	Simplified Problem
5!	$5 \cdot 4!$
4!	$4 \cdot 3!$
3!	$3 \cdot 2!$
2!	$2 \cdot 1!$
1!	$1 \cdot 0!$
0!	None

Once the problem of computing 5! has been reduced to solving subproblems, the solution to the simplest subproblem can be used to solve the next simplest subproblem, and so on, until the original problem has been solved. Table 3.4.2 shows how the subproblems are combined to compute 5!.

TABLE 3.4.2
Combining subproblems of the
factorial problem

Problem	Solution
0!	1
1!	$1 \cdot 0! = 1$
2!	$2 \cdot 1! = 2$
3!	$3 \cdot 2! = 3 \cdot 2 = 6$
4!	$4 \cdot 3! = 4 \cdot 6 = 24$
5!	$5 \cdot 4! = 5 \cdot 24 = 120$

$\square$

Next, we write a recursive algorithm that computes factorials. The algorithm is a direct translation of the equation

$$n! = n \cdot (n-1)!.$$

ALGORITHM 3.4.2 *Computing n Factorial*

This recursive algorithm computes $n!$.

Input: n, an integer greater than or equal to 0
Output: $n!$

```
1.  procedure factorial(n)
2.     if n = 0 then
3.        return(1)
4.     return(n * factorial(n − 1))
5.  end factorial
```

We show how Algorithm 3.4.2 computes $n!$ for several values of n. If $n = 0$, at line 3 the procedure correctly returns the value 1.

If $n = 1$, since $n \neq 0$, we proceed to line 4. We use this procedure to compute $0!$. We have just observed that the procedure computes 1 as the value of $0!$. At line 4, the procedure correctly computes the value $1!$

$$(n − 1)! \cdot n = 0! \cdot 1 = 1 \cdot 1 = 1.$$

If $n = 2$, since $n \neq 0$, we proceed to line 4. We use this procedure to compute $1!$. We have just observed that the procedure computes 1 as the value of $1!$. At line 4, the procedure correctly computes $2!$

$$(n − 1)! \cdot n = 1! \cdot 2 = 1 \cdot 2 = 2.$$

If $n = 3$, since $n \neq 0$, we proceed to line 4. We use this procedure to compute $2!$. We have just observed that the procedure computes 2 as the value of $2!$. At line 4, the procedure correctly computes the value $3!$

$$(n − 1)! \cdot n = 2! \cdot 3 = 2 \cdot 3 = 6.$$

The preceding arguments may be generalized using mathematical induction to *prove* that Algorithm 3.4.2 correctly outputs the value of $n!$ for any nonnegative integer n.

THEOREM 3.4.3

Algorithm 3.4.2 outputs the value of $n!$, $n \geq 0$.

Proof.

BASIS STEP $(n = 0)$. We have already observed that if $n = 0$, Algorithm 3.4.2 correctly outputs the value 1 of $0!$.

INDUCTIVE STEP. Assume that Algorithm 3.4.2 correctly outputs the value of $(n−1)!$, $n > 0$. Now suppose that n is input to Algorithm 3.4.2. Since $n \neq 0$, when we execute the procedure in Algorithm 3.4.2 we proceed to line 4. By the inductive assumption, the procedure correctly computes the value of $(n − 1)!$. At line 4, the procedure correctly computes the value $(n − 1)! \cdot n = n!$.

Therefore, Algorithm 3.4.2 correctly outputs the value of $n!$ for every integer $n \geq 0$. ∎

There must be some situations in which a recursive procedure does *not* invoke itself; otherwise, it would invoke itself forever. In Algorithm 3.4.2, if $n = 0$, the procedure does not invoke itself. We call the values for which a recursive procedure does not invoke itself the *base cases*. To summarize, every recursive procedure must have base cases.

We have shown how mathematical induction may be used to prove that a recursive algorithm computes the value it claims to compute. The link between mathematical induction and recursive algorithms runs deep. Often a proof by mathematical induction can be considered to be an algorithm to compute a value or to carry out a particular construction. The Basis Step of a proof by mathematical induction corresponds to the base cases of a recursive procedure and the Inductive Step of a proof by mathematical induction corresponds to the part of a recursive procedure where the procedure calls itself.

In Example 1.6.4, we gave a proof using mathematical induction that given a deficient $n \times n$ board (a board with one square removed) where n is a power of 2, we can tile the board with right trominoes (three squares that form an "el"; see Figure 1.6.3). We now translate the inductive proof into a recursive algorithm to construct a tiling by right trominoes of a deficient $n \times n$ board where n is a power of 2.

ALGORITHM 3.4.4   *Tiling a Deficient Board with Trominoes*

This algorithm constructs a tiling by right trominoes of a deficient $n \times n$ board where n is a power of 2.

Input: n, a power of 2 (the board size), and the location L of the missing square

Output: A tiling of a deficient $n \times n$ board

```
     procedure tile(n, L)
1.     if n = 2 then
            // the board is a right tromino T
2.          return(T)
3.     divide the board into four (n/2) × (n/2) boards
4.     rotate the board so that the missing square is in the upper left quadrant
5.     place one right tromino in the center // as in Figure 1.6.5
            // consider each of the squares covered by the center tromino as
            // missing and denote the missing squares as m₁, m₂, m₃, m₄
6.     call tile(n/2, m₁)
7.     call tile(n/2, m₂)
8.     call tile(n/2, m₃)
9.     call tile(n/2, m₄)
     end tile
```

We next give a recursive algorithm to compute the greatest common divisor of two nonnegative integers, not both zero.

Theorem 3.3.6 states that if a is a nonnegative integer, b is a positive integer, and

$$a = bq + r, \qquad 0 \le r < b,$$

then

$$\gcd(a, b) = \gcd(b, r). \tag{3.4.1}$$

[$\gcd(x, y)$ denotes the greatest common divisor of x and y.] Equation (3.4.1) is inherently recursive; it reduces the problem of computing the greatest common divisor of a and b to a smaller problem—that of computing the greatest common divisor of b and r. Recursive Algorithm 3.4.5, which computes the greatest common divisor, is based on equation (3.4.1).

ALGORITHM 3.4.5   *Recursively Computing the Greatest Common Divisor*

This algorithm recursively finds the greatest common divisor of the nonnegative integers a and b, where not both a and b are zero. (Algorithm 3.3.8 gives a nonrecursive algorithm for computing the greatest common divisor.)

Input: a and b (nonnegative integers, not both zero)

Output: Greatest common divisor of a and b

```
     procedure gcd_recurs(a, b)
        // make a largest
1.      if a < b then
2.         swap(a, b)
3.      if b = 0 then
4.         return(a)
5.      divide a by b to obtain a = bq + r, 0 ≤ r < b
6.      return(gcd_recurs(b, r))
     end gcd_recurs
```

We present one final example of a recursive algorithm.

EXAMPLE 3.4.6

A robot can take steps of 1 meter or 2 meters. We write an algorithm to calculate the number of ways the robot can walk n meters. As examples:

Distance	Sequence of Steps	Number of Ways to Walk
1	1	1
2	1, 1 or 2	2
3	1, 1, 1 or 1, 2 or 2, 1	3
4	1, 1, 1, 1 or 1, 1, 2	5
	or 1, 2, 1 or 2, 1, 1 or 2, 2	

Let walk(n) denote the number of ways the robot can walk n meters. We have observed that

$$\text{walk}(1) = 1, \qquad \text{walk}(2) = 2.$$

Now suppose that $n > 2$. The robot can begin by taking a step of 1 meter or a step of 2 meters. If the robot begins by taking a 1-meter step, a distance of $n - 1$ meters remains; but, by definition, the remainder of the walk can be completed in walk$(n - 1)$ ways. Similarly, if the robot begins by taking a 2-meter step, a distance of $n - 2$ meters remains and, in this case, the remainder of the walk can be completed in walk$(n - 2)$ ways. Since the walk must begin with either a 1-meter or a 2-meter step, all of the ways to walk n meters are accounted for. We obtain the formula

$$\text{walk}(n) = \text{walk}(n - 1) + \text{walk}(n - 2).$$

For example,

$$\text{walk}(4) = \text{walk}(3) + \text{walk}(2) = 3 + 2 = 5.$$

We can write a recursive algorithm to compute walk(n) by translating the equation

$$\text{walk}(n) = \text{walk}(n - 1) + \text{walk}(n - 2)$$

directly into an algorithm. The base cases are $n = 1$ and $n = 2$.   □

ALGORITHM 3.4.7   *Robot Walking*

This algorithm computes the function defined by

$$\text{walk}(n) = \begin{cases} 1, & n = 1 \\ 2, & n = 2 \\ \text{walk}(n - 1) + \text{walk}(n - 2), & n > 2 \end{cases}$$

Input: n
Output: walk(n)

procedure *robot_walk*(n)
  if $n = 1$ **or** $n = 2$ **then**
    return(n)
  return$(robot\_walk(n - 1) + robot\_walk(n - 2))$
end *robot_walk*

The sequence

$$\text{walk}(1), \ \text{walk}(2), \ \text{walk}(3), \ \ldots,$$

whose values begin

$$1, \quad 2, \quad 3, \quad 5, \quad 8, \quad 13, \quad \ldots,$$

is called the **Fibonacci sequence** in honor of Leonardo Fibonacci (ca. 1170–1250), an Italian merchant and mathematician. Subsequently, we denote the Fibonacci sequence as

$$f_1, \quad f_2, \quad \ldots.$$

This sequence is defined by the equations

$$f_1 = 1$$
$$f_2 = 2$$
$$f_n = f_{n-1} + f_{n-2}, \qquad n \geq 3.$$

The Fibonacci sequence originally arose in a puzzle about rabbits (see Exercises 13 and 14). After returning from the Orient in 1202, Fibonacci wrote his most famous work, *Liber Abaci*, which in addition to containing what we now call the Fibonacci sequence advocated the use of Hindu-Arabic numerals. This book was one of the main influences in bringing the decimal number system to Western Europe. Fibonacci signed much of his work "Leonardo Bigollo." *Bigollo* translates as "traveler" or "blockhead." There is some evidence that Fibonacci enjoyed having his contemporaries consider him a blockhead for advocating the new number system.

The Fibonacci sequence pops up in unexpected places. For example, the number of clockwise spirals and the number of counterclockwise spirals formed by the seeds of certain varieties of sunflowers are found in the Fibonacci sequence. Figure 3.4.1 shows a hypothetical sunflower with 13 clockwise spirals and 8 counterclockwise spirals. In Section 3.6, the Fibonacci sequence appears in the analysis of the Euclidean algorithm.

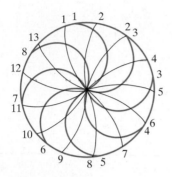

FIGURE 3.4.1
A hypothetical sunflower.

Exercises

1. Trace Algorithm 3.4.2 for $n = 4$.

2. Trace Algorithm 3.4.4 when $n = 4$ and the missing square is the upper left corner square.

3. Trace Algorithm 3.4.4 when $n = 8$ and the missing square is four from the left and six from the top.

4. Trace Algorithm 3.4.5 for $a = 5$ and $b = 0$.

5. Trace Algorithm 3.4.5 for $a = 55$ and $b = 20$.

6. (a) Use the formulas

$$s_1 = 1, \qquad s_n = s_{n-1} + n, \qquad n \geq 2,$$

to write a recursive algorithm that computes

$$s_n = 1 + 2 + 3 + \cdots + n.$$

(b) Give a proof using mathematical induction that your algorithm for part (a) is correct.

7. (a) Use the formulas

$$s_1 = 2, \qquad s_n = s_{n-1} + 2n, \qquad n \geq 2,$$

to write a recursive algorithm that computes

$$s_n = 2 + 4 + 6 + \cdots + 2n.$$

(b) Give a proof using mathematical induction that your algorithm for part (a) is correct.

8. (a) A robot can take steps of 1 meter, 2 meters, or 3 meters. Write a recursive algorithm to calculate the number of ways the robot can walk n meters.

(b) Give a proof using mathematical induction that your algorithm for part (a) is correct.

9. Write a recursive algorithm that computes the greatest common divisor of two non-negative integers, not both zero, that uses subtractions but no divisions (see Exercise 25, Section 3.3).

10. Write a nonrecursive algorithm to compute $n!$.

★ **11.** A robot can take steps of 1 meter or 2 meters. Write an algorithm to list all of the ways the robot can walk n meters.

★ **12.** A robot can take steps of 1 meter, 2 meters, or 3 meters. Write an algorithm to list all of the ways the robot can walk n meters.

Exercises 13–25 concern the Fibonacci sequence $\{f_n\}$.

13. Suppose that at the beginning of the year, there is one pair of rabbits and that every month each pair produces a new pair that becomes productive after one month. Suppose further that no deaths occur. Let a_n denote the number of pairs of rabbits at the end of the nth month. Show that $a_1 = 1$, $a_2 = 2$, and $a_n - a_{n-1} = a_{n-2}$. Explain why $a_n = f_n$, $n \geq 1$.

14. Fibonacci's original question was: Under the conditions of Exercise 13, how many pairs of rabbits are there after one year? Answer Fibonacci's question.

15. Use mathematical induction to show that

$$\sum_{k=1}^{n} f_k = f_{n+2} - 2, \qquad\qquad n \geq 1.$$

16. Use mathematical induction to show that

$$f_n^2 = f_{n-1} f_{n+1} + (-1)^n, \qquad\qquad n \geq 2.$$

17. Show that

$$f_{n+2}^2 - f_{n+1}^2 = f_n f_{n+3}, \qquad\qquad n \geq 1.$$

18. Use mathematical induction to show that

$$\sum_{k=1}^{n} f_k^2 = f_n f_{n+1} - 1, \qquad\qquad n \geq 1.$$

19. Use mathematical induction to show that f_n is even if and only if $n + 1$ is divisible by 3, $n \geq 1$.

20. Use mathematical induction to show that for $n \geq 5$,

$$f_n > \left(\frac{3}{2}\right)^n.$$

21. Use mathematical induction to show that for $n \geq 1$,

$$f_n < 2^n.$$

22. Use mathematical induction to show that for $n \geq 1$,

$$\sum_{k=1}^{n} f_{2k-1} = f_{2n} - 1, \qquad \sum_{k=1}^{n} f_{2k} = f_{2n+1} - 1.$$

★ **23.** Use mathematical induction to show that every integer $n \geq 1$ can be expressed as the sum of distinct Fibonacci numbers no two of which are consecutive.

★ **24.** Show that the representation in Exercise 23 is unique.

25. Show that for $n \geq 2$,

$$f_n = \frac{f_{n-1} + \sqrt{5f_{n-1}^2 + 4(-1)^n}}{2}.$$

Notice that this formula gives f_n in terms of one predecessor rather than two predecessors as in the original definition.

26. [Requires calculus.] Assume the formula for differentiating products:

$$\frac{d(fg)}{dx} = f\frac{dg}{dx} + g\frac{df}{dx}.$$

Use mathematical induction to prove that

$$\frac{dx^n}{dx} = nx^{n-1} \qquad \text{for } n = 1, 2, \ldots.$$

27. [Requires calculus.] Explain how the formula gives a recursive algorithm for integrating $\log^n |x|$:

$$\int \log^n |x| \, dx = x \log^n |x| - n \int \log^{n-1} |x| \, dx.$$

Give other examples of recursive integration formulas.

3.5 *COMPLEXITY OF ALGORITHMS*

A computer program, even though derived from a correct algorithm, might be useless for certain types of input because the time needed to run the program or the storage needed to hold the data, program variables, and so on, is too great. **Analysis of an algorithm** refers to the process of deriving estimates for the time and space needed to execute the algorithm. **Complexity of an algorithm** refers to the amount of time and space required to execute the algorithm. In this section we deal with the problem of estimating the time required to execute algorithms.

Suppose that we are given a set X of n elements, some labeled "red" and some labeled "black," and we want to find the number of subsets of X that contain at least one red item. Suppose we construct an algorithm that examines all subsets of X and counts those that contain at least one red item and then implement this algorithm as a computer program. Since a set that has n elements has 2^n subsets (Theorem 2.1.4), the program would require at least 2^n units of time to execute. It does not matter what the units of time are—2^n grows so fast as n increases (see Table 3.5.1) that, except for small values of n, it would be infeasible to run the program.

Determining the performance parameters of a computer program is a difficult task and depends on a number of factors such as the computer that is being used, the way the data are represented, and how the program is translated into machine instructions. Although precise estimates of the execution time of a program must take such factors into account, useful information can be obtained by analyzing the time complexity of the underlying algorithm.

The time needed to execute an algorithm is a function of the input. Usually, it is difficult to obtain an explicit formula for this function and we settle for less. Instead of dealing directly with the input, we use parameters that characterize the *size* of the input. We can ask for the minimum time needed to execute the algorithm among all inputs of size n. This time is called the **best-case time** for inputs of size n. We can also ask for the maximum time needed to execute the algorithm among all inputs of size n. This time is called the **worst-case time** for inputs of size n. Another important case is **average-case time**—the average time needed to execute the algorithm over some finite set of inputs all of size n.

We could measure the time required by an algorithm by counting the number of instructions executed. Alternatively, we could use a cruder time estimate such as the number of times each loop is executed. If the principal activity of an algorithm is making comparisons, as might happen in a sorting routine, we might count the number of comparisons. Usually, we are interested in general estimates since, as we have already observed, the actual performance of a program implementation of an algorithm is dependent on many factors.

EXAMPLE 3.5.1

A reasonable definition of the size of input for Algorithm 3.2.2 that finds the largest value in a finite sequence is the number of elements in the input sequence. A reasonable definition of the execution time is the number of iterations of the while loop. With these definitions, the worst-case, best-case, and average-case times for Algorithm 3.2.2 for input of size n are each $n - 1$ since the loop is always executed $n - 1$ times.   □

TABLE 3.5.1
Time to execute an algorithm if one step takes 1 microsecond to execute

Number of Steps to Termination for Input of Size n	Time to Execute if $n =$			
	3	**6**	**9**	**12**
1	10^{-6} sec	10^{-6} sec	10^{-6} sec	10^{-6} sec
$\lg \lg n$	10^{-6} sec	10^{-6} sec	2×10^{-6} sec	2×10^{-6} sec
$\lg n$	2×10^{-6} sec	3×10^{-6} sec	3×10^{-6} sec	4×10^{-6} sec
n	3×10^{-6} sec	6×10^{-6} sec	9×10^{-6} sec	10^{-5} sec
$n \lg n$	5×10^{-6} sec	2×10^{-5} sec	3×10^{-5} sec	4×10^{-5} sec
n^2	9×10^{-6} sec	4×10^{-5} sec	8×10^{-5} sec	10^{-4} sec
n^3	3×10^{-5} sec	2×10^{-4} sec	7×10^{-4} sec	2×10^{-3} sec
2^n	8×10^{-6} sec	6×10^{-5} sec	5×10^{-4} sec	4×10^{-3} sec
	50	**100**	**1000**	**10^5**	**10^6**
1	10^{-6} sec	10^{-6} sec	10^{-6} sec	10^{-6} sec	10^{-6} sec
$\lg \lg n$	2×10^{-6} sec	3×10^{-6} sec	3×10^{-6} sec	4×10^{-6} sec	4×10^{-6} sec
$\lg n$	6×10^{-6} sec	7×10^{-6} sec	10^{-5} sec	2×10^{-5} sec	2×10^{-5} sec
n	5×10^{-5} sec	10^{-4} sec	10^{-3} sec	0.1 sec	1 sec
$n \lg n$	3×10^{-4} sec	7×10^{-4} sec	10^{-2} sec	2 sec	20 sec
n^2	3×10^{-3} sec	0.01 sec	1 sec	3 hr	12 days
n^3	0.13 sec	1 sec	16.7 min	32 yr	31,710 yr
2^n	36 yr	4×10^{16} yr	3×10^{287} yr	3×10^{30089} yr	3×10^{301016} yr

Often we are less interested in the exact best-case or worst-case time required for an algorithm to execute than we are in how the best-case or worst-case time grows as the size of the input increases. For example, suppose that the worst-case time of an algorithm is

$$t(n) = 60n^2 + 5n + 1 \qquad (3.5.1)$$

for input of size n. For large n, the term $60n^2$ is approximately equal to $t(n)$ (see Table 3.5.2). In this sense, $t(n)$ grows like $60n^2$.

TABLE 3.5.2
Comparing growth of $t(n)$ with $60n^2$

n	$t(n) = 60n^2 + 5n + 1$	$60n^2$
10	6,051	6,000
100	600,501	600,000
1,000	60,005,001	60,000,000
10,000	6,000,050,001	6,000,000,000

If (3.5.1) measures the worst-case time for input of size n in seconds, then

$$T(n) = n^2 + \frac{5}{60}n + \frac{1}{60}$$

measures the worst-case time for input of size n in minutes. Now this change of units does not affect how the worst-case time grows as the size of the input increases but only the units in which we measure the worst-case time for input of size n. Thus when we describe how the best-case or worst-case time grows as the size of the input increases, we not only seek the dominant term [e.g., $60n^2$ in (3.5.1)], but we also may ignore constant coefficients. Under these assumptions, $t(n)$ grows like n^2 as n increases. We say that $t(n)$ is of **order** n^2 and write

$$t(n) = \Theta(n^2),$$

which is read "$t(n)$ is theta of n^2." The basic idea is to replace an expression such as $t(n) = 60n^2 + 5n + 1$ with a simpler expression such as n^2 that grows at the same rate as $t(n)$. The formal definitions follow.

DEFINITION 3.5.2

Let f and g be functions with domain $\{1, 2, 3, \ldots\}$.
    We write

$$f(n) = O(g(n))$$

and say that $f(n)$ *is of order at most* $g(n)$ if there exists a positive constant C_1 such that

$$|f(n)| \leq C_1|g(n)|$$

for all but finitely many positive integers n.
    We write

$$f(n) = \Omega(g(n))$$

and say that $f(n)$ is of *order at least* $g(n)$ if there exists a positive constant C_2 such that

$$|f(n)| \geq C_2|g(n)|$$

for all but finitely many positive integers n.
    We write

$$f(n) = \Theta(g(n))$$

and say that $f(n)$ is of *order* $g(n)$ if $f(n) = O(g(n))$ and $f(n) = \Omega(g(n))$.

Definition 3.5.2 can be loosely paraphrased as follows. $f(n) = O(g(n))$ if, except for constants and a finite number of exceptions, f is bounded above by g. $f(n) = \Omega(g(n))$ if, except for constants and a finite number of exceptions, f is bounded below by g. $f(n) = \Theta(g(n))$ if, except for constants and a finite number of exceptions, f is bounded above and below by g.

An expression of the form $f(n) = O(g(n))$ is sometimes referred to as a **big oh notation** for f. Similarly, $f(n) = \Omega(g(n))$ is sometimes referred to as an **omega notation** for f and $f(n) = \Theta(g(n))$ is sometimes referred to as a **theta notation** for f.

According to the definition, if $f(n) = O(g(n))$, all that one can conclude is that except for constants and a finite number of exceptions, f is bounded *above* by g, so g grows at least as fast as f. For example, if $f(n) = n$ and $g(n) = 2^n$, then $f(n) = O(g(n))$, but g grows considerably faster than f. The statement $f(n) = O(g(n))$ says nothing about a *lower* bound for f. On the other hand, if $f(n) = \Theta(g(n))$, one can draw the conclusion that, except for constants and a finite number of exceptions, f is bounded *above* and *below* by g, so f and g grow at the same rate. Notice that $n = O(2^n)$, but $n \neq \Theta(2^n)$. Unfortunately, it is not uncommon in the literature to find big oh notation used as if it were theta notation.

EXAMPLE 3.5.3

Since

$$60n^2 + 5n + 1 \leq 60n^2 + 5n^2 + n^2 = 66n^2 \qquad \text{for } n \geq 1,$$

we may take $C_1 = 66$ in Definition 3.5.2 to obtain

$$60n^2 + 5n + 1 = O(n^2).$$

Since

$$60n^2 + 5n + 1 \geq 60n^2 \qquad \text{for } n \geq 1,$$

we may take $C_2 = 60$ in Definition 3.5.2 to obtain

$$60n^2 + 5n + 1 = \Omega(n^2).$$

Since $60n^2 + 5n + 1 = O(n^2)$ and $60n^2 + 5n + 1 = \Omega(n^2)$,

$$60n^2 + 5n + 1 = \Theta(n^2). \qquad \square$$

The method of Example 3.5.3 can be used to show that a polynomial in n of degree k with nonnegative coefficients is $\Theta(n^k)$. [In fact, *any* polynomial in n of degree k is $\Theta(n^k)$, even if some of its coefficients are negative. To prove this more general result, the method of Example 3.5.3 has to be modified.]

THEOREM 3.5.4

Let

$$a_k n^k + a_{k-1} n^{k-1} + \cdots + a_1 n + a_0$$

be a polynomial in n of degree k, where each a_i is nonnegative. Then

$$a_k n^k + a_{k-1} n^{k-1} + \cdots + a_1 n + a_0 = \Theta(n^k).$$

Proof. Let

$$C = a_k + a_{k-1} + \cdots + a_1 + a_0.$$

Then

$$a_k n^k + a_{k-1} n^{k-1} + \cdots + a_1 n + a_0$$
$$\leq a_k n^k + a_{k-1} n^k + \cdots + a_1 n^k + a_0 n^k$$
$$= (a_k + a_{k-1} + \cdots + a_1 + a_0) n^k = C n^k.$$

Therefore,

$$a_k n^k + a_{k-1} n^{k-1} + \cdots + a_1 n + a_0 = O(n^k).$$

Since

$$a_k n^k + a_{k-1} n^{k-1} + \cdots + a_1 n + a_0 \geq a_k n^k,$$

$$a_k n^k + a_{k-1} n^{k-1} + \cdots + a_1 n + a_0 = \Omega(n^k).$$

Thus

$$a_k n^k + a_{k-1} n^{k-1} + \cdots + a_1 n + a_0 = \Theta(n^k). \qquad \blacksquare$$

EXAMPLE 3.5.5

In this book we let $\lg n$ denote $\log_2 n$ (the logarithm of n to the base 2). Since $\lg n < n$ for $n \geq 1$ (see Figure 3.5.1),

$$2n + 3\lg n < 2n + 3n = 5n \qquad \text{for } n \geq 1;$$

thus

$$2n + 3\lg n = O(n).$$

Also,

$$2n + 3\lg n \geq 2n \qquad \text{for } n \geq 1,$$

so

$$2n + 3\lg n = \Omega(n).$$

Therefore,

$$2n + 3\lg n = \Theta(n). \qquad \square$$

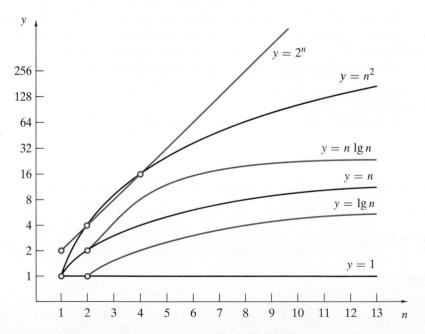

FIGURE 3.5.1   Growth of some common functions.

EXAMPLE 3.5.6

If we replace each integer $1, 2, \ldots, n$ by n in the sum $1 + 2 + \cdots + n$, the sum does not decrease and we have

$$1 + 2 + \cdots + n \leq n + n + \cdots + n = n \cdot n = n^2 \qquad (3.5.2)$$

for $n \geq 1$. It follows that

$$1 + 2 + \cdots + n = O(n^2).$$

To obtain a lower bound, we might imitate the preceding argument and replace each integer $1, 2, \ldots, n$ by 1 in the sum $1 + 2 + \cdots + n$ to obtain

$$1 + 2 + \cdots + n \geq 1 + 1 + \cdots + 1 = n.$$

In this case we conclude that

$$1 + 2 + \cdots + n = \Omega(n),$$

and while the preceding expression is true, we cannot deduce a Θ-estimate for $1 + 2 + \cdots + n$, since the upper bound n^2 and lower bound n are not equal. We must be craftier in deriving a lower bound.

One way to get a sharper lower bound is to argue as before, but first throw away the first half of the terms, to obtain

$$\begin{aligned}
1 + 2 + \cdots + n &\geq \lceil n/2 \rceil + \cdots + (n - 1) + n \\
&\geq \lceil n/2 \rceil + \cdots + \lceil n/2 \rceil + \lceil n/2 \rceil \\
&= \lceil (n + 1)/2 \rceil \lceil n/2 \rceil \geq (n/2)(n/2) = n^2/4. \quad (3.5.3)
\end{aligned}$$

We can now conclude that

$$1 + 2 + \cdots + n = \Omega(n^2).$$

Therefore,

$$1 + 2 + \cdots + n = \Theta(n^2). \qquad \square$$

EXAMPLE 3.5.7

If k is a positive integer and, as in Example 3.5.6, we replace each integer $1, 2, \ldots, n$ by n, we have

$$1^k + 2^k + \cdots + n^k \leq n^k + n^k + \cdots + n^k = n \cdot n^k = n^{k+1}$$

for $n \geq 1$; hence

$$1^k + 2^k + \cdots + n^k = O(n^{k+1}).$$

We can also obtain a lower bound as in Example 3.5.6:

$$1^k + 2^k + \cdots + n^k \geq \lceil n/2 \rceil^k + \cdots + (n-1)^k + n^k$$

$$\geq \lceil n/2 \rceil^k + \cdots + \lceil n/2 \rceil^k + \lceil n/2 \rceil^k$$

$$= \lceil (n+1)/2 \rceil \lceil n/2 \rceil^k \geq (n/2)(n/2)^k = n^{k+1}/2^{k+1}.$$

We conclude that

$$1^k + 2^k + \cdots + n^k = \Omega(n^{k+1}),$$

and hence

$$1^k + 2^k + \cdots + n^k = \Theta(n^{k+1}). \qquad \square$$

Notice the difference between the polynomial

$$a_k n^k + a_{k-1} n^{k-1} + \cdots + a_1 n + a_0$$

in Theorem 3.5.4 and the expression

$$1^k + 2^k + \cdots + n^k$$

in Example 3.5.7. A polynomial has a fixed number of terms, whereas the number of terms in the expression in Example 3.5.7 is dependent on the value of n. Furthermore, the polynomial in Theorem 3.5.4 is $\Theta(n^k)$, but the expression in Example 3.5.7 is $\Theta(n^{k+1})$.

Our next example gives a theta notation for $\lg n!$.

EXAMPLE 3.5.8

We show that

$$\lg n! = \Theta(n \lg n)$$

using an argument similar to that in Example 3.5.6.

By properties of logarithms, we have

$$\lg n! = \lg n + \lg(n-1) + \cdots + \lg 2 + \lg 1.$$

Since $\lg$ is an increasing function,

$$\lg n + \lg(n-1) + \cdots + \lg 2 + \lg 1 \leq \lg n + \lg n + \cdots + \lg n + \lg n = n \lg n.$$

We conclude that

$$\lg n! = O(n \lg n).$$

Now

$$\lg n + \lg(n-1) + \cdots + \lg 2 + \lg 1$$

$$\geq \lg n + \lg(n-1) + \cdots + \lg \lceil n/2 \rceil$$

$$\geq \lg \lceil n/2 \rceil + \cdots + \lg \lceil n/2 \rceil$$

$$= \lceil (n+1)/2 \rceil \lg \lceil n/2 \rceil \geq (n/2) \lg(n/2).$$

A proof by mathematical induction (see Exercise 46) shows that if $n \geq 4$,

$$(n/2) \lg(n/2) \geq (n \lg n)/4.$$

The last inequalities combine to give

$$\lg n + \lg(n-1) + \cdots + \lg 2 + \lg 1 \geq (n \lg n)/4$$

for $n \geq 4$. Therefore,

$$\lg n! = \Omega(n \lg n).$$

It follows that

$$\lg n! = \Theta(n \lg n). \qquad \square$$

We next define what it means for the best-case, worst-case, or average-case time of an algorithm to be of order at most $g(n)$.

DEFINITION 3.5.9

If an algorithm requires $t(n)$ units of time to terminate in the best case for an input of size n and

$$t(n) = O(g(n)),$$

we say that the *best-case time required by the algorithm is of order at most $g(n)$* or that the *best-case time required by the algorithm* is $O(g(n))$.

If an algorithm requires $t(n)$ units of time to terminate in the worst case for an input of size n and

$$t(n) = O(g(n)),$$

we say that the *worst-case time required by the algorithm is of order at most $g(n)$* or that the *worst-case time required by the algorithm* is $O(g(n))$.

If an algorithm requires $t(n)$ units of time to terminate in the average case for an input of size n and

$$t(n) = O(g(n)),$$

we say that the *average-case time required by the algorithm is of order at most $g(n)$* or that the *average-case time required by the algorithm* is $O(g(n))$.

By replacing O by Ω and "at most" by "at least" in Definition 3.5.9, we obtain the definition of what it means for the best-case, worst-case, or average-case time of an algorithm to be of order at least $g(n)$. If the best-case time required by an algorithm is $O(g(n))$ and $\Omega(g(n))$, we say that the best-case time required by the algorithm is $\Theta(g(n))$. An analogous definition applies to the worst-case and average-case time of an algorithm.

EXAMPLE 3.5.10

Suppose that an algorithm is known to take

$$60n^2 + 5n + 1$$

units of time to terminate in the worst case for inputs of size n. We showed in Example 3.5.3 that

$$60n^2 + 5n + 1 = \Theta(n^2).$$

Thus the worst-case time required by this algorithm is $\Theta(n^2)$. $\qquad \square$

EXAMPLE 3.5.11

Find a theta notation in terms of n for the number of times the statement $x := x + 1$ is executed.

```
1.    for i := 1 to n do
2.        for j := 1 to i do
3.            x := x + 1
```

First, i is set to 1 and, as j runs from 1 to 1, line 3 is executed one time. Next, i is set to 2 and, as j runs from 1 to 2, line 3 is executed two times, and so on. Thus the total number of times line 3 is executed is (see Example 3.5.6)

$$1 + 2 + \cdots + n = \Theta(n^2).$$

Thus a theta notation for the number of times the statement $x := x + 1$ is executed is $\Theta(n^2)$. □

EXAMPLE 3.5.12

Find a theta notation in terms of n for the number of times the statement $x := x + 1$ is executed.

```
1.    j := n
2.        while j ≥ 1 do
3.        begin
4.            for i := 1 to j do
5.                x := x + 1
6.            j := ⌊j/2⌋
7.        end
```

Let $t(n)$ denote the number of times we execute the statement $x := x+1$. The first time we arrive at the body of the while loop, the statement $x := x + 1$ is executed n times. Therefore $t(n) \geq n$ and $t(n) = \Omega(n)$.

Next we derive a big oh notation for $t(n)$. After j is set to n, we arrive at the while loop for the first time. The statement $x := x + 1$ is executed n times. At line 6, j is replaced by $\lfloor n/2 \rfloor$; hence $j \leq n/2$. If $j \geq 1$, we will execute $x := x + 1$ at most $n/2$ additional times in the next iteration of the while loop, and so on. If we let k denote the number of times we execute the body of the while loop, the number of times we execute $x := x + 1$ is at most

$$n + \frac{n}{2} + \frac{n}{4} + \cdots + \frac{n}{2^{k-1}}.$$

This geometric sum (see Example 1.6.2) is equal to

$$\frac{n \left(1 - \frac{1}{2^k}\right)}{1 - \frac{1}{2}}.$$

Now

$$t(n) \leq \frac{n \left(1 - \frac{1}{2^k}\right)}{1 - \frac{1}{2}} = 2n \left(1 - \frac{1}{2^k}\right) \leq 2n,$$

so $t(n) = O(n)$. Thus a theta notation for the number of times we execute $x := x + 1$ is $\Theta(n)$. □

EXAMPLE 3.5.13

Determine, in theta notation, the best-case, worst-case, and average-case times required to execute Algorithm 3.5.14, which follows. Assume that the input size is n and that the run time of the algorithm is the number of comparisons made at line 3. Also, assume that the $n + 1$ possibilities of *key* being at any particular position in the sequence or not being in the sequence are equally likely.

The best-case time can be analyzed as follows. If $s_1 = key$, line 3 is executed once. Thus the best-case time of Algorithm 3.5.14 is

$$\Theta(1).$$

The worst-case time of Algorithm 3.5.14 is analyzed as follows. If *key* is not in the sequence, line 3 will be executed n times, so the worst-case time of Algorithm 3.5.14 is

$$\Theta(n).$$

Finally, consider the average-case time of Algorithm 3.5.14. If *key* is found at the ith position, line 3 is executed i times and if *key* is not in the sequence, line 3 is executed n times. Thus the average number of times line 3 is executed is

$$\frac{(1 + 2 + \cdots + n) + n}{n + 1}.$$

Now

$$\frac{(1 + 2 + \cdots + n) + n}{n + 1} \leq \frac{n^2 + n}{n + 1} \qquad \text{by (3.5.2)}$$

$$= \frac{n(n + 1)}{n + 1} = n.$$

Therefore, the average-case time of Algorithm 3.5.14 is

$$O(n).$$

Also,

$$\frac{(1 + 2 + \cdots + n) + n}{n + 1} \geq \frac{n^2/4 + n}{n + 1} \qquad \text{by (3.5.3)}$$

$$\geq \frac{n^2/4 + n/4}{n + 1} = \frac{n}{4}.$$

Therefore the average-case time of Algorithm 3.5.14 is

$$\Omega(n).$$

Thus the average-case time of Algorithm 3.5.14 is

$$\Theta(n).$$

For this algorithm, the worst-case and average-case times are both $\Theta(n)$.

$\square$

> ALGORITHM 3.5.14   *Searching an Unordered Sequence*

Given the sequence

$$s_1, \quad s_2, \quad \cdots, \quad s_n,$$

and a value *key*, this algorithm finds the location of *key*. If *key* is not found, the algorithm outputs 0.

Input: $s_1, s_2, \ldots, s_n, n$, and *key* (the value to search for)

Output: The location of *key*, or if *key* is not found, 0

1.   **procedure** *linear_search*(s, n, key)
2.   **for** $i := 1$ **to** n **do**
3.   **if** $key = s_i$ **then**
4.   **return**(i) // successful search
5.   **return**(0) // unsuccessful search
6.   **end** *linear_search*

In Section 3.6 we consider a more involved example, the worst-case time of the Euclidean algorithm (Algorithm 3.3.8).

The constants that are suppressed in the theta notation may be important. Even if for any input of size n, algorithm A requires exactly $C_1 n$ time units and algorithm B requires exactly $C_2 n^2$ time units, for certain sizes of inputs algorithm B may be superior. For example, suppose that for any input of size n, algorithm A requires $300n$ units of time and algorithm B requires $5n^2$ units of time. For an input size of $n = 5$, algorithm A requires 1500 units of time and algorithm B requires 125 units of time, and thus algorithm B is faster. Of course, for sufficiently large inputs, algorithm A is considerably faster than algorithm B.

Certain forms occur so often that they are given special names, as shown in Table 3.5.3. The forms in Table 3.5.3, with the exception of $\Theta(n^m)$, are arranged so that if $\Theta(f(n))$ is above $\Theta(g(n))$, then $f(n) \le g(n)$ for all but finitely many positive integers n. Thus, if algorithms A and B have run times that are $\Theta(f(n))$ and $\Theta(g(n))$, respectively, and $\Theta(f(n))$ is above $\Theta(g(n))$ in Table 3.5.3, then algorithm A is more time efficient than algorithm B for sufficiently large inputs.

TABLE 3.5.3
Common growth functions

Theta Form[†]	Name
$\Theta(1)$	Constant
$\Theta(\lg \lg n)$	Log log
$\Theta(\lg n)$	Logarithmic
$\Theta(n)$	Linear
$\Theta(n \lg n)$	$n \log n$
$\Theta(n^2)$	Quadratic
$\Theta(n^3)$	Cubic
$\Theta(n^m)$	Polynomial
$\Theta(m^n), m \ge 2$	Exponential
$\Theta(n!)$	Factorial

[†] $\lg$ = log to the base 2;

m is a fixed nonnegative integer.

It is important to develop some feeling for the relative sizes of the functions in Table 3.5.3. In Figure 3.5.1 we have graphed some of these functions. Another way to develop some appreciation for the relative sizes of the functions $f(n)$ in Table 3.5.3 is to determine how long it would take an algorithm to terminate whose run time is exactly $f(n)$. For this purpose, let us assume that we have a computer that can execute one step in 1 microsecond (10^{-6} sec). Table 3.5.1 shows the execution times, under this assumption, for various input sizes. Notice that it is feasible to implement an algorithm that requires 2^n steps for an input of size n only for very small input sizes. Algorithms requiring n^2 or n^3 steps also become infeasible, but for relatively larger input sizes. Also, notice the dramatic improvement that results when we move from n^2 steps to $n \lg n$ steps.

A problem that has a worst-case polynomial time algorithm is considered to have a "good" algorithm; the interpretation is that such a problem has an efficient solution. Of course, if the worst-case time to solve the problem is proportional to a high degree polynomial, the problem can still take a long time to solve. Fortunately, in many important cases, the polynomial bound has small degree.

A problem that does not have a worst-case polynomial time algorithm is said to be **intractable**. Any algorithm, if there is one, that solves an intractable problem is guaranteed to take a long time to execute in the worst case even for modest sizes of the input.

Certain problems are so hard that they have no algorithms at all. A problem for which there is no algorithm is said to be an **unsolvable problem.** A large number of problems are known to be unsolvable, some of considerable practical importance. One of the earliest problems to be proved unsolvable is the **halting problem**: Given an arbitrary program and a set of inputs, will the program eventually halt?

A large number of solvable problems have an as yet undetermined status; they are thought to be intractable, but none of them have been proved intractable. (These problems belong to the class NP; see [Hopcroft] for details.) An example of a solvable problem thought to be intractable, but not known to be intractable, is:

> Given a collection $\mathcal{C}$ of finite sets and a positive integer $k \leq |\mathcal{C}|$, does $\mathcal{C}$ contain at least k mutually disjoint sets?

Other solvable problems thought to be intractable, but not known to be intractable, include the traveling salesperson problem and the Hamiltonian cycle problem (see Section 6.3).

ॐ ॐ ॐ

Exercises

Select a theta notation from Table 3.5.3 for each expression in Exercises 1–12.

1. $6n + 1$

2. $2n^2 + 1$

3. $6n^3 + 12n^2 + 1$

4. $3n^2 + 2n \lg n$

5. $2 \lg n + 4n + 3n \lg n$

6. $6n^6 + n + 4$

7. $2 + 4 + 6 + \cdots + 2n$

8. $(6n + 1)^2$

9. $(6n + 4)(1 + \lg n)$

10. $\dfrac{(n + 1)(n + 3)}{n + 2}$

11. $\dfrac{(n^2 + \lg n)(n + 1)}{n + n^2}$

12. $2 + 4 + 8 + 16 + \cdots + 2^n$

In Exercises 13–15, select a theta notation for $f(n) + g(n)$.

13. $f(n) = \Theta(1), \quad g(n) = \Theta(n^2)$

14. $f(n) = 6n^3 + 2n^2 + 4, \quad g(n) = \Theta(n \lg n)$

15. $f(n) = \Theta(n^{3/2}), \quad g(n) = \Theta(n^{5/2})$

In Exercises 16–26, select a theta notation from among

$$\Theta(1), \quad \Theta(\lg n), \quad \Theta(n), \quad \Theta(n \lg n), \quad \Theta(n^2), \quad \Theta(n^3), \quad \Theta(2^n), \text{ or } \Theta(n!)$$

for the number of times the statement $x := x + 1$ is executed.

16. for $i := 1$ **to** $2n$ **do**
    $x := x + 1$

17. $i := 1$
    while $i \le 2n$ **do**
      begin
        $x := x + 1$
        $i := i + 2$
      end

18. for $i := 1$ **to** n **do**
    for $j := 1$ **to** n **do**
      $x := x + 1$

19. for $i := 1$ **to** $2n$ **do**
    for $j := 1$ **to** n **do**
      $x := x + 1$

20. for $i := 1$ **to** n **do**
    for $j := 1$ **to** $\lfloor i/2 \rfloor$ **do**
      $x := x + 1$

21. for $i := 1$ **to** n **do**
    for $j := 1$ **to** n **do**
      for $k := 1$ **to** n **do**
        $x := x + 1$

22. for $i := 1$ **to** n **do**
    for $j := 1$ **to** n **do**
      for $k := 1$ **to** i **do**
        $x := x + 1$

23. for $i := 1$ **to** n **do**
    for $j := 1$ **to** i **do**
      for $k := 1$ **to** j **do**
        $x := x + 1$

24. $j := n$
    while $j \ge 1$ **do**
      begin
      for $i := 1$ **to** j **do**
       $x := x + 1$
      $j := \lfloor j/3 \rfloor$
      end

25. $i := n$
    while $i \ge 1$ **do**
      begin
      $x := x + 1$
      $i := \lfloor i/2 \rfloor$
      end

26. $i := n$
    while $i \ge 1$ **do**
      begin
      for $j := 1$ **to** n **do**
       $x := x + 1$
      $i := \lfloor i/2 \rfloor$
      end

27. Find a theta notation for the number of times the statement $x := x + 1$ is executed.

$i := 2$
while $i < n$ **do**
  begin
  $i := i^2$
  $x := x + 1$
  end

28. Find the exact number of comparisons (lines 12, 18, 20, 28, and 30) required by the following algorithm when n is even and when n is odd. Find a theta notation for this algorithm.

    Input: $s_1, s_2, \ldots, s_n, n$

   Output: *large* (the largest item in $s_1, s_2, \ldots, s_n$)
                    small (the smallest item in $s_1, s_2, \ldots, s_n$)

```
1.    procedure large_small(s, n, large, small)
2.      if n = 1 then
3.         begin
4.         large := s₁
5.         small := s₁
6.         return
7.         end
8.      m := 2⌊n/2⌋
9.      i := 1
10.     while i ≤ m − 1 do
11.        begin
12.        if sᵢ > sᵢ₊₁ then
13.           swap(sᵢ, sᵢ₊₁)
14.        i := i + 2
15.        end
16.     if n > m then
17.        begin
18.        if sₘ₋₁ > sₙ then
19.           swap(sₘ₋₁, sₙ)
20.        if sₙ > sₘ then
21.           swap(sₘ, sₙ)
22.        end
23.     small := s₁
24.     large := s₂
25.     i := 3
26.     while i ≤ m − 1 do
27.        begin
28.        if sᵢ < small then
29.           small := sᵢ
30.        if sᵢ₊₁ > large then
31.           large := sᵢ₊₁
32.        i := i + 2
33.        end
34.     end large_small
```

29. Suppose that $a > 1$ and that $f(n) = \Theta(\log_a n)$. Show that $f(n) = \Theta(\lg n)$.

30. Show that $n! = O(n^n)$.

31. Show that $2^n = O(n!)$.

32. Suppose that $g(n) > 0$ for $n = 1, 2, \ldots$. Show that $f(n) = \Theta(g(n))$ if and only if there exist positive constants c_1 and c_2 such that

$$c_1 g(n) \leq |f(n)| \leq c_2 g(n) \qquad \text{for all } n = 1, 2, \ldots.$$

Determine whether each statement in Exercises 33–42 is true or false. If the statement is false, give a counterexample. Assume that the functions f, g, and h take on only positive values.

33. If $f(n) = \Theta(g(n))$ and $g(n) = \Theta(h(n))$, then $f(n) = \Theta(h(n))$.

34. If $f(n) = \Theta(h(n))$ and $g(n) = \Theta(h(n))$, then $f(n) + g(n) = \Theta(h(n))$.

35. If $f(n) = \Theta(g(n))$, then $cf(n) = \Theta(g(n))$ for any $c \neq 0$.

36. If $f(n) = \Theta(g(n))$, then $2^{f(n)} = \Theta(2^{g(n)})$.

37. If $f(n) = \Theta(g(n))$, then $\lg f(n) = \Theta(\lg g(n))$. Assume that $f(n) \geq 1$ and $g(n) \geq 1$ for all $n = 1, 2, \ldots$.

38. If $f(n) = O(g(n))$, then $g(n) = O(f(n))$.

39. If $f(n) = O(g(n))$, then $g(n) = \Omega(f(n))$.

40. If $f(n) = \Theta(g(n))$, then $g(n) = \Theta(f(n))$.

41. $f(n) + g(n) = \Theta(h(n))$, where $h(n) = \max\{f(n), g(n)\}$

42. $f(n) + g(n) = \Theta(h(n))$, where $h(n) = \min\{f(n), g(n)\}$ ☐

★ **43.** Find functions f and g satisfying

$$f(n) \neq O(g(n)) \qquad \text{and} \qquad g(n) \neq O(f(n)).$$

44. Find functions f, g, h, and t satisfying

$$f(n) = \Theta(g(n)), \qquad h(n) = \Theta(t(n)), \qquad f(n) - h(n) \neq \Theta(g(n) - t(n)).$$

45. Where is the error in the following reasoning? Suppose that the worst-case time of an algorithm is $\Theta(n)$. Since $2n = \Theta(n)$, the worst-case time to run the algorithm with input of size $2n$ will be approximately the same as the worst-case time to run the algorithm with input of size n.

46. Show that if $n \geq 4$,

$$\frac{n}{2} \lg \frac{n}{2} \geq \frac{n \lg n}{4}.$$

47. Does

$$f(n) = O(g(n))$$

define an equivalence relation on the set of real-valued functions on $\{1, 2, \ldots\}$?

48. Does

$$f(n) = \Theta(g(n))$$

define an equivalence relation on the set of real-valued functions on $\{1, 2, \ldots\}$?

49. [Requires the integral.]

(a) Show, by consulting the figure, that

$$\frac{1}{2} + \frac{1}{3} + \cdots + \frac{1}{n} < \log_e n.$$

(b) Show, by consulting the figure, that

$$\log_e n < 1 + \frac{1}{2} + \cdots + \frac{1}{n-1}.$$

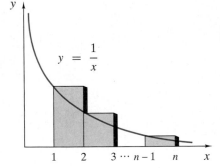

(c) Use parts (a) and (b) to show that

$$1 + \frac{1}{2} + \cdots + \frac{1}{n} = \Theta(\lg n).$$

50. [Requires the integral.] Use an argument like that in Exercise 49 to show that

$$\frac{n^{m+1}}{m+1} < 1^m + 2^m + \cdots + n^m < \frac{(n+1)^{m+1}}{m+1},$$

where m is a positive integer.

51. What is wrong with the following "proof" that any algorithm has a run time that is $O(n)$?

We must show that the time required for an input of size n is at most a constant times n.

BASIS STEP.   Suppose that $n = 1$. If the algorithm takes C units of time for an input of size 1, the algorithm takes at most $C \cdot 1$ units of time. Thus the assertion is true for $n = 1$.

INDUCTIVE STEP.   Assume that the time required for an input of size n is at most $C'n$ and that the time for processing an additional item is C''. Let C be the maximum of C' and C''. Then the total time required for an input of size $n + 1$ is at most

$$C'n + C'' \leq Cn + C = C(n + 1).$$

The Inductive Step has been verified.

By induction, for input of size n, the time required is at most Cn. Therefore, the run time is $O(n)$.

52. [Requires calculus.] Determine whether each statement is true or false. If the statement is false, give a counterexample. It is assumed that f and g are real-valued functions defined on the set of positive integers and that $g(n) \neq 0$ for $n \geq 1$.

   (a) If

$$\lim_{n \to \infty} \frac{f(n)}{g(n)}$$

   exists and is equal to some real number, then $f(n) = O(g(n))$.

   (b) If $f(n) = O(g(n))$, then

$$\lim_{n \to \infty} \frac{f(n)}{g(n)}$$

   exists and is equal to some real number.

   (c) If

$$\lim_{n \to \infty} \frac{f(n)}{g(n)}$$

   exists and is equal to some real number, then $f(n) = \Theta(g(n))$.

   (d) If

$$\lim_{n \to \infty} \frac{f(n)}{g(n)} = 1,$$

   then $f(n) = \Theta(g(n))$.

   (e) If $f(n) = \Theta(g(n))$, then

$$\lim_{n \to \infty} \frac{f(n)}{g(n)}$$

   exists and is equal to some real number.

★ 53. Use induction to prove that

$$\lg n! \geq \frac{n}{2} \lg \frac{n}{2}.$$

54. [Requires calculus.] Let $\ln x$ denote the natural logarithm ($\log_e x$) of x. Use the integral to obtain the estimate

$$n \ln n - n \leq \sum_{k=1}^{n} \ln k = \ln n!, \qquad n \geq 1.$$

55. Use the result of Exercise 54 and the change of base formula for logarithms to obtain the formula

$$n \lg n - n \lg e \leq \lg n!, \qquad n \geq 1.$$

56. Deduce

$$\lg n! \geq \frac{n}{2} \lg \frac{n}{2}$$

from the inequality of Exercise 55.

Problem

Develop and analyze an algorithm that outputs the maximum sum of consecutive values in the numerical sequence

$$s_1, \ldots, \quad s_n.$$

In mathematical notation, the problem is to find the maximum sum of the form $s_j + s_{j+1} + \cdots + s_i$. *Example*: If the sequence were

$$27 \quad 6 \quad -50 \quad 21 \quad -3 \quad 14 \quad 16 \quad -8 \quad 42 \quad 33 \quad -21 \quad 9,$$

the algorithm outputs 115—the sum of

$$21 \quad -3 \quad 14 \quad 16 \quad -8 \quad 42 \quad 33.$$

If all the numbers in a sequence are negative, the maximum sum of consecutive values is defined to be 0. (The idea is that the maximum of 0 is achieved by taking an "empty" sum.)

Attacking the Problem

In developing an algorithm, a good way to start is to ask the question, "How would I solve this problem by hand?" At least initially, take a straightforward approach. Here we might just list the sums of *all* consecutive values and pick the largest. For the example sequence, the sums are as shown below:

i	1	2	3	4	5	6	7	8	9	10	11	12
					j							
1	27											
2	33	6										
3	−17	−41	−50									
4	4	−2	−29	21								
5	1	−26	−32	18	−3							
6	15	−12	−18	32	11	14						
7	31	4	−2	48	27	30	16					
8	23	−4	−10	40	19	22	8	−8				
9	65	38	32	82	61	64	50	34	42			
10	98	71	65	115	94	97	83	67	75	33		
11	77	50	44	94	73	76	62	46	54	12	−21	
12	86	59	53	103	82	85	71	55	63	21	−12	9

The entry in column j, row i, is the sum

$$s_j + \cdots + s_i.$$

For example, the entry in column 4, row 7, is 48—the sum

$$s_4 + s_5 + s_6 + s_7 = 21 + -3 + 14 + 16 = 48.$$

By inspection, we find that 115 is the largest sum.

Finding a Solution

We began by writing pseudocode for the straightforward algorithm that computes all consecutive sums and finds the largest:

Input: $s_1, \ldots, s_n$
Output: *max*

```
procedure max_sum1(s, n)
// sum_ji is the sum s_j + ⋯ + s_i.
for i := 1 to n do
  begin
  for j := 1 to i − 1 do
    sum_ji := sum_{j,i−1} + s_i
  sum_ii := s_i
  end

// step through sum_ji and find the maximum
max := 0
for i := 1 to n do
  for j := 1 to i do
    if sum_ji > max then
      max := sum_ji
  return(max)
end max_sum1
```

The first nested for loops compute the sums

$$sum_{ji} = s_j + \cdots + s_i.$$

The computation relies on the fact that

$$sum_{ji} = s_j + \cdots + s_i = s_j + \cdots + s_{i-1} + s_i = sum_{j,i-1} + s_i.$$

The second nested for loops step through sum_{ji} and find the largest value.

Since each of the nested for loops takes time $\Theta(n^2)$, *max_sum1*'s time is $\Theta(n^2)$.

We can improve the actual time, but not the complexity of the algorithm, by computing the maximum within the same nested for loops in which we compute sum_{ji} :

Input: $s_1, \ldots, s_n$
Output: *max*

```
procedure max_sum2(s, n)
  // sum_ji is the sum s_j + ⋯ + s_i.
  max := 0
  for i := 1 to n do
    begin
    for j := 1 to i − 1 do
      begin
      sum_ji := sum_{j,i−1} + s_i
      if sum_ji > max then
        max := sum_ji
      end
    sum_ii := s_i
    if sum_ii > max then
      max := sum_ii
    end
  return(max)
end max_sum2
```

Since the nested for loops take time $\Theta(n^2)$, *max_sum2*'s time is $\Theta(n^2)$. To reduce the time complexity, we need to take a hard look at the pseudocode to see where it can be improved.

Two key observations lead to improved time. First, since we are looking only for the *maximum* sum, there is no need to record all of the sums; we will store only the maximum sum that ends at index i. Second, the line

$$sum_{ji} := sum_{j,i−1} + s_i$$

shows how a consecutive sum that ends at index $i − 1$ is related to a consecutive sum that ends at index i. The maximum can be computed by using a similar formula. If *sum* is the maximum consecutive sum that ends at index $i − 1$, the maximum consecutive sum that ends at index i is obtained by adding s_i to *sum* provided that $sum + s_i$ is positive. (If some sum of consecutive terms that ends at index i exceeds $sum + s_i$, we could remove the ith term and obtain a sum of consecutive terms ending at index $i − 1$ that exceeds *sum*, which is impossible.) If $sum + s_i \leq 0$, the maximum consecutive sum that ends at index i is obtained by taking no terms and has value 0. Thus we may compute the maximum consecutive sum that ends at index i by executing

```
if sum + s_i > 0 then
    sum := sum + s_i
else
    sum := 0
```

Formal Solution

Input: $s_1, \ldots, s_n$
Output: *max*

```
procedure max_sum3(s, n)
    // max is the maximum sum seen so far.
    // After the ith iteration of the for loop, sum is the
    // largest consecutive sum that ends at position i.
    max := 0
    sum := 0
    for i := 1 to n do
      begin
      if sum + s_i > 0 then
        sum := sum + s_i
      else
        sum := 0
      if sum > max then
        max := sum
      end
    return(max)
end max_sum3
```

Since this algorithm has a single for loop that runs from 1 to n, *max_sum3*'s time is $\Theta(n)$. The time complexity of this algorithm cannot be further improved. To solve this problem, we must at least look at each element in the sequence s, which takes time $\Theta(n)$.

Summary of Problem-Solving Techniques

- In developing an algorithm, a good way to start is to ask the question, "How would I solve this problem by hand?"

- In developing an algorithm, initially take a straightforward approach.

- After developing an algorithm, take a close look at the pseudocode to see where it can be improved. Look at the parts that perform key computations to gain insight into how to enhance the algorithm's efficiency.

- As in mathematical induction, extend a solution of a smaller problem to a larger problem. (In this problem, we extended a sum that ends at index $i - 1$ to a sum that ends at index i.)

- Don't repeat computations. (In this problem, we extended a sum that ends at index $i - 1$ to a sum that ends at index i by adding an additional term, rather than by computing the sum that ends at index i from scratch. This latter method would have meant recomputing the sum that ends at index $i - 1$.)

Comments

According to [Bentley], the problem discussed in this section is the one-dimensional version of the original two-dimensional problem that dealt with pattern matching in digital images. The original problem was to find the maximum sum in a rectangular submatrix of an $n \times n$ matrix of real numbers.

🙖 🙖 🙖

EXERCISE

1. Modify *max_sum3* so that it computes not only the maximum sum of consecutive values, but also the indexes of the first and last terms of a maximum-sum subsequence. If there is no maximum-sum subsequence (which would happen, for example, if all of the values of the sequence were negative), the algorithm should set the first and last indexes to zero.

3.6 ANALYSIS OF THE EUCLIDEAN ALGORITHM

In this section we analyze the worst-case performance of the Euclidean algorithm for finding the greatest common divisor of two nonnegative integers, not both zero (Algorithm 3.3.8). For reference, we summarize the algorithm:

Input: a and b (nonnegative integers, not both zero)

Output: Greatest common divisor of a and b

```
1.    procedure gcd(a, b)
2.       // make a largest
3.       if a < b then
4.          swap(a, b)
5.       while b ≠ 0 do
6.          begin
7.          divide a by b to obtain a = bq + r, 0 ≤ r < b
8.          a := b
9.          b := r
10.         end
11.      return(a)
12.   end gcd
```

We define the time required by the Euclidean algorithm as the number of divisions executed at line 7. Table 3.6.1 lists the number of divisions required for some small input values.

The worst case for the Euclidean algorithm occurs when the number of divisions is as large as possible. By referring to Table 3.6.1, we can determine the input pair $a, b, a > b$, with a as small as possible, that requires n divisions for $n = 0, \ldots, 4$. The results are given in Table 3.6.2.

Recall that the Fibonacci sequence $\{f_n\}$ (see Example 3.4.6) is defined by the equations

$$f_1 = 1; \qquad f_2 = 2; \qquad f_n = f_{n-1} + f_{n-2}, \qquad n \geq 3.$$

TABLE 3.6.1
Number of divisions required by the Euclidean algorithm for various values of the input

a \ b	0	1	2	3	4	5	6	7	8
0	—	0	0	0	0	0	0	0	0
1	0	1	1	1	1	1	1	1	1
2	0	1	1	2	1	2	1	2	1
3	0	1	2	1	2	3	1	2	3
4	0	1	1	2	1	2	2	3	1
5	0	1	2	3	2	1	2	3	4
6	0	1	1	1	2	2	1	2	2
7	0	1	2	2	3	3	2	1	2
8	0	1	1	3	1	4	2	2	1

The Fibonacci sequence begins

$$1, \quad 2, \quad 3, \quad 5, \quad 8, \quad 13, \quad \ldots.$$

A surprising pattern develops in Table 3.6.2: The a column is the beginning of the Fibonacci sequence and, except for the first value, the b column is the beginning of the Fibonacci sequence! We are led to conjecture that if the pair $a, b, a > b$, when input to the Euclidean algorithm requires $n \geq 1$ divisions, then $a \geq f_{n+1}$ and $b \geq f_n$. As further evidence of our conjecture, if we compute the smallest input pair that requires five divisions, we obtain $a = 13$ and $b = 8$. The values for six divisions are $a = 21$ and $b = 13$. Our next theorem confirms that our conjecture is correct. The proof of this theorem is illustrated in Figure 3.6.1.

TABLE 3.6.2
Smallest input pair that requires n divisions in the Euclidean algorithm

a	b	n (= number of divisions)
1	0	0
2	1	1
3	2	2
5	3	3
8	5	4

$$91 = 57 \cdot 1 + 34 \qquad \text{(1 division)}$$
$$34, 54 \text{ requires 4 divisions} \qquad \text{(to make a total of 5)}$$
$$54 \geq f_5 \text{ and } 34 \geq f_4 \qquad \text{(by inductive assumption)}$$
$$\therefore 91 = 57 \cdot 1 + 34 \geq 57 + 34 \geq f_5 + f_4 = f_6$$

FIGURE 3.6.1   The proof of Theorem 3.6.1. The pair 57, 91, which requires $n + 1 = 5$ divisions, is input to the Euclidean algorithm.

THEOREM 3.6.1

Suppose that the pair $a, b, a > b$, requires $n \geq 1$ divisions when input to the Euclidean algorithm. Then $a \geq f_{n+1}$ and $b \geq f_n$, where $\{f_n\}$ denotes the Fibonacci sequence.

Proof. The proof is by induction on n.

BASIS STEP ($n = 1$).   We have already observed that the theorem is true if $n = 1$.

Inductive Step. Assume that the theorem is true for $n \geq 1$. We must show that the theorem is true for $n + 1$.

Suppose that the pair $a, b, a > b$, requires $n + 1$ divisions when input to the Euclidean algorithm. At line 7, we divide a by b to obtain

$$a = bq + r, \qquad 0 \leq r < b. \qquad (3.6.1)$$

The algorithm then repeats using the values b and $r, b > r$. These values require n additional divisions. By the inductive assumption,

$$b \geq f_{n+1} \qquad \text{and} \qquad r \geq f_n. \qquad (3.6.2)$$

Combining (3.6.1) and (3.6.2), we obtain

$$a = bq + r \geq b + r \geq f_{n+1} + f_n = f_{n+2}. \qquad (3.6.3)$$

[The first inequality in (3.6.3) holds because $q > 0$; q cannot equal 0 because $a > b$.] Inequalities (3.6.2) and (3.6.3) give

$$a \geq f_{n+2} \qquad \text{and} \qquad b \geq f_{n+1}.$$

The Inductive Step is finished and the proof is complete. ■

We may use Theorem 3.6.1 to analyze the worst-case performance of the Euclidean algorithm.

THEOREM 3.6.2

If integers in the range 0 to m, $m \geq 8$, not both zero, are input to the Euclidean algorithm, then at most

$$\log_{3/2} \frac{2m}{3}$$

divisions are required.

Proof. Let n be the maximum number of divisions required by the Euclidean algorithm for integers in the range 0 to m, $m \geq 8$. Let a, b be an input pair in the range 0 to m that requires n divisions. Table 3.6.1 shows that $n \geq 4$ and that $a \neq b$. We may assume that $a > b$. (Interchanging the values of a and b does not alter the number of divisions required.) By Theorem 3.6.1, $a \geq f_{n+1}$. Thus

$$f_{n+1} \leq m.$$

By Exercise 20, Section 3.4, since $n + 1 \geq 5$,

$$\left(\frac{3}{2}\right)^{n+1} < f_{n+1}.$$

Combining these last inequalities, we obtain

$$\left(\frac{3}{2}\right)^{n+1} < m.$$

Taking the logarithm to the base $\frac{3}{2}$, we obtain

$$n + 1 < \log_{3/2} m.$$

Therefore,

$$n < \log_{3/2} m - 1 = \log_{3/2} m - \log_{3/2} 3/2 = \log_{3/2} \frac{2m}{3}. \qquad ■$$

Because the logarithm function grows so slowly, Theorem 3.6.2 tells us that the Euclidean algorithm is quite efficient, even for large values of the input. For example, since

$$\log_{3/2} \frac{2(1,000,000)}{3} = 33.07\ldots,$$

the Euclidean algorithm requires at most 33 divisions to compute the greatest common divisor of any pair of integers, not both zero, in the range 0 to 1,000,000.

୬ ୬ ୬

Exercises

1. Extend Tables 3.6.1 and 3.6.2 to the range 0 to 13.

2. Exactly how many divisions are required by the Euclidean algorithm in the worst case for numbers in the range 0 to 1,000,000?

3. How many subtractions are required by the algorithm of Exercise 26, Section 3.3, in the worst case for numbers in the range 0 to m? (This algorithm finds the greatest common divisor by using subtraction instead of division.)

4. Prove that when the pair f_{n+1}, f_n is input to the Euclidean algorithm, $n \geq 1$, exactly n divisions are required.

5. Show that for any integer $k > 1$, the number of divisions required by the Euclidean algorithm to compute $\gcd(a, b)$ is the same as the number of divisions required to compute $\gcd(ka, kb)$.

6. Show that $\gcd(f_n, f_{n+1}) = 1, n \geq 1$.

†3.7 THE RSA PUBLIC-KEY CRYPTOSYSTEM

Cryptology is the study of systems, called **cryptosystems**, for secure communications. In a cryptosystem, the sender transforms the message before transmitting it so that, hopefully, only authorized recipients can reconstruct the original message (i.e., the message before it was transformed). The sender is said to **encrypt** the message, and the recipient is said to **decrypt** the message. If the cryptosystem is secure, unauthorized persons will be unable to discover the decryption technique, so even if they read the encrypted message, they will be unable to decrypt it. Cryptosystems are important for large organizations (e.g., government and military) as well as for individuals. For example, if a credit card number is sent over a computer network, it is important for the number to be read only by the intended recipient. In this section, we look at some algorithms that support secure communication.

---

† This section can be omitted without loss of continuity.

In one of the oldest and simplest systems, the sender and receiver each have a key that defines a substitute character for each potential character to be sent. Moreover, the sender and receiver do not disclose the key. Such keys are said to be *private*.

EXAMPLE 3.7.1

If a key is defined as

    character: ABCDEFGHIJKLMNOPQRSTUVWXYZ
    replaced by: EIJFUAXVHWP GSRKOBTQYDMLZNC

the message SEND MONEY would be encrypted as QARUESKRAN. The encrypted message SKRANEKRELIN would be decrypted as MONEY ON WAY.   □

Simple systems such as that in Example 3.7.1 are easily broken since certain letters (e.g., E in English) and letter combinations (e.g., ER in English) appear more frequently than others. Also, a problem with private keys in general is that the keys have to be securely sent to the sender and recipient before messages can be sent. We devote the remainder of this section to the **RSA public key cryptosystem**, named after its inventors, Ronald L. Rivest, Adi Shamir, and Leonard M. Adleman, that is believed to be secure. In the RSA system, each participant makes public an encryption key and hides a decryption key. To send a message, all one needs to do is look up the recipient's encryption key in a publicly distributed table. The recipient then decrypts the message using the hidden decryption key.

In the RSA system, messages are represented as numbers. For example, each character might be represented as a number. If *blank* is represented as 1, A as 2, B as 3, and so on, the message SEND MONEY would be represented as 20, 6, 15, 5, 1, 14, 16, 15, 6, 26. If desired, the integers could be combined into the single integer

$$200615050111416150626.$$

We next describe how the RSA system works, present a concrete example, and then discuss why it works. Each prospective recipient chooses two primes p and q and computes $z = pq$. Since the security of the RSA system rests primarily on the inability of anyone knowing the value of z to discover the numbers p and q, p and q are typically chosen so that each has 100 or more digits. Next, the prospective recipient computes $\phi = (p - 1)(q - 1)$ and chooses an integer n such that $\gcd(n, \phi) = 1$. In practice, n is often chosen to be a prime. The pair z, n is then made public. Finally, the prospective recipient computes the unique number s, $0 < s < \phi$, satisfying $ns \bmod \phi = 1$. The number s is kept secret and used to decrypt messages.

To send the integer a, $0 \leq a \leq z - 1$, to the holder of public key z, n, the sender computes $c = a^n \bmod z$ and sends c. To decrypt the message, the recipient computes $c^s \bmod z$, which can be shown to be equal to a.

EXAMPLE 3.7.2

Suppose that we choose $p = 23$, $q = 31$, and $n = 29$. Then $z = pq = 713$ and $\phi = (p-1)(q-1) = 660$. Now $s = 569$ since $ns \bmod \phi = 29 \cdot 569 \bmod 660 = 16501 \bmod 660 = 1$. The pair $z, n = 713, 29$ is made publicly available.

To transmit $a = 572$ to the holder of public key 713, 29, the sender computes $c = a^n \bmod z = 572^{29} \bmod 713 = 113$ and sends 113. The receiver computes $c^s \bmod z = 113^{569} \bmod 713 = 572$ in order to decrypt the message.   □

It may appear that huge numbers must be computed in order to encrypt and decrypt messages using the RSA system. For example, the number 572^{29} in Example 3.7.2 has 80 digits, and if p and q have 100 or more digits, the numbers would be far larger. The key to simplifying the computation is to note that the arithmetic is done mod z. It can be shown that

$$ab \bmod z = [(a \bmod z)(b \bmod z)] \bmod z \qquad (3.7.1)$$

(see Exercise 10). We show how to use (3.7.1) to compute $572^{29} \bmod 713$.

EXAMPLE 3.7.3

We use (3.7.1) to compute $572^{29} \bmod 713$. We note that

$$29 = 16 + 8 + 4 + 1$$

(which is just the base 2 representation of 29), so we compute 572 to each of the powers 16, 8, 4, and 1, mod 713, by repeated squaring and then multiply them, mod 713:

$$572^2 \bmod 713 = 327184 \bmod 713 = 630$$

$$572^4 \bmod 713 = 630^2 \bmod 713 = 396900 \bmod 713 = 472$$

$$572^8 \bmod 713 = 472^2 \bmod 713 = 222784 \bmod 713 = 328$$

$$572^{16} \bmod 713 = 328^2 \bmod 713 = 107584 \bmod 713 = 634$$

$$572^{24} \bmod 713 = 572^{16} \cdot 572^8 \bmod 713 = 634 \cdot 328 \bmod 713$$
$$= 207952 \bmod 713 = 469$$

$$572^{28} \bmod 713 = 572^{24} \cdot 572^4 \bmod 713 = 469 \cdot 472 \bmod 713$$
$$= 221368 \bmod 713 = 338$$

$$572^{29} \bmod 713 = 572^{28} \cdot 572^1 \bmod 713 = 338 \cdot 572 \bmod 713$$
$$= 193336 \bmod 713 = 113.$$

The method is readily converted to an algorithm (see Exercise 11).   □

The Euclidean algorithm may be used by a prospective recipient to compute efficiently the unique number $s, 0 < s < \phi$, satisfying $ns \bmod \phi = 1$ (see Exercise 12). The main result that makes encryption and decryption work is that

$$a^u \bmod z = a \quad \text{for all } 0 \le a < z \text{ and } u \bmod \phi = 1$$

(for a proof, see [Cormen: Theorem 33.36, page 834]). Using this result and (3.7.1), we may show that decryption produces the correct result. Since $ns \bmod \phi = 1$,

$$c^s \bmod z = (a^n \bmod z)^s \bmod z = (a^n)^s \bmod z = a^{ns} \bmod z = a.$$

The security of the RSA encryption system relies mainly on the fact that at present there is no efficient algorithm known for factoring integers; that is, currently no algorithm is known for factoring d-digit integers in polynomial time, $O(d^k)$. Thus if the primes p and q are chosen large enough, it is impractical to compute the factorization $z = pq$. If the factorization could be found by a person who intercepts a message, the message could be decrypted just as the authorized recipient does. At this time, no practical method is known for factoring integers with 200 or more digits, so if p and q are chosen so that each has 100 or more digits, pq would then have about 200 or more digits, which seems to make RSA secure.

The first description of the RSA encryption system was in Martin Gardner's February 1977 *Scientific American* column (see [Gardner, 1977]). Included in this column was an encoded message using the key z, n, where z was the product of 64- and 65-digit primes, and $n = 9007$, and an offer of $100 to the first person to crack the code. At the time the article was written, it was estimated that it would take 40 quadrillion years to factor z. In fact, in April 1994, Arjen Lenstra, Paul Leyland, Michael Graff, and Derek Atkins, with the assistance of 600 volunteers from 25 countries using over 1600 computers, factored z (see [Taubes]). The work was coordinated on the Internet.

Another possible way a message could be intercepted and decrypted would be to take the nth root of c mod z, where c is the encrypted value sent. Since $c = a^n$ mod z, the nth root of c mod z would give a, the decrypted value. Again, at present there is no polynomial-time algorithm known for computing nth roots mod z. It is also conceivable that a message could be decrypted by some means other than factoring integers or taking nth roots mod z. For example, in 1996 Paul Kocher proposed a way to break RSA based on the time it takes to decrypt messages (see [English]). The idea is that different secret keys require different amounts of time to decrypt messages and, by using this timing information, an unauthorized person might be able to unveil the secret key and thus decrypt the message. Implementors of RSA have taken steps to alter the observed time to decrypt messages to thwart such attacks.

ॐ ॐ ॐ

Exercises

1. Encrypt the message COOL BEAVIS using the key of Example 3.7.1.
2. Decrypt the message UTWR ENKDTEKMIGYWRA using the key of Example 3.7.1.
3. Encrypt 333 using the public key 713, 29 of Example 3.7.2.
4. Decrypt 411 using $s = 569$ as in Example 3.7.2.

In Exercises 5–9, assume that we choose primes $p = 17$, $q = 23$, and $n = 31$.

5. Compute z.
6. Compute ϕ.
7. Verify that $s = 159$.
8. Encrypt 101 using the public key z, n.
9. Decrypt 250.

10. Prove equation (3.7.1).

11. Give an efficient algorithm to compute $a^n \bmod z$.

12. Show how to compute efficiently the value of s given n and ϕ; that is, given positive integers n and ϕ, with $\gcd(n, \phi) = 1$, give an efficient algorithm to compute positive integers s and t, with $0 < s < \phi$, such that $ns - t\phi = 1$ and, in particular, $ns \bmod \phi = 1$.

> *Hint*: Use the method of Exercise 21, Section 3.3, to compute efficiently integers s' and t' such that $s'n + t'\phi = 1$. If $s' > 0$, take $s = s'$. If $s' < 0$, take
> $$s = -s'(\phi - 1) \bmod \phi.$$

13. Show that the number s of Exercise 12 is unique.

14. Show how to use the method of Exercise 12 to compute the value s of Example 3.7.2.

15. Show how to use the method of Exercise 12 to compute the value s of Exercise 7.

❧ NOTES

The books by Knuth [1973, Vols. 1 and 3; 1981] are the first three books in a projected seven-volume set. The first half of Volume 1 introduces the concept of an algorithm and various mathematical topics, including mathematical induction. The second half of Volume 1 is devoted to data structures. These volumes are classics in the area of algorithms and are also among the finest examples of technical writing.

Most general references on computer science contain some discussion of algorithms. Books specifically on algorithms are [Aho; Baase; Brassard; Cormen; Knuth 1973, Vols. 1 and 3, 1981; Manber; Nievergelt; and Reingold]. [McNaughton] contains a very thorough discussion on an introductory level of what an algorithm is. Knuth's expository article about algorithms ([Knuth, 1977]) and his article about the role of algorithms in the mathematical sciences ([Knuth, 1985]) are also recommended. [Gardner, 1979] contains a chapter about the Fibonacci sequence.

Full details of the RSA cryptosystem may be found in [Cormen]. [Pfleeger] is devoted to computer security.

❧ CHAPTER REVIEW

Section 3.1

Algorithm

Properties of an algorithm: Precision, uniqueness, finiteness, input, output, generality

Assignment statement: $x := y$

Trace

Section 3.2

Pseudocode

Procedure

If-then structure:

if p **then**
  action

If-then-else structure:

if p **then**
  action 1
else
  action 2

Comment: Nonexecutable information. A comment starts with // and continues to the end of the line.

Return statements: **return** or
  return(x)

While loop:

while p **do**
  action

For loop:

> **for** *var* := *init* **to** *limit* **do**
>   *action*

Call statement:

> **call** *proc*($p_1, p_2, \ldots, p_k$)

Section 3.3

b divides $a : b \mid a$
b is a divisor of a
Quotient
Remainder
Common divisor
Greatest common divisor
Euclidean algorithm

Section 3.4

Recursive algorithm
Recursive procedure
Divide-and-conquer technique
n factorial, $n! : n(n-1) \cdots 2 \cdot 1$
Base cases: Situations where a recursive procedure does not invoke itself
Fibonacci sequence $\{f_n\} : f_1 = 1$, $f_2 = 2, f_n = f_{n-1} + f_{n-2}, n \geq 3$

Section 3.5

Analysis of algorithms
Complexity of algorithms
Worst-case time of an algorithm
Best-case time of an algorithm
Average-case time of an algorithm
Big oh notation: $f(n) = O(g(n))$
Omega notation: $f(n) = \Omega(g(n))$
Theta notation: $f(n) = \Theta(g(n))$

Section 3.6

If the pair $a, b, a > b$, requires $n \geq 1$ divisions when input to the Euclidean algorithm, then $a \geq f_{n+1}$ and $b \geq f_n$, where $\{f_n\}$ denotes the Fibonacci sequence.

If integers in the range 0 to m, $m \geq 8$, not both zero, are input to the Euclidean algorithm, then at most

$$\log_{3/2} \frac{2m}{3}$$

divisions are required.

Section 3.7

Cryptology
Cryptosystem
Encrypt a message
Decrypt a message
RSA public key cryptosystem: To encrypt a and send it to the holder of public key z, n, compute $c = a^n \bmod z$, and send c. To decrypt the message compute $c^s \bmod z$, which can be shown to be equal to a.
$ab \bmod z = [(a \bmod z)(b \bmod z)] \bmod z$
The security of the RSA encryption system relies mainly on the fact that at present there is no efficient algorithm known for factoring integers.

☙ CHAPTER SELF-TEST

Section 3.1

1. Trace the "find max" algorithm in Section 3.1 for the values $a = 12, b = 3, c = 0$.

2. Write an algorithm that receives as input the distinct numbers a, b, and c, and assigns the values a, b, and c to the variables x, y, and z so that

$$x < y < z.$$

3. Write an algorithm that outputs "Yes" if the values of a, b, and c are distinct, and "No" otherwise.

4. Which of the properties precision, uniqueness, finiteness, input, output, and generality, if any, are lacking in the following? Explain.

Input: S, a set of integers; m, an integer

Output: All subsets of S that sum to m

1.   List all subsets of S and their sums.
2.   Step through the subsets listed in 1 and output each whose sum is m.

Section 3.2

5. Trace Algorithm 3.2.2 for the input

$$s_1 = 7, \quad s_2 = 9, \quad s_3 = 17, \quad s_4 = 7.$$

6. Write an algorithm that receives as input the matrix of a relation R and tests whether R is symmetric.

7. Write an algorithm that receives as input the $n \times n$ matrix A and outputs the transpose A^T.

8. Write an algorithm that receives as input the sequence

$$s_1, \ldots, s_n$$

sorted in increasing order and prints all values that appear more than once. *Example*: If the sequence were

$$1 \quad 1 \quad 1 \quad 5 \quad 8 \quad 8 \quad 9 \quad 12$$

the output would be

$$1 \quad 8.$$

Section 3.3

9. If $a = 333$ and $b = 24$, find integers q and r so that $a = bq + r$, with $0 \le r < b$.

10. Using the Euclidean algorithm, find the greatest common divisor of the integers 396 and 480.

11. Using the Euclidean algorithm, find the greatest common divisor of the integers 2390 and 4326.

12. Fill in the blank to make a true statement: If a and b are integers satisfying $a > b > 0$ and $a = bq + r, 0 \le r < b$, then $\gcd(a, b) = $ ———.

Section 3.4

13. Trace Algorithm 3.4.4 (the tromino tiling algorithm) when $n = 8$ and the missing square is four from the left and two from the top.

Exercises 14–16 refer to the *tribonacci sequence* defined by the equations

$$t_1 = t_2 = t_3 = 1; \qquad t_n = t_{n-1} + t_{n-2} + t_{n-3}, \qquad n \ge 4.$$

14. Find t_4 and t_5.

15. Write a recursive algorithm to compute $t_n, n \ge 1$.

16. Give a proof using mathematical induction that your algorithm for Exercise 15 is correct.

Section 3.5

Select a theta notation from among $\Theta(1)$, $\Theta(n)$, $\Theta(n^2)$, $\Theta(n^3)$, $\Theta(n^4)$, $\Theta(2^n)$, or $\Theta(n!)$ for each of the expressions in Exercises 17 and 18.

17. $4n^3 + 2n - 5$ **18.** $1^3 + 2^3 + \cdots + n^3$

19. Select a theta notation from among $\Theta(1)$, $\Theta(n)$, $\Theta(n^2)$, $\Theta(n^3)$, $\Theta(2^n)$, or $\Theta(n!)$ for the number of times the line $x := x + 1$ is executed.

> **for** $i := 1$ **to** n **do**
>   **for** $j := 1$ **to** n **do**
> $x := x + 1$

20. Write an algorithm that tests whether two $n \times n$ matrices are equal and find a theta notation for its worst-case time.

Section 3.6

21. Exactly how many divisions are required by the Euclidean algorithm in the worst case for numbers in the range 0 to 1000?

22. Exactly how many divisions are required by the Euclidean algorithm to compute $\gcd(2, 76652913)$?

23. Exactly how many divisions are required by the Euclidean algorithm to compute $\gcd(f_{324}, f_{323})$? ($\{f_n\}$ denotes the Fibonacci sequence.)

24. Given that $\log_{3/2} 100 = 11.357747$, provide an upper bound for the number of divisions required by the Euclidean algorithm for integers in the range 0 to 100,000,000.

Section 3.7

In Exercises 25–28, assume that we choose primes $p = 13$, $q = 17$, and $n = 19$.

25. Compute z and ϕ.

26. Verify that $s = 91$.

27. Encrypt 144 using public key z, n.

28. Decrypt 28.

4

COUNTING METHODS
AND THE
PIGEONHOLE PRINCIPLE

There's only so many hands in a deck o' cards.

—from Shane

In many discrete problems, we are confronted with the problem of counting. For example, in Section 3.5 we saw that in order to estimate the run time of an algorithm, we needed to count the number of times certain steps or loops were executed. Counting also plays a crucial role in probability theory. Because of the importance of counting, a variety of useful aids, some quite sophisticated, have been developed. In this chapter we develop several tools for counting. These techniques can be used to derive the binomial theorem. The chapter concludes with a discussion of the Pigeonhole Principle, which often allows us to prove the existence of an object with certain properties.

4.1 BASIC PRINCIPLES

The menu for Kay's Quick Lunch is shown in Figure 4.1.1. As you can see, it features two appetizers, three main courses, and four beverages. How many different dinners consist of one main course and one beverage?

APPETIZERS	
Nachos	2.15
Salad	1.90

MAIN COURSES	
Hamburger	3.25
Cheeseburger	3.65
Fish Filet	3.15

BEVERAGES	
Tea	.70
Milk	.85
Cola	.75
Root Beer	.75

FIGURE 4.1.1
Kay's Quick Lunch menu.

If we list all possible dinners consisting of one main course and one beverage:

HT, HM, HC, HR, CT, CM, CC, CR, FT, FM, FC, FR,

we see that there are 12 different dinners. (The dinner consisting of a main course whose first letter is X and a beverage whose first letter is Y is denoted XY. For example, CR refers to the dinner consisting of a cheeseburger and root beer.) Notice that there are three main courses and four beverages and $12 = 3 \cdot 4$.

There are 24 possible dinners consisting of one appetizer, one main course, and one beverage:

NHT, NHM, NHC, NHR, NCT, NCM, NCC, NCR,

NFT, NFM, NFC, NFR, SHT, SHM, SHC, SHR,

SCT, SCM, SCC, SCR, SFT, SFM, SFC, SFR.

(The dinner consisting of an appetizer whose first letter is X, a main course whose first letter is Y, and a beverage whose first letter is Z is denoted XYZ.) Notice that there are two appetizers, three main courses, and four beverages and $24 = 2 \cdot 3 \cdot 4$.

In each of these examples, we found that the total number of dinners was equal to the product of numbers of each of the courses. These examples illustrate the **Multiplication Principle**.

MULTIPLICATION PRINCIPLE

If an activity can be constructed in t successive steps and step 1 can be done in n_1 ways; step 2 can then be done in n_2 ways; ...; and step t can then be done in n_t ways, then the number of different possible activities is $n_1 \cdot n_2 \cdots n_t$.

In the problem of counting the number of dinners consisting of one main course and one beverage, the first step is "select the main course" and the second step is "select the beverage." Thus $n_1 = 3$ and $n_2 = 4$ and, by the Multiplication Principle, the total number of dinners is $3 \cdot 4 = 12$. Figure 4.1.2 shows why we multiply 3 times 4—we have three groups of four objects.

We may summarize the Multiplication Principle by saying that we multiply together the numbers of ways of doing each step when an activity is constructed in successive steps.

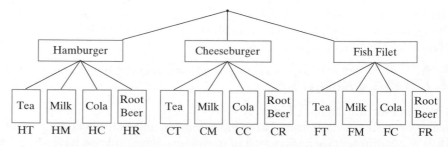

FIGURE 4.1.2   An illustration of the Multiplication Principle.

EXAMPLE 4.1.1

How many dinners are available from Kay's Quick Lunch consisting of one main course and an *optional* beverage?

We may construct a dinner consisting of one main course and an optional beverage by a two-step process. The first step is "select the main course" and the second step is "select an optional beverage." There are $n_1 = 3$ ways to select the main course (hamburger, cheeseburger, fish filet) and $n_2 = 5$ ways to select the optional beverage (tea, milk, cola, root beer, none). By the Multiplication Principle, there are $3 \cdot 5 = 15$ dinners. As confirmation, we list the 15 dinners:

$$HT, HM, HC, HR, HN, CT, CM,$$

$$CC, CR, CN, FT, FM, FC, FR, FN. \qquad \Box$$

EXAMPLE 4.1.2

(a) How many strings of length 4 can be formed using the letters $ABCDE$ if repetitions are not allowed?

(b) How many strings of part (a) begin with the letter B?

(c) How many strings of part (a) do not begin with the letter B?

(a) We use the Multiplication Principle. A string of length 4 can be constructed in four successive steps: Choose the first letter, choose the second letter, choose the third letter, and choose the fourth letter. The first letter can be selected in five ways. Once the first letter has been selected, the second letter can be selected in four ways. Once the second letter has been selected, the third letter can be selected in three ways. Once the third letter has been selected, the fourth letter can be selected in two ways. By the Multiplication Principle, there are

$$5 \cdot 4 \cdot 3 \cdot 2 = 120$$

strings.

(b) The strings that begin with the letter B can be constructed in four successive steps: Choose the first letter, choose the second letter, choose the third letter, and choose the fourth letter. The first letter (B) can be chosen in one way, the second letter in four ways, the third letter in three ways, and the fourth letter in two ways. Thus, by the Multiplication Principle, there are

$$1 \cdot 4 \cdot 3 \cdot 2 = 24$$

strings that start with the letter B.

(c) Part (a) shows that there are 120 strings of length 4 that can be formed using the letters $ABCDE$ and part (b) shows that 24 of these start with the letter B. It follows that there are

$$120 - 24 = 96$$

strings that do not begin with the letter B. $\qquad \Box$

EXAMPLE 4.1.3

In a digital picture, we wish to encode the amount of light at each point as an eight-bit string. How many values are possible at one point?

An eight-bit encoding can be constructed in eight successive steps: Select the first bit, select the second bit, ..., select the eighth bit. Since there are two ways to select each bit, by the Multiplication Principle the total number of eight-bit encodings is

$$2 \cdot 2 \cdot 2 \cdot 2 \cdot 2 \cdot 2 \cdot 2 \cdot 2 = 2^8 = 256. \qquad \square$$

We next give a proof using the Multiplication Principle that a set with n elements has 2^n subsets. We previously gave a proof of this result using mathematical induction (Theorem 2.1.4).

EXAMPLE 4.1.4

Use the Multiplication Principle to show that a set $\{x_1, \ldots, x_n\}$ containing n elements has 2^n subsets.

A subset can be constructed in n successive steps: Pick or do not pick x_1, pick or do not pick x_2, ..., pick or do not pick x_n. Each step can be done in two ways. Thus the number of possible subsets is

$$\underbrace{2 \cdot 2 \cdots 2}_{n \text{ factors}} = 2^n. \qquad \square$$

EXAMPLE 4.1.5

Let X be an n-element set. How many ordered pairs (A, B) satisfy $A \subseteq B \subseteq X$?

Given an ordered pair (A, B) satisfying $A \subseteq B \subseteq X$, each element in X is in exactly one of A, $B - A$, or $X - B$. Conversely, if we assign each element of X to one of the three sets A (and, by assumption, also to B and X), $B - A$ (and, by assumption, also to X), or $X - B$, we obtain a unique ordered pair (A, B) satisfying $A \subseteq B \subseteq X$. Thus the number of ordered pairs (A, B) satisfying $A \subseteq B \subseteq X$ is equal to the number of ways to assign the elements of X to the three sets A, $B - A$, and $X - B$. We can make such assignments by the following n-step process: Assign the first element of X to one of A, $B - A$, $X - B$; assign the second element of X to one of A, $B - A$, $X - B$; ...; assign the nth element of X to one of A, $B - A$, $X - B$. Since each step can be done in three ways, the number of ordered pairs (A, B) satisfying $A \subseteq B \subseteq X$ is

$$\underbrace{3 \cdot 3 \cdots 3}_{n \text{ factors}} = 3^n. \qquad \square$$

Next, we illustrate the Addition Principle by an example and then present the principle.

EXAMPLE 4.1.6

How many eight-bit strings begin either 101 or 111?

An eight-bit string that begins 101 can be constructed in five successive steps: select the fourth bit, select the fifth bit, ..., select the eighth bit. Since each of the five bits can be selected in two ways, by the Multiplication Principle, there are

$$2 \cdot 2 \cdot 2 \cdot 2 \cdot 2 = 2^5 = 32$$

eight-bit strings that begin 101. The same argument can be used to show that there are 32 eight-bit strings that begin 111. Since there are 32 eight-bit strings that begin 101 and 32 eight-bit strings that begin 111, there are $32 + 32 = 64$ eight-bit strings that begin either 101 or 111. □

In Example 4.1.6 we added the numbers of eight-bit strings (32 and 32) of each type to determine the final result. The **Addition Principle** tells us when to add to compute the total number of possibilities.

ADDITION PRINCIPLE

Suppose that $X_1, \ldots, X_t$ are sets and that the ith set X_i has n_i elements. If $\{X_1, \ldots, X_t\}$ is a pairwise disjoint family (i.e., if $i \neq j$, $X_i \cap X_j = \emptyset$), the number of possible elements that can be selected from X_1 or X_2 or ... or X_t is

$$n_1 + n_2 + \cdots + n_t.$$

(Equivalently, the union $X_1 \cup X_2 \cup \cdots \cup X_t$ contains $n_1 + n_2 + \cdots + n_t$ elements.)

In Example 4.1.6 we could let X_1 denote the set of eight-bit strings that begin 101 and X_2 denote the set of eight-bit strings that begin 111. Since X_1 is disjoint from X_2, according to the Addition Principle, the number of eight-bit strings of either type, which is the number of elements in $X_1 \cup X_2$, is $32 + 32 = 64$.

We may summarize the Addition Principle by saying that we add the numbers of elements in each subset when the elements being counted can be decomposed into disjoint subsets.

If we are counting objects that are constructed in successive steps, we use the Multiplication Principle. If we have disjoint sets of objects and we want to know the total number of objects, we use the Addition Principle. It is important to recognize when to apply each principle. This skill comes from practice and careful thinking about each problem.

We close this section with examples that illustrate both counting principles.

EXAMPLE 4.1.7

In how many ways can we select two books from different subjects from among five distinct computer science books, three distinct mathematics books, and two distinct art books?

Using the Multiplication Principle, we find that we can select two books, one from computer science and one from mathematics, in $5 \cdot 3 = 15$ ways.

Similarly, we can select two books, one from computer science and one from art, in $5 \cdot 2 = 10$ ways, and we can select two books, one from mathematics and one from art, in $3 \cdot 2 = 6$ ways. Since these sets of selections are pairwise disjoint, we may use the Addition Principle to conclude that there are

$$15 + 10 + 6 = 31$$

ways of selecting two books from different subjects from among the computer science, mathematics, and art books.   □

EXAMPLE 4.1.8

A six-person committee composed of Alice, Ben, Connie, Dolph, Egbert, and Francisco is to select a chairperson, secretary, and treasurer.

(a) In how many ways can this be done?

(b) In how many ways can this be done if either Alice or Ben must be chairperson?

(c) In how many ways can this be done if Egbert must hold one of the offices?

(d) In how many ways can this be done if both Dolph and Francisco must hold office?

(a) We use the Multiplication Principle. The officers can be selected in three successive steps: Select the chairperson, select the secretary, select the treasurer. The chairperson can be selected in six ways. Once the chairperson has been selected, the secretary can be selected in five ways. After selection of the chairperson and secretary, the treasurer can be selected in four ways. Therefore, the total number of possibilities is
$$6 \cdot 5 \cdot 4 = 120.$$

(b) Arguing as in part (a), if Alice is chairperson, there are $5 \cdot 4 = 20$ ways to select the remaining officers. Similarly, if Ben is chairperson, there are 20 ways to select the remaining officers. Since these cases are disjoint, by the Addition Principle, there are

$$20 + 20 = 40$$

possibilities.

(c) [First solution.] Arguing as in part (a), if Egbert is chairperson, there are 20 ways to select the remaining officers. Similarly, if Egbert is secretary, there are 20 possibilities, and if Egbert is treasurer, there are 20 possibilities. Since these three cases are pairwise disjoint, by the Addition Principle, there are

$$20 + 20 + 20 = 60$$

possibilities.

[Second solution.] Let us consider the activity of assigning Egbert and two others to offices to be made up of three successive steps: Assign Egbert an office, fill the highest remaining office, fill the last office. There are three ways to assign Egbert an office. Once Egbert has been assigned, there are five ways to fill the highest remaining office. Once Egbert has been assigned and the highest remaining office filled, there are four ways to fill the last office. By the Multiplication Principle, there are

$$3 \cdot 5 \cdot 4 = 60$$

possibilities.

(d) Let us consider the activity of assigning Dolph, Francisco, and one other person to offices to be made up of three successive steps: Assign Dolph, assign Francisco, fill the remaining office. There are three ways to assign Dolph. Once Dolph has been assigned, there are two ways to assign Francisco. Once Dolph and Francisco have been assigned, there are four ways to fill the remaining office. By the Multiplication Principle, there are

$$3 \cdot 2 \cdot 4 = 24$$

possibilities.   □

Exercises

Find the number of dinners at Kay's Quick Lunch (Figure 4.1.1) satisfying the conditions of Exercises 1–3.

1. One appetizer and one beverage

2. One appetizer, one main course, and an optional beverage

3. An optional appetizer, one main course, and an optional beverage

4. A man has eight shirts, four pairs of pants, and five pairs of shoes. How many different outfits are possible?

5. The options available on a particular model of a car are five interior colors, six exterior colors, two types of seats, three types of engines, and three types of radios. How many different possibilities are available to the consumer?

6. The Braille system of representing characters was developed early in the nineteenth century by Louis Braille. The characters, used by the blind, consist of raised dots. The positions for the dots are selected from two vertical columns of three dots each. At least one raised dot must be present. How many distinct Braille characters are possible?

In Exercises 7–15, two dice are rolled, one blue and one red.

7. How many outcomes are possible?

8. How many outcomes give the sum of 4?

9. How many outcomes are doubles? (A double occurs when both dice show the same number.)

10. How many outcomes give the sum of 7 or the sum of 11?

11. How many outcomes have the blue die showing 2?

12. How many outcomes have exactly one die showing 2?

13. How many outcomes have at least one die showing 2?

14. How many outcomes have neither die showing 2?

15. How many outcomes give an even sum?

In Exercises 16–18, suppose that there are 10 roads from Oz to Mid Earth and five roads from Mid Earth to Fantasy Island.

16. How many routes are there from Oz to Fantasy Island passing through Mid Earth?

17. How many round-trips are there of the form Oz–Mid Earth–Fantasy Island–Mid Earth–Oz?

18. How many round-trips are there of the form Oz–Mid Earth–Fantasy Island–Mid Earth–Oz in which on the return trip we do not reverse the original route from Oz to Fantasy Island?

19. How many different car license plates can be constructed if the licenses contain three letters followed by two digits if repetitions are allowed? if repetitions are not allowed?

20. How many eight-bit strings begin 1100?

21. How many eight-bit strings begin and end with 1?

22. How many eight-bit strings have either the second or the fourth bit 1 (or both)?

23. How many eight-bit strings have exactly one 1?

24. How many eight-bit strings have exactly two 1's?

25. How many eight-bit strings have at least one 1?.

26. How many eight-bit strings read the same from either end? (An example of such an eight-bit string is 01111110. Such strings are called *palindromes*.)

In Exercises 27–32, a six-person committee composed of Alice, Ben, Connie, Dolph, Egbert, and Francisco is to select a chairperson, secretary, and treasurer.

27. How many selections exclude Connie?

28. How many selections are there in which neither Ben nor Francisco is an officer?

29. How many selections are there in which both Ben and Francisco are officers?

30. How many selections are there in which Dolph is an officer and Francisco is not an officer?

31. How many selections are there in which either Dolph is chairperson or he is not an officer?

32. How many selections are there in which Ben is either chairperson or treasurer?

In Exercises 33–40, the letters $ABCDE$ are to be used to form strings of length 3.

33. How many strings can be formed if we allow repetitions?

34. How many strings can be formed if we do not allow repetitions?

35. How many strings begin with A, allowing repetitions?

36. How many strings begin with A if repetitions are not allowed?

37. How many strings do not contain the letter A, allowing repetitions?

38. How many strings do not contain the letter A if repetitions are not allowed?

39. How many strings contain the letter A, allowing repetitions?

40. How many strings contain the letter A if repetitions are not allowed?

Exercises 41–51 refer to the integers from 5 to 200, inclusive.

41. How many numbers are there?

42. How many are even?

43. How many are odd?

44. How many are divisible by 5?

45. How many are greater than 72?

46. How many consist of distinct digits?

47. How many contain the digit 7?

48. How many do not contain the digit 0?

49. How many are greater than 101 and do not contain the digit 6?

50. How many have the digits in strictly increasing order? (Examples are 13, 147, 8.)

51. How many are of the form xyz, where $0 \neq x < y$ and $y > z$?

52. (a) In how many ways can the months of the birthdays of five people be distinct?

(b) How many possibilities are there for the months of the birthdays of five people?

(c) In how many ways can at least two people from among five have their birthdays in the same month?

Exercises 53–57 refer to a set of five distinct computer science books, three distinct mathematics books, and two distinct art books.

53. In how many ways can these books be arranged on a shelf?

54. In how many ways can these books be arranged on a shelf if all five computer science books are on the left and both art books are on the right?

55. In how many ways can these books be arranged on a shelf if all five computer science books are on the left?

56. In how many ways can these books be arranged on a shelf if all books of the same discipline are grouped together?

★ 57. In how many ways can these books be arranged on a shelf if the two art books are not together?

58. In some versions of FORTRAN, an identifier consists of a string of one to six alphanumeric characters beginning with a letter. (An *alphanumeric* character is one of A to Z or 0 to 9.) How many valid FORTRAN identifiers are there?

59. If X is an n-element set and Y is an m-element set, how many functions are there from X to Y?

★ 60. There are 10 copies of one book and one copy each of 10 other books. In how many ways can we select 10 books?

61. How many terms are there in the expansion of

$$(x + y)(a + b + c)(e + f + g)(h + i)?$$

★ 62. How many subsets of a $(2n + 1)$-element set have n elements or less?

63. How many antisymmetric relations are there on an n-element set?

64. If X and Y are not disjoint subsets, we cannot add $|X|$ to $|Y|$ to compute the number of elements in $X \cup Y$. Prove that

$$|X \cup Y| = |X| + |Y| - |X \cap Y|$$

for arbitrary sets X and Y.

Use the result of Exercise 64 to solve Exercises 65–69.

65. How many eight-bit strings either begin 100 or have the fourth bit 1?

66. How many eight-bit strings either start with a 1 or end with a 1?

In Exercises 67 and 68, a six-person committee composed of Alice, Ben, Connie, Dolph, Egbert, and Francisco is to select a chairperson, secretary, and treasurer.

67. How many selections are there in which either Ben is chairperson or Alice is secretary?

68. How many selections are there in which either Connie is chairperson or Alice is an officer?

69. Two dice are rolled, one blue and one red. How many outcomes have either the blue die 3 or an even sum?

70. How many binary operators are there on $\{1, 2, \ldots, n\}$?

71. How many commutative binary operators are there on $\{1, 2, \ldots, n\}$?

Problem

Find the number of ordered triples of sets X_1, X_2, X_3 satisfying

$$X_1 \cup X_2 \cup X_3 = \{1, 2, 3, 4, 5, 6, 7, 8\} \quad \text{and} \quad X_1 \cap X_2 \cap X_3 = \emptyset.$$

By *ordered triple*, we mean that the order of the sets X_1, X_2, X_3 is taken into account. For example, the triples

$$\{1, 2, 3\}, \quad \{1, 4, 8\}, \quad \{2, 5, 6, 7\}$$

and

$$\{1, 4, 8\}, \quad \{1, 2, 3\}, \quad \{2, 5, 6, 7\}$$

are considered distinct.

Attacking the Problem

It would be nice to begin by enumerating triples, but there are so many it would be hard to gain much insight from staring at a few triples. Let's simplify the problem by making it smaller. Let's replace

$$\{1, 2, 3, 4, 5, 6, 7, 8\}$$

by $\{1\}$. What could be simpler than $\{1\}$? (Well, maybe $\emptyset$, but that's too simple!) We can now enumerate all ordered triples of sets X_1, X_2, X_3 satisfying $X_1 \cup X_2 \cup X_3 = \{1\}$ and $X_1 \cap X_2 \cap X_3 = \emptyset$. We must put 1 in at least one of the sets X_1, X_2, X_3 (so that the union will be $\{1\}$), but we must not put 1 in all three of the sets X_1, X_2, X_3 (otherwise, the intersection would not be empty). Thus 1 will be in exactly one or two of the sets X_1, X_2, X_3. The complete list of ordered triples is:

$$X_1 = \{1\}, \quad X_2 = \emptyset, \quad X_3 = \emptyset; \quad X_1 = \emptyset, \quad X_2 = \{1\}, \quad X_3 = \emptyset;$$
$$X_1 = \emptyset, \quad X_2 = \emptyset, \quad X_3 = \{1\}; \quad X_1 = \{1\}, \quad X_2 = \{1\}, \quad X_3 = \emptyset;$$
$$X_1 = \{1\}, \quad X_2 = \emptyset, \quad X_3 = \{1\}; \quad X_1 = \emptyset, \quad X_2 = \{1\}, \quad X_3 = \{1\}.$$

Thus there are six ordered triples of sets X_1, X_2, X_3 satisfying

$$X_1 \cup X_2 \cup X_3 = \{1\} \quad \text{and} \quad X_1 \cap X_2 \cap X_3 = \emptyset.$$

Let's step up one level and enumerate all ordered triples of sets X_1, X_2, X_3 satisfying $X_1 \cup X_2 \cup X_3 = \{1, 2\}$ and $X_1 \cap X_2 \cap X_3 = \emptyset$. As before, we must put 1 in at least one of the sets X_1, X_2, X_3 (so that 1 will be in the union), but we must not put 1 in all three of the sets X_1, X_2, X_3 (otherwise, the intersection would not be empty). This time we must also put 2 in at least one of the sets X_1, X_2, X_3 (so that 2 will also be in the union), but we must not put 2 in all three of the sets X_1, X_2, X_3 (otherwise, the intersection would not be empty). Thus each of 1 and 2 will be in exactly one or two of the sets X_1, X_2, X_3. We enumerate the sets in a systematic way so that we can recognize any patterns that appear. The complete list of ordered triples is:

1 is in	*2 is in*	*1 is in*	*2 is in*	*1 is in*	*2 is in*
X_1	X_1	X_1	X_2	X_1	X_3
X_2	X_1	X_2	X_2	X_2	X_3
X_3	X_1	X_3	X_2	X_3	X_3
X_1, X_2	X_1	X_1, X_2	X_2	X_1, X_2	X_3
X_1, X_3	X_1	X_1, X_3	X_2	X_1, X_3	X_3
X_2, X_3	X_1	X_2, X_3	X_2	X_2, X_3	X_3
X_1	X_1, X_2	X_1	X_1, X_3	X_1	X_2, X_3
X_2	X_1, X_2	X_2	X_1, X_3	X_2	X_2, X_3
X_2	X_1, X_2	X_3	X_1, X_3	X_3	X_2, X_3
X_1, X_2	X_1, X_2	X_1, X_2	X_1, X_3	X_1, X_2	X_2, X_3
X_1, X_3	X_1, X_2	X_1, X_3	X_1, X_3	X_1, X_3	X_2, X_3
X_2, X_3	X_1, X_2	X_2, X_3	X_1, X_3	X_2, X_3	X_2, X_3

For example, the top left entry, X_1 X_1, specifies that 1 is in X_1 and 2 is in X_1; therefore, this entry gives the ordered triple

$$X_1 = \{1, 2\}, \quad X_2 = \emptyset, \quad X_3 = \emptyset.$$

As shown, there are 36 ordered triples of sets X_1, X_2, X_3 satisfying

$$X_1 \cup X_2 \cup X_3 = \{1, 2\} \quad \text{and} \quad X_1 \cap X_2 \cap X_3 = \emptyset.$$

We see that there are six ways to assign 1 to the sets X_1, X_2, X_3, which accounts for six lines per block. Similarly, there are six ways to assign 2 to the sets X_1, X_2, X_3, which accounts for six blocks.

Before reading on, can you guess how many ordered triples of sets X_1, X_2, X_3 satisfy

$$X_1 \cup X_2 \cup X_3 = \{1, 2, 3\} \quad \text{and} \quad X_1 \cap X_2 \cap X_3 = \emptyset?$$

The pattern has emerged. If $X = \{1, 2, \ldots, n\}$, there are six ways to assign each of $1, 2, \ldots, n$ to the sets X_1, X_2, X_3. By the Multiplication Principle, the number of ordered triples is 6^n.

Finding Another Solution

We have just found a solution to the problem by starting with a simpler problem and then discovering and justifying the pattern that emerged.

Another approach is to look for a similar problem and imitate its solution. The problem of Example 4.1.5 is similar to the one at hand in that it also is a counting problem that deals with sets:

> Let X be an n-element set. How many ordered pairs
> (A, B) satisfy $A \subseteq B \subseteq X$?

(At this point, it would be a good idea to go back and reread Example 4.1.5.) The solution given in Example 4.1.5 counted the number of ways to assign elements of X to exactly one of the sets A, $B - A$, or $X - B$.

We can solve our problem by taking a similar approach. Each element of X is in exactly one of

$$\overline{X_1} \cap X_2 \cap X_3, \quad X_1 \cap \overline{X_2} \cap X_3, \quad X_1 \cap X_2 \cap \overline{X_3},$$
$$\overline{X_1} \cap \overline{X_2} \cap X_3, \quad \overline{X_1} \cap X_2 \cap \overline{X_3}, \quad X_1 \cap \overline{X_2} \cap \overline{X_3}.$$

Since each member of X can be assigned to one of these sets in six ways, by the Multiplication Principle, the number of ordered triples is 6^8.

Notice that while this approach to solving the problem is different than that of the preceding section, the final argument is essentially the same.

Formal Solution

Each element in X is in exactly one of

$$Y_1 = \overline{X_1} \cap X_2 \cap X_3, \quad Y_2 = X_1 \cap \overline{X_2} \cap X_3, \quad Y_3 = X_1 \cap X_2 \cap \overline{X_3},$$
$$Y_4 = \overline{X_1} \cap \overline{X_2} \cap X_3, \quad Y_5 = \overline{X_1} \cap X_2 \cap \overline{X_3}, \quad Y_6 = X_1 \cap \overline{X_2} \cap \overline{X_3}.$$

We can construct an ordered triple by the following eight-step process: Choose $j, 1 \le j \le 6$, and put 1 in Y_j; choose $j, 1 \le j \le 6$, and put 2 in Y_j; ...; choose $j, 1 \le j \le 6$, and put 8 in Y_j. For example, to construct the triple

$$\{1, 2, 3\}, \quad \{1, 4, 8\}, \quad \{2, 5, 6, 7\},$$

we first choose $j = 3$ and put 1 in $Y_3 = X_1 \cap X_2 \cap \overline{X_3}$. Next, we choose $j = 2$ and put 2 in $Y_2 = X_1 \cap \overline{X_2} \cap X_3$. The remaining choices for j are $j = 6, 5, 4, 4, 4, 5$.

Each choice for j can be made in six ways. By the Multiplication Principle, the number of ordered triples is

$$6 \cdot 6 \cdot 6 \cdot 6 \cdot 6 \cdot 6 \cdot 6 \cdot 6 = 6^8 = 1,679,616.$$

Summary of Problem-Solving Techniques

- Replace the original problem with a simpler problem. One way to do this is to reduce the size of the original problem.
- Directly enumerate the items to be counted.
- Enumerate items systematically so that patterns emerge.
- Look for patterns.
- Look for a similar problem and imitate its solution.

4.2 PERMUTATIONS AND COMBINATIONS

Four candidates, Zeke, Yung, Xeno, and Wilma, are running for the same office. So that the positions of the names on the ballot will not influence the voters, it is necessary to print ballots with the names listed in every possible order. How many distinct ballots will there be?

We can use the Multiplication Principle. A ballot can be constructed in four successive steps: Select the first name to be listed, select the second name to be listed, select the third name to be listed, select the fourth name to be listed. The first name can be selected in four ways. Once the first name has been selected, the second name can be selected in three ways. Once the second name has been selected, the third name can be selected in two ways. Once the third name has been selected, the fourth name can be selected in one way. By the Multiplication Principle, the number of ballots is

$$4 \cdot 3 \cdot 2 \cdot 1 = 24.$$

An ordering of objects, such as the names on the ballot, is called a **permutation**.

DEFINITION 4.2.1

A *permutation* of n distinct elements $x_1, \ldots, x_n$ is an ordering of the n elements $x_1, \ldots, x_n$.

EXAMPLE 4.2.2

There are six permutations of three elements. If the elements are denoted A, B, C, the six permutations are

$$ABC, \quad ACB, \quad BAC, \quad BCA, \quad CAB, \quad CBA. \qquad \square$$

We found that there are 24 ways to order four candidates on a ballot; thus there are 24 permutations of four objects. The method that we used to count the number of distinct ballots containing four names may be used to derive a formula for the number of permutations of n elements.

The proof of the following theorem for $n = 4$ is illustrated in Figure 4.2.1.

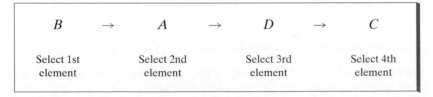

FIGURE 4.2.1   The proof of Theorem 4.2.3 for $n = 4$. A permutation of *ABCD* is constructed by successively selecting the first element, then the second element, then the third element, and, finally, the fourth element.

THEOREM 4.2.3

There are n! permutations of n elements.

Proof. We use the Multiplication Principle. A permutation of n elements can be constructed in n successive steps: Select the first element, select the second element, . . ., select the last element. The first element can be selected in n ways. Once the first element has been selected, the second element can be selected in $n - 1$ ways. Once the second element has been selected, the third element can be selected in $n - 2$ ways, and so on. By the Multiplication Principle, there are

$$n(n - 1)(n - 2) \cdots 2 \cdot 1 = n!$$

permutations of n elements.   ■

EXAMPLE 4.2.4

There are

$$10! = 10 \cdot 9 \cdot 8 \cdot 7 \cdot 6 \cdot 5 \cdot 4 \cdot 3 \cdot 2 \cdot 1 = 3,628,800$$

permutations of 10 elements.   □

EXAMPLE 4.2.5

How many permutations of the letters $ABCDEF$ contain the substring DEF?
    To guarantee the presence of the pattern DEF in the substring, these three letters must be kept together in this order. The remaining letters, A, B, and C, can be placed arbitrarily. We can think of constructing permutations of the letters $ABCDEF$ that contain the pattern DEF by permuting four tokens—one labeled DEF and the others labeled A, B, and C (see Figure 4.2.2). By Theorem 4.2.3, there are 4! permutations of four objects. Thus the number of permutations of the letters $ABCDEF$ that contain the substring DEF is

$$4! = 24.$$ □

FIGURE 4.2.2
Four tokens to permute.

EXAMPLE 4.2.6

How many permutations of the letters $ABCDEF$ contain the letters DEF together in any order?
    We can solve the problem by a two-step procedure: Select an ordering of the letters DEF, construct a permutation of $ABCDEF$ containing the given ordering of the letters DEF. By Theorem 4.2.3, the first step can be done in $3! = 6$ ways and, according to Example 4.2.5, the second step can be done in 24 ways. By the Multiplication Principle, the number of permutations of the letters $ABCDEF$ containing the letters DEF together in any order is

$$6 \cdot 24 = 144.$$ □

EXAMPLE 4.2.7

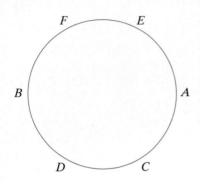

In how many ways can six persons be seated around a circular table? If a seating is obtained from another seating by having everyone move n seats clockwise, the seatings are considered identical.

Let us denote the persons as A, B, C, D, E, and F. Since seatings obtained by rotations are considered identical, we might as well seat A arbitrarily. To seat the remaining five persons, we can order them and then seat them in this order clockwise from A. For example, the permutation $CDBFE$ would define the seating in the adjacent figure. Since there are $5! = 120$ permutations of five elements, there are 120 ways that six persons can be seated around a circular table.

The same argument can be used to show that there are $(n-1)!$ ways that n persons can be seated around a circular table.   □

Sometimes we want to consider an ordering of r elements selected from n available elements. Such an ordering is called an **r-permutation**.

DEFINITION 4.2.8

An *r-permutation* of n (distinct) elements $x_1, \ldots, x_n$ is an ordering of an r-element subset of $\{x_1, \ldots, x_n\}$. The number of r-permutations of a set of n distinct elements is denoted $P(n, r)$.

EXAMPLE 4.2.9

Examples of 2-permutations of a, b, c are

$$ab, \quad ba, \quad ca.$$ □

If $r = n$ in Definition 4.2.8, we obtain an ordering of all n elements. Thus an n-permutation of n elements is what we previously called simply a permutation. Theorem 4.2.3 tells us that $P(n, n) = n!$. The number $P(n, r)$ of r-permutations of an n-element set when $r < n$ may be derived as in the proof of Theorem 4.2.3. The proof of the theorem for $n = 6$ and $r = 3$ is illustrated in Figure 4.2.3.

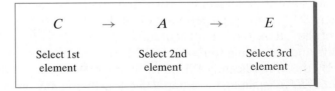

FIGURE 4.2.3   The proof of Theorem 4.2.10 for $n = 6$ and $r = 3$. An r-permutation of $ABCDEF$ is constructed by successively selecting the first element, then the second element, and, finally, the third element.

THEOREM 4.2.10

The number of r-permutations of a set of n distinct objects is

$$P(n, r) = n(n - 1)(n - 2) \cdots (n - r + 1), \quad r \le n.$$

Proof. We are to count the number of ways to order r elements selected from an n-element set. The first element can be selected in n ways. Once the first element has been selected, the second element can be selected in $n - 1$ ways. We continue selecting elements until having selected the $(r - 1)$st element, we select the rth element. This last element can be chosen in $n - r + 1$ ways. By the Multiplication Principle, the number of r-permutations of a set of n distinct objects is

$$n(n - 1)(n - 2) \cdots (n - r + 1).$$ ∎

EXAMPLE 4.2.11

According to Theorem 4.2.10, the number of 2-permutations of $X = \{a, b, c\}$ is

$$P(3, 2) = 3 \cdot 2 = 6.$$

These six 2-permutations are

$$ab, \ ac, \ ba, \ bc, \ ca, \ cb.$$ □

EXAMPLE 4.2.12

In how many ways can we select a chairperson, vice-chairperson, secretary, and treasurer from a group of 10 persons?

We need to count the number of orderings of four persons selected from a group of 10, since an ordering picks (uniquely) a chairperson (first pick), a vice-chairperson (second pick), a secretary (third pick), and a treasurer (fourth pick). By Theorem 4.2.10, the solution is

$$P(10, 4) = 10 \cdot 9 \cdot 8 \cdot 7 = 5040.$$ □

We could also have solved Example 4.2.12 by appealing directly to the Multiplication Principle.

We may also write $P(n, r)$ in terms of factorials:

$$P(n, r) = n(n - 1) \cdots (n - r + 1)$$

$$= \frac{n(n - 1) \cdots (n - r + 1)(n - r) \cdots 2 \cdot 1}{(n - r) \cdots 2 \cdot 1} = \frac{n!}{(n - r)!}. \quad (4.2.1)$$

EXAMPLE 4.2.13

Using (4.2.1), we may rewrite the solution of Example 4.2.12 as

$$P(10, 4) = \frac{10!}{(10 - 4)!} = \frac{10!}{6!}. \qquad \square$$

EXAMPLE 4.2.14

In how many ways can seven distinct Martians and five distinct Jovians wait in line if no two Jovians stand together?

We can line up the Martians and Jovians by a two-step process: Line up the Martians, line up the Jovians. The Martians can line up in $7! = 5040$ ways. Once we have lined up the Martians (e.g., in positions M_1–M_7), since no two Jovians can stand together, the Jovians have eight possible positions in which to stand (indicated by blanks):

$$\_ M_1 \_ M_2 \_ M_3 \_ M_4 \_ M_5 \_ M_6 \_ M_7 \_ .$$

Thus the Jovians can stand in $P(8, 5) = 8 \cdot 7 \cdot 6 \cdot 5 \cdot 4 = 6720$ ways. By the Multiplication Principle, the number of ways seven distinct Martians and five distinct Jovians can wait in line if no two Jovians stand together is

$$5040 \cdot 6720 = 33{,}868{,}800. \qquad \square$$

We turn next to combinations. A selection of objects without regard to order is called a **combination**.

DEFINITION 4.2.15

Given a set $X = \{x_1, \ldots, x_n\}$ containing n (distinct) elements,

(a) An *r-combination* of X is an unordered selection of *r*-elements of X (i.e., an *r*-element subset of X).

(b) The number of *r*-combinations of a set of n distinct elements is denoted $C(n, r)$ or $\binom{n}{r}$.

EXAMPLE 4.2.16

A group of five students, Mary, Boris, Rosa, Ahmad, and Nguyen, has decided to talk with the Mathematics Department chairperson about having the Mathematics Department offer more courses in discrete mathematics. The chairperson has said that she will speak with three of the students. In how many ways can these five students choose three of their group to talk with the chairperson?

In solving this problem, we must *not* take order into account. (For example, it will make no difference whether the chairperson talks to Mary, Ahmad, and Nguyen or to Nguyen, Mary, and Ahmad.) By simply listing the possibilities, we see that there are 10 ways that the five students can choose three of their group to talk to the chairperson:

MBR, MBA, MRA, BRA, MBN, MRN, BRN, MAN, BAN, RAN.

In the terminology of Definition 4.2.15, the number of ways the five students can choose three of their group to talk with the chairperson is $C(5, 3)$, the number of 3-combinations of five elements. We have found that

$$C(5, 3) = 10. \qquad \square$$

We next derive a formula for $C(n, r)$ by counting the number of r-permutations of an n-element set in two ways. The first way simply uses the formula $P(n, r)$. The second way of counting the number of r-permutations of an n-element set involves $C(n, r)$. Equating the two values will enable us to derive a formula for $C(n, r)$.

We can construct r-permutations of an n-element set X in two successive steps: First, select an r-combination of X (an unordered subset of r items), and then order it. For example, to construct a 2-permutation of $\{a, b, c, d\}$, we can first select a 2-combination and then order it. Figure 4.2.4 shows how all 2-permutations of $\{a, b, c, d\}$ are obtained in this way. The Multiplication Principle tells us that the number of r-permutations is the product of the number of r-combinations and the number of orderings of r elements. That is,

$$P(n, r) = C(n, r)r!.$$

Therefore,

$$C(n, r) = \frac{P(n, r)}{r!}.$$

Our next theorem states this result and gives some alternative ways to write $C(n, r)$.

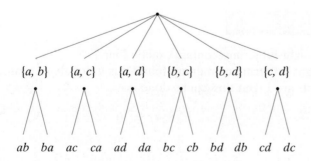

FIGURE 4.2.4   2-permutations of $\{a, b, c, d\}$.

THEOREM 4.2.17

The number of r-combinations of a set of n distinct objects is

$$C(n, r) = \frac{P(n, r)}{r!} = \frac{n(n-1) \cdots (n - r + 1)}{r!} = \frac{n!}{(n-r)! \, r!}, \quad r \leq n.$$

Proof. The proof of the first equation is given before the statement of the theorem. The other forms of the equation follow from Theorem 4.2.10 and equation (4.2.1). ∎

EXAMPLE 4.2.18

In how many ways can we select a committee of three from a group of 10 distinct persons?

Since a committee is an unordered group of people, the answer is

$$C(10, 3) = \frac{10 \cdot 9 \cdot 8}{3!} = 120.$$ □

EXAMPLE 4.2.19

In how many ways can we select a committee of two women and three men from a group of five distinct women and six distinct men?

As in Example 4.2.18, we find that the two women can be selected in

$$C(5, 2) = 10$$

ways and that the three men can be selected in

$$C(6, 3) = 20$$

ways. The committee can be constructed in two successive steps: Select the women, select the men. By the Multiplication Principle, the total number of committees is

$$10 \cdot 20 = 200.$$ □

EXAMPLE 4.2.20

How many eight-bit strings contain exactly four 1's?

An eight-bit string containing four 1's is uniquely determined once we tell which bits are 1. But this can be done in

$$C(8, 4) = 70$$

ways. □

EXAMPLE 4.2.21

An ordinary deck of 52 cards consists of four suits

> clubs, diamonds, hearts, spades

of 13 denominations each

> ace, 2–10, jack, queen, king.

(a) How many (unordered) five-card poker hands, selected from an ordinary 52-card deck, are there?

(b) How many poker hands contain cards all of the same suit?

(c) How many poker hands contain three cards of one denomination and two cards of a second denomination?

(a) The answer is given by the combination formula

$$C(52, 5) = 2,598,960.$$

(b) A hand containing cards all of the same suit can be constructed in two successive steps: Select a suit, select five cards from the chosen suit. The first step can be done in four ways and the second step can be done in $C(13, 5)$ ways. By the Multiplication Principle, the answer is

$$4 \cdot C(13, 5) = 5148.$$

(c) A hand containing three cards of one denomination and two cards of a second denomination can be constructed in four successive steps: Select the first denomination, select the second denomination, select three cards of the first denomination, select two cards of the second denomination. The first denomination can be chosen in 13 ways. Having selected the first denomination, we can choose the second denomination in 12 ways. We can select three cards of the first denomination in $C(4, 3)$ ways and we can select two cards of the second denomination in $C(4, 2)$ ways. By the Multiplication Principle, the answer is

$$13 \cdot 12 \cdot C(4, 3) \cdot C(4, 2) = 3744. \qquad \square$$

EXAMPLE 4.2.22

How many routes are there from the lower left corner of an $n \times n$ square grid to the upper right corner if we are restricted to traveling only to the right or upward? One such route is shown in a 4×4 grid in Figure 4.2.5a.

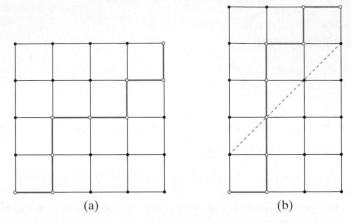

(a)   (b)

FIGURE 4.2.5   (a) A 4×4 grid with a route from the lower left corner to the upper right corner. (b) The route in (a) transformed to a route in a 5×3 grid.

Each route can be described by a string of n R's (right) and n U's (up). For example, the route shown in Figure 4.2.5a can be described by the string *RUURRURU*. Any such string can be obtained by selecting n positions for the R's, without regard to the order of selection, from among the $2n$ available positions in the string and then filling the remaining positions with U's. Thus there are $C(2n, n)$ possible routes.   □

EXAMPLE 4.2.23

How many routes are there from the lower left corner of an $n \times n$ square grid to the upper right corner if we are restricted to traveling only to the right or upward and if we are allowed to touch but not go above a diagonal line from the lower left corner to the upper right corner?

We call a route that touches but does not go above the diagonal a *good route*, and a route that goes above the diagonal a *bad route*. Our problem is to count the number of good routes. We also let G_n denote the number of good routes and B_n denote the number of bad routes. In Example 4.2.22 we showed that

$$G_n + B_n = C(2n, n);$$

thus, it suffices to compute the number of bad routes.

We call a route from the lower left corner of an $(n + 1) \times (n - 1)$ grid to the upper right corner (with no restrictions) an $(n + 1) \times (n - 1)$ route. A 5×3 route is shown in Figure 4.2.5b. We show that the number of bad routes is equal to the number of $(n + 1) \times (n - 1)$ routes by describing a one-to-one, onto function from the set of bad routes to the set of $(n + 1) \times (n - 1)$ routes.

Given a bad route, we find the first move (starting from the lower left) that takes it above the diagonal. Thereafter we replace each right move by an up move and each up move by a right move. For example, the route of Figure 4.2.5a is transformed to the route shown in Figure 4.2.5b. This transformation can also be effected by rotating the portion of the route following the first move above the diagonal about the dashed line shown in Figure 4.2.5b. We see that this transformation does indeed assign to each bad route an $(n + 1) \times (n - 1)$ route.

To show that our function is onto, consider any $(n + 1) \times (n - 1)$ route. Since this route ends above the diagonal, there is a first move where it goes above the diagonal. We may then rotate the remainder of the route about the dashed line shown in Figure 4.2.5b to obtain a bad route. The image of this bad route under our function is the $(n + 1) \times (n - 1)$ route with which we started. Therefore, our function is onto. Our function is also one-to-one, as we can readily verify that the function transforms distinct bad routes to distinct $(n + 1) \times (n - 1)$ routes. Therefore, there are equal numbers of bad routes and $(n + 1) \times (n - 1)$ routes.

An argument like that in Example 4.2.22 shows that the number of $(n + 1) \times (n - 1)$ routes is equal to $C(2n, n - 1)$. Thus the number of good routes is equal to

$$
C(2n, n) - B_n = C(2n, n) - C(2n, n - 1) = \frac{(2n)!}{n! \, n!} - \frac{(2n)!}{(n - 1)! \, (n + 1)!}
$$

$$
= \frac{(2n)!}{n! \, (n - 1)!} \left(\frac{1}{n} - \frac{1}{n + 1} \right) = \frac{(2n)!}{n! \, (n - 1)!} \cdot \frac{1}{n(n + 1)}
$$

$$
= \frac{(2n)!}{(n + 1)n! \, n!} = \frac{C(2n, n)}{n + 1}. \qquad \square
$$

The numbers $C(2n, n)/(n+1)$ are called **Catalan numbers** in honor of the Belgian mathematician Eugène-Charles Catalan (1814–1894), who discovered an elementary derivation of the formula $C(2n, n)/(n + 1)$. The problem first investigated by Catalan is given in Exercise 29, Section 5.1. Catalan published numerous papers in analysis, combinatorics, algebra, geometry, probability, and number theory. The truth of his conjecture that 0, 1 and 8, 9 are the only pairs of consecutive whole numbers that are powers (i.e., i^j where $j \geq 2$) was established relatively recently (in the 1970s).

In this book, we denote the Catalan number $C(2n, n)/(n + 1)$ as C_n, $n \geq 1$, and we define C_0 to be 1. The first few Catalan numbers are

$$
C_0 = 1, \quad C_1 = 1, \quad C_2 = 2, \quad C_3 = 5, \quad C_4 = 14, \quad C_5 = 42.
$$

Like the Fibonacci numbers, the Catalan numbers have a way of appearing in unexpected places (e.g., Exercises 28 and 29, Section 5.1).

We close this section by giving another proof of Theorem 4.2.17 that gives a formula for the number of r-element subsets of an n-element set. The proof is illustrated in Figure 4.2.6. Let X be an n-element set. We assume the formula $P(n, r) = n(n - 1) \cdots (n - r + 1)$ that counts the number of orderings of r-element subsets chosen from X. To count the number of r-element subsets of X, we do *not* want to take order into account—we want to consider permutations of the same subset *equivalent*. Formally, we define a relation R on the set S of r-permutations of X by the rule: $p_1 R p_2$ if p_1 and p_2 are permutations of the same r-element subset of X. It is straightforward to verify that R is an equivalence relation on S.

If p is an r-permutation of X, then p is a permutation of some r-element subset X_r of X; thus, the equivalence class containing p consists of all permutations of X_r. We see that each equivalence class has $r!$ elements. An equivalence class is determined by the r-element subset of X that is permuted to obtain its members. Therefore, there are $C(n, r)$ equivalence classes. Since the set S has $P(n, r)$ elements, by Theorem 2.5.16 $C(n, r) = P(n, r)/r!$.

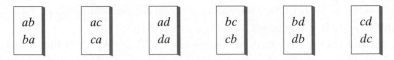

FIGURE 4.2.6   The alternative proof of Theorem 4.2.17 for $n = 4$ and $r = 2$. Each box contains an equivalence class for the relation R on the set of 2-permutations of $X = \{a, b, c, d\}$ defined by $p_1 R p_2$ if p_1 and p_2 are permutations of the same 2-element subset of X. There are $P(4, 2) = 12$ 2-permutations of X and 2 ways to permute each 2-permutation. Since each equivalence class corresponds to a subset of X, $12/2 = C(4, 2)$.

Exercises

1. How many permutations are there of a, b, c, d?

2. List the permutations of a, b, c, d.

3. How many 3-permutations are there of a, b, c, d?

4. List the 3-permutations of a, b, c, d.

5. How many permutations are there of 11 distinct objects?

6. How many 5-permutations are there of 11 distinct objects?

7. In how many ways can we select a chairperson, vice-chairperson, and recorder from a group of 11 persons?

8. In how many ways can we select a chairperson, vice-chairperson, secretary, and treasurer from a group of 12 persons?

9. In how many different ways can 12 horses finish in the order Win, Place, Show?

In Exercises 10–18, determine how many strings can be formed by ordering the letters $ABCDE$ subject to the conditions given.

10. Contains the substring ACE.

11. Contains the letters ACE together in any order.

12. Contains the substrings *DB* and *AE*.

13. Contains either the substring *AE* or the substring *EA*.

14. *A* appears before *D*. *Examples*: *BCAED*, *BCADE*.

15. Contains neither of the substrings *AB*, *CD*.

16. Contains neither of the substrings *AB*, *BE*.

17. *A* appears before *C* and *C* appears before *E*.

18. Contains either the substring *DB* or the substring *BE*.

19. In how many ways can five distinct Martians and eight distinct Jovians wait in line if no two Martians stand together?

20. In how many ways can five distinct Martians, ten distinct Vesuvians, and eight distinct Jovians wait in line if no two Martians stand together?

21. In how many ways can five distinct Martians and five distinct Jovians wait in line?

22. In how many ways can five distinct Martians and five distinct Jovians be seated at a circular table?

23. In how many ways can five distinct Martians and five distinct Jovians be seated at a circular table if no two Martians sit together?

24. In how many ways can five distinct Martians and eight distinct Jovians be seated at a circular table if no two Martians sit together?

In Exercises 25–27, let $X = \{a, b, c, d\}$.

25. Compute the number of 3-combinations of X.

26. List the 3-combinations of X.

27. Show the relationship between the 3-permutations and the 3-combinations of X by drawing a picture like that in Figure 4.2.4.

28. In how many ways can we select a committee of three from a group of 11 persons?

29. In how many ways can we select a committee of four from a group of 12 persons?

30. At one point in the Illinois state lottery Lotto game, a person was required to choose six numbers (in any order) from among 44 numbers. In how many ways can this be done? The state was considering changing the game so that a person would be required to choose six numbers from among 48 numbers. In how many ways can this be done?

Exercises 31–36 refer to a club consisting of six distinct men and seven distinct women.

31. In how many ways can we select a committee of five persons?

32. In how many ways can we select a committee of three men and four women?

33. In how many ways can we select a committee of four persons that has at least one woman?

34. In how many ways can we select a committee of four persons that has at most one man?

35. In how many ways can we select a committee of four persons that has persons of both sexes?

36. In how many ways can we select a committee of four persons so that Mabel and Ralph do not serve together?

37. In how many ways can we select a committee of four Republicans, three Democrats, and two Independents from a group of 10 distinct Republicans, 12 distinct Democrats, and four distinct Independents?

38. How many eight-bit strings contain exactly three 0's?

39. How many eight-bit strings contain three 0's in a row and five 1's?

★ **40.** How many eight-bit strings contain at least two 0's in a row?

In Exercises 41–49, find the number of (unordered) five-card poker hands, selected from an ordinary 52-card deck, having the properties indicated.

41. Containing four aces

42. Containing four of a kind, that is, four cards of the same denomination

43. Containing all spades

44. Containing cards of exactly two suits

45. Containing cards of all suits

46. Of the form A2345 of the same suit

47. Consecutive and of the same suit (Assume that the ace is the lowest denomination.)

48. Consecutive (Assume that the ace is the lowest denomination.)

49. Containing two of one denomination, two of another denomination, and one of a third denomination

50. Find the number of (unordered) 13-card bridge hands selected from an ordinary 52-card deck.

51. How many bridge hands are all of the same suit?

52. How many bridge hands contain exactly two suits?

53. How many bridge hands contain all four aces?

54. How many bridge hands contain five spades, four hearts, three clubs, and one diamond?

55. How many bridge hands contain five of one suit, four of another suit, three of another suit, and one of another suit?

56. How many bridge hands contain four cards of three suits and one card of the fourth suit?

57. How many bridge hands contain no face cards? (A face card is one of 10, J, Q, K, A.)

In Exercises 58–62, a coin is flipped 10 times.

58. How many outcomes are possible? (An *outcome* is a list of 10 H's and T's that gives the result of each of 10 tosses. For example, the outcome

$$H \ H \ T \ H \ T \ H \ H \ H \ T \ H$$

represents 10 tosses, where a head was obtained on the first two tosses, a tail was obtained on the third toss, a head was obtained on the fourth toss, etc.)

59. How many outcomes have exactly three heads?

60. How many outcomes have at most three heads?

61. How many outcomes have a head on the fifth toss?

62. How many outcomes have as many heads as tails?

Exercises 63–66 refer to a shipment of 50 microprocessors of which four are defective.

63. In how many ways can we select a set of four microprocessors?

64. In how many ways can we select a set of four nondefective microprocessors?

65. In how many ways can we select a set of four microprocessors containing exactly two defective microprocessors?

66. In how many ways can we select a set of four microprocessors containing at least one defective microprocessor?

★ 67. Show that the number of bit strings of length $n \geq 4$ that contain exactly two occurrences of 10 is $C(n + 1, 5)$.

★ 68. Show that the number of n-bit strings having exactly k 0's, with no two 0's consecutive, is $C(n - k + 1, k)$.

★ 69. Show that the product of any positive integer and its $k - 1$ successors is divisible by $k!$.

70. Show that there are $(2n - 1)(2n - 3) \cdots 3 \cdot 1$ ways to pick n pairs from $2n$ distinct items.

71. Suppose that we have n objects, r distinct and $n - r$ identical. Give another derivation of the formula

$$P(n, r) = r! \, C(n, r)$$

by counting the number of orderings of the n objects in two ways:

- Count the orderings by first choosing positions for the r distinct objects.

- Count the orderings by first choosing positions for the $n - r$ identical objects.

72. What is wrong with the following argument, which purports to show that $4C(39, 13)$ bridge hands contain three or fewer suits?

There are $C(39, 13)$ hands that contain only clubs, diamonds, and spades. In fact, for any three suits, there are $C(39, 13)$ hands that contain only those three suits. Since there are four 3-combinations of the suits, the answer is $4C(39, 13)$.

73. What is wrong with the following argument, which purports to show that there are $13^4 \cdot 48$ (unordered) five-card poker hands containing cards of all suits?

Pick one card of each suit. This can be done in $13 \cdot 13 \cdot 13 \cdot 13 = 13^4$ ways. Since the fifth card can be chosen in 48 ways, the answer is $13^4 \cdot 48$.

74. Let $s_{n,k}$ denote the number of ways to seat n persons at k round tables, with at least one person at each table. (The numbers $s_{n,k}$ are called *Stirling numbers of the first kind*.) The ordering of the tables is *not* taken into account. The seating arrangement at a table *is* taken into account except for rotations. *Examples*: The following pair is *not* distinct:

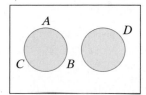
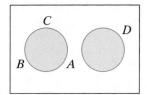

The following pair is *not* distinct:

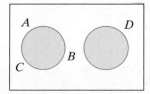
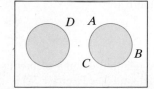

The following pair *is* distinct:

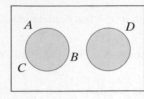 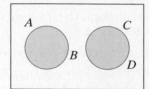

The following pair *is* distinct:

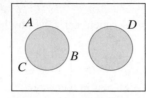 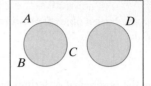

(a) Show that $s_{n,k} = 0$ if $k > n$.

(b) Show that $s_{n,n} = 1$ for all $n \geq 1$.

(c) Show that $s_{n,1} = (n-1)!$ for all $n \geq 1$.

(d) Show that $s_{n,n-1} = C(n, 2)$ for all $n \geq 2$.

(e) Show that

$$s_{n,2} = (n-1)! \left(1 + \frac{1}{2} + \frac{1}{3} + \cdots + \frac{1}{n-1}\right) \qquad \text{for all } n \geq 2.$$

(f) Show that

$$\sum_{k=1}^{n} s_{n,k} = n! \qquad \text{for all } n \geq 1.$$

(g) Find a formula for $s_{n,n-2}, n \geq 3$, and prove it.

75. Let $S_{n,k}$ denote the number of ways to partition an n-element set into exactly k nonempty subsets. The order of the subsets is not taken into account. (The numbers $S_{n,k}$ are called *Stirling numbers of the second kind*.)

(a) Show that $S_{n,k} = 0$ if $k > n$.

(b) Show that $S_{n,n} = 1$ for all $n \geq 1$.

(c) Show that $S_{n,1} = 1$ for all $n \geq 1$.

(d) Show that $S_{3,2} = 3$.

(e) Show that $S_{4,2} = 7$.

(f) Show that $S_{4,3} = 6$.

(g) Show that $S_{n,2} = 2^{n-1} - 1$ for all $n \geq 2$.

(h) Show that $S_{n,n-1} = C(n, 2)$ for all $n \geq 2$.

(i) Find a formula for $S_{n,n-2}, n \geq 3$, and prove it.

76. Show that there are

$$\sum_{k=1}^{n} S_{n,k}$$

equivalence relations on an n-element set. [The numbers $S_{n,k}$ are Stirling numbers of the second kind (see Exercise 75).]

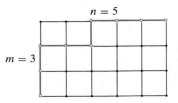

Problem

    (a) How many routes are there from the lower left corner to the upper right corner of an $m \times n$ grid in which we are restricted to traveling only to the right or upward? For example, the adjacent figure is a 3×5 grid and one route is shown.

    (b) Divide the routes into classes based on when the route first meets the top edge to derive the formula

$$\sum_{k=0}^{n} C(k + m - 1, k) = C(m + n, m).$$

Attacking the Problem

Example 4.2.22 was to count the number of paths from the lower left corner to the upper right corner of an $n \times n$ grid in which we are restricted to traveling only to the right or upward. The solution to that problem encoded each route as a string of n R's (right) and n U's (up). The problem then became one of counting the number of such strings. Any such string can be obtained by selecting n positions for the R's, without regard to the order of selection, from among the $2n$ available positions in the string and then filling the remaining positions with U's. Thus the number of strings and number of routes are equal to $C(2n, n)$.

    In the present problem, we can encode each route as a string of n R's (right) and m U's (up). As in the previous problem, we must count the number of such strings. Any such string can be obtained by selecting n positions for the R's, without regard to the order of selection, from among the $n + m$ available positions in the string and then filling the remaining positions with U's. Thus the number of strings and number of routes are equal to $C(n + m, n)$. We have answered part (a).

    In part (b) we are given a major hint: Divide the routes into classes based on when the route first meets the top edge. A route can first meet the top edge at any one of $n + 1$ positions. In the previous figure, the route shown first meets the top edge at the third position from the left. Before reading on, you might think about why we might divide the routes into classes.

    Notice that when we divide the routes into classes based on when the route first meets the top edge:

- The classes are *disjoint*.

(A route cannot first meet the top edge in two or more distinct positions.) Notice also that every route meets the top edge somewhere so

- Every route is in some class.

In the terminology of Section 2.1 (see Example 2.1.10 and the discussion that precedes it), the classes *partition* the set of routes. Because the classes partition the set of routes, the Addition Principle applies and the sum of the numbers of routes in each class is equal to the total number of routes. (No route is counted twice since the classes do not overlap, and every route is counted once since each route is in some class.) Evidently, the equation we're supposed to prove results from equating the sum of the number of routes in each class to the total number of routes.

Finding a Solution

We have already solved part (a). For part (b), let's look at the 3×5 grid. There is exactly one route that first meets the top edge at the first position from the left. There are three routes that first meet the top edge at the second position from the left:

 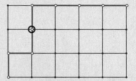 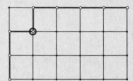

Notice that the only variation in the preceding figures occurs between the start and the circled dot. To put it another way, after a route meets the circled dot, there is only one way to finish the trip. Therefore, it suffices to count the number of routes from the lower left corner to the upper right corner of a 2×1 grid. But we already solved this problem in part (a)! The number of routes from the lower left corner to the upper right corner of a 2×1 grid is equal to $C(2 + 1, 1) = 3$. In a similar way, we find that the number of routes that first meet the top edge at the third position from the left is equal to the number of routes from the lower left corner to the upper right corner of a 2×2 grid—namely, $C(2 + 2, 2) = 6$. By summing we obtain all the routes

$$C(5 + 3, 5) = C(0 + 2, 0) + C(1 + 2, 1) + C(2 + 2, 2)$$
$$+ C(3 + 2, 3) + C(4 + 2, 4) + C(5 + 2, 5).$$

If we replace each term $C(k + 3 - 1, k)$ by its value, we obtain

$$56 = 1 + 3 + 6 + 10 + 15 + 21.$$

You should verify the preceding formula, find the six routes that first meet the top edge at the third position from the left, and see why the number of such routes is equal to the number of routes from the lower left corner to the upper right corner of a 2×2 grid.

Formal Solution

(a) We can encode each route as a string of n R's (right) and m U's (up). Any such string can be obtained by selecting n positions for the R's, without regard to the order of selection, from among the $n + m$ available positions in the string and then filling the remaining positions with U's. Thus the number of routes is equal to $C(n+m, n)$.

(b) Each route can be described as a string containing n R's and m U's. The last U in such a string marks the point at which the route first meets the top edge. We count the strings by dividing them into classes consisting of strings that end U, UR, URR, and so on. There are

$$C(n + m - 1, n)$$

strings that end U, since we must choose n slots from among the first $n + m - 1$ slots for the n R's. There are

$$C((n - 1) + m - 1, n - 1)$$

strings that end UR, since we must choose $n - 1$ slots from among the first $(n - 1) + m - 1$ slots for the $n - 1$ R's. In general, there are $C(k + m - 1, k)$ strings that end UR^{n-k}. Since there are $C(m + n, m)$ strings altogether, the formula follows.

Summary of Problem-Solving Techniques

- Look for a similar problem and imitate its solution.
- Counting the number of members of a set in two different ways leads to an equation. In particular, if $\{X_1, X_2, \ldots, X_n\}$ is a partition of X, the Addition Principle applies and

$$|X| = \sum_{i=1}^{n} |X_i|.$$

- Directly enumerate some of the items to be counted.
- Look for patterns.

Comments

It's important to verify that an alleged partition is truly a partition before using the Addition Principle. If X is the set of five-bit strings and X_i is the set of five-bit strings that contain i consecutive zeros, the Addition Principle does *not* apply; the sets X_i are *not* pairwise disjoint. For example, $00001 \in X_2 \cap X_3$. As an example of a partition of X, we could let X_i be the set of five-bit strings that contain exactly i zeros.

❧ ❧ ❧

Exercises

1. Divide the routes into classes based on when the route first crosses a vertical line and use the Addition Principle to derive a formula like that proved in this section.

2. Divide the routes into classes based on when the route crosses the slanted line shown.

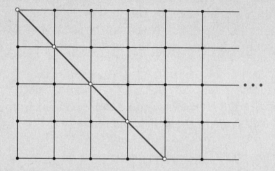

Then use the Addition Principle to derive a formula like that proved in this section.

4.3 ALGORITHMS FOR GENERATING PERMUTATIONS AND COMBINATIONS

The rock group Unhinged Universe has recorded n videos whose running times are

$$t_1, t_2, \ldots, t_n$$

seconds. A tape is to be released that can hold C seconds. Since this is the first tape by the Unhinged Universe, the group wants to include as much material as possible. Thus the problem is to choose a subset $\{i_1, \ldots, i_k\}$ of $\{1, 2, \ldots, n\}$ such that the sum

$$\sum_{j=1}^{k} t_{i_j} \qquad (4.3.1)$$

does not exceed C and is as large as possible. A straightforward approach is to examine all subsets of $\{1, 2, \ldots, n\}$ and choose a subset so that the sum (4.3.1) does not exceed C and is as large as possible. To implement this approach, we need an algorithm that generates all combinations of an n-element set. In this section we develop algorithms to generate combinations and permutations.

Since there are 2^n subsets of an n-element set, the running time of an algorithm that examines all subsets is $\Omega(2^n)$. As we saw in Section 3.5, such algorithms are impractical to run except for small values of n. Unfortunately, there are problems (an example of which is the tape filling problem described previously) for which no method much better than the "list all" approach is known.

Our algorithms list permutations and combinations in **lexicographic order**. Lexicographic order generalizes ordinary dictionary order.

Given two distinct words, to determine whether one precedes the other in the dictionary, we compare the letters in the words. There are two possibilities:

1. The words have different lengths, and each letter in the shorter word is identical to the corresponding letter in the longer word.
2. The words have the same or different lengths, and at some position, the letters in the words differ.   (4.3.2)

If 1 holds, the shorter word precedes the longer. (For example, "dog" precedes "doghouse" in the dictionary.) If 2 holds, we locate the leftmost position p at which the letters differ. The order of the words is determined by the order of the letters at position p. (For example, "gladiator" precedes "gladiolus" in the dictionary. At the leftmost position at which the letters differ we find "a" in "gladiator" and "o" in "gladiolus" and "a" precedes "o" in the alphabet.)

Lexicographic order generalizes ordinary dictionary order by replacing the alphabet by any set of symbols on which an order has been defined. We will be concerned with strings of integers.

DEFINITION 4.3.1

Let $\alpha = s_1 s_2 \cdots s_p$ and $\beta = t_1 t_2 \cdots t_q$ be strings over $\{1, 2, \ldots, n\}$. We say that α is *lexicographically less than* β and write $\alpha < \beta$ if either

(a) $p < q$ and $s_i = t_i$ for $i = 1, \ldots, p$

or

(b) for some i, $s_i \neq t_i$, and for the smallest such i, we have $s_i < t_i$.

In Definition 4.3.1, case (a) corresponds to possibility 1 of (4.3.2) and case (b) corresponds to possibility 2 of (4.3.2).

EXAMPLE 4.3.2

Let $\alpha = 132$ and $\beta = 1324$ be strings over $\{1, 2, 3, 4\}$. In the notation of Definition 4.3.1, $p = 3$, $q = 4$, $s_1 = 1$, $s_2 = 3$, $s_3 = 2$, $t_1 = 1$, $t_2 = 3$, $t_3 = 2$, and $t_4 = 4$. Since $p = 3 < 4 = q$ and $s_i = t_i$ for $i = 1, 2, 3$, condition (a) of Definition 4.3.1 is satisfied. Therefore, $\alpha < \beta$.   □

EXAMPLE 4.3.3

Let $\alpha = 13246$ and $\beta = 1342$ be strings over $\{1, 2, 3, 4, 5, 6\}$. In the notation of Definition 4.3.1, $p = 5$, $q = 4$, $s_1 = 1$, $s_2 = 3$, $s_3 = 2$, $s_4 = 4$, $s_5 = 6$, $t_1 = 1$, $t_2 = 3$, $t_3 = 4$, and $t_4 = 2$. The smallest i for which $s_i \neq t_i$ is $i = 3$. Since $s_3 < t_3$, by condition (b) of Definition 4.3.1, $\alpha < \beta$.   □

EXAMPLE 4.3.4

Let $\alpha = 1324$ and $\beta = 1342$ be strings over $\{1, 2, 3, 4\}$. In the notation of Definition 4.3.1, $p = q = 4$, $s_1 = 1$, $s_2 = 3$, $s_3 = 2$, $s_4 = 4$, $t_1 = 1$, $t_2 = 3$, $t_3 = 4$, and $t_4 = 2$. The smallest i for which $s_i \neq t_i$ is $i = 3$. Since $s_3 < t_3$, by condition (b) of Definition 4.3.1, $\alpha < \beta$.   □

EXAMPLE 4.3.5

Let $\alpha = 13542$ and $\beta = 21354$ be strings over $\{1, 2, 3, 4, 5\}$. In the notation of Definition 4.3.1, $s_1 = 1, s_2 = 3, s_3 = 5, s_4 = 4, s_5 = 2, t_1 = 2, t_2 = 1, t_3 = 3,$ $t_4 = 5,$ and $t_5 = 4$. The smallest i for which $s_i \neq t_i$ is $i = 1$. Since $s_1 < t_1$, by condition (b) of Definition 4.3.1, $\alpha < \beta$. $\square$

For strings of the same length over $\{1, 2, \ldots, 9\}$, lexicographic order is the same as numerical order on the positive integers if we interpret the strings as decimal numbers (see Examples 4.3.4 and 4.3.5). For strings of unequal length, lexicographic order may be different than numerical order (see Example 4.3.3). Throughout the remainder of this section, *order* will refer to lexicographic order.

First we consider the problem of listing all r-combinations of $\{1, 2, \ldots, n\}$. In our algorithm, we will list the r-combination $\{x_1, \ldots, x_r\}$ as the string $s_1 \cdots s_r$ where $s_1 < s_2 < \cdots < s_r$ and $\{x_1, \ldots, x_r\} = \{s_1, \ldots, s_r\}$. For example, the 3-combination $\{6, 2, 4\}$ will be listed 246.

We will list the r-combinations of $\{1, 2, \ldots, n\}$ in lexicographic order. Thus the first listed string will be $12 \cdots r$ and the last listed string will be $(n - r + 1) \cdots n$.

EXAMPLE 4.3.6

Consider the order in which the 5-combinations of $\{1, 2, 3, 4, 5, 6, 7\}$ will be listed. The first string is 12345, which is followed by 12346 and 12347. The next string is 12356 followed by 12457. The last string will be 34567. $\square$

EXAMPLE 4.3.7

Find the string that follows 13467 when we list the 5-combinations of $X = \{1, 2, 3, 4, 5, 6, 7\}$.

No string that begins 134 and represents a 5-combination of X exceeds 13467. Thus the string that follows 13467 must begin 135. Since 13567 is the smallest string that begins 135 and represents a 5-combination of X, the answer is 13567. $\square$

EXAMPLE 4.3.8

Find the string that follows 2367 when we list the 4-combinations of $X = \{1, 2, 3, 4, 5, 6, 7\}$.

No string that begins 23 and represents a 4-combination of X exceeds 2367. Thus the string that follows 2367 must begin 24. Since 2456 is the smallest string that begins 24 and represents a 4-combination of X, the answer is 2456. $\square$

A pattern is developing. Given a string $\alpha = s_1 \cdots s_r$, which represents the r-combination $\{s_1, \ldots, s_r\}$, to find the next string $\beta = t_1 \cdots t_r$, we find the rightmost element s_m that is not at its maximum value. (s_r may have the maximum value n, s_{r-1} may have the maximum value $n - 1$, etc.) Then

$$t_i = s_i \qquad \text{for } i = 1, \ldots, m - 1.$$

The element t_m is equal to $s_m + 1$. For the remainder of the string β we have

$$t_{m+1} \cdots t_r = (s_m + 2)(s_m + 3) \cdots.$$

The algorithm follows.

ALGORITHM 4.3.9   *Generating Combinations*

This algorithm lists all r-combinations of $\{1, 2, \ldots, n\}$ in increasing lexicographic order.

Input: r, n

Output: All r-combinations of $\{1, 2, \ldots, n\}$ in increasing lexicographic order

```
1.     procedure combination(r, n)
2.        for i := 1 to r do
3.           s_i := i
4.        print s_1, ..., s_r   // print the first r-combination
5.        for i := 2 to C(n, r) do
6.           begin
7.           m := r
8.           max_val := n
9.           while s_m = max_val do
10.             // find the rightmost element not at its maximum value
11.             begin
12.             m := m - 1
13.             max_val := max_val - 1
14.             end
15.          // the rightmost element is incremented
16.          s_m := s_m + 1
17.          // the rest of the elements are the successors of s_m
18.          for j := m + 1 to r do
19.             s_j := s_{j-1} + 1
20.          print s_1, ..., s_r   // print the ith combination
21.          end
22.    end combination
```

EXAMPLE 4.3.10

We will show how Algorithm 4.3.9 generates the 5-combination of $\{1, 2, 3, 4, 5, 6, 7\}$ following 23467. We are supposing that

$$s_1 = 2, \quad s_2 = 3, \quad s_3 = 4, \quad s_4 = 6, \quad s_5 = 7.$$

At line 15, we find that s_3 is the rightmost element not at its maximum value. At line 16, s_3 is set to 5. At lines 18 and 19, s_4 is set to 6 and s_5 is set to 7. At this point

$$s_1 = 2, \quad s_2 = 3, \quad s_3 = 5, \quad s_4 = 6, \quad s_5 = 7.$$

We have generated the 5-combination 23567 that follows 23467. ◻

EXAMPLE 4.3.11

The 4-combinations of $\{1, 2, 3, 4, 5, 6\}$ as listed by Algorithm 4.3.9 are

$$1234, \quad 1235, \quad 1236, \quad 1245, \quad 1246, \quad 1256, \quad 1345, \quad 1346,$$

$$1356, \quad 1456, \quad 2345, \quad 2346, \quad 2356, \quad 2456, \quad 3456.$$ ◻

Like the algorithm for generating r-combinations, the algorithm to generate permutations will list the permutations of $\{1, 2, \ldots, n\}$ in lexicographic order. (Exercise 16 asks for an algorithm that generates all r-permutations of an n-element set.)

EXAMPLE 4.3.12

To construct the permutation of $\{1, 2, 3, 4, 5, 6\}$ following 163542, we should keep as many digits as possible at the left the same.

Can the permutation following the given permutation have the form 1635__? Since the only permutation of the form 1635__ distinct from the given permutation is 163524 and 163524 is smaller than 163542, the permutation following the given permutation is not of the form 1635__.

Can the permutation following the given permutation have the form 163___? The last three digits must be a permutation of $\{2, 4, 5\}$. Since 542 is the largest permutation of $\{2, 4, 5\}$, any permutation that begins 163 is smaller than the given permutation. Thus the permutation following the given permutation is not of the form 163___.

The reason that the permutation following the given permutation cannot begin 1635 or 163 is that in either case the remaining digits in the given permutation (42 and 542, respectively) *decrease*. Therefore, working from the right, we must find the first digit d whose right neighbor r satisfies $d < r$. In our case, the third digit, 3, has this property. Thus the permutation following the given permutation will begin 16.

The digit following 16 must exceed 3. Since we want the next smallest permutation, the next digit is 4, the smallest available digit. Thus the desired permutation begins 164. The remaining digits 235 must be in increasing order to achieve the minimum value. Therefore, the permutation following the given permutation is 164235. ◻

We see that to generate all of the permutations of $\{1, 2, \ldots, n\}$, we can begin with the permutation $12 \cdots n$ and then repeatedly use the method of

Example 4.3.12 to generate the next permutation. We will end when the permutation $n(n - 1) \cdots 21$ is generated.

EXAMPLE 4.3.13

Using the method of Example 4.3.12, we can list the permutations of $\{1, 2, 3, 4\}$ in lexicographic order as

1234,	1243,	1324,	1342,	1423,	1432,	2134,	2143,
2314,	2341,	2413,	2431,	3124,	3142,	3214,	3241,
3412,	3421,	4123,	4132,	4213,	4231,	4312,	4321. □

The algorithm follows.

ALGORITHM 4.3.14   *Generating Permutations*

This algorithm lists all permutations of $\{1, 2, \ldots, n\}$ in increasing lexicographic order.

Input: n

Output: All permutations of $\{1, 2, \ldots, n\}$ in increasing lexicographic order

```
1.    procedure permutation(n)
2.        for i := 1 to n do
3.            s_i := i
4.        print s_1, ..., s_n    // print the first permutation
5.        for i := 2 to n! do
6.            begin
7.            m := n - 1
8.            while s_m > s_{m+1} do
9.                // find the first decrease working from the right
10.                m := m - 1
11.            k := n
12.            while s_m > s_k do
13.                // find the rightmost element s_k with s_m < s_k
14.                k := k - 1
15.            swap(s_m, s_k)
16.            p := m + 1
17.            q := n
18.            while p < q do
19.                // swap s_{m+1} and s_n, swap s_{m+2} and s_{n-1}, and so on
20.                begin
21.                swap(s_p, s_q)
22.                p := p + 1
23.                q := q - 1
24.                end
25.            print s_1, ..., s_n    // print the ith permutation
26.            end
27.    end permutation
```

> **EXAMPLE 4.3.15**

We will show how Algorithm 4.3.14 generates the permutation following 163542. Suppose that

$$s_1 = 1, \quad s_2 = 6, \quad s_3 = 3, \quad s_4 = 5, \quad s_5 = 4, \quad s_6 = 2$$

and that we are at line 7. The largest index m satisfying $s_m < s_{m+1}$ is 3. At lines 11–14, we find that the largest index k satisfying $s_k > s_m$ is 5. At line 15, we swap s_m and s_k. At this point, we have $s = 164532$. At lines 16–24, we reverse the order of the elements $s_4 s_5 s_6 = 532$. We obtain the desired permutation 164235. □

Exercises

In Exercises 1–3, find the r-combination that will be generated by Algorithm 4.3.9 with $n = 7$ after the r-combination given.

1. 1356   **2.** 12367   **3.** 14567

In Exercises 4–6, find the permutation that will be generated by Algorithm 4.3.14 after the permutation given.

4. 12354   **5.** 625431   **6.** 12876543

7. For each string in Exercises 1–3, explain (as in Example 4.3.10) exactly how Algorithm 4.3.9 generates the next r-combination.

8. For each string in Exercises 4–6, explain (as in Example 4.3.15) exactly how Algorithm 4.3.14 generates the next permutation.

9. Show the output from Algorithm 4.3.9 when $n = 6$ and $r = 3$.

10. Show the output from Algorithm 4.3.9 when $n = 6$ and $r = 2$.

11. Show the output from Algorithm 4.3.9 when $n = 7$ and $r = 5$.

12. Show the output from Algorithm 4.3.14 when $n = 2$.

13. Show the output from Algorithm 4.3.14 when $n = 3$.

14. Modify Algorithm 4.3.9 so that line 5

    5.   **for** $i := 2$ **to** $C(n, r)$ **do**

is eliminated. Base the terminating condition on the fact that the last r-combination has every element s_i equal to its maximum value.

15. Modify Algorithm 4.3.14 so that line 5

    5.   **for** $i := 2$ **to** $n!$ **do**

is eliminated. Base the terminating condition on the fact that the last permutation has the elements s_i in decreasing order.

16. Write an algorithm that generates all r-permutations of an n-element set.

★ **17.** Write a recursive algorithm that generates all r-combinations of the set $\{s_1, s_2, \ldots, s_n\}$. Divide the problem into two subproblems:

- List the r-combinations containing s_1.
- List the r-combinations not containing s_1.

★ **18.** Write a recursive algorithm that generates all permutations of the set $\{s_1, s_2, \ldots s_n\}$. Divide the problem into n subproblems:

- List the permutations that begin with s_1.
- List the permutations that begin with s_2.

$\vdots$

- List the permutations that begin with s_n.

4.4 GENERALIZED PERMUTATIONS AND COMBINATIONS

In Sections 4.2 and 4.3, we dealt with orderings and selections without allowing repetitions. In this section we consider orderings of sequences containing repetitions and unordered selections in which repetitions are allowed.

EXAMPLE 4.4.1

How many strings can be formed using the following letters?

$$M \; I \; S \; S \; I \; S \; S \; I \; P \; P \; I$$

Because of the duplication of letters, the answer is not 11!, but some number less than 11!.

Let us consider the problem of filling 11 blanks,

$$— \; — \; — \; — \; — \; — \; — \; — \; — \; — \; —,$$

with the letters given. There are $C(11, 2)$ ways to choose positions for the two P's. Once the positions for the P's have been selected, there are $C(9, 4)$ ways to choose positions for the four S's. Once the positions for the S's have been selected, there are $C(5, 4)$ ways to choose positions for the four I's. Once these selections have been made, there is one position left to be filled by the M. By the Multiplication Principle, the number of ways of ordering the letters is

$$C(11, 2)C(9, 4)C(5, 4) = \frac{11!}{2! \, 9!} \frac{9!}{4! \, 5!} \frac{5!}{4! \, 1!} = \frac{11!}{2! \, 4! \, 4! \, 1!} = 34{,}650. \qquad \square$$

The solution to Example 4.4.1 assumes a nice form. The number 11 that appears in the numerator is the total number of letters. The values in the denominator give the numbers of duplicates of each letter. The method can be used to establish a general formula.

THEOREM 4.4.2

Suppose that a sequence S of n items has n_1 identical objects of type 1, n_2 identical objects of type 2, $\ldots$, and n_t identical objects of type t. Then the number of orderings of S is

$$\frac{n!}{n_1! \, n_2! \cdots n_t!}.$$

Proof. We assign positions to each of the n items to create an ordering of S. We may assign positions to the n_1 items of type 1 in $C(n, n_1)$ ways. Having made these assignments, we may assign positions to the n_2 items of type 2 in $C(n - n_1, n_2)$ ways, and so on. By the Multiplication Principle, the number of orderings is

$$C(n, n_1)C(n - n_1, n_2)C(n - n_1 - n_2, n_3) \cdots C(n - n_1 - \cdots - n_{t-1}, n_t)$$

$$= \frac{n!}{n_1! \, (n - n_1)!} \frac{(n - n_1)!}{n_2! \, (n - n_1 - n_2)!} \cdots \frac{(n - n_1 - \cdots - n_{t-1})!}{n_t! \, 0!}$$

$$= \frac{n!}{n_1! \, n_2! \cdots n_t!}.$$
∎

EXAMPLE 4.4.3

In how many ways can eight distinct books be divided among three students if Bill gets four books and Shizuo and Marian each get two books?

Put the books in some fixed order. Now consider orderings of four B's, two S's, and two M's. An example is

$$B \; B \; B \; S \; M \; B \; M \; S.$$

Each such ordering determines a distribution of books. For the example ordering, Bill gets books 1, 2, 3, and 6, Shizuo gets books 4 and 8, and Marian gets books 5 and 7. Thus the number of ways of ordering $BBBBSSMM$ is the number of ways to distribute the books. By Theorem 4.4.2, this number is

$$\frac{8!}{4! \, 2! \, 2!} = 420. \qquad \square$$

We can give an alternate proof of Theorem 4.4.2 by using relations. Suppose that a sequence S of n items has n_i identical objects of type i for $i = 1, \ldots, t$. Let X denote the set of n elements obtained from S by considering the n_i objects of type i *distinct* for $i = 1, \ldots, t$. For example, if S is the sequence of letters

$$M \; I \; S \; S \; I \; S \; S \; I \; P \; P \; I,$$

X would be the set

$$\{M, I_1, S_1, S_2, I_2, S_3, S_4, I_3, P_1, P_2, I_4\}.$$

We define a relation R on the set of all permutations of X by the rule: $p_1 R p_2$ if p_2 is obtained from p_1 by permuting the order of the objects of type 1 (but not changing their location) and/or permuting the order of the objects of type 2 (but not changing their location) $\cdots$ and/or permuting the order of the objects of type t (but not changing their location); for example,

$$I_1 S_1 S_2 I_2 S_3 S_4 I_3 P_1 P_2 I_4 M R I_2 S_3 S_2 I_1 S_4 S_1 I_3 P_1 P_2 I_4 M.$$

It is straightforward to verify that R is an equivalence relation on the set of all permutations of X.

The equivalence class containing the permutation p consists of all permutations of X that are identical if we consider the objects of type i identical for $i = 1, \ldots, t$. Thus each equivalence class has $n_1! n_2! \cdots n_t!$ elements. Since an equivalence class is determined by an ordering of S, the number of orderings of S is equal to the number of equivalence classes. There are $n!$ permutations of X, so by Theorem 2.5.16 the number of orderings of S is

$$\frac{n!}{n_1! n_2! \cdots n_t!}.$$

Next, we turn to the problem of counting unordered selections where repetitions are allowed.

EXAMPLE 4.4.4

Consider three books: a computer science book, a physics book, and a history book. Suppose that the library has at least six copies of each of these books. In how many ways can we select six books?

The problem is to choose unordered, six-element selections from the set {computer science, physics, history}, repetitions allowed. A selection is uniquely determined by the number of each type of book selected. Let us denote a particular selection as

CS	Physics	History
$\times$ $\times$ $\times$	$\mid$ $\times$ $\times$ $\mid$	$\times$

Here we have designated the selection consisting of three computer science books, two physics books, and one history book. Another example of a selection is

CS	Physics	History
$\mid$	$\times$ $\times$ $\times$ $\times$ $\mid$	$\times$ $\times$

which denotes the selection consisting of no computer science books, four physics books, and two history books. We see that each ordering of six $\times$'s and two $\mid$'s denotes a selection. Thus our problem is to count the number of such orderings. But this is just the number of ways

$$C(8, 2) = 28$$

of selecting two positions for the $\mid$'s from eight possible positions. Thus there are 28 ways to select six books. $\square$

The method used in Example 4.4.4 can be used to derive a general result.

THEOREM 4.4.5

If X is a set containing t elements, the number of unordered, k-element selections from X, repetitions allowed, is

$$C(k + t - 1, t - 1) = C(k + t - 1, k).$$

Proof. Let $X = \{a_1, \ldots, a_t\}$. Consider the $k + t - 1$ slots

$$\underline{}\ \underline{}\ \underline{}\ \cdots\ \underline{}\ \underline{}$$

and $k + t - 1$ symbols consisting of k ×'s and $t - 1$ |'s. Each placement of these symbols into the slots determines a selection. The number n_1 of ×'s up to the first | represents the selection of n_1 a_1's; the number n_2 of ×'s between the first and second |'s represents the selection of n_2 a_2's; and so on. Since there are $C(k + t - 1, t - 1)$ ways to select the positions for the |'s, there are also $C(k + t - 1, t - 1)$ selections. This is equal to $C(k + t - 1, k)$, the number of ways to select the positions for the ×'s; hence there are

$$C(k + t - 1, t - 1) = C(k + t - 1, k)$$

unordered k-element selections from X, repetitions allowed.   ∎

EXAMPLE 4.4.6

Suppose that there are piles of red, blue, and green balls and that each pile contains at least eight balls.

   (a) In how many ways can we select eight balls?
   (b) In how many ways can we select eight balls if we must have at least one ball of each color?

By Theorem 4.4.5, the number of ways of selecting eight balls is

$$C(8 + 3 - 1, 3 - 1) = C(10, 2) = 45.$$

We can also use Theorem 4.4.5 to solve part (b) if we first select one ball of each color. To complete the selection, we must choose five additional balls. This can be done in

$$C(5 + 3 - 1, 3 - 1) = C(7, 2) = 21$$

ways.   □

EXAMPLE 4.4.7

In how many ways can 12 identical mathematics books be distributed among the students Anna, Beth, Candy, and Dan?

We can use Theorem 4.4.5 to solve this problem if we consider the problem to be that of labeling each book with the name of the student who receives it. This is the same as selecting 12 items (the names of the students) from the set {Anna, Beth, Candy, Dan}, repetitions allowed. By Theorem 4.4.5, the number of ways to do this is

$$C(12 + 4 - 1, 4 - 1) = C(15, 3) = 455.$$ □

> **EXAMPLE 4.4.8**

(a) How many solutions in nonnegative integers are there to the equation

$$x_1 + x_2 + x_3 + x_4 = 29? \tag{4.4.1}$$

(b) How many solutions in integers are there to (4.4.1) satisfying $x_1 > 0$, $x_2 > 1$, $x_3 > 2$, $x_4 \geq 0$?

(a) Each solution of (4.4.1) is equivalent to selecting 29 items, x_i of type i, $i = 1, 2, 3, 4$. According to Theorem 4.4.5, the number of selections is
$$C(29 + 4 - 1, 4 - 1) = C(32, 3) = 4960.$$

(b) Each solution of (4.4.1) satisfying the given conditions is equivalent to selecting 29 items, x_i of type i, $i = 1, 2, 3, 4$, where, in addition, we must have at least one item of type 1, at least two items of type 2, and at least three items of type 3. First, select one item of type 1, two items of type 2, and three items of type 3. Then, choose 23 additional items. By Theorem 4.4.5, this can be done in

$$C(23 + 4 - 1, 4 - 1) = C(26, 3) = 2600$$

ways. □

> **EXAMPLE 4.4.9**

How many times is the print statement executed?

```
for i_1 := 1 to n do
    for i_2 := 1 to i_1 do
        for i_3 := 1 to i_2 do
            .
              .
                .
                    for i_k := 1 to i_{k-1} do
                        print i_1, i_2, ..., i_k
```

Notice that each line of output consists of k integers

$$i_1 i_2 \cdots i_k, \tag{4.4.2}$$

where

$$n \geq i_1 \geq i_2 \geq \cdots \geq i_k \geq 1, \tag{4.4.3}$$

and that every sequence (4.4.2) satisfying (4.4.3) occurs. Thus the problem is to count the number of ways of choosing k integers, with repetitions allowed, from the set $\{1, 2, \ldots, n\}$. [Any such selection can be ordered to produce (4.4.3).] By Theorem 4.4.5, the total number of selections possible is

$$C(k + n - 1, k). \qquad \Box$$

ॐ ॐ ॐ

Exercises

In Exercises 1–3, determine the number of strings that can be formed by ordering the letters given.

1. *GUIDE*   **2.** *SCHOOL*   **3.** *SALESPERSONS*

4. In how many ways can 10 distinct books be divided among three students if the first student gets five books, the second three books, and the third two books?

Exercises 5–11 refer to piles of identical red, blue, and green balls where each pile contains at least 10 balls.

5. In how many ways can 10 balls be selected?

6. In how many ways can 10 balls be selected if at least one red ball must be selected?

7. In how many ways can 10 balls be selected if at least one red ball, at least two blue balls, and at least three green balls must be selected?

8. In how many ways can 10 balls be selected if exactly one red ball must be selected?

9. In how many ways can 10 balls be selected if exactly one red ball and at least one blue ball must be selected?

10. In how many ways can 10 balls be selected if at most one red ball is selected?

11. In how many ways can 10 balls be selected if twice as many red balls as green balls must be selected?

In Exercises 12–17, find the number of integer solutions of

$$x_1 + x_2 + x_3 = 15$$

subject to the conditions given.

12. $x_1 \geq 0, x_2 \geq 0, x_3 \geq 0$ **13.** $x_1 \geq 1, x_2 \geq 1, x_3 \geq 1$

14. $x_1 = 1, x_2 \geq 0, x_3 \geq 0$ **15.** $x_1 \geq 0, x_2 > 0, x_3 = 1$

16. $0 \leq x_1 \leq 6, x_2 \geq 0, x_3 \geq 0$ ★ **17.** $0 \leq x_1 < 6, 1 \leq x_2 < 9, x_3 \geq 0$

★ **18.** Find the number of solutions in integers to

$$x_1 + x_2 + x_3 + x_4 = 12$$

satisfying $0 \leq x_1 \leq 4, 0 \leq x_2 \leq 5, 0 \leq x_3 \leq 8$, and $0 \leq x_4 \leq 9$.

19. How many integers between 1 and 1,000,000 have the sum of the digits equal to 15?

★ **20.** How many integers between 1 and 1,000,000 have the sum of the digits equal to 20?

21. How many bridge deals are there? (A deal consists of partitioning a 52-card deck into four hands each containing 13 cards.)

22. In how many ways can three teams containing four, two, and two persons be selected from a group of eight persons?

23. A *domino* is a rectangle divided into two squares with each square numbered one of 0, 1, . . . , 6, repetitions allowed. How many distinct dominoes are there?

Exercises 24–29 refer to a bag containing 20 balls—six red, six green, and eight purple.

24. In how many ways can we select five balls if the balls are considered distinct?

25. In how many ways can we select five balls if balls of the same color are considered identical?

26. In how many ways can we draw two red, three green, and two purple balls if the balls are considered distinct?

27. We draw five balls, then replace the balls, and then draw five more balls. In how many ways can this be done if the balls are considered distinct?

28. We draw five balls without replacing them. We then draw five more balls. In how many ways can this be done if the balls are considered distinct?

29. We draw five balls and at least one is red, then replace them. We then draw five balls and at most one is green. In how many ways can this be done if the balls are considered distinct?

30. In how many ways can 15 identical mathematics books be distributed among six students?

31. In how many ways can 15 identical computer science books and 10 identical psychology books be distributed among five students?

32. In how many ways can we place 10 identical balls in 12 boxes if each box can hold one ball?

33. In how many ways can we place 10 identical balls in 12 boxes if each box can hold 10 balls?

34. Show that $(kn)!$ is divisible by $(n!)^k$.

35. By considering

```
for i₁ := 1 to n do
    for i₂ := 1 to i₁ do
        print i₁, i₂
```

and Example 4.4.9, deduce

$$1 + 2 + \cdots + n = \frac{n(n+1)}{2}.$$

★ **36.** Use Example 4.4.9 to prove the formula

$$C(k-1, k-1) + C(k, k-1) + \cdots + C(n+k-2, k-1) = C(k+n-1, k).$$

37. Write an algorithm that lists all solutions in nonnegative integers to

$$x_1 + x_2 + x_3 = n.$$

38. What is wrong with the following argument, which supposedly counts the number of partitions of a 10-element set into eight (nonempty) subsets?

List the elements of the set with blanks between them:

$$x_1 \underline{\quad} x_2 \underline{\quad} x_3 \underline{\quad} x_4 \underline{\quad} x_5 \underline{\quad} x_6 \underline{\quad} x_7 \underline{\quad} x_8 \underline{\quad} x_9 \underline{\quad} x_{10}.$$

Every time we fill seven of the nine blanks with seven vertical bars, we obtain a partition of $\{x_1, \ldots, x_{10}\}$ into eight subsets. For example, the partition $\{x_1\}$, $\{x_2\}$, $\{x_3, x_4\}$ $\{x_5\}$, $\{x_6\}$, $\{x_7, x_8\}$ $\{x_9\}$, $\{x_{10}\}$ would be represented as

$$x_1 \mid x_2 \mid x_3\, x_4 \mid x_5 \mid x_6 \mid x_7\, x_8 \mid x_9 \mid x_{10}.$$

Thus the solution to the problem is $C(9, 7)$.

4.5 BINOMIAL COEFFICIENTS AND COMBINATORIAL IDENTITIES

At first glance the expression $(a + b)^n$ does not have much to do with combinations; but as we will see in this section, we can obtain the formula for the expansion of $(a+b)^n$ by using the formula for the number of r-combinations of n objects. Frequently, we can relate an algebraic expression to some counting process. Several advanced counting techniques use such methods (see [Riordan; and Tucker]).

The **Binomial Theorem** gives a formula for the coefficients in the expansion of $(a + b)^n$. Since

$$(a + b)^n = \underbrace{(a + b)(a + b) \cdots (a + b)}_{n \text{ factors}}, \quad (4.5.1)$$

the expansion results from selecting either a or b from each of the n factors, multiplying the selections together, and then summing all such products obtained. For example, in the expansion of $(a+b)^3$, we select either a or b from the first factor $(a + b)$; either a or b from the second factor $(a + b)$; and either a or b from the third factor $(a + b)$; multiply the selections together; and then sum the products obtained. If we select a from all factors and multiply, we obtain the term aaa. If we select a from the first factor, b from the second factor, and a from the third factor and multiply, we obtain the term aba. Table 4.5.1 shows all the possibilities. If we sum the products of all the selections, we obtain

$$(a + b)^3 = (a + b)(a + b)(a + b)$$
$$= aaa + aab + aba + abb + baa + bab + bba + bbb$$
$$= a^3 + a^2b + a^2b + ab^2 + a^2b + ab^2 + ab^2 + b^3$$
$$= a^3 + 3a^2b + 3ab^2 + b^3.$$

In (4.5.1), a term of the form $a^{n-k}b^k$ arises from choosing b from k factors and a from the other $n - k$ factors. But this can be done in $C(n, k)$ ways, since $C(n, k)$ counts the number of ways of selecting k things from n items. Thus $a^{n-k}b^k$ appears $C(n, k)$ times. It follows that

$$(a + b)^n = C(n, 0)a^nb^0 + C(n, 1)a^{n-1}b^1 + C(n, 2)a^{n-2}b^2$$
$$+ \cdots + C(n, n - 1)a^1b^{n-1} + C(n, n)a^0b^n. \quad (4.5.2)$$

This result is known as the *Binomial Theorem*.

THEOREM 4.5.1 *Binomial Theorem*

If a and b are real numbers and n is a positive integer, then

$$(a + b)^n = \sum_{k=0}^{n} C(n, k)a^{n-k}b^k.$$

Proof. The proof precedes the statement of the theorem. ∎

The Binomial Theorem can also be proved using induction on n (see Exercise 16).

TABLE 4.5.1
Computing $(a + b)^3$

Selection from First Factor $(a + b)$	Selection from Second Factor $(a + b)$	Selection from Third Factor $(a + b)$	Product of Selections
a	a	a	$aaa = a^3$
a	a	b	$aab = a^2b$
a	b	a	$aba = a^2b$
a	b	b	$abb = ab^2$
b	a	a	$baa = a^2b$
b	a	b	$bab = ab^2$
b	b	a	$bba = ab^2$
b	b	b	$bbb = b^3$

The numbers $C(n, r)$ are known as **binomial coefficients** because they appear in the expansion (4.5.2) of the binomial $a + b$ raised to a power.

EXAMPLE 4.5.2

Taking $n = 3$ in Theorem 4.5.1, we obtain

$$(a + b)^3 = C(3, 0)a^3b^0 + C(3, 1)a^2b^1 + C(3, 2)a^1b^2 + C(3, 3)a^0b^3$$
$$= a^3 + 3a^2b + 3ab^2 + b^3.$$ □

EXAMPLE 4.5.3

Expand $(3x - 2y)^4$ using the Binomial Theorem.

If we take $a = 3x$, $b = -2y$, and $n = 4$ in Theorem 4.5.1, we obtain

$$(3x - 2y)^4 = (a + b)^4$$

$$= C(4, 0)a^4b^0 + C(4, 1)a^3b^1 + C(4, 2)a^2b^2$$
$$+ C(4, 3)a^1b^3 + C(4, 4)a^0b^4$$

$$= C(4, 0)(3x)^4(-2y)^0 + C(4, 1)(3x)^3(-2y)^1$$
$$+ C(4, 2)(3x)^2(-2y)^2 + C(4, 3)(3x)^1(-2y)^3$$
$$+ C(4, 4)(3x)^0(-2y)^4$$

$$= 3^4x^4 + 4 \cdot 3^3x^3(-2y) + 6 \cdot 3^2x^2(-2)^2y^2$$
$$+ 4(3x)(-2)^3y^3 + (-2)^4y^4$$

$$= 81x^4 - 216x^3y + 216x^2y^2 - 96xy^3 + 16y^4.$$ □

EXAMPLE 4.5.4

Find the coefficient of $a^5 b^4$ in the expansion of $(a + b)^9$.

The term involving $a^5 b^4$ arises in the Binomial Theorem by taking $n = 9$ and $k = 4$:

$$C(n, k)a^{n-k}b^k = C(9, 4)a^5 b^4 = 126 a^5 b^4.$$

Thus the coefficient of $a^5 b^4$ is 126. ◻

EXAMPLE 4.5.5

Find the coefficient of $x^2 y^3 z^4$ in the expansion of $(x + y + z)^9$.

Since

$$(x + y + z)^9 = (x + y + z)(x + y + z) \cdots (x + y + z) \quad \text{(nine terms)},$$

we obtain $x^2 y^3 z^4$ each time we multiply together x chosen from two of the nine terms, y chosen from three of the nine terms, and z chosen from four of the nine terms. We can choose two terms for the x's in $C(9, 2)$ ways. Having made this selection, we can choose three terms for the y's in $C(7, 3)$ ways. This leaves the remaining four terms for the z's. Thus the coefficient of $x^2 y^3 z^4$ in the expansion of $(x + y + z)^9$ is

$$C(9, 2)C(7, 3) = \frac{9!}{2! \, 7!} \frac{7!}{3! \, 4!} = \frac{9!}{2! \, 3! \, 4!} = 1260. \quad ◻$$

We can write the binomial coefficients in a triangular form known as **Pascal's triangle** (see Figure 4.5.1). The border consists of 1's and any interior value is the sum of the two numbers above it. This relationship is stated formally in the next theorem. The proof is a combinatorial argument. An identity that results from some counting process is called a **combinatorial identity** and the argument that leads to its formulation is called a **combinatorial argument**.

```
              1
            1   1
          1   2   1
        1   3   3   1
      1   4   6   4   1
    1   5  10  10   5   1
   .        .        .
  .        .        .
 .        .        .
```

FIGURE 4.5.1
Pascal's triangle.

THEOREM 4.5.6

$$C(n + 1, k) = C(n, k - 1) + C(n, k)$$

for $1 \le k \le n$.

Proof. Let X be a set with n elements. Choose $a \notin X$. Then $C(n + 1, k)$ is the number of k-element subsets of $Y = X \cup \{a\}$. Now the k-element subsets of Y can be divided into two disjoint classes:

1. Subsets of Y not containing a.
2. Subsets of Y containing a.

The subsets of class 1 are just k-element subsets of X and there are $C(n, k)$ of these. Each subset of class 2 consists of a $(k-1)$-element subset of X together with a and there are $C(n, k - 1)$ of these. Therefore,

$$C(n + 1, k) = C(n, k - 1) + C(n, k). \quad ∎$$

Theorem 4.5.6 can also be proved using Theorem 4.2.17 (Exercise 17).

We conclude by showing how the Binomial Theorem (Theorem 4.5.1) and Theorem 4.5.6 can be used to derive other combinatorial identities.

EXAMPLE 4.5.7

Use the Binomial Theorem to derive the equation

$$\sum_{k=0}^{n} C(n, k) = 2^n.$$

The sum is the same as the sum in the Binomial Theorem

$$\sum_{k=0}^{n} C(n, k) a^{n-k} b^k,$$

except that the expression $a^{n-k} b^k$ is missing. One way to "eliminate" this expression is to take $a = b = 1$, in which case the Binomial Theorem becomes

$$2^n = (1 + 1)^n = \sum_{k=0}^{n} C(n, k) 1^{n-k} 1^k = \sum_{k=0}^{n} C(n, k). \qquad (4.5.3)$$

$\square$

Equation (4.5.3) can also be proved by giving a combinatorial argument. Given an n-element set X, $C(n, k)$ counts the number of k-element subsets. Thus the right side of equation (4.5.3) counts the number of subsets of X. But the number of subsets of X is 2^n; we have reproved equation (4.5.3).

EXAMPLE 4.5.8

Use Theorem 4.5.6 to show that

$$\sum_{i=k}^{n} C(i, k) = C(n + 1, k + 1). \qquad (4.5.4)$$

We use Theorem 4.5.6 in the form

$$C(i, k) = C(i + 1, k + 1) - C(i, k + 1)$$

to obtain

$$C(k, k) + C(k + 1, k) + C(k + 2, k) + \cdots + C(n, k)$$
$$= 1 + C(k + 2, k + 1) - C(k + 1, k + 1) + C(k + 3, k + 1)$$
$$\quad - C(k + 2, k + 1) + \cdots + C(n + 1, k + 1) - C(n, k + 1)$$
$$= C(n + 1, k + 1). \qquad \square$$

Exercise 36, Section 4.4, shows another way to prove (4.5.4).

EXAMPLE 4.5.9

Use equation (4.5.4) to find the sum

$$1 + 2 + \cdots + n.$$

We may write

$$1 + 2 + \cdots + n = C(1, 1) + C(2, 1) + \cdots + C(n, 1)$$
$$= C(n + 1, 2) \qquad \text{by equation (4.5.4)}$$
$$= \frac{(n + 1)n}{2}.$$

$\square$

Exercises

1. Expand $(x + y)^4$ using the Binomial Theorem.
2. Expand $(2c - 3d)^5$ using the Binomial Theorem.

In Exercises 3–9, find the coefficient of the term when the expression is expanded.

3. $x^4 y^7$; $(x + y)^{11}$
4. $s^6 t^6$; $(2s - t)^{12}$
5. $x^2 y^3 z^5$; $(x + y + z)^{10}$
6. $w^2 x^3 y^2 z^5$; $(2w + x + 3y + z)^{12}$
7. $a^2 x^3$; $(a + x + c)^2 (a + x + d)^3$
8. $a^2 x^3$; $(a + ax + x)(a + x)^4$
9. $a^3 x^4$; $(a + \sqrt{ax} + x)^2 (a + x)^5$

In Exercises 10–12, find the number of terms in the expansion of each expression.

10. $(x + y + z)^{10}$
11. $(w + x + y + z)^{12}$
★ 12. $(x + y + z)^{10} (w + x + y + z)^2$

13. Find the next row of Pascal's triangle given the row

$$1 \quad 7 \quad 21 \quad 35 \quad 35 \quad 21 \quad 7 \quad 1.$$

14. (a) Show that $C(n, k) < C(n, k + 1)$ if and only if $k < (n - 1)/2$.
    (b) Use part (a) to deduce that the maximum of $C(n, k)$ for $k = 0, 1, \ldots, n$ is $C(n, \lfloor n/2 \rfloor)$.

15. Use the Binomial Theorem to show that

$$0 = \sum_{k=0}^{n} (-1)^k C(n, k).$$

16. Use induction on n to prove the Binomial Theorem.
17. Prove Theorem 4.5.6 by using Theorem 4.2.17.
18. Give a combinatorial argument to show that

$$C(n, k) = C(n, n - k).$$

★ 19. Prove equation (4.5.4) by giving a combinatorial argument.

20. Find the sum
$$1 \cdot 2 + 2 \cdot 3 + \cdots + (n-1)n.$$

★ **21.** Use equation (4.5.4) to derive a formula for
$$1^2 + 2^2 + \cdots + n^2.$$

22. Use the Binomial Theorem to show that
$$\sum_{k=0}^{n} 2^k C(n, k) = 3^n.$$

23. Suppose that n is even. Prove that
$$\sum_{k=0}^{n/2} C(n, 2k) = 2^{n-1} = \sum_{k=1}^{n/2} C(n, 2k - 1).$$

24. Prove
$$(a + b + c)^n = \sum_{0 \le i + j \le n} \frac{n!}{i!\, j!\, (n - i - j)!} a^i b^j c^{n-i-j}.$$

25. Use Exercise 24 to write the expansion of $(x + y + z)^3$.

26. Prove
$$3^n = \sum_{0 \le i + j \le n} \frac{n!}{i!\, j!\, (n - i - j)!}.$$

★ **27.** Give a combinatorial argument to prove that
$$\sum_{k=0}^{n} C(n, k)^2 = C(2n, n).$$

28. Prove
$$n(1 + x)^{n-1} = \sum_{k=1}^{n} C(n, k) k x^{k-1}.$$

29. Use the result of Exercise 28 to show that
$$n 2^{n-1} = \sum_{k=1}^{n} k C(n, k). \tag{4.5.5}$$

★ **30.** Prove equation (4.5.5) by induction.

31. A *smoothing sequence* $b_0, \ldots, b_{k-1}$ is a (finite) sequence satisfying $b_i \ge 0$ for $i = 0, \ldots, k - 1$, and $\sum_{i=0}^{k-1} b_i = 1$. A *smoothing of the (infinite) sequence* $a_1, a_2, \ldots$ *by the smoothing sequence* $b_0, \ldots, b_{k-1}$ is the sequence $\{a_j'\}$ defined by
$$a_j' = \sum_{i=0}^{k-1} a_{i+j} b_i.$$

The idea is that averaging smooths noisy data.

  The *binomial smoother of size k* is the sequence
$$\frac{B_0}{2^n}, \ldots, \frac{B_{k-1}}{2^n},$$
where $B_0, \ldots, B_{k-1}$ is row n of Pascal's triangle (row 0 being the top row).

  Let c_0, c_1 be the smoothing sequence defined by $c_0 = c_1 = \frac{1}{2}$. Show that if a sequence a is smoothed by c, the resulting sequence is smoothed by c, and so on k times; then, the sequence that results can be obtained by one smoothing of a by the binomial smoother of size $k + 1$.

32. In Example 4.1.5 we showed that there are 3^n ordered pairs (A, B) satisfying $A \subseteq B \subseteq X$, where X is an n-element set. Derive this result by considering the cases $|A| = 0, |A| = 1, \ldots, |A| = n$, and then using the Binomial Theorem.

4.6 *THE PIGEONHOLE PRINCIPLE*

The **Pigeonhole Principle** (also known as the *Dirichlet Drawer Principle* or the *Shoe Box Principle*) is sometimes useful in answering the question: Is there an item having a given property? When the Pigeonhole Principle is successfully applied, the Principle tells us only that the object exists; the Principle will not tell us how to find the object or how many there are.

The first version of the Pigeonhole Principle that we will discuss asserts that if n pigeons fly into k pigeonholes and $k < n$, some pigeonhole contains at least two pigeons (see Figure 4.6.1). The reason this statement is true can be seen by arguing by contradiction. If the conclusion is false, each pigeonhole contains at most one pigeon and, in this case, we can account for at most k pigeons. Since there are n pigeons and $n > k$, we have a contradiction.

PIGEONHOLE PRINCIPLE   *(First Form)*

If n pigeons fly into k pigeonholes and $k < n$, some pigeonhole contains at least two pigeons.

We note that the Pigeonhole Principle tells us nothing about how to locate the pigeonhole that contains two or more pigeons. It only asserts the *existence* of a pigeonhole containing two or more pigeons.

To apply the Pigeonhole Principle, we must decide which objects will play the roles of the pigeons and which objects will play the roles of the pigeonholes. Our first example illustrates one possibility.

FIGURE 4.6.1 $n = 6$ pigeons in $k = 4$ pigeonholes. Some pigeonhole contains at least two pigeons.

EXAMPLE 4.6.1

Ten persons have first names Alice, Bernard, and Charles and last names Lee, McDuff, and Ng. Show that at least two persons have the same first and last names.

There are nine possible names for the 10 persons. If we think of the persons as pigeons and the names as pigeonholes, we can consider the assignment of names to people to be that of assigning pigeonholes to the pigeons. By the Pigeonhole Principle some name (pigeonhole) is assigned to at least two persons (pigeons). □

We next restate the Pigeonhole Principle in an alternative form.

PIGEONHOLE PRINCIPLE   *(Second Form)*

If f is a function from a finite set X to a finite set Y and $|X| > |Y|$, then $f(x_1) = f(x_2)$ for some $x_1, x_2 \in X$, $x_1 \neq x_2$.

The second form of the Pigeonhole Principle can be reduced to the first form by letting X be the set of pigeons, and Y be the set of pigeonholes. We assign pigeon x to pigeonhole $f(x)$. By the first form of the Pigeonhole Principle, at least two pigeons, $x_1, x_2 \in X$, are assigned to the same pigeonhole; that is, $f(x_1) = f(x_2)$ for some $x_1, x_2 \in X$, $x_1 \neq x_2$.

Our next examples illustrate the use of the second form of the Pigeonhole Principle.

EXAMPLE 4.6.2

If 20 processors are interconnected, show that at least two processors are directly connected to the same number of processors.

Designate the processors $1, 2, \ldots, 20$. Let a_i be the number of processors to which processor i is directly connected. We are to show that $a_i = a_j$, for some $i \neq j$. The domain of the function a is $X = \{1, 2, \ldots, 20\}$ and the range Y is some subset of $\{0, 1, \ldots, 19\}$. Unfortunately, $|X| = |\{0, 1, \ldots, 19\}|$ and we cannot immediately use the second form of the Pigeonhole Principle.

Let us examine the situation more closely. Notice that we cannot have $a_i = 0$, for some i, *and* $a_j = 19$, for some j, for then we would have one processor (the ith processor) not connected to any other processor while, at the same time, some other processor (the jth processor) is connected to all the other processors (including the ith processor). Thus the range Y is a subset of either $\{0, 1, \ldots, 18\}$ or $\{1, 2, \ldots, 19\}$. In either case, $|Y| < 20 = |X|$. By the second form of the Pigeonhole Principle, $a_i = a_j$, for some $i \neq j$, as desired.

$\square$

EXAMPLE 4.6.3

Show that if we select 151 distinct computer science courses numbered between 1 and 300 inclusive, at least two are consecutively numbered.

Let the selected course numbers be

$$c_1, \quad c_2, \quad \ldots, \quad c_{151}. \qquad (4.6.1)$$

The 302 numbers consisting of (4.6.1) together with

$$c_1 + 1, \quad c_2 + 1, \quad \ldots, \quad c_{151} + 1 \qquad (4.6.2)$$

range in value between 1 and 301. By the second form of the Pigeonhole Principle, at least two of these values coincide. The numbers (4.6.1) are all distinct and hence the numbers (4.6.2) are also distinct. It must then be that one of (4.6.1) and one of (4.6.2) are equal. Thus we have

$$c_i = c_j + 1$$

and course c_i follows course c_j. $\square$

> ### EXAMPLE 4.6.4

An inventory consists of a list of 80 items, each marked "available" or "unavailable." There are 45 available items. Show that there are at least two available items in the list exactly nine items apart. (For example, available items at positions 13 and 22 or positions 69 and 78 satisfy the condition.)

Let a_i denote the position of the ith available item. We must show that $a_i - a_j = 9$ for some i and j. Consider the numbers

$$a_1, \quad a_2, \quad \ldots, \quad a_{45} \tag{4.6.3}$$

and

$$a_1 + 9, \quad a_2 + 9, \quad \ldots, \quad a_{45} + 9. \tag{4.6.4}$$

The 90 numbers in (4.6.3) and (4.6.4) have possible values only from 1 to 89. By the second form of the Pigeonhole Principle, two of the numbers must coincide. We cannot have two of (4.6.3) or two of (4.6.4) identical; thus some number in (4.6.3) is equal to some number in (4.6.4). Therefore, $a_i - a_j = 9$ for some i and j, as desired.   □

We next state yet another form of the Pigeonhole Principle.

> ### PIGEONHOLE PRINCIPLE   *(Third Form)*

Let f be a function from a finite set X into a finite set Y. Suppose that $|X| = n$ and $|Y| = m$. Let $k = \lceil n/m \rceil$. Then there are at least k values $a_1, \ldots, a_k \in X$ such that

$$f(a_1) = f(a_2) = \cdots = f(a_k).$$

To prove the third form of the Pigeonhole Principle, we argue by contradiction. Let $Y = \{y_1, \ldots, y_m\}$. Suppose that the conclusion is false. Then there are at most $k - 1$ values $x \in X$ with $f(x) = y_1$; there are at most $k - 1$ values $x \in X$ with $f(x) = y_2$; $\ldots$; there are at most $k - 1$ values $x \in X$ with $f(x) = y_m$. Thus there are at most $m(k-1)$ members in the domain of f. But

$$m(k - 1) < m\frac{n}{m} = n,$$

which is a contradiction. Therefore, there are at least k values, $a_1, \ldots, a_k \in X$ such that

$$f(a_1) = f(a_2) = \cdots = f(a_k).$$

Our last example illustrates the use of the third form of the Pigeonhole Principle.

> ### EXAMPLE 4.6.5

A useful feature of black-and-white pictures is the average brightness of the picture. Let us say that two pictures are similar if their average brightness differs by no more than some fixed value. Show that among six pictures, there are either three that are mutually similar or there are three that are mutually dissimilar.

Denote the pictures $P_1, P_2, \ldots, P_6$. Each of the five pairs

$$(P_1, P_2), \quad (P_1, P_3), \quad (P_1, P_4), \quad (P_1, P_5), \quad (P_1, P_6),$$

has the value "similar" or "dissimilar." By the third form of the Pigeonhole Principle, there are at least $\lceil 5/2 \rceil = 3$ pairs with the same value; that is, there are three pairs

$$(P_1, P_i), \quad (P_1, P_j), \quad (P_1, P_k)$$

all similar or all dissimilar. Suppose that each pair is similar. (The case that each pair is dissimilar is Exercise 8.) If any pair

$$(P_i, P_j), \quad (P_i, P_k), \quad (P_j, P_k) \tag{4.6.5}$$

is similar, then these two pictures together with P_1 are mutually similar and we have found three mutually similar pictures. Otherwise, each of the pairs (4.6.5) is dissimilar and we have found three mutually dissimilar pictures. $\qquad\square$

୬ ୬ ୬

Exercises

1. Thirteen persons have first names Dennis, Evita, and Ferdinand and last names Oh, Pietro, Quine, and Rostenkowski. Show that at least two persons have the same first and last names.

2. Eighteen persons have first names Alfie, Ben, and Cissi and last names Dumont and Elm. Show that at least three persons have the same first and last names.

3. Professor Euclid is paid every other week on Friday. Show that in some month she is paid three times.

4. Is it possible to interconnect five processors so that exactly two processors are directly connected to an identical number of processors? Explain.

5. An inventory consists of a list of 115 items, each marked "available" or "unavailable." There are 60 available items. Show that there are at least two available items in the list exactly four items apart.

6. An inventory consists of a list of 100 items, each marked "available" or "unavailable." There are 55 available items. Show that there are at least two available items in the list exactly nine items apart.

★ 7. An inventory consists of a list of 80 items, each marked "available" or "unavailable." There are 50 available items. Show that there are at least two unavailable items in the list either three or six items apart.

8. Complete Example 4.6.5 by showing that if the pairs $(P_1, P_i), (P_1, P_j), (P_1, P_k)$ are dissimilar, there are three pictures that are mutually similar or mutually dissimilar.

9. Does the conclusion to Example 4.6.5 necessarily follow if there are fewer than six pictures? Explain.

10. Does the conclusion to Example 4.6.5 necessarily follow if there are more than six pictures? Explain.

Answer Exercises 11–14 to give an argument that shows that if X is any $(n + 2)$-element subset of $\{1, 2, \ldots, 2n + 1\}$ and m is the greatest element in X, there exist distinct i and j in X with $m = i + j$.

For each element $k \in X - \{m\}$, let

$$a_k = \begin{cases} k & \text{if } k \le \dfrac{m}{2} \\ m - k & \text{if } k > \dfrac{m}{2}. \end{cases}$$

11. How many elements are in the domain of a?

12. Show that the range of a is contained in $\{1, 2, \ldots, n\}$.

13. Explain why Exercises 11 and 12 imply that $a_i = a_j$ for some $i \ne j$.

14. Explain why Exercise 13 implies that there exist distinct i and j in X with $m = i + j$.

15. Give an example of an $(n + 1)$-element subset X of $\{1, 2, \ldots, 2n + 1\}$ having the property: For no distinct $i, j \in X$ do we have $i + j \in X$.

Answer Exercises 16–19 to give an argument that proves the following result.

A sequence $a_1, a_2, \ldots, a_{n^2+1}$ of $n^2 + 1$ distinct numbers contains either an increasing subsequence of length $n + 1$ or a decreasing subsequence of length $n + 1$.

Suppose by way of contradiction that every increasing or decreasing subsequence has length n or less. Let b_i be the length of a longest increasing subsequence starting at a_i and let c_i be the length of a longest decreasing subsequence starting at a_i.

16. Show that the ordered pairs (b_i, c_i), $i = 1, \ldots, n^2 + 1$, are distinct.

17. How many ordered pairs (b_i, c_i) are there?

18. Explain why $1 \le b_i \le n$ and $1 \le c_i \le n$.

19. What is the contradiction?

Answer Exercises 20–23 to give an argument that shows that in a group of 10 persons there are at least two such that either the difference or sum of their ages is divisible by 16. Assume that the ages are given as whole numbers.

Let $a_1, \ldots, a_{10}$ denote the ages. Let $r_i = a_i \bmod 16$ and let

$$s_i = \begin{cases} r_i & \text{if } r_i \le 8 \\ 16 - r_i & \text{if } r_i > 8. \end{cases}$$

20. Show that $s_1, \ldots, s_{10}$ range in value from 0 to 8.

21. Explain why $s_j = s_k$ for some $j \ne k$.

22. Explain why if $s_j = r_j$ and $s_k = r_k$ or $s_j = 16 - r_j$ and $s_k = 16 - r_k$, then 16 divides $a_j - a_k$.

23. Show that if the conditions in Exercise 22 fail, then 16 divides $a_j + a_k$.

24. Show that in the decimal expansion of the quotient of two integers, eventually some block of digits repeats. *Examples:*

$$\frac{1}{6} = 0.1\underline{6}66\ldots, \qquad \frac{217}{660} = 0.32\underline{87}8787\ldots.$$

★ **25.** Twelve basketball players, whose uniforms are numbered 1 through 12, stand around the center ring on the court in an arbitrary arrangement. Show that some three consecutive players have the sum of their numbers at least 20.

★ **26.** For the situation of Exercise 25, find and prove an estimate for how large the sum of some four consecutive players' numbers must be.

★ **27.** Let f be a one-to-one function from $X = \{1, 2, \ldots, n\}$ onto X. Let $f^k = f \circ f \circ \cdots \circ f$ denote the k-fold composition of f with itself. Show that there are distinct positive integers i and j such that $f^i(x) = f^j(x)$ for all $x \in X$. Show that for some positive integer k, $f^k(x) = x$ for all $x \in X$.

★ **28.** A 3×7 rectangle is divided into 21 squares each of which is colored red or black. Prove that the board contains a nontrivial rectangle (not $1 \times k$ or $k \times 1$) whose four corner squares are all black or all red.

★ **29.** Prove that if p ones and q zeros are placed around a circle in an arbitrary manner, where p, q, and k are positive integers satisfying $p \geq kq$, the arrangement must contain at least k consecutive ones.

★ **30.** Write an algorithm that, given a sequence a, finds the length of a longest increasing subsequence of a.

✎ NOTES

An elementary book concerning counting methods is [Niven]. References on combinatorics are [Brualdi; Even, 1973; Liu, 1968; Riordan; and Roberts]. [Vilenkin] contains many worked-out combinatorial examples. The general discrete mathematical references [Liu, 1985; and Tucker] devote several sections to the topics of Chapter 4. [Even, 1973; Hu; and Reingold] treat combinatorial algorithms.

✎ CHAPTER REVIEW

Section 4.1

Multiplication Principle

Addition Principle

Section 4.2

Permutation of $x_1, \ldots, x_n$: Ordering of $x_1, \ldots, x_n$

$n!$ = number of permutations of an n-element set

r-permutation of $x_1, \ldots, x_n$: Ordering of r elements of $x_1, \ldots, x_n$

$P(n, r)$: Number of r-permutations of an n-element set; $P(n, r) = n(n - 1) \cdots (n - r + 1)$

r-combination of $\{x_1, \ldots, x_n\}$: (unordered) subset of $\{x_1, \ldots, x_n\}$ containing r elements

$C(n, r)$: Number of r-combinations of an n-element set; $C(n, r) = P(n, r)/r! = n!/[(n - r)!\, r!]$

Section 4.3

Lexicographic order

Algorithm for generating r-combinations: Algorithm 4.3.9

Algorithm for generating permutations: Algorithm 4.3.14

Section 4.4

Number of orderings of n items of t types with n_i identical objects of type $i = n!/[n_1! \cdots n_t!]$

Number of unordered, k-element selections from a t-element set, repetitions allowed $= C(k + t - 1, k)$

Section 4.5

Binomial Theorem: $(a + b)^n = \sum_{k=0}^n C(n, k) a^{n-k} b^k$

Pascal's triangle: $C(n + 1, k) = C(n, k - 1) + C(n, k)$

Section 4.6

Pigeonhole Principle (three forms)

↪ CHAPTER SELF-TEST

Section 4.1

1. How many eight-bit strings begin with 0 and end with 101?

2. How many ways can we select three books each from a different subject from a set of six distinct history books, nine distinct classics books, seven distinct law books, and four distinct education books?

3. How many functions are there from an n-element set onto $\{0, 1\}$?

4. A seven-person committee composed of Greg, Hwang, Isaac, Jasmine, Kirk, Lynn, and Manuel is to select a chairperson, vice-chairperson, social events chairperson, secretary, and treasurer. How many ways can the officers be chosen if either Greg is secretary or he is not an officer?

Section 4.2

5. How many 3-combinations are there of six objects?

6. How many strings can be formed by ordering the letters $ABCDEF$ if A appears before C and E appears before C?

7. How many six-card hands chosen from an ordinary 52-card deck contain three cards of one suit and three cards of another suit?

8. A shipment of 100 compact disks contains five defective disks. In how many ways can we select a set of four compact disks that contains more defective than nondefective disks?

Section 4.3

9. Find the 5-combination that will be generated by Algorithm 4.3.9 after 12467 if $n = 7$.

10. Find the 6-combination that will be generated by Algorithm 4.3.9 after 145678 if $n = 8$.

11. Find the permutation that will be generated by Algorithm 4.3.14 after 6427135.

12. Find the permutation that will be generated by Algorithm 4.3.14 after 625431.

Section 4.4

13. How many strings can be formed by ordering the letters $ILLINOIS$?

14. How many strings can be formed by ordering the letters $ILLINOIS$ if some I appears before some L?

15. In how many ways can 12 distinct books be divided among four students if each student gets three books?

16. How many integer solutions of

$$x_1 + x_2 + x_3 + x_4 = 17$$

satisfy $x_1 \geq 0, x_2 \geq 1, x_4 \geq 3$?

Section 4.5

17. Expand the expression $(s - r)^4$ using the Binomial Theorem.

18. Find the coefficient of x^3yz^4 in the expansion of $(2x + y + z)^8$.

19. Use the Binomial Theorem to prove that

$$\sum_{k=0}^{n} 2^{n-k}(-1)^k C(n, k) = 1.$$

20. Rotate Pascal's triangle counterclockwise so that the top row consists of 1's. Explain why the second row lists the positive integers in order $1, 2, \ldots$.

Section 4.6

21. Show that every set of 15 socks chosen from among 14 pairs of socks contains at least one matched pair.

22. Nineteen persons have first names Zeke, Wally, and Linda; middle names Lee and David; and last names Yu, Zamora, and Smith. Show that at least two persons have the same first, middle, and last names.

23. An inventory consists of a list of 200 items, each marked "available" or "unavailable." There are 110 available items. Show that there are at least two available items in the list exactly 19 items apart.

24. Let $P = \{p_1, p_2, p_3, p_4, p_5\}$ be a set of five (distinct) points in the ordinary Euclidean plane each of which has integer coordinates. Show that some pair has a midpoint that has integer coordinates.

RECURRENCE RELATIONS

*Y*ou want to tell me now?

Tell you what?

What it is you're trying to find out. You know, it's a funny thing. You're trying to find out what your father hired me to find out and I'm trying to find out why you want to find out.

You could go on forever, couldn't you?

—from The Big Sleep

This chapter offers an introduction to recurrence relations. Recurrence relations are useful in certain counting problems. A recurrence relation relates the nth element of a sequence to its predecessors. Because recurrence relations are closely related to recursive algorithms, recurrence relations arise naturally in the analysis of recursive algorithms.

5.1 INTRODUCTION

Consider the following instructions for generating a sequence:

1. Start with 5.
2. Given any term, add 3 to get the next term.

If we list the terms of the sequence, we obtain

$$5, \quad 8, \quad 11, \quad 14, \quad 17, \quad \ldots . \tag{5.1.1}$$

The first term is 5 because of instruction 1. The second term is 8 because instruction 2 says to add 3 to 5 to get the next term, 8. The third term is 11 because instruction 2 says to add 3 to 8 to get the next term, 11. By following instructions 1 and 2, we can compute any term in the sequence. Instructions 1 and 2 do not give an explicit formula for the nth term of the sequence in the sense of providing a formula that we can "plug n into" to obtain the value of the nth term, but by computing term by term we can eventually compute any term of the sequence.

If we denote the sequence (5.1.1) as $a_1, a_2, \ldots$, we may rephrase instruction 1 as

$$a_1 = 5 \qquad\qquad (5.1.2)$$

and we may rephrase instruction 2 as

$$a_n = a_{n-1} + 3, \qquad n \geq 2. \qquad\qquad (5.1.3)$$

Taking $n = 2$ in (5.1.3), we obtain

$$a_2 = a_1 + 3.$$

By (5.1.2), $a_1 = 5$; thus

$$a_2 = a_1 + 3 = 5 + 3 = 8.$$

Taking $n = 3$ in (5.1.3), we obtain

$$a_3 = a_2 + 3.$$

Since $a_2 = 8$,

$$a_3 = a_2 + 3 = 8 + 3 = 11.$$

By using (5.1.2) and (5.1.3), we can compute any term in the sequence just as we did using instructions 1 and 2. We see that (5.1.2) and (5.1.3) are equivalent to instructions 1 and 2.

Equation (5.1.3) furnishes an example of a **recurrence relation**. A recurrence relation defines a sequence by giving the nth value in terms of certain of its predecessors. In (5.1.3), the nth value is given in terms of the immediately preceding value. In order for a recurrence relation such as (5.1.3) to define a sequence, a "start-up" value or values, such as (5.1.2), must be given. These start-up values are called **initial conditions**. The formal definitions follow.

DEFINITION 5.1.1

A *recurrence relation* for the sequence $a_0, a_1, \ldots$ is an equation that relates a_n to certain of its predecessors $a_0, a_1, \ldots, a_{n-1}$.

Initial conditions for the sequence $a_0, a_1, \ldots$ are explicitly given values for a finite number of the terms of the sequence.

We have seen that it is possible to define a sequence by a recurrence relation together with certain initial conditions. We give several examples of recurrence relations.

EXAMPLE 5.1.2

The Fibonacci sequence (see the discussion following Algorithm 3.4.7) is defined by the recurrence relation

$$f_n = f_{n-1} + f_{n-2}, \qquad n \geq 3$$

and initial conditions

$$f_1 = 1, \qquad f_2 = 2. \qquad\qquad \square$$

EXAMPLE 5.1.3

A person invests \$1000 at 12 percent compounded annually. If A_n represents the amount at the end of n years, find a recurrence relation and initial conditions that define the sequence $\{A_n\}$.

At the end of $n - 1$ years, the amount is A_{n-1}. After one more year, we will have the amount A_{n-1} plus the interest. Thus

$$A_n = A_{n-1} + (0.12)A_{n-1} = (1.12)A_{n-1}, \qquad n \geq 1. \qquad (5.1.4)$$

To apply this recurrence relation for $n = 1$, we need to know the value of A_0. Since A_0 is the beginning amount, we have the initial condition

$$A_0 = 1000. \qquad (5.1.5)$$

□

The initial condition (5.1.5) and the recurrence relation (5.1.4) allow us to compute the value of A_n for any n. For example,

$$
\begin{aligned}
A_3 &= (1.12)A_2 = (1.12)(1.12)A_1 \\
&= (1.12)(1.12)(1.12)A_0 = (1.12)^3(1000) = 1404.93. \qquad (5.1.6)
\end{aligned}
$$

Thus, at the end of the third year, the amount is \$1404.93.

The computation (5.1.6) can be carried out for an arbitrary value of n to obtain

$$A_n = (1.12)A_{n-1}$$

$$\vdots$$

$$= (1.12)^n(1000).$$

We see that sometimes an explicit formula can be derived from a recurrence relation and initial conditions. Finding explicit formulas from recurrence relations is the topic of Section 5.2.

Although it is easy to obtain an explicit formula from the recurrence relation and initial condition for the sequence of Example 5.1.3, it is not immediately apparent how to obtain an explicit formula for the Fibonacci sequence. In Section 5.2 we give a method that yields an explicit formula for the Fibonacci sequence.

Recurrence relations, recursive algorithms, and mathematical induction are closely related. In all three, prior instances of the current case are assumed known. A recurrence relation uses prior values in a sequence to compute the current value. A recursive algorithm uses smaller instances of the current input to process the current input. The inductive step in a proof by mathematical induction assumes the truth of prior instances of the statement to prove the truth of the current statement.

A recurrence relation that defines a sequence can be directly converted to an algorithm to compute the sequence. For example, Algorithm 5.1.4, derived from recurrence relation (5.1.4) and initial condition (5.1.5), computes the sequence of Example 5.1.3.

ALGORITHM 5.1.4   *Computing Compound Interest*

This recursive algorithm computes the amount of money at the end of n years assuming an initial amount of \$1000 and an interest rate of 12 percent compounded annually.

Input: n, the number of years

Output: The amount of money at the end of n years

1.   **procedure** *compound_interest*(n)
2.   **if** $n = 0$ **then**
3.   **return**(1000)
4.   **return**$(1.12 * compound\_interest(n-1))$
5.   **end** *compound_interest*

Algorithm 5.1.4 is a direct translation of equations (5.1.4) and (5.1.5) that define the sequence $A_0, A_1, \dots$. Lines 2 and 3 correspond to initial condition (5.1.5) and line 4 corresponds to recurrence relation (5.1.4).

EXAMPLE 5.1.5

Let S_n denote the number of subsets of an n-element set. Since going from an $(n-1)$-element set to an n-element set doubles the number of subsets (see Theorem 2.1.4), we obtain the recurrence relation

$$S_n = 2S_{n-1}.$$

The initial condition is
$$S_0 = 1. \qquad \square$$

One of the main reasons for using recurrence relations is that sometimes it is easier to determine the nth term of a sequence in terms of its predecessors than it is to find an explicit formula for the nth term in terms of n. The next examples are intended to illustrate this thesis.

EXAMPLE 5.1.6

Let S_n denote the number of n-bit strings that do not contain the pattern 111. Develop a recurrence relation for $S_1, S_2, \dots$, and initial conditions that define the sequence S.

We will count the number of n-bit strings that do not contain the pattern 111

(a) that begin with 0;
(b) that begin with 10;
(c) that begin with 11.

Since the sets of strings of types (a), (b), and (c) are disjoint, by the Addition Principle S_n will equal the sum of the numbers of strings of types (a), (b), and (c). Suppose that an n-bit string begins with 0 and does not contain the pattern 111. Then the $(n-1)$-bit string following the initial 0 does not contain the pattern 111. Since any $(n-1)$-bit string not containing 111 can follow the initial 0, there are S_{n-1} strings of type (a). If an n-bit string begins with 10 and does not contain the pattern 111, then the $(n-2)$-bit string following the initial 10 cannot contain the pattern 111; therefore, there are S_{n-2} strings of type (b). If an n-bit string begins with 11 and does not contain the pattern 111, then the third bit must be 0. The $(n-3)$-bit string following the initial 110 cannot contain the pattern 111; therefore, there are S_{n-3} strings of type (c). Thus

$$S_n = S_{n-1} + S_{n-2} + S_{n-3}, \qquad n \geq 4.$$

By inspection, we find the initial conditions

$$S_1 = 2, \quad S_2 = 4, \quad S_3 = 7. \qquad\qquad \square$$

EXAMPLE 5.1.7

Recall (see Example 4.2.23) that the Catalan number C_n is equal to the number of routes from the lower left corner of an $n \times n$ square grid to the upper right corner if we are restricted to traveling only to the right or upward and if we are allowed to touch but not go above a diagonal line from the lower left corner to the upper right corner. We call such a route a *good route*. We give a recurrence relation for the Catalan numbers.

We divide the good routes into classes based on when they first meet the diagonal after leaving the lower left corner. For example, the route in Figure 5.1.1 first meets the diagonal at the point $(3, 3)$. We regard the routes that first meet the diagonal at (k, k) as constructed by a two-step process: First,

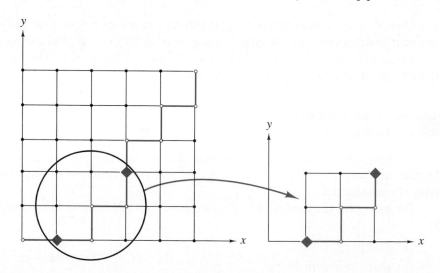

FIGURE 5.1.1   Decomposition of a good route.

construct the part from $(0, 0)$ to (k, k). Second, construct the part from (k, k) to (n, n). A good route always leaves $(0, 0)$ by moving right to $(1, 0)$ and it always arrives at (k, k) by moving up from $(k, k - 1)$. The moves from $(1, 0)$ to $(k, k - 1)$ give a good route in the $(k - 1) \times (k - 1)$ grid with corners at $(1, 0)$, $(1, k - 1)$, $(k, k - 1)$, and $(k, 0)$. [In Figure 5.1.1, we have marked the points $(1, 0)$ and $(k, k - 1)$, $k = 3$, with diamonds and we have isolated the $(k - 1) \times (k - 1)$ subgrid.] Thus there are C_{k-1} routes from $(0, 0)$ to (k, k) that first meet the diagonal at (k, k). The part from (k, k) to (n, n) is a good route in the $(n - k) \times (n - k)$ grid with corners at (k, k), (k, n), (n, n), and (n, k) (see Figure 5.1.1). There are C_{n-k} such routes. By the Multiplication Principle, there are $C_{k-1}C_{n-k}$ good routes in an $n \times n$ grid that first meet the diagonal at (k, k). The good routes that first meet the diagonal at (k, k) are distinct from those that first meet the diagonal at (k', k'), $k \neq k'$. Thus we may use the Addition Principle to obtain a recurrence relation for the total number of good routes in an $n \times n$ grid:

$$C_n = \sum_{k=1}^{n} C_{k-1}C_{n-k}.$$ $\square$

EXAMPLE 5.1.8   *Tower of Hanoi*

The Tower of Hanoi is a puzzle consisting of three pegs mounted on a board and n disks of various sizes with holes in their centers (see Figure 5.1.2). It is assumed that if a disk is on a peg, only a disk of smaller diameter can be placed on top of the first disk. Given all the disks stacked on one peg as in Figure 5.1.2, the problem is to transfer the disks to another peg by moving one disk at a time.

We will give a solution and then find a recurrence relation and an initial condition for the sequence $c_1, c_2, \ldots$, where c_n denotes the number of moves our solution takes to solve the n-disk puzzle. We will then show that our solution is optimal; that is, we will show that no other solution uses fewer moves.

We give a recursive algorithm. If there is only one disk, we simply move it to the desired peg. If we have $n > 1$ disks on peg 1 as in Figure 5.1.2, we begin by recursively invoking our algorithm to move the top $n - 1$ disks to peg 2 (see Figure 5.1.3). During these moves, the bottom disk on peg 1 stays fixed. Next, we move the remaining disk on peg 1 to peg 3. Finally, we again recursively invoke our algorithm to move the $n - 1$ disks on peg 2 to peg 3. We have succeeded in moving n disks from peg 1 to peg 3.

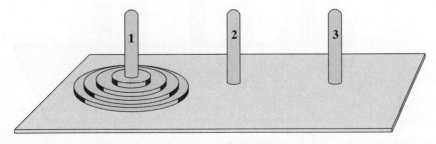

FIGURE 5.1.2   Tower of Hanoi.

If $n > 1$, we solve the $(n-1)$-disk problem twice and we explicitly move one disk. Therefore,

$$c_n = 2c_{n-1} + 1, \quad n > 1.$$

The initial condition is

$$c_1 = 1.$$

In Section 5.2, we will show that $c_n = 2^n - 1$.

We next show that our solution is optimal. We let d_n be the number of moves required by an optimal solution. We use mathematical induction to show that

$$c_n = d_n, \qquad n \geq 1. \tag{5.1.7}$$

BASIS STEP ($n = 1$).   By inspection

$$c_1 = 1 = d_1;$$

thus (5.1.7) is true when $n = 1$.

INDUCTIVE STEP.   Assume that (5.1.7) is true for $n - 1$. Consider the point in an optimal solution to the n-disk problem when the largest disk is moved for the first time. The largest disk must be on a peg by itself (so that it can be removed from the peg) and another peg must be empty (so that this peg can receive the largest disk). Thus the $n-1$ smaller disks must be stacked on a third peg (see Figure 5.1.3). In other words, the $n-1$ disk problem must have been solved, which required at least d_{n-1} moves. The largest disk was then moved, which required one additional move. Finally, at some point the $n-1$ disks were moved on top of the largest disk, which required at least d_{n-1} additional moves. It follows that

$$d_n \geq 2d_{n-1} + 1.$$

By the inductive assumption, $c_{n-1} = d_{n-1}$. Thus

$$d_n \geq 2d_{n-1} + 1 = 2c_{n-1} + 1 = c_n. \tag{5.1.8}$$

The last equality follows from the recurrence relation for the sequence c_1, $c_2, \ldots$. By definition, no solution can take fewer moves than an optimal solution, so

$$c_n \geq d_n. \tag{5.1.9}$$

Inequalities (5.1.8) and (5.1.9) combine to give

$$c_n = d_n.$$

The inductive step is complete. Therefore, our solution is optimal.   □

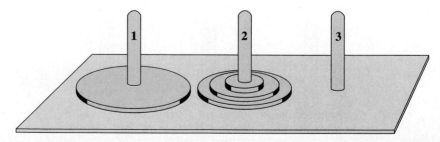

FIGURE 5.1.3   After recursively moving the top $n - 1$ disks from peg 1 to peg 2 in the Tower of Hanoi.

The Tower of Hanoi puzzle was invented by the French mathematician Édouard Lucas in the late nineteenth century. (Lucas was the first person to call the sequence $1, 2, 3, 5, \ldots$ the Fibonacci sequence.) The following myth was also created to accompany the puzzle (and, one assumes, to help market the puzzle). The puzzle was said to be derived from a mythical gold tower that consisted of 64 disks. The 64 disks were to be transferred by monks according to the rules set forth previously. It was said that before the monks finished moving the tower, the tower would collapse and the world would end in a clap of thunder. Since at least $2^{64} - 1 = 18,446,744,073,709,551,615$ moves are required to solve the 64-disk Tower of Hanoi puzzle, we can be fairly certain that something would happen to the tower before it was completely moved.

EXAMPLE 5.1.9   *The Cobweb in Economics*

We assume an economics model in which the supply and demand are given by linear equations (see Figure 5.1.4). Specifically, the demand is given by the equation

$$p = a - bq,$$

where p is the price, q is the quantity, and a and b are positive parameters. The idea is that as the price increases, the consumers demand less of the product. The supply is given by the equation

$$p = kq,$$

where p is the price, q is the quantity, and k is a positive parameter. The idea is that as the price increases, the manufacturer is willing to supply greater quantities.

We assume further that there is a time lag as the supply reacts to changes. (For example, it takes time to manufacture goods and it takes time to grow crops.) We denote the discrete time intervals as $n = 0, 1, \ldots$. We assume that the demand is given by the equation

$$p_n = a - bq_n;$$

that is, at time n, the quantity q_n of the product will be sold at price p_n. We assume that the supply is given by the equation

$$p_n = kq_{n+1}; \tag{5.1.10}$$

that is, one unit of time is required for the manufacturer to adjust the quantity q_{n+1}, at time $n + 1$, to the price p_n, at the prior time n.

If we solve equation (5.1.10) for q_{n+1} and substitute into the demand equation for time $n + 1$,

$$p_{n+1} = a - bq_{n+1},$$

we obtain the recurrence relation

$$p_{n+1} = a - \frac{b}{k}p_n$$

for the price. We will solve this recurrence relation in Section 5.2.

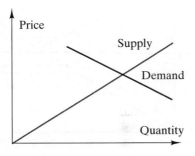

FIGURE 5.1.4
An economics model.

The price changes through time may be viewed graphically. If the initial price is p_0, the manufacturer will be willing to supply the quantity q_1, at time $n = 1$. We locate this quantity by moving horizontally to the supply curve (see Figure 5.1.5). However, the market forces drive the price down to p_1, as we can see by moving vertically to the demand curve. At price p_1, the manufacturer will be willing to supply the quantity q_2 at time $n = 2$, as we can see by moving horizontally to the supply curve. Now the market forces drive the price up to p_2, as we can see by moving vertically to the demand curve. By continuing this process, we obtain the "cobweb" shown in Figure 5.1.5.

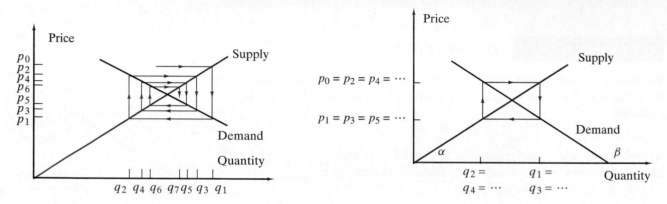

FIGURE 5.1.5 A cobweb with a stabilizing price.   **FIGURE 5.1.6** A cobweb with a fluctuating price.

For the supply and demand functions of Figure 5.1.5, the price approaches that given by the intersection of the supply and demand curves. This is not always the case, however. For example, in Figure 5.1.6, the price fluctuates between p_0 and p_1, whereas in Figure 5.1.7, the price swings become more and more pronounced. The behavior is determined by the slopes of the supply and demand lines. To produce the fluctuating behavior of Figure 5.1.6, the angles α and β must add to $180°$. The slopes of the supply and demand curves are $\tan \alpha$ and $\tan \beta$, respectively; thus in Figure 5.1.6, we have

$$k = \tan \alpha = - \tan \beta = b.$$

We have shown that the price fluctuates between two values when $k = b$. A similar analysis shows that the price tends to that given by the intersection of the supply and demand curves (Figure 5.1.5) when $b < k$ and the increasing

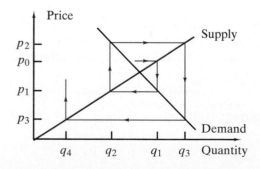

FIGURE 5.1.7 A cobweb with increasing price swings.

price swings case (Figure 5.1.7) occurs when $b > k$ (see Exercises 35 and 36). In Section 5.2 we discuss the behavior of the price through time by analyzing an explicit formula for the price p_n.   □

It is possible to extend the definition of recurrence relation to include functions indexed over n-tuples of positive integers. Our last example is of this form.

> **EXAMPLE 5.1.10**   *Ackermann's Function*

Ackermann's function can be defined by the recurrence relations

$$A(m, 0) = A(m - 1, 1), \qquad\qquad m = 1, 2, \ldots, \qquad (5.1.11)$$
$$A(m, n) = A(m - 1, A(m, n - 1)), \qquad m = 1, 2, \ldots,$$
$$n = 1, 2, \ldots, \qquad (5.1.12)$$

and the initial conditions

$$A(0, n) = n + 1, \qquad n = 0, 1, \ldots. \qquad (5.1.13)$$

Ackermann's function is of theoretical importance because of its rapid rate of growth. Functions related to Ackermann's function appear in the time complexity of certain algorithms such as the time to execute union/find algorithms (see [Tarjan, pp. 22–29]).

The computation

$$
\begin{aligned}
A(1, 1) &= A(0, A(1, 0)) & &\text{by (5.1.12)} \\
&= A(0, A(0, 1)) & &\text{by (5.1.11)} \\
&= A(0, 2) & &\text{by (5.1.13)} \\
&= 3 & &\text{by (5.1.13)}
\end{aligned}
$$

illustrates the use of equations (5.1.11)–(5.1.13).   □

෨ ෨ ෨
Exercises

In Exercises 1–3, find a recurrence relation and initial conditions that generate a sequence that begins with the given terms.

1. $3, 7, 11, 15, \ldots$

2. $3, 6, 9, 15, 24, 39, \ldots$

3. $1, 1, 2, 4, 16, 128, 4096, \ldots$

In Exercises 4–8, assume that a person invests $2000 at 14 percent compounded annually. Let A_n represent the amount at the end of n years.

4. Find a recurrence relation for the sequence $A_0, A_1, \ldots$.

5. Find an initial condition for the sequence $A_0, A_1, \ldots$.

6. Find $A_1, A_2,$ and A_3.

7. Find an explicit formula for A_n.

8. How long will it take for a person to double the initial investment?

If a person invests in a tax-sheltered annuity, the money invested, as well as the interest earned, is not subject to taxation until withdrawn from the account. In Exercises 9–12, assume that a person invests $2000 *each* year in a tax-sheltered annuity at 10 percent compounded annually. Let A_n represent the amount at the end of n years.

9. Find a recurrence relation for the sequence $A_0, A_1, \dots$.

10. Find an initial condition for the sequence $A_0, A_1, \dots$.

11. Find A_1, A_2, and A_3.

12. Find an explicit formula for A_n.

In Exercises 13–17, assume that a person invests $3000 at 12 percent annual interest compounded quarterly. Let A_n represent the amount at the end of n years.

13. Find a recurrence relation for the sequence $A_0, A_1, \dots$.

14. Find an initial condition for the sequence $A_0, A_1, \dots$.

15. Find A_1, A_2, and A_3.

16. Find an explicit formula for A_n.

17. How long will it take for a person to double the initial investment?

18. Let S_n denote the number of n-bit strings that do not contain the pattern 000. Find a recurrence relation and initial conditions for the sequence $\{S_n\}$.

Exercises 19–21 refer to the sequence S where S_n denotes the number of n-bit strings that do not contain the pattern 00.

19. Find a recurrence relation and initial conditions for the sequence$\{S_n\}$.

20. Show that $S_n = f_{n+1}, n = 1, 2, \dots$, where f denotes the Fibonacci sequence.

21. By considering the number of n-bit strings with exactly i 0's and Exercise 20, show that

$$f_{n+1} = \sum_{i=0}^{\lfloor (n+1)/2 \rfloor} C(n+1-i, i), \qquad n = 1, 2, \dots,$$

where f denotes the Fibonacci sequence.

Exercises 22–24 refer to the sequence $S_1, S_2, \dots$, where S_n denotes the number of n-bit strings that do not contain the pattern 010.

22. Compute S_1, S_2, S_3, and S_4.

23. By considering the number of n-bit strings that do not contain the pattern 010 that have no leading 0's (i.e., that begin with 1); that have one leading 0 (i.e., that begin 01); that have two leading 0's; and so on, derive the recurrence relation

$$S_n = S_{n-1} + S_{n-3} + S_{n-4} + S_{n-5} + \dots + S_1 + 3. \qquad (5.1.14)$$

24. By replacing n by $n - 1$ in (5.1.14), write a formula for S_{n-1}. Subtract the formula for S_{n-1} from the formula for S_n and use the result to derive the recurrence relation

$$S_n = 2S_{n-1} - S_{n-2} + S_{n-3}.$$

In Exercises 25–30, $C_0, C_1, C_2, \dots$ denotes the sequence of Catalan numbers.

25. Given that $C_0 = C_1 = 1$ and $C_2 = 2$, compute C_3, C_4, and C_5 by using the recurrence relation of Example 5.1.7.

26. Show that the Catalan numbers are given by the recurrence relation

$$(n + 2)C_{n+1} = (4n + 2)C_n, \qquad n \geq 0,$$

and initial condition $C_0 = 1$.

27. Prove that

$$C_n \geq \frac{4^{n-1}}{n^2} \qquad \text{for all } n \geq 1.$$

28. Derive a recurrence relation and an initial condition for the number of ways to parenthesize the product

$$a_1 * a_2 * \cdots * a_n, \qquad n \geq 2.$$

Examples: There is one way to parenthesize $a_1 * a_2$, namely, $(a_1 * a_2)$. There are two ways to parenthesize $a_1 * a_2 * a_3$, namely, $((a_1 * a_2) * a_3)$ and $(a_1 * (a_2 * a_3))$. Deduce that the number of ways to parenthesize the product of n elements is $C_{n-1}, n \geq 2$.

★ **29.** This is the problem originally investigated by Catalan.

Derive a recurrence relation and an initial condition for the number of ways to divide a convex $(n+2)$-sided polygon, $n \geq 1$, into triangles by drawing $n-1$ lines through the corners that do not intersect in the interior of the polygon. (A polygon is *convex* if any line joining two points in the polygon lies wholly in the polygon.) For example, there are five ways to divide a convex pentagon into triangles by drawing two nonintersecting lines through the corners:

Deduce that the number of ways to divide a convex $(n + 2)$-sided polygon into triangles by drawing $n - 1$ nonintersecting lines through the corners is $C_n, n \geq 1$.

30. By dividing the routes from the lower left corner to the upper right corner in an $(n + 1) \times (n + 1)$ grid in which we are restricted to traveling only to the right or upward into classes based on when, after leaving the lower left corner, the route first meets the diagonal line from the lower left corner to the upper right corner, derive the recurrence relation

$$C_n = \frac{1}{2}C(2(n + 1), n + 1) - \sum_{k=0}^{n-1} C_k C(2(n - k), n - k).$$

In Exercises 31 and 32, let S_n denote the number of routes from the lower left corner of an $n \times n$ grid to the upper right corner in which we are restricted to traveling to the right, upward, or diagonally northeast [i.e., from (i, j) to $(i + 1, j + 1)$] and in which we are allowed to touch but not go above a diagonal line from the lower left corner to the upper right corner. The numbers $S_0, S_1, \ldots$ are called the *Schröder numbers*.

31. Show that $S_0 = 1, S_1 = 2, S_2 = 6,$ and $S_3 = 22$.

32. Derive a recurrence relation for the sequence of Schröder numbers.

33. Write explicit solutions for the Tower of Hanoi puzzle for $n = 3, 4$.

34. To what values do the price and quantity tend in Example 5.1.9 when $b < k$?

35. Show that when $b < k$ in Example 5.1.9, the price tends to that given by the intersection of the supply and demand curves.

36. Show that when $b > k$ in Example 5.1.9, the differences between successive prices increase.

Exercises 37–43 refer to Ackermann's function $A(m, n)$.

37. Compute $A(2, 2)$ and $A(2, 3)$.

38. Use induction to show that

$$A(1, n) = n + 2, \qquad n = 0, 1, \ldots.$$

39. Use induction to show that

$$A(2, n) = 3 + 2n, \qquad n = 0, 1 \ldots.$$

40. Guess a formula for $A(3, n)$ and prove it by using induction.

★ **41.** Prove that $A(m, n) > n$ for all $m \geq 0, n \geq 0$ by induction on m. The inductive step will use induction on n.

42. By using Exercise 41 or otherwise, prove that $A(m, n) > 1$ for all $m \geq 1$, $n \geq 0$.

43. By using Exercise 41 or otherwise, prove that $A(m, n) < A(m, n + 1)$ for all $m \geq 0, n \geq 0$.

What we and others have called Ackermann's function is actually derived from Ackermann's original function defined by

$$
\begin{aligned}
AO(0, y, z) &= z + 1, & y, z &\geq 0 \\
AO(1, y, z) &= y + z, & y, z &\geq 0 \\
AO(2, y, z) &= yz, & y, z &\geq 0 \\
AO(x + 3, y, 0) &= 1, & x, y &\geq 0 \\
AO(x + 3, y, z + 1) &= AO(x + 2, y, AO(x + 3, y, z)), & x, y, z &\geq 0.
\end{aligned}
$$

Exercises 44–47 refer to the function AO and to Ackermann's function A.

44. Show that $A(x, y) = AO(x, 2, y + 3) - 3$ for $y \geq 0$ and $x = 0, 1, 2$.

45. Show that $AO(x, 2, 1) = 2$ for $x \geq 2$.

46. Show that $AO(x, 2, 2) = 4$ for $x \geq 2$.

★ **47.** Show that $A(x, y) = AO(x, 2, y + 3) - 3$ for $x, y \geq 0$.

48. A network consists of n nodes. Each node has communications facilities and local storage. Periodically, all files must be shared. A *link* consists of two nodes sharing files. Specifically, when nodes A and B are linked, A transmits all its files to B and B transmits all its files to A. Only one link exists at a time, and after a link is established and the files are shared, the link is deleted. Let a_n be the minimum number of links required by n nodes so that all files are known to all nodes.

  (a) Show that $a_2 = 1, a_3 \leq 3, a_4 \leq 4$.

  (b) Show that $a_n \leq a_{n-1} + 2, n \geq 3$.

49. If P_n denotes the number of permutations of n distinct objects, find a recurrence relation and an initial condition for the sequence $P_1, P_2, \ldots$.

50. Suppose that we have n dollars and that each day we buy either orange juice (\$1), milk (\$2), or beer (\$2). If R_n is the number of ways of spending all the money, show that

$$R_n = R_{n-1} + 2R_{n-2}.$$

Order is taken into account. For example, there are 11 ways to spend four dollars: *MB, BM, OOM, OOB, OMO, OBO, MOO, BOO, OOOO, MM, BB.*

51. Suppose that we have n dollars and that each day we buy either tape (\$1), paper (\$1), pens (\$2), pencils (\$2), or binders (\$3). If R_n is the number of ways of spending all the money, derive a recurrence relation for the sequence $R_1, R_2, \ldots$.

52. Let R_n denote the number of regions into which the plane is divided by n lines. Assume that each pair of lines meets in a point, but that no three lines meet in a point. Derive a recurrence relation for the sequence $R_1, R_2, \ldots$.

Exercises 53 and 54 refer to the sequence S_n defined by

$$S_1 = 0, \qquad S_2 = 1, \qquad S_n = \frac{S_{n-1} + S_{n-2}}{2}, \qquad n = 3, 4, \ldots.$$

53. Compute S_3 and S_4.

★ 54. Guess a formula for S_n and show that it is correct by using induction.

★ 55. Let F_n denote the number of functions f from $X = \{1, \ldots, n\}$ into X having the property that if i is in the range of f, then $1, 2, \ldots, i - 1$ are also in the range of f. (Set $F_0 = 1$.) Show that the sequence $F_0, F_1, \ldots$ satisfies the recurrence relation

$$F_n = \sum_{j=0}^{n-1} C(n, j) F_j.$$

56. If α is a bit string, let $C(\alpha)$ be the maximum number of consecutive 0's in α. [*Examples*: $C(10010) = 2$, $C(00110001) = 3$.] Let S_n be the number of n-bit strings α with $C(\alpha) \le 2$. Develop a recurrence relation for $S_1, S_2, \ldots$.

57. Derive a recurrence relation for $C(n, k)$, the number of k-element subsets of an n-element set. Specifically, write $C(n + 1, k)$ in terms of $C(n, i)$ for appropriate i.

58. Derive a recurrence relation for $S(k, n)$, the number of ways of choosing k items, allowing repetitions, from n available types. Specifically, write $S(k, n)$ in terms of $S(k - 1, i)$ for appropriate i.

59. Let $S(n, k)$ denote the number of functions from $\{1, \ldots, n\}$ onto $\{1, \ldots, k\}$. Show that $S(n, k)$ satisfies the recurrence relation

$$S(n, k) = k^n - \sum_{i=1}^{k-1} C(k, i) S(n, i).$$

60. The *Lucas sequence* $L_1, L_2, \ldots$ (named after Édouard Lucas, the inventor of the Tower of Hanoi puzzle) is defined by the recurrence relation

$$L_n = L_{n-1} + L_{n-2}, \qquad n \ge 3,$$

and the initial conditions
$$L_1 = 1, \qquad L_2 = 3.$$

(a) Find the values of L_3, L_4, and L_5.

(b) Show that
$$L_{n+2} = f_n + f_{n+2}, \qquad n \ge 1,$$

where $f_1, f_2, \ldots$ denotes the Fibonacci sequence.

61. Establish the recurrence relation

$$s_{n+1,k} = s_{n,k-1} + n s_{n,k}$$

for Stirling numbers of the first kind (see Exercise 74, Section 4.2).

62. Establish the recurrence relation

$$S_{n+1,k} = S_{n,k-1} + k S_{n,k}$$

for Stirling numbers of the second kind (see Exercise 75, Section 4.2).

★ **63.** Show that

$$S_{n,k} = \frac{1}{k!} \sum_{i=0}^{k} (-1)^i (k-i)^n C(k,i),$$

where $S_{n,k}$ denotes a Stirling number of the second kind (see Exercise 75, Section 4.2).

64. Assume that a person invests a sum of money at r percent compounded annually. Explain the rule of thumb: To estimate the time to double the investment, divide 70 by r.

65. Derive a recurrence relation for the number of multiplications needed to evaluate an $n \times n$ determinant by the cofactor method.

A *rise/fall permutation* is a permutation p of $1, 2, \ldots, n$ satisfying

$$p(i) < p(i+1) \qquad \text{for } i = 1, 3, 5, \ldots$$

and

$$p(i) > p(i+1) \qquad \text{for } i = 2, 4, 6, \ldots.$$

For example, there are five rise/fall permutations of 1, 2, 3, 4:

$$1, 3, 2, 4; \qquad 1, 4, 2, 3; \qquad 2, 3, 1, 4; \qquad 2, 4, 1, 3; \qquad 3, 4, 1, 2.$$

Let E_n denote the number of rise/fall permutations of $1, 2, \ldots, n$. (Define $E_0 = 1$.) The numbers $E_0, E_1, E_2, \ldots$ are called the *Euler numbers*.

66. List all rise/fall permutations of 1, 2, 3. What is the value of E_3?

67. List all rise/fall permutations of 1, 2, 3, 4, 5. What is the value of E_5?

68. Show that in a rise/fall permutation of $1, 2, \ldots, n$, n must occur in position $2i$, for some i.

★ **69.** Use Exercise 68 to derive the recurrence relation

$$E_n = \sum_{j=1}^{\lfloor n/2 \rfloor} C(n-1, 2j-1) E_{2j-1} E_{n-2j}.$$

★ **70.** By considering where 1 must occur in a rise/fall permutation, derive the recurrence relation

$$E_n = \sum_{j=0}^{\lfloor (n-1)/2 \rfloor} C(n-1, 2j) E_{2j} E_{n-2j-1}.$$

★ **71.** Prove that

$$E_n = \frac{1}{2} \sum_{j=1}^{n-1} C(n-1, j) E_j E_{n-j-1}.$$

5.2 SOLVING RECURRENCE RELATIONS

To solve a recurrence relation involving the sequence $a_0, a_1, \ldots$ is to find an explicit formula for the general term a_n. In this section we discuss two methods of solving recurrence relations: **iteration** and a special method that applies to **linear homogeneous recurrence relations with constant coefficients**. For more powerful methods, such as methods that make use of generating functions, consult [Brualdi].

To solve a recurrence relation involving the sequence $a_0, a_1, \ldots$ by iteration, we use the recurrence relation to write the nth term a_n in terms of certain of its predecessors $a_{n-1}, \ldots, a_0$. We then successively use the recurrence relation to replace each of $a_{n-1}, \ldots$ by certain of their predecessors. We continue until an explicit formula is obtained. The iterative method was used to solve the recurrence relation of Example 5.1.3.

EXAMPLE 5.2.1

We can solve the recurrence relation

$$a_n = a_{n-1} + 3, \qquad\qquad (5.2.1)$$

subject to the initial condition

$$a_1 = 2,$$

by iteration. Replacing n by $n - 1$ in (5.2.1), we obtain

$$a_{n-1} = a_{n-2} + 3.$$

If we substitute this expression for a_{n-1} into (5.2.1), we obtain

$$a_n = \boxed{a_{n-1}} \qquad + 3$$
$$\downarrow$$
$$= \boxed{a_{n-2} + 3} \qquad + 3$$
$$= a_{n-2} + 2 \cdot 3. \qquad\qquad (5.2.2)$$

Replacing n by $n - 2$ in (5.2.1), we obtain

$$a_{n-2} = a_{n-3} + 3.$$

If we substitute this expression for a_{n-2} into (5.2.2), we obtain

$$a_n = \boxed{a_{n-2}} \qquad + 2 \cdot 3$$
$$\downarrow$$
$$= \boxed{a_{n-3} + 3} \qquad + 2 \cdot 3$$
$$= a_{n-3} + 3 \cdot 3.$$

In general, we have

$$a_n = a_{n-k} + k \cdot 3.$$

If we set $k = n - 1$ in this last expression, we have

$$a_n = a_1 + (n - 1) \cdot 3.$$

Since $a_1 = 2$, we obtain the explicit formula

$$a_n = 2 + 3(n - 1)$$

for the sequence a.   □

EXAMPLE 5.2.2

We can solve the recurrence relation

$$S_n = 2S_{n-1}$$

of Example 5.1.5, subject to the initial condition

$$S_0 = 1,$$

by iteration:

$$S_n = 2S_{n-1} = 2(2S_{n-2}) = \cdots = 2^n S_0 = 2^n. \qquad \square$$

EXAMPLE 5.2.3   *Population Growth*

Assume that the deer population of Rustic County is 1000 at time $n = 0$ and that the increase from time $n - 1$ to time n is 10 percent of the size at time $n - 1$. Write a recurrence relation and an initial condition that define the deer population at time n and then solve the recurrence relation.

Let d_n denote the deer population at time n. We have the initial condition

$$d_0 = 1000.$$

The increase from time $n - 1$ to time n is $d_n - d_{n-1}$. Since this increase is 10 percent of the size at time $n - 1$, we obtain the recurrence relation

$$d_n - d_{n-1} = 0.1d_{n-1},$$

which may be rewritten

$$d_n = 1.1d_{n-1}.$$

The recurrence relation may be solved by iteration:

$$d_n = 1.1d_{n-1} = 1.1(1.1d_{n-2}) = (1.1)^2(d_{n-2})$$
$$= \cdots = (1.1)^n d_0 = (1.1)^n 1000.$$

The assumptions imply exponential population growth. $\qquad \square$

EXAMPLE 5.2.4

Find an explicit formula for c_n, the minimum number of moves in which the n-disk Tower of Hanoi puzzle can be solved (see Example 5.1.8).

In Example 5.1.8 we obtained the recurrence relation

$$c_n = 2c_{n-1} + 1 \tag{5.2.3}$$

and initial condition

$$c_1 = 1.$$

Applying the iterative method to (5.2.3), we obtain

$$
\begin{aligned}
c_n &= 2c_{n-1} + 1 \\
&= 2(2c_{n-2} + 1) + 1 \\
&= 2^2 c_{n-2} + 2 + 1 \\
&= 2^2(2c_{n-3} + 1) + 2 + 1 \\
&= 2^3 c_{n-3} + 2^2 + 2 + 1 \\
&\;\;\vdots \\
&= 2^{n-1} c_1 + 2^{n-2} + 2^{n-3} + \cdots + 2 + 1 \\
&= 2^{n-1} + 2^{n-2} + 2^{n-3} + \cdots + 2 + 1 \\
&= 2^n - 1.
\end{aligned}
$$

The last step results from the formula for the geometric sum (see Example 1.6.2). □

EXAMPLE 5.2.5

We can solve the recurrence relation

$$
p_n = a - \frac{b}{k} p_{n-1}
$$

for the price p_n in the economics model of Example 5.1.9 by iteration. To simplify the notation, we set $s = -b/k$.

$$
\begin{aligned}
p_n &= a + s p_{n-1} \\
&= a + s(a + s p_{n-2}) \\
&= a + as + s^2 p_{n-2} \\
&= a + as + s^2(a + s p_{n-3}) \\
&= a + as + as^2 + s^3 p_{n-3} \\
&\;\;\vdots \\
&= a + as + as^2 + \cdots + as^{n-1} + s^n p_0 \\
&= \frac{a - as^n}{1 - s} + s^n p_0 \\
&= s^n\left(\frac{-a}{1-s} + p_0\right) + \frac{a}{1-s} \\
&= \left(-\frac{b}{k}\right)^n \left(\frac{-ak}{k+b} + p_0\right) + \frac{ak}{k+b}, \tag{5.2.4}
\end{aligned}
$$

We see that if $b/k < 1$, the term

$$
\left(-\frac{b}{k}\right)^n \left(\frac{-ak}{k+b} + p_0\right)
$$

becomes small as n gets large so that the price tends to stabilize at approximately $ak/(k+b)$. If $b/k = 1$, (5.2.4) shows that p_n oscillates between p_0 and p_1. If $b/k > 1$, (5.2.4) shows that the differences between successive prices increase. Previously, we observed these properties graphically (see Example 5.1.9).   □

We turn next to a special class of recurrence relations.

DEFINITION 5.2.6

A linear homogeneous recurrence relation of order k with constant coefficients is a recurrence relation of the form

$$a_n = c_1 a_{n-1} + c_2 a_{n-2} + \cdots + c_k a_{n-k}, \quad c_k \neq 0. \tag{5.2.5}$$

Notice that a linear homogeneous recurrence relation of order k with constant coefficients (5.2.5), together with the k initial conditions

$$a_0 = C_0, \quad a_1 = C_1, \ldots, \quad a_{k-1} = C_{k-1},$$

uniquely defines a sequence $a_0, a_1, \ldots$.

EXAMPLE 5.2.7

The recurrence relations

$$S_n = 2S_{n-1} \tag{5.2.6}$$

of Example 5.2.2 and

$$f_n = f_{n-1} + f_{n-2}, \tag{5.2.7}$$

which defines the Fibonacci sequence, are both linear homogeneous recurrence relations with constant coefficients. The recurrence relation (5.2.6) is of order 1 and (5.2.7) is of order 2.   □

EXAMPLE 5.2.8

The recurrence relation

$$a_n = 3a_{n-1}a_{n-2} \tag{5.2.8}$$

is not a linear homogeneous recurrence relation with constant coefficients. In a linear homogeneous recurrence relation with constant coefficients, each term is of the form ca_k. Terms such as $a_{n-1}a_{n-2}$ are not permitted. Recurrence relations like (5.2.8) are said to be *nonlinear*.   □

EXAMPLE 5.2.9

The recurrence relation

$$a_n - a_{n-1} = 2n$$

is not a linear homogeneous recurrence relation with constant coefficients because the expression on the right side of the equation is not zero. (Such an equation is said to be *inhomogeneous*. Linear inhomogeneous recurrence relations with constant coefficients are discussed in Exercises 33–39.)   □

EXAMPLE 5.2.10

The recurrence relation

$$a_n = 3na_{n-1}$$

is not a linear homogeneous recurrence relation with constant coefficients because the coefficient $3n$ is not constant. It is a linear homogeneous recurrence relation with nonconstant coefficients.   □

We will illustrate the general method of solving linear homogeneous recurrence relations with constant coefficients by finding an explicit formula for the sequence defined by the recurrence relation

$$a_n = 5a_{n-1} - 6a_{n-2} \qquad (5.2.9)$$

and initial conditions

$$a_0 = 7, \qquad a_1 = 16. \qquad (5.2.10)$$

Often in mathematics, when trying to solve a more difficult instance of some problem, we begin with an expression that solved a simpler version. For the first-order recurrence relation (5.2.6), we found in Example 5.2.2 that the solution was of the form

$$S_n = t^n;$$

thus for our first attempt at finding a solution of the second-order recurrence relation (5.2.9), we will search for a solution of the form $V_n = t^n$.

If $V_n = t^n$ is to solve (5.2.9), we must have

$$V_n = 5V_{n-1} - 6V_{n-2}$$

or

$$t^n = 5t^{n-1} - 6t^{n-2}$$

or

$$t^n - 5t^{n-1} + 6t^{n-2} = 0.$$

Dividing by t^{n-2}, we obtain the equivalent equation

$$t^2 - 5t + 6 = 0. \qquad (5.2.11)$$

Solving (5.2.11), we find the solutions

$$t = 2, \qquad t = 3.$$

At this point, we have two solutions S and T of (5.2.9), given by

$$S_n = 2^n, \qquad T_n = 3^n. \qquad (5.2.12)$$

We can verify (see Theorem 5.2.11) that if S and T are solutions of (5.2.9), then $bS + dT$, where b and d are any numbers whatever, is also a solution of (5.2.9). In our case, if we define the sequence U by the equation

$$U_n = bS_n + dT_n$$

$$= b2^n + d3^n,$$

U is a solution of (5.2.9).

To satisfy the initial conditions (5.2.10), we must have

$$7 = U_0 = b2^0 + d3^0 = b + d, \qquad 16 = U_1 = b2^1 + d3^1 = 2b + 3d.$$

Solving these equations for b and d, we obtain

$$b = 5, \qquad d = 2.$$

Therefore, the sequence U defined by

$$U_n = 5 \cdot 2^n + 2 \cdot 3^n$$

satisfies the recurrence relation (5.2.9) and the initial conditions (5.2.10). We conclude that

$$a_n = U_n = 5 \cdot 2^n + 2 \cdot 3^n, \qquad \text{for } n = 0, 1, \dots.$$

At this point we will summarize and justify the techniques used to solve the preceding recurrence relation.

THEOREM 5.2.11

Let

$$a_n = c_1 a_{n-1} + c_2 a_{n-2} \tag{5.2.13}$$

be a second-order linear homogeneous recurrence relation with constant coefficients.

If S and T are solutions of (5.2.13), then $U = bS + dT$ is also a solution of (5.2.13).

If r is a root of

$$t^2 - c_1 t - c_2 = 0, \tag{5.2.14}$$

then the sequence r^n, $n = 0, 1, \dots$, is a solution of (5.2.13).

If a is the sequence defined by (5.2.13),

$$a_0 = C_0, \qquad a_1 = C_1, \tag{5.2.15}$$

and r_1 and r_2 are roots of (5.2.14) with $r_1 \neq r_2$, then there exist constants b and d such that

$$a_n = br_1^n + dr_2^n, \qquad n = 0, 1, \dots.$$

Proof. Since S and T are solutions of (5.2.13),

$$S_n = c_1 S_{n-1} + c_2 S_{n-2}, \qquad T_n = c_1 T_{n-1} + c_2 T_{n-2}.$$

If we multiply the first equation by b and the second by d and add, we obtain

$$U_n = bS_n + dT_n = c_1(bS_{n-1} + dT_{n-1}) + c_2(bS_{n-2} + dT_{n-2})$$

$$= c_1 U_{n-1} + c_2 U_{n-2}.$$

Therefore, U is a solution of (5.2.13).

Since r is a root of (5.2.14),

$$r^2 = c_1 r + c_2.$$

Now

$$c_1 r^{n-1} + c_2 r^{n-2} = r^{n-2}(c_1 r + c_2) = r^{n-2} r^2 = r^n;$$

thus the sequence $r^n, n = 0, 1, \ldots$, is a solution of (5.2.13).

If we set $U_n = b r_1^n + d r_2^n$, then U is a solution of (5.2.13). To meet the initial conditions (5.2.15), we must have

$$U_0 = b + d = C_0, \qquad U_1 = b r_1 + d r_2 = C_1.$$

If we multiply the first equation by r_1 and subtract, we obtain

$$d(r_1 - r_2) = r_1 C_0 - C_1.$$

Since $r_1 - r_2 \neq 0$, we can solve for d. Similarly, we can solve for b. With these choices for b and d, we have

$$U_0 = C_0, \qquad U_1 = C_1.$$

Let a be the sequence defined by (5.2.13) and (5.2.15). Since U also satisfies (5.2.13) and (5.2.15), it follows that $U_n = a_n, n = 0, 1, \ldots$. ∎

> **EXAMPLE 5.2.12**   *More Population Growth*

Assume that the deer population of Rustic County is 200 at time $n = 0$ and 220 at time $n = 1$ and that the increase from time $n - 1$ to time n is twice the increase from time $n - 2$ to time $n - 1$. Write a recurrence relation and an initial condition that define the deer population at time n and then solve the recurrence relation.

Let d_n denote the deer population at time n. We have the initial conditions

$$d_0 = 200, \qquad d_1 = 220.$$

The increase from time $n - 1$ to time n is $d_n - d_{n-1}$ and the increase from time $n - 2$ to time $n - 1$ is $d_{n-1} - d_{n-2}$. Thus we obtain the recurrence relation

$$d_n - d_{n-1} = 2(d_{n-1} - d_{n-2}),$$

which may be rewritten

$$d_n = 3 d_{n-1} - 2 d_{n-2}.$$

To solve this recurrence relation, we first solve the quadratic equation

$$t^2 - 3t + 2 = 0$$

to obtain roots 1 and 2. The sequence d is of the form

$$d_n = b \cdot 1^n + c \cdot 2^n = b + c 2^n.$$

To meet the initial conditions, we must have

$$200 = d_0 = b + c, \qquad 220 = d_1 = b + 2c.$$

Solving for b and c, we find that $b = 180$ and $c = 20$. Thus d_n is given by

$$d_n = 180 + 20 \cdot 2^n.$$

As in Example 5.2.3, the growth is exponential. □

> **EXAMPLE 5.2.13**

Find an explicit formula for the Fibonacci sequence.

The Fibonacci sequence is defined by the linear, homogeneous, second-order recurrence relation

$$f_n - f_{n-1} - f_{n-2} = 0, \qquad n \geq 3,$$

and initial conditions

$$f_1 = 1, \qquad f_2 = 2.$$

We begin by using the quadratic formula to solve

$$t^2 - t - 1 = 0.$$

The solutions are

$$t = \frac{1 \pm \sqrt{5}}{2}.$$

Thus the solution is of the form

$$f_n = b \left(\frac{1 + \sqrt{5}}{2} \right)^n + d \left(\frac{1 - \sqrt{5}}{2} \right)^n.$$

To satisfy the initial conditions, we must have

$$b \left(\frac{1 + \sqrt{5}}{2} \right) + d \left(\frac{1 - \sqrt{5}}{2} \right) = 1$$

$$b \left(\frac{1 + \sqrt{5}}{2} \right)^2 + d \left(\frac{1 - \sqrt{5}}{2} \right)^2 = 2.$$

Solving these equations for b and d, we obtain

$$b = \frac{1}{\sqrt{5}} \left(\frac{1 + \sqrt{5}}{2} \right), \qquad d = -\frac{1}{\sqrt{5}} \left(\frac{1 - \sqrt{5}}{2} \right).$$

Therefore, an explicit formula for the Fibonacci sequence is

$$f_n = \frac{1}{\sqrt{5}} \left(\frac{1 + \sqrt{5}}{2} \right)^{n+1} - \frac{1}{\sqrt{5}} \left(\frac{1 - \sqrt{5}}{2} \right)^{n+1}.$$

Surprisingly, even though f_n is an integer, the preceding formula involves the irrational number $\sqrt{5}$. □

Theorem 5.2.11 states that any solution of (5.2.13) may be given in terms of two basic solutions r_1^n and r_2^n. However, in case (5.2.14) has two equal roots r, we obtain only one basic solution r^n. The next theorem shows that in this case, nr^n furnishes the other basic solution.

THEOREM 5.2.14

Let

$$a_n = c_1 a_{n-1} + c_2 a_{n-2} \qquad (5.2.16)$$

be a second-order linear homogeneous recurrence relation with constant coefficients.

Let a be the sequence satisfying (5.2.16) and

$$a_0 = C_0, \qquad a_1 = C_1.$$

If both roots of

$$t^2 - c_1 t - c_2 = 0 \qquad (5.2.17)$$

are equal to r, then there exist constants b and d such that

$$a_n = b r^n + d n r^n, \qquad n = 0, 1, \ldots.$$

Proof. The proof of Theorem 5.2.11 shows that the sequence r^n, $n = 0, 1, \ldots$, is a solution of (5.2.16). We show that the sequence $n r^n$, $n = 0, 1, \ldots$, is also a solution of (5.2.16).

Since r is the only solution of (5.2.17), we must have

$$t^2 - c_1 t - c_2 = (t - r)^2.$$

It follows that

$$c_1 = 2r, \qquad c_2 = -r^2.$$

Now

$$c_1 \left[(n-1) r^{n-1} \right] + c_2 \left[(n-2) r^{n-2} \right] = 2r(n-1) r^{n-1} - r^2 (n-2) r^{n-2}$$
$$= r^n \left[2(n-1) - (n-2) \right] = n r^n.$$

Therefore, the sequence $n r^n$, $n = 0, 1, \ldots$, is a solution of (5.2.16).

By Theorem 5.2.11, the sequence U defined by $U_n = b r^n + d n r^n$ is a solution of (5.2.16).

The proof that there are constants b and d such that $U_0 = C_0$ and $U_1 = C_1$ is similar to the argument given in Theorem 5.2.11 and is left as an exercise (Exercise 41). It follows that $U_n = a_n$, $n = 0, 1, \ldots$. ∎

EXAMPLE 5.2.15

Solve the recurrence relation

$$d_n = 4(d_{n-1} - d_{n-2}) \qquad (5.2.18)$$

subject to the initial conditions

$$d_0 = 1 = d_1.$$

According to Theorem 5.2.11, $S_n = r^n$ is a solution of (5.2.18), where r is a solution of

$$t^2 - 4t + 4 = 0. \qquad (5.2.19)$$

Thus we obtain the solution

$$S_n = 2^n$$

of (5.2.18). Since 2 is the only solution of (5.2.19), by Theorem 5.2.14,

$$T_n = n2^n$$

is also a solution of (5.2.18). Thus the general solution of (5.2.18) is of the form

$$U = aS + bT.$$

We must have

$$U_0 = 1 = U_1.$$

These last equations become

$$aS_0 + bT_0 = a + 0b = 1, \qquad aS_1 + bT_1 = 2a + 2b = 1.$$

Solving for a and b, we obtain

$$a = 1, \qquad b = -\frac{1}{2}.$$

Therefore, the solution of (5.2.18) is

$$d_n = 2^n - n2^{n-1}. \qquad \square$$

For the general linear homogeneous recurrence relation of order k with constant coefficients (5.2.5), if r is a root of

$$t^k - c_1 t^{k-1} - c_2 t^{k-2} - \ldots - c_k = 0$$

of multiplicity m, it can be shown that

$$r^n, \quad nr^n, \quad \ldots, \quad n^{m-1}r^n$$

are solutions of (5.2.5). This fact can be used, just as in the previous examples for recurrence relations of order 2, to solve a linear homogeneous recurrence relation of order k with constant coefficients. For a precise statement and a proof of the general result, see [Brualdi].

Exercises

Tell whether or not each recurrence relation in Exercises 1–10 is a linear homogeneous recurrence relation with constant coefficients. Give the order of each linear homogeneous recurrence relation with constant coefficients.

1. $a_n = -3a_{n-1}$ **2.** $a_n = 2na_{n-1}$

3. $a_n = 2na_{n-2} - a_{n-1}$ **4.** $a_n = a_{n-1} + n$

5. $a_n = 7a_{n-2} - 6a_{n-3}$ **6.** $a_n = a_{n-1} + 1 + 2^{n-1}$

7. $a_n = (\lg 2n)a_{n-1} - [\lg(n-1)]a_{n-2}$ **8.** $a_n = 6a_{n-1} - 9a_{n-2}$

9. $a_n = -a_{n-1} - a_{n-2}$ **10.** $a_n = -a_{n-1} + 5a_{n-2} - 3a_{n-3}$

In Exercises 11–25, solve the given recurrence relation for the initial conditions given.

11. Exercise 1; $a_0 = 2$

12. Exercise 2; $a_0 = 1$

13. Exercise 4; $a_0 = 0$

14. $a_n = 6a_{n-1} - 8a_{n-2}$; $a_0 = 1$, $a_1 = 0$

15. $a_n = 7a_{n-1} - 10a_{n-2}$; $a_0 = 5$, $a_1 = 16$

16. $a_n = 2a_{n-1} + 8a_{n-2}$; $a_0 = 4$, $a_1 = 10$

17. $2a_n = 7a_{n-1} - 3a_{n-2}$; $a_0 = a_1 = 1$

18. Exercise 6; $a_0 = 0$

19. Exercise 8; $a_0 = a_1 = 1$

20. $a_n = -8a_{n-1} - 16a_{n-2}$; $a_0 = 2$, $a_1 = -20$

21. $9a_n = 6a_{n-1} - a_{n-2}$; $a_0 = 6$, $a_1 = 5$

22. The Lucas sequence

$$L_n = L_{n-1} + L_{n-2}, \quad n \geq 3; \qquad L_1 = 1, \quad L_2 = 3$$

23. Exercise 50, Section 5.1

24. Exercise 52, Section 5.1

25. The recurrence relation preceding Exercise 53, Section 5.1

26. Assume that the deer population of Rustic County is 0 at time $n = 0$. Suppose that at time n, $100n$ deer are introduced into Rustic County and that the population increases 20 percent each year. Write a recurrence relation and an initial condition that define the deer population at time n and then solve the recurrence relation. The following formula may be of use:

$$\sum_{i=1}^{n-1} ix^{i-1} = \frac{(n-1)x^n - nx^{n-1} + 1}{(x-1)^2}.$$

Sometimes a recurrence relation that is not a linear homogeneous equation with constant coefficients can be transformed into a linear homogeneous equation with constant coefficients. In Exercises 27 and 28, make the given substitution and solve the resulting recurrence relation, then find the solution to the original recurrence relation.

27. Solve the recurrence relation

$$\sqrt{a_n} = \sqrt{a_{n-1}} + 2\sqrt{a_{n-2}}$$

with initial conditions $a_0 = a_1 = 1$ by making the substitution $b_n = \sqrt{a_n}$.

28. Solve the recurrence relation

$$a_n = \sqrt{\frac{a_{n-2}}{a_{n-1}}}$$

with initial conditions $a_0 = 8$, $a_1 = 1/(2\sqrt{2})$ by taking the logarithm of both sides and making the substitution $b_n = \lg a_n$.

In Exercises 29–31, solve the recurrence relation for the initial conditions given.

29. $a_n = -2na_{n-1} + 3n(n-1)a_{n-2}$; $a_0 = 1$, $a_1 = 2$

★ **30.** $c_n = 2 + \sum_{i=1}^{n-1} c_i$, $n \geq 2$; $c_1 = 1$

★ **31.** $A(n, m) = 1 + A(n - 1, m - 1) + A(n - 1, m); \quad n - 1 \geq m \geq 1, \quad n \geq 2;$
$A(n, 0) = A(n, n) = 1, \quad n \geq 0$

32. Show that

$$f_n \geq \left(\frac{1 + \sqrt{5}}{2} \right)^{n-1}, \qquad n \geq 1,$$

where f denotes the Fibonacci sequence.

33. The equation

$$a_n = c_1 a_{n-1} + c_2 a_{n-2} + f(n) \tag{5.2.20}$$

is called a **second-order linear inhomogeneous recurrence relation with constant coefficients.**

Let $g(n)$ be a solution of (5.2.20). Show that any solution U of (5.2.20) is of the form

$$U_n = V_n + g(n), \tag{5.2.21}$$

where V is a solution of the homogeneous equation (5.2.13).

If $f(n) = C$ in (5.2.20), it can be shown that $g(n) = C'$ in (5.2.21). Also, if $f(n) = Cn, g(n) = C_1'n + C_0'$; if $f(n) = Cn^2, g(n) = C_2'n^2 + C_1'n + C_0'$; and if $f(n) = C^n, g(n) = C'C^n$. Use these facts together with Exercise 33 to find the general solutions of the recurrence relations of Exercises 34–39.

34. $a_n = 6a_{n-1} - 8a_{n-2} + 3$

35. $a_n = 7a_{n-1} - 10a_{n-2} + 16n$

36. $a_n = 2a_{n-1} + 8a_{n-2} + 81n^2$

37. $2a_n = 7a_{n-1} - 3a_{n-2} + 2^n$

38. $a_n = -8a_{n-1} - 16a_{n-2} + 3n$

39. $9a_n = 6a_{n-1} - a_{n-2} + 5n^2$

40. The equation

$$a_n = f(n)a_{n-1} + g(n)a_{n-2} \tag{5.2.22}$$

is called a **second-order linear homogeneous recurrence relation**. The coefficients $f(n)$ and $g(n)$ are not necessarily constant. Show that if S and T are solutions of (5.2.22), then $bS + dT$ is also a solution of (5.2.22).

41. Suppose that both roots of

$$t^2 - c_1 t - c^2 = 0$$

are equal to r and suppose that a_n satisfies

$$a_n = c_1 a_{n-1} + c_2 a_{n-2}, \qquad a_0 = C_0, \qquad a_1 = C_1.$$

Show that there exist constants b and d such that

$$a_n = br^n + dnr^n, \qquad n = 0, 1, \ldots,$$

thus completing the proof of Theorem 5.2.14.

42. Let a_n be the minimum number of links required to solve the n-node communication problem (see Exercise 48, Section 5.1). Use iteration to show that $a_n \leq 2n - 4$, $n \geq 4$.

The n-disk, four-peg Tower of Hanoi puzzle has the same rules as the three-peg puzzle; the only difference is that there is an extra peg. Exercises 43–46 refer to the following algorithm to solve the n-disk, four-peg Tower of Hanoi puzzle.

Assume that the pegs are numbered 1, 2, 3, 4 and that the problem is to move the disks, which are initially stacked on peg 1, to peg 4. If $n = 1$, move the disk to peg 4 and stop. If $n > 1$, let k_n be the largest integer satisfying

$$\sum_{i=1}^{k_n} i \leq n.$$

Fix k_n disks at the bottom of peg 1. Recursively invoke this algorithm to move the $n - k_n$ disks at the top of peg 1 to peg 2. During this part of the algorithm, the k_n bottom disks on peg 1 remain fixed. Next, move the k_n disks on peg 1 to peg 4 by invoking the optimal three-peg algorithm (see Example 5.1.8) and using only pegs 1, 3, and 4. Finally, again recursively invoke this algorithm to move the $n - k_n$ disks on peg 2 to peg 4. During this part of the algorithm, the k_n disks on peg 4 remain fixed. Let $T(n)$ denote the number of moves required by this algorithm.

This algorithm, although not known to be optimal, uses as few moves as any other algorithm that has been proposed for the four-peg problem.

43. Derive the recurrence relation

$$T(n) = 2T(n - k_n) + 2^{k_n} - 1.$$

44. Compute $T(n)$ for $n = 1, \ldots, 10$. Compare these values with the optimal number of moves to solve the three-peg problem.

★ **45.** Let

$$r_n = n - \frac{k_n(k_n + 1)}{2}.$$

Using induction or otherwise, prove that

$$T(n) = (k_n + r_n - 1)2^{k_n} + 1.$$

★ **46.** Show that $T(n) = O(4^{\sqrt{n}})$.

Problem

(a) At a dinner, n persons check their coats. When they leave, the coats are returned randomly; unfortunately, no one receives the correct coat. Let D_n denote the number of ways that n persons can all receive the wrong coats. Show that the sequence $D_1, D_2, \ldots$ satisfies the recurrence relation

$$D_n = (n-1)(D_{n-1} + D_{n-2}).$$

(b) Solve the recurrence relation of part (a) by making the substitution $C_n = D_n - nD_{n-1}$.

Attacking the Problem

Before attacking the problem, consider what is required for part (a). To prove the recurrence relation, we must reduce the n-person, wrong-coat problem to the $(n-1)$- and $(n-2)$-person, wrong-coat problems (since in the formula D_n is given in terms of D_{n-1} and D_{n-2}). Thus as we look systematically at some examples, we should try to see how the case for n persons relates to the cases for $n-1$ and $n-2$ persons. The situation is similar to that of mathematical induction and recursive algorithms in which a given instance of a problem is related to smaller instances of the same problem.

Now for some examples. The smallest case is $n = 1$. A single person must get the correct coat so $D_1 = 0$. For $n = 2$, there is one way for everyone to get the wrong coat: person 1 gets coat 2, and person 2 gets coat 1. Thus $D_2 = 1$. Before continuing, let's develop some notation for the distribution of coats. Carefully chosen notation can help in solving a problem.

We'll write

$$c_1, \quad c_2, \ldots, \quad c_n$$

to mean that person 1 got coat c_1, person 2 got coat $c_2, \ldots$, and so on. The one way for two persons to get the wrong coats is denoted 2, 1.

If $n = 3$, person 1 gets either coat 2 or 3, so the possibilities are 2, ?, ? and 3, ?, ?. Let's fill in the missing numbers. Suppose that person 1 gets coat 2. Person 2 can't get coat 1 (for then person 3 would get the correct coat); thus, person 2 gets coat 3. This leaves coat 1 for person 3. Thus if person 1 gets coat 2, the only possibility is

$$2, 3, 1.$$

Suppose that person 1 gets coat 3. Person 3 can't get coat 1 (for then person 2 would get the correct coat); thus, person 3 gets coat 2. This leaves coat 1 for person 2. Thus if person 1 gets coat 3, the only possibility is

$$3, 1, 2.$$

Thus $D_3 = 2$.

Let's check that the recurrence relation holds for $n = 3$:

$$D_3 = 2 = 2(1 + 0) = (3 - 1)(D_2 + D_1).$$

If $n = 4$, person 1 gets either coat 2, 3, or 4, so the possibilities are 2, ?, ?, ?; 3, ?, ?, ?; and 4, ?, ?, ?. (If there are n persons, person 1 gets either coat 2 or 3 or ... or $n-1$. These $n-1$ possibilities account for the leading $n-1$ factor in the recurrence relation.) Let's fill in the missing numbers. Suppose that person 1 gets coat 2. If person 2 gets coat 1, then person 3 gets coat 4 and person 4 gets coat 3, which gives 2, 1, 4, 3. If person 2 does not get coat 1, the possibilities are 2, 3, 4, 1 and 2, 4, 1, 3. Thus if person 1 gets coat 2, there are three possibilities. Similarly, if person 1 gets coat 3, there are three possibilities, and if person 1 gets coat 4, there are three possibilities. You should list the possibilities to confirm this last statement. Thus $D_4 = 9$.

Let's check that the recurrence relation also holds for $n = 4$:

$$D_4 = 9 = 3(2 + 1) = (4 - 1)(D_3 + D_2).$$

Before reading on, work through the case $n = 5$. List only the possibilities when person 1 gets coat 2. (There are too many possibilities to list them all.) Also verify the recurrence relation for $n = 5$.

Notice that if person 1 gets coat 2 and person 2 gets coat 1 (i.e., if persons 1 and 2 trade coats), the number of ways for the other persons to get the wrong coats is D_{n-2} (the remaining $n-2$ persons must all get the wrong coats). This accounts for the presence of D_{n-2} in the recurrence relation. We'll have a solution, provided that the D_{n-1} term appears in the remaining case: person 1 gets coat 2, but person 2 does not get coat 1 (i.e., persons 1 and 2 do not trade coats).

Finding a Solution

Suppose that there are n persons. Let's summarize the argument we've developed through our examples. Person 1 gets either coat 2, or 3, ..., or n; so, there are $n-1$ possible ways for person 1 to get the wrong coat. Suppose that person 1 gets coat 2. There are two possibilities: person 2 gets coat 1, person 2 does not get coat 1. If person 2 gets coat 1, the number of ways for the other persons to get the wrong coats is D_{n-2}. The remaining case is that person 2 does not get coat 1.

Let's write out carefully what it is we have to count. Persons 2, 3, ..., n have among themselves coats 1, 3, 4, ..., n (coat 2 is missing because person 1 has it). We want to find the number of ways for persons 2, 3, ..., n to each get the wrong coat *and* for person 2 to not get coat 1. This is almost the problem of $n-1$ persons all not getting the right coats. We can turn it into this problem if we tell persons 2, 3, ..., n that coat 1 is coat 2. (Just sew in a temporary label!). Now person 2 will not get coat 1 because person 2 thinks it's coat 2. Since there are $n-1$ persons, there are D_{n-1} ways for persons 2, 3, ..., n to each get the wrong coat *and* for person 2 to not get coat 1. It follows that there are $D_{n-1} + D_{n-2}$ ways for person 1 to get coat 2 and for all the others to get the wrong coats. Since there are $n-1$ possible ways for person 1 to get the wrong coat, the recurrence relation now follows.

The recurrence relation defines D_n in terms of D_{n-1} and D_{n-2}, so it can't be solved using iteration. Also, the recurrence relation does not have constant coefficients (although it is linear), so it can't be solved using Theorem 5.2.11 or 5.2.14. This explains the need to make the substitution in part (b). Evidently, after making the substitution, the recurrence relation for C_n can be solved by the methods of Section 5.2.

If we expand

$$D_n = (n-1)(D_{n-1} + D_{n-2}),$$

we obtain

$$D_n = nD_{n-1} - D_{n-1} + (n-1)D_{n-2}.$$

If we then move nD_{n-1} to the left side of the equation (to obtain an expression equal to C_n), we obtain

$$D_n - nD_{n-1} = -D_{n-1} + (n-1)D_{n-2}.$$

Now the left side of the equation is equal to C_n and the right side is equal to $-C_{n-1}$. Thus we obtain the recurrence relation

$$C_n = -C_{n-1}.$$

This equation may be solved by iteration.

Formal Solution

Part (a): Suppose that n persons all have the wrong coats. We consider the coat that one person p has. Suppose that p has q's coat. We consider two cases: q has p's coat and q does not have p's coat.

There are D_{n-2} distributions in which q has p's coat since the remaining $n-2$ coats are in possession of the remaining $n-2$ persons, but each has the wrong coat.

We show that there are D_{n-1} distributions in which q does not have p's coat. Note that the set of coats C that the $n-1$ persons (excluding p) possess includes all except q's (which p has). Temporarily assign ownership of p's coat to q. Then any distribution of C among the $n-1$ persons in which no one has his own coat yields a distribution in which q does not have the coat that is really p's. Since there are D_{n-1} such distributions, there are D_{n-1} distributions in which q does not have p's coat.

It follows that there are $D_{n-1} + D_{n-2}$ distributions in which p has q's coat. Since p can have any one of $n-1$ coats, the recurrence relation follows.

Part (b): Making the given substitution, we obtain

$$C_n = -C_{n-1}.$$

Using iteration we obtain

$$C_n = (-1)^1 C_{n-1} = (-1)^2 C_{n-2} = \cdots$$
$$= (-1)^{n-2} C_2 = (-1)^n C_2 = (-1)^n (D_2 - 2D_1) = (-1)^n.$$

Therefore,

$$D_n - nD_{n-1} = (-1)^n.$$

Solving this last recurrence relation by iteration, we obtain

$$
\begin{aligned}
D_n &= (-1)^n + nD_{n-1} \\
&= (-1)^n + n\left[(-1)^{n-1} + (n-1)D_{n-2}\right] \\
&= (-1)^n + n(-1)^{n-1} + n(n-1)\left[(-1)^{n-2} + (n-2)D_{n-3}\right] \\
&\vdots \\
&= (-1)^n + n(-1)^{n-1} + n(n-1)(-1)^{n-2} + \cdots \\
&\quad - [n(n-1)\cdots 4] + [n(n-1)\cdots 3].
\end{aligned}
$$

Summary of Problem-Solving Techniques

- When examples get too wordy, develop a notation for concisely describing the examples. Carefully chosen notation can help enormously in solving a problem.
- When looking at examples, try to see how the current problem relates to smaller instances of the same problem.
- It often helps to write out carefully what is to be counted.
- It is sometimes possible to convert a recurrence relation that is not a linear homogeneous equation with constant coefficients into a linear homogeneous equation with constant coefficients. Such a recurrence relation can then be solved by the methods of Section 5.2.

Comment

The technical name of a permutation in which no element is in its original position is a **derangement**.

5.3 APPLICATIONS TO THE ANALYSIS OF ALGORITHMS

In this section we use recurrence relations to analyze the time required by algorithms. The technique is to develop a recurrence relation and initial conditions that define a sequence $a_1, a_2, \ldots$, where a_n is the time (best-case, average-case, or worst-case time) required for an algorithm to execute an input of size n. By solving the recurrence relation, we can determine the time needed by the algorithm.

Our first algorithm is a version of the selection sorting algorithm. This algorithm selects the largest item and places it last, then recursively repeats this process.

ALGORITHM 5.3.1   *Selection Sort*

This algorithm sorts the sequence

$$s_1, \quad s_2, \quad \ldots, \quad s_n$$

in increasing order by first selecting the largest item and placing it last and then recursively sorting the remaining elements.

Input: $s_1, s_2, \ldots, s_n$ and the length n of the sequence

Output: $s_1, s_2, \ldots, s_n$, arranged in increasing order

```
1.    procedure selection_sort(s, n)
2.      // base case
3.      if n = 1 then
4.         return
5.      // find largest
6.      max_index := 1    // assume initially that s₁ is largest
7.      for i := 2 to n do
8.         if sᵢ > s_max_index then    // found larger, so update
9.            max_index := i
10.     // move largest to end
11.     swap(sₙ, s_max_index)
12.     call selection_sort(s, n − 1)
13.   end selection_sort
```

As a measure of the time required by this algorithm, we count the number of comparisons b_n at line 8 required to sort n items. (Notice that the best-case, average-case, and worst-case times are all the same for this algorithm.) We immediately obtain the initial condition

$$b_1 = 0.$$

To obtain a recurrence relation for the sequence $b_1, b_2, \ldots,$ we simulate the execution of the algorithm for arbitrary input of size $n > 1$. We count the number of comparisons at each line and then sum these numbers to obtain the total number of comparisons b_n. At lines 1–7, there are zero comparisons (of the type we are counting). At line 8, there are $n - 1$ comparisons (since line 7 causes line 8 to be executed $n - 1$ times). There are zero comparisons at lines 9–11. The recursive call occurs at line 12, where we invoke this algorithm with input of size $n - 1$. But, by definition, this algorithm requires b_{n-1} comparisons for input of size $n - 1$. Thus there are b_{n-1} comparisons at line 12. Therefore, the total number of comparisons is

$$b_n = n - 1 + b_{n-1},$$

which yields the desired recurrence relation.

Our recurrence relation can be solved by iteration:

$$\begin{aligned}
b_n &= b_{n-1} + n - 1 \\
&= (b_{n-2} + n - 2) + (n - 1) \\
&= (b_{n-3} + n - 3) + (n - 2) + (n - 1) \\
&\ \vdots \\
&= b_1 + 1 + 2 + \cdots + (n - 2) + (n - 1) \\
&= 0 + 1 + 2 + \cdots + (n - 1) = \frac{(n-1)n}{2} = \Theta(n^2).
\end{aligned}$$

Thus the time required by Algorithm 5.3.1 is $\Theta(n^2)$.

Our next algorithm (Algorithm 5.3.2) is **binary search**. Binary search looks for a value in a *sorted* sequence and returns the index of the value if it is found, or 0, if it is not found. The algorithm uses the divide-and-conquer approach. The sequence is divided into two nearly equal parts (line 4). If the item is found at the dividing point (line 5), the algorithm terminates. If the item is not found, because the sequence is sorted, an additional comparison (line 7) will locate the half of the sequence in which the item appears if it is present. We then recursively invoke binary search (line 11) to continue the search.

ALGORITHM 5.3.2   *Binary Search*

This algorithm looks for a value in an increasing sequence and returns the index of the value if it is found, or 0, if it is not found.

Input: A sequence $s_i, s_{i+1}, \ldots, s_j, i \geq 1$, sorted in increasing order, a value *key*, i, and j

Output: The output is an index k for which $s_k = key$, or if *key* is not in the sequence, the output is the value 0.

1. **procedure** *binary_search*(s, i, j, key)
2.   **if** $i > j$ **then**   // not found
3.   **return**(0)
4. $k := \lfloor (i + j)/2 \rfloor$
5.   **if** $key = s_k$ **then**   // found
6.   **return**(k)
7.   **if** $key < s_k$ **then**   // search left half
8. $j := k - 1$
9.   **else**   // search right half
10. $i := k + 1$
11.   **return**(*binary_search*(s, i, j, key))
12. **end** *binary_search*

EXAMPLE 5.3.3

We illustrate Algorithm 5.3.2 for the input

$$s_1 = \text{'}B\text{'}, \qquad s_2 = \text{'}D\text{'}, \qquad s_3 = \text{'}F\text{'}, \qquad s_4 = \text{'}S\text{'},$$

and $key = \text{'}S\text{'}$. At line 2, since $i > j$ ($1 > 4$) is false, we proceed to line 4, where we set k to 2. At line 5, since key ('S') is not equal to s_2 ('D'), we proceed to line 7. At line 7, $key < s_k$ ('S' < 'D') is false, so at line 10, we set i to 3. We then invoke this algorithm with $i = 3, j = 4$ to search for *key* in

$$s_3 = \text{'}F\text{'}, \qquad s_4 = \text{'}S\text{'}.$$

At line 2, since $i > j$ ($3 > 4$) is false, we proceed to line 4, where we set k to 3. At line 5, since *key* ('S') is not equal to s_3 ('F'), we proceed to line 7. At line 7, $key < s_k$ ('S' < 'F') is false, so at line 10, we set i to 4. We then invoke this algorithm with $i = j = 4$ to search for key in

$$s_4 = \text{'S'}.$$

At line 2, since $i > j$ (4 > 4) is false, we proceed to line 4, where we set k to 4. At line 5, since key ('S') is equal to s_4 ('S'), we return 4, the index of key in the sequence s. $\square$

Next we turn to the worst-case analysis of binary search. We define the worst-case time required by binary search to be the number of times the algorithm is invoked in the worst case for a sequence containing n items. We let a_n denote the worst-case time.

Suppose that n is 1; that is, suppose that the sequence consists of one element s_i and $i = j$. In the worst case, the item will not be found at line 5, so the algorithm will be invoked a second time at line 11. However, at the second invocation we will have $i > j$ and the algorithm will terminate unsuccessfully at line 3. We have shown that if n is 1, the algorithm is invoked twice. We obtain the initial condition

$$a_1 = 2. \tag{5.3.1}$$

Now suppose that $n > 1$. In this case, $i < j$, so the condition in line 2 is false. In the worst case, the item will not be found at line 5, so the algorithm will be invoked at line 11. By definition, the invocation at line 11 will require a total of a_m invocations, where m is the size of the sequence that is input at line 11. Since the sizes of the left and right sides of the original sequence are $\lfloor n/2 \rfloor$ and $\lfloor (n-1)/2 \rfloor$ and the worst case occurs with the larger sequence, the total number of invocations at line 11 will be $a_{\lfloor n/2 \rfloor}$. The original invocation together with the invocations at line 11 gives all the invocations; thus we obtain the recurrence relation

$$a_n = 1 + a_{\lfloor n/2 \rfloor}. \tag{5.3.2}$$

The recurrence relation (5.3.2) is typical of those that result from divide-and-conquer algorithms. Such recurrence relations are usually not easily solved explicitly (see, however, Exercise 6). Rather, one estimates the growth of the sequence involved using theta notation. Our method of deriving a theta notation for the sequence defined by (5.3.1) and (5.3.2) illustrates a general method of handling such recurrence relations. First we *explicitly solve* (5.3.2) in case n is a power of 2. When n is not a power of 2, n lies between two powers of 2, say 2^{k-1} and 2^k, and a_n lies between $a_{2^{k-1}}$ and a_{2^k}. Since explicit formulas are known for $a_{2^{k-1}}$ and a_{2^k}, we can *estimate* a_n and thereby derive a theta notation for a_n.

First we solve the recurrence relation (5.3.2) in case n is a power of 2. If $n = 2^k$, (5.3.2) becomes

$$a_{2^k} = 1 + a_{2^{k-1}}, \qquad k = 1, 2, \ldots.$$

If we let $b_k = a_{2^k}$, we obtain the recurrence relation

$$b_k = 1 + b_{k-1}, \qquad k = 1, 2, \ldots, \tag{5.3.3}$$

and the initial condition

$$b_0 = 2.$$

The recurrence relation (5.3.3) can be solved by the iterative method:

$$b_k = 1 + b_{k-1} = 2 + b_{k-2} = \cdots = k + b_0 = k + 2.$$

Thus, if $n = 2^k$,

$$a_n = 2 + \lg n. \qquad (5.3.4)$$

An arbitrary value of n falls between two powers of 2, say

$$2^{k-1} < n \leq 2^k. \qquad (5.3.5)$$

Since the sequence a is increasing (a fact that can be *proved* using induction—see Exercise 5),

$$a_{2^{k-1}} \leq a_n \leq a_{2^k}. \qquad (5.3.6)$$

Notice that (5.3.5) gives

$$k - 1 < \lg n \leq k. \qquad (5.3.7)$$

From (5.3.4), (5.3.6), and (5.3.7), we deduce that

$$\lg n < 1 + k = a_{2^{k-1}} \leq a_n \leq a_{2^k} = 2 + k < 3 + \lg n = O(\lg n).$$

Therefore $a_n = \Theta(\lg n)$, so binary search is $\Theta(\lg n)$ in the worst case. This result is important enough to highlight as a theorem.

THEOREM 5.3.4

The worst-case time for binary search for input of size n is $\Theta(\lg n)$.

Proof. The proof precedes the statement of the theorem.   ∎

For our last example, we present and analyze another sorting algorithm known as **merge sort** (Algorithm 5.3.8). We will show that merge sort has worst-case run time $\Theta(n \lg n)$, so for large input, merge sort is much faster than selection sort (Algorithm 5.3.1), which has worst-case run time $\Theta(n^2)$. In Section 7.7 we will show that *any* sorting algorithm that compares elements and, based on the result of a comparison, moves items around in an array is $\Omega(n \lg n)$ in the worst case; thus merge sort is optimal within this class of sorting algorithms.

In merge sort the sequence to be sorted,

$$s_i, \ldots, s_j,$$

is divided into two nearly equal sequences,

$$s_i, \ldots, s_m, \qquad s_{m+1}, \ldots, s_j,$$

where $m = \lfloor (i + j)/2 \rfloor$. Each of these sequences is recursively sorted, after which they are combined to produce a sorted arrangement of the original sequence. The process of combining two sorted sequences is called **merging**.

> ### ALGORITHM 5.3.5   *Merging Two Sequences*

This algorithm combines two increasing sequences into a single increasing sequence.

Input: Two increasing sequences: $s_i, \ldots, s_m$ and $s_{m+1}, \ldots, s_j$, and indexes i, m, and j

Output: The sequence $c_i, \ldots, c_j$ consisting of the elements $s_i, \ldots, s_m$ and $s_{m+1}, \ldots, s_j$ combined into one increasing sequence

```
1.   procedure merge(s, i, m, j, c)
2.      // p is the position in the sequence s_i, ..., s_m
3.      // q is the position in the sequence s_{m+1}, ..., s_j
4.      // r is the position in the sequence c_i, ..., c_j
5.      p := i
6.      q := m + 1
7.      r := i
8.      // copy smaller of s_p and s_q
9.      while p ≤ m and q ≤ j do
10.        begin
11.        if s_p < s_q then
12.           begin
13.           c_r := s_p
14.           p := p + 1
15.           end
16.        else
17.           begin
18.           c_r := s_q
19.           q := q + 1
20.           end
21.        r := r + 1
22.        end
23.      // copy remainder of first sequence
24.      while p ≤ m do
25.        begin
26.        c_r := s_p
27.        p := p + 1
28.        r := r + 1
29.        end
30.      // copy remainder of second sequence
31.      while q ≤ j do
32.        begin
33.        c_r := s_q
34.        q := q + 1
35.        r := r + 1
36.        end
37.   end merge
```

EXAMPLE 5.3.6

Figure 5.3.1 shows how Algorithm 5.3.5 merges the sequences

$$1, 3, 4 \qquad 2, 4, 5, 6.$$ □

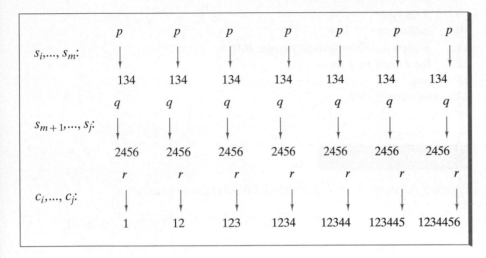

FIGURE 5.3.1   Merging $s_i, \ldots, s_m$ and $s_{m+1}, \ldots, s_j$. The result is $c_i, \ldots, c_j$.

Theorem 5.3.7 shows that in the worst case, $n - 1$ comparisons are needed to merge two sequences the sum of whose lengths is n.

In the worst case, Algorithm 5.3.5 requires $j - i$ comparisons. In particular, in the worst case, $n - 1$ comparisons are needed to merge two sequences the sum of whose lengths is n.

Proof. In Algorithm 5.3.5, the comparison of elements in the sequences occurs in the while loop at line 11. The while loop will execute as long as $p \leq m$ and $q \leq j$. Thus, in the worst case, Algorithm 5.3.5 requires $j - i$ comparisons. ■

We next use Algorithm 5.3.5 (merging) to construct merge sort.

ALGORITHM 5.3.8   *Merge Sort*

This recursive algorithm sorts a sequence into increasing order by using Algorithm 5.3.5, which merges two increasing sequences.

   Input: $s_i, \ldots, s_j, i$, and j

  Output: $s_i, \ldots, s_j$, arranged in increasing order

```
1.  procedure merge_sort(s, i, j)
2.      // base case: i = j
3.      if  i = j then
4.        return
5.      // divide sequence and sort
6.      m := ⌊(i + j)/2⌋
7.      call merge_sort(s, i, m)
8.      call merge_sort(s, m + 1, j)
9.      // merge
10.     call merge (s, i, m, j, c)
11.     // copy c, the output of merge, into s
12.     for k := i to j do
13.       s_k := c_k
14.  end merge_sort
```

EXAMPLE 5.3.9

Figure 5.3.2 shows how Algorithm 5.3.8 sorts the sequence

$$12, \quad 30, \quad 21, \quad 8, \quad 6, \quad 9, \quad 1, \quad 7. \qquad \square$$

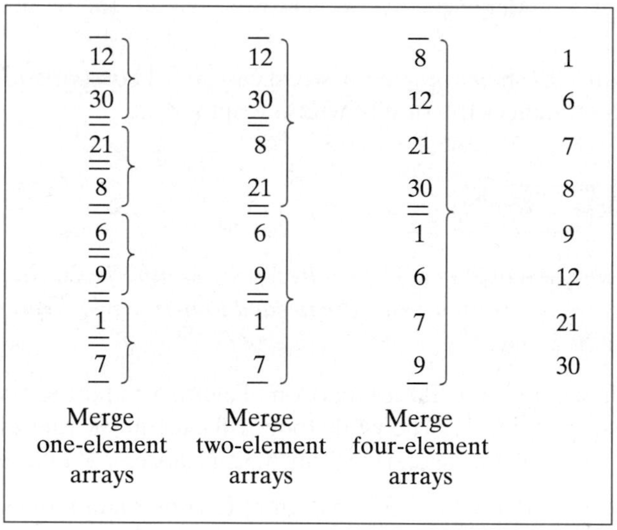

Merge one-element arrays	Merge two-element arrays	Merge four-element arrays	
12	12	8	1
30	30	12	6
21	8	21	7
8	21	30	8
6	6	1	9
9	9	6	12
1	1	7	21
7	7	9	30

FIGURE 5.3.2   Sorting by merge sort.

We conclude by showing that merge sort (Algorithm 5.3.8) is $\Theta(n \lg n)$ in the worst case. The method of proof is the same as we used to show that binary search is $\Theta(\lg n)$ in the worst case.

THEOREM 5.3.10

Merge sort (Algorithm 5.3.8) is $\Theta(n \lg n)$ in the worst case.

Proof. Let a_n be the number of comparisons required by Algorithm 5.3.8 to sort n items in the worst case. Then $a_1 = 0$. If $n > 1$, a_n is at most the sum of the numbers of comparisons in the worst case resulting from the recursive calls at lines 7 and 8, and the number of comparisons in the worst case required by merge at line 10. That is,

$$a_n \leq a_{\lfloor n/2 \rfloor} + a_{\lfloor (n+1)/2 \rfloor} + n - 1.$$

In fact, this upper bound is achievable (see Exercise 11), so that

$$a_n = a_{\lfloor n/2 \rfloor} + a_{\lfloor (n+1)/2 \rfloor} + n - 1.$$

First we solve the preceding recurrence relation in case n is a power of 2, say $n = 2^k$. The equation becomes

$$a_{2^k} = 2a_{2^{k-1}} + 2^k - 1.$$

We may solve this last equation by using iteration (see Section 5.2):

$$
\begin{aligned}
a_{2^k} &= 2a_{2^{k-1}} + 2^k - 1 \\
&= 2[2a_{2^{k-2}} + 2^{k-1} - 1] + 2^k - 1 \\
&= 2^2 a_{2^{k-2}} + 2 \cdot 2^k - 1 - 2 \\
&= 2^2 [2a_{2^{k-3}} + 2^{k-2} - 1] + 2 \cdot 2^k - 1 - 2 \\
&= 2^3 a_{2^{k-3}} + 3 \cdot 2^k - 1 - 2 - 2^2 \\
&\;\;\vdots \\
&= 2^k a_{2^0} + k \cdot 2^k - 1 - 2 - 2^2 - \cdots - 2^{k-1} \\
&= k \cdot 2^k - (2^k - 1) \\
&= (k - 1)2^k + 1. \quad\quad\quad\quad\quad\quad\quad\quad (5.3.8)
\end{aligned}
$$

An arbitrary value of n falls between two powers of 2, say,

$$2^{k-1} < n \leq 2^k. \quad\quad\quad\quad\quad\quad (5.3.9)$$

Since the sequence a is increasing (see Exercise 14),

$$a_{2^{k-1}} \leq a_n \leq a_{2^k}. \quad\quad\quad\quad\quad\quad (5.3.10)$$

Notice that (5.3.9) gives
$$k - 1 < \lg n \leq k. \quad\quad\quad\quad\quad\quad (5.3.11)$$

From (5.3.8), (5.3.10), and (5.3.11), we deduce that

$$\Omega(n \lg n) = (-2 + \lg n)\frac{n}{2} < (k-2)2^{k-1} + 1 = a_{2^{k-1}}$$

$$\leq a_n \leq a_{2^k} \leq k2^k + 1 \leq (1 + \lg n)2n + 1 = O(n \lg n).$$

Therefore $a_n = \Theta(n \lg n)$, so merge sort is $\Theta(n \lg n)$ in the worst case.   ∎

As remarked previously, in Section 7.7 we will show that any comparison-based sorting algorithm is $\Omega(n \lg n)$ in the worst case. This result implies, in particular, that merge sort is $\Omega(n \lg n)$ in the worst case. If we had already proved this result, to prove that merge sort is $\Theta(n \lg n)$ in the worst case, it would have been sufficient to prove that merge sort is $O(n \lg n)$ in the worst case.

Even though merge sort, Algorithm 5.3.8, is optimal, it may not be the algorithm of choice for a particular sorting problem. Factors such as the average-case time of the algorithm, the number of items to be sorted, available memory, the data structures to be used, whether the items to be sorted are in memory or reside on peripheral storage devices such as disks or tapes, whether the items to be sorted are already "nearly" sorted, and the hardware to be used must be taken into account.

৩৲ ৩৲ ৩৲

Exercises

Exercises 1–4 refer to the sequence

$$s_1 = \text{'C'}, \quad s_2 = \text{'G'}, \quad s_3 = \text{'J'}, \quad s_4 = \text{'M'}, \quad s_5 = \text{'X'}.$$

1. Show how Algorithm 5.3.2 executes in case $key = $ 'G'.
2. Show how Algorithm 5.3.2 executes in case $key = $ 'P'.
3. Show how Algorithm 5.3.2 executes in case $key = $ 'C'.
4. Show how Algorithm 5.3.2 executes in case $key = $ 'Z'.
5. Let a_n denote the worst-case time of binary search (Algorithm 5.3.2). Prove that $a_n \le a_{n+1}$ for $n \ge 1$.

★ 6. Prove that if a_n is the number of times the binary search algorithm (Algorithm 5.3.2) is invoked in the worst case for a sequence containing n items, then

$$a_n = 2 + \lfloor \lg n \rfloor$$

for every positive integer n.

7. Suppose that algorithm A requires $\lceil n \lg n \rceil$ comparisons to sort n items and algorithm B requires $\lceil n^2/4 \rceil$ comparisons to sort n items. For which n is algorithm B superior to algorithm A?
8. Show how merge sort (Algorithm 5.3.8) sorts the sequence 1, 9, 7, 3.
9. Show how merge sort (Algorithm 5.3.8) sorts the sequence 2, 3, 7, 2, 8, 9, 7, 5, 4.
10. Suppose that we have two sequences each of size n sorted in increasing order.
    (a) Under what conditions does the maximum number of comparisons occur in Algorithm 5.3.5?
    (b) Under what conditions does the minimum number of comparisons occur in Algorithm 5.3.5?
11. Let a_n be as in the proof of Theorem 5.3.10. Describe input for which

$$a_n = a_{\lfloor n/2 \rfloor} + a_{\lfloor (n+1)/2 \rfloor} + n - 1.$$

12. What is the minimum number of comparisons required by Algorithm 5.3.8 to sort an array of size 6?
13. What is the maximum number of comparisons required by Algorithm 5.3.8 to sort an array of size 6?

14. Let a_n be as in the proof of Theorem 5.3.10. Show that $a_n \leq a_{n+1}$ for all $n \geq 1$.

15. Let a_n denote the number of comparisons required by merge sort in the worst case. Show that $a_n \leq 3n \lg n$ for $n = 1, 2, 3, \ldots$.

16. Show that in the best case, merge sort requires $\Theta(n \lg n)$ comparisons.

Exercises 17–21 refer to Algorithm 5.3.11.

ALGORITHM 5.3.11   *Computing an Exponential*

This algorithm computes a^n recursively, where a is a real number and n is a positive integer.

   Input: a (a real number), n (a positive integer)

   Output: a^n

```
1.     procedure exp1(a, n)
2.       if n = 1 then
3.         return(a)
4.       m := ⌊n/2⌋
5.       return(exp1(a, m) · exp1(a, n − m))
6.     end exp1
```

Let b_n be the number of multiplications (line 5) required to compute a^n.

17. Explain how Algorithm 5.3.11 computes a^n.

18. Find a recurrence relation and initial conditions for the sequence $\{b_n\}$.

19. Compute b_2, b_3, and b_4.

20. Solve the recurrence relation of Exercise 18 in case n is a power of 2.

21. Prove that $b_n = n - 1$ for every positive integer n.

Exercises 22–27 refer to Algorithm 5.3.12.

ALGORITHM 5.3.12   *Computing an Exponential*

This algorithm computes a^n recursively, where a is a real number and n is a positive integer.

   Input: a (a real number), n (a positive integer)

   Output: a^n

```
1.     procedure exp2(a, n)
2.       if n = 1 then
3.         return(a)
4.       m := ⌊n/2⌋
5.       power := exp2(a, m)
6.       power := power · power
7.       if n is even then
8.         return(power)
9.       else
10.        return(power · a)
11.    end exp2
```

Let b_n be the number of multiplications (lines 6 and 10) required to compute a^n.

22. Explain how Algorithm 5.3.12 computes a^n.

23. Show that
$$b_n = \begin{cases} b_{(n-1)/2} + 2, & \text{if } n \text{ is odd;} \\ b_{n/2} + 1, & \text{if } n \text{ is even.} \end{cases}$$

24. Find b_1, b_2, b_3, and b_4.

25. Solve the recurrence relation of Exercise 23 in case n is a power of 2.

26. Show, by an example, that b is not increasing.

★ 27. Prove that $b_n = \Theta(\lg n)$.

Exercises 28–33 refer to Algorithm 5.3.13.

| ALGORITHM 5.3.13 | *Finding the Largest and Smallest Elements in a Sequence* |

This recursive algorithm finds the largest and smallest elements in a sequence.

Input: $s_i, \ldots, s_j, i$, and j

Output: *large* (the largest element in the sequence), *small* (the smallest element in the sequence)

```
1.    procedure large_small(s, i, j, large, small)
2.      if i = j then
3.        begin
4.        large := sᵢ
5.        small := sᵢ
6.        return
7.        end
8.      m := ⌊(i + j)/2⌋
9.      large_small(s, i, m, large_left, small_left)
10.     large_small(s, m + 1, j, large_right, small_right)
11.     if large_left > large_right then
12.       large := large_left
13.     else
14.       large := large_right
15.     if small_left > small_right then
16.       small := small_right
17.     else
18.       small := small_left
19.   end large_small
```

Let b_n be the number of comparisons (lines 11 and 15) required for an input of size n.

28. Explain how Algorithm 5.3.13 finds the largest and smallest elements.

29. Show that $b_1 = 0$ and $b_2 = 2$.

30. Find b_3.

31. Establish the recurrence relation

$$b_n = b_{\lfloor n/2 \rfloor} + b_{\lfloor (n+1)/2 \rfloor} + 2 \tag{5.3.12}$$

for $n > 1$.

32. Solve the recurrence relation (5.3.12) in case n is a power of 2 to obtain

$$b_n = 2n - 2, \qquad n = 1, 2, 4, \ldots .$$

33. Use induction to show that

$$b_n = 2n - 2$$

for every positive integer n.

Exercises 34–37 refer to Algorithm 5.3.13, with the following inserted after line 7.

```
7a.   if j = i + 1 then
7b.      begin
7c.      if s_i > s_j then
7d.         begin
7e.         large := s_i
7f.         small := s_j
7g.         end
7h.      else
7i.         begin
7j.         small := s_i
7k.         large := s_j
7l.         end
7m.      return
7n.   end
```

Let b_n be the number of comparisons (lines 7c, 11, and 15) for an input of size n.

34. Show that $b_1 = 0$ and $b_2 = 1$.

35. Compute b_3 and b_4.

36. Show that the recurrence relation (5.3.12) holds for $n > 2$.

37. Solve the recurrence relation (5.3.12) in case n is a power of 2 to obtain

$$b_n = \frac{3n}{2} - 2, \quad n = 2, 4, 8, \ldots .$$

★ **38.** Modify Algorithm 5.3.13 by inserting the lines preceding Exercise 34 after line 7 and replacing line 8 with the following.

```
8a.   if j - i is odd and (1 + j - i)/2 is odd then
8b.      m := ⌊(i + j)/2⌋ - 1
8c.   else
8d.      m := ⌊(i + j)/2⌋
```

Show that in the worse case, this modified algorithm requires at most $\lceil (3n/2) - 2 \rceil$ comparisons to find the largest and smallest elements in an array of size n.

Exercises 39–41 refer to Algorithm 5.3.14.

ALGORITHM 5.3.14   *Insertion Sort*

This algorithm sorts the sequence

$$s_1, \quad s_2, \quad \ldots, \quad s_n$$

in increasing order by recursively sorting the first $n - 1$ elements and then inserting s_n in the correct position.

Input: $s_1, s_2, \ldots, s_n$ and the length n of the sequence

Output: $s_1, s_2, \ldots, s_n$ arranged in increasing order

```
1.      procedure insertion_sort(s, n)
2.         if n = 1 then
3.            return
4.         insertion_sort(s, n − 1)
5.         i := n − 1
6.         temp := s_n
7.         while i ≥ 1 and s_i > temp do
8.            begin
9.               s_{i+1} := s_i
10.              i := i − 1
11.           end
12.        s_{i+1} := temp
13.     end insertion_sort
```

Let b_n be the number of times the comparison $s_i > temp$ in line 7 is made in the worst case. Assume that if $i < 1$, the comparison $s_i > temp$ is not made.

39. Explain how Algorithm 5.3.14 sorts the sequence.

40. Which input produces the worst-case behavior for Algorithm 5.3.14?

41. Find b_1, b_2, and b_3.

42. Find a recurrence relation for the sequence $\{b_n\}$.

43. Solve the recurrence relation of Exercise 42.

Exercises 44–46 refer to Algorithm 5.3.15.

ALGORITHM 5.3.15

Input: $s_1, \ldots, s_n, n$
Output: $s_1, \ldots, s_n$

```
procedure algor(s, n)
   i := n
   while i ≥ 1 do
      begin
         s_i := s_i + 1
         i := ⌊i/2⌋
      end
   n := ⌊n/2⌋
   if n ≥ 1 then
      algor(s, n)
end algor
```

Let b_n be the number of times the statement $s_i := s_i + 1$ is executed.

44. Find a recurrence relation for the sequence $\{b_n\}$ and compute b_1, b_2, and b_3.

45. Solve the recurrence relation of Exercise 44 in case n is a power of 2.

46. Prove that $b_n = \Theta((\lg n)^2)$.

47. Solve the recurrence relation

$$a_n = 3a_{\lfloor n/2 \rfloor} + n, \qquad n > 1,$$

in case n is a power of 2. Assume that $a_1 = 1$.

48. Show that $a_n = \Theta(n^{\lg 3})$, where a_n is as in Exercise 47.

Exercises 49–56 refer to an algorithm that accepts as input the sequence

$$s_i, \ldots, s_j.$$

If $j > i$, the subproblems

$$s_i, \ldots, s_{\lfloor (i+j)/2 \rfloor} \quad \text{and} \quad s_{\lfloor (i+j)/2+1 \rfloor}, \ldots, s_j$$

are solved recursively. Solutions to subproblems of sizes m and k can be combined in time $c_{m,k}$ to solve the original problem. Let b_n be the time required by the algorithm for an input of size n.

49. Write a recurrence relation for b_n assuming that $c_{m,k} = 3$.

50. Write a recurrence relation for b_n assuming that $c_{m,k} = m + k$.

51. Solve the recurrence relation of Exercise 49 in case n is a power of 2, assuming that $b_1 = 0$.

52. Solve the recurrence relation of Exercise 49 in case n is a power of 2, assuming that $b_1 = 1$.

53. Solve the recurrence relation of Exercise 50 in case n is a power of 2, assuming that $b_1 = 0$.

54. Solve the recurrence relation of Exercise 50 in case n is a power of 2, assuming that $b_1 = 1$.

★ **55.** Assume that if $m_1 \geq m_2$ and $k_1 \geq k_2$, then $c_{m_1,k_1} \geq c_{m_2,k_2}$. Show that the sequence $b_1, b_2, \ldots$ is increasing.

★ **56.** Assuming that $c_{m,k} = m + k$ and $b_1 = 0$, show that $b_n \leq 4n \lg n$.

Exercises 57–62 refer to the following situation. We let P_n denote a particular problem of size n. If P_n is divided into subproblems of sizes i and j, there is an algorithm that combines the solutions of these two subproblems into a solution to P_n in time at most $2 + \lg(ij)$. Assume that a problem of size 1 is already solved.

57. Write a recursive algorithm to solve P_n similar to Algorithm 5.3.8.

58. Let a_n be the worst-case time to solve P_n by the algorithm of Exercise 57. Show that

$$a_n \leq a_{\lfloor n/2 \rfloor} + a_{\lfloor (n+1)/2 \rfloor} + 2 \lg n.$$

59. Let b_n be the recurrence relation obtained from Exercise 58 by replacing "$\leq$" by "$=$". Assume that $b_1 = a_1 = 0$. Show that if n is a power of 2,

$$b_n = 4n - 2 \lg n - 4.$$

60. Show that $a_n \leq b_n$ for $n = 1, 2, 3, \ldots$.

61. Show that $b_n \leq b_{n+1}$ for $n = 1, 2, 3, \ldots$.

62. Show that $a_n \leq 8n$ for $n = 1, 2, 3, \ldots$.

63. Suppose that $\{a_n\}$ is an increasing sequence and that whenever m divides n,

$$a_n = a_{n/m} + d,$$

where d is a positive real number and m is an integer satisfying $m > 1$. Show that $a_n = \Theta(\lg n)$.

★ **64.** Suppose that $\{a_n\}$ is an increasing sequence and that whenever m divides n,

$$a_n = ca_{n/m} + d,$$

where c and d are real numbers satisfying $c > 1$ and $d > 0$, and m is an integer satisfying $m > 1$. Show that $a_n = \Theta(n^{\log_m c})$.

65. [Project] Investigate other sorting algorithms. Consider specifically complexity, empirical studies, and special features of the algorithms (see [Knuth, 1973, Vol. 3]).

⤳ NOTES

Recurrence relations are treated more fully in [Liu, 1985; Roberts; and Tucker]. Several applications to the analysis of algorithms are presented in [Cormen].

[Cull] gives algorithms for solving certain Tower of Hanoi problems with minimum space and time complexity. [Hinz] is a comprehensive discussion of the Tower of Hanoi with 50 references.

The cobweb in economics first appeared in [Ezekiel].

All data structures and algorithms books have extended discussions of searching and sorting (see, e.g., [Brassard; Cormen; Knuth, 1973, Vol. 3; Kruse; and Nyhoff]).

Recurrence relations are also called *difference equations*. [Goldberg] contains a discussion of difference equations and applications.

⤳ CHAPTER REVIEW

Section 5.1

Recurrence relation
Initial condition
Compound interest
Tower of Hanoi
Cobweb in economics
Ackermann's function

Section 5.2

Solving a recurrence relation by iteration
nth-order linear homogeneous recurrence relation with constant coefficients and how to solve a second-order recurrence relation
Population growth

Section 5.3

How to find a recurrence relation that describes the time required by a recursive algorithm
Selection sort
Binary search
Merging sequences
Merge sort

⤳ CHAPTER SELF-TEST

Section 5.1

1. Answer parts (a)–(c) for the sequence defined by the rules:

    1. The first term is 3.

    2. The nth term is n plus the previous term.

    (a) Write the first four terms of the sequence.

    (b) Find an initial condition for the sequence.

    (c) Find a recurrence relation for the sequence.

2. Assume that a person invests $4000 at 17 percent compounded annually. Let A_n represent the amount at the end of n years. Find a recurrence relation and an initial condition for the sequence $A_0, A_1, \ldots$.

3. Let P_n be the number of partitions of an n-element set. Show that the sequence $P_0, P_1, \ldots$ satisfies the recurrence relation

$$P_n = \sum_{k=0}^{n-1} C(n-1, k) P_k.$$

4. Suppose that we have a $2 \times n$ rectangular board divided into $2n$ squares. Let a_n denote the number of ways to exactly cover this board by 1×2 dominoes. Show that the sequence $\{a_n\}$ satisfies the recurrence relation

$$a_n = a_{n-1} + a_{n-2}.$$

Show that $a_n = f_n$, where $\{f_n\}$ is the Fibonacci sequence.

Section 5.2

5. Is the recurrence relation

$$a_n = a_{n-1} + a_{n-3}$$

a linear homogeneous recurrence relation with constant coefficients?

In Exercises 6 and 7, solve the recurrence relation subject to the initial conditions.

6. $a_n = -4a_{n-1} - 4a_{n-2};$ $a_0 = 2,$ $a_1 = 4$

7. $a_n = 3a_{n-1} + 10a_{n-2};$ $a_0 = 4,$ $a_1 = 13$

8. Let c_n denote the number of strings over $\{0, 1, 2\}$ of length n that contain an even number of 1's. Write a recurrence relation and initial condition that define the sequence $c_1, c_2, \ldots$. Solve the recurrence relation to obtain an explicit formula for c_n.

Section 5.3

Exercises 9–12 refer to the following algorithm.

ALGORITHM   *Polynomial Evaluation*

This algorithm evaluates the polynomial

$$p(x) = \sum_{k=0}^{n} c_k x^{n-k}$$

at the point t.

   Input: The sequence of coefficients $c_0, c_1, \ldots, c_n$, the value t, and n
 Output: $p(t)$

```
procedure poly(c, n, t)
  if n = 0 then
    return(c₀)
  return(t· poly(c, n − 1, t) + cₙ)
end poly
```

Let b_n be the number of multiplications required to compute $p(t)$.

9. Find a recurrence relation and an initial condition for the sequence $\{b_n\}$.

10. Compute $b_1, b_2,$ and b_3.

11. Solve the recurrence relation of Exercise 9.

12. Suppose that we compute $p(t)$ by a straightforward technique that requires $n - k$ multiplications to compute $c_k t^{n-k}$. How many multiplications would be required to compute $p(t)$? Would you prefer this method or the preceding algorithm? Explain.

6

GRAPH THEORY

Well, I got on the road, and I went north to Providence. Met the Mayor.
The Mayor of Providence!
He was sitting in the hotel lobby. What'd he say?
He said, "Morning!" And I said, "You got a fine city here, Mayor."
And then he had coffee with me. And then I went to Waterbury. Waterbury is a fine city. Big clock city, the famous Waterbury clock. Sold a nice bill there. And then Boston—Boston is the cradle of the Revolution. A fine city. And a couple of other towns in Mass., and on to Portland and Bangor and straight home!

—from Death of a Salesman

Although the first paper in graph theory goes back to 1736 (see Example 6.2.16) and several important results in graph theory were obtained in the nineteenth century, it is only since the 1920s that there has been a sustained, widespread, intense interest in graph theory. Indeed, the first text on graph theory ([König]) appeared in 1936. Undoubtedly, one of the reasons for the recent interest in graph theory is its applicability in many diverse fields, including computer science, chemistry, operations research, electrical engineering, linguistics, and economics.

We begin with some basic graph terminology and examples. We then discuss some important concepts in graph theory, including paths and cycles. A shortest-path algorithm is presented that efficiently finds a shortest path between two given points. Two classical graph problems, the existence of Hamiltonian cycles and the traveling salesperson

† This section can be omitted without loss of continuity.

problem, are then considered. After presenting ways of representing graphs, we study the question of when two graphs are essentially the same (i.e., when two graphs are isomorphic) and when a graph can be drawn in the plane without having any of its edges cross. We conclude by presenting a solution based on a graph model to the Instant Insanity puzzle.

6.1 *INTRODUCTION*

Figure 6.1.1 shows the highway system in Wyoming that a particular person is responsible for inspecting. Specifically, this road inspector must travel all of these roads and file reports on road conditions, visibility of lines on the roads, status of traffic signs, and so on. Since the road inspector lives in Greybull, the most economical way to inspect all of the roads would be to start in Greybull, travel each of the roads exactly once, and return to Greybull. Is this possible? See if you can decide before reading on.

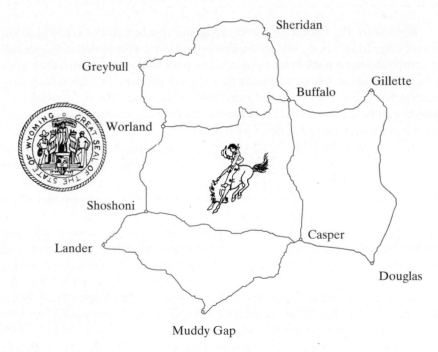

FIGURE 6.1.1   Part of the Wyoming highway system.

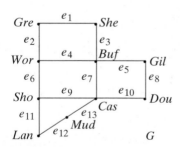

FIGURE 6.1.2
A graph model of the highway system shown in Figure 6.1.1.

The problem can be modeled as a **graph**. In fact, since graphs are drawn with dots and lines, they look like road maps. In Figure 6.1.2, we have drawn a graph G that models the map of Figure 6.1.1. The dots in Figure 6.1.2 are called **vertices** and the lines that connect the vertices are called **edges**. (Later in this section we will define all of these terms carefully.) We have labeled each vertex with the first three letters of the city to which it corresponds. We have labeled the edges $e_1, \ldots, e_{13}$. In drawing a graph, the only information of importance is which vertices are connected by which edges. For this reason, the graph of Figure 6.1.2 could just as well be drawn as in Figure 6.1.3.

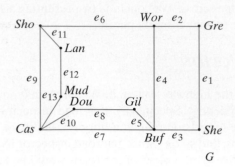

FIGURE 6.1.3   An alternative, but equivalent, graph model of the highway system shown in Figure 6.1.1.

If we start at a vertex v_0, travel along an edge to vertex v_1, travel along another edge to vertex v_2, and so on, and eventually arrive at vertex v_n, we call the complete tour a **path** from v_0 to v_n. The path that starts at *She*, then goes to *Buf*, and ends at *Gil* corresponds to a trip on the map of Figure 6.1.1 that begins in Sheridan, goes to Buffalo, and ends at Gillette. The road inspector's problem can be rephrased for the graph model G in the following way: Is there a path from vertex *Gre* to vertex *Gre* that traverses every edge exactly once?

We can show that the road inspector cannot start in Greybull, travel each of the roads exactly once, and return to Greybull. To put the answer in graph terms, there is no path from vertex *Gre* to vertex *Gre* in Figure 6.1.2 that traverses every edge exactly once. To see this, suppose that there is such a path and consider vertex *Wor*. Each time we arrive at *Wor* on some edge, we must leave *Wor* on a different edge. Furthermore, every edge that touches *Wor* must be used. Thus the edges at *Wor* occur in pairs. It follows that an even number of edges must touch *Wor*. Since three edges touch *Wor*, we have a contradiction. Therefore, there is no path from vertex *Gre* to vertex *Gre* in Figure 6.1.2 that traverses every edge exactly once. The argument applies to an arbitrary graph G. If G has a path from vertex v to v that traverses every edge exactly once, an even number of edges must touch each vertex. We discuss this problem in greater detail in Section 6.2.

At this point we give some formal definitions.

DEFINITION 6.1.1

A *graph* (or *undirected graph*) G consists of a set V of *vertices* (or *nodes*) and a set E of *edges* (or *arcs*) such that each edge $e \in E$ is associated with an unordered pair of vertices. If there is a unique edge e associated with the vertices v and w, we write $e = (v, w)$ or $e = (w, v)$. In this context, (v, w) denotes an edge between v and w in an undirected graph and *not* an ordered pair.

A *directed graph* (or *digraph*) G consists of a set V of *vertices* (or *nodes*) and a set E of *edges* (or *arcs*) such that each edge $e \in E$ is associated with an ordered pair of vertices. If there is a unique edge e associated with the ordered pair (v, w) of vertices, we write $e = (v, w)$, which denotes an edge from v to w.

An edge e in a graph (undirected or directed) that is associated with the pair of vertices v and w is said to be *incident on* v and w, and v and w are said to be *incident on* e and to be *adjacent vertices*.

If G is a graph (undirected or directed) with vertices V and edges E, we write $G = (V, E)$.

Unless specified otherwise, the sets E and V are assumed to be finite and V is assumed to be nonempty.

EXAMPLE 6.1.2

In Figure 6.1.2 the (undirected) graph G consists of the set

$$V = \{Gre,\ She,\ Wor,\ Buf,\ Gil,\ Sho,\ Cas,\ Dou,\ Lan,\ Mud\}$$

of vertices and the set

$$E = \{e_1, e_2, \ldots, e_{13}\}$$

of edges. Edge e_1 is associated with the unordered pair $\{Gre, She\}$ of vertices and edge e_{10} is associated with the unordered pair $\{Cas, Dou\}$ of vertices. Edge e_1 is denoted (Gre, She) or (She, Gre) and edge e_{10} is denoted (Cas, Dou) or (Dou, Cas). Edge e_4 is incident on Wor and Buf and the vertices Wor and Buf are adjacent. $\qquad\square$

EXAMPLE 6.1.3

A directed graph is shown in Figure 6.1.4. The directed edges are indicated by arrows. Edge e_1 is associated with the ordered pair (v_2, v_1) of vertices and edge e_7 is associated with the ordered pair (v_6, v_6) of vertices. Edge e_1 is denoted (v_2, v_1) and edge e_7 is denoted (v_6, v_6). $\qquad\square$

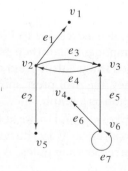

FIGURE 6.1.4   A directed graph.

Definition 6.1.1 allows distinct edges to be associated with the same pair of vertices. For example, in Figure 6.1.5 below, edges e_1 and e_2 are both associated with the vertex pair $\{v_1, v_2\}$. Such edges are called **parallel edges**. An edge incident on a single vertex is called a **loop**. For example, in Figure 6.1.5 edge $e_3 = (v_2, v_2)$ is a loop. A vertex, such as vertex v_4 in Figure 6.1.5, that is not incident on any edge is called an **isolated vertex**. A graph with neither loops nor parallel edges is called a **simple graph**.

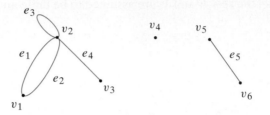

FIGURE 6.1.5   A graph with parallel edges and loops.

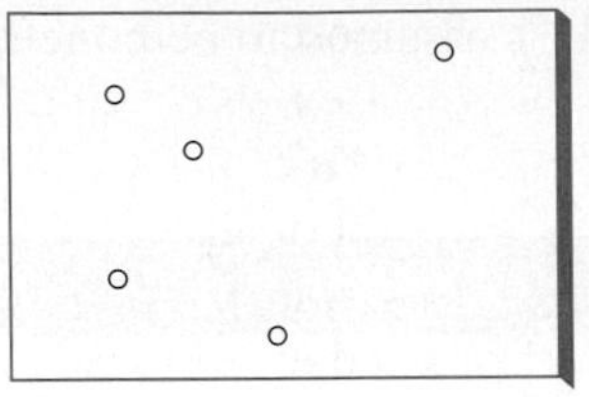

FIGURE 6.1.6   A sheet of metal with holes for bolts.

EXAMPLE 6.1.4

Since the graph of Figure 6.1.2 has neither parallel edges nor loops, it is a simple graph.   □

Some authors do not permit loops and parallel edges when they define graphs. One would expect that if agreement has not been reached on the definition of graph, most other terms in graph theory would also not have standard definitions. This is indeed the case. In reading articles and books about graphs, it is necessary to check on the definitions being used.

We turn next to an example that shows how a graph model can be used to analyze a manufacturing problem.

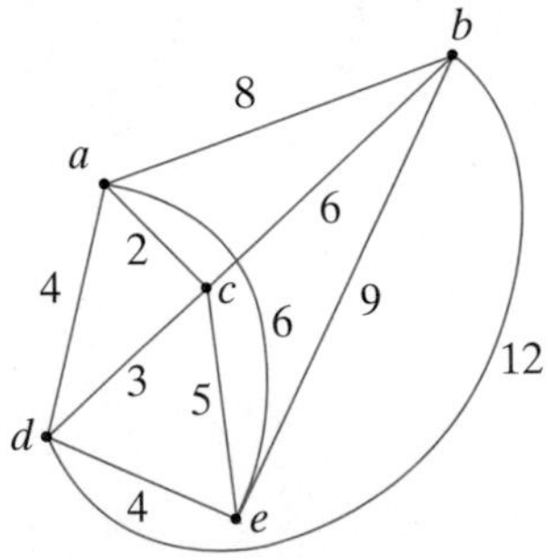

EXAMPLE 6.1.5

Frequently in manufacturing, it is necessary to bore many holes in sheets of metal (see Figure 6.1.6). Components can then be bolted to these sheets of metal. The holes can be drilled using a drill press under the control of a computer. To save time and money, the drill press should be moved as quickly as possible. We model the situation as a graph.

The vertices of the graph correspond to the holes (see Figure 6.1.7). Every pair of vertices is connected by an edge. We write on each edge the time to move the drill press between the corresponding holes. A graph with numbers on the

FIGURE 6.1.7
A graph model of sheet metal in Figure 6.1.6. The edge weight is the time to move the drill press.

edges (such as the graph of Figure 6.1.7) is called a **weighted graph**. If edge e is labeled k, we say that the **weight of edge** e is k. For example, in Figure 6.1.7 the weight of edge (c, e) is 5. In a weighted graph, the **length of a path** is the sum of the weights of the edges in the path. For example, in Figure 6.1.7 the length of the path that starts at a, visits c, and terminates at b is 8. In this problem, the length of a path that starts at vertex v_1 and then visits $v_2, v_3, \ldots$, in this order, and terminates at v_n represents the time it takes the drill press to start at hole h_1 and then move to $h_2, h_3, \ldots$, in this order, and terminate at h_n, where hole h_i corresponds to vertex v_i. A path of minimum length that visits every vertex exactly one time represents the optimal path for the drill press to follow.

Suppose that in this problem the path is required to begin at vertex a and end at vertex e. We can find the minimum-length path by listing all possible paths from a to e that pass through every vertex exactly one time and choose the shortest one (see Table 6.1.1). We see that the path that visits the vertices a, b, c, d, e, in this order, has minimum length. Of course, a different pair of starting and ending vertices might produce an even shorter path.   □

TABLE 6.1.1
Paths in the graph of Figure 6.1.7 from a to e that pass through every vertex exactly one time, and their lengths

Path	Length
a, b, c, d, e	21
a, b, d, c, e	28
a, c, b, d, e	24
a, c, d, b, e	26
a, d, b, c, e	27
a, d, c, b, e	22

Listing all paths from vertex v to vertex w, as we did in Example 6.1.5, is a rather time-consuming way to find a minimum-length path from v to w that visits every vertex exactly one time. Unfortunately, no one knows a method that is much more practical for arbitrary graphs. This problem is a version of the **traveling salesperson problem**. We discuss that problem in Section 6.3.

EXAMPLE 6.1.6   *Similarity Graphs*

This example deals with the problem of grouping "like" objects into classes based on properties of the objects. For example, suppose that a particular algorithm is implemented in C by a number of persons and we want to group "like" programs into classes based on certain properties of the programs (see Table 6.1.2). Suppose that we select as properties

1. The number of lines in the program.

2. The number of `return` statements in the program.

3. The number of function calls in the program.

TABLE 6.1.2
C programs that implement the same algorithm

Program	*Number of Program Lines*	*Number of* return *Statements*	*Number of Function Calls*
1	66	20	1
2	41	10	2
3	68	5	8
4	90	34	5
5	75	12	14

A **similarity graph** G is constructed as follows. The vertices correspond to programs. A vertex is denoted (p_1, p_2, p_3), where p_i is the value of property i. We define a **dissimilarity function** s as follows. For each pair of vertices $v = (p_1, p_2, p_3)$ and $w = (q_1, q_2, q_3)$, we set

$$s(v, w) = |p_1 - q_1| + |p_2 - q_2| + |p_3 - q_3|.$$

If we let v_i be the vertex corresponding to program i, we obtain

$$s(v_1, v_2) = 36, \quad s(v_1, v_3) = 24, \quad s(v_1, v_4) = 42, \quad s(v_1, v_5) = 30,$$
$$s(v_2, v_3) = 38, \quad s(v_2, v_4) = 76, \quad s(v_2, v_5) = 48, \quad s(v_3, v_4) = 54,$$
$$s(v_3, v_5) = 20, \quad s(v_4, v_5) = 46.$$

If v and w are vertices corresponding to two programs, $s(v, w)$ is a measure of how dissimilar the programs are. A large value of $s(v, w)$ indicates dissimilarity, while a small value indicates similarity.

For a fixed number S, we insert an edge between vertices v and w if $s(v, w) < S$. (In general, there will be different similarity graphs for different values of S.) We say that v and w are **in the same class** if $v = w$ or there is a path from v to w. In Figure 6.1.8 we show the graph corresponding to the programs of Table 6.1.2 with $S = 25$. In this graph, the programs are grouped into three classes: $\{1, 3, 5\}$, $\{2\}$, and $\{4\}$. In a real problem, an appropriate value for S might be selected by trial and error or the value of S might be selected automatically according to some predetermined criteria.   □

Example 6.1.6 belongs to the subject called **pattern recognition**. Pattern recognition is concerned with grouping data into classes based on properties of the data. Pattern recognition by computer has much practical significance. For example, computers have been programmed to detect cancer from X-rays, to select tax returns to be audited, to analyze satellite pictures, to recognize text, and to forecast weather.

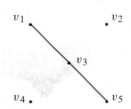

FIGURE 6.1.8
A similarity graph corresponding to the programs of Table 6.1.2 with $S = 25$.

| EXAMPLE 6.1.7 | *The n-Cube (Hypercube)* |

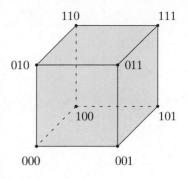

FIGURE 6.1.9
The 3-cube.

The traditional computer, often called a **serial computer**, executes one instruction at a time. Our definition of algorithm also assumes that one instruction is executed at a time. Such algorithms are called **serial algorithms**. Recently, as hardware costs have declined, it has become feasible to build **parallel computers** with many processors that are capable of executing several instructions at a time. Graphs are often convenient models to describe these machines. The associated algorithms are known as **parallel algorithms**. Many problems can be solved much faster using parallel computers rather than serial computers. We discuss one model for parallel computation known as the ***n*-cube** or **hypercube**.

The n-cube has 2^n processors, $n \geq 1$, which are represented by vertices (see Figure 6.1.9) labeled $0, 1, \ldots, 2^n - 1$. Each processor has its own local memory. An edge connects two vertices if the binary representation of their labels differs in exactly one bit. During one time unit, all processors in the n-cube may execute an instruction simultaneously and then communicate with an adjacent processor. If a processor needs to communicate with a nonadjacent processor, the first processor sends a message that includes the route to and ultimate destination of the recipient. It may take several time units for a processor to communicate with a nonadjacent processor.

The n-cube may also be described recursively. The 1-cube has two processors, labeled 0 and 1, and one edge. Let H_1 and H_2 be two $(n - 1)$-cubes whose vertices are labeled in binary $0, \ldots, 2^{n-1} - 1$ (see Figure 6.1.10). We place an edge between each pair of vertices, one from H_1 and one from H_2, provided that the vertices have identical labels. We then change the label L on each vertex in H_1 to $0L$ and we change the label L on each vertex in H_2 to $1L$. We obtain an n-cube (Exercise 32).

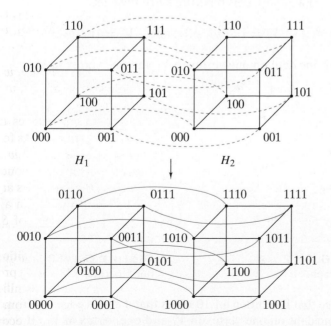

FIGURE 6.1.10   Combining two 3-cubes to obtain a 4-cube.

The n-cube is an important model of computation because several such machines have been built and are running. Furthermore, several other parallel computation models can be simulated by the hypercube. The latter point is considered in more detail in Examples 6.3.5 and 6.6.3.   □

We conclude this introductory section by defining some special graphs that appear frequently in graph theory.

DEFINITION 6.1.8

The *complete graph on n vertices*, denoted K_n, is the simple graph with n vertices in which there is an edge between every pair of distinct vertices.

EXAMPLE 6.1.9

The complete graph on four vertices, K_4, is shown in Figure 6.1.11.   □

DEFINITION 6.1.10

A graph $G = (V, E)$ is *bipartite* if the vertex set V can be partitioned into two subsets V_1 and V_2 such that each edge in E is incident on one vertex in V_1 and one vertex in V_2.

EXAMPLE 6.1.11

The graph in Figure 6.1.12 is bipartite since if we let

$$V_1 = \{v_1, v_2, v_3\} \quad \text{and} \quad V_2 = \{v_4, v_5\},$$

each edge is incident on one vertex in V_1 and one vertex in V_2.   □

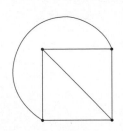

FIGURE 6.1.11
The complete graph K_4.

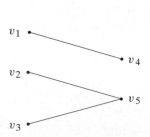

FIGURE 6.1.12   A bipartite graph.

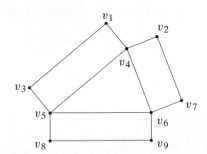

FIGURE 6.1.13   A graph that is not bipartite.

Notice that Definition 6.1.10 states that if e is an edge in a bipartite graph, then e is incident on one vertex in V_1 and one vertex in V_2. It does *not* state that if v_1 is a vertex in V_1 and v_2 is a vertex in V_2, then there is an edge between v_1 and v_2. For example, the graph of Figure 6.1.12 is bipartite since each edge is incident on one vertex in $V_1 = \{v_1, v_2, v_3\}$ and one vertex in $V_2 = \{v_4, v_5\}$. However, not all edges between vertices in V_1 and V_2 are in the graph. For example, the edge (v_1, v_5) is absent.

EXAMPLE 6.1.12

The graph in Figure 6.1.13 is *not* bipartite. It is often easiest to prove that a graph is not bipartite by arguing by contradiction.

Suppose that the graph in Figure 6.1.13 is bipartite. Then the vertex set can be partitioned into two subsets V_1 and V_2 such that each edge is incident on one vertex in V_1 and one vertex in V_2. Now consider the vertices v_4, v_5, and v_6. Since v_4 and v_5 are adjacent, one is in V_1 and the other in V_2. We may assume that v_4 is in V_1 and that v_5 is in V_2. Since v_5 and v_6 are adjacent and v_5 is in V_2, v_6 is in V_1. Since v_4 and v_6 are adjacent and v_4 is in V_1, v_6 is in V_2. But now v_6 is in both V_1 and V_2, which is a contradiction since V_1 and V_2 are disjoint. Therefore the graph in Figure 6.1.13 is not bipartite.   □

DEFINITION 6.1.13

The *complete bipartite graph on m and n vertices*, denoted $K_{m,n}$, is the simple graph whose vertex set is partitioned into sets V_1 with m vertices and V_2 with n vertices in which there is an edge between each pair of vertices v_1 and v_2 where v_1 is in V_1 and v_2 is in V_2.

EXAMPLE 6.1.14

The complete bipartite graph on two and four vertices, $K_{2,4}$, is shown in Figure 6.1.14.

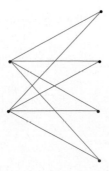

FIGURE 6.1.14   The complete bipartite graph $K_{2,4}$.   □

ᔓ ᔓ ᔓ

Exercises

Explain why none of the graphs in Exercises 1–3 has a path from *a* to *a* that passes through each edge exactly one time.

1.

2.

3.

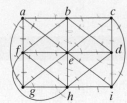

Show that each graph in Exercises 4–6 has a path from *a* to *a* that passes through each edge exactly one time by finding such a path by inspection.

4.

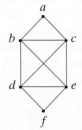

5.

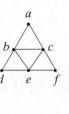

6.

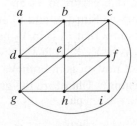

For each graph $G = (V, E)$ in Exercises 7–9, find V, E, all parallel edges, all loops, all isolated vertices, and tell whether G is a simple graph. Also, tell on which vertices edge e_1 is incident.

7.

8.

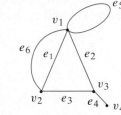

9.

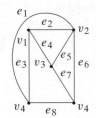

10. Draw K_3 and K_5.

11. Find a formula for the number of edges in K_n.

12. Give an example of a bipartite graph different from that in Figure 6.1.12. Specify the disjoint vertex sets.

State which graphs in Exercises 13–19 are bipartite graphs. If the graph is bipartite, specify the disjoint vertex sets.

13.

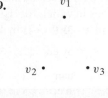

14.

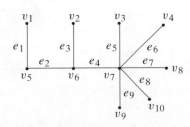

15. Figure 6.1.2

16. Figure 6.1.5

17. Exercise 7

18. Exercise 8

19. Exercise 9

20. Draw $K_{2,3}$ and $K_{3,3}$.

21. Find a formula for the number of edges in $K_{m,n}$.

In Exercises 22–24, find a path of minimum length from v to w in the graph of Figure 6.1.7 that passes through each vertex exactly one time.

22. $v = b, w = e$ **23.** $v = c, w = d$ **24.** $v = a, w = b$

25. Draw the similarity graph that results from setting $S = 40$ in Example 6.1.6. How many classes are there?

26. Draw the similarity graph that results from setting $S = 50$ in Example 6.1.6. How many classes are there?

27. In general, is "is similar to" an equivalence relation?

28. Suggest additional properties for Example 6.1.6 that might be useful in comparing programs.

29. How might one automate the selection of S to group data into classes using a similarity graph?

30. Draw a 2-cube.

31. Draw a picture like that in Figure 6.1.10 to show how a 3-cube may be constructed from two 2-cubes.

32. Prove that the recursive construction in Example 6.1.7 actually yields an n-cube.

33. How many edges are incident on a vertex in an n-cube?

34. How many edges are in an n-cube?

★ **35.** In how many ways can the vertices of an n-cube be labeled $0, \ldots, 2^n - 1$ so that there is an edge between two vertices if and only if the binary representation of their labels differs in exactly one bit?

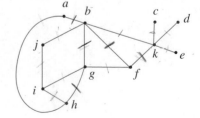

Exercises 36–38 refer to the adjacent graph. The vertices represent offices. An edge connects two offices if there is a communication link between the two. Notice that any office can communicate with any other either directly through a communication link or by having others relay the message.

36. Show, by giving an example, that communication among all offices is still possible even if some communication links are broken.

37. What is the maximum number of communication links that can be broken with communication among all offices still possible?

38. Show a configuration in which the maximum number of communication links are broken with communication among all offices still possible.

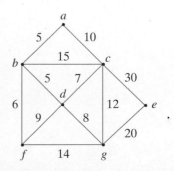

39. In the adjacent graph the vertices represent cities and the numbers on the edges represent the costs of building the indicated roads. Find a least expensive road system that connects all the cities.

6.2 PATHS AND CYCLES

If we think of the vertices in a graph as cities and the edges as roads, a path corresponds to a trip beginning at some city, passing through several cities, and terminating at some city. We begin by giving a formal definition of path.

DEFINITION 6.2.1

Let v_0 and v_n be vertices in a graph. A *path* from v_0 to v_n of length n is an alternating sequence of $n + 1$ vertices and n edges beginning with vertex v_0 and ending with vertex v_n,

$$(v_0, e_1, v_1, e_2, v_2, \ldots, v_{n-1}, e_n, v_n),$$

in which edge e_i is incident on vertices v_{i-1} and v_i for $i = 1, \ldots, n$.

The formalism in Definition 6.2.1 means: Start at vertex v_0; go along edge e_1 to v_1; go along edge e_2 to v_2; and so on.

EXAMPLE 6.2.2

In the graph of Figure 6.2.1,

$$(1, e_1, 2, e_2, 3, e_3, 4, e_4, 2) \tag{6.2.1}$$

is a path of length 4 from vertex 1 to vertex 2.   □

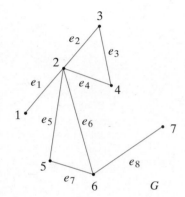

FIGURE 6.2.1
A connected graph with paths $(1, e_1, 2, e_2, 3, e_3, 4, e_4, 2)$ of length 4 and (6) of length 0.

EXAMPLE 6.2.3

In the graph of Figure 6.2.1, the path (6) consisting solely of vertex 6 is a path of length 0 from vertex 6 to vertex 6.   □

In the absence of parallel edges, in denoting a path, we may suppress the edges. For example, the path (6.2.1) may also be written

$$(1, 2, 3, 4, 2).$$

A **connected graph** is a graph in which we can get from any vertex to any other vertex on a path. The formal definition follows.

DEFINITION 6.2.4

A graph G is *connected* if given any vertices v and w in G, there is a path from v to w.

EXAMPLE 6.2.5

The graph G of Figure 6.2.1 is connected since, given any vertices v and w in G, there is a path from v to w.   □

FIGURE 6.2.2
A graph that is not connected.

EXAMPLE 6.2.6

The graph G of Figure 6.2.2 is not connected since, for example, there is no path from vertex v_2 to vertex v_5.   □

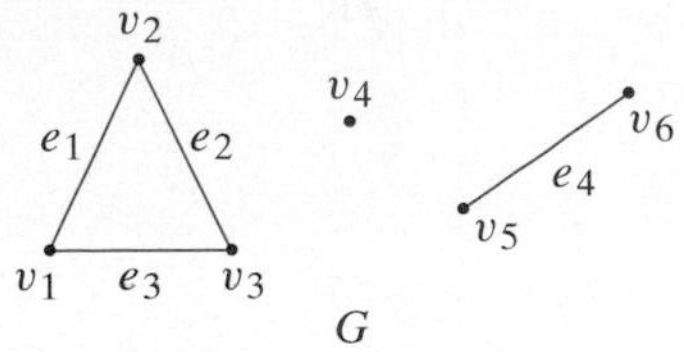

FIGURE 6.2.2
A graph that is not connected.

EXAMPLE 6.2.7

Let G be the graph whose vertex set consists of the 50 states of the United States. Put an edge between states v and w if v and w share a border. For example, there is an edge between California and Oregon and between Illinois and Missouri. There is no edge between Georgia and New York, nor is there an edge between Utah and New Mexico. (Touching does not count; the states must share a border.) The graph G is not connected because there is no path from Hawaii to California (or from Hawaii to any other state).   □

As we can see from Figures 6.2.1 and 6.2.2, a connected graph consists of one "piece," while a graph that is not connected consists of two or more "pieces." These "pieces" are **subgraphs** of the original graph and are called **components**. We give the formal definitions beginning with subgraph.

A subgraph G' of a graph G is obtained by selecting certain edges and vertices from G subject to the restriction that if we select an edge e in G that is incident on vertices v and w, we must include v and w in G'. The restriction is to ensure that G' is actually a graph. The formal definition follows.

DEFINITION 6.2.8

Let $G = (V, E)$ be a graph. We call (V', E') a *subgraph* of G if

   (a) $V' \subseteq V$ and $E' \subseteq E$.

   (b) For every edge $e' \in E'$, if e' is incident on v' and w', then $v', w' \in V'$.

EXAMPLE 6.2.9

The graph $G' = (V', E')$ of Figure 6.2.3 is a subgraph of the graph $G = (V, E)$ of Figure 6.2.4 since $V' \subseteq V$ and $E' \subseteq E$.

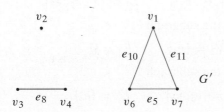

FIGURE 6.2.3   A subgraph of the graph of Figure 6.2.4.

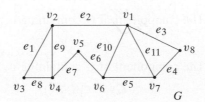

FIGURE 6.2.4   A graph, one of whose subgraphs is shown in Figure 6.2.3.

□

EXAMPLE 6.2.10

Find all subgraphs of the graph G of Figure 6.2.5 having at least one vertex.

If we select no edges, we may select one or both vertices yielding the subgraphs G_1, G_2, and G_3 shown in Figure 6.2.6. If we select the one available edge e_1, we must select the two vertices on which e_1 is incident. In this case, we obtain the subgraph G_4 shown in Figure 6.2.6. Thus G has the four subgraphs shown in Figure 6.2.6.

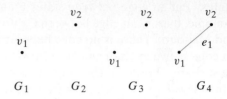

FIGURE 6.2.5   The graph for Example 6.2.10.

FIGURE 6.2.6   The four subgraphs of the graph of Figure 6.2.5.

We can now define component.

DEFINITION 6.2.11

Let G be a graph and let v be a vertex in G. The subgraph G' of G consisting of all edges and vertices in G that are contained in some path beginning at v is called the *component* of G containing v.

EXAMPLE 6.2.12

The graph G of Figure 6.2.1 has one component, namely itself. Indeed, a graph is connected if and only if it has exactly one component.

EXAMPLE 6.2.13

Let G be the graph of Figure 6.2.2. The component of G containing v_3 is the subgraph

$$G_1 = (V_1, E_1), \qquad V_1 = \{v_1, v_2, v_3\}, \qquad E_1 = \{e_1, e_2, e_3\}.$$

The component of G containing v_4 is the subgraph

$$G_2 = (V_2, E_2), \qquad V_2 = \{v_4\}, \qquad E_2 = \emptyset.$$

The component of G containing v_5 is the subgraph

$$G_3 = (V_3, E_3), \qquad V_3 = \{v_5, v_6\}, \qquad E_3 = \{e_4\}.$$

Another characterization of the components of a graph $G = (V, E)$ is obtained by defining a relation R on the set of vertices V by the rule

$$v_1 R v_2, \text{ if there is a path from } v_1 \text{ to } v_2.$$

It can be shown (Exercise 68) that R is an equivalence relation on V and that if $v \in V$, the set of vertices in the component containing v is the equivalence class

$$[v] = \{w \in V \mid wRv\}.$$

Notice that the definition of path allows repetitions of vertices or edges or both. In the path (6.2.1), vertex 2 appears twice.

Subclasses of paths are obtained by prohibiting duplicate vertices or edges or by making the vertices v_0 and v_n of Definition 6.2.1 identical.

DEFINITION 6.2.14

Let v and w be vertices in a graph G.

A *simple path* from v to w is a path from v to w with no repeated vertices.

A *cycle* (or *circuit*) is a path of nonzero length from v to v with no repeated edges.

A *simple cycle* is a cycle from v to v in which, except for the beginning and ending vertices that are both equal to v, there are no repeated vertices.

EXAMPLE 6.2.15

For the graph of Figure 6.2.1 we have

Path	Simple Path?	Cycle?	Simple Cycle?
$(6, 5, 2, 4, 3, 2, 1)$	No	No	No
$(6, 5, 2, 4)$	Yes	No	No
$(2, 6, 5, 2, 4, 3, 2)$	No	Yes	No
$(5, 6, 2, 5)$	No	Yes	Yes
(7)	Yes	No	No

$\square$

We next reexamine the problem introduced in Section 6.1 of finding a cycle in a graph that traverses each edge exactly one time.

EXAMPLE 6.2.16   *Königsberg Bridge Problem*

The first paper in graph theory was Leonhard Euler's in 1736. The paper presented a general theory that included a solution to what is now called the Königsberg bridge problem.

Two islands lying in the Pregel River in Königsberg (now Kaliningrad in Russia) were connected to each other and the river banks by bridges, as shown in Figure 6.2.7. The problem is to start at any location—*A*, *B*, *C*, or *D*; walk over each bridge exactly once; then return to the starting location.

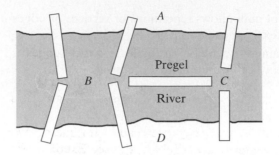

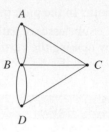

FIGURE 6.2.7   The bridges of Königsberg.

FIGURE 6.2.8   A graph model of the bridges of Königsberg.

The bridge configuration can be modeled as a graph, as shown in Figure 6.2.8. The vertices represent the locations and the edges represent the bridges. The Königsberg bridge problem is now reduced to finding a cycle in the graph of Figure 6.2.8 that includes all the edges and all the vertices. In honor of Euler, a cycle in a graph *G* that includes all the edges and all the vertices of *G* is called an **Euler cycle**.[†] From the discussion of Section 6.1, we see that there is no Euler cycle in the graph of Figure 6.2.8 because there are an odd number of edges incident on vertex *A*. (In fact, in the graph of Figure 6.2.8, every vertex is incident on an odd number of edges.)   □

The solution to the existence of Euler cycles is nicely stated by introducing the degree of a vertex. The **degree of a vertex** v, $\delta(v)$, is the number of edges incident on v. (By definition, each loop on v contributes 2 to the degree of v.) In Section 6.1 we found that if a graph *G* has an Euler cycle, then every vertex in *G* has even degree. We can also prove that *G* is connected.

THEOREM 6.2.17

If a graph G has Euler cycle, then G is connected and every vertex has even degree.

Proof. Suppose that *G* has an Euler cycle. We argued in Section 6.1 that every vertex in *G* has even degree. If v and w are vertices in *G*, the portion of the Euler cycle that takes us from v to w serves as a path from v to w. Therefore, *G* is connected.   ■

The converse of Theorem 6.2.17 is also true. We give a proof by mathematical induction due to [Fowler].

[†]   For technical reasons, if *G* consists of one vertex v and no edges, we call the path (v) an Euler cycle for *G*.

THEOREM 6.2.18

If G is a connected graph and every vertex has even degree, then G has an Euler cycle.

Proof. The proof is by induction on the number n of edges in G.

BASIS STEP $(n = 0)$. Since G is connected, if G has no edges, G consists of a single vertex. An Euler cycle consists of the single vertex and no edges.

INDUCTIVE STEP. Suppose that G has n edges, $n > 0$, and that any connected graph with k edges, $k < n$, in which every vertex has even degree, has an Euler cycle.

It is straightforward to verify that a connected graph with one or two vertices, each of which has even degree, has an Euler cycle (see Exercise 69); thus, we assume that the graph has at least three vertices.

Since G is connected, there are vertices v_1, v_2, and v_3 in G with edge e_1 incident on v_1 and v_2 and edge e_2 incident on v_2 and v_3. We delete the edges e_1 and e_2, but no vertices, and add an edge e incident on v_1 and v_3 to obtain the graph G' [see Figure 6.2.9(a)]. Notice that each component of the graph G' has less than n edges and that in each component of the graph G', every vertex has even degree. We show that G' has either one or two components.

Let v be a vertex. Since G is connected, there is a path P in G from v to v_1. Let P' be the portion of the path P starting at v whose edges are also in G'. Now P' ends at either v_1, v_2, or v_3 because the only way that P could fail to be a path in G' is that P contains one of the deleted edges e_1 or e_2. If P' ends at v_1, then v is in the same component as v_1 in G'. If P' ends at v_3 [see Figure 6.2.9(b)], then v is in the same component as v_3 in G', which is in the same

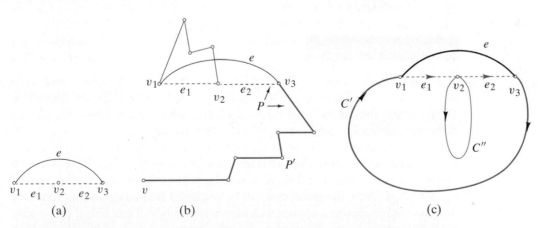

(a)   (b)   (c)

FIGURE 6.2.9 The proof of Theorem 6.2.18. In (a), the edges e_1 and e_2 are deleted and the edge e is added. In (b), P (shown in color) is a path in G from v to v_1, and P' (shown in heavy color) is the portion of P starting at v whose edges are also in G'. As shown, P' ends at v_3. Since edge e is in G', there is a path in G' from v to v_1. Thus v and v_1 are in the same component. In (c), C' (shown with a heavy line) is an Euler cycle for one component, and C'' (shown with a light, solid line) is an Euler cycle for the other component. If we replace e in C' by e_1, C'', e_2, we obtain an Euler cycle (shown in color) for G.

component as v_1 in G' (since edge e in G' is incident on v_1 and v_3). If P' ends at v_2, then v_2 is in the same component as v. Therefore, any vertex in G' is in the same component as either v_1 or v_2. Thus G' has one or two components.

If G' has one component, that is, if G' is connected, we may apply the inductive hypothesis to conclude that G' has an Euler cycle C'. This Euler cycle may be modified to produce an Euler cycle in G: we simply replace the occurrence of the edge e in C' by the edges e_1 and e_2.

Suppose that G' has two components [see Figure 6.2.9(c)]. By the inductive hypothesis, the component containing v_1 has an Euler cycle C' and the component containing v_2 has an Euler cycle C'' beginning and ending at v_2. An Euler cycle in G is obtained by modifying C' by replacing (v_1, v_3) in C' by (v_1, v_2) followed by C'' followed by (v_2, v_3) or by replacing (v_3, v_1) in C' by (v_3, v_2) followed by C'' followed by (v_2, v_1). The Inductive Step is complete; G has an Euler cycle. ∎

If G is a connected graph and every vertex has even degree and G has only a few edges, we can usually find an Euler cycle by inspection.

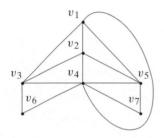

FIGURE 6.2.10
The graph for Example 6.2.19.

EXAMPLE 6.2.19

Let G be the graph of Figure 6.2.10. Use Theorem 6.2.18 to verify that G has an Euler cycle. Find an Euler cycle for G.

We observe that G is connected and that

$$\delta(v_1) = \delta(v_2) = \delta(v_3) = \delta(v_5) = 4, \qquad \delta(v_4) = 6, \qquad \delta(v_6) = \delta(v_7) = 2.$$

Since the degree of every vertex is even, by Theorem 6.2.18, G has an Euler cycle. By inspection, we find the Euler cycle

$$(v_6, v_4, v_7, v_5, v_1, v_3, v_4, v_1, v_2, v_5, v_4, v_2, v_3, v_6).$$ □

EXAMPLE 6.2.20

A domino is a rectangle divided into two squares with each square numbered one of $0, 1, \ldots, 6$. Two squares on a single domino can have the same number. We show that the distinct dominoes can be arranged in a circle so that touching dominoes have adjacent squares with identical numbers.

We model the situation as a graph G with seven vertices labeled $0, 1, \ldots, 6$. The edges represent the dominoes: there is one edge between each distinct pair of vertices and there is one loop at each vertex. Notice that G is connected. Now the dominoes can be arranged in a circle so that touching dominoes have adjacent squares with identical numbers if and only if G contains an Euler cycle. Since the degree of each vertex is 8 (remember that a loop contributes 2 to the degree), each vertex has even degree. By Theorem 6.2.18, G has an Euler cycle. Therefore, the dominoes can be arranged in a circle so that touching dominoes have adjacent squares with identical numbers. □

What can be said about a connected graph in which not all the vertices have even degree? The first observation (Corollary 6.2.22) is that there are an even number of vertices of odd degree. This follows from the fact (Theorem 6.2.21) that the sum of all of the degrees in a graph is an even number.

THEOREM 6.2.21

If G is a graph with m edges and vertices $\{v_1, v_2, \ldots, v_n\}$, then

$$\sum_{i=1}^{n} \delta(v_i) = 2m.$$

In particular, the sum of the degrees of all the vertices in a graph is even.

Proof. When we sum over the degrees of all the vertices, we count each edge (v_i, v_j) twice—once when we count it as (v_i, v_j) in the degree of v_i and again when we count it as (v_j, v_i) in the degree of v_j. The conclusion follows. ■

COROLLARY 6.2.22

In any graph, there are an even number of vertices of odd degree.

Proof. Let us divide the vertices into two groups: those with even degree $x_1, \ldots, x_m$ and those with odd degree $y_1, \ldots, y_n$. Let

$$S = \delta(x_1) + \delta(x_2) + \cdots + \delta(x_m), \qquad T = \delta(y_1) + \delta(y_2) + \cdots + \delta(y_n).$$

By Theorem 6.2.21, $S + T$ is even. Since S is the sum of even numbers, S is even. Thus T is even. But T is the sum of n odd numbers, and therefore n is even. ■

Suppose that a connected graph G has exactly two vertices v and w of odd degree. Let us temporarily insert an edge e from v to w. The resulting graph G' is connected and every vertex has even degree. By Theorem 6.2.18, G' has an Euler cycle. If we delete e from this Euler cycle, we obtain a path with no repeated edges from v to w containing all the edges and vertices of G. We have shown that if a graph has exactly two vertices v and w of odd degree, there is a path with no repeated edges containing all the edges and vertices from v to w. The converse can be proved similarly.

THEOREM 6.2.23

A graph has a path with no repeated edges from v to w ($v \neq w$) containing all the edges and vertices if and only if it is connected and v and w are the only vertices having odd degree.

Proof. Suppose that a graph has a path P with no repeated edges from v to w containing all the edges and vertices. The graph is surely connected. If we add an edge from v to w, the resulting graph has an Euler cycle, namely, the path P together with the added edge. By Theorem 6.2.17, every vertex has even degree. Removing the added edge affects only the degrees of v and w that are each reduced by 1. Thus in the original graph, v and w have odd degree and all other vertices have even degree.

The converse was discussed just before the statement of the theorem. ■

Generalizations of Theorem 6.2.23 are given as Exercises 42 and 44.

We conclude by proving a rather special result that we will use in Section 7.2.

THEOREM 6.2.24

If a graph G contains a cycle from v to v, G contains a simple cycle from v to v.

Proof. Let

$$C = (v_0, e_1, v_1, \ldots, e_i, v_i, e_{i+1}, \ldots, e_j, v_j, e_{j+1}, v_{j+1}, \ldots, e_n, v_n)$$

be a cycle from v to v where $v = v_0 = v_n$ (see Figure 6.2.11). If C is not a simple cycle, then $v_i = v_j$, for some $i < j < n$. We can replace C by the cycle

$$C' = (v_0, e_1, v_1, \ldots, e_i, v_i, e_{j+1}, v_{j+1}, \ldots, e_n, v_n).$$

If C' is not a simple cycle from v to v, we repeat the previous procedure. Eventually we obtain a simple cycle from v to v. ∎

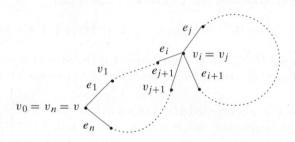

FIGURE 6.2.11   A cycle that is either a simple cycle or can be reduced to a simple cycle.

Exercises

In Exercises 1–9, tell whether the given path in the graph is

(a) A simple path   (b) A cycle   (c) A simple cycle

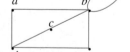

1. (b, b) **2.** (e, d, c, b)

3. (a, d, c, d, e) **4.** (d, c, b, e, d)

5. $(b, c, d, a, b, e, d, c, b)$ **6.** (b, c, d, e, b, b)

7. (a, d, c, b, e) **8.** (d)

9. (d, c, b)

In Exercises 10–18, draw a graph having the given properties or explain why no such graph exists.

10. Six vertices each of degree 3

11. Five vertices each of degree 3

12. Four vertices each of degree 1
13. Six vertices; four edges
14. Four edges; four vertices having degrees 1, 2, 3, 4
15. Four vertices having degrees 1, 2, 3, 4
16. Simple graph; six vertices having degrees 1, 2, 3, 4, 5, 5
17. Simple graph; five vertices having degrees 2, 3, 3, 4, 4
18. Simple graph; five vertices having degrees 2, 2, 4, 4, 4
19. Find all the simple cycles in the adjacent graph.

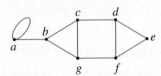

20. Find all simple paths from a to e in the graph of Exercise 19.
21. Find all connected subgraphs of the adjacent graph of Exercise 21 containing all of the vertices of the original graph and having as few edges as possible. Which are simple paths? Which are cycles? Which are simple cycles?

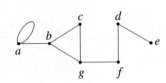

Find the degree of each vertex for the following graphs.

22.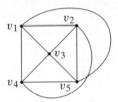

23.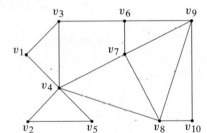

In Exercises 24–27, find all subgraphs having at least one vertex of the graph given.

24.

25.

26.

★ 27.

In Exercises 28–33, decide whether the graph has an Euler cycle. If the graph has an Euler cycle, exhibit one.

28. Exercise 21
30. Exercise 23
29. Exercise 22
31. Figure 6.2.4

32.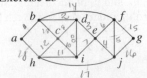

33.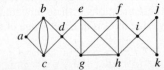

34. The following graph is continued to an arbitrary, finite depth. Does the graph contain an Euler cycle? If the answer is yes, describe one.

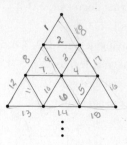

35. When does the complete graph K_n contain an Euler cycle?

36. When does the complete bipartite graph $K_{m,n}$ contain an Euler cycle?

37. For which values of m and n does the graph contain an Euler cycle?

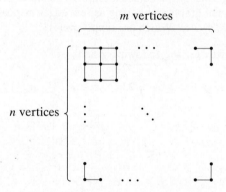

38. For which values of n does the n-cube contain an Euler cycle?

In Exercises 39 and 40, verify that there are an even number of vertices of odd degree in the graph.

39.

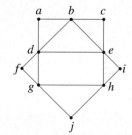

40.

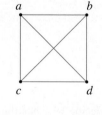

41. For the graph of Exercise 39, find a path with no repeated edges from d to e containing all the edges.

42. Let G be a connected graph with four vertices v_1, v_2, v_3, and v_4 of odd degree. Show that there are paths with no repeated edges from v_1 to v_2 and from v_3 to v_4 such that every edge in G is in exactly one of the paths.

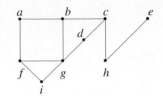

43. Illustrate Exercise 42 using the adjacent graph.

44. State and prove a generalization of Exercise 42 where there are an arbitrary number of vertices of odd degree.

In Exercises 45 and 46, tell whether each assertion is true or false. If false, give a counterexample and if true, explain.

45. Let G be a graph and let v and w be distinct vertices. If there is a path from v to w, there is a simple path from v to w.

46. If a graph contains a cycle that includes all the edges, the cycle is an Euler cycle.

47. Let G be a connected graph. Suppose that an edge e is in a cycle. Show that G with e removed is still connected.

48. Give an example of a connected graph such that the removal of any edge results in a graph that is not connected. (Assume that removing an edge does not remove any vertices.)

★ **49.** Can a knight move around a chessboard and return to its original position making every move exactly once? (A move is considered to be made when the move is made in either direction.)

50. Show that if G' is a connected subgraph of a graph G, then G' is contained in a component.

51. Show that if a graph G is partitioned into connected subgraphs so that each edge and each vertex in G belong to one of the subgraphs, the subgraphs are components.

52. Let G be a directed graph and let G' be the undirected graph obtained from G by ignoring the direction of edges in G. Assume that G is connected. If v is a vertex in G, we say the *parity of* v is *even* if the number of edges of the form (v, w) is even; *odd parity* is defined similarly. Prove that if v and w are vertices in G having odd parity, it is possible to change the orientation of certain edges in G so that v and w have even parity and the parity of all other vertices in G is unchanged.

★ **53.** Show that the maximum number of edges in a simple, disconnected graph with n vertices is $(n - 1)(n - 2)/2$.

★ **54.** Show that the maximum number of edges in a simple, bipartite graph with n vertices is $\lfloor n^2/4 \rfloor$.

A vertex v in a connected graph G is an *articulation point* if the removal of v and all edges incident on v disconnects G.

55. Give an example of a graph with six vertices that has exactly two articulation points.

56. Give an example of a graph with six vertices that has no articulation points.

57. Show that a vertex v in a connected graph G is an articulation point if and only if there are vertices w and x in G having the property that every path from w to x passes through v.

Let G be a directed graph and let v be a vertex in G. The *indegree* of v, in(v), is the number of edges of the form (w, v). The *outdegree* of v, out(v), is the number of edges of the form (v, w). A *directed Euler cycle* in G is a sequence of edges of the form

$$(v_0, v_1), (v_1, v_2), \ldots, (v_{n-1}, v_n),$$

where $v_0 = v_n$, every edge in G occurs exactly one time, and all vertices appear.

58. Show that a directed graph G contains a directed Euler cycle if and only if the undirected graph obtained by ignoring the directions of the edges of G is connected and in(v) = out(v) for every vertex v in G.

A *de Bruijn sequence* for n (in 0's and 1's) is a sequence

$$a_1, \quad \ldots, \quad a_{2^n}$$

of 2^n bits having the property that if s is a bit string of length n, for some m,

$$s = a_m a_{m+1} \cdots a_{m+n-1}. \tag{6.2.2}$$

In (6.2.2), we define $a_{2^n+i} = a_i$ for $i = 1, \ldots, 2^n - 1$.

59. Verify that 00011101 is a de Bruijn sequence for $n = 3$.

60. Let G be a directed graph with vertices corresponding to all bit strings of length $n-1$. A directed edge exists from vertex $x_1 \cdots x_{n-1}$ to $x_2 \cdots x_n$. Show that a directed Euler cycle in G corresponds to a de Bruijn sequence.

★ **61.** Show that there is a de Bruijn sequence for every $n = 1, 2, \ldots$.

★ **62.** A *closed path* is a path from v to v. Show that a connected graph G is bipartite if and only if every closed path in G has even length.

63. How many paths of length $k \geq 1$ are there in K_n?

64. Show that there are

$$\frac{n(n-1)[(n-1)^k - 1]}{n-2}$$

paths whose lengths are between 1 and k, inclusive, in $K_n, n > 2$.

65. Let v and w be distinct vertices in K_n. Let p_m denote the number of paths of length m from v to w in $K_n, 1 \leq m \leq n$.

(a) Derive a recurrence relation for p_m.

(b) Find an explicit formula for p_m.

66. Let v and w be distinct vertices in $K_n, n \geq 2$. Show that the number of simple paths from v to w is

$$(n-2)! \sum_{k=0}^{n-2} \frac{1}{k!}.$$

★ **67.** [Requires calculus.]   Show that there are $\lfloor n!e - 1 \rfloor$ simple paths in K_n. ($e = 2.71828\ldots$ is the base of the natural logarithm.)

68. Let G be a graph. Define a relation R on the set V of vertices of G as vRw if there is a path from v to w. Prove that R is an equivalence relation on V.

69. Prove that a connected graph with one or two vertices, each of which has even degree, has an Euler cycle.

Let G be a connected graph. The *distance* between vertices v and w in G, $\mathrm{dist}(v, w)$, is the length of a shortest path from v to w. The *diameter* of G is

$$d(G) = \max\{\mathrm{dist}(v, w) \mid v \text{ and } w \text{ are vertices in } G\}.$$

70. Find the diameter of the graph of Figure 6.2.10.

71. Find the diameter of the n-cube. In the context of parallel computation, what is the meaning of this value?

72. Find the diameter of K_n, the complete graph on n vertices.

73. Show that the number of paths in the adjacent graph from v_1 to v_1 of length n is equal to the nth Fibonacci number f_n.

$v_1 \qquad\qquad v_2$

74. Let G be a simple graph with n vertices in which every vertex has degree k and

$$k \geq \frac{n-3}{2} \qquad \text{if } n \bmod 4 = 1,$$

$$k \geq \frac{n-1}{2} \qquad \text{if } n \bmod 4 \neq 1.$$

Show that G is connected.

A *cycle* in a simple directed graph [i.e., a directed graph in which there is at most one edge of the form (v, w) and no edges of the form (v, v)] is a sequence of three or more vertices

$$(v_0, v_1, \ldots, v_n)$$

in which (v_{i-1}, v_i) is an edge for $i = 1, \ldots, n$ and $v_0 = v_n$. A *directed acyclic graph (dag)* is a simple directed graph with no cycles.

75. Show that a dag has at least one vertex with no out edges [i.e., there is at least one vertex v such that there are no edges of the form (v, w)].

76. Show that the maximum number of edges in an n-vertex dag is $n(n-1)/2$.

77. An *independent set* in a graph G is a subset S of the vertices of G having the property that no two vertices in S are adjacent. (Note that $\emptyset$ is an independent set for any graph.) Prove the following result due to [Prodinger].

Let P_n be the graph that is a simple path with n vertices. Prove that the number of independent sets in P_n is equal to f_{n+1}, $n = 1, 2, \ldots$, where $\{f_n\}$ is the Fibonacci sequence.

PROBLEM-SOLVING CORNER:

Graphs

Problem

Is it possible in a department of 25 persons, racked by dissension, for each person to get along with exactly five others?

Attacking the Problem

Where do we start? Since this problem is in Chapter 6, which deals with graphs, it would probably be a good idea to try to model the problem as a graph. If this problem were not associated with a particular section or chapter in the book, we might try several approaches—one of which might be to model the problem as a graph. Many discrete problems can be solved by modeling them using graphs. This is not to say that this is the only approach possible. Most of the time by taking different approaches, a single problem can be solved in many ways. (A nice example is [Wagon].)

Finding a Solution

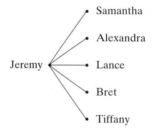

A fundamental issue in building a graph model is to figure out what the graph is—what are the vertices, and what are the edges? In this problem, there's not much choice; we have persons and dissension. Let's try letting the vertices be the people. It's very common in a graph model for the edges to indicate a *relationship* between the vertices. Here the relationship is "gets along with," so we'll put an edge between two vertices (people) if they get along.

Now suppose that each person gets along with exactly five others. For example, in the adjacent figure, which shows part of our graph, Jeremy gets along with Samantha, Alexandra, Lance, Bret, and Tiffany, and no others.

It follows that the degree of every vertex is 5. Now let's take stock of the situation: We have 25 vertices and each vertex has degree 5. Before reading on, try to determine whether this is possible.

Corollary 6.2.22 says that there are an even number of vertices of odd degree. We have a contradiction because there are an *odd* number of vertices of odd degree. Therefore, it is not possible in a department of 25 persons racked by dissension for each person to get along with exactly five others.

Formal Solution

No. It is not possible in a department of 25 persons racked by dissension for each person to get along with exactly five others. Suppose by way of contradiction that it is possible. Consider a graph where the vertices are the persons and an edge connects two vertices (people) if the people get along. Since every vertex has odd degree, there are an odd number of vertices of odd degree, which is a contradiction.

Summary of Problem-Solving Techniques

- Many discrete problems can be solved by modeling them using graphs.
- To build a graph model, determine what the vertices represent and what the edges represent.
- It's very common in a graph model for the edges to indicate a relationship between the vertices.

6.3 HAMILTONIAN CYCLES AND THE TRAVELING SALESPERSON PROBLEM

Sir William Rowan Hamilton marketed a puzzle in the mid-1800s in the form of a dodecahedron (see Figure 6.3.1). Each corner bore the name of a city and the problem was to start at any city, travel along the edges, visit each city exactly one time, and return to the initial city. The graph of the edges of the dodecahedron is given in Figure 6.3.2. We can solve Hamilton's puzzle if we can find a cycle in the graph of Figure 6.3.2 that contains each vertex exactly once (except for the starting and ending vertex that appears twice). See if you can find a solution before looking at a solution given in Figure 6.3.3.

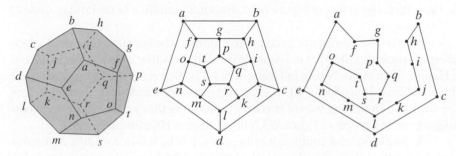

FIGURE 6.3.1
Hamilton's puzzle.

FIGURE 6.3.2
The graph of Hamilton's puzzle.

FIGURE 6.3.3
Visiting each vertex once in the graph of Figure 6.3.2.

In honor of Hamilton, we call a cycle in a graph G that contains each vertex in G exactly once, except for the starting and ending vertex that appears twice, a **Hamiltonian cycle**.

Hamilton (1805–1865) was one of Ireland's greatest scholars. He was professor of astronomy at the University of Dublin, where he published articles in physics and mathematics. In mathematics, Hamilton is most famous for inventing the quaternions, a generalization of the complex number system. The quaternions provided inspiration for the development of modern abstract algebra. In this connection, Hamilton introduced the term *vector*.

EXAMPLE 6.3.1

The cycle (a, b, c, d, e, f, g, a) is a Hamiltonian cycle for the graph of Figure 6.3.4. □

The problem of finding a Hamiltonian cycle in a graph sounds similar to the problem of finding an Euler cycle in a graph. An Euler cycle visits each edge once, whereas a Hamiltonian cycle visits each vertex once; however, the problems are actually quite distinct. For example, the graph G of Figure 6.3.4 does not have an Euler cycle since there are vertices of odd degree, yet Example 6.3.1 showed that G has a Hamiltonian cycle. Furthermore, unlike the situation for Euler cycles (see Theorems 6.2.17 and 6.2.18), no easily verified necessary and sufficient conditions are known for the existence of a Hamiltonian cycle in a graph.

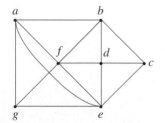

FIGURE 6.3.4
A graph with a Hamiltonian cycle.

The following examples show that sometimes we can argue that a graph does not contain a Hamiltonian cycle.

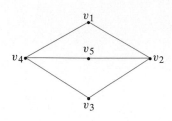

FIGURE 6.3.5
A graph with no Hamiltonian cycle.

EXAMPLE 6.3.2

Show that the graph of Figure 6.3.5 does not contain a Hamiltonian cycle.

Since there are five vertices, a Hamiltonian cycle must have five edges. Suppose that we could eliminate edges from the graph leaving just a Hamiltonian cycle. We would have to eliminate one edge incident at v_2 and one edge incident at v_4, since each vertex in a Hamiltonian cycle has degree 2. But this leaves only four edges—not enough for a Hamiltonian cycle of length 5. Therefore, the graph of Figure 6.3.5 does not contain a Hamiltonian cycle.□

We must be careful not to count an eliminated edge more than once when using an argument like that in Example 6.3.2 to show that a graph does not have a Hamiltonian cycle. Notice in Example 6.3.2 (which refers to Figure 6.3.5) that if we eliminate one edge incident at v_2 and one edge incident at v_4, these edges are distinct. Therefore, we are correct in reasoning that we must eliminate two edges from the graph of Figure 6.3.5 to produce a Hamiltonian cycle.

As an example of double counting, consider the following *faulty* argument that purports to show that the graph of Figure 6.3.6 has no Hamiltonian cycle. Since there are five vertices, a Hamiltonian cycle must have five edges. Suppose that we could eliminate edges from the graph to produce a Hamiltonian cycle. We would have to eliminate two edges incident at c and one edge incident at each of a, b, d, and e. This leaves two edges—not enough for a Hamiltonian cycle. Therefore, the graph of Figure 6.3.6 does not contain a Hamiltonian cycle. The error in this argument is that if we eliminate two edges incident at c (as we must do), we also eliminate edges incident at two of a, b, d, or e. We must not count the two eliminated edges incident at the two vertices again. Notice that the graph of Figure 6.3.6 does have a Hamiltonian cycle.

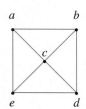

FIGURE 6.3.6
A graph *with* a Hamiltonian cycle.

EXAMPLE 6.3.3

Show that the graph G of Figure 6.3.7 does not contain a Hamiltonian cycle.

Suppose that G has a Hamiltonian cycle H. The edges (a, b), (a, g), (b, c), and (c, k) must be in H since each vertex in a Hamiltonian cycle has degree 2. Thus edges (b, d) and (b, f) are not in H. Therefore, edges (g, d), (d, e), (e, f), and (f, k) are in H. The edges now known to be in H form a cycle C. Adding an additional edge to C will give some vertex in H degree greater than 2. This contradiction shows that G does not have a Hamiltonian cycle. □

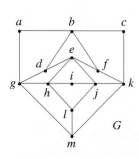

FIGURE 6.3.7
A graph with no Hamiltonian cycle.

The **traveling salesperson problem** is related to the problem of finding a Hamiltonian cycle in a graph. (We referred briefly to a variant of the traveling salesperson problem in Section 6.1.) The problem is: Given a weighted graph G, find a minimum-length Hamiltonian cycle in G. If we think of the vertices in a weighted graph as cities and the edge weights as distances, the traveling salesperson problem is to find a shortest route in which the salesperson can visit each city one time, starting and ending at the same city.

EXAMPLE 6.3.4

The cycle $C = (a, b, c, d, a)$ is a Hamiltonian cycle for the graph G of Figure 6.3.8. Replacing any of the edges in C by either of the edges labeled 11 would increase the length of C; thus C is a minimum-length Hamiltonian cycle for G. Thus C solves the traveling salesperson problem for G.   □

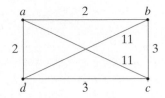

FIGURE 6.3.8
A graph for the traveling salesperson problem.

Although there are algorithms (see, e.g., [Even, 1979]) for finding an Euler cycle, if there is one, in time $\Theta(n)$ for a graph having n edges, every known algorithm for finding Hamiltonian cycles requires either exponential or factorial time in the worst case. For this reason, methods that produce near-minimum-length cycles are often used for problems that ask for a solution to the traveling salesperson problem. Instant fame awaits the discoverer of a polynomial-time algorithm for solving the Hamiltonian cycle problem (or the traveling salesperson problem) or of a proof that there is no polynomial-time algorithm for these problems.

We conclude this section by looking at Hamiltonian cycles in the n-cube.

EXAMPLE 6.3.5   *Gray Codes and Hamiltonian Cycles in the n-Cube*

Consider a **ring model** for parallel computation that, when represented as a graph, is a simple cycle (see Figure 6.3.9). The vertices represent processors. An edge between processors p and q indicates that p and q can communicate directly with one another. We see that each processor can communicate directly with exactly two other processors. Nonadjacent processors communicate by sending messages.

FIGURE 6.3.9
The ring model for parallel computation.

The n-cube (see Example 6.1.7) is another model for parallel computation. The n-cube has a greater degree of connectivity among its processors. We consider the question of when an n-cube can simulate a ring model with 2^n processors. In graph terminology, we are asking when the n-cube contains a simple cycle with 2^n vertices as a subgraph or, since the n-cube has 2^n processors, when the n-cube contains a Hamiltonian cycle. [We leave to the exercises the question of when an n-cube can simulate a ring model with an arbitrary number of processors (see Exercise 18).]

We first observe that if the n-cube contains a Hamiltonian cycle, we must have $n \geq 2$ since the 1-cube has no cycles at all.

Recall (see Example 6.1.7) that we may label the vertices of the n-cube $0, 1, \ldots, 2^n - 1$ in such a way that an edge connects two vertices if and only if the binary representation of their labels differs in exactly one bit. Thus the n-cube has a Hamiltonian cycle if and only if $n \geq 2$ and there is a sequence,

$$s_1, \quad s_2, \quad \ldots, \quad s_{2^n} \tag{6.3.1}$$

where each s_i is a string of n bits, satisfying:

- Every n-bit string appears somewhere in the sequence.
- s_i and s_{i+1} differ in exactly one bit, $i = 1, \ldots, 2^n - 1$.
- s_{2^n} and s_1 differ in exactly one bit.

A sequence (6.3.1) is called a **Gray code**. When $n \geq 2$, a Gray code (6.3.1) corresponds to the Hamiltonian cycle

$$s_1, \quad s_2, \quad \ldots, \quad s_{2^n}, \quad s_1$$

since every vertex appears and the edges (s_i, s_{i+1}), $i = 1, \ldots, 2^n - 1$, and (s_{2^n}, s_1) are distinct. When $n = 1$, the Gray code, 0, 1, corresponds to the path $(0, 1, 0)$, which is not a cycle because the edge $(0, 1)$ is repeated.

Gray codes have been extensively studied in other contexts. For example, Gray codes have been used in converting analog information to digital form (see [Deo]). We show how to construct a Gray code for each positive integer n, thus proving that the n-cube has a Hamiltonian cycle for every positive integer $n \geq 2$. ☐

THEOREM 6.3.6

Let G_1 denote the sequence $0, 1$. We define G_n in terms of G_{n-1} by the following rules:

(a) Let G_{n-1}^R denote the sequence G_{n-1} written in reverse.

(b) Let G_{n-1}' denote the sequence obtained by prefixing each member of G_{n-1} with 0.

(c) Let G_{n-1}'' denote the sequence obtained by prefixing each member of G_{n-1}^R with 1.

(d) Let G_n be the sequence consisting of G_{n-1}' followed by G_{n-1}''.

Then G_n is a Gray code for every positive integer n.

Proof. We prove the theorem by induction on n.

BASIS STEP $(n = 1)$.   Since the sequence $0, 1$ is a Gray code, the theorem is true when n is 1.

INDUCTIVE STEP.   Assume that G_{n-1} is a Gray code. Each string in G_{n-1}' begins with 0, so any difference between consecutive strings must result from differing bits in the corresponding strings in G_{n-1}. But since G_{n-1} is a Gray code, each consecutive pair of strings in G_{n-1} differs in exactly one bit. Therefore, each consecutive pair of strings in G_{n-1}' differs in exactly one bit. Similarly, each consecutive pair of strings in G_{n-1}'' differs in exactly one bit.

Let α denote the last string in G_{n-1}', and let β denote the first string in G_{n-1}''. If we delete the first bit from α and the first bit from β, the resulting strings are identical. Since the first bit in α is 0 and the first bit in β is 1, the last string in G_{n-1}' and the first string in G_{n-1}'' differ in exactly one bit. Similarly, the first string in G_{n-1}' and the last string in G_{n-1}'' differ in exactly one bit. Therefore, G_n is a Gray code. ■

COROLLARY 6.3.7

The n-cube has a Hamiltonian cycle for every positive integer $n \geq 2$.

EXAMPLE 6.3.8

We use Theorem 6.3.6 to construct the Gray code G_3 beginning with G_1.

G_1:	**0**	**1**						
G_1^R:	1	0						
G_1':	00	01						
G_1'':	11	10						
G_2:	00	01	11	10				
G_2^R:	10	11	01	00				
G_2':	000	001	011	010				
G_2'':	110	111	101	100				
G_3:	000	001	011	010	110	111	101	100

◻

We close this section by examining a problem that goes back some 200 years.

EXAMPLE 6.3.9   *The Knight's Tour*

In chess, the knight's move consists of moving two squares horizontally or vertically and then moving one square in the perpendicular direction. For example, in Figure 6.3.10 a knight on the square marked K can move to any of the squares marked X. A **knight's tour of an $n \times n$ board** begins at some square, visits each square exactly once making legal moves, and returns to the initial square. The problem is to determine for which n a knight's tour exists.

We can use a graph to model this problem. We let the squares of the board, alternately colored black and white in the usual way, be the vertices of the graph and we place an edge between two vertices if the corresponding squares on the board represent a legal move for the knight (see Figure 6.3.11). We denote the graph as GK_n. Then there is a knight's tour on the $n \times n$ board if and only if GK_n has a Hamiltonian cycle.

We show that if GK_n has a Hamiltonian cycle, n is even. To see this, note that GK_n is bipartite. We can partition the vertices into sets V_1, those corresponding to the white squares, and V_2, those corresponding to the black squares; each edge is incident on a vertex in V_1 and V_2. Since any cycle must alternate between a vertex in V_1 and one in V_2, any cycle in GK_n must have even length. But since a Hamiltonian cycle must visit each vertex exactly once, a Hamiltonian cycle in GK_n must have length n^2. Thus n must be even.

In view of the preceding result, the smallest possible board that might have a knight's tour is the 2×2 board, but it does not have a knight's tour because the board is so small the knight has no legal moves. The next smallest board that might have a knight's tour is the 4×4 board, although, as we shall show, it too does not have a knight's tour.

We argue by contradiction to show that GK_4 does not have a Hamiltonian cycle. Suppose that GK_4 has a Hamiltonian cycle $C = (v_1, v_2, \ldots, v_{17})$. We

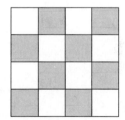

FIGURE 6.3.10
The knight's legal moves in chess.

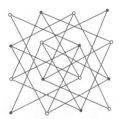

FIGURE 6.3.11
A 4×4 chessboard and the graph GK_4.

assume that v_1 corresponds to the upper left square. We call the eight squares across the top and bottom *outside squares*, and we call the other eight squares *inside squares*. Notice that the knight must arrive at an outside square from an inside square and that the knight must move from an outside square to an inside square. Thus in the cycle C, each vertex corresponding to an outside square must be preceded and followed by a vertex corresponding to an inside square. Since there are equal numbers of outside and inside squares, vertices v_i where i is odd correspond to outside squares, and vertices v_i where i is even correspond to inside squares. But looking at the moves the knight makes, we see that vertices v_i where i is odd correspond to white squares, and vertices v_i where i is even correspond to black squares. Therefore the only outside squares visited are white and the only inside squares visited are black. Thus C is not a Hamiltonian cycle. This contradiction completes the proof that GK_4 has no Hamiltonian cycle. This argument was given by Louis Pósa when he was a teenager.

The graph GK_6 has a Hamiltonian cycle. This fact can be proved by simply exhibiting one (see Exercise 21). It can be shown using elementary methods that GK_n has a Hamiltonian cycle for all even $n \geq 6$ (see [Schwenk]). The proof explicitly constructs Hamiltonian cycles for certain smaller boards and then pastes smaller boards together to obtain Hamiltonian cycles for the larger boards.   □

ॐ ॐ ॐ

Exercises

Find a Hamiltonian cycle in each graph.

1.

2.

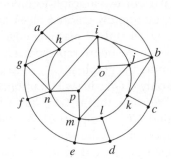

Show that none of the graphs contains a Hamiltonian cycle.

3.   **4.**   **5.**

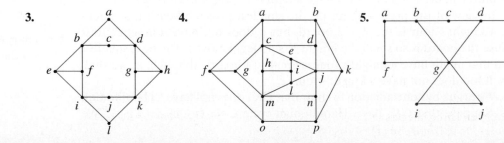

Determine whether or not each graph contains a Hamiltonian cycle. If there is a Hamiltonian cycle, exhibit it; otherwise, give an argument that shows there is no Hamiltonian cycle.

6. 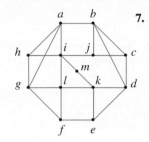   **7.** 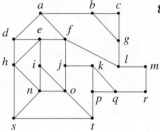   **8.**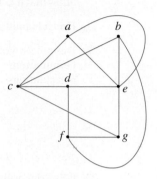

9. Give an example of a graph that has an Euler cycle but contains no Hamiltonian cycle.

10. Given an example of a graph that has an Euler cycle that is also a Hamiltonian cycle.

11. Give an example of a graph that has an Euler cycle and a Hamiltonian cycle that are not identical.

★ **12.** For which values of m and n does the graph of Exercise 37, Section 6.2, contain a Hamiltonian cycle?

13. Modify the graph of Exercise 37, Section 6.2, by inserting an edge between the vertex in row i, column 1, and the vertex in row i, column m, for $i = 1, \ldots, n$. Show that the resulting graph always has a Hamiltonian cycle.

14. Show that if $n \geq 3$, the complete graph on n vertices K_n contains a Hamiltonian cycle.

15. When does the complete bipartite graph $K_{m,n}$ contain a Hamiltonian cycle?

16. Show that the cycle (e, b, a, c, d, e) provides a solution to the traveling salesperson problem for the graph shown.

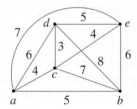

17. Solve the traveling salesperson problem for the graph.

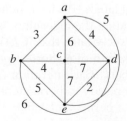

★ **18.** Let m and n be integers satisfying $1 \leq m \leq 2^n$. Prove that the n-cube has a simple cycle of length m if and only if $m \geq 4$ and m is even.

19. Use Theorem 6.3.6 to compute the Gray code G_4.

20. Let G be a bipartite graph with disjoint vertex sets V_1 and V_2, as in Definition 6.1.10. Show that if G has a Hamiltonian cycle, V_1 and V_2 have the same number of elements.

21. Find a Hamiltonian cycle in GK_6 (see Example 6.3.9).

22. Describe a graph model appropriate for solving the following problem: Can the permutations of $\{1, 2, \ldots, n\}$ be arranged in a sequence so that adjacent permutations

$$p: \quad p_1, \quad \ldots, \quad p_n \quad \text{and} \quad q: \quad q_1, \quad \ldots, \quad q_n$$

satisfy $p_i \neq q_i$ for $i = 1, \ldots, n$?

23. Solve the problem of Exercise 22 for $n = 1, 2, 3, 4$. (The answer to the question is "yes" for $n \geq 5$; see [Problem 1186] in the References.)

6.4 *A SHORTEST-PATH ALGORITHM*

Recall (see Section 6.1) that a weighted graph is a graph in which values are assigned to the edges and that the length of a path in a weighted graph is the sum of the weights of the edges in the path. We let $w(i, j)$ denote the weight of edge (i, j). In weighted graphs, we often want to find a **shortest path** (i.e., a path having minimum length) between two given vertices. Algorithm 6.4.1, due to E. W. Dijkstra, which efficiently solves this problem, is the topic of this section.

Edsger W. Dijkstra (1930–) was born in The Netherlands. He was an early proponent of programming as a science. So dedicated to programming was he that when he was married in 1957, he listed his profession as a programmer. However, the Dutch authorities said that there was no such profession, and he had to change the entry to "theoretical physicist." He won the prestigious Turing Award from the Association for Computing Machinery in 1972. He was appointed to the Schlumberger Centennial Chair in Computer Science at the University of Texas at Austin in 1984.

Throughout this section, G denotes a connected, weighted graph. We assume that the weights are positive numbers and that we want to find a shortest path from vertex a to vertex z. The assumption that G is connected can be dropped (see Exercise 9).

Dijkstra's algorithm involves assigning labels to vertices. We let $L(v)$ denote the label of vertex v. At any point, some vertices have temporary labels and the rest have permanent labels. We let T denote the set of vertices having temporary labels. In illustrating the algorithm, we will circle vertices having permanent labels. We will show later that if $L(v)$ is the permanent label of vertex v, then $L(v)$ is the length of a shortest path from a to v. Initially, all vertices have temporary labels. Each iteration of the algorithm changes the status of one label from temporary to permanent; thus we may terminate the algorithm when z receives a permanent label. At this point $L(z)$ gives the length of a shortest path from a to z.

ALGORITHM 6.4.1   *Dijkstra's Shortest-Path Algorithm*

This algorithm finds the length of a shortest path from vertex a to vertex z in a connected, weighted graph. The weight of edge (i, j) is $w(i, j) > 0$ and the label of vertex x is $L(x)$. At termination, $L(z)$ is the length of a shortest path from a to z.

Input: A connected, weighted graph in which all weights are positive. Vertices a and z.

Output: $L(z)$, the length of a shortest path from a to z.

```
1.  procedure dijkstra(w, a, z, L)
2.    L(a) := 0
3.    for all vertices x ≠ a do
4.       L(x) := ∞
5.    T := set of all vertices
6.    // T is the set of vertices whose shortest distance from a has
7.    // not been found
8.    while z ∈ T do
9.      begin
10.       choose v ∈ T with minimum L(v)
11.       T := T − {v}
12.       for each x ∈ T adjacent to v do
13.          L(x) := min{L(x), L(v) + w(v, x)}
14.      end
15.  end dijkstra
```

EXAMPLE 6.4.2

We show how Algorithm 6.4.1 finds a shortest path from a to z in the graph of Figure 6.4.1. (The vertices in T are uncircled and have temporary labels. The circled vertices have permanent labels.) Figure 6.4.2 shows the result of executing lines 2–5. At line 8, z is not circled. We proceed to line 10, where we select vertex a, the uncircled vertex with the smallest label, and circle it

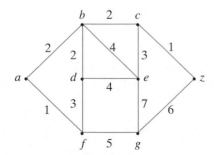

FIGURE 6.4.1   The graph for Example 6.4.2.

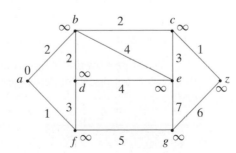

FIGURE 6.4.2   Initialization in Dijkstra's shortest-path algorithm.

(see Figure 6.4.3). At lines 12 and 13 we update each of the uncircled vertices, b and f, adjacent to a. We obtain the new labels

$$L(b) = \min\{\infty, 0 + 2\} = 2, \qquad L(f) = \min\{\infty, 0 + 1\} = 1$$

(see Figure 6.4.3). At this point, we return to line 8.

Since z is not circled, we proceed to line 10, where we select vertex f, the uncircled vertex with the smallest label, and circle it (see Figure 6.4.4). At lines 12 and 13 we update each label of the uncircled vertices, d and g, adjacent to f. We obtain the labels shown in Figure 6.4.4.

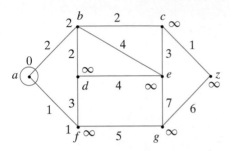

FIGURE 6.4.3   The first iteration of Dijkstra's shortest-path algorithm.

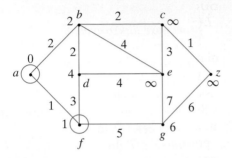

FIGURE 6.4.4   The second iteration of Dijkstra's shortest-path algorithm.

You should verify that the next iteration of the algorithm produces the labeling shown in Figure 6.4.5 and that at the termination of the algorithm, z is labeled 5 indicating that the length of a shortest path from a to z is 5. A shortest path is given by (a, b, c, z).

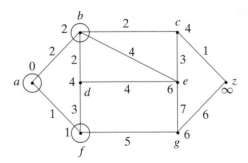

FIGURE 6.4.5   The third iteration of Dijkstra's shortest-path algorithm.   □

We next show that Algorithm 6.4.1 is correct. The proof hinges on the fact that Dijkstra's algorithm finds the lengths of shortest paths from a in increasing order.

THEOREM 6.4.3

Dijkstra's shortest-path algorithm (Algorithm 6.4.1) correctly finds the length of a shortest path from a to z.

Proof. We use mathematical induction on i to prove that the ith time we arrive at line 10, $L(v)$ is the length of a shortest path from a to v. When this is proved, correctness of the algorithm follows since when z is chosen at line 10, $L(z)$ will give the length of a shortest path from a to z.

BASIS STEP $(i = 1)$.   The first time we arrive at line 10, because of the initialization steps (lines 2–4), $L(a)$ is zero and all other L values are ∞. Thus a is chosen the first time we arrive at line 10. Since $L(a)$ is zero, $L(a)$ is the length of a shortest path from a to a.

INDUCTIVE STEP.   Assume that for all $k < i$, the kth time we arrive at line 10, $L(v)$ is the length of a shortest path from a to v.

Suppose that we are at line 10 for the ith time and we choose v in T with minimum value $L(v)$.

First we show that if there is a path from a to a vertex w whose length is less than $L(v)$, then w is not in T (i.e., w was previously selected at line 10). Suppose by way of contradiction that w is in T. Let P be a shortest path from a to w, let x be the vertex nearest a on P that is in T, and let u be the predecessor of x on P (see Figure 6.4.6). Then u is not in T, so u was chosen at line 10 during a previous iteration of the while loop. By the inductive assumption, $L(u)$ is the length of a shortest path from a to u. Now

$$L(x) \leq L(u) + w(u, x) \leq \text{ length of } P < L(v).$$

But this inequality shows that v is not the vertex in T with minimum $L(v)$ [$L(x)$ is smaller]. This contradiction completes the proof that if there is a path from a to a vertex w whose length is less than $L(v)$, then w is not in T.

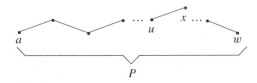

FIGURE 6.4.6   The proof of Theorem 6.4.3. P is a shortest path from a to w, x is the vertex nearest a on P that is in T, and u is the predecessor of x on P.

The preceding result shows, in particular, that if there were a path from a to v whose length is less than $L(v)$, v would already have been selected at line 10 and removed from T. Therefore, every path from a to v has length at least $L(v)$. By construction, there is a path from a to v of length $L(v)$, so this is a shortest path from a to v. The proof is complete.   ■

Algorithm 6.4.1 finds the length of a shortest path from a to z. In most applications, we would also want to identify a shortest path. A slight modification of Algorithm 6.4.1 allows us to find a shortest path.

EXAMPLE 6.4.4

Find a shortest path from a to z and its length for the graph of Figure 6.4.7.

We will apply Algorithm 6.4.1 with a slight modification. In addition to circling a vertex, we will also label it with the name of the vertex from which it was labeled.

Figure 6.4.7 shows the result of executing lines 2–4 of Algorithm 6.4.1. First, we circle a (see Figure 6.4.8). Next, we label the vertices b and d adjacent to a. Vertex b is labeled "$a, 2$" to indicate its value and the fact that it was labeled from a. Similarly, vertex d is labeled "$a, 1$".

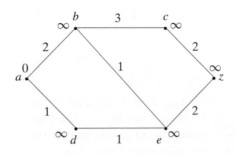

FIGURE 6.4.7   Initialization in Dijkstra's shortest-path algorithm.

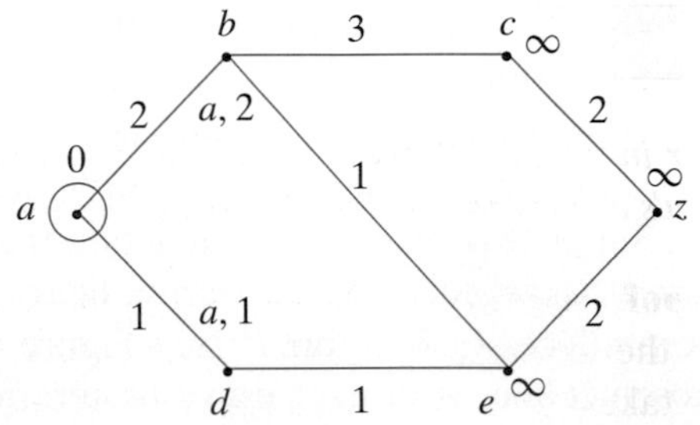

FIGURE 6.4.8   The first iteration of Dijkstra's shortest-path algorithm.

Next, we circle vertex d and update the label of the vertex e adjacent to d (see Figure 6.4.9). Then we circle vertex b and update the labels of vertices c and e (see Figure 6.4.10). Next, we circle vertex e and update the label of vertex z (see Figure 6.4.11). At this point, we may circle z, so the algorithm terminates. The length of a shortest path from a to z is 4. Starting at z, we can retrace the labels to find the shortest path

$$(a, d, e, z).$$

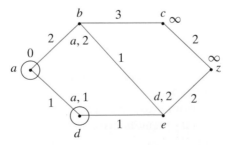

FIGURE 6.4.9   The second iteration of Dijkstra's shortest-path algorithm.

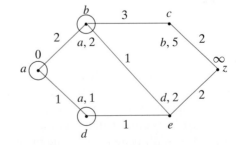

FIGURE 6.4.10   The third iteration of Dijkstra's shortest-path algorithm.

Our next theorem shows that Dijkstra's algorithm is $\Theta(n^2)$ in the worst case.

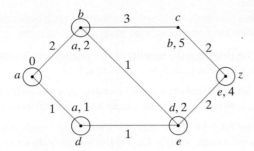

FIGURE 6.4.11   The conclusion of Dijkstra's shortest-path algorithm.

THEOREM 6.4.5

For input consisting of an n-vertex, simple, connected, weighted graph, Dijkstra's algorithm (Algorithm 6.4.1) has worst-case run time $\Theta(n^2)$.

Proof. We consider the time spent in the loops, which provides an upper bound on the total time. Line 4 is executed $O(n)$ times. Within the while loop, line 10 takes time $O(n)$ (we could find the minimum $L(v)$ by examining all the vertices in T). The body of the for loop (line 13) takes time $O(n)$. Since lines 10 and 13 are nested within a while loop, which takes time $O(n)$, the total time for lines 10 and 13 is $O(n^2)$. Thus Dijkstra's algorithm runs in time $O(n^2)$.

In fact, for an appropriate choice of z, the time is $\Omega(n^2)$ for K_n, the complete graph on n vertices, because every vertex is adjacent to every other. Thus the worst-case run time is $\Theta(n^2)$.   ∎

Any shortest-path algorithm that receives as input K_n, the complete graph on n vertices, must examine all of the edges of K_n at least once. Since K_n has $n(n-1)/2$ edges (see Exercise 11, Section 6.1), its worst-case run time must be at least $n(n-1)/2 = \Omega(n^2)$. It follows from Theorem 6.4.5 that Algorithm 6.4.1 is optimal.

॰ॐ ॰ॐ ॰ॐ

Exercises

In Exercises 1–5, find the length of a shortest path and a shortest path between each pair of vertices in the weighted graph.

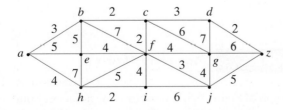

1. a, f **2.** a, g **3.** a, z **4.** b, j **5.** h, d

6. Write an algorithm that finds the length of a shortest path between two given vertices in a connected, weighted graph and also finds a shortest path.

7. Write an algorithm that finds the lengths of the shortest paths from a given vertex to every other vertex in a connected, weighted graph G.

★ 8. Write an algorithm that finds the lengths of the shortest paths between all vertex pairs in a simple, connected, weighted graph having n vertices in time $O(n^3)$.

9. Modify Algorithm 6.4.1 so that it accepts a weighted graph that is not necessarily connected. At termination, what is $L(z)$ if there is no path from a to z?

10. True or false? When a connected, weighted graph and vertices a and z are input to the following algorithm, it returns the length of a shortest path from a to z. If the algorithm is correct, prove it; otherwise, give an example of a connected, weighted graph and vertices a and z for which it fails.

ALGORITHM 6.4.6

```
procedure algor(w, a, z)
    length := 0
    v := a
    T := set of all vertices
    while v ≠ z do
      begin
      T := T − {v}
      choose x ∈ T with minimum w(v, x)
      length := length + w(v, x)
      v := x
      end
    return(length)
end algor
```

11. True or false? Algorithm 6.4.1 finds the length of a shortest path in a connected, weighted graph even if some weights are negative. If true, prove it; otherwise, provide a counterexample.

6.5 REPRESENTATIONS OF GRAPHS

In the preceding sections we represented a graph by drawing it. Sometimes, as for example in using a computer to analyze a graph, we need a more formal representation. Our first method of representing a graph uses the **adjacency matrix**.

EXAMPLE 6.5.1   *Adjacency Matrix*

Consider the graph of Figure 6.5.1. To obtain the adjacency matrix of this graph, we first select an ordering of the vertices, say a, b, c, d, e. Next, we label the rows and columns of a matrix with the ordered vertices. The entry in this matrix

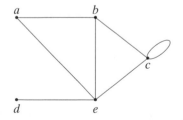

FIGURE 6.5.1
The graph for Example 6.5.1.

is 1 if the row and column vertices are adjacent and 0 otherwise. The adjacency matrix for this graph is

$$
\begin{array}{c}
\begin{array}{ccccc} a & b & c & d & e \end{array} \\
\begin{array}{c} a \\ b \\ c \\ d \\ e \end{array}
\left(\begin{array}{ccccc}
0 & 1 & 0 & 0 & 1 \\
1 & 0 & 1 & 0 & 1 \\
0 & 1 & 1 & 0 & 1 \\
0 & 0 & 0 & 0 & 1 \\
1 & 1 & 1 & 1 & 0
\end{array}\right).
\end{array}
$$
□

Notice that we can obtain the degree of a vertex v in a simple graph G by summing row v or column v in G's adjacency matrix. Also, note that while the adjacency matrix allows us to represent loops, it does not allow us to represent parallel edges; however, if we modify the definition of adjacency matrix to allow arbitrary nonnegative integers as entries in the adjacency matrix, we can represent parallel edges. In the modified adjacency matrix, we interpret the ijth entry as specifying the number of edges between i and j.

The adjacency matrix is not a very efficient way to represent a graph. Since the matrix is symmetric about the main diagonal (the elements on a line from the upper left corner to the lower right corner), the information, except that on the main diagonal, appears twice.

EXAMPLE 6.5.2

The adjacency matrix of the simple graph of Figure 6.5.2 is

$$
A = \begin{array}{c}
\begin{array}{ccccc} & a & b & c & d & e \end{array} \\
\begin{array}{c} a \\ b \\ c \\ d \\ e \end{array}
\left(\begin{array}{ccccc}
0 & 1 & 0 & 1 & 0 \\
1 & 0 & 1 & 0 & 1 \\
0 & 1 & 0 & 1 & 1 \\
1 & 0 & 1 & 0 & 0 \\
0 & 1 & 1 & 0 & 0
\end{array}\right).
\end{array}
$$

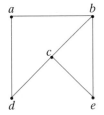

□ **FIGURE 6.5.2**
The graph for Example 6.5.2.

We will show that if A is the adjacency matrix of a simple graph G, the powers of A,

$$A, A^2, A^3, \ldots,$$

count the number of paths of various lengths. More precisely, if the vertices of G are labeled $1, 2, \ldots$, the ijth entry in the matrix A^n is equal to the number of paths from i to j of length n. For example, suppose that we square the matrix A of Example 6.5.2 to obtain

$$
A^2 = \left(\begin{array}{ccccc}
0 & 1 & 0 & 1 & 0 \\
1 & 0 & 1 & 0 & 1 \\
0 & 1 & 0 & 1 & 1 \\
1 & 0 & 1 & 0 & 0 \\
0 & 1 & 1 & 0 & 0
\end{array}\right)
\left(\begin{array}{ccccc}
0 & 1 & 0 & 1 & 0 \\
1 & 0 & 1 & 0 & 1 \\
0 & 1 & 0 & 1 & 1 \\
1 & 0 & 1 & 0 & 0 \\
0 & 1 & 1 & 0 & 0
\end{array}\right)
= \begin{array}{c}
\begin{array}{ccccc} & a & b & c & d & e \end{array} \\
\begin{array}{c} a \\ b \\ c \\ d \\ e \end{array}
\left(\begin{array}{ccccc}
2 & 0 & 2 & 0 & 1 \\
0 & 3 & 1 & 2 & 1 \\
2 & 1 & 3 & 0 & 1 \\
0 & 2 & 0 & 2 & 1 \\
1 & 1 & 1 & 1 & 2
\end{array}\right).
\end{array}
$$

Consider the entry for row a, column c in A^2, obtained by multiplying pairwise the entries in row a by the entries in column c of the matrix A and summing:

$$a \begin{pmatrix} 0 & 1 & 0 & 1 & 0 \end{pmatrix} \begin{matrix} {}_b \quad {}_d \end{matrix} \quad \overset{c}{\begin{pmatrix} 0 \\ 1 \\ 0 \\ 1 \\ 1 \end{pmatrix}} \begin{matrix} b \\ \\ d \end{matrix} = 0 \cdot 0 + 1 \cdot 1 + 0 \cdot 0 + 1 \cdot 1 + 0 \cdot 1 = 2.$$

The only way a nonzero product appears in this sum is if both entries to be multiplied are 1. This happens if there is a vertex v whose entry in row a is 1 and whose entry in column c is 1. In other words, there must be edges of the form (a, v) and (v, c). Such edges form a path (a, v, c) of length 2 from a to c and each path increases the sum by 1. In this example, the sum is 2 because there are two paths

$$(a, b, c), \quad (a, d, c)$$

of length 2 from a to c. In general, the entry in row x and column y of the matrix A^2 is the number of paths of length 2 from vertex x to vertex y.

The entries on the main diagonal of A^2 give the degrees of the vertices (when the graph is a simple graph). Consider, for example, vertex c. The degree of c is 3 since c is incident on the three edges (c, b), (c, d), and (c, e). But each of these edges can be converted to a path of length 2 from c to c:

$$(c, b, c), \quad (c, d, c), \quad (c, e, c).$$

Similarly, a path of length 2 from c to c defines an edge incident on c. Thus the number of paths of length 2 from c to c is 3, the degree of c.

We now use induction to show that the entries in the nth power of an adjacency matrix give the number of paths of length n.

THEOREM 6.5.3

If A is the adjacency matrix of a simple graph, the ijth entry of A^n is equal to the number of paths of length n from vertex i to vertex j, $n = 1, 2, \ldots$.

Proof. We will use induction on n.

In case $n = 1$, A^1 is simply A. The ijth entry is 1 if there is an edge from i to j, which is a path of length 1, and 0 otherwise. Thus the theorem is true in case $n = 1$. The basis step has been verified.

Assume that the theorem is true for n. Now

$$A^{n+1} = A^n A$$

so that the ikth entry in A^{n+1} is obtained by multiplying pairwise the elements in the ith row of A^n by the elements in the kth column of A and summing:

$$k\text{th column of } A$$

$$i\text{th row of } A^n \quad (s_1, \; s_2, \; \ldots, \; s_j, \; \ldots, \; s_m) \begin{pmatrix} t_1 \\ t_2 \\ \vdots \\ t_j \\ \vdots \\ t_m \end{pmatrix}$$

$$= s_1 t_1 + s_2 t_2 + \cdots + s_j t_j + \cdots + s_m t_m$$

$$= ik\text{th entry in } A^{n+1}.$$

By induction, s_j gives the number of paths of length n from i to j in the graph G. Now t_j is either 0 or 1. If t_j is 0, there is no edge from j to k, so there are $s_j t_j = 0$ paths of length $n+1$ from i to k, where the last edge is (j, k). If t_j is 1, there is an edge from vertex j to vertex k (see Figure 6.5.3). Since there are s_j paths of length n from vertex i to vertex j, there are $s_j t_j = s_j$ paths of length $n+1$ from i to k, where the last edge is (j, k) (see Figure 6.5.3). Summing over all j, we will count all paths of length $n+1$ from i to k. Thus the ikth entry in A^{n+1} gives the number of paths of length $n+1$ from i to k and the inductive step is verified.

By the Principle of Mathematical Induction, the theorem is established. ∎

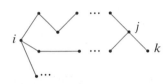

FIGURE 6.5.3
The proof of Theorem 6.5.3. A path from i to k of length $n+1$ whose next-to-last vertex is j consists of a path of length n from i to j followed by edge (j, k). If there are s_j paths of length n from i to j and t_j is 1 if edge (j, k) exists and 0 otherwise, the sum of $s_j t_j$ over all j gives the number of paths of length $n+1$ from i to k.

EXAMPLE 6.5.4

After Example 6.5.2, we showed that if A is the matrix of the graph of Figure 6.5.2,

$$A^2 = \begin{array}{c} \\ a \\ b \\ c \\ d \\ e \end{array} \begin{array}{c} \begin{array}{ccccc} a & b & c & d & e \end{array} \\ \begin{pmatrix} 2 & 0 & 2 & 0 & 1 \\ 0 & 3 & 1 & 2 & 1 \\ 2 & 1 & 3 & 0 & 1 \\ 0 & 2 & 0 & 2 & 1 \\ 1 & 1 & 1 & 1 & 2 \end{pmatrix} \end{array}.$$

By multiplying,

$$A^4 = A^2 A^2 = \begin{pmatrix} 2 & 0 & 2 & 0 & 1 \\ 0 & 3 & 1 & 2 & 1 \\ 2 & 1 & 3 & 0 & 1 \\ 0 & 2 & 0 & 2 & 1 \\ 1 & 1 & 1 & 1 & 2 \end{pmatrix} \begin{pmatrix} 2 & 0 & 2 & 0 & 1 \\ 0 & 3 & 1 & 2 & 1 \\ 2 & 1 & 3 & 0 & 1 \\ 0 & 2 & 0 & 2 & 1 \\ 1 & 1 & 1 & 1 & 2 \end{pmatrix},$$

we find that

$$A^4 = \begin{array}{c} \\ a \\ b \\ c \\ d \\ e \end{array} \begin{pmatrix} \begin{array}{ccccc} a & b & c & d & e \end{array} \\ 9 & 3 & 11 & 1 & 6 \\ 3 & 15 & 7 & 11 & 8 \\ 11 & 7 & 15 & 3 & 8 \\ 1 & 11 & 3 & 9 & 6 \\ 6 & 8 & 8 & 6 & 8 \end{pmatrix}.$$

The entry from row d, column e is 6, which means that there are six paths of length 4 from d to e. By inspection, we find them to be

$$(d, a, d, c, e), \quad (d, c, d, c, e), \quad (d, a, b, c, e),$$

$$(d, c, e, c, e), \quad (d, c, e, b, e), \quad (d, c, b, c, e). \qquad \square$$

Another useful matrix representation of a graph is known as the **incidence matrix**.

EXAMPLE 6.5.5   *Incidence Matrix*

To obtain the incidence matrix of the graph in Figure 6.5.4, we label the rows with the vertices and the columns with the edges (in some arbitrary order). The entry for row v and column e is 1 if e is incident on v and 0 otherwise. Thus the incidence matrix for the graph of Figure 6.5.4 is

$$\begin{array}{c} \\ v_1 \\ v_2 \\ v_3 \\ v_4 \\ v_5 \end{array} \begin{pmatrix} \begin{array}{ccccccc} e_1 & e_2 & e_3 & e_4 & e_5 & e_6 & e_7 \end{array} \\ 1 & 1 & 1 & 0 & 0 & 0 & 0 \\ 0 & 0 & 1 & 1 & 1 & 0 & 1 \\ 0 & 0 & 0 & 0 & 0 & 1 & 0 \\ 1 & 1 & 0 & 1 & 0 & 0 & 0 \\ 0 & 0 & 0 & 0 & 1 & 1 & 0 \end{pmatrix}.$$

A column such as e_7 is understood to represent a loop. $\qquad \square$

The incidence matrix allows us to represent both parallel edges and loops. Notice that in a graph without loops each column has two 1's and that the sum of a row gives the degree of the vertex identified with that row.

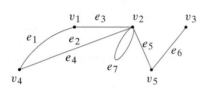

FIGURE 6.5.4
The graph for Example 6.5.5.

ᕽ ᕽ ᕽ

Exercises

In Exercises 1–6, write the adjacency matrix of each graph.

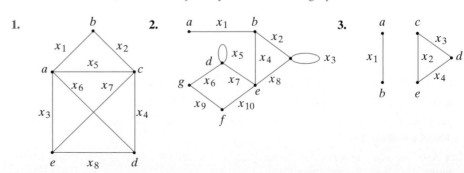

4. The graph of Figure 6.2.2

5. The complete graph on five vertices K_5

6. The complete bipartite graph $K_{2,3}$

In Exercises 7–12, write the incidence matrix of each graph.

7. The graph of Exercise 1

8. The graph of Exercise 2

9. The graph of Exercise 3

10. The graph of Figure 6.2.1

11. The complete graph on five vertices K_5

12. The complete bipartite graph $K_{2,3}$

In Exercises 13–17, draw the graph represented by each adjacency matrix.

13.
$$
\begin{array}{c c}
 & \begin{array}{c c c c c} a & b & c & d & e \end{array} \\
\begin{array}{c} a \\ b \\ c \\ d \\ e \end{array} &
\begin{pmatrix}
1 & 0 & 0 & 1 & 0 \\
0 & 0 & 1 & 0 & 1 \\
0 & 1 & 1 & 1 & 1 \\
1 & 0 & 1 & 0 & 0 \\
0 & 1 & 1 & 0 & 0
\end{pmatrix}
\end{array}
$$

14.
$$
\begin{array}{c c}
 & \begin{array}{c c c c c} a & b & c & d & e \end{array} \\
\begin{array}{c} a \\ b \\ c \\ d \\ e \end{array} &
\begin{pmatrix}
0 & 1 & 0 & 0 & 0 \\
1 & 0 & 0 & 0 & 0 \\
0 & 0 & 0 & 1 & 1 \\
0 & 0 & 1 & 0 & 1 \\
0 & 0 & 1 & 1 & 1
\end{pmatrix}
\end{array}
$$

15.
$$
\begin{array}{c c}
 & \begin{array}{c c c c c c} a & b & c & d & e & f \end{array} \\
\begin{array}{c} a \\ b \\ c \\ d \\ e \\ f \end{array} &
\begin{pmatrix}
0 & 0 & 1 & 0 & 0 & 1 \\
0 & 1 & 0 & 1 & 1 & 0 \\
1 & 0 & 0 & 0 & 0 & 1 \\
0 & 1 & 0 & 0 & 1 & 0 \\
0 & 1 & 0 & 1 & 0 & 0 \\
1 & 0 & 1 & 0 & 0 & 0
\end{pmatrix}
\end{array}
$$

16.
$$
\begin{array}{c c}
 & \begin{array}{c c c c c c} a & b & c & d & e & f \end{array} \\
\begin{array}{c} a \\ b \\ c \\ d \\ e \\ f \end{array} &
\begin{pmatrix}
1 & 1 & 1 & 1 & 0 & 1 \\
1 & 0 & 1 & 1 & 1 & 0 \\
1 & 1 & 0 & 1 & 1 & 1 \\
1 & 1 & 1 & 0 & 1 & 1 \\
0 & 1 & 1 & 1 & 0 & 1 \\
1 & 0 & 1 & 1 & 1 & 0
\end{pmatrix}
\end{array}
$$

17. The 7×7 matrix whose ijth entry is 1 if $i + 1$ divides $j + 1$ or $j + 1$ divides $i + 1$ and whose ijth entry is 0 otherwise.

18. Write the adjacency matrices of the components of the graphs given by the adjacency matrices of Exercises 13–17.

19. Compute the squares of the adjacency matrices of K_5 and the graphs of Exercises 1 and 3.

20. Let A be the adjacency matrix for the graph of Exercise 1. What is the entry in row a, column d of A^5?

21. Suppose that a graph has an adjacency matrix of the form

$$
A = \left(\begin{array}{c|c} & A' \\ \hline A'' & \end{array} \right),
$$

where all entries of the submatrices A' and A'' are 0. What must the graph look like?

22. Repeat Exercise 21 with "adjacency" replaced by "incidence."

23. How might the definition of adjacency matrix be changed to allow for the representation of parallel edges?

24. Let A be an adjacency matrix of a graph. Why is A^n symmetric about the main diagonal for every positive integer n?

In Exercises 25 and 26, draw the graphs represented by the incidence matrices.

25. $\begin{array}{c} a \\ b \\ c \\ d \\ e \end{array} \begin{pmatrix} 1 & 0 & 0 & 0 & 0 & 1 \\ 0 & 1 & 1 & 0 & 1 & 0 \\ 1 & 0 & 0 & 1 & 0 & 0 \\ 0 & 1 & 0 & 1 & 0 & 0 \\ 0 & 0 & 1 & 0 & 1 & 1 \end{pmatrix}$

26. $\begin{array}{c} a \\ b \\ c \\ d \\ e \end{array} \begin{pmatrix} 0 & 1 & 0 & 0 & 1 & 1 \\ 0 & 1 & 1 & 0 & 1 & 0 \\ 0 & 0 & 0 & 0 & 0 & 1 \\ 1 & 0 & 0 & 1 & 0 & 0 \\ 1 & 0 & 0 & 1 & 0 & 0 \end{pmatrix}$

27. What must a graph look like if some row of its incidence matrix consists only of 0's?

28. Let A be the adjacency matrix of a graph G with n vertices. Let

$$Y = A + A^2 + \cdots + A^{n-1}.$$

If some off-diagonal entry in the matrix Y is zero, what can you say about the graph G?

Exercises 29–32 refer to the adjacency matrix A of K_5.

29. Let n be a positive integer. Explain why all the diagonal elements of A^n are equal and all the off-diagonal elements of A^n are equal.

Let d_n be the common value of the diagonal elements of A^n and let a_n be the common value of the off-diagonal elements of A^n.

★ 30. Show that

$$d_{n+1} = 4a_n; \qquad a_{n+1} = d_n + 3a_n; \qquad a_{n+1} = 3a_n + 4a_{n-1}.$$

★ 31. Show that

$$a_n = \frac{1}{5}[4^n + (-1)^{n+1}].$$

32. Show that

$$d_n = \frac{4}{5}[4^{n-1} + (-1)^n].$$

★ 33. Derive results similar to those of Exercises 30–32 for the adjacency matrix A of the graph K_m.

★ 34. Let A be the adjacency matrix of the graph $K_{m,n}$. Find a formula for the entries in A^j.

6.6 ISOMORPHISMS OF GRAPHS

The following instructions are given to two persons who cannot see each other's paper: "Draw and label five vertices a, b, c, d, and e. Connect a and b, b and c, c and d, d and e, and a and e." The graphs produced are shown in Figure 6.6.1. Surely these figures define the same graph even though they appear dissimilar. Such graphs are said to be **isomorphic**.

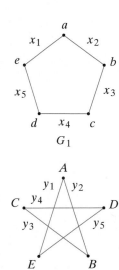

FIGURE 6.6.1
Isomorphic graphs.

> DEFINITION 6.6.1

Graphs G_1 and G_2 are *isomorphic* if there is a one-to-one, onto function f from the vertices of G_1 to the vertices of G_2 and a one-to-one, onto function g from the edges of G_1 to the edges of G_2, so that an edge e is incident on v and w in G_1 if and only if the edge $g(e)$ is incident on $f(v)$ and $f(w)$ in G_2. The pair of functions f and g is called an *isomorphism* of G_1 onto G_2.

EXAMPLE 6.6.2

An isomorphism for the graphs G_1 and G_2 of Figure 6.6.1 is defined by

$$f(a) = A, \quad f(b) = B, \quad f(c) = C, \quad f(d) = D, \quad f(e) = E$$
$$g(x_i) = y_i, \quad i = 1, \dots, 5. \qquad \square$$

If we define a relation R on a set of graphs by the rule $G_1 R G_2$ if G_1 and G_2 are isomorphic, R is an equivalence relation. Each equivalence class consists of a set of mutually isomorphic graphs.

EXAMPLE 6.6.3   *The Mesh Model for Parallel Computation*

Previously, we considered the problem of when the n-cube could simulate a ring model for parallel computation (see Example 6.3.5). We now consider when the n-cube can simulate the **mesh model for parallel computation**.

The two-dimensional mesh model for parallel computation when described as a graph consists of a rectangular array of vertices connected as shown (see Figure 6.6.2). The problem "When can an n-cube simulate a two-dimensional mesh?" can be rephrased in graph terminology as "When does an n-cube contain a subgraph isomorphic to a two-dimensional mesh?" We show that if M is a mesh p vertices by q vertices, where $p \leq 2^i$ and $q \leq 2^j$, then the $(i + j)$-cube contains a subgraph isomorphic to M. (In Figure 6.6.2, we may take $p = 6$, $q = 4$, $i = 3$, and $j = 2$. Thus our result shows that the 5-cube contains a subgraph isomorphic to the graph in Figure 6.6.2.)

Let M be a mesh p vertices by q vertices, where $p \leq 2^i$ and $q \leq 2^j$. We consider M to be a rectangular array in ordinary 2-space with p vertices in the horizontal direction and q vertices in the vertical direction (see Figure 6.6.2). As coordinates for the vertices, we use elements of Gray codes. (Gray codes are discussed in Example 6.3.5.) The coordinates in the horizontal direction are the first p members of an i-bit Gray code and the coordinates in the vertical direction are the first q members of a j-bit Gray code (see Example 6.6.2). If a vertex v is in the mesh, we let v_x denote the horizontal coordinate of v and v_y denote the vertical coordinate of v. We then define a function f on the vertices of M by

$$f(v) = v_x v_y.$$

(The string $v_x v_y$ is the string v_x followed by the string v_y.) Notice that f is one-to-one.

If (v, w) is an edge in M, the bit strings $v_x v_y$ and $w_x w_y$ differ in exactly one bit. Thus $(v_x v_y, w_x w_y)$ is an edge in the $(i + j)$-cube. We define a function g on the edges of M by

$$g((v, w)) = (v_x v_y, w_x w_y).$$

Notice that g is one-to-one. The pair f, g of functions is an isomorphism of M onto the subgraph (V, E) of the $(i + j)$-cube where

$$V = \{f(v) \mid v \text{ is a vertex in } M\}, \qquad E = \{g(e) \mid e \text{ is an edge in } M\}.$$

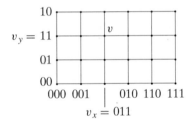

FIGURE 6.6.2
Mesh model for parallel computation.

Therefore, if M is a mesh p vertices by q vertices, where $p \leq 2^i$ and $q \leq 2^j$, the $(i + j)$-cube contains a subgraph isomorphic to M.

The argument given extends to an arbitrary number of dimensions (see Exercise 11); that is, if M is a $p_1 \times p_2 \times \cdots \times p_k$ mesh, where $p_i \leq 2^{t_i}$ for $i = 1, \ldots, k$, then the $(t_1 + t_2 + \cdots + t_k)$-cube contains a subgraph isomorphic to M.   □

Suppose that graphs G_1 and G_2 are isomorphic. If f is the function of Definition 6.6.1, it follows from the definition that vertices v and w in G_1 are adjacent if and only if the vertices $f(v)$ and $f(w)$ are adjacent in G_2. The converse is true for *simple* graphs.

THEOREM 6.6.4

Let G_1 and G_2 be simple graphs. The following are equivalent.

(a) *G_1 and G_2 are isomorphic.*

(b) *There is a one-to-one, onto function f from the vertex set of G_1 to the vertex set of G_2 satisfying: Vertices v and w are adjacent in G_1 if and only if the vertices $f(v)$ and $f(w)$ are adjacent in G_2.*

Proof. It follows immediately from Definition 6.6.1 that (a) implies (b).

We prove that (b) implies (a). Suppose that there is a one-to-one, onto function f from the vertex set of G_1 to the vertex set of G_2 satisfying: Vertices v and w are adjacent in G_1 if and only if the vertices $f(v)$ and $f(w)$ are adjacent in G_2.

We define a function g from the edges of G_1 to the edges of G_2 by the rule

$$g\left((v, w)\right) = (f(v), f(w)).$$

Because G_1 and G_2 are simple graphs, neither has parallel edges and so the notation (v', w') unambiguously designates an edge. Notice that the range of g is a subset of edges of G_2 since, if (v, w) is an edge in G_1, v and w are adjacent, which, in turn, implies that $f(v)$ and $f(w)$ are adjacent, that is, that $(f(v), f(w))$ is an edge in G_2.

We leave to the reader the details of checking that g is a one-to-one and onto and that an edge e is incident on v and w in G_1 if and only if the edge $g(e)$ is incident on $f(v)$ and $f(w)$ in G_2. It follows that G_1 and G_2 are isomorphic.
■

It follows immediately from Theorem 6.6.4 that simple graphs are isomorphic if and only if for some orderings of their vertices their adjacency matrices are identical.

THEOREM 6.6.5

Simple graphs G_1 and G_2 are isomorphic if and only if for some orderings of their vertices, their adjacency matrices are equal.

Proof. See Exercise 25.
■

EXAMPLE 6.6.6

The adjacency matrix of graph G_1 in Figure 6.6.1 relative to the vertex ordering $a, b, c, d, e,$

$$
\begin{array}{c@{\quad}ccccc}
 & a & b & c & d & e \\
\begin{array}{c} a \\ b \\ c \\ d \\ e \end{array} &
\left(\begin{array}{ccccc}
0 & 1 & 0 & 0 & 1 \\
1 & 0 & 1 & 0 & 0 \\
0 & 1 & 0 & 1 & 0 \\
0 & 0 & 1 & 0 & 1 \\
1 & 0 & 0 & 1 & 0
\end{array}\right),
\end{array}
$$

is equal to the adjacency matrix of graph G_2 in Figure 6.6.1 relative to the vertex ordering $A, B, C, D, E,$

$$
\begin{array}{c@{\quad}ccccc}
 & A & B & C & D & E \\
\begin{array}{c} A \\ B \\ C \\ D \\ E \end{array} &
\left(\begin{array}{ccccc}
0 & 1 & 0 & 0 & 1 \\
1 & 0 & 1 & 0 & 0 \\
0 & 1 & 0 & 1 & 0 \\
0 & 0 & 1 & 0 & 1 \\
1 & 0 & 0 & 1 & 0
\end{array}\right).
\end{array}
$$

We see again that G_1 and G_2 are isomorphic. □

An interesting problem is to determine whether two graphs are isomorphic. Although every known algorithm to test whether two graphs are isomorphic requires exponential or factorial time in the worst case, there are algorithms that can determine whether a pair of graphs is isomorphic in linear time in the average case (see [Read] and [Babai]).

The following is one way to show that two simple graphs G_1 and G_2 are *not* isomorphic. Find a property of G_1 that G_2 does *not* have, but that G_2 *would* have if G_1 and G_2 were isomorphic. Such a property is called an **invariant**. More precisely, a property P is an invariant, if whenever G_1 and G_2 are isomorphic graphs:

If G_1 has property P, G_2 also has property P.

By Definition 6.6.1, if graphs G_1 and G_2 are isomorphic, there are one-to-one, onto functions from the edges (respectively, vertices) of G_1 to the edges (respectively, vertices) of G_2. Thus, if G_1 and G_2 are isomorphic, then G_1 and G_2 have the same number of edges and the same number of vertices. Therefore, if e and n are nonnegative integers, the properties "has e edges" and "has n vertices" are invariants.

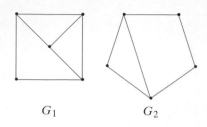

G_1 G_2

FIGURE 6.6.3
Nonisomorphic graphs. G_1 has seven edges and G_2 has six edges.

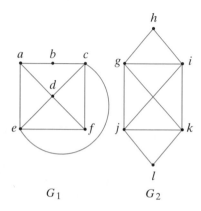

G_1 G_2

FIGURE 6.6.4
Nonisomorphic graphs. G_1 has vertices of degree 3, but G_2 has no vertices of degree 3.

EXAMPLE 6.6.7

The graphs G_1 and G_2 in Figure 6.6.3 are not isomorphic, since G_1 has seven edges and G_2 has six edges and "has seven edges" is an invariant. □

EXAMPLE 6.6.8

Show that if k is a positive integer, "has a vertex of degree k" is an invariant.

Suppose G_1 and G_2 are isomorphic graphs and f (respectively, g) is a one-to-one, onto function from the vertices (respectively, edges) of G_1 onto the vertices (respectively, edges) of G_2. Suppose that G_1 has a vertex v of degree k. Then there are k edges $e_1, \ldots, e_k$ incident on v. By Definition 6.6.1, $g(e_1), \ldots, g(e_k)$ are incident on $f(v)$. Because g is one-to-one, $\delta(f(v)) \geq k$.

Let E be an edge that is incident on $f(v)$ in G_2. Since g is onto, there is an edge e in G_1 with $g(e) = E$. Since $g(e)$ is incident on $f(v)$ in G_2, by Definition 6.6.1, e is incident on v in G_1. Since $e_1, \ldots, e_k$ are the only edges in G_1 incident on $v, e = e_i$ for some $i \in \{1, \ldots, k\}$. Now $g(e_i) = g(e) = E$. Thus $\delta(f(v)) = k$, so G_2 has a vertex, namely $f(v)$, of degree k. □

EXAMPLE 6.6.9

Since "has a vertex of degree 3" is an invariant, the graphs G_1 and G_2 of Figure 6.6.4 are not isomorphic; G_1 has vertices (a and f) of degree 3, but G_2 does not have a vertex of degree 3. Notice that G_1 and G_2 have the same numbers of edges and vertices. □

Another invariant that is sometimes useful is "has a simple cycle of length k." We leave the proof that this property is an invariant to the exercises (Exercise 12).

EXAMPLE 6.6.10

Since "has a simple cycle of length 3" is an invariant, the graphs G_1 and G_2 of Figure 6.6.5 are not isomorphic; the graph G_2 has a simple cycle of length 3, but all simple cycles in G_1 have length at least 4. Notice that G_1 and G_2 have the same numbers of edges and vertices and that every vertex in G_1 or G_2 has degree 4. □

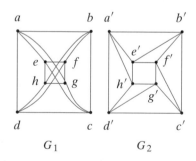

G_1 G_2

FIGURE 6.6.5
Nonisomorphic graphs. G_2 has a simple cycle of length 3, but G_1 has no simple cycles of length 3.

It would be easy to test whether a pair of graphs is isomorphic if we could find a small number of easily checked invariants that isomorphic graphs and *only* isomorphic graphs share. Unfortunately, no one has succeeded in finding such a set of invariants.

Exercises

In Exercises 1–10, determine whether the graphs G_1 and G_2 are isomorphic. If the graphs are isomorphic, find functions f and g for Definition 6.6.1; otherwise, give an invariant that the graphs do not share.

1.

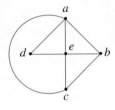

G_1 G_2

2.

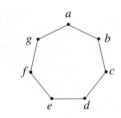

G_1 G_2

3.

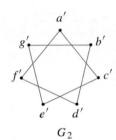

G_1 G_2

4.

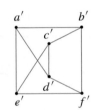

G_1 G_2

5.

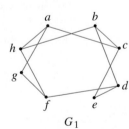

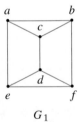

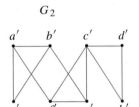

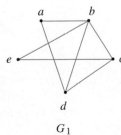

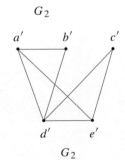

G_1 G_2

6.

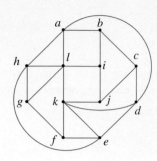

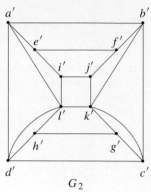

G_1

G_2

7.

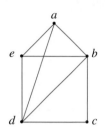

G_1

G_2

★ **8.**

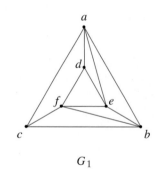

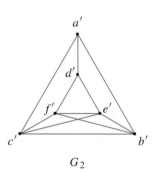

G_1

G_2

★ **9.**

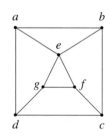

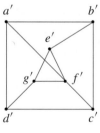

G_1

G_2

★ **10.**

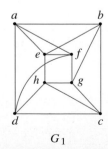

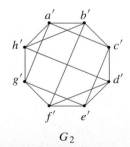

G_1

G_2

11. Show that if M is a $p_1 \times p_2 \times \cdots \times p_k$ mesh, where $p_i \le 2^{t_i}$ for $i = 1, \ldots, k$, then the $(t_1 + t_2 + \cdots + t_k)$-cube contains a subgraph isomorphic to M.

In Exercises 12–16, show that the property given is an invariant.

12. Has a simple cycle of length k

13. Has n vertices of degree k

14. Is connected

15. Has n simple cycles of length k

16. Has an edge (v, w), where $\delta(v) = i$ and $\delta(w) = j$

17. Find an invariant not given in this section or in Exercises 12–16. Prove that your property is an invariant.

In Exercises 18–20, tell whether each property is an invariant or not. If the property is an invariant, prove that it is; otherwise, give a counterexample.

18. Has an Euler cycle

19. Has a vertex inside some simple cycle

20. Is bipartite

21. Draw all nonisomorphic simple graphs having three vertices.

22. Draw all nonisomorphic simple graphs having four vertices.

23. Draw all nonisomorphic, cycle-free, connected graphs having five vertices.

24. Draw all nonisomorphic, cycle-free, connected graphs having six vertices.

25. Show that simple graphs G_1 and G_2 are isomorphic if and only if their vertices may be ordered so that their adjacency matrices are identical.

The *complement* of a simple graph G is the simple graph $\overline{G}$ with the same vertices as G. An edge exists in $\overline{G}$ if and only if it does not exist in G.

26. Draw the complement of the graph G_1 of Exercise 1.

27. Draw the complement of the graph G_2 of Exercise 1.

★ 28. Show that if G is a simple graph, either G or $\overline{G}$ is connected.

29. A simple graph G is **self-complementary** if G and $\overline{G}$ are isomorphic.

   (a) Find a self-complementary graph having five vertices.

   (b) Find another self-complementary graph.

30. Let G_1 and G_2 be simple graphs. Show that G_1 and G_2 are isomorphic if and only if $\overline{G_1}$ and $\overline{G_2}$ are isomorphic.

31. Given two graphs G_1 and G_2, suppose that there is a one-to-one, onto function f from the vertices of G_1 to the vertices of G_2 and a one-to-one, onto function g from the edges of G_1 to the edges of G_2, so that if an edge e is incident on v and w in G_1, the edge $g(e)$ is incident on $f(v)$ and $f(w)$ in G_2. Are G_1 and G_2 isomorphic?

A *homomorphism* from a graph G_1 to a graph G_2 is a function f from the vertex set of G_1 to the vertex set of G_2 with the property that if v and w are adjacent in G_1, then $f(v)$ and $f(w)$ are adjacent in G_2.

32. Suppose that G_1 and G_2 are simple graphs. Show that if f is a homomorphism of G_1 to G_2 and f is one-to-one and onto, G_1 and G_2 are isomorphic.

In Exercises 33–37, for each pair of graphs, give an example of a homomorphism from G_1 to G_2.

33.

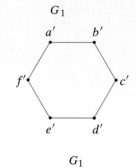

G_1 G_2

34.

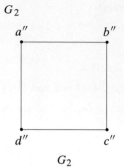

G_1 G_2

35. $G_1 = G_1$ of Exercise 34; $G_2 = G_1$ of Exercise 33

36. $G_1 = G_1$ of Exercise 33

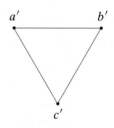

G_2

37.

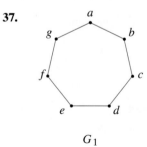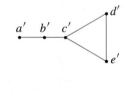

G_1 G_2

★ **38.** [Hell] Show that the only homomorphism from the graph to itself is the identity function.

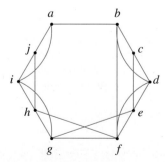

6.7 *PLANAR GRAPHS*

Three cities, C_1, C_2, and C_3, are to be directly connected by expressways to each of three other cities, C_4, C_5, and C_6. Can this road system be designed so that the expressways do not cross? A system in which the roads do cross is illustrated in Figure 6.7.1. If you try drawing a system in which the roads do not cross, you will soon be convinced that it cannot be done. Later in this section we explain carefully why it cannot be done.

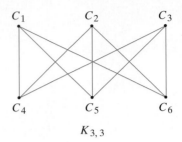

FIGURE 6.7.1
Cities connected by expressways.

DEFINITION 6.7.1

A graph is *planar* if it can be drawn in the plane without its edges crossing.

In designing printed circuits it is desirable to have as few lines cross as possible; thus the designer of printed circuits faces the problem of planarity.

If a connected, planar graph is drawn in the plane, the plane is divided into contiguous regions called **faces**. A face is characterized by the cycle that forms its boundary. For example, in the graph of Figure 6.7.2, face A is bounded by the cycle $(5, 2, 3, 4, 5)$ and face C is bounded by the cycle $(1, 2, 5, 1)$. The outer face D is considered to be bounded by the cycle $(1, 2, 3, 4, 6, 1)$. The graph of Figure 6.7.2 has $f = 4$ faces, $e = 8$ edges, and $v = 6$ vertices. Notice that f, e, and v satisfy the equation

$$f = e - v + 2. \tag{6.7.1}$$

In 1752, Euler proved that equation (6.7.1) holds for any connected, planar graph. At the end of this section we will show how to prove (6.7.1), but for now let us show how (6.7.1) can be used to show that certain graphs are not planar.

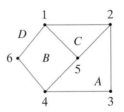

FIGURE 6.7.2
A connected, planar graph with $f = 4$ faces (A, B, C, D), $e = 8$ edges, and $v = 6$ vertices; $f = e - v + 2$.

EXAMPLE 6.7.2

Show that the graph $K_{3,3}$ of Figure 6.7.1 is not planar.

Suppose that $K_{3,3}$ is planar. Since every cycle has at least four edges, each face is bounded by at least four edges. Thus the number of edges that bound faces is at least $4f$. In a planar graph, each edge belongs to at most two bounding cycles. Therefore,

$$2e \geq 4f.$$

Using (6.7.1), we find that

$$2e \geq 4(e - v + 2). \tag{6.7.2}$$

For the graph of Figure 6.7.1, $e = 9$ and $v = 6$, so (6.7.2) becomes

$$18 = 2 \cdot 9 \geq 4(9 - 6 + 2) = 20,$$

which is a contradiction. Therefore, $K_{3,3}$ is not planar. □

By a similar kind of argument (see Exercise 15), we can show that the graph K_5 of Figure 6.7.3 is not planar.

Obviously, if a graph contains $K_{3,3}$ or K_5 as a subgraph, it cannot be planar. The converse is almost true. To state the situation precisely, we must introduce some new terms.

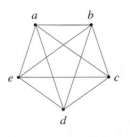

FIGURE 6.7.3
The nonplanar graph K_5.

DEFINITION 6.7.3

If a graph G has a vertex v of degree 2 and edges (v, v_1) and (v, v_2) with $v_1 \neq v_2$, we say that the edges (v, v_1) and (v, v_2) are in *series*. A *series reduction* consists of deleting the vertex v from the graph G and replacing the edges (v, v_1) and (v, v_2) by the edge (v_1, v_2). The resulting graph G' is said to be *obtained from G by a series reduction*. By convention, G is said to be obtainable from itself by a series reduction.

EXAMPLE 6.7.4

In the graph G of Figure 6.7.4, the edges (v, v_1) and (v, v_2) are in series. The graph G' of Figure 6.7.4 is obtained from G by a series reduction.

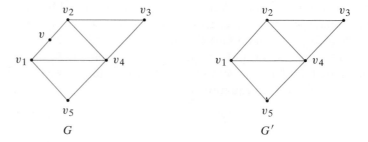

FIGURE 6.7.4 G' is obtained from G by a series reduction.   □

DEFINITION 6.7.5

Graphs G_1 and G_2 are *homeomorphic* if G_1 and G_2 can be reduced to isomorphic graphs by performing a sequence of series reductions.

According to Definitions 6.7.3 and 6.7.5, any graph is homeomorphic to itself. Also, graphs G_1 and G_2 are homeomorphic if G_1 can be reduced to a graph isomorphic to G_2 or if G_2 can be reduced to a graph isomorphic to G_1.

EXAMPLE 6.7.6

The graphs G_1 and G_2 of Figure 6.7.5 are homeomorphic, since they can both be reduced to the graph G' of Figure 6.7.5 by a sequence of series reductions.   □

If we define a relation R on a set of graphs by the rule $G_1 R G_2$ if G_1 and G_2 are homeomorphic, R is an equivalence relation. Each equivalence class consists of a set of mutually homeomorphic graphs.

We now state a necessary and sufficient condition for a graph to be planar. The theorem was first stated and proved by Kuratowski in 1930. The proof appears in [Even, 1979].

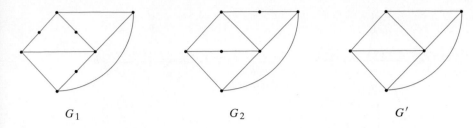

FIGURE 6.7.5 G_1 and G_2 are homeomorphic; each can be reduced to G'.

| THEOREM 6.7.7 | *Kuratowski's Theorem* |

A graph G is planar if and only if G does not contain a subgraph homeomorphic to K_5 or $K_{3,3}$. ∎

| EXAMPLE 6.7.8 |

Show that the graph G of Figure 6.7.6 is not planar by using Kuratowski's Theorem.

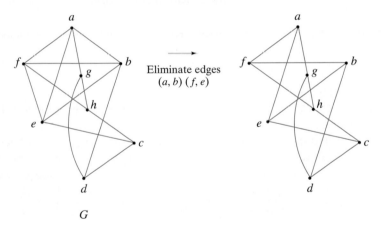

FIGURE 6.7.6   Eliminating edges to obtain a subgraph.

Let us try to find $K_{3,3}$ in the graph G of Figure 6.7.6. We first note that the vertices a, b, f, and e each have degree 4. In $K_{3,3}$ each vertex has degree 3, so let us eliminate the edges (a, b) and (f, e) so that all vertices have degree 3 (see Figure 6.7.6). We note that if we eliminate one more edge, we will obtain two vertices of degree 2 and we can then carry out two series reductions. The resulting graph will have nine edges and since $K_{3,3}$ has nine edges, this approach looks promising. Using trial and error, we finally see that if we eliminate edge (g, h) and carry out the series reductions, we obtain an isomorphic copy of $K_{3,3}$ (see Figure 6.7.7). Therefore, the graph G of Figure 6.7.6 is not planar, since it contains a subgraph homeomorphic to $K_{3,3}$. □

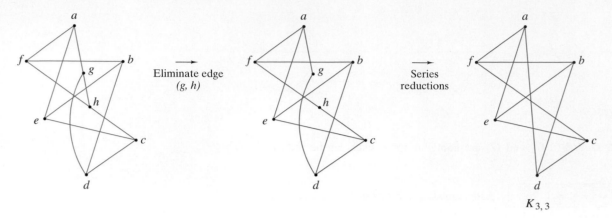

FIGURE 6.7.7   Elimination of an edge to obtain a subgraph, followed by series reductions.

Although Theorem 6.7.7 does give an elegant characterization of planar graphs, it does not lead to an efficient algorithm for recognizing planar graphs. However, algorithms are known that can determine whether a graph having n vertices is planar in time $O(n)$ (see [Even, 1979]).

We will conclude this section by proving Euler's formula.

| THEOREM 6.7.9 | *Euler's Formula for Graphs* |

If G is a connected, planar graph with e edges, v vertices, and f faces, then

$$f = e - v + 2. \tag{6.7.3}$$

Proof. We will use induction on the number of edges.

Suppose that $e = 1$. Then G is one of the two graphs shown in Figure 6.7.8. In either case, the formula holds. We have verified the Basis Step.

Suppose that the formula holds for connected, planar graphs with n edges. Let G be a graph with $n + 1$ edges. First, suppose that G contains no cycles. Pick a vertex v and trace a path starting at v. Since G is cycle-free, every time we trace an edge, we arrive at a new vertex. Eventually, we will reach a vertex a, with degree 1, that we cannot leave (see Figure 6.7.9). We delete a and the

$f = 1, e = 1, v = 2$

$f = 2, e = 1, v = 1$

FIGURE 6.7.8
The Basis Step of Theorem 6.7.9.

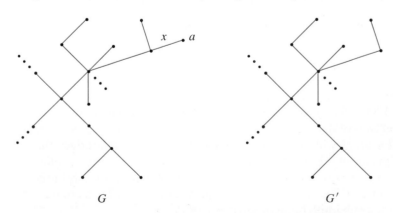

FIGURE 6.7.9   The proof of Theorem 6.7.9 for the case that G has no cycles. We find a vertex a of degree 1 and delete a and the edge x incident on it.

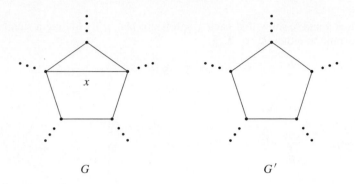

FIGURE 6.7.10   The proof of Theorem 6.7.9 for the case that G has a cycle. We delete edge x in a cycle.

edge x incident on a from the graph G. The resulting graph G' has n edges; hence, by the inductive assumption, (6.7.3) holds for G'. Since G has one more edge than G', one more vertex than G', and the same number of faces as G', it follows that (6.7.3) also holds for G.

Now suppose that G contains a cycle. Let x be an edge in a cycle (see Figure 6.7.10). Now x is part of a boundary for two faces. This time we delete the edge x but no vertices to obtain the graph G' (see Figure 6.7.10). Again G' has n edges; hence, by the inductive assumption, (6.7.3) holds for G'. Since G has one more face than G', one more edge than G', and the same number of vertices as G', it follows that (6.7.3) also holds for G.

Since we have verified the Inductive Step, by the Principle of Mathematical Induction, the theorem is proved.   ∎

ॐ ॐ ॐ

Exercises

In Exercises 1–3, show that each graph is planar by redrawing it so that no edges cross.

1.

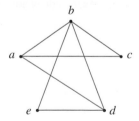

2.

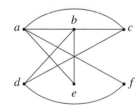

3.

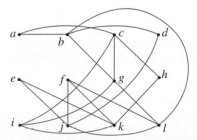

In Exercises 4 and 5, show that each graph is not planar by finding a subgraph homeomorphic to either K_5 or $K_{3,3}$.

4.

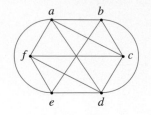

5.

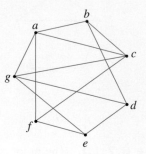

In Exercises 6–8, determine whether each graph is planar. If the graph is planar, redraw it so that no edges cross; otherwise, find a subgraph homeomorphic to either K_5 or $K_{3,3}$.

6.

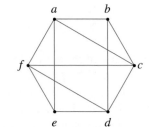

7.

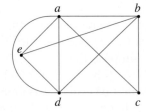

8.

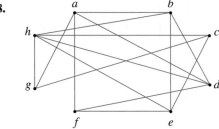

9. A connected, planar graph has nine vertices having degrees 2, 2, 2, 3, 3, 3, 4, 4, and 5. How many edges are there? How many faces are there?

10. Show that adding or deleting loops, parallel edges, or edges in series does not affect the planarity of a graph.

11. Show that any graph having four or fewer vertices is planar.

12. Show that any graph having five or fewer vertices and a vertex of degree 2 is planar.

13. Show that in any simple, connected, planar graph $e \leq 3v - 6$.

14. Give an example of a simple, connected, nonplanar graph for which $e \leq 3v - 6$.

15. Use Exercise 13 to show that K_5 is not planar.

★ 16. Show that if a simple graph G has 11 or more vertices, then either G or its complement $\overline{G}$ is not planar.

★ 17. Prove that if a planar graph has an Euler cycle, it has an Euler cycle with no crossings. A path P in a planar graph has a *crossing* if a vertex v appears at least twice in P and P crosses itself at v; that is,

$$P = (\ldots, w_1, v, w_2, \ldots, w_3, v, w_4, \ldots),$$

where the vertices are arranged so that w_1, v, w_2 crosses w_3, v, w_4 at v as in the adjacent figure.

A *coloring* of a graph G by the colors $C_1, C_2, \ldots, C_n$ assigns to each vertex a color C_i so that any vertex has a color different from that of any adjacent vertex. For example, the accompanying graph is colored with three colors. The rest of the exercises deal with coloring planar graphs.

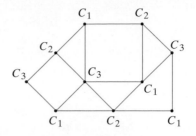

A *planar map* is a planar graph where the faces are interpreted as countries, the edges are interpreted as borders between countries, and the vertices represent the intersections of borders. The problem of coloring a planar map G, so that no countries with adjoining boundaries have the same color, can be reduced to the problem of coloring a graph by first constructing the *dual graph* G' of G in the following way. The vertices of the dual graph G' consist of one point in each face of G, including the unbounded face. An edge in G' connects two vertices if the corresponding faces in G are separated by a boundary. Coloring the map G is equivalent to coloring the vertices of the dual graph G'.

18. Find the dual of the adjacent map.

19. Show that the dual of a planar map is a planar graph.

20. Show that any coloring of the map of Exercise 18, excluding the unbounded region, requires at least three colors.

21. Color the map of Exercise 18, excluding the unbounded region, using three colors.

22. Find the dual of the adjacent map.

23. Show that any coloring of the map of Exercise 22, excluding the unbounded region, requires at least four colors.

24. Color the map of Exercise 22, excluding the unbounded region, using four colors.

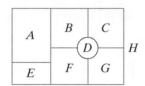

Exercise 18

A *triangulation* of a simple, planar graph G is obtained from G by connecting as many vertices as possible while maintaining planarity and not introducing loops or parallel edges.

25. Find a triangulation of the adjacent graph.

26. Show that if a triangulation G' of a simple, planar graph G can be colored with n colors, so can G.

27. Show that in a triangulation of a simple, planar graph, $3f = 2e$.

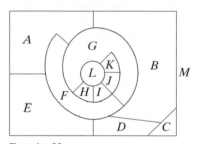

Exercise 22

Exercise 25

Appel and Haken proved (see [Appel]) that every simple, planar graph can be colored with four colors. The problem had been posed in the mid-1800s and for years no one had succeeded in giving a proof. Those working on the four-color problem in recent years had one advantage their predecessors did not—the use of fast electronic computers. The following exercises show how the proof begins.

Suppose there is a simple, planar graph that requires more than four colors to color. Among all such graphs, there is one with the fewest number of vertices. Let G be a triangulation of this graph. Then G also has a minimal number of vertices and by Exercise 26, G requires more than four colors to color.

28. If the dual of a map has a vertex of degree 3, what must the original map look like?

29. Show that G cannot have a vertex of degree 3.

★ 30. Show that G cannot have a vertex of degree 4.

★ 31. Show that G has a vertex of degree 5.

The contribution of Appel and Haken was to show that only a finite number of cases involving the vertex of degree 5 needed to be considered and to analyze all of these cases and show that all could be colored using four colors. The reduction to a finite number of cases was facilitated by using the computer to help find the cases to be analyzed. The computer was then used again to analyze the resulting cases.

★ 32. Show that any simple, planar graph can be colored using five colors.

†6.8 *INSTANT INSANITY*

Instant Insanity is a puzzle consisting of four cubes each of whose faces is painted one of the four colors, red, white, blue, or green (see Figure 6.8.1). (There are different versions of the puzzle, depending on which faces are painted which colors.) The problem is to stack the cubes, one on top of the other, so that whether viewed from front, back, left, or right, one sees all four colors (see Figure 6.8.2). Since 331,776 different stacks are possible (see Exercise 12), a solution by hand by trial and error is impractical. We present a solution, using a graph model, that makes it possible to discover a solution, if there is one, in a few minutes!

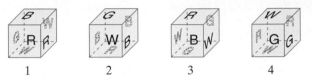

FIGURE 6.8.1   An Instant Insanity puzzle.

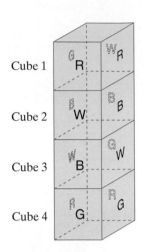

Cube 1

Cube 2

Cube 3

Cube 4

FIGURE 6.8.2
A solution to the Instant Insanity puzzle of Figure 6.8.1.

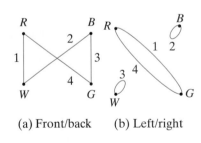

(a) Front/back   (b) Left/right

FIGURE 6.8.3
Graphs that represent the stacking of Figure 6.8.2.

Cube 1

FIGURE 6.8.4
Rotating a cube to obtain a left/right orientation, without changing the front/back colors.

First, notice that any particular stacking can be represented by two graphs, one representing the front/back colors and the other representing the left/right colors. For example, in Figure 6.8.3 we have represented the stacking of Figure 6.8.2. The vertices represent the colors, and an edge connects two vertices if the opposite faces have those colors. For example, in the front/back graph, the edge labeled 1 connects R and W, since the front and back faces of cube 1 are red and white. As another example, in the left/right graph, W has a loop, since both the left and right faces of cube 3 are white.

We can also construct a stacking from a pair of graphs such as those in Figure 6.8.3, which represent a solution of the Instant Insanity puzzle. Begin with the front/back graph. Cube 1 is to have red and white opposing faces. Arbitrarily assign one of these colors, say red, to the front. Then cube 1 has a white back face. The other edge incident on W is 2, so make cube 2's front face white. This gives cube 2 a blue back face. The other edge incident on B is 3, so make cube 3's front face blue. This gives cube 3 a green back face. The other edge incident on G is 4. Cube 4 then gets a green front face and a red back face. The front and back faces are now properly aligned. At this point, the left and right faces are randomly arranged; however, we will show how to correctly orient the left and right faces without altering the colors of the front and back faces.

Cube 1 is to have red and green opposing left and right faces. Arbitrarily assign one of these colors, say green, to the left. Then cube 1 has a red right face. Notice that by rotating the cube, we can obtain this left/right orientation without changing the colors of the front and back (see Figure 6.8.4). We can similarly orient cubes 2, 3, and 4. Notice that cubes 2 and 3 have the same colors on opposing sides. The stacking of Figure 6.8.2 has been reconstructed.

It is apparent from the preceding discussion that a solution to the Instant Insanity puzzle can be obtained if we can find two graphs like those of Figure 6.8.3. The properties needed are

† This section can be omitted without loss of continuity.

- Each vertex should have degree 2.   (6.8.1)
- Each cube should be represented by an edge exactly
  once in each graph.   (6.8.2)
- The two graphs should not have any edges in common.   (6.8.3)

Property (6.8.1) assures us that each color can be used twice, once on the front
(or left) and once on the back (or right). Property (6.8.2) assures us that each
cube is used exactly once. Property (6.8.3) assures us that, after orienting the
front and back sides, we can successfully orient the left and right sides.

To obtain a solution, we first draw a graph G that represents all of the
faces of all of the cubes. The vertices of G represent the four colors and an
edge labeled i connects two vertices (colors) if the opposing faces of cube i
have those colors. In Figure 6.8.5 we have drawn the graph that represents
the cubes of Figure 6.8.1. Then, by inspection, we find two subgraphs of G
satisfying properties (6.8.1)–(6.8.3). Try your hand at the method by finding
another solution to the puzzle represented by Figure 6.8.5.

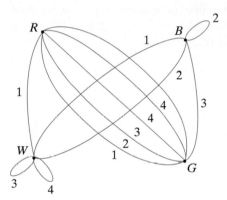

FIGURE 6.8.5   A graph representation of the Instant Insanity puzzle of Figure 6.8.1.

EXAMPLE 6.8.1

Find a solution to the Instant Insanity puzzle of Figure 6.8.6.

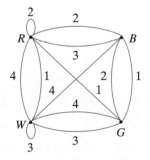

FIGURE 6.8.6   The Instant Insanity puzzle for Example 6.8.1.

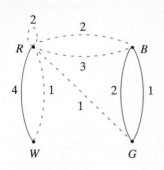

FIGURE 6.8.7
Trying to find a subgraph of Figure
6.8.6 satisfying (6.8.1) and (6.8.2).

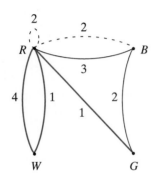

FIGURE 6.8.8
Another attempt to find
a subgraph of Figure 6.8.6 satisfying
(6.8.1) and (6.8.2).

We begin by trying to construct one subgraph having properties (6.8.1) and (6.8.2). We arbitrarily choose a vertex, say B, and choose two edges incident on vertex B. Suppose that we select the two edges shown as solid lines in Figure 6.8.7. Now consider the problem of picking two edges incident on vertex R. We cannot select any edges incident on B or G since B and G must each have degree 2. Since each cube must appear in each subgraph exactly once, we cannot select any of the edges labeled 1 or 2 since we already have selected edges with these labels. Edges incident on R that cannot be selected are shown dashed in Figure 6.8.7. This leaves only the edge labeled 4. Since we need two edges incident on R, our initial selection of edges incident on B must be revised.

For our next attempt at choosing two edges incident on vertex B, let us select the edges labeled 2 and 3, as shown in Figure 6.8.8. Since this choice includes one edge incident on R, we must choose one additional edge incident on R. We have three possibilities for selecting the additional edge (shown in color in Figure 6.8.8). (The loop incident at R counts as two edges and so cannot be chosen.) If we select the edge labeled 1 incident on R and G, we would need a loop at W labeled 4. Since there is no such loop, we do not select this edge. If we select the edge labeled 1 incident on R and W, we can then select the edge labeled 4 incident on W and G (see Figure 6.8.9). We have obtained one of the graphs.

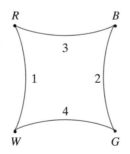

FIGURE 6.8.9   A
subgraph of Figure 6.8.6
satisfying (6.8.1) and (6.8.2).

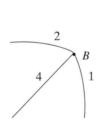

FIGURE 6.8.10

Edges incident on
B not used in Figure 6.8.9.

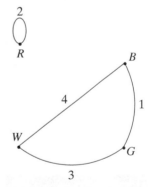

FIGURE 6.8.11   Another subgraph of Figure
6.8.6, with no edges in common with Figure 6.8.9, satisfying (6.8.1) and (6.8.2). This
graph and that of Figure 6.8.9
solve the Instant Insanity puzzle of Figure 6.8.6.

We now look for a second graph having no edges in common with the graph just chosen. Let us again begin by picking two edges incident on B. Because we cannot reuse edges, our choices are limited to three edges (see Figure 6.8.10). Choosing the edges labeled 1 and 4 leads to the graph of Figure 6.8.11. The graphs of Figures 6.8.9 and 6.8.11 solve the Instant Insanity puzzle of Figure 6.8.6.   □

Exercises

Find solutions to the following Instant Insanity puzzles.

1.
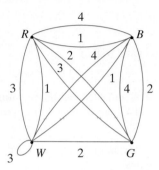

2.
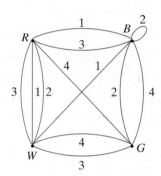

3.
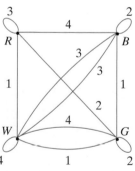

4.
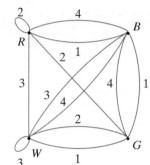

5.
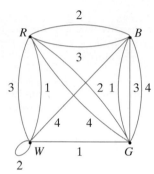

6.
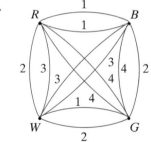

7. (a) Find all subgraphs of Figure 6.8.5 satisfying properties (6.8.1) and (6.8.2).

   (b) Find all the solutions to the Instant Insanity puzzle of Figure 6.8.5.

8. (a) Represent the Instant Insanity puzzle by a graph.

   (b) Find a solution to the puzzle.

   (c) Find all subgraphs of your graph of part (a) satisfying properties (6.8.1) and (6.8.2).

   (d) Use (c) to show that the puzzle has a unique solution.

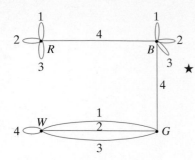

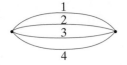

9. Show that the adjacent Instant Insanity puzzle has no solution by giving an argument to show that no subgraph satisfies properties (6.8.1) and (6.8.2). Notice that there is no solution even though each cube contains all four colors.

★ **10.** Give an example of an Instant Insanity puzzle satisfying:

(a) There is no solution.

(b) Each cube contains all four colors.

(c) There is a subgraph satisfying properties (6.8.1) and (6.8.2).

11. Show that there are 24 orientations of a cube.

12. Number the cubes of an Instant Insanity puzzle 1, 2, 3, and 4. Show that the number of stackings in which the cubes are stacked 1, 2, 3, and 4, reading from bottom to top, is 331,776.

★ **13.** How many Instant Insanity graphs are there; that is, how many graphs are there with four vertices and 12 edges—three of each of four types?

Exercises 14–21 refer to a modified version of Instant Insanity where a solution is defined to be a stacking that, when viewed from the front, back, left, or right, shows one color. (The front, back, left, and right are of different colors.)

14. Give an argument that shows that if we graph the puzzle as in regular Instant Insanity, a solution to modified Instant Insanity consists of two subgraphs of the form shown in the figure, with no edges or vertices in common.

Find solutions to modified Instant Insanity for the following puzzles.

15.

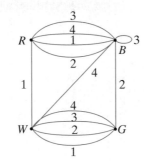

16.

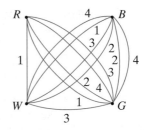

17.

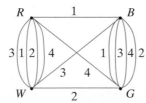

18. Graph of Exercise 6

19. Show that for Figure 6.8.5, Instant Insanity, as given in the text, has a solution, but the modified version does not have a solution.

★ **20.** Show that if modified Instant Insanity has a solution, the version given in the text must also have a solution.

21. Is it possible for neither version of Instant Insanity to have a solution even if each cube contains all four colors? If the answer is yes, prove it; otherwise, give a counterexample.

⤙ NOTES

Virtually any reference on discrete mathematics contains one or more chapters on graph theory. Books specifically on graph theory are [Berge; Bondy; Chartrand; Deo; Even, 1979; Gibbons; Harary; König; Ore; and Wilson]. [Deo; Even, 1979; and Gibbons] emphasize graph algorithms. [Brassard and Cormen] also treat graphs and graph algorithms.

[Fukunaga; Gose; and Nadler] are texts on pattern recognition.

[Akl; Leighton; Lester; Lewis; and Quinn] discuss parallel computers and algorithms for parallel computers. Our results on subgraphs of the hypercube are from [Saad].

Euler's original paper on the Königsberg bridges, edited by J. R. Newman, was reprinted as [Newman].

In [Gardner, 1959], Hamiltonian cycles are related to the Tower of Hanoi puzzle.

In many cases, so-called *branch-and-bound* methods (see, e.g., [Tucker]) often give solutions to the traveling salesperson problem more efficiently than will exhaustive search.

Dijkstra's shortest-path algorithm appeared in [Dijkstra, 1959].

The complexity of the graph isomorphism problem is discussed in [Köbler].

Appel and Haken published their solution to the four-color problem in [Appel].

⤙ CHAPTER REVIEW

Section 6.1

Graph $G = (V, E)$
   (undirected and directed)
Vertex
Edge
Edge e is incident on vertex v
Vertex v is incident on edge e
v and w are adjacent vertices
Parallel edges
Loop
Isolated vertex
Simple graph
Weighted graph
Weight of an edge
Length of a path in a weighted graph
Similarity graph
Dissimilarity function
Pattern recognition
n-cube (hypercube)
Serial computer
Serial algorithm
Parallel computer
Parallel algorithm

Section 6.2

Path
Simple path

Cycle
Simple cycle
Connected graph
Subgraph
Component of a graph
Degree of a vertex $\delta(v)$
Königsberg bridge problem
Euler cycle
A graph G has an Euler cycle if and only if G is connected and every vertex has even degree.
The sum of the degrees of all vertices in a graph is equal to twice the number of edges.
In any graph, there are an even number of vertices of odd degree.
A graph has a path with no repeated edges from v to w ($v \neq w$) containing all the edges and vertices if and only if it is connected and v and w are the only vertices having odd degree.
If a graph G contains a cycle from v to v, G contains a simple cycle from v to v.

Section 6.3

Hamiltonian cycle

✍ CHAPTER SELF-TEST

Section 6.1

1. For the graph $G = (V, E)$, find V, E, all parallel edges, all loops, all isolated vertices, and state whether G is a simple graph. Also state on which vertices edge e_3 is incident and on which edges vertex v_2 is incident.

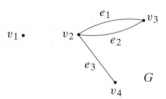

2. Explain why the graph does not have a path from a to a that passes through each edge exactly one time.

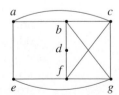

3. Draw $K_{2,5}$ the complete bipartite graph on 2 and 5 vertices.

4. Prove that the n-cube is bipartite for all $n \geq 1$.

Section 6.2

5. Tell whether the path $(v_2, v_3, v_4, v_2, v_6, v_1, v_2)$ in the graph is a simple path, a cycle, a simple cycle, or none of these.

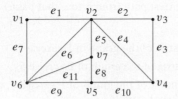

6. Draw all subgraphs containing exactly two edges of the graph

7. Find a connected subgraph of the graph of Exercise 5 containing all of the vertices of the original graph and having as few edges as possible.

8. Does the graph of Exercise 5 contain an Euler cycle? Explain.

Section 6.3

9. Find a Hamiltonian cycle in the graph of Exercise 5.

10. Find a Hamiltonian cycle in the 3-cube.

11. Show that the graph has no Hamiltonian cycle.

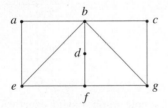

12. Show that the cycle $(a, b, c, d, e, f, g, h, i, j, a)$ provides a solution to the traveling salesperson problem for the graph shown.

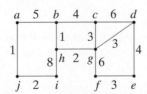

Section 6.4

Exercises 13–16 refer to the following graph.

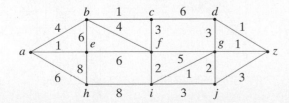

13. Find the length of a shortest path from a to i.

14. Find the length of a shortest path from a to z.

15. Find a shortest path from a to z.

16. Find the length of a shortest path from a to z that passes through c.

Section 6.5

17. Write the adjacency matrix of the graph of Exercise 5.

18. Write the incidence matrix of the graph of Exercise 5.

19. If A is the adjacency matrix of the graph of Exercise 5, what does the entry in row v_2 and column v_3 of A^3 represent?

20. Can a column of an incidence matrix consist only of zeros? Explain.

Section 6.6

In Exercises 21 and 22, determine whether the graphs G_1 and G_2 are isomorphic. If the graphs are isomorphic, give orderings of their vertices that produce equal adjacency matrices. If the graphs are not isomorphic, give an invariant that the graphs do not share.

21.

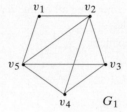
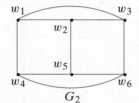

22.

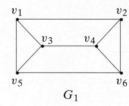

23. Draw all nonisomorphic, simple graphs having exactly five vertices and two edges.

24. Draw all nonisomorphic, simple graphs having exactly five vertices, two components, and no cycles.

Section 6.7

In Exercises 25 and 26, determine whether the graph is planar. If the graph is planar, redraw it so that no edges cross; otherwise, find a subgraph homeomorphic to either K_5 or $K_{3,3}$.

25.

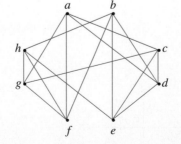

26.

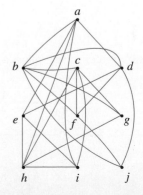

27. Show that any simple, connected graph with 31 edges and 12 vertices is not planar.

28. Show that the n-cube is planar if $n \leq 3$ and not planar if $n > 3$.

Section 6.8

29. Represent the Instant Insanity puzzle by a graph.

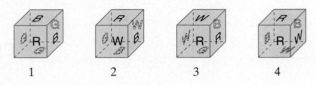

        1   2   3   4

30. Find a solution to the puzzle of Exercise 29.

31. Find all subgraphs of the graph of Exercise 29 satisfying properties (6.8.1) and (6.8.2).

32. Use Exercise 31 to determine how many solutions the puzzle of Exercise 29 has.

7

TREES

I talk to the trees,
but they don't listen to me.

—from Paint Your Wagon

Trees form one of the most widely used subclasses of graphs. Computer science, in particular, makes extensive use of trees. In computer science, trees are useful in organizing and relating data in a database (see Example 7.1.6). Trees also arise in theoretical problems such as the optimal time for sorting (see Section 7.7).

In this chapter we begin by giving the requisite terminology. We look at subclasses of trees (e.g., rooted trees and binary trees) and many applications of trees (e.g., spanning trees, decision trees, and game trees). Our discussion of tree isomorphisms extends the discussion of Section 6.6 on graph isomorphisms.

[†] This section can be omitted without loss of continuity.

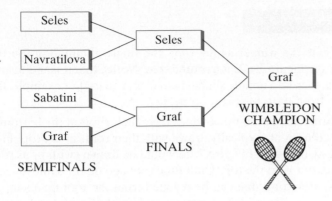

FIGURE 7.1.1   Semifinals and finals at Wimbledon.

7.1 INTRODUCTION

Figure 7.1.1 shows the results of the semifinals and finals of a tennis competition at Wimbledon, which featured four of the greatest players in the history of tennis. At Wimbledon, when a player loses, she is out of the tournament. Winners continue to play until only one person, the champion, remains. (Such a competition is called a *single-elimination tournament*.) Figure 7.1.1 shows that in the semifinals Monica Seles defeated Martina Navratilova and Steffi Graf defeated Gabriela Sabatini. The winners, Seles and Graf, then played, and Graf defeated Seles. Steffi Graf, being the sole undefeated player, became Wimbledon champion.

If we regard the single-elimination tournament of Figure 7.1.1 as a graph (see Figure 7.1.2), we obtain a **tree**. If we rotate Figure 7.1.2, it looks like a natural tree (see Figure 7.1.3). The formal definition follows.

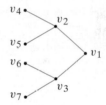

FIGURE 7.1.2
The tournament of Figure 7.1.1 as a tree.

DEFINITION 7.1.1

A (*free*) *tree* T is a simple graph satisfying: If v and w are vertices in T, there is a unique simple path from v to w.

A *rooted tree* is a tree in which a particular vertex is designated the root.

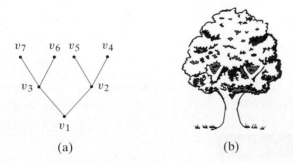

FIGURE 7.1.3   The tree of Figure 7.1.2 rotated (a) compared with a natural tree (b).

EXAMPLE 7.1.2

If we designate the winner as the root, the single-elimination tournament of Figure 7.1.1 (or Figure 7.1.2) is a rooted tree. Notice that if v and w are vertices in this graph, there is a unique simple path from v to w. For example, the unique simple path from v_2 to v_7 is (v_2, v_1, v_3, v_7). □

In contrast to natural trees, which have their roots at the bottom, in graph theory rooted trees are typically drawn with their roots at the top. Figure 7.1.4 shows the way the tree of Figure 7.1.2 would be drawn (with v_1 as root). First, we place the root v_1 at the top. Under the root and on the same level, we place the vertices, v_2 and v_3, that can be reached from the root on a simple path of length 1. Under each of these vertices and on the same level, we place the vertices, v_4, v_5, v_6, and v_7, that can be reached from the root on a simple path of length 2. We continue in this way until the entire tree is drawn. Since the simple path from the root to any given vertex is unique, each vertex is on a uniquely determined level. We call the level of the root level 0. The vertices under the root are said to be on level 1, and so on. Thus the **level of a vertex** v is the length of the simple path from the root to v. The **height** of a rooted tree is the maximum level number that occurs.

FIGURE 7.1.4
The tree Figure 7.1.3(a) with the root at the top.

EXAMPLE 7.1.3

The vertices v_1, v_2, v_3, v_4, v_5, v_6, v_7 in the rooted tree of Figure 7.1.4 are on (respectively) levels 0, 1, 1, 2, 2, 2, 2. The height of the tree is 2. □

EXAMPLE 7.1.4

If we designate e as the root in the tree T of Figure 7.1.5, we obtain the rooted tree T' shown in Figure 7.1.5. The vertices $a, b, c, d, e, f, g, h, i, j$ are on (respectively) levels 2, 1, 2, 1, 0, 1, 1, 2, 2, 3. The height of T' is 3. □

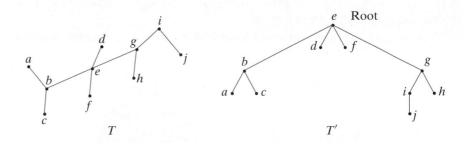

FIGURE 7.1.5   A tree T and a rooted tree T'. T' is obtained from T by designating e as the root.

EXAMPLE 7.1.5

A rooted tree is often used to specify hierarchical relationships. When a tree is used in this way, if vertex a is on a level one less than the level of vertex b

and *a* and *b* are adjacent, then *a* is "just above" *b* and a logical relationship exists between *a* and *b*: *a* dominates *b* or *b* is subordinate to *a* in some way. An example of such a tree, which is the administrative organizational chart of a hypothetical university, is given in Figure 7.1.6.   □

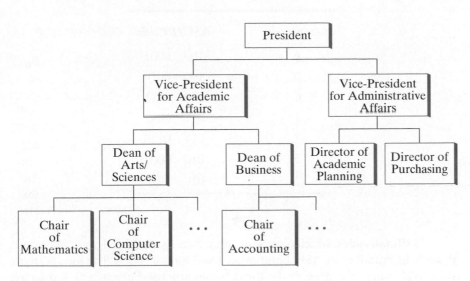

FIGURE 7.1.6   An administrative organizational chart.

EXAMPLE 7.1.6   *Hierarchical Definition Trees*

Figure 7.1.7 is an example of a **hierarchical definition tree**. Such trees are used to show logical relationships among records in a database. [Recall (see Section 2.7) that a database is a collection of records that are manipulated by a computer.] The tree of Figure 7.1.7 might be used as a model for setting up a database to maintain records about books housed in several libraries.   □

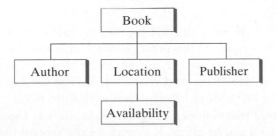

FIGURE 7.1.7   A hierarchical definition tree.

EXAMPLE 7.1.7   *Huffman Codes*

The most common way to represent characters internally in a computer is by using fixed-length bit strings. For example, ASCII (American Standard Code for Information Interchange) represents each character by a string of seven bits. Examples are given in Table 7.1.1.

TABLE 7.1.1
A portion of the ASCII table

Character	ASCII Code
A	100 0001
B	100 0010
C	100 0011
1	011 0001
2	011 0010
!	010 0001
*	010 1010

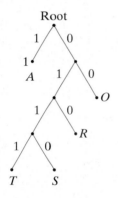

FIGURE 7.1.8
A Huffman code.

Huffman codes, which represent characters by variable-length bit strings, provide alternatives to ASCII and other fixed-length codes. The idea is to use short bit strings to represent the most frequently used characters and to use longer bit strings to represent less frequently used characters. In this way it is generally possible to represent strings of characters, such as text and programs, in less space than if ASCII were used. VCR Plus+, a device that automatically programs a videocassette recorder, uses a Huffman code to generate numbers that the user then enters to choose which programs to record. The numbers are published in many television listings.

A Huffman code is most easily defined by a rooted tree (see Figure 7.1.8). To decode a bit string, we begin at the root and move down the tree until a character is encountered. The bit, 0 or 1, tells us whether to move right or left. As an example, let us decode the string

$$01010111. \qquad (7.1.1)$$

We begin at the root. Since the first bit is 0, the first move is right. Next, we move left and then right. At this point, we encounter the first character R. To decode the next character, we begin again at the root. The next bit is 1, so we move left and encounter the next character A. The last bits 0111 decode as T. Therefore, the bit string (7.1.1) represents the word RAT.

Given a tree that defines a Huffman code, such as Figure 7.1.8, any bit string [e.g., (7.1.1)] can be uniquely decoded even though the characters are represented by variable-length bit strings. For the Huffman code defined by the tree of Figure 7.1.8, the character A is represented by a bit string of length 1, whereas S and T are represented by bit strings of length 4. (A is represented as 1, S is represented as 0110, and T is represented as 0111.)   □

Huffman gave an algorithm (Algorithm 7.1.8) to construct a Huffman code from a table giving the frequency of occurrence of the characters to be represented so that the code constructed represents strings of characters in minimal space, provided that the strings to be represented have character frequencies identical to the character frequencies in the table. A proof that the code constructed is optimal may be found in [Cormen].

ALGORITHM 7.1.8   *Constructing an Optimal Huffman Code*

This algorithm constructs an optimal Huffman code from a table giving the frequency of occurrence of the characters to be represented. The output is a rooted tree with the vertices at the lowest levels labeled with the frequencies and with the edges labeled with bits as in Figure 7.1.8. The coding tree is obtained by replacing each frequency by a character having that frequency.

Input: A sequence of n frequencies, $n \geq 2$

Output: A rooted tree that defines an optimal Huffman code

```
procedure huffman(f, n)
   if n = 2 then
      begin
         let f₁ and f₂ denote the frequencies
         let T be as in Figure 7.1.9
         return(T)
      end
   let fᵢ and fⱼ denote the smallest frequencies
   replace fᵢ and fⱼ in the list f by fᵢ + fⱼ
   T' := huffman(f, n − 1)
   replace a vertex in T' labeled fᵢ + fⱼ by the tree shown in
      Figure 7.1.10 to obtain the tree T
   return(T)
end huffman
```

FIGURE 7.1.9
The case $n = 2$ for Algorithm 7.1.8.

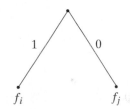

FIGURE 7.1.10
The case $n > 2$ for Algorithm 7.1.8.

EXAMPLE 7.1.9

We show how Algorithm 7.1.8 constructs an optimal Huffman code using Table 7.1.2.

TABLE 7.1.2
Input for Example 7.1.9

Character	Frequency
!	2
@	3
#	7
$	8
%	12

The algorithm begins by repeatedly replacing the smallest two frequencies with the sum until a two-element sequence is obtained:

$$2, 3, 7, 8, 12 \rightarrow 2 + 3, 7, 8, 12$$

$$5, 7, 8, 12 \rightarrow 5 + 7, 8, 12$$

$$8, 12, 12 \rightarrow 8 + 12, 12$$

$$12, 20$$

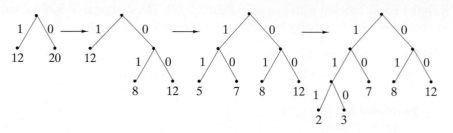

FIGURE 7.1.11   Constructing an optimal Huffman code.

The algorithm then constructs trees working backward beginning with the two-element sequence 12, 20 as shown in Figure 7.1.11. For example, the second tree is obtained from the first by replacing the vertex labeled 20 by the tree of Figure 7.1.12 since 20 arose as the sum of 8 and 12. Finally, to obtain the optimal Huffman coding tree, we replace each frequency by a character having that frequency (see Figure 7.1.13).

FIGURE 7.1.12
The tree that replaces the vertex labeled 20 in Figure 7.1.11.

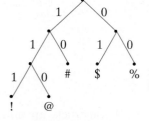

FIGURE 7.1.13
The final tree of Figure 7.1.11 with each frequency replaced by a character having that frequency.

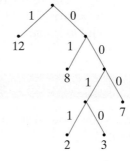

FIGURE 7.1.14
Another optimal Huffman tree for Example 7.1.9.

Notice that the Huffman tree for Table 7.1.2 is not unique. When 12 is replaced by 5, 7, because there are two vertices labeled 12, there is a choice. In Figure 7.1.11, we arbitrarily chose one of the vertices labeled 12. If we choose the other vertex labeled 12, we will obtain the tree of Figure 7.1.14. Either of the Huffman trees gives an optimal code; that is, either will encode text having the frequencies of Table 7.1.2 in exactly the same (optimal) space.   □

Exercises

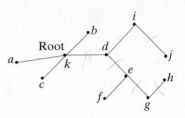

1. Find the level of each vertex in the tree shown in the adjacent figure.
2. Find the height of the tree shown in the adjacent figure.
3. Draw the tree T of Figure 7.1.5 as a rooted tree with a as root. What is the height of the resulting tree?
4. Draw the tree T of Figure 7.1.5 as a rooted tree with b as root. What is the height of the resulting tree?
5. Give an example similar to Example 7.1.5 of a tree that is used to specify hierarchical relationships.
6. Give an example different from Example 7.1.6 of a hierarchical definition tree.

Decode each bit string using the Huffman code given in the adjacent figure.

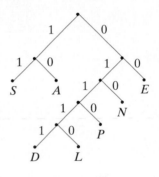

7. 011000010
8. 01110100110
9. 01111001001110
10. 1110011101001111

Encode each word using the preceding Huffman code.

11. DEN   12. NEED   13. LEADEN   14. PENNED

15. What factors in addition to the amount of memory used should be considered when choosing a code, such as ASCII or a Huffman code, to represent characters in a computer?
16. What techniques in addition to the use of Huffman codes might be used to save memory when storing text?
17. Construct an optimal Huffman code for the set of letters in the table.

Letter	Frequency
α	5
β	6
γ	6
δ	11
ε	20

18. Construct an optimal Huffman code for the set of letters in the table.

Letter	Frequency
I	7.5
U	20.0
B	2.5
S	27.5
C	5.0
H	10.0
M	2.5
P	25.0

19. Use the code developed in Exercise 18 to encode the following words (which have frequencies consistent with the table of Exercise 18):

BUS, CUPS, MUSH, PUSS, SIP, PUSH, CUSS, HIP, PUP, PUPS, HIPS.

20. Construct two optimal Huffman coding trees for the table of Exercise 17 of different heights.

21. Professor Gig A. Byte needs to store text made up of the characters A, B, C, D, E, which occur with the following frequencies:

Character	Frequency
A	6
B	2
C	3
D	2
E	8

Professor Byte suggests using the variable-length codes

Character	Code
A	1
B	00
C	01
D	10
E	0

which, he argues, store the text in less space than that used by an optimal Huffman code. Is the professor correct? Explain.

22. Show that any tree with two or more vertices has a vertex of degree 1.

23. Show that a tree is a planar graph.

24. Show that a tree is a bipartite graph.

25. Show that the vertices of a tree can be colored with two colors so that each edge is incident on vertices of different colors.

The *eccentricity* of a vertex v in a tree T is the maximum length of a simple path that begins at v.

26. Find the eccentricity of each vertex in the tree of Figure 7.1.5.

A vertex v in a tree T is a *center* for T if the eccentricity of v is minimal.

27. Find the center(s) of the tree of Figure 7.1.5.

★ 28. Show that a tree has either one or two centers.

★ 29. Show that if a tree has two centers they are adjacent.

30. Define the radius r of a tree using the concepts of eccentricity and center. The diameter d of any graph was defined before Exercise 70, Section 6.2. Is it always true, according to your definition of radius, that $2r = d$? Explain.

31. Give an example of a tree T that does not satisfy the property: If v and w are vertices in T, there is a unique path from v to w.

7.2 *TERMINOLOGY AND CHARACTERIZATIONS OF TREES*

A portion of the family tree of the ancient Greek gods is shown in Figure 7.2.1. (Not all children are listed.) As shown, we can regard a family tree as a rooted tree. The vertices adjacent to a vertex v and on the next-lower level are the children of v. For example, Kronos's children are Zeus, Poseidon, Hades, and Ares. The terminology adapted from a family tree is used routinely for any rooted tree. The formal definitions follow.

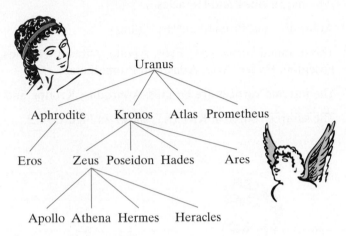

FIGURE 7.2.1   A portion of the family tree of ancient Greek gods.

DEFINITION 7.2.1

Let T be a tree with root v_0. Suppose that x, y, and z are vertices in T and that $(v_0, v_1, \ldots, v_n)$ is a simple path in T. Then

(a) v_{n-1} is the *parent* of v_n.

(b) $v_0, \ldots, v_{n-1}$ are *ancestors* of v_n.

(c) v_n is a *child* of v_{n-1}.

(d) If x is an ancestor of y, y is a *descendant* of x.

(e) If x and y are children of z, x and y are *siblings*.

(f) If x has no children, x is a *terminal vertex* (or a *leaf*).

(g) If x is not a terminal vertex, x is an *internal* (or *branch*) *vertex*.

(h) The *subtree of T rooted at x* is the graph with vertex set V and edge set E, where V is x together with the descendants of x and

$$E = \{e \mid e \text{ is an edge on a simple path from } x \text{ to some vertex in } V\}.$$

EXAMPLE 7.2.2

In the rooted tree of Figure 7.2.1,

(a) The parent of Eros is Aphrodite.

(b) The ancestors of Hermes are Zeus, Kronos, and Uranus.

(c) The children of Zeus are Apollo, Athena, Hermes, and Heracles.

(d) The descendants of Kronos are Zeus, Poseidon, Hades, Ares, Apollo, Athena, Hermes, and Heracles.

(e) Aphrodite and Prometheus are siblings.

(f) The terminal vertices are Eros, Apollo, Athena, Hermes, Heracles, Poseidon, Hades, Ares, Atlas, and Prometheus.

(g) The internal vertices are Uranus, Aphrodite, Kronos, and Zeus.

(h) The subtree rooted at Kronos is shown in Figure 7.2.2.   □

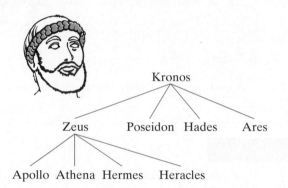

FIGURE 7.2.2   The subtree rooted at Kronos of the tree of Figure 7.2.1.

The remainder of this section is devoted to providing alternative characterizations of trees. Let T be a tree. We note that T is connected since there is a simple path from any vertex to any other vertex. Further, we can show that T does not contain a cycle. To see this, suppose that T contains a cycle C'. By Theorem 6.2.24, T contains a simple cycle (see Figure 7.2.3)

$$C = (v_0, \ldots, v_n),$$

$v_0 = v_n$. Since T is a simple graph, C cannot be a loop; so C contains at least two distinct vertices v_i and v_j, $i < j$. Now

$$(v_i, v_{i+1}, \ldots, v_j), \qquad (v_i, v_{i-1}, \ldots, v_0, v_{n-1}, \ldots, v_j)$$

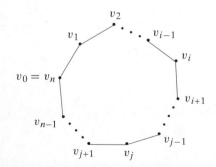

FIGURE 7.2.3
A simple cycle.

are distinct simple paths from v_i to v_j, which contradicts the definition of tree. Therefore, a tree cannot contain a cycle.

A graph with no cycles is called an **acyclic graph**. We just showed that a tree is a connected, acyclic graph. The converse is also true; every connected, acyclic graph is a tree. The next theorem gives this characterization of trees as well as others.

THEOREM 7.2.3

Let T be a graph with n vertices. The following are equivalent.

    (a) T is a tree.

    (b) T is connected and acyclic.

    (c) T is connected and has $n - 1$ edges.

    (d) T is acyclic and has $n - 1$ edges.

Proof. To show that (a)–(d) are equivalent, we will prove four results: if (a), then (b); if (b), then (c); if (c), then (d); and, if (d), then (a).

[If (a), then (b).] The proof of this result was given before the statement of the theorem.

[If (b), then (c).] Suppose that T is connected and acyclic. We will prove that T has $n - 1$ edges by induction on n.

If $n = 1$, T consists of one vertex and zero edges, so the result is true if $n = 1$.

Now suppose that the result holds for a connected, acyclic graph with n vertices. Let T be a connected, acyclic graph with $n + 1$ vertices. Choose a simple path P of maximum length. Since T is acyclic, P is not a cycle. Therefore, P contains a vertex v of degree 1 (see Figure 7.2.4). Let T^* be T with v and the edge incident on v removed. Then T^* is connected and acyclic, and because T^* contains n vertices, by the inductive hypothesis T^* contains $n - 1$ edges. Therefore, T contains n edges. The inductive argument is complete and this portion of the proof is complete.

[If (c), then (d).] Suppose that T is connected and has $n - 1$ edges. We must show that T is acyclic.

Suppose that T contains at least one cycle. Since removing an edge from a cycle does not disconnect a graph, we may remove edges, but no vertices, from cycle(s) in T until the resulting graph T^* is connected and acyclic. Now T^* is an acyclic, connected graph with n vertices. We may use our just proven result, (b) implies (c), to conclude that T^* has $n - 1$ edges. But now T has more than $n - 1$ edges. This is a contradiction. Therefore, T is acyclic. This portion of the proof is complete.

[If (d), then (a).] Suppose that T is acyclic and has $n - 1$ edges. We must show that T is a tree, that is, that T is a simple graph and that T has a unique simple path from any vertex to any other vertex.

The graph T cannot contain any loops because loops are cycles and T is acyclic. Similarly, T cannot contain distinct edges e_1 and e_2 incident on v and w because we would then have the cycle (v, e_1, w, e_2, v). Therefore T is a simple graph.

FIGURE 7.2.4
The proof of Theorem 7.2.3 [If (b), then (c).]. P is a simple path. v and the edge incident on v are removed so that the inductive hypothesis can be invoked.

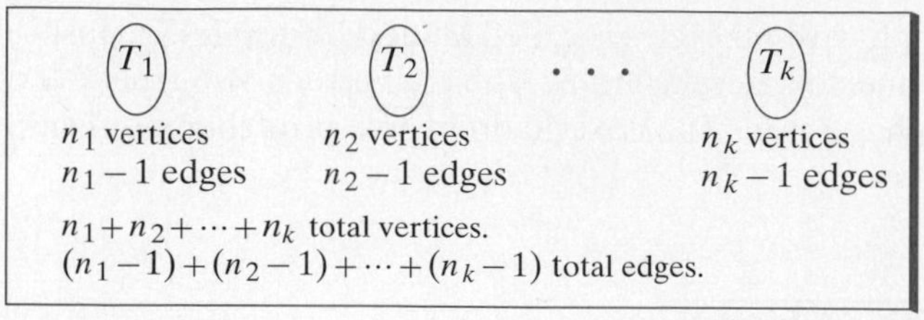

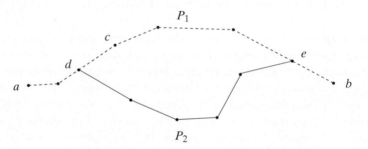

FIGURE 7.2.5 The proof of Theorem 7.2.3 [If (d), then (a).]. The T_i are components of T. T_i has n_i vertices and $n_i - 1$ edges. A contradiction results from the fact that the total number of edges must equal $n - 1$.

Suppose, by way of contradiction, that T is not connected (see Figure 7.2.5). Let

$$T_1, \quad T_2, \quad \ldots, \quad T_k$$

be the components of T. Since T is not connected, $k > 1$. Suppose that T_i has n_i vertices. Each T_i is connected and acyclic, so we may use our previously proven result, (b) implies (c), to conclude that T_i has $n_i - 1$ edges. Now

$$
\begin{aligned}
n - 1 &= (n_1 - 1) + (n_2 - 1) + \cdots + (n_k - 1) &&\text{(counting edges)} \\
&< (n_1 + n_2 + \cdots + n_k) - 1 &&\text{(since } k > 1\text{)} \\
&= n - 1, &&\text{(counting vertices)}
\end{aligned}
$$

which is impossible. Therefore, T is connected.

FIGURE 7.2.6 The proof of Theorem 7.2.3 [If (d), then (a).]. P_1 (shown dashed) and P_2 (shown in color) are distinct simple paths from a to b. c is the first vertex after a on P_1 not in P_2. d is the vertex preceding c on P_1. e is the first vertex after d on P_1 that is also on P_2. As shown, a cycle results, which gives a contradiction.

Suppose that there are distinct simple paths P_1 and P_2 from a to b in T (see Figure 7.2.6). Let c be the first vertex after a on P_1 that is not in P_2; let d be the vertex preceding c on P_1; and let e be the first vertex after d on P_1 that is also on P_2. Let

$$(v_0, v_1, \ldots, v_{n-1}, v_n)$$

be the portion of P_1 from $d = v_0$ to $e = v_n$. Let

$$(w_0, w_1, \ldots, w_{m-1}, w_m)$$

be the portion of P_2 from $d = w_0$ to $e = w_m$. Now

$$(v_0, \ldots, v_n = w_m, w_{m-1}, \ldots, w_1, w_0) \tag{7.2.1}$$

is a cycle in T, which is a contradiction. [In fact, (7.2.1) is a simple cycle since no vertices are repeated except for v_0 and w_0.] Thus there is a unique simple path from any vertex to any other vertex in T. Therefore, T is a tree. This completes the proof. ∎

Exercises

Answer the questions in Exercises 1–6 for the tree in Figure 7.2.1.

1. Find the parent of Poseidon.
2. Find the ancestors of Eros.
3. Find the children of Uranus.
4. Find the descendants of Zeus.
5. Find the siblings of Ares.
6. Draw the subtree rooted at Aphrodite.

Answer the questions in Exercises 7–15 for the adjacent tree.

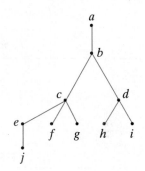

7. Find the parents of c and of h.
8. Find the ancestors of c and of j.
9. Find the children of d and of e.
10. Find the descendants of c and of e.
11. Find the siblings of f and of h.
12. Find the terminal vertices.
13. Find the internal vertices.
14. Draw the subtree rooted at j.
15. Draw the subtree rooted at e.
16. Answer the questions in Exercises 7–15 for the adjacent tree.
17. What can you say about two vertices in a rooted tree that have the same parent?
18. What can you say about two vertices in a rooted tree that have the same ancestors?
19. What can you say about a vertex in a rooted tree that has no ancestors?
20. What can you say about two vertices in a rooted tree that have a descendant in common?
21. What can you say about a vertex in a rooted tree that has no descendants?

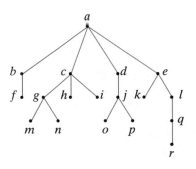

In Exercises 22–26, draw a graph having the given properties or explain why no such graph exists.

22. Six edges; eight vertices
23. Acyclic; four edges, six vertices
24. Tree; all vertices of degree 2
25. Tree; six vertices having degrees 1, 1, 1, 1, 3, 3
26. Tree; four internal vertices; six terminal vertices
27. Explain why if we allow cycles of length 0, a graph consisting of a single vertex and no edges is not acyclic.
28. Explain why if we allow cycles to repeat edges, a graph consisting of a single edge and two vertices is not acyclic.

v_2

v_1

29. The connected graph shown here has a unique simple path from any vertex to any other vertex but it is not a tree. Explain.

A *forest* is a simple graph with no cycles.

30. Explain why a forest is a union of trees.

31. If a forest F consists of m trees and has n vertices, how many edges does F have?

32. If $P_1 = (v_0, \ldots, v_n)$ and $P_2 = (w_0, \ldots, w_m)$ are distinct simple paths from a to b in a simple graph G, is

$$(v_0, \ldots, v_n = w_m, w_{m-1}, \ldots, w_1, w_0)$$

necessarily a cycle? Explain. (This exercise is relevant to the last paragraph of the proof of Theorem 7.2.3.)

33. Show that a graph G with n vertices and fewer than $n - 1$ edges is not connected.

★ 34. Prove that T is a tree if and only if T is connected and when an edge is added between any two vertices, exactly one cycle is created.

35. Show that if G is a tree, every vertex of degree 2 or more is an articulation point. ("Articulation point" is defined before Exercise 55, Section 6.2.)

36. Give an example to show that the converse of Exercise 35 is false, even if G is assumed to be connected.

PROBLEM-SOLVING CORNER:

TREES

Problem

Let T be a simple graph. Prove that T is a tree if and only if T is connected but the removal of any edge (but no vertices) from T disconnects T.

Attacking the Problem

Let's be clear about what we have to prove. Since the statement is "if and only if," we must prove two statements:

If T is a tree, then T is connected but the removal of any edge

(but no vertices) from T disconnects T.   (1)

If T is connected but the removal of any edge (but no vertices)

from T disconnects T, then T is a tree.   (2)

In (1), from the assumption that T is a tree, we must deduce that T is connected but the removal of any edge (but no vertices) from T disconnects T. In (2), from the assumption that T is connected but the removal of any edge (but no vertices) from T disconnects T, we must deduce that T is a tree.

In developing a proof, it is usually helpful to review definitions and other results related to the statement to be proved. Here of direct relevance are the definition of a tree and Theorem 7.2.3, which gives equivalent conditions for a graph to be a tree.

Definition 7.1.1 states:

A tree T is a simple graph satisfying: If v and w are vertices

in T, there is a unique simple path from v to w.   (3)

Theorem 7.2.3 states that the following are equivalent for an n-vertex graph T:

T is a tree.   (4)

T is connected and acyclic.   (5)

T is connected and has $n - 1$ edges.   (6)

T is acyclic and has $n - 1$ edges.   (7)

Finding a Solution

Let's first try to prove (1). We assume that T is a tree. We must prove two things: T is connected and the removal of any edge (but no vertices) from T disconnects T.

Statements (5) and (6) immediately tell us that T is connected. None of the statements (3) through (7) directly tell us anything about removal of edges or about a graph that's not connected. However, if we argue by contradiction and assume that the removal of some edge (but no vertices) from T does not disconnect T, then when we remove that edge from T, the graph T' that results is connected. In this case, for the graph T', (5) is true, but (6) and (7) are false, which is a contradiction since either (5), (6), and (7) are all true (and the graph is a tree), or (5) , (6) , and (7) are all false (and the graph is a not a tree).

Now consider proving (2). We assume that T is connected and that the removal of any edge (but no vertices) from T disconnects T. We must show that T is a tree. Let's try to show that T is connected and acyclic. We can then appeal to (5) to conclude that T is a tree.

Since T is connected, all we have to do is show that T is acyclic. Again, we'll approach this by contradiction. Suppose that T has a cycle. Keeping in mind what we're assuming (the removal of any edge from T disconnects T), try to figure out how to deduce a contradiction from the assumption that T contains a cycle before reading on.

If we delete an edge from T's cycle, T will remain connected. This contradiction shows that T is acyclic. By (5), T is a tree.

Formal Solution

Suppose that T has n vertices.

Suppose that T is a tree. Then by Theorem 7.2.3, T is connected and has $n - 1$ edges. Suppose that we can remove an edge from T to obtain T' so that T' is connected. Since T contains no cycles, T' also contains no cycles. By Theorem 7.2.3, T' is a tree. Again by Theorem 7.2.3, T' has $n - 1$ edges. This is a contradiction. Therefore, T is connected but the removal of any edge (but no vertices) from T disconnects T.

If T is connected and the removal of any edge (but no vertices) from T disconnects T, then T contains no cycles. By Theorem 7.2.3, T is a tree.

Summary of Problem-Solving Techniques

- When trying to construct a proof, write out carefully what is assumed and what is to be proved.
- When trying to construct a proof, consider using closely related definitions and theorems.
- When trying to construct a proof, review the proofs of similar and related theorems.
- If none of the conditions of potentially useful definitions and theorems apply, try proof by contradiction. When you assume the negation of the hypotheses, additional statements become available that might make some of the conditions of the definitions and theorems apply.

7.3 SPANNING TREES

In this section we consider the problem of finding a subgraph T of a graph G such that T is a tree containing all of the vertices of G. Such a tree is called a **spanning tree**. We will see that the methods of finding spanning trees may be applied to other problems as well.

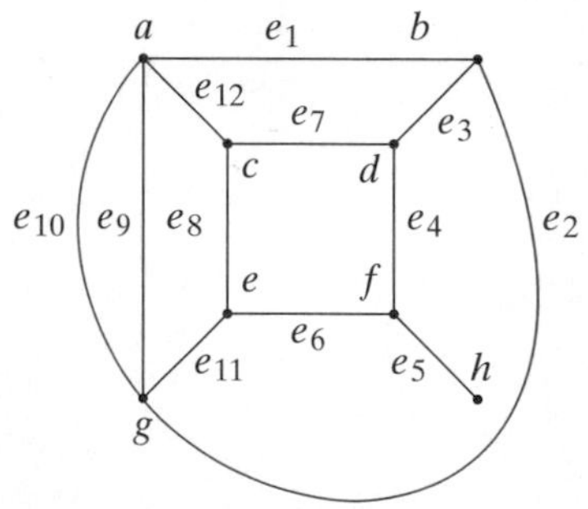

DEFINITION 7.3.1

A tree T is a *spanning tree* of a graph G if T is a subgraph of G that contains all of the vertices of G.

EXAMPLE 7.3.2

A spanning tree of the graph G of Figure 7.3.1 is shown in black.   □

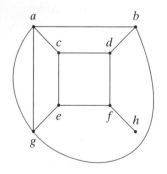

FIGURE 7.3.1
A graph and a spanning tree shown in black.

FIGURE 7.3.2
Another spanning tree (in black) of the graph of Figure 7.3.1.

EXAMPLE 7.3.3

In general, a graph will have several spanning trees. Another spanning tree of the graph G of Figure 7.3.1 is shown in Figure 7.3.2.   □

Suppose that a graph G has a spanning tree T. Let a and b be vertices of G. Since a and b are also vertices in T and T is a tree, there is a path P from a to b. However, P also serves as a path from a to b in G; thus G is connected. The converse is also true.

THEOREM 7.3.4

A graph G has a spanning tree if and only if G is connected.

Proof. We have already shown that if G has a spanning tree, then G is connected. Suppose that G is connected. If G is acyclic, by Theorem 7.2.3, G is a tree.

Suppose that G contains a cycle. We remove an edge (but no vertices) from this cycle. The graph produced is still connected. If it is acyclic, we stop. If it contains a cycle, we remove an edge from this cycle. Continuing in this way, we eventually produce an acyclic, connected subgraph T. By Theorem 7.2.3, T is a tree. Since T contains all the vertices of G, T is a spanning tree of G. ∎

An algorithm for finding a spanning tree based on the proof of Theorem 7.3.4 would not be very efficient; it would involve the time-consuming process of finding cycles. We can do much better. We shall illustrate the first algorithm for finding a spanning tree by an example and then state the algorithm.

EXAMPLE 7.3.5

Find a spanning tree for the graph G of Figure 7.3.1.

We will use a method called **breadth-first search** (Algorithm 7.3.6). The idea of breadth-first search is to process all the vertices on a given level before moving to the next higher level.

First, select an ordering, say $abcdefgh$, of the vertices of G. Select the first vertex a and label it the root. Let T consist of the single vertex a and no edges. Add to T all edges (a, x) and vertices on which they are incident, for $x = b$ to h, that do not produce a cycle when added to T. We would add to T edges (a, b), (a, c), and (a, g). (We could use either of the parallel edges incident on a and g.) Repeat this procedure with the vertices on level 1 by examining each in order:

  b: Include (b, d).
  c: Include (c, e).
  g: None.

Repeat this procedure with the vertices on level 2:

  d: Include (d, f).
  e: None.

Repeat this procedure with the vertices on level 3:

  f: Include (f, h).

Since no edges can be added to the single vertex h on level 4, the procedure ends. We have found the spanning tree shown in Figure 7.3.1. □

We formalize the method of Example 7.3.5 as Algorithm 7.3.6.

ALGORITHM 7.3.6   *Breadth-First Search for a Spanning Tree*

This algorithm finds a spanning tree using the breadth-first search method.

Input: A connected graph G with vertices ordered

$$v_1, \quad v_2, \quad \ldots, \quad v_n$$

Output: A spanning tree T

```
procedure bfs(V, E)
  // V = vertices ordered v_1, ..., v_n;  E = edges
  // V' = vertices of spanning tree T;  E' = edges of spanning tree T
  // v_1 is the root of the spanning tree
  // S is an ordered list
  S := (v_1)
  V' := {v_1}
  E' := ∅
  while true do
    begin
    for each x ∈ S, in order, do
      for each y ∈ V − V', in order, do
        if (x, y) is an edge then
          add edge (x, y) to E' and y to V'
      if no edges were added then
        return (T)
      S := children of S ordered consistently, with the original
          vertex ordering
    end
  end bfs
```

Exercise 16 is to give an argument to show that Algorithm 7.3.6 correctly finds a spanning tree.

Breadth-first search can be used to test whether an arbitrary graph G with n vertices is connected (see Exercise 26). We use the method of Algorithm 7.3.6 to produce a tree T. Then G is connected if and only if T has n vertices.

Breadth-first search can also be used to find minimum-length paths in an unweighted graph from a fixed vertex v to all other vertices (see Exercise 20). We use the method of Algorithm 7.3.6 to generate a spanning tree rooted at v. We note that the length of a shortest path from v to a vertex on level i of the spanning tree is i. Dijkstra's shortest-path algorithm for weighted graphs (Algorithm 6.4.1) can be considered as a generalization of breadth-first search (see Exercise 21).

An alternative to breadth-first search is **depth-first search**, which proceeds to successive levels in a tree at the earliest possible opportunity.

ALGORITHM 7.3.7   *Depth-First Search for a Spanning Tree*

This algorithm finds a spanning tree using the depth-first search method.

Input: A connected graph G with vertices ordered

$$v_1, \quad v_2, \quad \ldots, \quad v_n$$

Output: A spanning tree T

```
procedure dfs(V, E)
  // V' = vertices of spanning tree T;  E' = edges of spanning tree T
  // v₁ is the root of the spanning tree
  V' := {v₁}
  E' := ∅
  w := v₁
  while true do
    begin
    while there is an edge (w, v) that when added to T does not
      create a cycle in T do
      begin
      choose the edge (w, vₖ) with minimum k that when added to
        T does not create a cycle in T
      add (w, vₖ) to E'
      add vₖ to V'
      w := vₖ
      end
    if w = v₁ then
      return(T)
    w := parent of w in T // backtrack
    end
  end dfs
```

Exercise 17 is to give an argument to show that Algorithm 7.3.7 correctly finds a spanning tree.

EXAMPLE 7.3.8

Use depth-first search (Algorithm 7.3.7) to find a spanning tree for the graph of Figure 7.3.2 with the vertex ordering $abcdefgh$.

We select the first vertex a and call it the root (see Figure 7.3.2). Next, we add the edge (a, x), with minimal x, to our tree. In our case we add the edge (a, b).

We repeat this process. We add the edges $(b, d), (d, c), (c, e), (e, f)$, and (f, h). At this point, we cannot add an edge of the form (h, x), so we backtrack to the parent f of h and try to add an edge of the form (f, x). Again, we cannot add an edge of the form (f, x), so we backtrack to the parent e of f. This time we succeed in adding the edge (e, g). At this point, no more edges can be added, so we finally backtrack to the root and the procedure ends.   □

Because of the line in Algorithm 7.3.7, where we retreat along an edge toward the initially chosen root, depth-first search is also called **backtracking**. In the following example, we use backtracking to solve a puzzle.

EXAMPLE 7.3.9   *Four-Queens Problem*

The four-queens problem is to place four tokens on a 4×4 grid so that no two tokens are on the same row, column, or diagonal. Construct a backtracking algorithm to solve the four-queens problem. (To use chess terminology, this is the problem of placing four queens on a 4×4 board so that no queen attacks another queen.)

The idea of the algorithm is to place tokens successively in the columns. When it is impossible to place a token in a column, we backtrack and adjust the token in the preceding column. ☐

ALGORITHM 7.3.10   *Solving the Four-Queens Problem Using Backtracking*

This algorithm uses backtracking to search for an arrangement of four tokens on a 4×4 grid so that no two tokens are on the same row, column, or diagonal.

Input: An array *row* of size 4

Output: **true**, if there is a solution
         false, if there is no solution
         [If there is a solution, the kth queen is in column k, row $row(k)$.]

```
procedure four_queens (row)
  k := 1    // start in column 1
  // row(k) is incremented prior to use, so we'll start in row 1
  row(1) := 0
  while k > 0 do
    begin
    row(k) := row(k) + 1
    // look for a legal move in column k
    while row(k) ≤ 4 and column k, row(k) conflicts do
      // try next row
      row(k) := row(k) + 1
    if row(k) ≤ 4 then    // found legal move in column k
      if k = 4 then    // solution complete
        return(true)
      else    // next column
        begin
        k := k + 1
        row(k) := 0
        end
    else    // backtrack to previous column
      k := k − 1
    end
  return(false)    // no solution
end four_queens
```

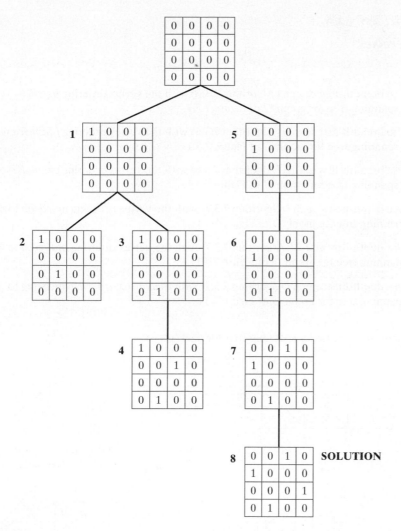

FIGURE 7.3.3   The tree generated by the backtracking algorithm (Algorithm 7.3.10) in the search for a solution to the four-queens problem.

The tree that Algorithm 7.3.10 generates is shown in Figure 7.3.3. The numbering indicates the order in which the vertices are generated. The solution is found at vertex 8.

The n-queens problem is to place n tokens on an $n \times n$ grid so that no two tokens are on the same row, column, or diagonal. It is straightforward to check that there is no solution to the two- or three-queens problem (see Exercise 10). We have just seen that Algorithm 7.3.10 generates a solution to the four-queens problem. Many constructions have been given to generate solutions to the n-queens problem for all $n \geq 4$ (see, e.g., [Erbas]).

Backtracking or depth-first search is especially attractive in a problem such as that in Example 7.3.9, where all that is desired is one solution. Since a solution, if one exists, is found at a terminal vertex, by moving to the terminal vertices as rapidly as possible, in general we can avoid generating some unnecessary vertices.

🌿 🌿 🌿
Exercises

1. Use breadth-first search (Algorithm 7.3.6) with the vertex ordering *hgfedcba* to find a spanning tree for graph *G* of Figure 7.3.1.

2. Use breadth-first search (Algorithm 7.3.6) with the vertex ordering *hfdbgeca* to find a spanning tree for graph *G* of Figure 7.3.1.

3. Use breadth-first search (Algorithm 7.3.6) with the vertex ordering *chbgadfe* to find a spanning tree for graph *G* of Figure 7.3.1.

4. Use depth-first search (Algorithm 7.3.7) with the vertex ordering *hgfedcba* to find a spanning tree for graph *G* of Figure 7.3.1.

5. Use depth-first search (Algorithm 7.3.7) with the vertex ordering *hfdbgeca* to find a spanning tree for graph *G* of Figure 7.3.1.

6. Use depth-first search (Algorithm 7.3.7) with the vertex ordering *dhcbefag* to find a spanning tree for graph *G* of Figure 7.3.1.

In Exercises 7–9, find a spanning tree for each graph.

7.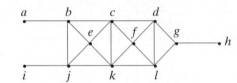

8. 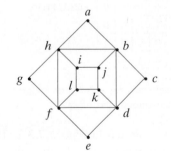   9.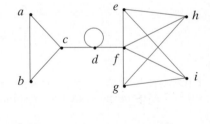

10. Show that there is no solution to the two-queens or the three-queens problem.

11. Find a solution to the five-queens and six-queens problems.

12. True or false? If *G* is a connected graph and *T* is a spanning tree for *G*, there is an ordering of the vertices of *G* such that Algorithm 7.3.6 produces *T* as a spanning tree. If true, prove it; otherwise, give a counterexample.

13. True or false? If *G* is a connected graph and *T* is a spanning tree for *G*, there is an ordering of the vertices of *G* such that Algorithm 7.3.7 produces *T* as a spanning tree. If true, prove it; otherwise, give a counterexample.

14. Show, by an example, that Algorithm 7.3.6 can produce identical spanning trees for a connected graph *G* from two distinct vertex orderings of *G*.

15. Show, by an example, that Algorithm 7.3.7 can produce identical spanning trees for a connected graph *G* from two distinct vertex orderings of *G*.

16. Prove that Algorithm 7.3.6 is correct.

17. Prove that Algorithm 7.3.7 is correct.

18. Under what conditions is an edge in a connected graph G contained in every spanning tree of G?

19. Let T and T' be two spanning trees of a connected graph G. Suppose that an edge x is in T but not in T'. Show that there is an edge y in T' but not in T such that $(T - \{x\}) \cup \{y\}$ and $(T' - \{y\} \cup \{x\}$ are spanning trees of G.

20. Write an algorithm based on breadth-first search that finds the minimum length of each path in an unweighted graph from a fixed vertex v to all other vertices.

21. Let G be a weighted graph in which the weight of each edge is a positive integer. Let G' be the graph obtained from G by replacing each edge

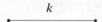

$$k$$

in G of weight k by k unweighted edges in series:

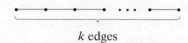

$$k \text{ edges}$$

Show that Dijkstra's algorithm for finding the minimum length of each path in the weighted graph G from a fixed vertex v to all other vertices (Algorithm 6.4.1) and performing a breadth-first search in the unweighted graph G' starting with vertex v are, in effect, the same process.

22. Let T be a spanning tree for a graph G. Show that if an edge in G, but not in T, is added to T, a unique cycle is produced.

A cycle as described in Exercise 22 is called a *fundamental cycle*. The *fundamental cycle matrix* of a graph G has its rows indexed by the fundamental cycles of G relative to a spanning tree T for G and its columns indexed by the edges of G. The ijth entry is 1 if edge j is in the ith fundamental cycle and 0 otherwise. For example, the fundamental cycle matrix of the graph G of Figure 7.3.1 relative to the spanning tree shown in Figure 7.3.1 is

	e_7	e_6	e_{11}	e_{10}	e_2	e_1	e_3	e_4	e_5	e_8	e_9	e_{12}
(abdca)	1	0	0	0	0	1	1	0	0	0	0	1
(efdbace)	0	1	0	0	0	1	1	1	0	1	0	1
(ageca)	0	0	1	0	0	0	0	0	0	1	1	1
(aga)	0	0	0	1	0	0	0	0	0	0	1	0
(abga)	0	0	0	0	1	1	0	0	0	0	1	0

Find the fundamental cycle matrix of each graph. The spanning tree to be used is drawn in black.

23.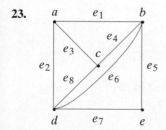

24.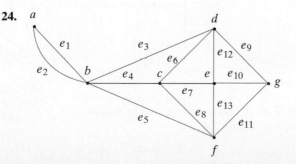

25.

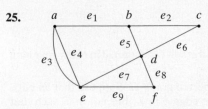

26. Write a breadth-first search algorithm to test whether a graph is connected.

27. Write a depth-first search algorithm to test whether a graph is connected.

28. Write a depth-first search algorithm that finds all solutions to the four-queens problem.

7.4 MINIMAL SPANNING TREES

The weighted graph G of Figure 7.4.1 shows six cities and the costs of building roads between certain pairs of cities. We want to build the lowest-cost road system that will connect the six cities. The solution can be represented by a subgraph. This subgraph must be a spanning tree since it must contain all the vertices (so that each city is in the road system), it must be connected (so that any city can be reached from any other), and it must have a unique simple path between each pair of vertices (since a graph containing multiple simple paths between a vertex pair could not represent a minimum-cost system). Thus what is needed is a spanning tree the sum of whose weights is a minimum. Such a tree is called a **minimal spanning tree**.

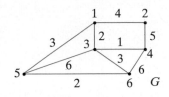

FIGURE 7.4.1
Six cities (a, b, c, d, e, f) and the costs of building roads between certain pairs of them.

DEFINITION 7.4.1

Let G be a weighted graph. A *minimal spanning tree* of G is a spanning tree of G with minimum weight.

EXAMPLE 7.4.2

The tree T' shown in Figure 7.4.2 is a spanning tree for graph G of Figure 7.4.1. The weight of T' is 20. This tree is not a minimal spanning tree since spanning tree T shown in Figure 7.4.3 has weight 12. We will see later that T is a minimal spanning tree for G. □

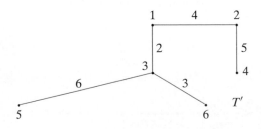

FIGURE 7.4.2 A spanning tree of weight 20 of the graph of Figure 7.4.1.

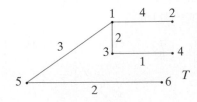

FIGURE 7.4.3 A spanning tree of weight 12 of the graph of Figure 7.4.1.

The algorithm to find a minimal spanning tree that we will discuss is known as **Prim's Algorithm** (Algorithm 7.4.3). This algorithm builds a tree by iteratively adding edges until a minimal spanning tree is obtained. At each iteration, we add a minimum-weight edge that does not complete a cycle to the current tree. Another algorithm to find a minimal spanning tree, known as **Kruskal's Algorithm**, is presented in the exercises (see Exercises 20–22).

ALGORITHM 7.4.3   *Prim's Algorithm*

This algorithm finds a minimal spanning tree in a connected, weighted graph.

Input: A connected, weighted graph with vertices $1, \ldots, n$ and start vertex s. If (i, j) is an edge, $w(i, j)$ is equal to the weight of (i, j); if (i, j) is not an edge, $w(i, j)$ is equal to ∞ (a value greater than any actual weight).

Output: The set of edges E in a minimal spanning tree.

```
        procedure prim(w, n, s)
            // v(i) = 1 if vertex i has been added to mst
            // v(i) = 0 if vertex i has not been added to mst
 1.     for i := 1 to n do
 2.        v(i) := 0
            // add start vertex to mst
 3.        v(s) := 1
            // begin with an empty edge set
 4.        E := ∅
            // put n − 1 edges in the minimal spanning tree
 5.     for i := 1 to n − 1 do
 6.        begin
               // add edge of minimum weight with one vertex in mst and one
               // vertex not in mst
 7.           min := ∞
 8.           for j := 1 to n do
 9.              if v(j) = 1 then    // j is a vertex in mst
10.                 for k = 1 to n do
11.                    if v(k) = 0 and w(j, k) < min then
12.                       begin
13.                       add_vertex := k
14.                       e := (j, k)
15.                       min := w(j, k)
16.                       end
               // put vertex and edge in mst
17.           v(add_vertex) := 1
18.           E := E ∪ {e}
19.        end
20.     return(E)
21.     end prim
```

EXAMPLE 7.4.4

Show how Prim's Algorithm finds a minimal spanning tree for the graph of Figure 7.4.1. Assume that the start vertex s is 1.

At line 3 we add vertex 1 to the minimal spanning tree. The first time we execute the for loop in lines 8–16, the edges with one vertex in the tree and one vertex not in the tree are

Edge	Weight
(1, 2)	4
(1, 3)	2
(1, 5)	3

The edge (1, 3) with minimum weight is selected. At lines 17 and 18, vertex 3 is added to the minimal spanning tree and edge (1, 3) is added to E.

The next time we execute the for loop in lines 8–16, the edges with one vertex in the tree and one vertex not in the tree are

Edge	Weight
(1, 2)	4
(1, 5)	3
(3, 4)	1
(3, 5)	6
(3, 6)	3

The edge (3, 4) with minimum weight is selected. At lines 17 and 18, vertex 4 is added to the minimal spanning tree and edge (3, 4) is added to E.

The next time we execute the for loop in lines 8–16, the edges with one vertex in the tree and one vertex not in the tree are

Edge	Weight
(1, 2)	4
(1, 5)	3
(2, 4)	5
(3, 5)	6
(3, 6)	3
(4, 6)	6

This time two edges have minimum weight 3. A minimal spanning tree will be constructed whichever is selected. In this version, edge (1, 5) is selected. At lines 17 and 18, vertex 5 is added to the minimal spanning tree and edge (1, 5) is added to E.

The next time we execute the for loop in lines 8–16, the edges with one vertex in the tree and one vertex not in the tree are

Edge	Weight
(1, 2)	4
(2, 4)	5
(3, 6)	3
(4, 6)	6
(5, 6)	2

The edge (5, 6) with minimum weight is selected. At lines 17 and 18, vertex 6 is added to the minimal spanning tree and edge (5, 6) is added to E.

The last time we execute the for loop in lines 8–16, the edges with one vertex in the tree and one vertex not in the tree are

Edge	Weight
(1, 2)	4
(2, 4)	5

The edge (1, 2) with minimum weight is selected. At lines 17 and 18, vertex 2 is added to the minimal spanning tree and edge (1, 2) is added to E. The minimal spanning tree constructed is shown in Figure 7.4.3.   □

Prim's Algorithm furnishes an example of a **greedy algorithm**. A greedy algorithm is an algorithm that optimizes the choice at each iteration without regard to previous choices. The principle can be summarized as "doing the best locally." In Prim's Algorithm, since we want a minimal spanning tree, at each iteration we simply add an available edge with minimum weight.

Optimizing at each iteration does not necessarily give an optimal solution to the original problem. We will show shortly (Theorem 7.4.5) that Prim's Algorithm is correct—we do obtain a minimal spanning tree. As an example of a greedy algorithm that does not lead to an optimal solution, consider a "shortest-path algorithm" in which at each step we select an available edge having minimum weight incident on the most recently added vertex. If we apply this algorithm to the weighted graph of Figure 7.4.4 to find a shortest path from a to z, we would select the edge (a, c) and then the edge (c, z). Unfortunately, this is not the shortest path from a to z.

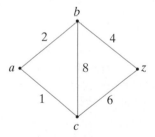

FIGURE 7.4.4
A graph that shows that selecting an edge having minimum weight incident on the most recently added vertex does *not* necessarily yield a shortest path. Starting at a, we obtain (a, c, z), but the shortest path from a to z is (a, b, z).

We next show that Prim's Algorithm is correct.

> **THEOREM 7.4.5**

Prim's Algorithm (Algorithm 7.4.3) is correct; that is, at the termination of Algorithm 7.4.3, T is a minimal spanning tree.

Proof. We let T_i denote the graph constructed by Algorithm 7.4.3 after the ith iteration of the for loop, lines 5–19. More precisely, the edge set of T_i is the set E constructed after the ith iteration of the for loop, lines 5–19, and the vertex set of T_i is the set of vertices on which the edges in E are incident. We let T_0 be the graph constructed by Algorithm 7.4.3 just before the for loop at line 5 is entered for the first time; T_0 consists of the single vertex s and no edges. Subsequently in this proof, we suppress the vertex set and refer to a graph by specifying its edge set.

By construction, at the termination of Algorithm 7.4.3, the graph constructed, T_{n-1}, is a connected, acyclic subgraph of the given graph G containing all the vertices of G; hence T_{n-1} is a spanning tree of G.

We use induction to show that for all $i = 0, \ldots, n - 1$, T_i is contained in a minimal spanning tree. It then follows that at termination, T_{n-1} is a minimal spanning tree.

If $i = 0$, T_0 consists of a single vertex. In this case T_0 is contained in every minimal spanning tree. We have verified the Basis Step.

Next, assume that T_i is contained in a minimal spanning tree T'. Let V be the set of vertices in T_i. Algorithm 7.4.3 selects an edge (j, k) of minimum weight where $j \in V$ and $k \notin V$ and adds it to T_i to produce T_{i+1}. If (j, k) is in T', then T_{i+1} is contained in the minimal spanning tree T'. If (j, k) is not in T', $T' \cup \{(j, k)\}$ contains a cycle C. Choose an edge (x, y) in C, different from (j, k) with $x \in V$ and $y \notin V$. Then

$$w(x, y) \geq w(j, k). \tag{7.4.1}$$

Because of (7.4.1), the graph $T'' = [T' \cup \{(j, k)\}] - \{(x, y)\}$ has weight less than or equal to the weight of T'. Since T'' is a spanning tree, T'' is a minimal spanning tree. Since T_{i+1} is contained in T'', the Inductive Step has been verified. The proof is complete. ∎

Our version of Prim's Algorithm examines $\Theta(n^3)$ edges in the worst case (see Exercise 6) to find a minimal spanning tree for a graph having n vertices. It is possible (see Exercise 8) to implement Prim's Algorithm so that only $\Theta(n^2)$ edges are examined in the worst case. Since K_n has $\Theta(n^2)$ edges, the latter version is optimal.

⁕ ⁕ ⁕
Exercises

In Exercises 1–5, find the minimal spanning tree given by Algorithm 7.4.3 for each graph.

1.

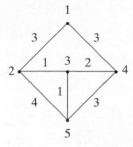

2.

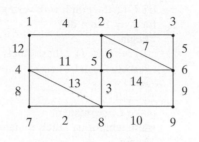

3.

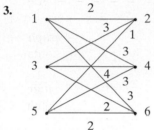

4.

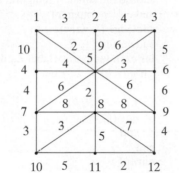

5.

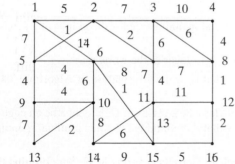

6. Show that Algorithm 7.4.3 examines $\Theta(n^3)$ edges in the worst case.

Exercises 7–9 refer to an alternate version of Prim's Algorithm (Algorithm 7.4.6).

ALGORITHM 7.4.6   *Alternate Version of Prim's Algorithm*

This algorithm finds a minimal spanning tree in a connected, weighted graph G. At each step, some vertices have temporary labels and some have permanent labels. The label of vertex i is denoted L_i.

Input: A connected, weighted graph with vertices $1, \ldots, n$ and start vertex s. If (i, j) is an edge, $w(i, j)$ is equal to the weight of (i, j); if (i, j) is not an edge, $w(i, j)$ is equal to ∞ (a value greater than any actual weight).

Output: A minimal spanning tree T.

```
procedure prim_alternate(w, n, s)
  let T be the graph with vertex s and no edges
  for j := 1 to n do
    begin
    L_j := w(s, j) // these labels are temporary
    back(j) := s
    end
  L_s := 0
  make L_s permanent
  while temporary labels remain do
    begin
    choose the smallest temporary label L_i
    make L_i permanent
    add edge (i, back(i)) to T
    add vertex i to T
    for each temporary label L_k do
      if w(i, k) < L_k then
        begin
        L_k := w(i, k)
        back(k) := i
        end
    end
  return(T)
end prim_alternate
```

7. Show how Algorithm 7.4.6 finds a minimal spanning tree for the graphs of Exercises 1–5.

8. Show that Algorithm 7.4.6 examines $O(n^2)$ edges in the worst case.

9. Prove that Algorithm 7.4.6 is correct; that is, at the termination of Algorithm 7.4.6, T is a minimal spanning tree.

10. Let G be a connected, weighted graph, let v be a vertex in G, and let e be an edge of minimum weight incident on v. Show that e is contained in some minimal spanning tree.

11. Let G be a connected, weighted graph and let v be a vertex in G. Suppose that the weights of the edges incident on v are distinct. Let e be the edge of minimum weight incident on v. Must e be contained in every minimal spanning tree?

12. Show that any algorithm that finds a minimal spanning tree in K_n, when all the weights are the same, must examine every edge in K_n.

13. Show that if all weights in a connected graph G are distinct, G has a unique minimal spanning tree.

In Exercises 14–16, decide if the statement is true or false. If the statement is true, prove it; otherwise, give a counterexample. In each exercise, G is a connected, weighted graph.

14. If all the weights in G are distinct, distinct spanning trees of G have distinct weights.

15. If e is an edge in G whose weight is less than the weight of every other edge, e is in every minimal spanning tree of G.

16. If T is a minimal spanning tree of G, there is a labeling of the vertices of G so that Algorithm 7.4.3 produces T.

17. Let G be a connected, weighted graph. Show that if, as long as possible, we remove an edge from G having maximum weight whose removal does not disconnect G, the result is a minimal spanning tree for G.

★ 18. Write an algorithm that finds a maximal spanning tree in a connected, weighted graph.

19. Prove that your algorithm in Exercise 18 is correct.

Kruskal's Algorithm finds a minimal spanning tree in a connected, weighted graph G having n vertices as follows. The graph T initially consists of the vertices of G and no edges. At each iteration, we add an edge e to T having minimum weight that does not complete a cycle in T. When T has $n-1$ edges, we stop.

20. Formally state Kruskal's Algorithm.

21. Show how Kruskal's Algorithm finds minimal spanning trees for the graphs of Exercises 1–5.

22. Show that Kruskal's Algorithm is correct; that is, at the termination of Kruskal's Algorithm, T is a minimal spanning tree.

23. Let V be a set of n vertices and let s be a "dissimilarity function" on $V \times V$ (see Example 6.1.6). Let G be the complete, weighted graph having vertices V and weights $w(v_i, v_j) = s(v_i, v_j)$. Modify Kruskal's Algorithm so that it groups data into classes. This modification is known as the **method of nearest neighbors** (see [Gose]).

Exercises 24–30 refer to the following situation. Suppose that we have stamps of various denominations and that we want to choose the minimum number of stamps to make a given amount of postage. Consider a greedy algorithm that selects stamps by choosing as many of the largest denomination as possible, then as many of the second largest denomination as possible, and so on.

24. Show that if the available denominations are 1, 8, and 10 cents, the algorithm does not always produce the fewest number of stamps to make a given amount of postage.

★ 25. Show that if the available denominations are 1, 5, and 25 cents, the algorithm produces the fewest number of stamps to make any given amount of postage.

26. Find positive integers a_1 and a_2 such that $a_1 > 2a_2 > 1$, a_2 does not divide a_1, and the algorithm, with available denominations 1, a_1, a_2, does not always produce the fewest number of stamps to make a given amount of postage.

★ 27. Find positive integers a_1 and a_2 such that $a_1 > 2a_2 > 1$, a_2 does not divide a_1, and the algorithm, with available denominations 1, a_1, a_2, produces the fewest number of stamps to make any given amount of postage. Prove that your values do give an optimal solution.

★ 28. Suppose that the available denominations are

$$1 = a_1 < a_2 < \cdots < a_n.$$

Show, by giving counterexamples, that the condition

$$a_i \geq 2a_{i-1} - a_{i-2}, \qquad 3 \leq i \leq n$$

is neither necessary nor sufficient for the greedy algorithm to be optimal.

★ 29. Show that the greedy algorithm is optimal for denominations

$$1 = a_1 < a_2 < a_3$$

if and only if the greedy algorithm is optimal for $n = 1, 2, \ldots, a_2 a_3 - 1$.

★ **30.** Show that the greedy algorithm is optimal for denominations

$$1 = a_1 < a_2 < \cdots < a_m$$

if and only if the greedy algorithm is optimal for $n = 1, 2, \ldots, k$, where

$$k = \sum_{i=1}^{m-1} \left(\frac{a_{i+1}}{\gcd(a_{i+1}, a_i)} - 1 \right) a_i.$$

7.5 BINARY TREES

Binary trees are among the most important special types of rooted trees. Every vertex in a binary tree has at most two children (see Figure 7.5.1). Moreover, each child is designated as either a **left child** or a **right child**. In drawing a binary tree, a left child is drawn to the left and a right child is drawn to the right. The formal definition follows.

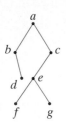

FIGURE 7.5.1
A binary tree.

DEFINITION 7.5.1

A *binary tree* is a rooted tree in which each vertex has either no children, one child, or two children. If a vertex has one child, that child is designated as either a left child or a right child (but not both). If a vertex has two children, one child is designated a left child and the other child is designated a right child.

EXAMPLE 7.5.2

In the binary tree of Figure 7.5.1, vertex b is the left child of vertex a and vertex c is the right child of vertex a. Vertex d is the right child of vertex b; vertex b has no left child. Vertex e is the left child of vertex c; vertex c has no right child. □

EXAMPLE 7.5.3

A tree that defines a Huffman code is a binary tree. For example, in the Huffman coding tree of Figure 7.1.8, moving from a vertex to a left child corresponds to using the bit 1 and moving from a vertex to a right child corresponds to using the bit 0. □

A **full binary tree** is a binary tree in which each vertex has either two children or zero children. A fundamental result about full binary trees is our next theorem.

THEOREM 7.5.4

If T is a full binary tree with i internal vertices, then T has $i+1$ terminal vertices and $2i+1$ total vertices.

Proof. The vertices of T consist of the vertices that are children (of some parent) and the vertices that are not children (of any parent). There is one nonchild—the root. Since there are i internal vertices, each having two children, there are $2i$ children. Thus the total number of vertices of T is $2i + 1$ and the number of terminal vertices is

$$(2i + 1) - i = i + 1.$$ ∎

EXAMPLE 7.5.5

A single-elimination tournament is a tournament in which a contestant is eliminated after one loss. The graph of a single-elimination tournament is a full binary tree (see Figure 7.5.2). The contestants' names are listed on the left. Winners progress to the right. Eventually, there is a single winner at the root. If the number of contestants is not a power of 2, some contestants receive byes. In Figure 7.5.2, contestant 7 has a first-round bye.

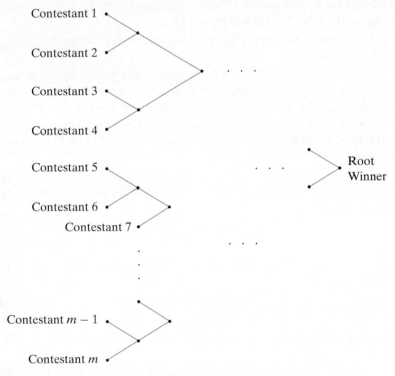

FIGURE 7.5.2   The graph (full binary tree) of a single-elimination tournament.

We show that if there are n contestants in a single-elimination tournament, a total of $n - 1$ matches are played.

The number of contestants is the same as the number of terminal vertices and the number of matches i is the same as the number of internal vertices. Thus, by Theorem 7.5.4,

$$n + i = 2i + 1$$

so that $i = n - 1$. □

Our next result about binary trees relates the number of terminal vertices to the height.

> **THEOREM 7.5.6**

If a binary tree of height h has t terminal vertices, then

$$\lg t \le h. \tag{7.5.1}$$

Proof. We will prove the equivalent inequality

$$t \le 2^h \tag{7.5.2}$$

by induction on h. Inequality (7.5.1) is obtained from (7.5.2) by taking the logarithm to the base 2 of both sides of (7.5.2).

If $h = 0$, the binary tree consists of a single vertex. In this case, $t = 1$ and thus (7.5.2) is true.

Assume that the result holds for a binary tree whose height is less than h. Let T be a binary tree of height $h > 0$ with t terminal vertices. Suppose first that the root of T has only one child. If we eliminate the root and the edge incident on the root, the resulting tree has height $h - 1$ and the same number of terminals as T. By induction, $t \le 2^{h-1}$. Since $2^{h-1} < 2^h$, (7.5.2) is established for this case.

Now suppose that the root of T has children v_1 and v_2. Let T_i be the subtree rooted at v_i and suppose that T_i has height h_i and t_i terminal vertices, $i = 1, 2$. By induction,

$$t_i \le 2^{h_i}, \qquad i = 1, 2. \tag{7.5.3}$$

The terminal vertices of T consist of the terminal vertices of T_1 and T_2. Hence

$$t = t_1 + t_2. \tag{7.5.4}$$

Combining (7.5.3) and (7.5.4), we obtain

$$t = t_1 + t_2 \le 2^{h_1} + 2^{h_2} \le 2^{h-1} + 2^{h-1} = 2^h.$$

The inductive step has been verified and the proof is complete.   ∎

> **EXAMPLE 7.5.7**

The binary tree in Figure 7.5.3 has height $h = 3$ and the number of terminals $t = 8$. For this tree, the inequality (7.5.1) becomes an equality.   □

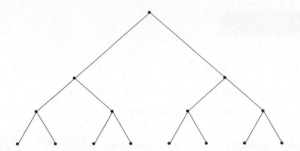

FIGURE 7.5.3   A binary tree of height $h = 3$ with $t = 8$ terminals. For this binary tree, $\lg t = h$.

Suppose that we have a set S whose elements can be ordered. For example, if S consists of numbers, we can use ordinary ordering defined on numbers, and if S consists of strings of alphabetic characters, we can use lexicographic order. Binary trees are used extensively in computer science to store elements from an ordered set such as a set of numbers or a set of strings. If data item $d(v)$ is stored in vertex v and data item $d(w)$ is stored in vertex w, then if v is a left child (or right child) of w, some ordering relationship will be guaranteed to exist between $d(v)$ and $d(w)$. One example is a **binary search tree**.

DEFINITION 7.5.8

A *binary search tree* is a binary tree T in which data are associated with the vertices. The data are arranged so that, for *each* vertex v in T, each data item in the left subtree of v is less than the data item in v and each data item in the right subtree of v is greater than the data item in v.

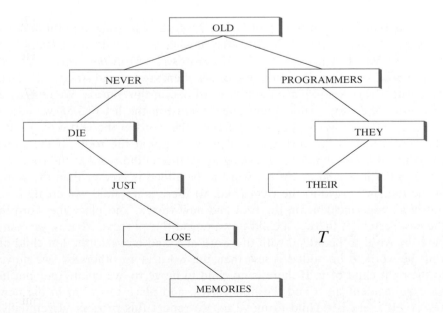

FIGURE 7.5.4   A binary search tree.

EXAMPLE 7.5.9

The words

OLD PROGRAMMERS NEVER DIE

THEY JUST LOSE THEIR MEMORIES   (7.5.5)

may be placed in a binary search tree as shown in Figure 7.5.4. Notice that for any vertex v, each data item in the left subtree of v is less than (i.e., precedes alphabetically) the data item in v and each data item in the right subtree of v is greater than the data item in v.   □

In general, there will be many ways to place data into a binary search tree. Figure 7.5.5 shows another binary search tree that stores the words (7.5.5).

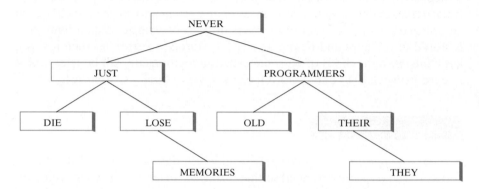

FIGURE 7.5.5   Another binary search tree that stores the same words as the tree in Figure 7.5.4.

The binary search tree T of Figure 7.5.4 was constructed in the following way. We begin with an **empty tree**, that is, a tree with no vertices and no edges. We then inspect each of the words (7.5.5) *in the order in which they appear*, OLD first, then PROGRAMMERS, then NEVER, and so on. To start, we create a vertex and place the first word OLD in this vertex. We designate this vertex the root. Thereafter, given a word in the list (7.5.5), we add a vertex v and an edge to the tree and place the word in the vertex v. To decide where to add the vertex and edge, we begin at the root. If the word to be added is less than (using lexicographic order) the word at the root, we move to the left child and if the word to be added is greater than the word at the root, we move to the right child. If there is no child, we create one, put in an edge incident on the root and new vertex, and place the word in the new vertex. If there is a child v, we repeat this process. That is, we compare the word to be added with the word at v and move to the left child of v if the word to be added is less than the word at v; otherwise, we move to the right child of v. If there is no child to move to, we create one, put in an edge incident on v and the new vertex, and place the word in the new vertex. If there is a child to move to, we repeat this process. Eventually, we place the word in the tree. We then get the next word in the list, compare it with the root, move left or right, compare it with the new vertex, move left or right, and so on, and eventually store it in the tree. In this way, we store all of the words in the tree and thus create a binary search tree. We formally state this method of constructing a binary search tree as Algorithm 7.5.10.

ALGORITHM 7.5.10   *Constructing a Binary Search Tree*

This algorithm constructs a binary search tree. The input is read in the order submitted. After each word is read, it is inserted into the tree.

Input: A sequence $w_1, \ldots, w_n$ of distinct words and the length n of the sequence

Output: A binary search tree T

```
procedure make_bin_search_tree(w, n)
  let T be the tree with one vertex, root
  store w₁ in root
  for i := 2 to n do
    begin
    v := root
    search := true      // find spot for wᵢ
    while search do
      begin
      s := word in v
      if wᵢ < s then
        if v has no left child then
          begin
          add a left child l to v
          store wᵢ in l
          search := false      // end search
          end
        else
          v := left child of v
      else      // wᵢ > s
        if v has no right child then
          begin
          add a right child r to v
          store wᵢ in r
          search := false      // end search
          end
        else
          v := right child of v
      end      // while
    end      // for
  return(T)
  end make_bin_search_tree
```

Binary search trees are useful for locating data. That is, given a data item D, we can easily determine if D is in a binary search tree and, if it is present, where it is located. To determine if a data item D is in a binary search tree, we would begin at the root. We would then repeatedly compare D with the data item at the current vertex. If D is equal to the data item at the current vertex, we have found D, so we stop. If D is less than the data item at the current vertex v, we move to v's left child and repeat this process. If D is greater than the data item at the current vertex v, we move to v's right child and repeat this process. If at any point the child to move to is missing, we conclude that D is not in the tree. (Exercise 2 asks for a formal statement of this process.)

The time spent searching for an item in a binary search tree is longest when the item is not present and we follow a longest path from the root. Thus the maximum time to search for an item in a binary search tree is approximately proportional to the height of the tree. Therefore, if the height of a binary search tree is small, searching the tree will always be very fast (see Exercise 21). Many ways are known to minimize the height of a binary search tree (see, e.g., [Cormen]).

We make more precise statements about worst-case searching in a binary search tree. Let T be a binary search tree with n vertices and let T^* be the full binary tree obtained from T by adding left and right children to existing vertices in T wherever possible. In Figure 7.5.6, we show the full binary tree that results from modifying the binary search tree of Figure 7.5.4. The added vertices are drawn as boxes. An unsuccessful search in T corresponds to arriving at an added (box) vertex in T^*. Let us define the worst-case time needed to execute the search procedure as the height h of the tree T^*. By Theorem 7.5.6, $\lg t \leq h$, where t is the number of terminal vertices in T^*. The full binary tree T^* has n internal vertices, so by Theorem 7.5.4, $t = n + 1$. Thus in the worst case, the time will be equal to at least $\lg t = \lg(n + 1)$. Exercise 3 shows that if the height of T is minimized, the worst case requires time equal to $\lceil \lg(n + 1) \rceil$. For example, since

$$\lceil \lg(2{,}000{,}000 + 1) \rceil = 21,$$

it is possible to store 2 million items in a binary search tree and find an item, or determine that it is not present, in at most 21 steps.

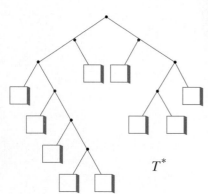

T^*

FIGURE 7.5.6
Expanding a binary search tree to a full binary tree.

Exercises

1. Place the words FOUR SCORE AND SEVEN YEARS AGO OUR FOREFA-THERS BROUGHT FORTH, in the order in which they appear, in a binary search tree.

2. Write a formal algorithm for searching in a binary search tree.

3. Write an algorithm that stores n distinct words in a binary search tree T of minimal height. Show that the derived tree T^*, as described in the text, has height $\lceil \lg(n + 1) \rceil$.

4. True or false? Let T be a binary tree. If for every vertex v in T the data item in v is greater than the data item in the left child of v and the data item in v is less than the data item in the right child of v, then T is a binary search tree. Explain.

In Exercises 5–7, draw a graph having the given properties or explain why no such graph exists.

5. Full binary tree; four internal vertices; five terminal vertices

6. Full binary tree; height $= 3$; nine terminal vertices

7. Full binary tree; height $= 4$; nine terminal vertices

8. A **full m-ary tree** is a rooted tree such that every parent has m ordered children. If T is a full m-ary tree with i internal vertices, how many vertices does T have? How many terminal vertices does T have? Prove your results.

9. Give an algorithm for constructing a full binary tree with $n > 1$ terminal vertices.

10. Give a recursive algorithm to insert a word in a binary search tree.

11. Find the maximum height of a full binary tree having t terminal vertices.

12. Write an algorithm that tests whether a binary tree in which data are stored in the vertices is a binary search tree.

13. Let T be a full binary tree. Let I be the sum of the lengths of the simple paths from the root to the internal vertices. We call I the *internal path length*. Let E be the sum of the lengths of the simple paths from the root to the terminal vertices. We call E the *external path length*. Prove that if T has n internal vertices, then $E = I - 2n$.

A binary tree T is *balanced* if for every vertex v in T, the heights of the left and right subtrees of v differ by at most 1. (Here the height of an empty tree is defined to be -1.)

State whether each tree in Exercises 14–17 is balanced or not.

14.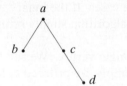

15.

16.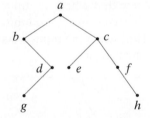

17.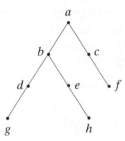

In Exercises 18–20, N_h is defined as the minimum number of vertices in a balanced binary tree of height h and $f_1, f_2, \ldots$ denotes the Fibonacci sequence.

18. Show that $N_0 = 1$, $N_1 = 2$, and $N_2 = 4$.

19. Show that $N_h = 1 + N_{h-1} + N_{h-2}$, for $h \geq 0$.

20. Show that $N_h = f_{h+2} - 1$, for $h \geq 0$.

★ 21. Show that the height h of an n-vertex balanced binary tree satisfies $h = O(\lg n)$. This result shows that the worst-case time to search in an n-vertex balanced binary search tree is $O(\lg n)$.

★ 22. Prove that if a binary tree of height h has $n \geq 1$ vertices, then $\lg n < h+1$. This result, together with Exercise 21, shows that the worst-case time to search in an n-vertex balanced binary search tree is $\Theta(\lg n)$.

7.6 TREE TRAVERSALS

Breadth-first search and depth-first search provide ways to "walk" a tree, that is, to traverse a tree in a systematic way so that each vertex is visited exactly once. In this section we consider three additional tree traversal methods. We define these traversals recursively.

ALGORITHM 7.6.1   *Preorder Traversal*

This recursive algorithm processes the vertices of a binary tree using preorder traversal.

Input: PT, the root of a binary tree
Output: Dependent on how "process" is interpreted in line 3

> **procedure** *preorder*(PT)
> 1.   **if** PT is empty **then**
> 2.   **return**
> 3.   process PT
> 4. $l :=$ left child of PT
> 5.   *preorder*(l)
> 6. $r :=$ right child of PT
> 7.   *preorder*(r)
> **end** *preorder*

Let us examine Algorithm 7.6.1 for some simple cases. If the binary tree is empty, nothing is processed since, in this case, the algorithm simply returns at line 2.

Suppose that the input consists of tree with a single vertex. We set PT to the root and call *preorder*(PT). Since PT is not empty, we proceed to line 3, where we process the root. At line 5, we call *preorder* with PT equal to the (empty) left child of the root. However, we just saw that when we input an empty tree to *preorder* nothing is processed. Similarly at line 7, we input an empty tree to *preorder* and nothing is processed. Thus when the input consists of a tree with a single vertex, we process the root and return.

FIGURE 7.6.1
Input for Algorithm 7.6.1.

FIGURE 7.6.2
At line 5 of Algorithm 7.6.1, where the input is the tree of Figure 7.6.1.

Now suppose that the input is the tree of Figure 7.6.1. We set PT to the root and call *preorder*(PT). Since PT is not empty, we proceed to line 3, where we process the root. At line 5 we call *preorder* with PT equal to the left child of the root (see Figure 7.6.2). We just saw that if the tree input to *preorder* consists of a single vertex, *preorder* processes that vertex. Thus we next process vertex B. Similarly, at line 7, we process vertex C. Thus the vertices are processed in the order ABC.

FIGURE 7.6.3
A binary tree. Preorder is
$ABCDEFGHIJ$. Inorder is
$CBDEAFIHJG$. Postorder is
$CEDBIJHGFA$.

EXAMPLE 7.6.2

In what order are the vertices of the tree of Figure 7.6.3 processed if preorder traversal is used?

Following lines 3–7 (root/left/right) of Algorithm 7.6.1, the traversal proceeds as shown in Figure 7.6.4. Thus the order of processing is

$$ABCDEFGHIJ. \qquad \square$$

Inorder traversal and postorder traversal are obtained by changing the position of line 3 (root) in Algorithm 7.6.1. "Pre," "in," and "post" refer to the position of the root in the traversal; that is, "preorder" means root first, "inorder" means root second, and "postorder" means root last.

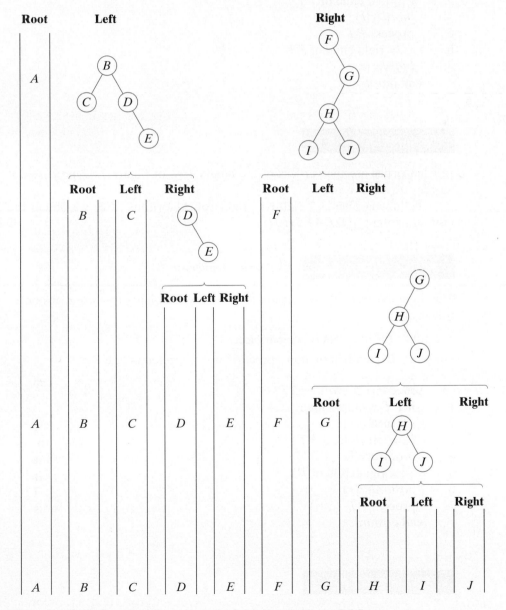

FIGURE 7.6.4   Preorder traversal of the tree in Figure 7.6.3.

ALGORITHM 7.6.3   *Inorder Traversal*

This recursive algorithm processes the vertices of a binary tree using inorder traversal.

   Input: PT, the root of a binary tree

Output: Dependent on how "process" is interpreted in line 5

```
     procedure inorder(PT)
1.      if PT is empty then
2.         return
3.         l := left child of PT
4.         inorder(l)
5.         process PT
6.         r := right child of PT
7.         inorder(r)
     end inorder
```

EXAMPLE 7.6.4

In what order are the vertices of the binary tree of Figure 7.6.3 processed if inorder traversal is used?

   Following lines 3–7 (left/root/right) of Algorithm 7.6.3, we obtain the inorder listing $CBDEAFIHJG$.   □

ALGORITHM 7.6.5   *Postorder Traversal*

This recursive algorithm processes the vertices of a binary tree using postorder traversal.

   Input: PT, the root of a binary tree

Output: Dependent on how "process" is interpreted in line 7

```
     procedure postorder(PT)
1.      if PT is empty then
2.         return
3.         l := left child of PT
4.         postorder(l)
5.         r := right child of PT
6.         postorder(r)
7.         process PT
     end postorder
```

EXAMPLE 7.6.6

In what order are the vertices of the binary tree of Figure 7.6.3 processed if postorder traversal is used?

Following lines 3–7 (left/right/root) of Algorithm 7.6.5, we obtain the postorder listing $CEDBIJHGFA$. □

Notice that preorder traversal may be obtained by following the route shown in Figure 7.6.5, and that reverse postorder traversal may be obtained by following the route shown in Figure 7.6.6.

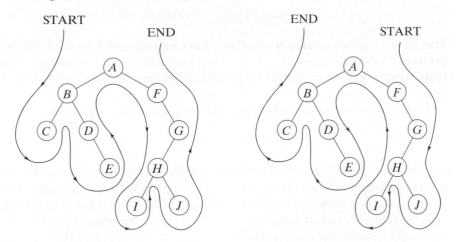

FIGURE 7.6.5
Preorder traversal.

FIGURE 7.6.6
Reverse postorder traversal.

If data are stored in a binary search tree, as described in Section 7.5, inorder traversal will process the data in order, since the sequence left/root/right agrees with the ordering of the data in the tree.

In the remainder of this section we consider binary tree representations of arithmetic expressions. Such representations facilitate the computer evaluation of expressions.

We will restrict our operators to $+$, $-$, $*$, and $/$. An example of an expression involving these operators is

$$(A + B) * C - D/E. \qquad (7.6.1)$$

This standard way of representing expressions is called the **infix form of an expression**. The variables A, B, C, D, and E are referred to as **operands**. The **operators** $+$, $-$, $*$, and $/$ operate on pairs of operands or expressions. In the infix form of an expression, an operator appears between its operands.

An expression such as (7.6.1) can be represented as a binary tree. The terminal vertices correspond to the operands, and the internal vertices correspond to the operators. The expression (7.6.1) would be represented as shown in Figure 7.6.7. In the binary tree representation of an expression, an operator operates on its left and right subtrees. For example, in the subtree whose root is $/$ in Figure 7.6.7, the divide operator operates on the operands D and E; that is, D is to be divided by E. In the subtree whose root is $*$ in Figure 7.6.7, the multiplication operator operates on the subtree headed by $+$, which itself represents an expression, and C.

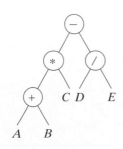

FIGURE 7.6.7
The binary tree representation of the expression $(A + B) * C - D/E$.

In a binary tree we distinguish the left and right subtrees of a vertex. The left and right subtrees of a vertex correspond to the left and right operands or expressions. This left/right distinction is important in expressions. For example, $4 - 6$ and $6 - 4$ are different.

If we traverse the binary tree of Figure 7.6.7 using inorder, and insert a pair of parentheses for each operation, we obtain

$$(((A + B) * C) - (D/E)) .$$

This form of an expression is called the **fully parenthesized form of the expression**. In this form we do not need to specify which operations (such as multiplication) are to be performed before others (such as addition), since the parentheses unambiguously dictate the order of operations.

If we traverse the tree of Figure 7.6.7 using postorder, we obtain

$$A B + C * D E / - .$$

This form of the expression is called the **postfix form of the expression** (or **reverse Polish notation**). In postfix, the operator follows its operands. For example, the first three symbols $A B +$ indicate that A and B are to be added. Advantages of the postfix form over the infix form are that in postfix no parentheses are needed and no conventions are necessary regarding the order of operations. The expression will be unambiguously evaluated. For these reasons and others, many compilers translate infix expressions to postfix form. Also, some calculators require expressions to be entered in postfix form.

A third form of an expression can be obtained by applying preorder traversal to a binary tree representation of an expression. In this case, the result is called the **prefix form of the expression** (or **Polish notation**). As in postfix, no parentheses are needed and no conventions are necessary regarding the order of operations. The prefix form of (7.6.1), obtained by applying preorder traversal to the tree of Figure 7.6.7, is

$$- * + A B C / D E.$$

Exercises

In Exercises 1–5, list the order in which the vertices are processed using preorder, inorder, and postorder traversal.

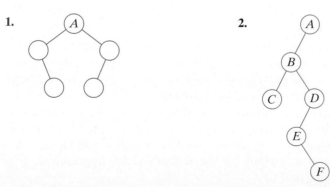

1.

2.

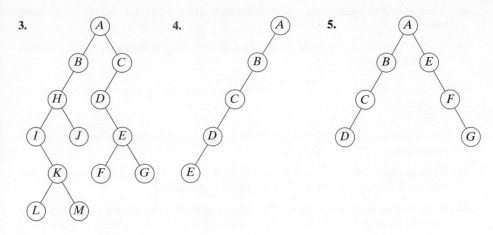

3.   **4.**   **5.**

In Exercises 6–10, represent the expression as a binary tree and write the prefix and postfix forms of the expression.

6. $(A + B) * (C - D)$

7. $((A - C) * D) / (A + (B + D))$

8. $(A * B + C * D) - (A/B - (D + E))$

9. $(((A + B) * C + D) * E) - ((A + B) * C - D)$

10. $(A * B - C/D + E) + (A - B - C - D * D)/(A + B + C)$

In Exercises 11–15, represent the postfix expression as a binary tree and write the prefix form, the usual infix form, and the fully parenthesized infix form of the expression.

11. $AB+C-$ **12.** $ABC+-$

13. $ABCD+*/E-$ **14.** $ABC**CDE+/-$

15. $AB+CD*EF/--A*$

In Exercises 16–21, find the value of the postfix expression if $A = 1$, $B = 2$, $C = 3$, and $D = 4$.

16. $ABC+-$ **17.** $AB+C-$

18. $AB+CD*AA/--B*$ **19.** $ABC**ABC++-$

20. $ABAB*+*D*$ **21.** $ADBCD*-+*$

22. Show, by example, that distinct binary trees with vertices A, B, and C can have the same preorder listing ABC.

23. Show that there is a unique binary tree with six vertices whose preorder vertex listing is $ABCEFD$ and whose inorder vertex listing is $ACFEBD$.

★ **24.** Write an algorithm that reconstructs the binary tree given its preorder and inorder vertex orderings.

25. Give examples of distinct binary trees, B_1 and B_2, each with two vertices, with the preorder vertex listing of B_1 equal to the preorder listing of B_2 and the postorder vertex listing of B_1 equal to the postorder listing of B_2.

26. Let P_1 and P_2 be permutations of $ABCDEF$. Is there a binary tree with vertices A, B, C, D, E, and F whose preorder listing is P_1 and whose inorder listing is P_2? Explain.

27. Write a recursive algorithm that prints the contents of the terminal vertices of a binary tree from left to right.

28. Write a recursive algorithm that interchanges all left and right children of a binary tree.

29. Write a recursive algorithm that initializes each vertex of a binary tree to the number of its descendants.

In Exercises 30 and 31, every expression involves only the operands $A, B, \ldots, Z$ and the operators $+, -, *, /$.

★ 30. Give a necessary and sufficient condition for a string of symbols to be a valid postfix expression.

31. Write an algorithm that, given the binary tree representation of an expression, outputs the fully parenthesized infix form of the expression.

32. Write an algorithm that prints the characters and their codes given a Huffman coding tree (see Example 7.1.7). Assume that each terminal vertex stores a character and its frequency.

★ 33. A *vertex cover* of a graph $G = (V, E)$ is a subset W of V such that for each edge $(v, w) \in E$, either $v \in W$ or $w \in W$. The *size* of a vertex cover W is the number of vertices in W. Write an algorithm that finds a vertex cover of minimal size for a tree $T = (V, E)$ whose worst-case time is $\Theta(|E|)$.

7.7 DECISION TREES AND THE MINIMUM TIME FOR SORTING

The binary tree of Figure 7.7.1 gives an algorithm for choosing a restaurant. Each internal vertex asks a question. If we begin at the root, answer each question, and follow the appropriate edge, we will eventually arrive at a terminal vertex that chooses a restaurant. Such a tree is called a **decision tree**. In this section we use decision trees to specify algorithms and to obtain lower bounds on the worst-case time for sorting as well as solving certain coin puzzles. We begin with coin puzzles.

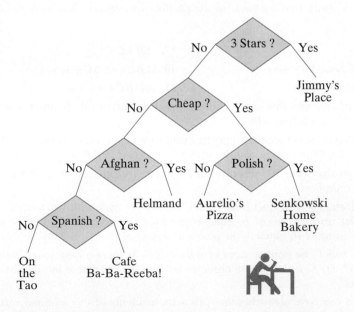

FIGURE 7.7.1   A decision tree.

| EXAMPLE 7.7.1 | *Five-Coins Puzzle* |

Five coins are identical in appearance, but one coin is either heavier or lighter than the others, which all weigh the same. The problem is to identify the bad coin and determine whether it is heavier or lighter than the others using only a pan balance (see Figure 7.7.2), which compares the weights of two sets of coins.

An algorithm to solve the puzzle is given in Figure 7.7.3 as a decision tree. The coins are labeled C_1, C_2, C_3, C_4, C_5. As shown, we begin at the root and place coin C_1 in the left pan and coin C_2 in the right pan. An edge labeled ⬈ means that the left side of the pan balance is heavier than the right side. Similarly, an edge labeled ⬊ means that the right side of the pan balance is heavier than the left side and an edge labeled ⬌ means that the two sides balance. For example, at the root when we compare C_1 with C_2, if the left side is heavier than the right side, we know that either C_1 is the heavy coin or C_2 is the light coin. In this case, as shown in the decision tree, we next compare C_1 with C_5 (which is known to be a good coin) and immediately determine whether the bad coin is C_1 or C_2 and whether it is heavy or light. The terminal vertices give the solution. For example, when we compare C_1 with C_5 and the pans balance, we follow the edge to the terminal vertex labeled C_2, L, which tells us that the bad coin is C_2 and that it is lighter than the others.   □

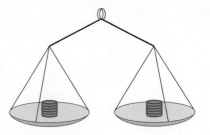

FIGURE 7.7.2
A pan balance for comparing weights of coins.

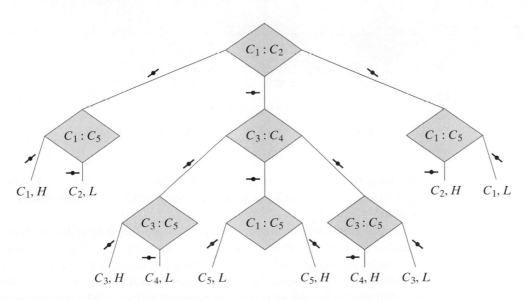

FIGURE 7.7.3   An algorithm to solve the five-coins puzzle.

If we define the worst-case time to solve a coin-weighing problem to be the number of weighings required in the worst case, it is easy to determine the worst-case time from the decision tree; the worst-case time is equal to the height of the tree. For example, the height of the decision tree of Figure 7.7.3 is 3, so the worst-case time for this algorithm is equal to 3.

We can use decision trees to show that the algorithm given in Figure 7.7.3 to solve the five-coins puzzle is optimal, that is, that *no* algorithm that solves the five-coins puzzle has worst-case time less than 3.

We argue by contradiction to show that no algorithm that solves the five-coins puzzle has worst-case time less than 3. Suppose that there is an algorithm that solves the five-coins puzzle in the worst-case in two or fewer weighings. The algorithm can be described by a decision tree, and since the worst-case time is 2 or less, the height of the decision tree is two or less. Since each internal vertex has at most three children, such a tree can have at most nine terminal vertices (see Figure 7.7.4). Now the terminal vertices correspond to possible outcomes. Thus a decision tree of height 2 or less can account for at most nine outcomes. But the five-coins puzzle has 10 outcomes:

$$C_1, L, \qquad C_1, H, \qquad C_2, L, \qquad C_2, H, \qquad C_3, L,$$
$$C_3, H, \qquad C_4, L, \qquad C_4, H, \qquad C_5, L, \qquad C_5, H.$$

This is a contradiction. Therefore, no algorithm that solves the five-coins puzzle has worst-case time less than 3, and the algorithm of Figure 7.7.3 is optimal.

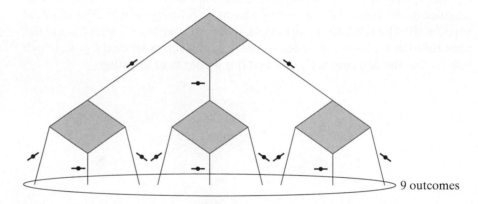

9 outcomes

FIGURE 7.7.4   A five-coins puzzle algorithm that uses at most two weighings.

We have seen how a decision tree can be used to give a lower bound for the worst-case time to solve a problem. Sometimes, the lower bound is unattainable.

Consider the four-coins puzzle (all the rules are the same as for the five-coins puzzle except that the number of coins is reduced by one). Since there are now eight outcomes rather than 10, we can conclude that any algorithm to solve the four-coins puzzle requires at least two weighings in the worst case. (This time we *cannot* conclude that at least three weighings are required in the worst case.) However, closer inspection shows that, in fact, three weighings are required.

The first weighing either compares two coins against two coins or one coin against one coin. Figure 7.7.5 shows that if we begin by comparing two coins against two coins, the decision tree can account for at most six outcomes. Since there are eight outcomes, no algorithm that begins by comparing two coins against two coins can solve the problem in two weighings or less in the worst case. Similarly, Figure 7.7.6 shows that if we begin by comparing one coin against one coin and the coins balance, the decision tree can account for only three outcomes. Since four outcomes are possible after identifying two good coins, no algorithm that begins by comparing one coin against one coin can solve the problem in two weighings or less in the worst case. Therefore, any algorithm that solves the four-coins puzzle requires at least three weighings in the worst case.

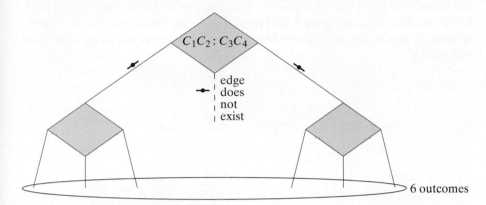

FIGURE 7.7.5   A four-coins puzzle algorithm that begins by comparing two coins against two coins.

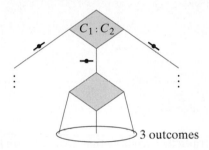

FIGURE 7.7.6   A four-coins puzzle algorithm that begins by comparing one coin against one coin.

If we modify the four-coins puzzle by requiring only that we identify the bad coin (without determining whether it is heavy or light), we can solve the puzzle in two weighings in the worst case (see Exercise 1).

We turn now to sorting. We can use decision trees to estimate the worst-case time to sort.

The sorting problem is easily described: Given n items

$$x_1, \quad \ldots, \quad x_n,$$

arrange them in ascending (or descending) order. We restrict our attention to sorting algorithms that repeatedly compare two elements, and based on the result of the comparison, modify the original list.

EXAMPLE 7.7.2

An algorithm to sort a_1, a_2, a_3 is given by the decision tree of Figure 7.7.7. Each edge is labeled with the arrangement of the list based on the answer to the question at an internal vertex. The terminal vertices give the sorted order.

Let us define the worst-case time to sort to be the number of comparisons in the worst case. Just as in the case of the decision trees that solve coin puzzle problems, the height of a decision tree that solves a sorting problem is equal to the worst-case time. For example, the worst-case time for the algorithm given by the decision tree of Figure 7.7.7 is equal to 3. We show that this algorithm is optimal, that is, that *no* algorithm that sorts three items has worst-case time less than 3.

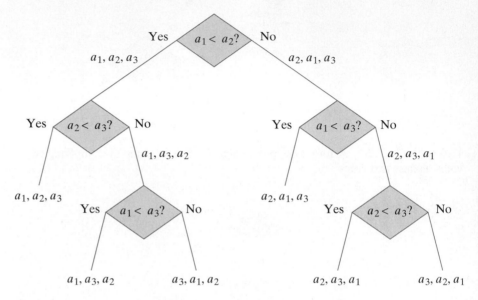

FIGURE 7.7.7   An algorithm to sort a_1, a_2, a_3.

We argue by contradiction to show that no algorithm that sorts three items has worst-case time less than 3. Suppose that there is an algorithm that sorts three items in the worst case in two or fewer comparisons. The algorithm can be described by a decision tree, and since the worst-case time is 2 or less, the height of the decision tree is two or less. Since each internal vertex has at most two children, such a tree can have at most four terminal vertices (see Figure 7.7.8). Now the terminal vertices correspond to possible outcomes. Thus a decision tree of height 2 or less can account for at most four outcomes. But the problem

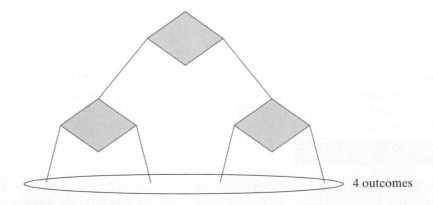

FIGURE 7.7.8   A sorting algorithm that makes at most two comparisons.

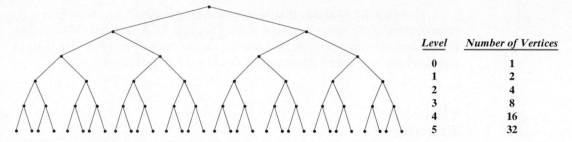

Level	Number of Vertices
0	1
1	2
2	4
3	8
4	16
5	32

FIGURE 7.7.9   Level compared with the maximum number of vertices in that level in a binary tree.

of sorting three items has six possible outcomes, corresponding to the $3! = 6$ ways that three items can be arranged:

$$s_1, s_2, s_3, \quad s_1, s_3, s_2, \quad s_2, s_1, s_3, \quad s_2, s_3, s_1, \quad s_3, s_1, s_2, \quad s_3, s_2, s_1.$$

This is a contradiction. Therefore, no algorithm that sorts three items has worst-case time less than 3 and the algorithm of Figure 7.7.7 is optimal. $\square$

Since $4! = 24$, there are 24 possible outcomes to the problem of sorting four items. To accommodate 24 terminal vertices, we must have a tree of height at least five (see Figure 7.7.9). Therefore, any algorithm that sorts four items requires at least five comparisons in the worst case. Exercise 9 is to give an algorithm that sorts four items using five comparisons in the worst case.

The method of Example 7.7.2 can be used to give a lower bound on the number of comparisons required in the worst case to sort an arbitrary number of items.

THEOREM 7.7.3

If $f(n)$ is the number of comparisons needed to sort n items in the worst case by a sorting algorithm, then $f(n) = \Omega(n \lg n)$.

Proof. Let T be the decision tree that represents the algorithm for input of size n and let h denote the height of T. Then the algorithm requires h comparisons in the worst case, so

$$h = f(n). \tag{7.7.1}$$

The tree T has at least $n!$ terminal vertices, so by Theorem 7.5.6,

$$\lg n! \le h. \tag{7.7.2}$$

Example 3.5.8 shows that $\lg n! = \Theta(n \lg n)$; thus, for some positive constant C,

$$Cn \lg n \le \lg n! \tag{7.7.3}$$

for all but finitely many integers n. Combining (7.7.1) through (7.7.3), we obtain

$$Cn \lg n \le f(n)$$

for all but finitely many integers n. Therefore

$$f(n) = \Omega(n \lg n). \qquad \blacksquare$$

Theorem 5.3.10 states that merge sort (Algorithm 5.3.8) uses $\Theta(n \lg n)$ comparisons in the worst case and is, by Theorem 7.7.3, optimal. Many other sorting algorithms are known that also attain the optimal number $\Theta(n \lg n)$ of comparisons; one, tournament sort, is described before Exercise 12.

୬ ୬ ୬

Exercises

1. Four coins are identical in appearance, but one coin is either heavier or lighter than the others, which all weigh the same. Draw a decision tree that gives an algorithm that identifies in at most two weighings the bad coin (but not necessarily determines whether it is heavier or lighter than the others) using only a pan balance.

2. Show that at least two weighings are required to solve the problem of Exercise 1.

3. Eight coins are identical in appearance, but one coin is either heavier or lighter than the others, which all weigh the same. Draw a decision tree that gives an algorithm that identifies in at most three weighings the bad coin and determines whether it is heavier or lighter than the others using only a pan balance.

4. Twelve coins are identical in appearance, but one coin is either heavier or lighter than the others, which all weigh the same. Draw a decision tree that gives an algorithm that identifies in at most three weighings the bad coin and determines whether it is heavier or lighter than the others using only a pan balance.

5. What is wrong with the following argument, which supposedly shows that the twelve-coins puzzle requires at least four weighings in the worst case if we begin by weighing four coins against four coins?

    If we weigh four coins against four coins and they balance, we must then determine the bad coin from the remaining four coins. But the discussion in this section showed that determining the bad coin from among four coins requires at least three weighings in the worst case. Therefore, in the worst case, if we begin by weighing four coins against four coins, the twelve-coins puzzle requires at least four weighings.

★ 6. Thirteen coins are identical in appearance, but one coin is either heavier or lighter than the others, which all weigh the same. How many weighings in the worst case are required to find the bad coin and determine whether it is heavier or lighter than the others using only a pan balance? Prove your answer.

7. Solve Exercise 6 for the fourteen-coins puzzle.

8. $(3^n - 3)/2, n \geq 2$, coins are identical in appearance, but one coin is either heavier or lighter than the others, which all weigh the same. [Kurosaka] gave an algorithm to find the bad coin and determine whether it is heavier or lighter than the others using only a pan balance in n weighings in the worst case. Prove that the coin cannot be found and identified as heavy or light in fewer than n weighings.

9. Give an algorithm that sorts four items using five comparisons in the worst case.

10. Use decision trees to find a lower bound on the number of comparisons required to sort five items in the worst case. Give an algorithm that uses this number of comparisons to sort five items in the worst case.

11. Use decision trees to find a lower bound on the number of comparisons required to sort six items in the worst case. Give an algorithm that uses this number of comparisons to sort six items in the worst case.

Exercises 12–18 refer to *tournament sort*.

    Tournament Sort. We are given a sequence

$$s_1, \quad \dots, \quad s_{2^k}$$

to sort in increasing order.

    We will build a binary tree with terminal vertices labeled $s_1, \dots, s_{2^k}$. An example is shown in the margin.

    Working left to right, create a parent for each pair and label it with the maximum of the children. Continue in this way until you reach the root. At this point, the largest value, m, has been found.

    To find the second largest value, first pick a value v less than all the items in the sequence. Replace the terminal vertex w containing m with v. Relabel the vertices by following the path from w to the root, as shown in the figure in the margin. At this point, the second largest value is found. Continue until the sequence is ordered.

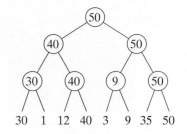

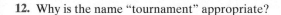

12. Why is the name "tournament" appropriate?

13. Draw the two trees that would be created after the adjacent tree when tournament sort is applied.

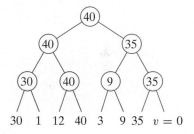

14. How many comparisons does tournament sort require to find the largest element?

15. Show that any algorithm that finds the largest value among n items requires at least $n - 1$ comparisons.

16. How many comparisons does tournament sort require to find the second largest element?

17. Write tournament sort as a formal algorithm.

18. Show that if n is a power of 2, tournament sort requires $\Theta(n \lg n)$ comparisons.

19. Give an example of a real situation (like that of Figure 7.7.1) that can be modeled as a decision tree. Draw the decision tree.

20. Draw a decision tree that can be used to determine who must file a federal tax return.

21. Draw a decision tree that gives a reasonable strategy for playing blackjack (see, e.g., [Ainslie]).

7.8 ISOMORPHISMS OF TREES

In Section 6.6 we defined what it means for two graphs to be isomorphic. (You might want to review Section 6.6 before continuing.) In this section we discuss isomorphic trees, isomorphic rooted trees, and isomorphic binary trees.

Theorem 6.6.4 states that simple graphs G_1 and G_2 are isomorphic if and only if there is a one-to-one, onto function f from the vertex set of G_1 to the vertex set of G_2 that preserves the adjacency relation in the sense that vertices v_i and v_j are adjacent in G_1 if and only if the vertices $f(v_i)$ and $f(v_j)$ are adjacent in G_2. Since a (free) tree is a simple graph, trees T_1 and T_2 are isomorphic if and only if there is a one-to-one, onto function f from the vertex set of T_1 to the vertex set of T_2 that preserves the adjacency relation; that is, vertices v_i and v_j are adjacent in T_1 if and only if the vertices $f(v_i)$ and $f(v_j)$ are adjacent in T_2.

EXAMPLE 7.8.1

The function f from the vertex set of the tree T_1 shown in Figure 7.8.1 to the vertex set of the tree T_2 shown in Figure 7.8.2 defined by

$$f(a) = 1, \qquad f(b) = 3, \qquad f(c) = 2, \qquad f(d) = 4, \qquad f(e) = 5$$

is a one-to-one, onto function that preserves the adjacency relation. Thus the trees T_1 and T_2 are isomorphic. $\square$

FIGURE 7.8.1   A tree.

FIGURE 7.8.2   A tree isomorphic to the tree in Figure 7.8.1.

As in the case of graphs, we can show that two trees are not isomorphic if we can exhibit an invariant that the trees do not share.

EXAMPLE 7.8.2

FIGURE 7.8.3
Nonisomorphic trees. T_2 has a vertex of degree 3, but T_2 does not.

The trees T_1 and T_2 of Figure 7.8.3 are not isomorphic because T_2 has a vertex (x) of degree 3, but T_1 does not have a vertex of degree 3. $\square$

We can show that there are three nonisomorphic trees with five vertices. The three nonisomorphic trees are shown in Figures 7.8.1 and 7.8.3.

THEOREM 7.8.3

There are three nonisomorphic trees with five vertices.

Proof. We will give an argument to show that any tree with five vertices is isomorphic to one of the trees in Figure 7.8.1 or 7.8.3.

If T is a tree with five vertices, by Theorem 7.2.3 T has four edges. If T had a vertex v of degree greater than 4, v would be incident on more than four edges. It follows that each vertex in T has degree at most 4.

We will first find all nonisomorphic trees with five vertices in which the maximum vertex degree that occurs is 4. We will next find all nonisomorphic trees with five vertices in which the maximum vertex degree that occurs is 3, and so on.

Let T be a tree with five vertices and suppose that T has a vertex v of degree 4. Then there are four edges incident on v and, because of Theorem 7.2.3, these are all the edges. It follows that in this case T is isomorphic to the tree in Figure 7.8.1.

Suppose that T is a tree with five vertices and the maximum vertex degree that occurs is 3. Let v be a vertex of degree 3. Then v is incident on three edges, as shown in Figure 7.8.4. The fourth edge cannot be incident on v since then v would have degee 4. Thus the fourth edge is incident on one of v_1, v_2, or v_3. Adding an edge incident on any of v_1, v_2, or v_3 gives a tree isomorphic to the tree T_2 of Figure 7.8.3.

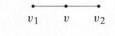

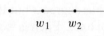

FIGURE 7.8.4
Vertex v has degree 3.

FIGURE 7.8.5
Vertex v has degree 2.

FIGURE 7.8.6
Adding a third edge
to the graph of Figure
7.8.5.

Now suppose that T is a tree with five vertices and the maximum vertex degree that occurs is 2. Let v be a vertex of degree 2. Then v is incident on two edges, as shown in Figure 7.8.5. A third edge cannot be incident on v; thus it must be incident on either v_1 or v_2. Adding the third edge gives the graph of Figure 7.8.6. For the same reason, the fourth edge cannot be incident on either of the vertices w_1 or w_2 of Figure 7.8.6. Adding the last edge gives a tree isomorphic to the tree T_1 of Figure 7.8.3.

Since a tree with five vertices must have a vertex of degree 2, we have found all nonisomorphic trees with five vertices. ■

For two *rooted* trees T_1 and T_2 to be isomorphic, there must be a one-to-one, onto function f from T_1 to T_2 that preserves the adjacency relation and that preserves the root. The latter condition means that f(root of T_1) = root of T_2. The formal definition follows.

DEFINITION 7.8.4

Let T_1 be a rooted tree with root r_1 and let T_2 be a rooted tree with root r_2. The rooted trees T_1 and T_2 are *isomorphic* if there is a one-to-one, onto function f from the vertex set of T_1 to the vertex set of T_2 satisfying

(a) Vertices v_i and v_j are adjacent in T_1 if and only if the vertices $f(v_i)$ and $f(v_j)$ are adjacent in T_2.

(b) $f(r_1) = r_2$.

We call the function f an *isomorphism*.

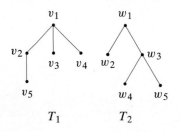

FIGURE 7.8.7
Isomorphic rooted trees.

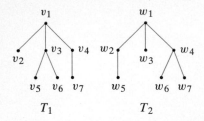

FIGURE 7.8.8
Nonisomorphic rooted trees. (The trees are isomorphic as *free* trees.)

EXAMPLE 7.8.5

The rooted trees T_1 and T_2 in Figure 7.8.7 are isomorphic. An isomorphism is

$$f(v_1) = w_1, \qquad f(v_2) = w_3, \qquad f(v_3) = w_4, \qquad f(v_4) = w_2,$$

$$f(v_5) = w_7, \qquad f(v_6) = w_6, \qquad f(v_7) = w_5. \qquad \square$$

The isomorphism of Example 7.8.5 is not unique. Can you find another isomorphism of the rooted trees of Figure 7.8.7?

EXAMPLE 7.8.6

The rooted trees T_1 and T_2 of Figure 7.8.8 are not isomorphic since the root of T_1 has degree 3 but the root of T_2 has degree 2. These trees are isomorphic as *free* trees. Each is isomorphic to the tree T_2 of Figure 7.8.3. $\qquad \square$

Arguing as in the proof of Theorem 7.8.3, we can show that there are four nonisomorphic rooted trees with four vertices.

THEOREM 7.8.7

There are four nonisomorphic rooted trees with four vertices. These four rooted trees are shown in Figure 7.8.9.

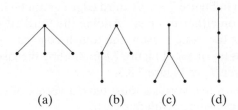

FIGURE 7.8.9   The four nonisomorphic rooted trees with four vertices.

Proof. We first find all nonisomorphic rooted trees with four vertices in which the root has degree 3; we then find all nonisomorphic rooted trees with four vertices in which the root has degree 2; and so on. We note that the root of a rooted tree with four vertices cannot have degree greater than 3.

A rooted tree with four vertices in which the root has degree 3 must be isomorphic to the tree in Figure 7.8.9a.

A rooted tree with four vertices in which the root has degree 2 must be isomorphic to the tree in Figure 7.8.9b.

Let T be a rooted tree with four vertices in which the root has degree 1. Then the root is incident on one edge. The two remaining edges may be added in one of two ways (see Figure 7.8.9c and d). Therefore, all nonisomorphic rooted trees with four vertices are shown in Figure 7.8.9. $\qquad \blacksquare$

Binary trees are special kinds of rooted trees; thus an isomorphism of binary trees must preserve the adjacency relation and must preserve the roots.

However, in binary trees a child is designated a left child or a right child. We require that an isomorphism of binary trees preserve the left and right children. The formal definition follows.

DEFINITION 7.8.8

Let T_1 be a binary tree with root r_1 and let T_2 be a binary tree with root r_2. The binary trees T_1 and T_2 are *isomorphic* if there is a one-to-one, onto function f from the vertex set of T_1 to the vertex set of T_2 satisfying

(a) Vertices v_i and v_j are adjacent in T_1 if and only if the vertices $f(v_i)$ and $f(v_j)$ are adjacent in T_2.

(b) $f(r_1) = r_2$.

(c) v is a left child of w in T_1 if and only if $f(v)$ is a left child of $f(w)$ in T_2.

(d) v is a right child of w in T_1 if and only if $f(v)$ is a right child of $f(w)$ in T_2.

We call the function f an *isomorphism*.

EXAMPLE 7.8.9

The binary trees T_1 and T_2 in Figure 7.8.10 are isomorphic. The isomorphism is $f(v_i) = w_i$ for $i = 1, \ldots, 4$.   □

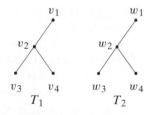

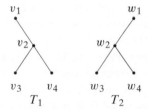

FIGURE 7.8.10   Isomorphic binary trees.

FIGURE 7.8.11   Nonisomorphic binary trees. (The trees are isomorphic as *rooted* trees and as *free* trees.)

EXAMPLE 7.8.10

The binary trees T_1 and T_2 in Figure 7.8.11 are not isomorphic. The root v_1 in T_1 has a right child, but the root w_1 in T_2 has no right child.   □

The trees T_1 and T_2 in Figure 7.8.11 *are* isomorphic as rooted trees and as free trees. As *rooted* trees, either of the trees of Figure 7.8.11 is isomorphic to the rooted tree T of Figure 7.8.9c.

Arguing as in the proofs of Theorems 7.8.3 and 7.8.7, we can show that there are five nonisomorphic binary trees with three vertices.

THEOREM 7.8.11

There are five nonisomorphic binary trees with three vertices. These five binary trees are shown in Figure 7.8.12.

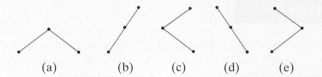

(a)   (b)   (c)   (d)   (e)

FIGURE 7.8.12   The five nonisomorphic binary trees with three vertices.

Proof. We first find all nonisomorphic binary trees with three vertices in which the root has degree 2. We then find all nonisomorphic binary trees with three vertices in which the root has degree 1. We note that the root of any binary tree cannot have degree greater than 2.

A binary tree with three vertices in which the root has degree 2 must be isomorphic to the tree in Figure 7.8.12a. In a binary tree with three vertices in which the root has degree 1, the root either has a left child and no right child or a right child and no left child. If the root has a left child, the child itself has either a left or a right child. We obtain the two binary trees in Figure 7.8.12b and c. Similarly, if the root has a right child, the child itself has either a left or a right child. We obtain the two binary trees in Figure 7.8.12d and e. Therefore, all nonisomorphic binary trees with three vertices are shown in Figure 7.8.12. ∎

If S is a set of trees of a particular type (e.g., S is a set of free trees or S is a set of rooted trees or S is a set of binary trees) and we define a relation R on S by the rule $T_1 R T_2$ if T_1 and T_2 are isomorphic, R is an equivalence relation. Each equivalence class consists of a set of mutually isomorphic trees.

In Theorem 7.8.3 we showed that there are three nonisomorphic free trees having five vertices. In Theorem 7.8.7 we showed that there are four nonisomorphic rooted trees having four vertices. In Theorem 7.8.11 we showed that there are five nonisomorphic binary trees having three vertices. You might have wondered if there are formulas for the number of nonisomorphic n-vertex trees of a particular type. There are formulas for the number of nonisomorphic n-vertex free trees, for the number of nonisomorphic n-vertex rooted trees, and for the number of nonisomorphic n-vertex binary trees. The formulas for the number of nonisomorphic free trees and for the number of nonisomorphic rooted trees with n vertices are quite complicated. Furthermore, the derivations of these formulas require techniques beyond those that we develop in this book. The formulas and proofs appear in [Deo, Sec. 10–3]. We derive a formula for the number of binary trees with n vertices.

THEOREM 7.8.12

There are $C(2n, n)/(n + 1)$ nonisomorphic binary trees with n vertices.

Proof. Let a_n denote the number of binary trees with n vertices. For example, $a_0 = 1$ since there is one binary tree having no vertices; $a_1 = 1$ since there is one binary tree having one vertex; $a_2 = 2$ since there are two binary trees having two vertices (see Figure 7.8.13); and $a_3 = 5$ since there are five binary trees having three vertices (see Figure 7.8.12).

We derive a recurrence relation for the sequence $a_0, a_1, \ldots$. Consider the construction of a binary tree with n vertices, $n > 0$. One vertex must be the root. Since there are $n - 1$ vertices remaining, if the left subtree has k vertices, the right subtree must have $n - k - 1$ vertices. We construct an n-vertex binary tree whose left subtree has k vertices and whose right subtree has $n - k - 1$ vertices by a two-step process: Construct the left subtree, construct the right subtree. (Figure 7.8.14 shows this construction for $n = 6$ and $k = 2$.) By the Multiplication Principle, this construction can be carried out in $a_k a_{n-k-1}$ ways. Different values of k give distinct n-vertex binary trees, so by the Addition Principle, the total number of n-vertex binary trees is

$$\sum_{k=0}^{n-1} a_k a_{n-k-1}.$$

We obtain the recurrence relation

$$a_n = \sum_{k=0}^{n-1} a_k a_{n-k-1}, \qquad n \geq 1.$$

But this recurrence relation and initial condition $a_0 = 1$ define the sequence of Catalan numbers (see Examples 4.2.23 and 5.1.7). Thus a_n is equal to the Catalan number $C(2n, n)/(n + 1)$. ∎

FIGURE 7.8.13
The two nonisomorphic binary trees with two vertices.

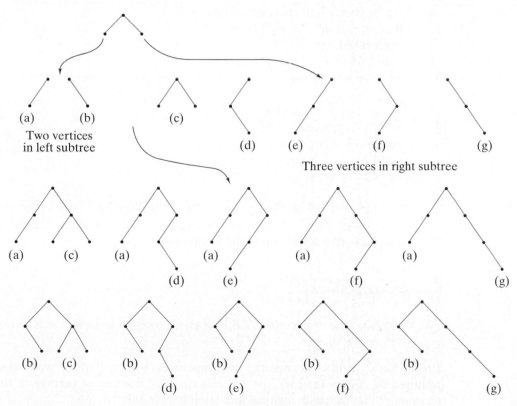

FIGURE 7.8.14   The proof of Theorem 7.8.12 for the case $n = 6$ vertices and $k = 2$ vertices in the left subtree.

When discussing graph isomorphisms in Section 6.6, we remarked that there is no fast method of deciding whether two arbitrary graphs are isomorphic. The situation is different for trees. It is possible to determine in polynomial time whether two arbitrary trees are isomorphic. As a special case, we give a linear time algorithm to determine whether two binary trees T_1 and T_2 are isomorphic. The algorithm is based on preorder traversal (see Section 7.6). We first check that each of T_1 and T_2 is nonempty, after which we check that the left subtrees of T_1 and T_2 are isomorphic and that the right subtrees of T_1 and T_2 are isomorphic.

| ALGORITHM 7.8.13 | *Testing Whether Two Binary Trees Are Isomorphic* |

Input: The roots r_1 and r_2 of two binary trees. (If the first tree is empty, r_1 has the special value *null*. If the second tree is empty, r_2 has the special value *null*.)

Output: **true**, if the trees are isomorphic
            false, if the trees are not isomorphic

```
        procedure bin_tree_isom(r₁, r₂)
1.        if r₁ = null and r₂ = null then
2.          return(true)
          // now one or both of r₁ or r₂ is not null
3.        if r₁ = null or r₂ = null then
4.          return(false)
          // now neither of r₁ or r₂ is null
5.        lc_r₁ := left child of r₁
6.        lc_r₂ := left child of r₂
7.        rc_r₁ := right child of r₁
8.        rc_r₂ := right child of r₂
9.        return(bin_tree_isom(lc_r₁, lc_r₂)
            and bin_tree_isom(rc_r₁, rc_r₂))
        end bin_tree_isom
```

As a measure of the time required by Algorithm 7.8.13, we count the number of comparisons with *null* in lines 1 and 3. We show that Algorithm 7.8.13 is a linear time algorithm in the worst case.

| THEOREM 7.8.14 |

The worst-case time of Algorithm 7.8.13 is $\Theta(n)$, where n is the total number of vertices in the two trees.

Proof. Let a_n denote the number of comparisons with *null* in the worst case required by Algorithm 7.8.13, where n is the total number of vertices in the trees input. We use mathematical induction to prove that

$$a_n \leq 3n + 2 \qquad \text{for } n \geq 0.$$

Basis Step ($n = 0$). If $n = 0$, the trees input to Algorithm 7.8.13 are both empty. In this case, there are two comparisons with *null* at line 1, after which the procedure returns. Thus $a_0 = 2$ and the inequality holds when $n = 0$.

Inductive Step. Assume that

$$a_k \leq 3k + 2,$$

when $k < n$. We must show that

$$a_n \leq 3n + 2.$$

We first find an upper bound for the number of comparisons in the worst case when the total number of vertices in the trees input to the procedure is $n > 0$ and neither tree is empty. In this case, there are four comparisons at lines 1 and 3. Let L denote the sum of the numbers of vertices in the two left subtrees of the trees input and let R denote the sum of the numbers of vertices in the two right subtrees of the trees input. Then at line 9 there are at most $a_L + a_R$ additional comparisons. Therefore, at most $4 + a_L + a_R$ comparisons are required in the worst case. By the inductive assumption,

$$a_L \leq 3L + 2 \qquad \text{and} \qquad a_R \leq 3R + 2. \tag{7.8.1}$$

Now

$$2 + L + R = n \tag{7.8.2}$$

because the vertices comprise the two roots, the vertices in the left subtrees, and the vertices in the right subtrees. Combining (7.8.1) and (7.8.2), we obtain

$$4 + a_L + a_R \leq 4 + (3L + 2) + (3R + 2) = 3(2 + L + R) + 2 = 3n + 2.$$

If either tree is empty, four comparisons are required at lines 1 and 3, after which the procedure returns. Thus, whether one of the trees is empty or not, at most $3n + 2$ comparisons are required in the worst case. Therefore,

$$a_n \leq 3n + 2$$

and the Inductive Step is complete. We conclude that the worst-case time of Algorithm 7.8.13 is $O(n)$.

If n is even, say $n = 2k$, one can use induction to show (see Exercise 24) that when two k-vertex isomorphic binary trees are input to Algorithm 7.8.13, the number of comparisons is equal to $3n + 2$. Using this result, one can show (see Exercise 25) that if n is odd, say $n = 2k + 1$, when the two binary trees shown in Figure 7.8.15 are input to Algorithm 7.8.13, the number of comparisons is equal to $3n + 1$. Thus the worst-case time of Algorithm 7.8.13 is $\Omega(n)$.

Since the worst-case time is $O(n)$ and $\Omega(n)$, the worst-case time of Algorithm 7.8.13 is $\Theta(n)$. ∎

[Aho] gives an algorithm whose worst-case time is linear in the number of vertices that determines whether two arbitrary (not necessarily binary) rooted trees are isomorphic.

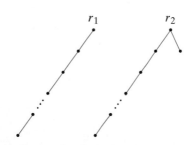

k vertices $k + 1$ vertices

Figure 7.8.15
Two binary trees that give worst-case run time $3n + 1$ for Algorithm 7.8.13 when $n = 2k + 1$ is odd.

☙ ☙ ☙
Exercises

In Exercises 1–6, determine whether each pair of free trees is isomorphic. If the pair is isomorphic, specify an isomorphism. If the pair is not isomorphic, give an invariant that one tree satisfies but the other does not.

1.

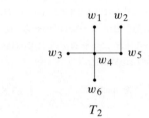

T_1 T_2

2. T_1 as in Exercise 1

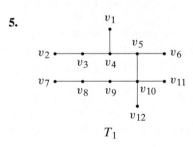

T_2

3.

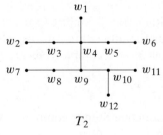

T_1 T_2

4.

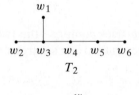

T_1 T_2

5.

T_1 T_2

6.

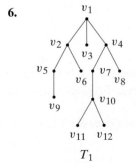

T_1

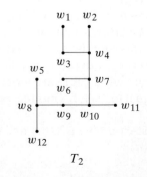

T_2

In Exercises 7–9, determine whether each pair of rooted trees is isomorphic. If the pair is isomorphic, specify an isomorphism. If the pair is not isomorphic, give an invariant that one tree satisfies but the other does not. Also, determine whether the trees are isomorphic as free trees.

7.

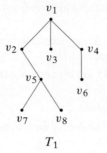
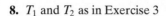

T_1 T_2

8. T_1 and T_2 as in Exercise 3

9.

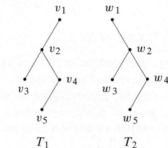

T_1 T_2

In Exercises 10–12, determine whether each pair of binary trees is isomorphic. If the pair is isomorphic, specify an isomorphism. If the pair is not isomorphic, give an invariant that one tree satisfies but the other does not. Also, determine whether the trees are isomorphic as free trees or as rooted trees.

10. T_1 and T_2 as in Exercise 9

11.

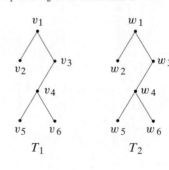

T_1 T_2

12.

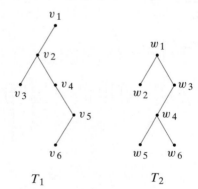

T_1 T_2

13. Draw all nonisomorphic free trees having three vertices.
14. Draw all nonisomorphic free trees having four vertices.
15. Draw all nonisomorphic free trees having six vertices.
16. Draw all nonisomorphic rooted trees having three vertices.
17. Draw all nonisomorphic rooted trees having five vertices.
18. Draw all nonisomorphic binary trees having two vertices.
19. Draw all nonisomorphic binary trees having four vertices
20. Draw all nonisomorphic full binary trees having seven vertices. (A full binary tree is a binary tree in which each internal vertex has two children.)

21. Draw all nonisomorphic full binary trees having nine vertices.

22. Find a formula for the number of nonisomorphic n-vertex full binary trees.

23. Find all nonisomorphic (as free trees and not as rooted trees) spanning trees for each graph in Exercises 7–9, Section 7.3.

24. Use induction to show that when two k-vertex isomorphic binary trees are input to Algorithm 7.8.13, the number of comparisons with *null* is equal to $6k + 2$.

25. Show that when the two binary trees shown in Figure 7.8.15 are input to Algorithm 7.8.13, the number of comparisons with *null* is equal to $6k + 4$.

26. Write an algorithm to generate an n-vertex random binary tree.

27. [*Project*] Report on the formulas for the number of nonisomorphic free trees and for the number of nonisomorphic rooted trees with n vertices (see [Deo]).

†7.9 GAME TREES

Trees are useful in the analysis of games such as tic-tac-toe, chess, and checkers, in which players alternate moves. In this section we show how trees can be used to develop game-playing strategies. This kind of approach is used in the development of many computer programs that allow human beings to play against computers or even computers against computers.

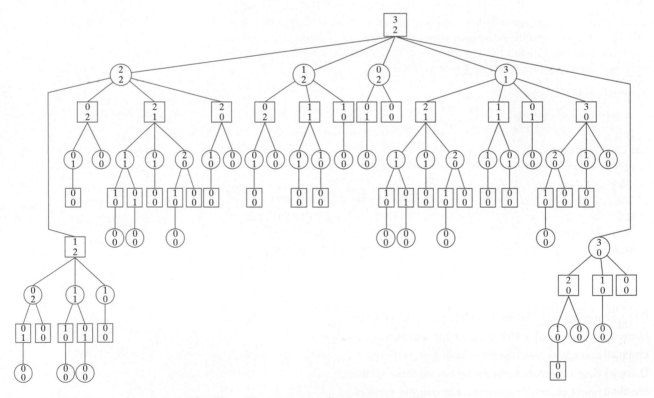

FIGURE 7.9.1 A game tree for nim. The initial distribution is two piles of three and two tokens, respectively.

---

†This section can be omitted without loss of continuity.

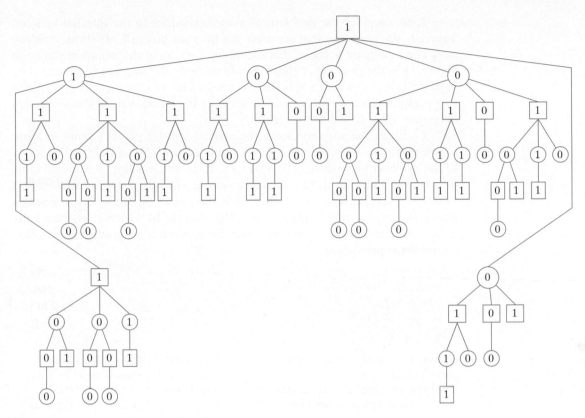

FIGURE 7.9.2   The game tree of Figure 7.9.1 showing the values of all vertices.

As an example of the general approach, consider a version of the game of nim. Initially, there are n piles, each containing a number of identical tokens. Players alternate moves. A move consists of removing one or more tokens from any one pile. The player who removes the last token loses. As a specific case, consider an initial distribution consisting of two piles: one containing three tokens and one containing two tokens. All possible move sequences can be listed in a **game tree** (see Figure 7.9.1). The first player is represented by a box and the second player is represented by a circle. Each vertex shows a particular position in the game. In our game, the initial position is shown as $\binom{3}{2}$. A path represents a sequence of moves. If a position is shown in a square, it is the first player's move; if a position is shown in a circle, it is the second player's move. A terminal vertex represents the end of the game. In nim, if the terminal vertex is a circle, the first player removed the last token and lost the game. If the terminal vertex is a box, the second player lost.

The analysis begins with the terminal vertices. We label each terminal vertex with the value of the position to the first player. If the terminal vertex is a circle, since the first player lost, this position is worthless to the first player and we assign it the value 0 (see Figure 7.9.2). If the terminal vertex is a box, since the first player won, this position is valuable to the first player and we label it with a value greater than 0, say 1 (see Figure 7.9.2). At this point, all terminal vertices have been assigned values.

Now, consider the problem of assigning values to the internal vertices. Suppose, for example, that we have an internal box, all of whose children have been assigned a value. For example, if we have the situation shown in Figure 7.9.3, the first player (box) should move to the position represented by vertex B, since this position is the most valuable. In other words, box moves to a position represented by a child with the maximum value. We assign this maximum value to the box vertex.

Consider the situation from the second (circle) player's point of view. Suppose that we have the situation shown in Figure 7.9.4. Circle should move to the position represented by vertex C, since this position is least valuable to box and therefore most valuable to circle. In other words, circle moves to a position represented by a child with the minimum value. We assign this minimum value to the circle vertex. The process by which circle seeks the minimum of its children and box seeks the maximum of its children is called the **minimax procedure**.

FIGURE 7.9.3   The first player (box) should move to position B since it is most valuable. This maximum value (1) is assigned to the box.

FIGURE 7.9.4   The second player (circle) should move to position C since it is least valuable (to box). This minimum value (0) is assigned to the circle.

Working upward from the terminal vertices and using the minimax procedure, we can assign values to all of the vertices in the game tree (see Figure 7.9.2). These numbers represent the value of the game, at any position, to the first player. Notice that the root in Figure 7.9.2, which represents the original position, has a value of 1. This means that the first player can always win the game by using an optimal strategy. This optimal strategy is contained in the game tree: The first player always moves to a position that maximizes the value of the children. No matter what the second player does, the first player can always move to a vertex having value 1. Ultimately, a terminal vertex having value 1 is reached where the first player wins the game.

Many interesting games, such as chess, have game trees so large that it is not feasible to use a computer to generate the entire tree. Nevertheless, the concept of a game tree is still useful for analyzing such games.

When using a game tree we should use a depth-first search. If the game tree is so large that it is not feasible to reach a terminal vertex, we limit the level to which depth-first search is carried out. The search is said to be an ***n*-level search** if we limit the search to n levels below the given vertex. Since the vertices at the lowest level may not be terminal vertices, some method must be found to assign them a value. This is where the specifics of the game must be dealt with. An **evaluation function** E is constructed that assigns each possible game position P the value $E(P)$ of the position to the first player. After the vertices at the lowest level are assigned values by using the function E, the minimax procedure can be applied to generate the values of the other vertices. We illustrate these concepts with an example.

EXAMPLE 7.9.1

Apply the minimax procedure to find the value of the root in tic-tac-toe using a two-level, depth-first minimax search. Use the evaluation function E, which assigns a position the value

$$NX - NO$$

where NX (respectively, NO) is the number of rows, columns, or diagonals containing an X (respectively, O) that X (respectively, O) might complete. For example, position P of Figure 7.9.5 has $NX = 2$, since X might complete the column or the diagonal, and $NO = 1$, since O can only complete a column. Therefore,

$$E(P) = 2 - 1 = 1.$$

FIGURE 7.9.5
The value of position P is
$E(P) = NX - NO = 2 - 1 = 1.$

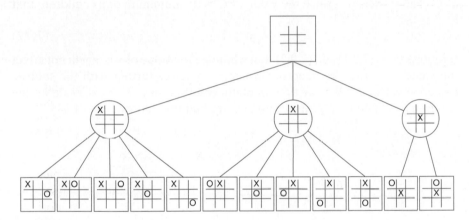

FIGURE 7.9.6   The game tree for tic-tac-toe to level 2 with symmetric positions omitted.

In Figure 7.9.6, we have drawn the game tree for tic-tac-toe to level 2. We have omitted symmetric positions. We first assign the vertices at level 2 the values given by E (see Figure 7.9.7). Next, we compute circle's values by minimizing over the children. Finally, we compute the value of the root by maximizing over the children. Using this analysis, the first move by the first player would be to the center square.   □

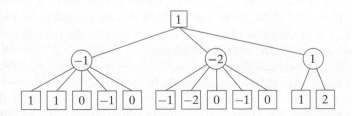

FIGURE 7.9.7   The game tree of Figure 7.9.6 with the values of all vertices shown.

Evaluation of a game tree, or even a part of a game tree, can be a time-consuming task, so any technique that reduces the effort is welcomed. The most general technique is called **alpha-beta pruning**. In general, alpha-beta pruning allows us to bypass many vertices in a game tree, yet still find the value of a vertex. The value obtained is the same as if we had evaluated all the vertices.

As an example, consider the game tree in Figure 7.9.8. Suppose that we want to evaluate vertex A using a two-level, depth-first search. We evaluate children left to right. We begin at the lower left by evaluating the vertices E, F, and G. The values shown are obtained from an evaluation function. Vertex B is 2, the minimum of its children. At this point, we know that the value x of A must be at least 2, since the value of A is the maximum of its children; that is,

$$x \geq 2. \tag{7.9.1}$$

This lower bound for A is called an **alpha value** of A. The next vertices to be evaluated are H, I, and J. When I evaluates to 1, we know that the value y of C cannot exceed 1, since the value of C is the minimum of its children; that is

$$y \leq 1. \tag{7.9.2}$$

It follows from (7.9.1) and (7.9.2) that whatever the value of y is, it will not affect the value of x; thus we need not concern ourselves further with the subtree rooted at vertex C. We say that an **alpha cutoff** occurs. We next evaluate the children of D and then D itself. Finally, we find that the value of A is 3.

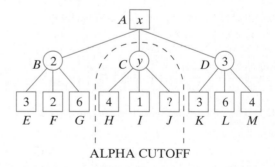

FIGURE 7.9.8   Evaluating vertex A using a two-level, depth-first search with alpha-beta pruning. An alpha cutoff occurs at vertex C when vertex I is evaluated since I's value (1) is less than or equal to the current lower bound estimate (2) for vertex A.

To summarize, an alpha cutoff occurs at a box vertex v when a grandchild w of v has a value less than or equal to the alpha value of v. The subtree whose root is the parent of w may be deleted (pruned). This deletion will not affect the value of v. An alpha value for a vertex v is only a lower bound for the value of v. The alpha value of a vertex is dependent on the current state of the search and changes as the search progresses.

Similarly, a **beta cutoff** occurs at a circle vertex v when a grandchild w of v has a value greater than or equal to the beta value of v. The subtree whose root is the parent of w may be pruned. This deletion will not affect the value of v. A **beta value** for a vertex v is only an upper bound for the value of v. The beta value of a vertex is dependent on the current state of the search and changes as the search progresses.

EXAMPLE 7.9.2

Evaluate the root of the tree of Figure 7.9.9 using depth-first search with alpha-beta pruning. Assume that children are evaluated left to right. For each vertex whose value is computed, write the value in the vertex. Place a check by the root of each subtree that is pruned. The value of each terminal vertex is written under the vertex.

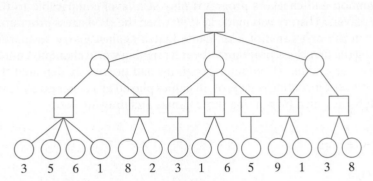

FIGURE 7.9.9   The game tree for Example 7.9.2.

We begin by evaluating vertices A, B, C, and D (see Figure 7.9.10). Next, we find that the value of E is 6. This results in a beta value of 6 for F. Next, we evaluate vertex G. Since its value is 8 and 8 exceeds the beta value of F, we obtain a beta cutoff and prune the subtree with root H. The value of F is 6. This results in an alpha value of 6 for I. Next, we evaluate vertices J and K. Since the value 3 of K is less than the alpha value 6 of I, an alpha cutoff occurs and the subtree with root L may be pruned. Next, we evaluate M, N, O, P, Q, R, and S. No further pruning is possible. Finally, we determine that the root I has value 8.   □

It has been shown (see [Pearl]) that for game trees in which every parent has n children and in which the terminal values are randomly ordered, for a given amount of time, the alpha-beta procedure permits a search depth $\frac{4}{3}$ greater than the pure minimax procedure, which evaluates every vertex. Pearl also shows that for such game trees, the alpha-beta procedure is optimal.

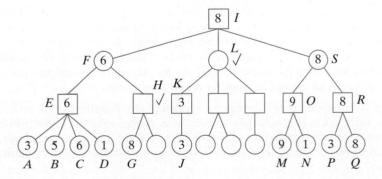

FIGURE 7.9.10   Evaluating the root of the game tree of Figure 7.9.9 using depth-first search with alpha-beta pruning. Checked vertices are roots of subtrees that are pruned. The values of vertices that are evaluated are written inside the vertices.

Other techniques have been combined with alpha-beta pruning to facilitate the search of a game tree. One idea is to order the children of the vertices to be evaluated so that the most promising moves are examined first (see Exercises 23–26). Another idea is to allow a variable-depth search in which the search backtracks when it reaches an unpromising position as measured by some function.

Some game-playing programs have been quite successful. The best backgammon and checkers programs play at a level comparable to the best human players. History was made in 1996 when the IBM chess program, Deep Blue, won the first game of a six-game match against Garry Kasparov, thus becoming the first chess program to beat a reigning world champion under regular time constraints. (Computer programs had previously defeated the best human players, including Kasparov, in games played at a faster rate.) Kasparov ultimately prevailed by winning three games and drawing two.

ॐ ॐ ॐ

Exercises

1. Draw the complete game tree for a version of nim in which the initial position consists of one pile of six tokens and a turn consists of taking one, two, or three tokens. Assign values to all vertices so that the resulting tree is analogous to Figure 7.9.2. Assume that the last player to take a token loses. Will the first or second player, playing an optimal strategy, always win? Describe an optimal strategy for the winning player.

2. Draw the complete game tree for nim in which the initial position consists of two piles of three tokens each. Omit symmetric positions. Assume that the last player to take a token loses. Assign values to all vertices so that the resulting tree is analogous to Figure 7.9.2. Will the first or second player, playing an optimal strategy, always win? Describe an optimal strategy for the winning player.

3. Draw the complete game tree for nim in which the initial position consists of two piles, one containing three tokens and the other containing two tokens. Assume that the last player to take a token wins. Assign values to all vertices so that the resulting tree is analogous to Figure 7.9.2. Will the first or second player, playing an optimal strategy, always win? Describe an optimal strategy for the winning player.

4. Draw the complete game tree for nim in which the initial position consists of two piles of three tokens each. Omit symmetric positions. Assume that the last player to take a token wins. Assign values to all vertices so that the resulting tree is analogous to Figure 7.9.2. Will the first or second player, playing an optimal strategy, always win? Describe an optimal strategy for the winning player.

5. Draw the complete game tree for the version of nim described in Exercise 1. Assume that the last person to take a token wins. Assign values to all vertices so that the resulting tree is analogous to Figure 7.9.2. Will the first or second player, playing an optimal strategy, always win? Describe an optimal strategy for the winning player.

6. Give an example of a (possibly hypothetical) complete game tree in which a terminal vertex is 1 if the first player won and 0 if the first player lost having the following

properties: There are more 0's than 1's among the terminal vertices, but the first player can always win by playing an optimal strategy.

Exercises 7 and 8 refer to nim and nim'. Nim is the game using n piles of tokens as described in this section in which the last player to move loses. Nim' is the game using n piles of tokens as described in this section except that the last player to move wins. We fix n piles each with a fixed number of tokens. We assume that at least one pile has at least two tokens.

★ **7.** Show that the first player can always win nim if and only if the first player can always win nim'.

★ **8.** Given a winning strategy for a particular player for nim, describe a winning strategy for this player for nim'.

Evaluate each vertex in each game tree. The values of the terminal vertices are given.

9.

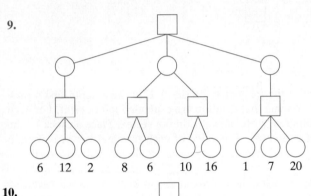

10.

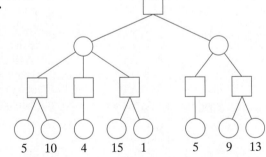

11.

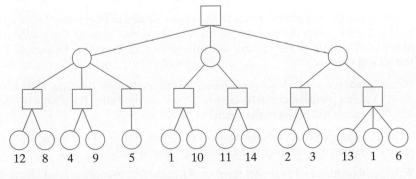

12.

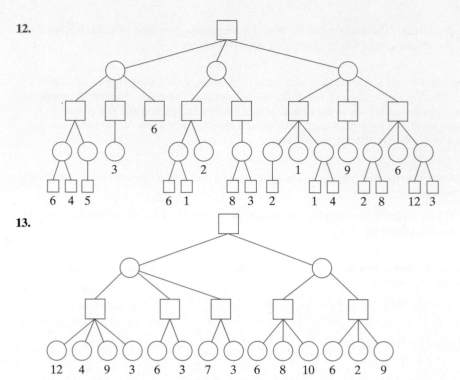

6 4 5   3   6 1   8 3 2   1 4   2 8   12 3

13.

12 4 9 3 6 3 7 3 6 8 10 6 2 9

14. Evaluate the root of each of the trees of Exercises 9–13 using a depth-first search with alpha-beta pruning. Assume that children are evaluated left to right. For each vertex whose value is computed, write the value in the vertex. Place a check by the root of each subtree that is pruned. The value of each terminal vertex is written under the vertex.

In Exercises 15–18, determine the value of the tic-tac-toe position using the evaluation function of Example 7.9.1.

15.

16.

17.

18.

19. Assume that the first player moves to the center square in tic-tac-toe. Draw a two-level game tree, with the root having an X in the center square. Omit symmetric positions. Evaluate all the vertices using the evaluation function of Example 7.9.1. Where will O move?

★ **20.** Would a two-level search program based on the evaluation function E of Example 7.9.1 play a perfect game of tic-tac-toe? If not, can you alter E so that a two-level search program will play a perfect game of tic-tac-toe?

21. Write an algorithm that evaluates vertices of a game tree to level n using depth-first search. Assume the existence of an evaluation function E.

★ **22.** Write an algorithm that evaluates the root of a game tree using an n-level, depth-first search with alpha-beta pruning. Assume the existence of an evaluation function E.

The following approach often leads to more pruning than pure alpha-beta minimax. First, perform a two-level search. Evaluate children from left to right. At this point, all the children of the root will have values. Next, order the children of the root with the most promising moves to the left. Now, use an n-level depth-first search with alpha-beta pruning. Evaluate children from left to right.

Carry out this procedure for $n = 4$ for each game tree of Exercises 23–25. Place a check by the root of each subtree that is pruned. The value of each vertex, as given by the evaluation function, is given under the vertex.

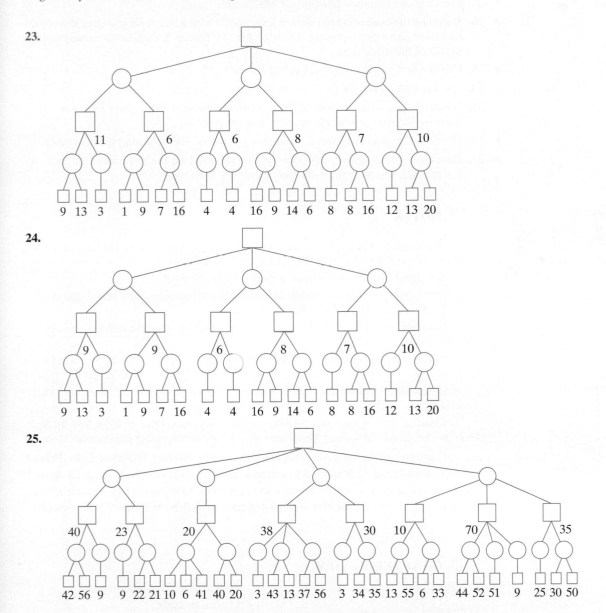

23.

24.

25.

26. Write an algorithm to carry out the procedure described before Exercise 23.

Mu Torere is a two-person game played by the Maoris (see [Bell]). The board is an eight-pointed star (see the figure) with a circular area in the center known as the *putahi*. The first player has four black tokens and the second player has four

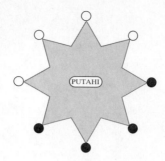

white tokens. The initial position is shown. A player who cannot make a move loses. Players alternate moves. At most one token can occupy a point of the star or the putahi. A move consists of:

(a) Moving to an adjacent point

(b) Moving from the putahi to a point

(c) Moving from a point to the putahi provided that one or both of the adjacent points contain the opponent's pieces

★ **27.** Develop an evaluation function for Mu Torere.

★ **28.** Combine the evaluation function in Exercise 27 with a two-level search of the game tree to obtain a game-playing algorithm for Mu Torere. Evaluate the game-playing ability of this algorithm.

★ **29.** Can the first player always win in Mu Torere?

★ **30.** Can the first player always tie in Mu Torere?

31. [*Project*] According to [Nilsson], the complete game tree for chess has over 10^{100} vertices. Report on how this estimate was obtained.

★ **32.** [*Project*] Develop an evaluation function for Kalah. (See [Ainslie] for the rules.)

★ **33.** Develop a game-playing algorithm for Kalah based on the evaluation function of Exercise 32. Evaluate the game-playing ability of this algorithm.

ᔑ NOTES

The following are recommended references on trees: [Berge; Bondy; Deo; Even, 1979; Gibbons; Harary; Knuth, 1973, Vol. 1; Liu, 1985; and Ore].

See [Date] for the use of trees in hierarchical databases.

[Cormen] has additional information on Huffman codes and a proof that Algorithm 7.1.8 constructs an optimal Huffman tree.

[Golomb, 1965] describes backtracking and contains several examples and applications.

Minimal spanning tree algorithms and their implementation can be found in [Tarjan].

[Baase] discusses the minimal time for sorting as well as lower bounds for other problems.

Classical sorting algorithms are thoroughly covered in [Knuth, 1973, Vol. 3]. See [Akl; Baase; Leighton; Lester; Lewis; and Quinn] for sorting using parallel machines.

Good references on game trees are [Nievergelt; Nilsson; and Slagle]. In [Frey], the minimax procedure is applied to a simple game. Various methods to speed up the search of the game tree are discussed and compared. Computer programs are given. [Berlekamp] contains a general theory of games as well as analyses of many specific games.

ᔑ CHAPTER REVIEW

Section 7.1

Free tree

Rooted tree

Level of a vertex in a rooted tree

Height of a rooted tree

Hierarchical definition tree

Huffman code

Section 7.2

Parent

Ancestor

Child
Descendant
Sibling
Terminal vertex
Internal vertex
Subtree
Acyclic graph
Alternative characterizations of trees
   (Theorem 7.2.3)

Section 7.3

Spanning tree
A graph has a spanning tree if and
   only if it is connected.
Breadth-first search
Depth-first search
Backtracking

Section 7.4

Minimal spanning tree
Prim's Algorithm to find a minimal
   spanning tree
Greedy algorithm

Section 7.5

Binary tree
Left child in a binary tree
Right child in a binary tree
Full binary tree
If T is full binary tree with i
   internal vertices, then T has $i + 1$
   terminal vertices and $2i + 1$ total
   vertices.
If a binary tree of height h has t ter-
   minal vertices, then $\lg t \leq h$.
Binary search tree
Algorithm to construct a binary search
   tree

Section 7.6

Preorder traversal

Inorder traversal
Postorder traversal
Prefix form of an expression
   (Polish notation)
Infix form of an expression
Postfix form of an expression
   (reverse Polish notation)
Tree representation of an expression

Section 7.7

Decision tree
The height of a decision tree that
   represents an algorithm is proportional to the worst-case time of the
   algorithm.
Any sorting algorithm requires at
   least $\Omega(n \lg n)$ comparisons in the
   worst case to sort n items.

Section 7.8

Isomorphic free trees
Isomorphic rooted trees
Isomorphic binary trees
The Catalan number
   $C(2n, n)/(n + 1)$ is equal to the
   number of nonisomorphic binary
   trees with n vertices.
Linear time algorithm (Algorithm
   7.8.13) to test whether two binary
   trees are isomorphic

Section 7.9

Game tree
Minimax procedure
n-level search
Evaluation function
Alpha-beta pruning
Alpha value
Alpha cutoff
Beta value
Beta cutoff

✌ CHAPTER SELF-TEST

Section 7.1

1. Draw the adjacent free tree as a rooted tree with root c.
2. Find the level of every vertex in the adjacent tree rooted at c.
3. Find the height of the adjacent tree rooted at c.

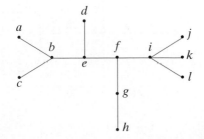

4. Construct an optimal Huffman code for the set of letters in the table.

Letter	Frequency
A	5
B	8
C	5
D	12
E	20
F	10

Section 7.2

5. Draw the free tree of Exercise 1 as a rooted tree with root f. Find

 (a) The parent of a.

 (b) The children of b.

 (c) The terminal vertices.

 (d) The subtree rooted at e.

Answer true or false in Exercises 6–8 and explain your answer.

6. If T is a tree with six vertices, T must have five edges.

7. If T is a rooted tree with six vertices, the height of T is at most 5.

8. An acyclic graph with eight vertices has seven edges.

Section 7.3

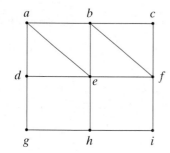

9. Use breadth-first search (Algorithm 7.3.6) with the vertex ordering *eachgbdfi* to find a spanning tree for the adjacent graph.

10. Use depth-first search (Algorithm 7.3.7) with the vertex ordering *eachgbdfi* to find a spanning tree for the adjacent graph.

11. Use breadth-first search (Algorithm 7.3.6) with the vertex ordering *fdehagbci* to find a spanning tree for the adjacent graph.

12. Use depth-first search (Algorithm 7.3.7) with the vertex ordering *fdehagbci* to find a spanning tree for the adjacent graph.

Section 7.4

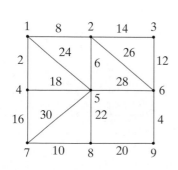

13. Find a minimal spanning tree for the adjacent graph.

14. In what order are the edges added by Prim's Algorithm for the adjacent graph if the initial vertex is 1?

15. In what order are the edges added by Prim's Algorithm for the adjacent graph if the initial vertex is 6?

16. Give an example of the use of the greedy method that does not lead to an optimal algorithm.

Section 7.5

17. Draw a binary tree with exactly two left children and one right child.

18. A full binary tree has 15 internal vertices. How many terminal vertices does it have?

19. Place the words

WORD PROCESSING PRODUCES CLEAN MANUSCRIPTS

BUT NOT NECESSARILY CLEAR PROSE,

in the order in which they appear, in a binary search tree.

20. Explain how we would look for MORE in the binary search tree of Exercise 19.

Section 7.6

Exercises 21–23 refer to the adjacent binary tree.

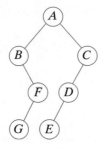

21. List the order in which the vertices are processed using preorder traversal.

22. List the order in which the vertices are processed using inorder traversal.

23. List the order in which the vertices are processed using postorder traversal.

24. Represent the prefix expression $- * E/BD - CA$ as a binary tree. Also write the postfix form and the fully parenthesized infix form of the expression.

Section 7.7

25. Six coins are identical in appearance, but one coin is either heavier or lighter than the others, which all weigh the same. Prove that at least three weighings are required in the worst case to identify the bad coin and determine whether it is heavy or light using only a pan balance.

26. Draw a decision tree that gives an algorithm to solve the coin puzzle of Exercise 25 in at most three weighings in the worst case.

27. Professor E. Sabic claims to have discovered an algorithm that uses at most $100n$ comparisons in the worst case to sort n items, for all $n \geq 1$. The Professor's algorithm repeatedly compares two elements and, based on the result of the comparison, modifies the original list. Give an argument that shows that the Professor must be mistaken.

28. The *binary insertion sort* algorithm sorts an array of size n as follows. If $n = 1, 2,$ or 3, the algorithm uses an optimal sort. If $n > 3$, the algorithm sorts $s_1, \ldots, s_n$ in the following way. First, $s_1, \ldots, s_{n-1}$ is recursively sorted. Then binary search is used to determine the correct position for s_n, after which s_n is inserted in its correct position. Determine the number of comparisons used by binary insertion sort in the worst case for $n = 4, 5, 6$. Does any algorithm use fewer comparisons for $n = 4, 5, 6$?

Section 7.8

Answer true or false in Exercises 29 and 30 and explain your answer.

29. If T_1 and T_2 are isomorphic as rooted trees, then T_1 and T_2 are isomorphic as free trees.

30. If T_1 and T_2 are rooted trees that are isomorphic as free trees, then T_1 and T_2 are isomorphic as rooted trees.

31. Determine whether the free trees are isomorphic. If the trees are isomorphic, give an isomorphism. If the trees are not isomorphic, give an invariant the trees do not share.

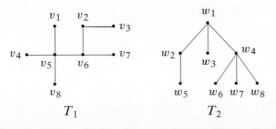

32. Determine whether the rooted trees are isomorphic. If the trees are isomorphic, give an isomorphism. If the trees are not isomorphic, give an invariant the trees do not share.

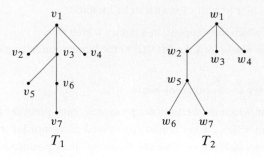

Section 7.9

33. Find the value of the tic-tac-toe position using the evaluation function of Example 7.9.1.

34. Give an evaluation function for a tic-tac-toe position different from that of Example 7.9.1. Attempt to discriminate more among the positions than does the evaluation function of Example 7.9.1.

35. Evaluate each vertex in the game tree. The values of the terminal vertices are given.

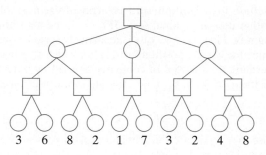

36. Evaluate the root of the tree of Exercise 35 using the minimax procedure with alpha-beta pruning. Assume that the children are evaluated left to right. For each vertex whose value is computed, write the value in the vertex. Place a check by the root of each subtree that is pruned.

8

NETWORK MODELS AND PETRI NETS

*M*arry me,
and I'll never look at any
other horse!

—from A Day at the Races

In this chapter we discuss two topics, network models and Petri nets, which make use of directed graphs. The major portion of the chapter is devoted to the problem of maximizing the flow through a network. The network might be a transportation network through which commodities flow, a pipeline network through which oil flows, a computer network through which data flow, or any number of other possibilities. In each case the problem is to find a maximal flow. Many other problems, which on the surface seem not to be flow problems, can, in fact, be modeled as network flow problems.

Maximizing the flow in a network is a problem that belongs both to graph theory and to operations research. The traveling salesperson problem furnishes another example of a problem in graph theory and operations research. **Operations research** studies the very broad category of problems of optimizing the performance of a system. Typical problems studied in operations research are network problems, allocation of resources problems, and personnel assignment problems.

Petri nets model systems in which processing can occur concurrently. The model provides a framework for dealing with questions such as whether the system will deadlock and whether the capacities of components of the system will be exceeded.

8.1 NETWORK MODELS

Consider the directed graph in Figure 8.1.1, which represents an oil pipeline network. Oil is unloaded at the dock a and pumped through the network to the refinery z. The vertices b, c, d, and e represent intermediate pumping stations. The directed edges represent subpipelines of the system and show the direction the oil can flow. The labels on the edges show the capacities of the subpipelines. The problem is to find a way to maximize the flow from the dock to the refinery and to compute the value of this maximum flow. Figure 8.1.1 provides an example of a **transport network**.

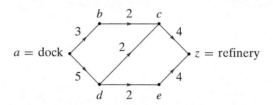

FIGURE 8.1.1   A transport network.

DEFINITION 8.1.1

A *transport network* (or more simply *network*) is a simple, weighted, directed graph satisfying:

   (a) A designated vertex, the *source*, has no incoming edges.

   (b) A designated vertex, the *sink*, has no outgoing edges.

   (c) The weight C_{ij} of the directed edge (i, j), called the *capacity* of (i, j), is a nonnegative number.

EXAMPLE 8.1.2

The graph of Figure 8.1.1 is a transport network. The source is vertex a and the sink is vertex z. The capacity of edge (a, b), C_{ab}, is 3 and the capacity of edge (b, c), C_{bc}, is 2.   □

Throughout this chapter, if G is a network, we will denote the source by a and the sink by z.

A **flow** in a network assigns a flow in each directed edge that does not exceed the capacity of that edge. Moreover, it is assumed that the flow into a vertex v, which is neither the source nor the sink, is equal to the flow out of v. The next definition makes these ideas precise.

DEFINITION 8.1.3

Let G be a transport network. Let C_{ij} denote the capacity of the directed edge (i, j). A *flow* F in G assigns each directed edge (i, j) a nonnegative number F_{ij} such that:

(a) $F_{ij} \leq C_{ij}$.

(b) For each vertex j, which is neither the source nor the sink,

$$\sum_i F_{ij} = \sum_i F_{ji}. \tag{8.1.1}$$

[In a sum such as (8.1.1), unless specified otherwise, the sum is assumed to be taken over all vertices i. Also, if (i, j) is not an edge, we set $F_{ij} = 0$.]

We call F_{ij} the *flow in edge* (i, j). For any vertex j, we call

$$\sum_i F_{ij}$$

the *flow into j* and we call

$$\sum_i F_{ji}$$

the *flow out of j.*

The property expressed by equation (8.1.1) is called **conservation of flow**. In the oil-pumping example of Figure 8.1.1, conservation of flow means that oil is neither used nor supplied at pumping stations b, c, d, and e.

EXAMPLE 8.1.4

The assignments,

$$F_{ab} = 2, \quad F_{bc} = 2, \quad F_{cz} = 3, \quad F_{ad} = 3,$$
$$F_{dc} = 1, \quad F_{de} = 2, \quad F_{ez} = 2,$$

define a flow for the network of Figure 8.1.1. For example, the flow into vertex d,

$$F_{ad} = 3,$$

is the same as the flow out of vertex d,

$$F_{dc} + F_{de} = 1 + 2 = 3. \qquad \square$$

In Figure 8.1.2 we have redrawn the network of Figure 8.1.1 to show the flow of Example 8.1.4. An edge e is labeled "x, y" if the capacity of e is x and the flow in e is y. This notation will be used throughout this chapter.

Notice that in Example 8.1.4, the flow out of the source a,

$$F_{ab} + F_{ad},$$

is the same as the flow into the sink z,

$$F_{cz} + F_{ez};$$

both values are 5. The next theorem shows that it is always true that the flow out of the source equals the flow into the sink.

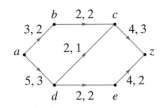

FIGURE 8.1.2
Flow in a network. Edges are labeled x, y to indicate capacity x and flow y.

THEOREM 8.1.5

Given a flow F in a network, the flow out of the source a equals the flow into the sink z; that is,

$$\sum_i F_{ai} = \sum_i F_{iz}.$$

Proof. Let V be the set of vertices. We have

$$\sum_{j \in V} \left(\sum_{i \in V} F_{ij} \right) = \sum_{j \in V} \left(\sum_{i \in V} F_{ji} \right),$$

since each double sum is

$$\sum_{e \in E} F_e,$$

where E is the set of edges. Now

$$
\begin{aligned}
0 &= \sum_{j \in V} \left(\sum_{i \in V} F_{ij} - \sum_{i \in V} F_{ji} \right) \\
&= \left(\sum_{i \in V} F_{iz} - \sum_{i \in V} F_{zi} \right) + \left(\sum_{i \in V} F_{ia} - \sum_{i \in V} F_{ai} \right) \\
&\quad + \sum_{\substack{j \in V \\ j \neq a, z}} \left(\sum_{i \in V} F_{ij} - \sum_{i \in V} F_{ji} \right) \\
&= \sum_{i \in V} F_{iz} - \sum_{i \in V} F_{ai}
\end{aligned}
$$

since $F_{zi} = 0 = F_{ia}$, for all $i \in V$, and (Definition 8.1.3b)

$$\sum_{i \in V} F_{ij} - \sum_{i \in V} F_{ji} = 0 \qquad \text{if } j \in V - \{a, z\}. \qquad \blacksquare$$

In light of Theorem 8.1.5, we can state the following definition.

DEFINITION 8.1.6

Let F be a flow in a network G. The value

$$\sum_i F_{ai} = \sum_i F_{iz}$$

is called the *value of the flow F*.

EXAMPLE 8.1.7

The value of the flow in the network of Figure 8.1.2 is 5.   □

The problem for a transport network G may be stated: Find a maximal flow in G; that is, among all possible flows in G, find a flow F so that the value of F is a maximum. In the next section we will give an algorithm that efficiently solves this problem. We will conclude this section by giving additional examples.

EXAMPLE 8.1.8   *A Pumping Network*

Figure 8.1.3 represents a pumping network in which water for two cities, A and B, is delivered from three wells, w_1, w_2, and w_3. The capacities of the intermediate systems are shown on the edges. Vertices b, c, and d represent intermediate pumping stations. Model this system as a transport network.

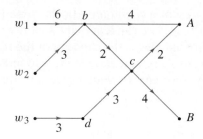

FIGURE 8.1.3   A pumping network. Water for cities A and B is delivered from wells w_1, w_2, and w_3. Capacities are shown on the edges.

To obtain a designated source and sink, we can obtain an equivalent transport network by tying together the sources into a **supersource** and tying together the sinks into a **supersink** (see Figure 8.1.4). In Figure 8.1.4, ∞ represents an unlimited capacity.

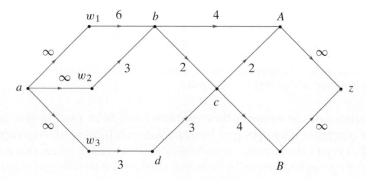

FIGURE 8.1.4   The network of Figure 8.1.3 with a designated source and sink.   □

EXAMPLE 8.1.9   *A Traffic Flow Network*

It is possible to go from city A to city C directly or by going through city B. During the period 6:00 to 7:00 P.M., the average trip times are

A to B (15 minutes)

B to C (30 minutes)

A to C (30 minutes).

The maximum capacities of the routes are

A to B (3000 vehicles)

B to C (2000 vehicles)

A to C (4000 vehicles).

Represent the flow of traffic from A to C during the period 6:00 to 7:00 P.M. as a network.

A vertex will represent a city at a particular time (see Figure 8.1.5). An edge connects X, t_1 to Y, t_2 if we can leave city X at t_1 P.M. and arrive at city Y at t_2 P.M. The capacity of an edge is the capacity of the route. Edges of infinite capacity connect A, t_1, to A, t_2, and B, t_1 to B, t_2 to indicate that any number of cars can wait at city A or city B. Finally, we introduce a supersource and supersink.   □

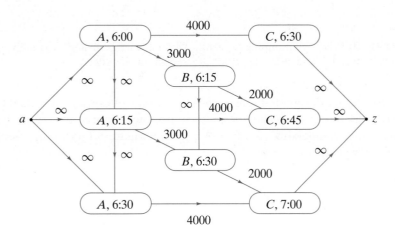

FIGURE 8.1.5   A network that represents the flow of traffic from city A to city C during the period 6:00 to 7:00 P.M.

Variants of the network flow problem have been used in the design of efficient computer networks (see [Jones; Kleinrock]). In modeling a computer network, a vertex is a message or switching center, an edge represents a channel on which data can be transmitted between vertices, a flow is the average number of bits per second being transmitted on a channel, and the capacity of an edge is the capacity of the corresponding channel.

ॐ ॐ ॐ
Exercises

In Exercises 1–3, fill in the missing edge flows so that the result is a flow in the given network. Determine the values of the flows.

1.

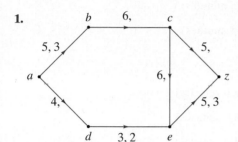

2.

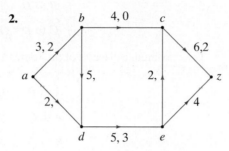

3.

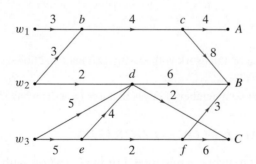

4. The accompanying graph represents a pumping network in which oil for three refineries, A, B, and C, is delivered from three wells, w_1, w_2, and w_3. The capacities of the intermediate systems are shown on the edges. Vertices b, c, d, e, and f represent intermediate pumping stations. Model this system as a network.

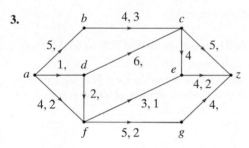

5. Model the system of Exercise 4 as a network assuming that well w_1 can pump at most 2 units, well w_2 at most 4 units, and well w_3 at most 7 units.

6. Model the system of Exercise 5 as a network assuming, in addition to the limitations on the wells, that city A requires 4 units, city B requires 3 units, and city C requires 4 units.

7. Model the system of Exercise 6 as a network assuming, in addition to the limitations on the wells and the requirements by the cities, that the intermediate pumping station d can pump at most 6 units.

8. There are two routes from city A to city D. One route passes through city B and the other route passes through city C. During the period 7:00 to 8:00 A.M., the average trip times are

$$A \text{ to } B \quad (30 \text{ minutes})$$

$$A \text{ to } C \quad (15 \text{ minutes})$$

$$B \text{ to } D \quad (15 \text{ minutes})$$

$$C \text{ to } D \quad (15 \text{ minutes}).$$

The maximum capacities of the routes are

$$A \text{ to } B \quad (1000 \text{ vehicles})$$

$$A \text{ to } C \quad (3000 \text{ vehicles})$$

$$B \text{ to } D \quad (4000 \text{ vehicles})$$

$$C \text{ to } D \quad (2000 \text{ vehicles}).$$

Represent the flow of traffic from A to D during the period 7:00 to 8:00 A.M. as a network.

9. In the system shown, we want to maximize the flow from a to z. The capacities are shown on the edges. The flow between two vertices, neither of which is a or z, can be in either direction. Model this system as a network.

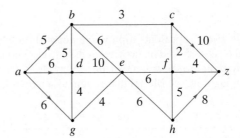

10. Give an example of a network with exactly two maximal flows, where each F_{ij} is a nonnegative integer.

11. What is the maximum number of edges that a network can have?

8.2 A MAXIMAL FLOW ALGORITHM

If G is a transport network, a **maximal flow** in G is a flow with maximum value. In general, there will be several flows having the same maximum value. In this section we give an algorithm for finding a maximal flow. The basic idea is simple—start with some initial flow and iteratively increase the value of the flow until no more improvement is possible. The resulting flow will then be a maximal flow.

We can take the initial flow to be the one in which the flow in each edge is zero. To increase the value of a given flow, we must find a path from the source to the sink and increase the flow along this path.

It is helpful at this point to introduce some terminology. Throughout this section, G denotes a network with source a, sink z, and capacity C. Momentarily, consider the edges of G to be undirected and let

$$P = (v_0, v_1, \ldots, v_n), \qquad v_0 = a, \qquad v, = z,$$

be a path from a to z in this undirected graph. (All paths in this section are with reference to the underlying undirected graph.) If an edge e in P is directed from v_{i-1} to v_i, we say that e is **properly oriented (with respect to P)**; otherwise, we say that e is **improperly oriented (with respect to P)** (see Figure 8.2.1).

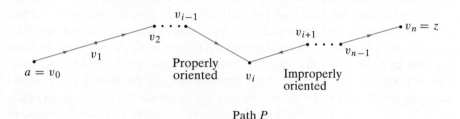

Path P

FIGURE 8.2.1   Properly and improperly oriented edges. Edge (v_{i-1}, v_i) is properly oriented because it is oriented in the a-to-z direction. Edge (v_i, v_{i+1}) is improperly oriented because it is *not* in the a-to-z direction.

If we can find a path P from the source to the sink in which every edge in P is properly oriented and the flow in each edge is less than the capacity of the edge, it is possible to increase the value of the flow.

EXAMPLE 8.2.1

Consider the path from a to z in Figure 8.2.2. All the edges in P are properly oriented. The value of the flow in this network can be increased by 1, as shown in Figure 8.2.3.

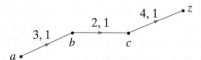

FIGURE 8.2.2   A path all of whose edges are properly oriented.

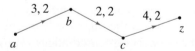

FIGURE 8.2.3   After increasing the flow of Figure 8.2.2 by 1.   □

It is also possible to increase the flow in certain paths from the source to the sink in which we have properly and improperly oriented edges. Let P be a path from a to z and let x be a vertex in P that is neither a nor z

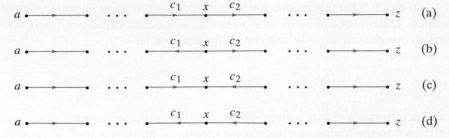

FIGURE 8.2.4   The four possible orientations of the edges incident on x.

(see Figure 8.2.4). There are four possibilities for the orientations of the edges e_1 and e_2 incident on x. In case (a), both edges are properly oriented. In this case, if we increase the flow in each edge by Δ, the flow into x will still equal the flow out of x. In case (b), if we increase the flow in e_2 by Δ, we must *decrease* the flow in e_1 by Δ so that the flow into x will still equal the flow out of x. Case (c) is similar to case (b), except that we increase the flow in e_1 by Δ and decrease the flow in e_2 by Δ. In case (d), we decrease the flow in both edges by Δ. In every case, the resulting edge assignments give a flow. Of course, to carry out these alterations, we must have flow less than capacity in a properly oriented edge and a nonzero flow in an improperly oriented edge.

EXAMPLE 8.2.2

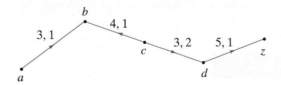

FIGURE 8.2.5   A path with an improperly oriented edge: (c, b).

Consider the path from a to z in Figure 8.2.5. Edges (a, b), (c, d), and (d, z) are properly oriented and edge (c, b) is improperly oriented. We decrease the flow by 1 in the improperly oriented edge (c, b) and increase the flow by 1 in the properly oriented edges (a, b), (c, d), and (d, z) (see Figure 8.2.6). The value of the new flow is 1 greater than that of the original flow.

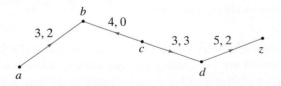

FIGURE 8.2.6   After increasing the flow of Figure 8.2.5 by 1.   □

We summarize the method of Examples 8.2.1 and 8.2.2 as a theorem.

THEOREM 8.2.3

Let P be a path from a to z in a network G satisfying:

(a) For each properly oriented edge (i, j) in P,

$$F_{ij} < C_{ij}.$$

(b) For each improperly oriented edge (i, j) in P,

$$0 < F_{ij}.$$

Let

$$\Delta = \min X,$$

where X consists of the numbers $C_{ij} - F_{ij}$, for properly oriented edges (i, j) in P, and F_{ij}, for improperly oriented edges (i, j) in P. Define

$$F_{ij}^* = \begin{cases} F_{i,j} & \text{if } (i, j) \text{ is not in P} \\ F_{ij} + \Delta & \text{if } (i, j) \text{ is properly oriented in P} \\ F_{ij} - \Delta & \text{if } (i, j) \text{ is not properly oriented in P.} \end{cases}$$

Then F^* is a flow whose value is Δ greater than the value of F.

Proof. (See Figures 8.2.2, 8.2.3, 8.2.5, and 8.2.6.) The argument that F^* is a flow is given just before Example 8.2.2. Since the edge (a, v) in P is increased by Δ, the value of F^* is Δ greater than the value of F. ∎

In the next section we will show that if there are no paths satisfying the conditions of Theorem 8.2.3, the flow is maximal. Thus it is possible to construct an algorithm based on Theorem 8.2.3. The outline is

1. Start with a flow (e.g., the flow in which the flow in each edge is 0).

2. Search for a path satisfying the conditions of Theorem 8.2.3. If no such path exists, stop; the flow is maximal.

3. Increase the flow through the path by Δ, where Δ is defined as in Theorem 8.2.3, and go to line 2.

In the formal algorithm, we search for a path satisfying the conditions of Theorem 8.2.3 while simultaneously keeping track of the quantities

$$C_{ij} - F_{ij}, F_{ij}.$$

ALGORITHM 8.2.4   *Finding a Maximal Flow in a Network*

This algorithm finds a maximal flow in a network. The capacity of each edge is a nonnegative integer.

Input: A network with source a, sink z, capacity C, vertices $a = v_0, \ldots, v_n = z$, and n

Output: A maximal flow F

```
        procedure max_flow(a, z, C, v, n)
          // v's label is (predecessor(v), val(v))
          // start with zero flow
1.        for each edge (i, j) do
2.           F_ij := 0
3.        while true do
4.          begin
            // remove all labels
5.          for i := 0 to n do
6.            begin
7.              predecessor(v_i) := null
8.              val(vi) := null
9.            end
            // label a
10.         predecessor(a) := −
11.         val(a) := ∞
            // U is the set of unexamined, labeled vertices
12.         U := {a}
            // continue until z is labeled
13.         while val(z) = null do
14.           begin
15.           if U = ∅ then // flow is maximal
16.              return(F)
17.           choose v in U
18.           U := U − {v}
19.           Δ := val(v)
20.           for each edge (v, w) with val(w) = null do
21.             if F_vw < C_uw then
22.               begin
23.                 predecessor(w) := v
24.                 val(w) := min{Δ, C_vw − F_vw}
25.                 U := U ∪ {w}
26.               end
27.           for each edge (w, v) with val(w) = null do
28.             if F_wv > 0 then
29.               begin
30.                 predecessor(w) := v
31.                 val(w) := min{Δ, F_wv}
32.                 U := U ∪ {w}
33.               end
34.           end
```

// find path P from a to z on which to revise flow

35. $w_0 := z$

36. $k := 0$

37.   **while** $w_k \neq a$ **do**

38.   **begin**

39. $w_{k+1} := predecessor(w_k)$

40. $k := k + 1$

41.   **end**

42. $P := (w_k, w_{k+1}, \ldots, w_1, w_0)$

43. $\Delta := val(z)$

44.   **for** $i = 1$ **to** k **do**

45.   **begin**

46. $e := (w_i, w_{i-1})$

47.   **if** e is properly oriented in P **then**

48. $F_e := F_e + \Delta$

49.   **else**

50. $F_e := F_e - \Delta$

51.   **end**

52.   **end**

     end *max_flow*

A proof that Algorithm 8.2.4 terminates is left as an exercise (Exercise 19). If the capacities are allowed to be nonnegative rational numbers, the algorithm also terminates; however, if arbitrary nonnegative real capacities are allowed and we permit the edges in line 20 to be examined in any order, the algorithm may not terminate (see [Ford, pp. 21–22]).

Algorithm 8.2.4 is often referred to as the **labeling procedure**. We will illustrate the algorithm with two examples.

EXAMPLE 8.2.5

In this discussion, if vertex v satisfies

$$predecessor(v) = p \qquad \text{and} \qquad val(v) = t,$$

we show v's label on the graph as (p, t).

At lines 1 and 2, we initialize the flow to 0 in each edge (see Figure 8.2.7). Next, at lines 5–9 we set all labels to *null*. Then, at lines 10 and 11 we label vertex a as $(-, \infty)$. At line 12 we set $U = \{a\}$. We then enter the while loop (line 13).

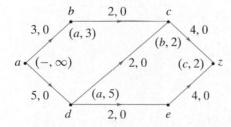

FIGURE 8.2.7   After the first labeling. Vertex v is labeled $(predecessor(v), val(v))$.

Since z is not labeled and U is not empty, we move to line 17, where we choose vertex a in U and remove it from U at line 18. At this point, $U = \emptyset$. We set Δ to ∞ [$= val(a)$] at line 19. At line 20 we examine the edges (a, b) and (a, d) since neither b nor d is labeled. For edge (a, b) we have

$$F_{ab} = 0 < C_{ab} = 3.$$

At lines 23 and 24, we label vertex b as $(a, 3)$ since

$$predecessor(b) = a$$

and

$$val(b) = \min\{\Delta, 3 - 0\} = \min\{\infty, 3 - 0\} = 3.$$

At line 25, we add b to U. Similarly, we label vertex d as $(a, 5)$ and add d to U. At this point, $U = \{b, d\}$.

We then return to the top of the while loop (line 13). Since z is not labeled and U is not empty, we move to line 17, where we choose a vertex in U. Suppose that we choose b. We remove b from U at line 18. We set Δ to 3 [$= val(b)$] at line 19. At line 20 we examine the edge (b, c). At lines 23 and 24 we label vertex c as $(b, 2)$ since

$$predecessor(c) = b$$

and

$$val(c) = \min\{\Delta, 2 - 0\}, = \min\{3, 2 - 0\} = 2.$$

At line 25 we add c to U. At this point, $U = \{c, d\}$.

We then return to the top of the while loop (line 13). Since z is not labeled and U is not empty, we move to line 17, where we choose a vertex in U. Suppose that we choose c. We remove c from U at line 18. We set Δ to 2 [$= val(c)$] at line 19. At line 20 we examine the edge (c, z). At lines 23 and 24 we label vertex z as $(c, 2)$. At line 25, we add z to U. At this point, $U = \{d, z\}$.

We then return to the top of the while loop (line 13). Since z is labeled, we proceed to line 35. At lines 35–42, by following predecessors from z, we find the path

$$P = (a, b, c, z)$$

from a to z. At line 42 we set Δ to 2. Since each edge in P is properly oriented, at line 48 we increase the flow in each edge in P by $\Delta = 2$ to obtain Figure 8.2.8.

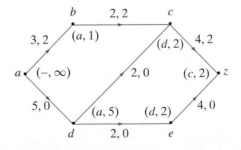

FIGURE 8.2.8   After increasing the flow on path (a, b, c, z) by 2 and the second labeling.

We then return to the top of the while loop (line 3). Next, at lines 5–9 we set all labels to *null*. Then, at lines 10 and 11 we label vertex a as $(-, \infty)$ (see Figure 8.2.8). At line 12 we set $U = \{a\}$. We then enter the while loop (line 13).

Since z is not labeled and U is not empty, we move to line 17, where we choose vertex a in U and remove it from U at line 18. At lines 23 and 24 we label vertex b as $(a, 1)$, and we label vertex d as $(a, 5)$. We add b and d to U so that $U = \{b, d\}$.

We then return to the top of the while loop (line 13). Since z is not labeled and U is not empty, we move to line 17, where we choose a vertex in U. Suppose that we choose b. We remove b from U at line 18. At line 20 we examine the edge (b, c). Since $F_{bc} = C_{bc}$, we do not label vertex c at this point. Now $U = \{d\}$.

We then return to the top of the while loop (line 13). Since z is not labeled and U is not empty, we move to line 17, where we choose vertex d in U and remove it from U at line 18. At lines 23 and 24 we label vertex c as $(d, 2)$ and we label vertex e as $(d, 2)$. We add c and e to U so that $U = \{c, e\}$.

We then return to the top of the while loop (line 13). Since z is not labeled and U is not empty, we move to line 17, where we choose a vertex in U. Suppose that we choose c in U and remove it from U at line 18. At lines 23 and 24 we label vertex z as $(c, 2)$. We add z to U so that $U = \{z, e\}$.

We then return to the top of the while loop (line 13). Since z is labeled, we proceed to line 35. At line 42 we find that

$$P = (a, d, c, z).$$

Since each edge in P is properly oriented, at line 48 we increase the flow in each edge in P by $\Delta = 2$ to obtain Figure 8.2.9.

You should check that the next iteration of the algorithm produces the labeling shown in Figure 8.2.9. Increasing the flow by $\Delta = 2$ produces Figure 8.2.10.

We then return to the top of the while loop (line 3). Next, at lines 5–9 we set all labels to *null*. Then, at lines 10 and 11 we label vertex a as $(-, \infty)$ (see Figure 8.2.10). At line 12 we set $U = \{a\}$. We then enter the while loop (line 13).

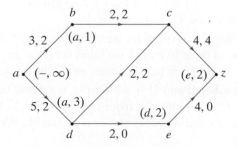

FIGURE 8.2.9   After increasing the flow on path (a, d, c, z) by 2 and the third labeling.

Since z is not labeled and U is not empty, we move to line 17, where we choose vertex a in U and remove it from U at line 18. At lines 23 and 24 we label vertex b as $(a, 1)$ and we label vertex d as $(a, 1)$. We add b and d to U so that $U = \{b, d\}$.

We then return to the top of the while loop (line 13). Since z is not labeled and U is not empty, we move to line 17, where we choose a vertex in U. Suppose that we choose b. We remove b from U at line 18. At line 20, we examine the edge (b, c). Since $F_{bc} = C_{bc}$, we do not label vertex c. Now $U = \{d\}$.

We then return to the top of the while loop (line 13). Since z is not labeled and U is not empty, we move to line 17, where we choose vertex d in U and remove it from U at line 18. At line 20 we examine edges (d, c) and (d, e). Since $F_{dc} = C_{dc}$ and $F_{de} = C_{de}$, we do not label either vertex c or vertex e. Now $U = \emptyset$.

We then return to the top of the while loop (line 13). Since z is not labeled, we move to line 15. Since U is empty, the algorithm terminates. The flow of Figure 8.2.10 is maximal.

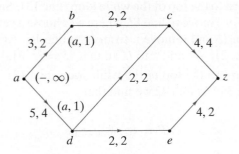

FIGURE 8.2.10   After increasing the flow on path (a, d, e, z) by 2 and the fourth and final labeling. The flow is maximal.   □

Our last example shows how to modify Algorithm 8.2.4 to generate a maximal flow from a given flow.

EXAMPLE 8.2.6

Replace the zero flow in lines 1 and 2 of Algorithm 8.2.4 with the flow of Figure 8.2.11 and then find a maximal flow.

After initializing the given flow, we move to lines 5–9, where we set all labels to *null*. Then, at lines 10 and 11 we label vertex a as $(-, \infty)$ (see Figure 8.2.11). At line 12 we set $U = \{a\}$. We then enter the while loop (line 13).

Since z is not labeled and U is not empty, we move to line 17, where we choose vertex a in U and remove it from U at line 18. At lines 23 and 24, we label vertex b as $(a, 1)$ and we label vertex d as $(a, 1)$. We add b and d to U so that $U = \{b, d\}$.

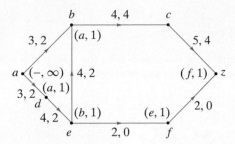

FIGURE 8.2.11   After labeling.

We then return to the top of the while loop (line 13). Since z is not labeled and U is not empty, we move to line 17, where we choose a vertex in U. Suppose that we choose b. We remove b from U at line 18. At line 20 we examine edges (b, c) and (e, b). Since $F_{bc} = C_{bc}$, we do not label vertex c. At lines 30 and 31, vertex e is labeled $(b, 1)$ since

$$val(e) = \min\{val(b), F_{eb}\} = \min\{1, 2\} = 1.$$

We then return to the top of the while loop (line 13). We ultimately label z (see Figure 8.2.11) and at line 42 we find the path

$$P = (a, b, e, f, z).$$

Edges (a, b), (e, f), and (f, z) are properly oriented, so the flow in each is increased by 1. Since edge (e, b) is improperly oriented, its flow is decreased by 1. We obtain the flow of Figure 8.2.12.

Another iteration of the algorithm produces the maximal flow shown in Figure 8.2.13.   □

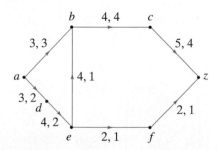

FIGURE 8.2.12   After increasing the flow on path (a, b, e, f, z) by 1. Notice that edge (e, b) is improperly oriented and so has its flow *decreased* by 1.

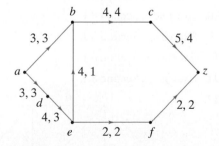

FIGURE 8.2.13   After increasing the flow on path (a, d, e, f, z) by 1. The flow is maximal.

ॐ ॐ ॐ

Exercises

In Exercises 1–3, a path from the source a to the sink z in a network is given. Find the maximum possible increase in the flow obtainable by altering the flows in the edges in the path.

1.

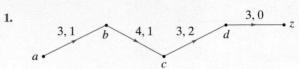

2.

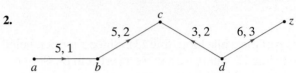

3.

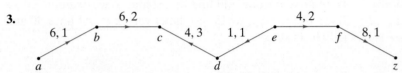

In Exercises 4–12, use Algorithm 8.2.4 to find a maximal flow in each network.

4. Figure 8.1.4

5. Figure 8.1.5

6.

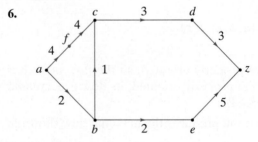

7. Exercise 5, Section 8.1

8. Exercise 6, Section 8.1

9. Exercise 7, Section 8.1

10. Exercise 8, Section 8.1

11. Exercise 9, Section 8.1

12.

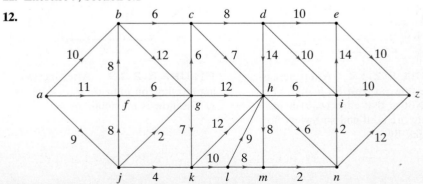

In Exercises 13–18, find a maximal flow in each network starting with the flow given.

13. Figure 8.1.2

14. Exercise 1, Section 8.1

15. Exercise 2, Section 8.1

16. Exercise 3, Section 8.1

17. Figure 8.1.4 with flows

$$F_{a,w_1} = 2, \qquad F_{w_1,b} = 2, \qquad F_{bA} = 0, \qquad F_{cA} = 0, \qquad F_{Az} = 0,$$
$$F_{a,w_2} = 0, \qquad F_{w_2,b} = 0, \qquad F_{bc} = 2, \qquad F_{cB} = 4, \qquad F_{Bz} = 4,$$
$$F_{a,w_3} = 2, \qquad F_{w_3,d} = 2, \qquad F_{dc} = 2.$$

18. Figure 8.1.4 with flows

$$F_{a,w_1} = 1, \qquad F_{w_1,b} = 1, \qquad F_{bA} = 4, \qquad F_{cA} = 2, \qquad F_{Az} = 6,$$
$$F_{a,w_2} = 3, \qquad F_{w_2,b} = 3, \qquad F_{bc} = 0, \qquad F_{cB} = 1, \qquad F_{Bz} = 1,$$
$$F_{a,w_3} = 3, \qquad F_{w_3,d} = 3, \qquad F_{dc} = 3.$$

19. Show that Algorithm 8.2.4 terminates.

8.3 *THE MAX FLOW, MIN CUT THEOREM*

In this section we show that at the termination of Algorithm 8.2.4, the flow in the network is maximal. Along the way we will define and discuss cuts in networks.

Let G be a network and consider the flow F at the termination of Algorithm 8.2.4. Some vertices are labeled and some are unlabeled. Let P ($\overline{P}$) denote the set of labeled (unlabeled) vertices. (Recall that $\overline{P}$ denotes the complement of P.) Then the source a is in P and the sink z is in $\overline{P}$. The set S of edges (v, w), with $v \in P$ and $w \in \overline{P}$, is called a **cut** and the sum of the capacities of the edges in S is called the **capacity of the cut**. We will see that this cut has a minimum capacity and, since a minimal cut corresponds to a maximal flow (Theorem 8.3.9), the flow F is maximal. We begin with the formal definition of cut.

Throughout this section, G is a network with source a and sink z. The capacity of edge (i, j) is C_{ij}.

DEFINITION 8.3.1

A *cut* $(P, \overline{P})$ in G consists of a set P of vertices and the complement $\overline{P}$ of P, with $a \in P$ and $z \in \overline{P}$.

> **EXAMPLE 8.3.2**

Consider the network G of Figure 8.3.1. If we let $P = \{a, b, d\}$, then $\overline{P} = \{c, e, f, z\}$ and $(P, \overline{P})$ is a cut in G. As shown, we sometimes indicate a cut by drawing a dashed line to partition the vertices.

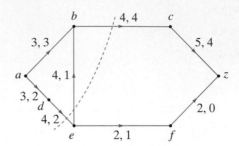

FIGURE 8.3.1   A cut in a network. The dashed line divides the vertices into sets $P = \{a, b, d\}$ and $\overline{P} = \{c, e, f, z\}$ producing the cut $(P, \overline{P})$.   □

> **EXAMPLE 8.3.3**

Figure 8.2.10 shows the labeling at the termination of Algorithm 8.2.4 for a particular network. If we let P ($\overline{P}$) denote the set of labeled (unlabeled) vertices, we obtain the cut shown in Figure 8.3.2.

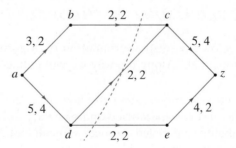

FIGURE 8.3.2   A network at termination of Algorithm 8.2.4. The cut $(P, \overline{P})$, $P = \{a, b, d\}$, is obtained by letting P be the set of labeled vertices.   □

We next define the capacity of a cut.

> **DEFINITION 8.3.4**

The *capacity of the cut* $(P, \overline{P})$ is the number

$$C(P, \overline{P}) = \sum_{i \in P} \sum_{j \in \overline{P}} C_{ij}.$$

EXAMPLE 8.3.5

The capacity of the cut of Figure 8.3.1 is

$$C_{bc} + C_{de} = 8. \qquad \square$$

EXAMPLE 8.3.6

The capacity of the cut of Figure 8.3.2 is

$$C_{bc} + C_{dc} + C_{de} = 6. \qquad \square$$

The next theorem shows that the capacity of any cut is always greater than or equal to the value of any flow.

THEOREM 8.3.7

Let F be a flow in G and let $(P, \overline{P})$ be a cut in G. Then the capacity of $(P, \overline{P})$ is greater than or equal to the value of F; that is,

$$\sum_{i \in P} \sum_{j \in \overline{P}} C_{ij} \geq \sum_{i} F_{ai}. \qquad (8.3.1)$$

(The notation $\displaystyle\sum_{i}$ means the sum over all vertices i.)

Proof. Note that

$$\sum_{j \in P} \sum_{i \in P} F_{ji} = \sum_{j \in P} \sum_{i \in P} F_{ij},$$

since either side of the equation is merely the sum of F_{ij} over all $i, j \in P$.

Now

$$\sum_{i} F_{ai} = \sum_{j \in P} \sum_{i} F_{ji} - \sum_{j \in P} \sum_{i} F_{ij}$$

$$= \sum_{j \in P} \sum_{i \in P} F_{ji} + \sum_{j \in P} \sum_{i \in \overline{P}} F_{ji} - \sum_{j \in P} \sum_{i \in P} F_{ij} - \sum_{j \in P} \sum_{i \in \overline{P}} F_{ij}$$

$$= \sum_{j \in P} \sum_{i \in \overline{P}} F_{ji} - \sum_{j \in P} \sum_{i \in \overline{P}} F_{ij} \leq \sum_{j \in P} \sum_{i \in \overline{P}} F_{ji} \leq \sum_{j \in P} \sum_{i \in \overline{P}} C_{ji}. \qquad \blacksquare$$

EXAMPLE 8.3.8

In Figure 8.3.1, the value 5 of the flow is less than the capacity 8 of the cut.   □

A **minimal cut** is a cut having minimum capacity.

| THEOREM 8.3.9 | *Max Flow, Min Cut Theorem* |

Let F be a flow in G and let $(P, \overline{P})$ be a cut in G. If equality holds in (8.3.1), then the flow is maximal and the cut is minimal. Moreover, equality holds in (8.3.1) if and only if

(a) $F_{ij} = C_{ij}$ for $i \in P$, $j \in \overline{P}$

and

(b) $F_{ij} = 0$ for $i \in \overline{P}$, $j \in P$.

Proof. The first statement follows immediately.

The proof of Theorem 8.3.7 shows that equality holds precisely when

$$\sum_{j \in P} \sum_{i \in \overline{P}} F_{ij} = 0 \qquad \text{and} \qquad \sum_{j \in P} \sum_{i \in \overline{P}} F_{ji} = \sum_{j \in P} \sum_{i \in \overline{P}} C_{ji};$$

thus the last statement is also true.   ∎

| EXAMPLE 8.3.10 |

In Figure 8.3.2, the value of the flow and the capacity of the cut are both 6; therefore, the flow is maximal and the cut is minimal.   □

We can use Theorem 8.3.9 to show that Algorithm 8.2.4 produces a maximal flow.

| THEOREM 8.3.11 |

At termination, Algorithm 8.2.4 produces a maximal flow. Moreover, if P (respectively, $\overline{P}$) is the set of labeled (respectively, unlabeled) vertices at the termination of Algorithm 8.2.4, the cut $(P, \overline{P})$ is minimal.

Proof. Let P ($\overline{P}$) be the set of labeled (unlabeled) vertices of G at the termination of Algorithm 8.2.4. Consider an edge (i, j), where $i \in P$, $j \in \overline{P}$. Since i is labeled, we must have

$$F_{ij} = C_{ij};$$

otherwise, we would have labeled j at lines 23 and 24. Now consider an edge (j, i), where $j \in \overline{P}$, $i \in P$. Since i is labeled, we must have

$$F_{ji} = 0;$$

otherwise, we would have labeled j at lines 30 and 31. By Theorem 8.3.9, the flow at the termination of Algorithm 8.2.4 is maximal and the cut $(P, \overline{P})$ is minimal.   ∎

ॐ ॐ ॐ

Exercises

In Exercises 1–3, find the capacity of the cut $(P, \overline{P})$. Also, determine whether the cut is minimal.

1. $P = \{a, d\}$ for Exercise 1, Section 8.1

2. $P = \{a, d, e\}$ for Exercise 2, Section 8.1

3. $P = \{a, b, c, d\}$ for Exercise 3, Section 8.1

In Exercises 4–16, find a minimal cut in each network.

4. Figure 8.1.1   **5.** Figure 8.1.4   **6.** Figure 8.1.5

7. Exercise 1, Section 8.1   **8.** Exercise 2, Section 8.1

9. Exercise 3, Section 8.1   **10.** Exercise 4, Section 8.1

11. Exercise 5, Section 8.1   **12.** Exercise 6, Section 8.1

13. Exercise 7, Section 8.1   **14.** Exercise 8, Section 8.1

15. Exercise 9, Section 8.1   **16.** Exercise 12, Section 8.2

Exercises 17–22 refer to a network G that, in addition to having nonnegative integer capacities C_{ij}, has nonnegative integer minimal edge flow requirements m_{ij}. That is, a flow F must satisfy

$$m_{ij} \leq F_{ij} \leq C_{ij}$$

for all edges (i, j).

17. Give an example of a network G, in which $m_{ij} \leq C_{ij}$ for all edges (i, j), for which no flow exists.

Define

$$C(\overline{P}, P) = \sum_{i \in \overline{P}} \sum_{j \in P} C_{ij},$$

$$m(P, \overline{P}) = \sum_{i \in P} \sum_{j \in \overline{P}} m_{ij}, \qquad m(\overline{P}, P) = \sum_{i \in \overline{P}} \sum_{j \in P} m_{ij}.$$

18. Show that the value V of any flow satisfies

$$m(P, \overline{P}) - C(\overline{P}, P) \leq V \leq C(P, \overline{P}) - m(\overline{P}, P)$$

for any cut $(P, \overline{P})$.

19. Show that if a flow exists in G, a maximal flow exists in G with value

$$\min\{C(P, \overline{P}) - m(\overline{P}, P) \mid (P, \overline{P}) \text{ is a cut in } G\}.$$

20. Assume that G has a flow F. Develop an algorithm for finding a maximal flow in G.

21. Show that if a flow exists in G, a minimal flow exists in G with value

$$\max\{m(P, \overline{P}) - C(\overline{P}, P) \mid (P, \overline{P}) \text{ is a cut in } G\}.$$

22. Assume that G has a flow F. Develop an algorithm for finding a minimal flow in G.

23. True or false? If F is a flow in a network G and $(P, \overline{P})$ is a cut in G and the capacity of $(P, \overline{P})$ exceeds the value of the flow, F, then the cut $(P, \overline{P})$ is not minimal and the flow F is not maximal. If true, prove it; otherwise, give a counterexample.

8.4 MATCHING

In this section we consider the problem of matching elements in one set to elements in another set. We will see that this problem can be reduced to finding a maximal flow in a network. We begin with an example.

EXAMPLE 8.4.1

Suppose that four persons A, B, C, and D apply for five jobs J_1, J_2, J_3, J_4, and J_5. Suppose that applicant A is qualified for jobs J_2 and J_5; applicant B is qualified for jobs J_2, and J_5; applicant C is qualified for jobs J_1, J_3, J_4, and J_5; and applicant D is qualified for jobs J_2 and J_5. Is it possible to find a job for each applicant?

The situation can be modeled by the graph of Figure 8.4.1. The vertices represent the applicants and the jobs. An edge connects an applicant to a job for which the applicant is qualified. We can show that it is not possible to match a job to each applicant by considering applicants A, B, and D, who are qualified for jobs J_2 and J_5. If A and B are assigned a job, none remains for D. Therefore, no assignments exist for A, B, C, and D. □

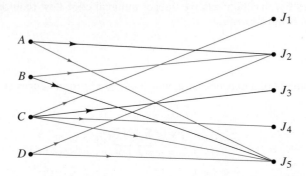

FIGURE 8.4.1   Applicants (A, B, C, D) and jobs $(J_1, J_2, J_3, J_4, J_5)$. An edge connects an applicant to a job for which the applicant is qualified. The black lines show a maximal matching (i.e., the maximum number of applicants have jobs).

In Example 8.4.1 a matching consists of finding jobs for qualified persons. A maximal matching finds jobs for the maximum number of persons. A maximal matching for the graph of Figure 8.4.1 is shown with black lines. A complete matching finds jobs for everyone. We showed that the graph of Figure 8.4.1 has no complete matching. The formal definitions follow.

DEFINITION 8.4.2

Let G be a directed, bipartite graph with disjoint vertex sets V and W in which the edges are directed from vertices in V to vertices in W. (Any vertex in G is either in V or in W.) A *matching* for G is a set of edges E with no vertices in common. A *maximal matching* for G is a matching E in which E contains the maximum number of edges. A *complete matching* for G is a matching E having the property that if $v \in V$, then $(v, w) \in E$ for some $w \in W$.

EXAMPLE 8.4.3

The matching for the graph of Figure 8.4.2, shown with black lines, is a maximal matching and a complete matching.   □

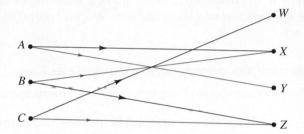

FIGURE 8.4.2   The black lines show a maximal matching (the maximum number of edges are used) and a complete matching (each of A, B, and C is matched).

In the next example we illustrate how the matching problem can be modeled as a network problem.

EXAMPLE 8.4.4   *A Matching Network*

Model the matching problem of Example 8.4.1 as a network.

We first assign each edge in the graph of Figure 8.4.1 capacity 1 (see Figure 8.4.3). Next we add a supersource a and edges of capacity 1 from a to each of A, B, C, and D. Finally, we introduce a supersink z and edges of capacity 1 from each of J_1, J_2, J_3, J_4, and J_5 to z. We call a network such as that of Figure 8.4.3 a **matching network**.

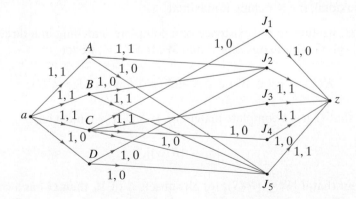

FIGURE 8.4.3   The matching problem (Figure 8.4.1) as a matching network.   □

The next theorem relates matching networks and flows.

THEOREM 8.4.5

Let G be a directed, bipartite graph with disjoint vertex sets V and W in which the edges are directed from vertices in V to vertices in W. (Any vertex in G is either in V or in W.)

(a) *A flow in the matching network gives a matching in G. The vertex $v \in V$ is matched with the vertex $w \in W$ if and only if the flow in edge (v, w) is 1.*

(b) *A maximal flow corresponds to a maximal matching.*

(c) *A flow whose value is $|V|$ corresponds to a complete matching.*

Proof. Let a (z) represent the source (sink) in the matching network and suppose that a flow is given.

Suppose that the edge (v, w), $v \in V$, $w \in W$, has flow 1. The only edge into vertex v is (a, v). This edge must have flow 1; thus the flow into vertex v is 1. Since the flow out of v is also 1, the only edge of the form (v, x) having flow 1 is (v, w). Similarly, the only edge of the form (x, w) having flow 1 is (v, w). Therefore, if E is the set of edges of the form (v, w) having flow 1, the members of E have no vertices in common; thus E is a matching for G.

Parts (b) and (c) follow from the fact that the number of vertices in V matched is equal to the value of the corresponding flow. ■

Since a maximal flow gives a maximal matching, Algorithm 8.2.4 applied to a matching network produces a maximal matching. In practice, the implementation of Algorithm 8.2.4 can be simplified by using the adjacency matrix of the graph (see Exercise 11).

EXAMPLE 8.4.6

The matching of Figure 8.4.1 is represented as a flow in Figure 8.4.3. Since the flow is maximal, the matching is maximal. □

Next, we turn to the existence of a complete matching in a directed, bipartite graph G with vertex sets V and W. If $S \subseteq V$, we let

$$R(S) = \{w \in W \mid v \in S \text{ and } (v, w) \text{ is an edge in } G\}.$$

Suppose that G has a complete matching. If $S \subseteq V$ we must have

$$|S| \leq |R(S)|.$$

It turns out that if $|S| \leq |R(S)|$ for all subsets S of V, then G has a complete matching. This result was first given by the English mathematician Philip Hall and is known as **Hall's Marriage Theorem**, since if V is a set of men and W is a set of women and edges exist from $v \in V$ to $w \in W$ if v and w are compatible, the theorem gives a condition under which each man can marry a compatible woman.

| THEOREM 8.4.7 | *Hall's Marriage Theorem* |

Let G be a directed, bipartite graph with disjoint vertex sets V and W in which the edges are directed from vertices in V to vertices in W. (Any vertex in G is either in V or in W.) There exists a complete matching in G if and only if

$$|S| \leq |R(S)| \qquad \text{for all } S \subseteq V. \tag{8.4.1}$$

Proof. We have already pointed out that if there is a complete matching in G, condition (8.4.1) holds.

Suppose that condition (8.4.1) holds. Let $n = |V|$ and let $(P, \overline{P})$ be a minimal cut in the matching network. If we can show that the capacity of this cut is n, a maximal flow would have value n. The matching corresponding to this maximal flow would be a complete matching.

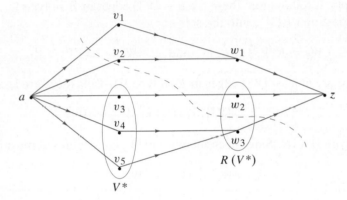

FIGURE 8.4.4   The proof of Theorem 8.4.7. $V = \{v_1, v_2, v_3, v_4, v_5\}$; $n = |V| = 5$; $W = \{w_1, w_2, w_3\}$; the cut is $(P, \overline{P})$, where $P = \{a, v_3, v_4, v_5, w_3\}$; $V^* = V \cap P = \{v_3, v_4, v_5\}$; $R(V^*) = \{w_2, w_3\}$; $W_1 = R(V^*) \cap P = \{w_3\}$; $W_2 = R(V^*) \cap \overline{P} = \{w_2\}$; $E = \{(a, v_1), (a, v_2), (v_3, w_2), (w_3, z)\}$. The capacity of the cut is $|E| = 4 < n$. The type I edges are (a, v_1) and (a, v_2). Edge (v_3, w_2) is the only type II edge, and edge (w_3, z) is the only type III edge.

The argument is by contradiction. Assume that the capacity of the minimal cut $(P, \overline{P})$ is less than n. The capacity of this cut is the number of edges in the set

$$E = \{(x, y) \mid x \in P, y \in \overline{P}\}$$

(see Figure 8.4.4). A member of E is one of the three types

Type I: $(a, v), v \in V$.
Type II: $(v, w), v \in V, w \in W$.
Type III: $(w, z), w \in W$.

We will estimate the number of edges of each type.

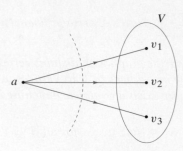

FIGURE 8.4.5   The proof of Theorem 8.4.7 for $n = 3$. If $V \subseteq \overline{P}$, as shown the capacity of the cut is n. Since we are assuming that the capacity of the cut is less than n, this case cannot occur. Therefore, $V \cap P$ is nonempty.

If $V \subseteq \overline{P}$, the capacity of the cut is n (see Figure 8.4.5); thus

$$V^* = V \cap P$$

is nonempty. It follows that there are $n - |V^*|$ edges in E of type I.

We partition $R(V^*)$ into the sets

$$W_1 = R(V^*) \cap P \qquad \text{and} \qquad W_2 = R(V^*) \cap \overline{P}.$$

Then there are at least $|W_1|$ edges in E of type III. Thus there are less than

$$n - (n - |V^*|) - |W_1| = |V^*| - |W_1|$$

edges of type II in E. Since each member of W_2 contributes at most one type II edge,

$$|W_2| < |V^*| - |W_1|.$$

Thus

$$|R(V^*)| = |W_1| + |W_2| < |V^*|,$$

which contradicts (8.4.1). Therefore, a complete matching exists.   ■

EXAMPLE 8.4.8

For the graph in Figure 8.4.1, if $S = \{A, B, D\}$, we have $R(S) = \{J_2, J_5\}$ and

$$|S| = 3 > 2 = |R(S)|.$$

By Theorem 8.4.7, there is not a complete matching for the graph of Figure 8.4.1.   □

EXAMPLE 8.4.9

There are n computers and n disk drives. Each computer is compatible with $m > 0$ disk drives and each disk drive is compatible with m computers. Is it possible to match each computer with a compatible disk drive?

Let V be the set of computers and W be the set of disk drives. An edge exists from $v \in V$ to $w \in W$ if v and w are compatible. Notice that every vertex has degree m. Let $S = \{v_1, \ldots, v_k\}$ be a subset of V. Then there are km edges from the set S. If $R(S) = \{w_1, \ldots, w_j\}$, then $R(S)$ receives at most jm edges from S. Therefore,

$$km \leq jm.$$

Now

$$|S| = k \leq j = |R(S)|.$$

By Theorem 8.4.7 there is a complete matching. Thus it is possible to match each computer with a compatible disk drive. □

Exercises

1. Show that the flow in Figure 8.4.3 is maximal by exhibiting a minimal cut whose capacity is 3.

2. Find the flow that corresponds to the matching of Figure 8.4.2. Show that this flow is maximal by exhibiting a minimal cut whose capacity is 3.

3. Applicant A is qualified for jobs J_1 and J_4; B is qualified for jobs J_2, J_3, and J_6; C is qualified for jobs J_1, J_3, J_5, and J_6; D is qualified for jobs J_1, J_3, and J_4; and E is qualified for jobs J_1, J_3, and J_6.

   (a) Model this situation as a matching network.

   (b) Use Algorithm 8.2.4 to find a maximal matching.

   (c) Is there a complete matching?

4. Applicant A is qualified for jobs J_1, J_2, J_4, and J_5; B is qualified for jobs J_1, J_4, and J_5; C is qualified for jobs J_1, J_4, and J_5; D is qualified for jobs J_1 and J_5; E is qualified for jobs J_2, J_3, and J_5; and F is qualified for jobs J_4 and J_5. Answer parts (a)–(c) of Exercise 3 for this situation.

5. Applicant A is qualified for jobs J_1, J_2, and J_4; B is qualified for jobs J_3, J_4, J_5, and J_6; C is qualified for jobs J_1 and J_5; D is qualified for jobs J_1, J_3, J_4, and J_8; E is qualified for jobs J_1, J_2, J_4, J_6, and J_8; F is qualified for jobs J_4 and J_6; and G is qualified for jobs J_3, J_5, and J_7. Answer parts (a)–(c) of Exercise 3 for this situation.

6. Five students, V, W, X, Y, and Z, are members of four committees, C_1, C_2, C_3, and C_4. The members of C are V, X, and Y; the members of C_2 are X and Z; the members of C_3 are V, Y, and Z; and the members of C_4 are V, W, X, and Z. Each committee is to send a representative to the administration. No student can represent two committees.

   (a) Model this situation as a matching network.

   (b) What is the interpretation of a maximal matching?

   (c) What is the interpretation of a complete matching?

   (d) Use Algorithm 8.2.4 to find a maximal matching.

   (e) Is there a complete matching?

7. Show that by suitably ordering the vertices, the adjacency matrix of a bipartite graph can be written

$$\begin{pmatrix} 0 & A \\ A^T & 0 \end{pmatrix},$$

where 0 is a matrix consisting only of 0's and A^T is the transpose of the matrix A.

In Exercises 8–10, G is a bipartite graph, A is the matrix of Exercise 7, and F is a flow in the associated matching network. Label each entry in A that represents an edge with flow 1.

8. What kind of labeling corresponds to a matching?

9. What kind of labeling corresponds to a complete matching?

10. What kind of labeling corresponds to a maximal matching?

11. Restate Algorithm 8.2.4, applied to a matching network, in terms of operations on the matrix A of Exercise 7.

Let G be a directed, bipartite graph with disjoint vertex sets V and W in which the edges are directed from vertices in V to vertices in W. (Any vertex in G is either in V or in W.) We define the *deficiency of G* as

$$\delta(G) = \max\{|S| - |R(S)| \mid S \subseteq V\}.$$

12. Show that G has a complete matching if and only if $\delta(G) = 0$.

★ **13.** Show that the maximum number of vertices in V that can be matched with vertices in W is $|V| - \delta(G)$.

14. True or false? Any matching is contained in a maximal matching. If true, prove it; if false, give a counterexample.

PROBLEM-SOLVING CORNER:

MATCHING

Problem

Let G be a directed, bipartite graph with disjoint vertex sets V and W in which the edges are directed from vertices in V to vertices in W. (Any vertex in G is either in V or in W.) Let M_W denote the maximum degree that occurs among vertices in W, and let m_V denote the minimum degree that occurs among vertices in V. Show that if $0 < M_W \le m_V$, then G has a complete matching.

Attacking the Problem

Hall's Marriage Theorem (Theorem 8.4.7) says that a directed, bipartite graph with disjoint vertex sets V and W has a complete matching if and only if $|S| \le |R(S)|$ for all $S \subseteq V$. Thus a possible way to solve the problem is to show that the given condition $M_W \le m_V$ implies that $|S| \le |R(S)|$ for all $S \subseteq V$.

Finding a Solution

Our goal is to prove that if $M_W \le m_V$, then $|S| \le |R(S)|$ for all $S \subseteq V$. Let's start with an example graph G for which $M_W \le m_V$; in fact, let's arrange to have $M_W = 2$ and $m_V = 3$:

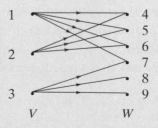

Consider an example subset $S = \{1, 3\}$ of V and the edges incident on vertices in S:

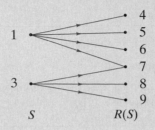

$$S \qquad\qquad R(S)$$

The fact that the minimum degree of vertices in V is 3 means that for *any* subset S of V, *at least* three edges are incident on each vertex in S. In general there will be at least $3|S| = m_V|S|$ edges incident on vertices in S. In our example, $3|S| = 6$, but there are actually seven edges incident on vertices in S. The expression $m_V|S|$ is always a *lower bound* for the number of edges incident on vertices in S.

The fact that the maximum degree of vertices in W is 2 means that for *any* subset S of V, *at most* two edges are incident on each vertex in $R(S)$. In general there will be at most $2|R(S)| = M_W|R(S)|$ edges incident on vertices in $R(S)$. In our example, $2|R(S)| = 12$, but there are actually 10 edges incident on vertices in $R(S)$. Since the edges incident on vertices in S are a subset of the edges incident on vertices in $R(S)$, the expression $M_W|R(S)|$ is always an *upper bound* for the number of edges incident on vertices in S.

We have two ways of estimating the number of edges incident on vertices in S. The first way, using S, gives us a lower bound $m_V|S|$ on the number of edges, and the second way, using $R(S)$, gives us an upper bound $M_W|R(S)|$ on the number of edges. Comparing these estimates gives us the inequality

$$m_V|S| \leq M_W|R(S)|.$$

We can't deduce $|S| \leq |R(S)|$, but we haven't used the hypothesis

$$M_W \leq m_V$$

yet! Combining the last two inequalities, we have

$$m_V|S| \leq M_W|R(S)| \leq m_V|R(S)|.$$

If we now cancel m_V from both ends of the inequality, we obtain

$$|S| \leq |R(S)|,$$

which is exactly the inequality we wanted to prove.

Formal Solution

Let $S \subseteq V$. Each vertex in S is incident on at least $m_V|S|$ edges; thus, there are at least $m_V|S|$ edges incident on vertices in S. Each vertex in $R(S)$ is incident on at most M_W edges; thus, there are at most $M_W|R(S)|$ edges incident on vertices in $R(S)$. It follows that $m_V|S| \leq M_W|R(S)|$. Since $M_W \leq m_V$, $|R(S)|M_W \leq |R(S)|m_V$. Therefore $m_V|S| \leq m_V|R(S)|$, and $|S| \leq |R(S)|$. By Theorem 8.4.7, G has a complete matching.

Summary of Problem-Solving Techniques

- Look at example graphs.
- When looking at examples, it's a good idea to assign distinct values to the parameters in the problem so you can keep them straight. (In our example we set $M_W = 2$ and $m_V = 3$.)
- Try to reduce given conditions to those in a useful theorem. (We reduced the conditions given in this problem to the conditions given in Hall's Marriage Theorem.)
- An inequality can sometimes be proved by estimating the size of some set in two different ways. If one estimate gives an upper bound M and another gives a lower bound m, it follows that $m \leq M$.

Comments

The last summarized problem-solving technique provides a method of proving an inequality. In a similar way, an equality can sometimes be proved by counting the number of elements in some set in two different ways. If one way of counting gives c_1 and the other way of counting gives c_2, it follows that $c_1 = c_2$. These techniques are widely used and their usefulness cannot be overemphasized. For example, in Section 4.2 we derived a formula for $C(n, r)$ by counting the number of r-permutations of an n-element set in two different ways.

ৡ৵ ৡ৵ ৡ৵

EXERCISE

1. Give an example of a bipartite graph G that has a complete matching but does not satisfy the condition $M_W \leq m_V$.

$A = 1$

$B = 2$

$C = 3$

$A = A+1$

$C = B+C$

$B = A+C$

FIGURE 8.5.1
A computer program.

8.5 PETRI NETS

Consider the computer program shown in Figure 8.5.1. Normally, the instructions would be processed sequentially—first, $A = 1$, then $B = 2$, and so on. However, notice that there is no logical reason that prevents the first three instructions—$A = 1$; $B = 2$; $C = 3$—from being processed in any order or concurrently. With the continuing decline of the cost of computer hardware, and processors in particular, there is increasing interest in concurrent processing to achieve greater speed and efficiency. The use of **Petri nets**, graph models of concurrent processing, is one method of modeling and studying concurrent processing.

DEFINITION 8.5.1

A *Petri net* is a directed graph $G = (V, E)$, where $V = P \cup T$ and $P \cap T = \emptyset$. Any edge e in E is incident on one member of P and one member of T. The set P is called the set of *places* and the set T is called the set of *transitions*.

Less formally, a Petri net is a directed, bipartite graph where the two classes of vertices are called places and transitions. In general, parallel edges are allowed in Petri nets; however, for simplicity, we will not permit parallel edges.

EXAMPLE 8.5.2

An example of a Petri net is given in Figure 8.5.2. Places are typically drawn as circles and transitions as bars or rectangular boxes.

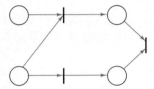

FIGURE 8.5.2   A Petri net. Circles are *places* and bars are *transitions*.   □

DEFINITION 8.5.3

A *marking* of a Petri net assigns each place a nonnegative integer. A Petri net with a marking is called a *marked Petri net* (or sometimes just a Petri net).

If a marking assigns the nonnegative integer n to place p, we say that there are n *tokens* on p. The tokens are represented as black dots.

EXAMPLE 8.5.4

An example of a marked Petri net is given in Figure 8.5.3.

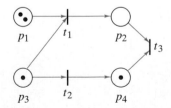

FIGURE 8.5.3   A marked Petri net.   □

In modeling, the places represent **conditions**, the transitions represent **events**, and the presence of at least one token in a place (condition) indicates that that condition is met.

EXAMPLE 8.5.5   *Petri Net Model of a Computer Program*

In Figure 8.5.4 we have modeled the computer program of Figure 8.5.1. Here the events (transitions) are the instructions, and the places represent the conditions under which an instruction can be executed.   □

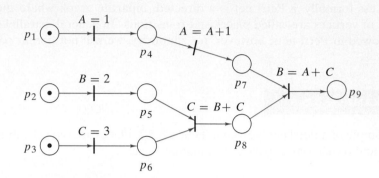

FIGURE 8.5.4   The program of Figure 8.5.1 as a Petri net. The tokens indicate that the conditions for executing $A = 1$, $B = 2$, and $C = 3$ are met.

DEFINITION 8.5.6

In a Petri net, if an edge is directed from place p to transition t, we say that p is an *input place* for transition t. An *output place* is defined similarly. If every input place for a transition t has at least one token, we say that t is *enabled*. A *firing* of an enabled transition removes one token from each input place and adds one token to each output place.

EXAMPLE 8.5.7

In the Petri net of Figure 8.5.3, places p_1 and p_3 are input places for transition t_1. Transitions t_1 and t_2 are enabled, but transition t_3 is not enabled. If we fire transition t_1, we obtain the marked Petri net of Figure 8.5.5. Transition t_3 is now enabled. If we then fire transition t_3, we obtain the net shown. At this point no transition is enabled and thus none may be fired.   □

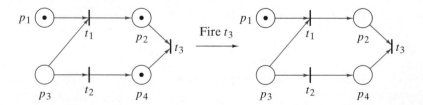

FIGURE 8.5.5   Firing transition t_3.

DEFINITION 8.5.8

If a sequence of firings transforms a marking M to a marking M', we say that M' is *reachable* from M.

EXAMPLE 8.5.9

In Figure 8.5.6, M'' is reachable from M by first firing transition t_1 and then firing t_2. □

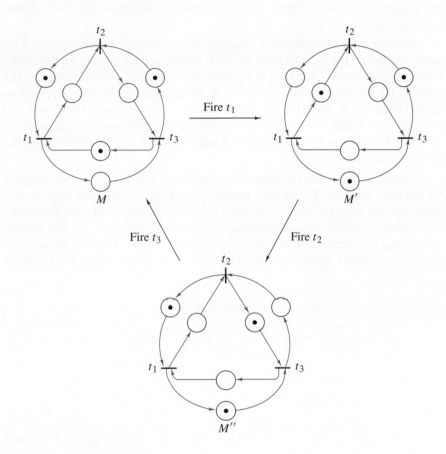

FIGURE 8.5.6   Marking M'' is reachable from M by firing t_1 and then t_2.

In modeling, the firing of a transition simulates the occurrence of that event. Of course, an event can take place only if all of the conditions for its execution have been met; that is, the transition can be fired only if it is enabled. By putting tokens in places p_1, p_2, and p_3 in Figure 8.5.4, we show that the conditions for executing the instructions $A = 1$, $B = 2$, and $C = 3$ are met. The program is ready to be executed. Since the transitions $A = 1$, $B = 2$, and $C = 3$ are enabled, they can be fired in any order or concurrently. Transition $C = B + C$ is enabled only if places p_5 and p_6 have tokens. But these places will have tokens only if transitions $B = 2$ and $C = 3$ have been fired. In other words, the condition under which the event $C = B + C$ can occur is that $B = 2$ and $C = 3$ must have been executed. In this way we model the legal execution sequences of Figure 8.5.1 and the implicit concurrency within this program.

Among the most important properties studied in Petri net theory are **liveness** and **safeness**. Liveness is related to the absence of deadlocks and safeness is related to bounded memory capacity.

EXAMPLE 8.5.10   *Petri Net Model of a Shared Computer System*

Two persons are sharing a computer system that has a disk drive D and a printer P. Each person needs both D and P. A possible Petri net model of this situation is shown in Figure 8.5.7. The marking indicates that both D and P are available.

Now suppose that person 1 requests D and then P (while person 2 requests neither). The occurrences of these events are simulated by first firing transition "request D" and then firing transition "request P" for person 1. The resulting Petri net is shown in Figure 8.5.8. When person 1 finishes the processing and releases D and P, simulated by firing the transitions "process" and then "release D and P," we return to the Petri net of Figure 8.5.7. If person 2 requests D and then P (while person 1 requests neither), we obtain a similar firing sequence.

Again, assume that we have the situation of Figure 8.5.7. Now suppose that person 1 requests D and then person 2 requests P. After the appropriate transitions are fired to simulate the occurrences of these events, we obtain the Petri net of Figure 8.5.9. Notice that at this point, no transition can fire. Person 1 is waiting for person 2 to release P and person 2 is waiting for person 1 to release D. Activity within the system stops. We say that a deadlock occurs.

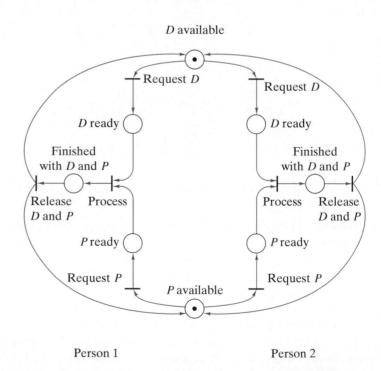

FIGURE 8.5.7   A Petri net model of a shared computer system. Each person needs disk drive D and printer P. The marking indicates that D and P are available.

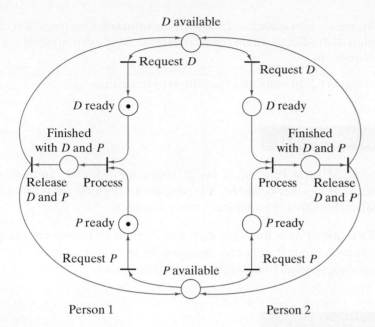

FIGURE 8.5.8 The Petri net of Figure 8.5.7 after firing "request D" and then "request P" for person 1. After person 1 finishes processing and releases D and P, simulated by firing "process" and then "release D and P," we obtain the Petri net of Figure 8.5.7 again.

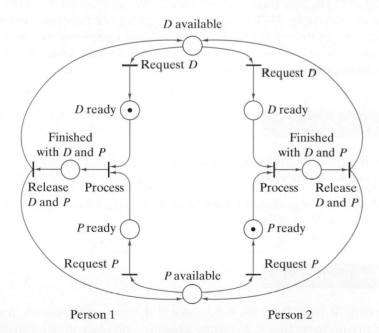

FIGURE 8.5.9 The Petri net of Figure 8.5.7 after firing "request D" for person 1 and "request P" for person 2. At this point the Petri net is deadlocked; that is, no transition can fire.

Formally, we say that a marked Petri net is **deadlocked** if no transition can fire. Prevention of deadlocks within concurrent processing environments is a major practical concern. □

Example 8.5.10 motivates the following definition.

DEFINITION 8.5.11

A marking M for a Petri net is *live* if, beginning from M, no matter what sequence of firings has occurred, it is possible to fire any given transition by proceeding through some additional firing sequence.

If a marking M is live for a Petri net P, then no matter what sequence of transitions is fired, P will never deadlock. Indeed, we can fire any transition by proceeding through some additional firing sequence.

EXAMPLE 8.5.12

The marking M of the net of Figure 8.5.6 is live. To see this, notice that the only transition for marking M that can be fired is t_1, which produces marking M'. The only transition for marking M' that can be fired is t_2, which produces marking M''. The only transition for marking M'' that can be fired is t_3, which returns us to marking M. Thus any firing sequence, starting with marking M, produces one of the markings M, M', or M'' and from there we can fire any transition t_1, t_2 or t_3 by proceeding as in Figure 8.5.6. Therefore, the marking M for the net of Figure 8.5.6 is live. □

EXAMPLE 8.5.13

The marking shown in Figure 8.5.4 is not live since after transition $A = 1$ is fired, it can never fire again. □

If a place is regarded as having limited capacity, **boundedness** assures us that no place will overflow.

DEFINITION 8.5.14

A marking M for a Petri net is *bounded* if there is some positive integer n having the property that in any firing sequence, no place ever receives more than n tokens. If a marking M is bounded and in any firing sequence no place ever receives more than one token, we call M a *safe* marking.

If each place represents a register capable of holding one computer word and if an initial marking is safe, we are guaranteed that the memory capacity of the registers will not be exceeded.

EXAMPLE 8.5.15

The markings of Figure 8.5.6 are safe. The marking M of Figure 8.5.10 is not safe, since as shown, if transition t_1 is fired, place p_2 then has two tokens. By listing all the markings reachable from M, it can be verified that M is bounded and live (see Exercise 7). $\qquad\qquad$ □

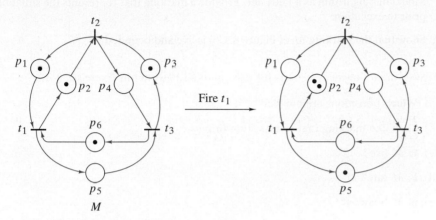

FIGURE 8.5.10   Marking M is not safe. After t_1 is fired, p_2 holds two tokens.

ॐ ॐ ॐ

Exercises

In Exercises 1–3, model each program by a Petri net. Provide a marking that represents the situation prior to execution of the program.

1. $A = 1$

   $B = 2$

   $C = A + B$

   $C = C + 1$

2. $A = 2$

   $B = A + A$

   $C = 3$

   $D = A + A$

   $C = A + B + C$

3. $A = 1$

      $S = 0$

  10 $S = S + A$

      $A = A + 1$

      GOTO 10

4. Describe three situations involving concurrency that might be modeled as Petri nets.

5. Give an example of a marked Petri net in which two transitions are enabled, but firing either one disables the other.

6. Consider the following algorithm for washing a lion.

  1. Get lion.

  2. Get soap.

  3. Get tub.

  4. Put water in tub.

  5. Put lion in tub.

6. Wash lion with soap.

7. Rinse lion.

8. Remove lion from tub.

9. Dry lion.

Model this algorithm as a Petri net. Provide a marking that represents the situation prior to execution.

7. Show that the marking M of Figure 8.5.10 is live and bounded.

Answer the following questions for each marked Petri net in Exercises 8–12.

(a) Which transitions are enabled?

(b) Show the marking that results from firing t_1.

(c) Is M live?

(d) Is M safe?

(e) Is M bounded?

(f) Show or describe all markings reachable from M.

(g) Exhibit a marking (other than the marking that puts zero tokens in each place) not reachable from M.

8.

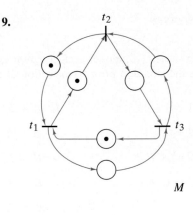

9.

10.

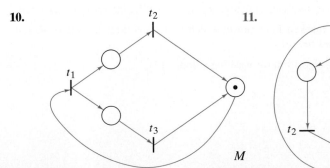

11.

★ **12.**

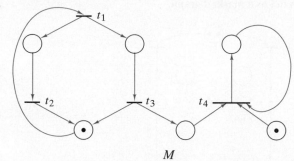

M

13. Give an example of a Petri net with a marking that is safe, but not live.

14. Give an example of a Petri net with a marking that is bounded, but not safe.

15. The **Dining Philosophers' Problem** (see [Dijkstra, 1968]) concerns five philosophers seated at a round table. Each philosopher either eats or meditates. The table is set alternately with one plate and one chopstick. Eating requires two chopsticks so that if each philosopher picks up the chopstick to the right of the plate, none can eat—the system will deadlock. Model this situation as a Petri net. Your model should be live so that the system will not deadlock and so that, at any point, any philosopher can potentially either eat or meditate.

16. Develop an alternative Petri net model for the situation of Example 8.5.10 that prevents deadlock.

If each place in a marked Petri net P has one incoming and one outgoing edge, then P can be redrawn as a directed graph where vertices correspond to transitions and edges to places. The tokens are placed on the edges. Such a graph is called a *marked graph*. Here we show a marked Petri net and its representation as a marked graph.

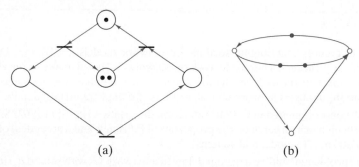

(a)   (b)

17. Which Petri nets in Exercises 8–17 can be redrawn as marked graphs?

18. Redraw the marked Petri net as a marked graph.

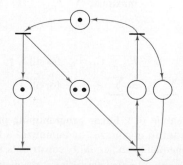

19. Redraw the marked Petri net as a marked graph.

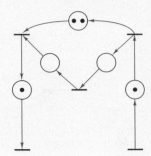

The *token count* of a simple directed cycle in a marked graph is the number of tokens on all the edges in the cycle.

20. Show that the token count of a simple directed cycle does not change during any firing sequence.

★ **21.** Show that a marking M for a marked graph G is live if and only if M places at least one token in each simple directed cycle in G.

★ **22.** Show that a live marking is safe for a marked graph G if and only if every edge in G belongs to a simple directed cycle with token count one.

23. Give an example of a marked graph with a nonlive marking in which every edge belongs to a simple directed cycle with token count one.

24. Let G be a marked graph. Show that each edge in G is contained in a simple directed cycle if and only if every marking for G is bounded.

★ **25.** Let G be a directed graph where, if we ignore the direction of the edges in G, G is connected as an undirected graph. Show that G has a live and safe marking if and only if given any two vertices v and w in G there is a directed path from v to w.

✺ NOTES

General references that contain sections on network models are [Berge; Deo; Liu, 1968, 1985; and Tucker]. The classic work on networks is [Ford]; many of the results on networks, especially the early results, are due to Ford and Fulkerson, the authors of this book. [Tarjan] discusses network flow algorithms and implementation details.

Petri nets originated in C. Petri's doctoral dissertation [Petri] in 1962. Since then there has been much research on the properties of Petri nets and a great deal of interest in using Petri nets to model real systems.

The problem of finding a maximal flow in a network G, with source a, sink z, and capacities C_{ij}, may be rephrased as follows:

$$\text{maximize} \sum_{j} F_{aj}$$

subject to

$$0 \le F_{ij} \le C_{ij} \qquad \text{for all } i, j;$$
$$\sum_{i} F_{ij} = \sum_{i} F_{ji} \qquad \text{for all } j.$$

Such a problem is an example of a **linear programming problem**. In a linear programming problem, we want to maximize (or minimize) a linear expression, such as $\sum_{j} F_{aj}$, subject to linear inequality and equality constraints, such as $0 \le F_{ij} \le C_{ij}$ and $\sum_{i} F_{ij} = \sum_{i} F_{ji}$. Although the **simplex algorithm** is normally an efficient way to solve a general linear programming problem, network transport problems are usually more efficiently solved using Algorithm 8.2.4. See [Hillier] for an exposition of the simplex algorithm.

Suppose that for each edge (i, j) in a network G, c_{ij} represents the cost of the flow of one unit through edge (i, j). Suppose that we want a maximal flow, with minimal cost

$$\sum_i \sum_j c_{ij} F_{ij}.$$

This problem, called the **transportation problem**, is again a linear programming problem and, as with the maximal flow problem, a specific algorithm can be used to obtain a solution that is, in general, more efficient than the simplex algorithm (see [Hillier]).

CHAPTER REVIEW

Section 8.1

(Transport) network
Source
Sink
Capacity
Flow in a network
Flow in an edge
Flow into a vertex
Flow out of a vertex
Conservation of flow
Given a flow F in a network, the flow out of the source equals the flow into the sink. This common value is called the value of the flow F.
Supersource
Supersink

Section 8.2

Maximal flow
Properly oriented edge with respect to a path
Improperly oriented edge with respect to a path
How to increase the flow in a path from the source to the sink when:

    (a) for each properly oriented edge the flow is less than the capacity and

    (b) each improperly oriented edge has positive flow (see Theorem 8.2.3)

How to find a maximal flow in a network (Algorithm 8.2.4)

Section 8.3

Cut in a network

Capacity of a cut
The capacity of any cut is greater than or equal to the value of any flow (Theorem 8.3.7).
Minimal cut
Max flow, min cut theorem (Theorem 8.3.9)
At the termination of the maximal flow algorithm, Algorithm 8.2.4, the set of labeled vertices defines a minimal cut.

Section 8.4

Matching
Maximal matching
Complete matching
Matching network
Relationship between flows and matchings (Theorem 8.4.5)
Hall's marriage theorem (Theorem 8.4.7)

Section 8.5

(Marked) Petri net
Place
Transition
Marking
Conditions and events
Input place
Output place
Enabled transition
Firing of an enabled transition
Reachable marking
Deadlock
Live marking
Bounded marking
Safe marking

CHAPTER SELF-TEST

Section 8.1

Exercises 1–4 refer to the following network. The capacities are shown on the edges.

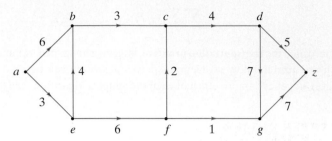

1. Explain why

$$F_{a,e} = 2, \quad F_{e,b} = 2, \quad F_{b,c} = 3, \quad F_{c,d} = 3, \quad F_{d,z} = 3, \quad F_{a,b} = 1,$$

with all other $F_{x,y} = 0$, is a flow.

2. What is the flow into b?

3. What is the flow out of c?

4. What is the value of the flow F?

Section 8.2

5. For the flow of Exercise 1, find a path from a to z satisfying: (a) for each properly oriented edge the flow is less than the capacity and (b) each improperly oriented edge has positive flow.

6. By modifying only the flows in the edges of the path of Exercise 5, find a flow with a larger value than F.

7. Use Algorithm 8.2.4 to find a maximal flow in the network of Exercise 1 (beginning with the flow in which the flow in each edge is equal to zero).

8. Use Algorithm 8.2.4 to find a maximal flow in the following network (beginning with the flow in which the flow in each edge is equal to zero).

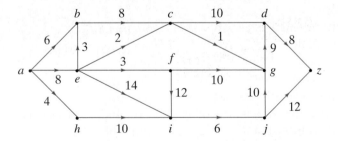

Section 8.3

9. In each of parts (a)–(d), answer true if the statement is true for every network; otherwise, answer false.

    (a) If the capacity of a cut in a network is equal to Ca, then the value of any flow is less than or equal to Ca.

    (b) If the capacity of a cut in a network is equal to Ca, then the value of any flow is greater than or equal to Ca.

(c) If the capacity of a cut in a network is equal to Ca, then the value of some flow is greater than or equal to Ca.

(d) If the capacity of a cut in a network is equal to Ca, then the value of some flow is less than or equal to Ca.

10. Find the capacity of the cut $(P, \bar{P})$ in the network of Exercise 1, where $P = \{a, b, e, f\}$.

11. Is the cut $(P, \bar{P})$, $P = \{a, b, e, f\}$, in the network of Exercise 1 minimal? Explain.

12. Find a minimal cut in the network of Exercise 8.

Section 8.4

Exercises 13–16 refer to the following situation. Applicant A is qualified for jobs J_2, J_4, and J_5; applicant B is qualified for jobs J_1 and J_3; applicant C is qualified for jobs J_1, J_3, and J_5; and applicant D is qualified for jobs J_3 and J_5.

13. Model the situation as a matching network.

14. Use Algorithm 8.2.4 to find a maximal matching.

15. Is there a complete matching?

16. Find a minimal cut in the matching network.

Section 8.5

Exercises 17–20 refer to the following Petri net.

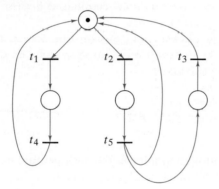

17. Which transitions are enabled?

18. Show the marking that results from firing t_2.

19. Is the marking safe?

20. Is the marking live?

9

BOOLEAN ALGEBRAS AND COMBINATORIAL CIRCUITS

He's contemptible, dishonest, selfish, deceitful, vicious— and yet he's out there, and I'm in here. He's called normal, and I'm not. Well, if that's normal, I don't want it.

—from Miracle on 34th Street

Several definitions honor the nineteenth-century mathematician George Boole—Boolean algebra, Boolean function, Boolean expression, and Boolean ring—to name a few. Boole is one of the persons in a long historical chain who were concerned with formalizing and mechanizing the process of logical thinking. In fact, in 1854 Boole wrote a book entitled *The Laws of Thought*. Boole's contribution was the development of a theory of logic using symbols instead of words. For a discussion of Boole's work, see [Hailperin].

Almost a century after Boole's work, it was observed, especially by C. E. Shannon in 1938 (see [Shannon]), that Boolean algebra could be used to analyze electrical circuits. Thus Boolean algebra became an indispensable tool for the analysis and design of electronic computers in the succeeding decades. We explore the relationship of Boolean algebra to circuits throughout this chapter.

9.1 COMBINATORIAL CIRCUITS

In a digital computer, there are only two possibilities, written 0 and 1, for the smallest, indivisible object. All programs and data are ultimately reducible to combinations of bits. A variety of devices have been used throughout the years in digital computers to store bits. Electronic circuits allow these storage devices to communicate with each other. A bit in one part of a circuit is transmitted to another part of the circuit as a voltage. Thus two voltage levels are needed—for example, a high voltage can communicate 1 and a low voltage can communicate 0.

In this section we discuss **combinatorial circuits**. The output of a combinatorial circuit is uniquely defined for every combination of inputs. A combinatorial circuit has no memory; previous inputs and the state of the system do not affect the output of a combinatorial circuit. Circuits for which the output is a function, not only of the inputs, but also of the state of the system, are called **sequential circuits** and are considered in Chapter 10.

Combinatorial circuits can be constructed using solid-state devices, called **gates**, which are capable of switching voltage levels (bits). We will begin by discussing AND, OR, and NOT gates.

DEFINITION 9.1.1

An *AND gate* receives inputs x_1 and x_2, where x_1 and x_2 are bits, and produces output denoted $x_1 \wedge x_2$, where

$$x_1 \wedge x_2 = \begin{cases} 1 & \text{if } x_1 = 1 \text{ and } x_2 = 1 \\ 0 & \text{otherwise.} \end{cases}$$

An AND gate is drawn as shown in Figure 9.1.1.

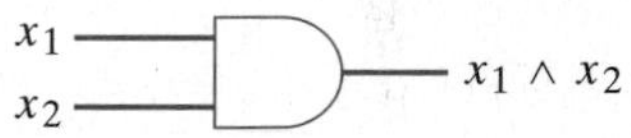

FIGURE 9.1.1
AND gate.

DEFINITION 9.1.2

An *OR gate* receives inputs x_1 and x_2, where x_1 and x_2 are bits, and produces output denoted $x_1 \vee x_2$, where

$$x_1 \vee x_2 = \begin{cases} 1 & \text{if } x_1 = 1 \text{ or } x_2 = 1 \\ 0 & \text{otherwise.} \end{cases}$$

An OR gate is drawn as shown in Figure 9.1.2.

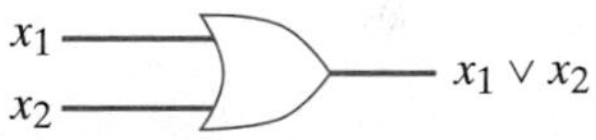

FIGURE 9.1.2
OR gate.

DEFINITION 9.1.3

A *NOT gate* (or *inverter*) receives input x, where x is a bit, and produces output denoted $\overline{x}$, where

$$\overline{x} = \begin{cases} 1 & \text{if } x = 0 \\ 0 & \text{if } x = 1. \end{cases}$$

A NOT gate is drawn as shown in Figure 9.1.3.

FIGURE 9.1.3
NOT gate.

The **logic table** of a combinatorial circuit lists all possible inputs together with the resulting outputs.

EXAMPLE 9.1.4

Following are the logic tables for the basic AND, OR, and NOT circuits (Figures 9.1.1–9.1.3).

x_1	x_2	$x_1 \wedge x_2$	x_1	x_2	$x_1 \vee x_2$	x	$\bar{x}$
1	1	1	1	1	1	1	0
1	0	0	1	0	1	0	1
0	1	0	0	1	1		
0	0	0	0	0	0		

We note that performing the operation AND (OR) is the same as taking the minimum (maximum) of the two bits x_1 and x_2.   □

EXAMPLE 9.1.5

The circuit of Figure 9.1.4 is an example of a combinatorial circuit since the output y is uniquely defined for each combination of inputs x_1, x_2, and x_3.

FIGURE 9.1.4   A combinatorial circuit.

The logic table for this combinatorial circuit follows.

x_1	x_2	x_3	y
1	1	1	0
1	1	0	0
1	0	1	0
1	0	0	1
0	1	1	0
0	1	0	1
0	0	1	0
0	0	0	1

Notice that all possible combinations of values for the inputs x_1, x_2, and x_3 are listed. For a given set of inputs, we can compute the value of the output y by tracing the flow through the circuit. For example, the fourth line of the table gives the value of the output y for the input values

$$x_1 = 1, \qquad x_2 = 0, \qquad x_3 = 0.$$

If $x_1 = 1$ and $x_2 = 0$, the output from the AND gate is 0 (see Figure 9.1.5). Since $x_3 = 0$, the inputs to the OR gate are both 0. Therefore, the output of the OR gate is 0. Since the input to the NOT gate is 0, it produces output $y = 1$.

□

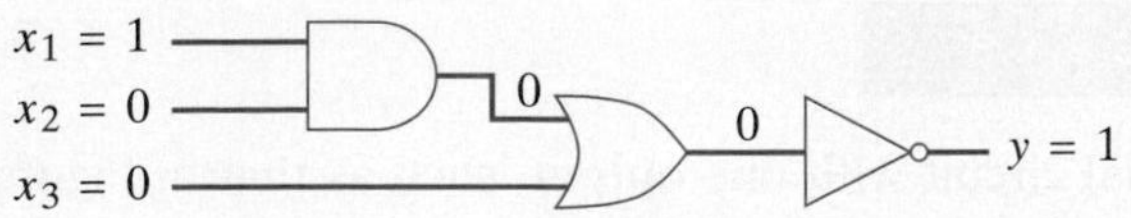

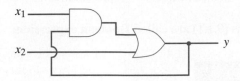

FIGURE 9.1.5   The circuit of Figure 9.1.4 when $x_1 = 1$ and $x_2 = x_3 = 0$.

EXAMPLE 9.1.6

The circuit of Figure 9.1.6 is not a combinatorial circuit because the output y is not uniquely defined for each combination of inputs x_1 and x_2. For example, suppose that $x_1 = 1$ and $x_2 = 0$. If the output of the AND gate is 0, then $y = 0$. On the other hand, if the output of the AND gate is 1, then $y = 1$. Such a circuit might be used to store one bit.

FIGURE 9.1.6   A circuit that is not a combinatorial circuit.   □

EXAMPLE 9.1.7

Individual combinatorial circuits may be interconnected. The combinatorial circuits C_1, C_2, and C_3 of Figure 9.1.7 may be combined, as shown, to obtain the combinatorial circuit C.

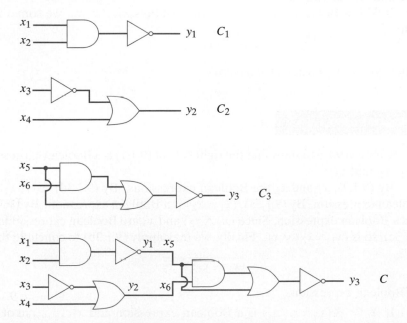

FIGURE 9.1.7   Combinatorial circuit C is obtained by interconnecting the combinatorial circuits C_1, C_2, and C_3.   □

EXAMPLE 9.1.8

A combinatorial circuit with one output, such as that in Figure 9.1.4, can be represented by an expression using the symbols $\wedge$, $\vee$, and $^-$. We follow the flow of the circuit symbolically. First, x_1 and x_2 are ANDed (see Figure 9.1.8), which produces output $x_1 \wedge x_2$. This output is then ORed with x_3 to produce output $(x_1 \wedge x_2) \vee x_3$. This output is then NOTed. Thus the output y may be

$$y = \overline{(x_1 \wedge x_2) \vee x_3}. \tag{9.1.1}$$

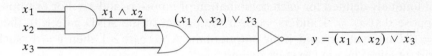

FIGURE 9.1.8   Boolean expression representation of a combinatorial circuit.

Expressions such as (9.1.1) are called **Boolean expressions**.   □

DEFINITION 9.1.9

Boolean expressions in the symbols $x_1, \ldots, x_n$ are defined recursively as follows.

$$0, \quad 1, \quad x_1, \ldots, \quad x_n \tag{9.1.2}$$

are Boolean expressions. If X_1 and X_2 are Boolean expressions, then

$$\text{(a) } (X_1), \qquad \text{(b) } \overline{X}_1, \qquad \text{(c) } X_1 \vee X_2, \qquad \text{(d) } X_1 \wedge X_2 \tag{9.1.3}$$

are Boolean expressions.

If X is a Boolean expression in the symbols $x_1, \ldots, x_n$, we sometimes write

$$X = X(x_1, \ldots, x_n).$$

Either symbol x or $\overline{x}$ is called a *literal*.

EXAMPLE 9.1.10

Use definition 9.1.9 to show that the right side of (9.1.1) is a Boolean expression in x_1, x_2, and x_3.

By (9.1.2), x_1 and x_2 are Boolean expressions. By (9.1.3d), $x_1 \wedge x_2$ is a Boolean expression. By (9.1.3a), $(x_1 \wedge x_2)$ is a Boolean expression. By (9.1.2), x_3 is a Boolean expression. Since $(x_1 \wedge x_2)$ and x_3 are Boolean expressions, by (9.1.3c), so is $(x_1 \wedge x_2) \vee x_3$. Finally, we may apply (9.1.3b) to conclude that

$$\overline{(x_1 \wedge x_2) \vee x_3}$$

is a Boolean expression.   □

If $X = X(x_1, \ldots, x_n)$ is a Boolean expression and $x_1, \ldots, x_n$ are assigned values $a_1, \ldots, a_n$ in $\{0, 1\}$, we may use Definitions 9.1.1–9.1.3 to compute a value for X. We denote this value $X(a_1, \ldots, a_n)$ or $X(x_i = a_i)$.

EXAMPLE 9.1.11

For $x_1 = 1$, $x_2 = 0$, and $x_3 = 0$, the Boolean expression $X(x_1, x_2, x_3) = \overline{(x_1 \wedge x_2) \vee x_3}$ of (9.1.1) becomes

$$
\begin{aligned}
X(1, 0, 0) &= \overline{(1 \wedge 0) \vee 0} \\
&= \overline{0 \vee 0} \qquad \text{since } 1 \wedge 0 = 0 \\
&= \overline{0} \qquad \text{since } 0 \vee 0 = 0 \\
&= 1 \qquad \text{since } \overline{0} = 1.
\end{aligned}
$$

We have again computed the fourth row of the table in Example 9.1.5.   □

In a Boolean expression in which parentheses are not used to specify the order of operations, we assume that $\wedge$ is evaluated before $\vee$.

EXAMPLE 9.1.12

For $x_1 = 0$, $x_2 = 0$, and $x_3 = 1$ the value of the Boolean expression $x_1 \wedge x_2 \vee x_3$ is

$$x_1 \wedge x_2 \vee x_3 = 0 \wedge 0 \vee 1 = 0 \vee 1 = 1.$$ □

Example 9.1.8 showed how to represent a combinatorial circuit with one output as a Boolean expression. The following example shows how to construct a combinatorial circuit that represents a Boolean expression.

EXAMPLE 9.1.13

Find the combinatorial circuit corresponding to the Boolean expression

$$(x_1 \wedge (\overline{x}_2 \vee x_3)) \vee x_2$$

and write the logic table for the circuit obtained.

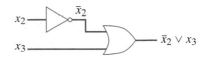

FIGURE 9.1.9
The combinatorial circuit corresponding to the Boolean expression $\overline{x}_2 \vee x_3$

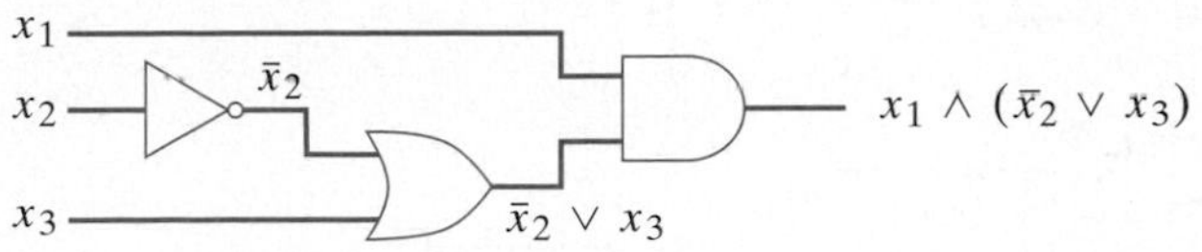

We begin with the expression $\overline{x}_2 \vee x_3$ in the innermost parentheses. This expression is converted to a combinatorial circuit, as shown in Figure 9.1.9. The output of this circuit is ANDed with x_1 to produce the circuit drawn in Figure 9.1.10. Finally, the output of this circuit is ORed with

FIGURE 9.1.10   The combinatorial circuit corresponding to the Boolean expression $x_1 \wedge (\overline{x}_2 \vee x_3)$.

x_2 to give the desired circuit drawn in Figure 9.1.11. The logic table is as follows:

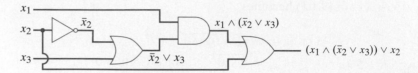

FIGURE 9.1.11   The combinatorial circuit corresponding to the Boolean expression $(x_1 \wedge (\overline{x}_2 \vee x_3)) \vee x_2$.

x_1	x_2	x_3	$(x_1 \wedge (\overline{x}_2 \vee x_3)) \vee x_2$
1	1	1	1
1	1	0	1
1	0	1	1
1	0	0	1
0	1	1	1
0	1	0	1
0	0	1	0
0	0	0	0

☐

ぺ ぺ ぺ

Exercises

In Exercises 1–6, write the Boolean expression that represents the combinatorial circuit, write the logic table, and write the output of each gate symbolically as in Figure 9.1.8.

1.

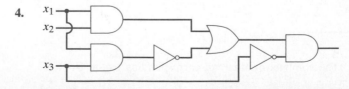

3.

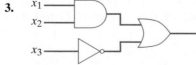

4.

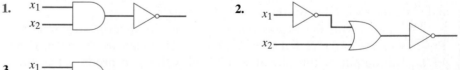

5.

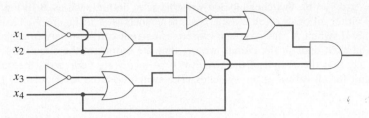

6. The circuit at the bottom of Figure 9.1.7.

Exercises 7–9 refer to the circuit

7. Show that this circuit is not a combinatorial circuit.

8. Show that if $x = 0$, the output y is uniquely determined.

9. Show that if $x = 1$, the output y is undetermined.

In Exercises 10–14, find the value of the Boolean expressions for

$$x_1 = 1, \quad x_2 = 1, \quad x_3 = 0, \quad x_4 = 1.$$

10. $\overline{x_1 \wedge x_2}$ **11.** $x_1 \vee (\overline{x}_2 \wedge x_3)$ **12.** $(x_1 \wedge \overline{x}_2) \vee (x_1 \vee \overline{x}_3)$

13. $(x_1 \wedge (x_2 \vee (x_1 \wedge \overline{x}_2))) \vee ((x_1 \wedge \overline{x}_2) \vee (x_1 \wedge \overline{x}_3))$

14. $(((x_1 \wedge x_2) \vee (x_3 \wedge \overline{x}_4)) \vee (\overline{(x_1 \vee x_3)} \wedge (\overline{x}_2 \vee x_3))) \vee (x_1 \wedge \overline{x}_3)$

15. Using Definition 9.1.9, show that each expression in Exercises 10–14 is a Boolean expression.

In Exercises 16–20, tell whether the given expression is a Boolean expression. If it is a Boolean expression, show that it is using Definition 9.1.9.

16. $x_1 \wedge (x_2 \vee x_3)$ **17.** $x_1 \wedge \overline{x}_2 \vee x_3$ **18.** (x_1)

19. $((x_1 \wedge x_2) \vee \overline{x}_3$ **20.** $((x_1))$

21. Find the combinatorial circuit corresponding to each Boolean expression in Exercises 10–14 and write the logic table.

A *switching circuit* is an electrical network consisting of switches each of which is open or closed. An example is given in Figure 9.1.12. If switch X is open (closed),

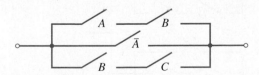

FIGURE 9.1.12   A switching circuit.

we write $X = 0$ ($X = 1$). Switches labeled with the same letter, such as B in Figure 9.1.12, are either all open or all closed. Switch X, such as A in Figure 9.1.12, is open if and only if switch $\overline{X}$, such as $\overline{A}$, is closed. If current can flow between the extreme left and right ends of the circuit, we say that the output of the circuit is 1; otherwise, we say that the output of the circuit is 0. A *switching table* gives the output of the circuit for all values of the switches. The switching table for Figure 9.1.12 is as follows:

A	B	C	Circuit Output
1	1	1	1
1	1	0	1
1	0	1	0
1	0	0	0
0	1	1	1
0	1	0	1
0	0	1	1
0	0	0	1

22. Draw a circuit with two switches A and B having the property that the circuit output is 1 precisely when both A and B are closed. This configuration is labeled $A \wedge B$ and is called a **series circuit.**

23. Draw a circuit with two switches A and B having the property that the circuit output is 1 precisely when either A or B is closed. This configuration is labeled $A \vee B$ and is called a **parallel circuit.**

24. Show that the circuit of Figure 9.1.12 can be represented symbolically as

$$(A \wedge B) \vee \overline{A} \vee (B \wedge C).$$

Represent each circuit in Exercises 25–29 symbolically and give its switching table.

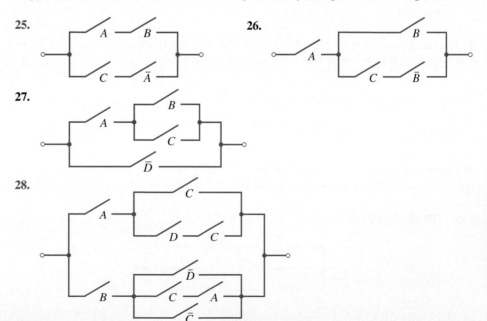

25.

26.

27.

28.

29.

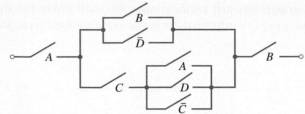

Represent the expressions in Exercises 30–34 as switching circuits and write the switching tables.

30. $(A \vee \overline{B}) \wedge A$ **31.** $A \vee (\overline{B} \wedge C)$ **32.** $(\overline{A} \wedge B) \vee (C \wedge A)$

33. $(A \wedge ((B \wedge \overline{C}) \vee (\overline{B} \wedge C))) \vee (\overline{A} \wedge B \wedge C)$

34. $A \wedge ((B \wedge C \wedge \overline{D}) \vee ((\overline{B} \wedge C) \vee D) \vee (\overline{B} \wedge \overline{C} \wedge D)) \wedge (B \vee \overline{D})$

9.2 PROPERTIES OF COMBINATORIAL CIRCUITS

In the preceding section we defined two binary operators $\wedge$ and $\vee$ on $Z_2 = \{0, 1\}$ and a unary operator $^{-}$ on Z_2. (Throughout the remainder of this chapter we let Z_2 denote the set $\{0, 1\}$.) We saw that these operators could be implemented in circuits as gates. In this section we discuss some properties of the system consisting of Z_2 and the operators $\wedge$, $\vee$, and $^{-}$.

> **THEOREM 9.2.1**
>
> If $\wedge$, $\vee$, and $^{-}$ are as in Definitions 9.1.1–9.1.3, then the following properties hold.
>
> (a) *Associative laws:*
>
> $$(a \vee b) \vee c = a \vee (b \vee c)$$
> $$(a \wedge b) \wedge c = a \wedge (b \wedge c) \qquad \text{for all } a, b, c \in Z_2.$$
>
> (b) *Commutative laws:*
>
> $$a \vee b = b \vee a, \qquad a \wedge b = b \wedge a \qquad \text{for all } a, b \in Z_2.$$
>
> (c) *Distributive laws:*
>
> $$a \wedge (b \vee c) = (a \wedge b) \vee (a \wedge c)$$
> $$a \vee (b \wedge c) = (a \vee b) \wedge (a \vee c) \qquad \text{for all } a, b, c \in Z_2.$$
>
> (d) *Identity laws:*
>
> $$a \vee 0 = a, \qquad a \wedge 1 = a \qquad \text{for all } a \in Z_2.$$
>
> (e) *Complement laws:*
>
> $$a \vee \overline{a} = 1, \qquad a \wedge \overline{a} = 0 \qquad \text{for all } a \in Z_2.$$

Proof. The proofs are straightforward verifications. We shall prove the first distributive law only and leave the other equations as exercises (see Exercises 16 and 17).

We must show that

$$a \wedge (b \vee c) = (a \wedge b) \vee (a \wedge c) \qquad \text{for all } a, b, c \in Z_2. \qquad (9.2.1)$$

We simply evaluate both sides if (9.2.1) for all possible values of a, b, and c in Z_2 and verify that in each case we obtain the same result. The table gives the details.

a	b	c	$a \wedge (b \vee c)$	$(a \wedge b) \vee (a \wedge c)$
1	1	1	1	1
1	1	0	1	1
1	0	1	1	1
1	0	0	0	0
0	1	1	0	0
0	1	0	0	0
0	0	1	0	0
0	0	0	0	0

■

EXAMPLE 9.2.2

By using Theorem 9.2.1, show that the combinatorial circuits of Figure 9.2.1 have identical outputs for given identical inputs.

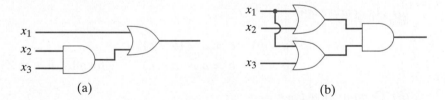

(a)   (b)

FIGURE 9.2.1   The combinatorial circuits (a) and (b) have identical outputs for given identical inputs and are said to be equivalent.

The Boolean expressions representing the circuits are, respectively,

$$x_1 \vee (x_2 \wedge x_3), \qquad (x_1 \vee x_2) \wedge (x_1 \vee x_3).$$

By Theorem 9.2.1c,

$$a \vee (b \wedge c) = (a \vee b) \wedge (a \vee c) \qquad \text{for all } a, b, c \in Z_2. \qquad (9.2.2)$$

But (9.2.2) says that the combinatorial circuits of Figure 9.2.1 have identical outputs for given identical inputs.   □

Arbitrary Boolean expressions are defined to be equal if they have the same values for all possible assignments of bits to the literals.

DEFINITION 9.2.3

Let

$$X_1 = X_1(x_1, \ldots, x_n) \qquad \text{and} \qquad X_2 = X_2(x_1, \ldots, x_n)$$

be Boolean expressions. We define X_1 to be *equal* to X_2 and write

$$X_1 = X_2$$

if

$$X_1(a_1, \ldots, a_n) = X_2(a_1, \ldots, a_n) \qquad \text{for all } a_i \in Z_2.$$

EXAMPLE 9.2.4

Show that

$$\overline{(x \vee y)} = \overline{x} \wedge \overline{y}. \tag{9.2.3}$$

According to Definition 9.2.3, (9.2.3) holds if the equation is true for all choices of x and y in Z_2. Thus we may simply construct a table listing all possibilities to verify (9.2.3).

x	y	$\overline{(x \vee y)}$	$\overline{x} \wedge \overline{y}$
1	1	0	0
1	0	0	0
0	1	0	0
0	0	1	1

□

If we define a relation R on a set of Boolean expressions by the rule $X_1 R X_2$ if $X_1 = X_2$, R is an equivalence relation. Each equivalence class consists of a set of Boolean expressions any one of which is equal to any other.

Because of the associative laws, Theorem 9.2.1a, we can unambiguously write

$$a_1 \vee a_2 \vee \cdots \vee a_n \tag{9.2.4}$$

or

$$a_1 \wedge a_2 \wedge \cdots \wedge a_n \tag{9.2.5}$$

for $a_i \in Z_2$. The combinatorial circuit corresponding to (9.2.4) is drawn as in Figure 9.2.2 and the combinatorial circuit corresponding to (9.2.5) is drawn as in Figure 9.2.3.

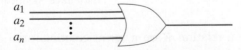

FIGURE 9.2.2   An *n*-input OR gate.

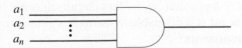

FIGURE 9.2.3   An *n*-input AND gate.

The properties listed in Theorem 9.2.1 hold for a variety of systems. Any system satisfying these properties is called a **Boolean algebra**. Abstract Boolean algebras are examined in Section 9.3.

Having defined equality of Boolean expressions, we define equivalence of combinatorial circuits.

DEFINITION 9.2.5

We say that two combinatorial circuits, each having inputs $x_1, \ldots, x_n$ and a single output, are *equivalent* if, whenever the circuits receive the same inputs, they produce the same outputs.

EXAMPLE 9.2.6

The combinatorial circuits of Figures 9.2.4 and 9.2.5 are equivalent since, as shown, they have identical logic tables.

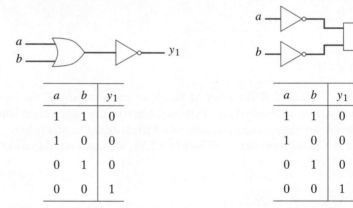

a	b	y_1
1	1	0
1	0	0
0	1	0
0	0	1

a	b	y_1
1	1	0
1	0	0
0	1	0
0	0	1

FIGURE 9.2.4   A combinatorial circuit and its logic table.

FIGURE 9.2.5   A combinatorial circuit and its logic table, which is identical to the logic table of Figure 9.2.4. The circuits of Figures 9.2.4 and 9.2.5 are said to be equivalent because they have identical logic tables.   □

If we define a relation R on a set of combinatorial circuits by the rule $C_1 R C_2$ if C_1 and C_2 are equivalent (in the sense of Definition 9.2.5), R is an equivalence relation. Each equivalence class consists of a set of mutually equivalent combinatorial circuits.

Example 9.2.6 shows that equivalent circuits may not have the same number of gates. In general, it is desirable to use as few gates as possible to minimize the cost of the components.

It follows immediately from the definitions that combinatorial circuits are equivalent if and only if the Boolean expressions that represent them are equal.

THEOREM 9.2.7

Let C_1 and C_2 be combinatorial circuits represented, respectively, by the Boolean expressions $X_1 = X_1(x_1, \ldots, x_n)$ and $X_2 = X_2(x_1, \ldots, x_n)$. Then C_1 and C_2 are equivalent if and only if $X_1 = X_2$.

Proof. The value $X_1(a_1, \ldots, a_n)$ [respectively, $X_2(a_1, \ldots, a_n)$] for $a_i \in Z_2$ is the output for circuit C_1 (respectively, C_2) for inputs $a_1, \ldots, a_n$.

According to Definition 9.2.5, circuits C_1 and C_2 are equivalent if and only if they have the same outputs $X_1(a_1, \ldots, a_n)$ and $X_2(a_1, \ldots, a_n)$ for all possible inputs $a_1, \ldots, a_n$. Thus circuits C_1 and C_2 are equivalent if and only if

$$X_1(a_1, \ldots, a_n) = X_2(a_1, \ldots, a_n) \qquad \text{for all values } a_i \in Z_2. \qquad (9.2.6)$$

But by Definition 9.2.3, (9.2.6) holds if and only if $X_1 = X_2$. ∎

EXAMPLE 9.2.8

In Example 9.2.4 we showed that

$$\overline{(x \vee y)} = \overline{x} \wedge \overline{y}.$$

By Theorem 9.2.7, the combinatorial circuits (Figures 9.2.4 and 9.2.5) corresponding to these expressions are equivalent. □

Exercises

Show that the combinatorial circuits of Exercises 1–5 are equivalent.

1.

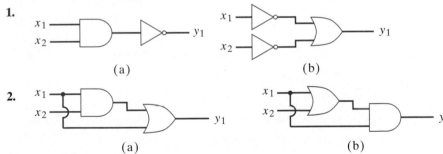

(a)   (b)

2.

(a)   (b)

3.

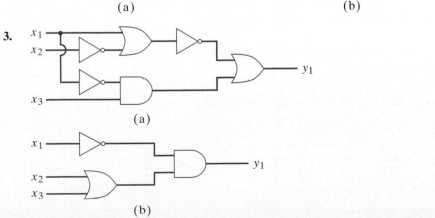

(a)

(b)

4.

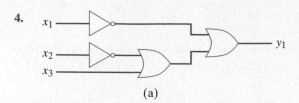

(a)

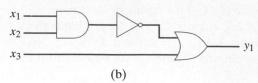

(b)

5.

(a)   (b)

Verify the equations in Exercises 6–10.

6. $x_1 \vee x_1 = x_1$

7. $x_1 \vee (x_1 \wedge x_2) = x_1$

8. $x_1 \wedge \overline{x}_2 = \overline{(\overline{x}_1 \vee x_2)}$

9. $x_1 \wedge \overline{(x_2 \wedge x_3)} = (x_1 \wedge \overline{x}_2) \vee (x_1 \wedge \overline{x}_3)$

10. $(x_1 \vee x_2) \wedge (x_3 \vee x_4) = (x_3 \wedge x_1) \vee (x_3 \wedge x_2) \vee (x_4 \wedge x_1) \vee (x_4 \wedge x_2)$

Prove or disprove the equations in Exercises 11–15.

11. $\overline{\overline{x}} = x$

12. $\overline{x}_1 \wedge \overline{x}_2 = x_1 \vee x_2$

13. $\overline{x}_1 \wedge ((x_2 \wedge x_3) \vee (x_1 \wedge x_2 \wedge x_3)) = x_2 \wedge x_3$

14. $\overline{((\overline{x}_1 \wedge x_2) \vee (x_1 \wedge \overline{x}_3))} = (x_1 \vee \overline{x}_2) \wedge (x_1 \vee \overline{x}_3)$

15. $(x_1 \vee x_2) \wedge (\overline{x}_3 \vee x_4) \wedge (x_3 \wedge \overline{x}_2) = 0$

16. Prove the second statement of Theorem 9.2.1c.

17. Prove Theorem 9.2.1, parts (a), (b), (d), and (e).

We say that two switching circuits are *equivalent* if the Boolean expressions that represent them are equal.

18. Show that the switching circuits are equivalent.

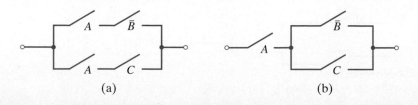

(a)   (b)

19. For each switching circuit in Exercises 25–29, Section 9.1, find an equivalent switching circuit using parallel and series circuits having as few switches as you can.

20. For each Boolean expression in Exercises 30–34, Section 9.1, find a switching circuit using parallel and series circuits having as few switches as you can.

A *bridge circuit* is a switching circuit, such as the one shown here, that uses nonparallel and nonseries circuits.

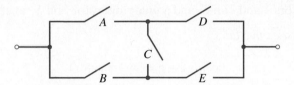

For each switching circuit, find an equivalent switching circuit using bridge circuits having as few switches as you can.

21.

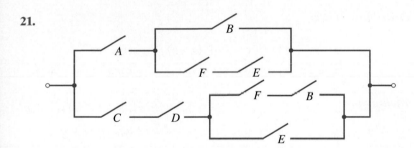

22.

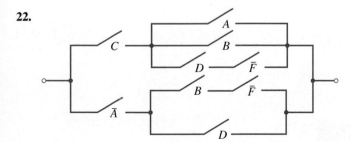

★ 23.

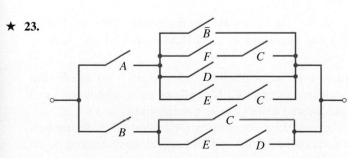

24. For each Boolean expression in Exercises 30–34, Section 9.1, find a switching circuit using bridge circuits having as few switches as you can.

9.3 *BOOLEAN ALGEBRAS*

In this section we consider general systems that have properties like those given in Theorem 9.2.1. We will see that apparently diverse systems obey these same laws. We call such systems Boolean algebras.

DEFINITION 9.3.1

A *Boolean algebra* B consists of a set S containing distinct elements 0 and 1, binary operators $+$ and $\cdot$ on S, and a unary operator $'$ on S satisfying

(a) *Associative laws:*

$$(x + y) + z = x + (y + z)$$
$$(x \cdot y) \cdot z = x \cdot (y \cdot z) \qquad \textit{for all } x, y, z \in S.$$

(b) *Commutative laws:*

$$x + y = y + x, \qquad x \cdot y = y \cdot x \qquad \textit{for all } x, y \in S.$$

(c) *Distributive laws:*

$$x \cdot (y + z) = (x \cdot y) + (x \cdot z)$$
$$x + (y \cdot z) = (x + y) \cdot (x + z) \qquad \textit{for all } x, y, z \in S.$$

(d) *Identity laws:*

$$x + 0 = x, \qquad x \cdot 1 = x \qquad \textit{for all } x \in S.$$

(e) *Complement laws:*

$$x + x' = 1, \qquad x \cdot x' = 0 \qquad \textit{for all } x \in S.$$

If B is a Boolean algebra, we write $B = (S, +, \cdot, ', 0, 1)$.

EXAMPLE 9.3.2

By Theorem 9.2.1, $(Z_2, \vee, \wedge, ^-, 0, 1)$ is a Boolean algebra. (We are letting Z_2 denote the set $\{0, 1\}$.) The operators $+, \cdot, '$ in Definition 9.3.1 are $\vee, \wedge, ^-$, respectively. □

As is the standard custom, we will usually abbreviate $a \cdot b$ as ab. We also assume that $\cdot$ is evaluated before $+$. This allows us to eliminate some parentheses. For example, we can write $(xy) + z$ more simply as $xy + z$.

Several comments are in order concerning Definition 9.3.1. In the first place, 0 and 1 are merely symbolic names and, in general, have nothing to do with the numbers 0 and 1. This same comment applies to $+$ and $\cdot$, which merely denote binary operators and, in general, have nothing to do with ordinary addition and multiplication.

EXAMPLE 9.3.3

Let U be a universal set and let $S = \mathcal{P}(U)$, the power set of U. If we define the following operations

$$X + Y = X \cup Y, \qquad X \cdot Y = X \cap Y, \qquad X' = \overline{X}$$

on S, then $(S, \cup, \cap, ^-, \emptyset, U)$ is a Boolean algebra. The empty set $\emptyset$ plays the role of 0 and the universal set U plays the role of 1. If we let X, Y, and Z be subsets of S, properties (a)–(e) of Definition 9.3.1 become the following properties of sets (see Theorem 2.1.8):

(a') $(X \cup Y) \cup Z = X \cup (Y \cup Z)$
  $(X \cap Y) \cap Z = X \cap (Y \cap Z)$ for all $X, Y, Z \in \mathcal{P}(U)$.
(b') $X \cup Y = Y \cup X, \qquad X \cap Y = Y \cap X$ for all $X, Y \in \mathcal{P}(U)$.
(c') $X \cap (Y \cup Z) = (X \cap Y) \cup (X \cap Z)$
  $X \cup (Y \cap Z) = (X \cup Y) \cap (X \cup Z)$ for all $X, Y, Z \in \mathcal{P}(U)$.
(d') $X \cup \emptyset = X, \qquad X \cap U = X$ for every $X \in \mathcal{P}(U)$.
(e') $X \cup \overline{X} = U, \qquad X \cap \overline{X} = \emptyset$ for every $X \in \mathcal{P}(U)$. $\square$

At this point we will deduce several other properties of Boolean algebras. We begin by showing that the element x' in Definition 9.3.1e is unique.

THEOREM 9.3.4

In a Boolean algebra, the element x' of Definition 9.3.1e is unique. Specifically, if $x + y = 1$ and $xy = 0$, then $y = x'$.

Proof.

$y = y1$	Definition 9.3.1d
$= y(x + x')$	Definition 9.3.1e
$= yx + yx'$	Definition 9.3.1c
$= xy + yx'$	Definition 9.3.1b
$= 0 + yx'$	Given
$= xx' + yx'$	Definition 9.3.1e
$= x'x + x'y$	Definition 9.3.1b
$= x'(x + y)$	Definition 9.3.1c
$= x'1$	Given
$= x'$	Definition 9.3.1d

∎

DEFINITION 9.3.5

In a Boolean algebra, we call the element x' the *complement* of x.

We can now derive several additional properties of Boolean algebras.

> ### THEOREM 9.3.6

Let $B = (S, +, \cdot, ', 0, 1)$ *be a Boolean algebra. The following properties hold.*

  (a) Idempotent laws:

$$x + x = x, \qquad xx = x \qquad\qquad \text{for all } x \in S.$$

  (b) Bound laws:

$$x + 1 = 1, \qquad x0 = 0 \qquad\qquad \text{for all } x \in S.$$

  (c) Absorption laws:

$$x + xy = x, \qquad x(x + y) = x \qquad\qquad \text{for all } x, y \in S.$$

  (d) Involution law:

$$(x')' = x \qquad\qquad \text{for all } x \in S.$$

  (e) 0 and 1 laws:
$$0' = 1, \qquad 1' = 0.$$

  (f) De Morgan's laws for Boolean algebras:

$$(x + y)' = x'y', \qquad (xy)' = x' + y', \qquad \text{for all } x, y \in S.$$

Proof. We will prove (b) and the first statement of parts (a), (c), and (f) and leave the others as exercises (see Exercises 18–20).

$$
\begin{aligned}
\text{(a)} \quad x &= x + 0 & &\text{Definition 9.3.1d} \\
&= x + (xx') & &\text{Definition 9.3.1e} \\
&= (x + x)(x + x') & &\text{Definition 9.3.1c} \\
&= (x + x)1 & &\text{Definition 9.3.1e} \\
&= x + x & &\text{Definition 9.3.1d} \\
\text{(b)} \quad x + 1 &= (x + 1)1 & &\text{Definition 9.3.1d} \\
&= (x + 1)(x + x') & &\text{Definition 9.3.1e} \\
&= x + 1x' & &\text{Definition 9.3.1c} \\
&= x + x'1 & &\text{Definition 9.3.1b} \\
&= x + x' & &\text{Definition 9.3.1d} \\
&= 1 & &\text{Definition 9.3.1e} \\
x0 &= x0 + 0 & &\text{Definition 9.3.1d} \\
&= x0 + xx' & &\text{Definition 9.3.1e} \\
&= x(0 + x') & &\text{Definition 9.3.1c} \\
&= x(x' + 0) & &\text{Definition 9.3.1b} \\
&= xx' & &\text{Definition 9.3.1d} \\
&= 0 & &\text{Definition 9.3.1e} \\
\text{(c)} \quad x + xy &= x1 + xy & &\text{Definition 9.3.1d} \\
&= x(1 + y) & &\text{Definition 9.3.1c} \\
&= x(y + 1) & &\text{Definition 9.3.1b} \\
&= x1 & &\text{Part (b)} \\
&= x & &\text{Definition 9.3.1d}
\end{aligned}
$$

(f) If we show that

$$(x + y)(x'y') = 0 \tag{9.3.1}$$

and

$$(x + y) + x'y' = 1, \tag{9.3.2}$$

it will follow from Theorem 9.3.4 that $x'y' = (x + y)'$. Now

$$
\begin{aligned}
(x + y)(x'y') &= (x'y')(x + y) & \text{Definition 9.3.1b} \\
&= (x'y')x + (x'y')y & \text{Definition 9.3.1c} \\
&= x(x'y') + (x'y')y & \text{Definition 9.3.1b} \\
&= (xx')y' + x'(y'y) & \text{Definition 9.3.1a} \\
&= (xx')y' + x'(yy') & \text{Definition 9.3.1b} \\
&= 0y' + x'0 & \text{Definition 9.3.1e} \\
&= y'0 + x'0 & \text{Definition 9.3.1b} \\
&= 0 + 0 & \text{Part (b)} \\
&= 0 & \text{Definition 9.3.1d}
\end{aligned}
$$

Therefore, (9.3.1) holds.

Next we verify (9.3.2).

$$
\begin{aligned}
(x + y) + x'y' &= ((x + y) + x')((x + y) + y') & \text{Definition 9.3.1c} \\
&= ((y + x) + x')((x + y) + y') & \text{Definition 9.3.1b} \\
&= (y + (x + x'))(x + (y + y')) & \text{Definition 9.3.1a} \\
&= (y + 1)(x + 1) & \text{Definition 9.3.1e} \\
&= 1 \cdot 1 & \text{Part (b)} \\
&= 1 & \text{Definition 9.3.1d}
\end{aligned}
$$

By Theorem 9.3.4, $x'y' = (x + y)'$.   ∎

EXAMPLE 9.3.7

As explained in Example 9.3.3, if U is a set, $\mathcal{P}(U)$ can be considered a Boolean algebra. Therefore, De Morgan's laws, which for sets may be stated

$$(\overline{X \cup Y}) = \overline{X} \cap \overline{Y}, \quad (\overline{X \cap Y}) = \overline{X} \cup \overline{Y} \qquad \text{for all } X, Y \in \mathcal{P}(U),$$

hold. These equations may be verified directly (see Theorem 2.1.8), but Theorem 9.3.6 shows that they are a consequence of other laws.   □

The reader has surely noticed that equations involving elements of a Boolean algebra come in pairs. For example, the identity laws (Definition 9.3.1d) are

$$x + 0 = x, \qquad x1 = x.$$

Such pairs are said to be **dual**.

DEFINITION 9.3.8

The *dual* of a statement involving Boolean expressions is obtained by replacing 0 by 1, 1 by 0, + by ·, and · by +.

EXAMPLE 9.3.9

The dual of

$$(x + y)' = x'y'$$

is

$$(xy)' = x' + y'. \qquad \square$$

Each condition in the definition of a Boolean algebra (Definition 9.3.1) includes its dual. Therefore, we have the following result.

THEOREM 9.3.10

The dual of a theorem about Boolean algebras is also a theorem.

Proof. Suppose that T is a theorem about Boolean algebras. Then there is a proof P of T involving only the definitions of a Boolean algebra (Definition 9.3.1). Let P' be the sequence of statements obtained by replacing every statement in P by its dual. Then P' is a proof of the dual of T. ■

EXAMPLE 9.3.11

The dual of

$$x + x = x \qquad (9.3.3)$$

is

$$xx = x. \qquad (9.3.4)$$

We proved (9.3.3) earlier (see the proof of Theorem 9.3.6a). If we write the dual of each statement in the proof of (9.3.3), we obtain the following proof of (9.3.4):

$$x = x1$$
$$= x(x + x')$$
$$= xx + xx'$$
$$= xx + 0$$
$$= xx. \qquad \square$$

EXAMPLE 9.3.12

The proofs given in Theorem 9.3.6 of the two statements of part (b) are dual to each other. $\square$

ॐ ॐ ॐ

Exercises

1. Verify properties (a′)–(e′) of Example 9.3.3.

2. Let $S = \{1, 2, 3, 6\}$. Define

$$x + y = \text{lcm}(x, y), \qquad x \cdot y = \text{gcd}(x, y), \qquad x' = \frac{6}{x}$$

for $x, y \in S$ (lcm and gcd denote, respectively, the least common multiple and the greatest common divisor.) Show that $(S, +, \cdot, ', 1, 6)$ is a Boolean algebra.

3. $S = \{1, 2, 4, 8\}$. Define $+$ and $\cdot$ as in Exercise 2 and define $x' = 8/x$. Show that $(S, +, \cdot, ', 1, 8)$ is not a Boolean algebra.

Let $S_n = \{1, 2, \ldots, n\}$. Define

$$x + y = \max\{x, y\}, \qquad x \cdot y = \min\{x, y\}.$$

4. Show that parts (a)–(c) of Definition 9.3.1 hold for S_n.

5. Show that it is possible to define 0,1 and $'$ so that $(S_n, +, \cdot, ', 0, 1)$ is a Boolean algebra if and only if $n = 2$.

6. Rewrite the conditions of Theorem 9.3.6 for sets as in Example 9.3.3.

7. Interpret Theorem 9.3.4 for sets as in Example 9.3.3.

Write the dual of each statement in Exercises 8–14.

8. $(x + y)(x + 1) = x + xy + y$

9. $(x' + y')' = xy$

10. If $x + y = x + z$ and $x' + y = x' + z$, then $y = z$.

11. $xy' = 0$ if and only if $xy = x$.

12. If $x + y = 0$, then $x = 0 = y$.

13. $x = 0$ if and only if $y = xy' + x'y$ for all y.

14. $x + x(y + 1) = x$

15. Prove the statements of Exercises 8–14.

16. Prove the duals of the statements of Exercises 8–14.

17. Write the dual of Theorem 9.3.4. How does the dual relate to Theorem 9.3.4 itself?

18. Prove the second statements of parts (a), (c), and (f) of Theorem 9.3.6.

19. Prove the second statements of parts (a), (c), and (f) of Theorem 9.3.6 by dualizing the proofs of the first statements given in the text.

20. Prove Theorem 9.3.6, parts (d) and (e).

★ 21. Deduce part (a) of Definition 9.3.1 from parts (b)–(e) of Definition 9.3.1.

22. Let U be the set of positive integers. Let S be the collection of subsets X of U with either X or $\overline{X}$ finite. Show that $(S, \cup, \cap, ^-, \emptyset, U)$ is a Boolean algebra.

★ 23. Let n be a positive integer. Let S be the set of all divisors of n, including 1 and n. Define $+$ and $\cdot$ as in Exercise 2 and define $x' = n/x$. What conditions must n satisfy so that $(S, +, \cdot, ', 1, n)$ is a Boolean algebra?

Problem

Let $(S, +, \cdot, {}', 0, 1)$ be a Boolean algebra and let A be subset of S. Show that $(A, +, \cdot, {}', 0, 1)$ is a Boolean algebra if and only if $1 \in A$ and $xy' \in A$ for all $x, y \in A$.

Attacking the Problem

Since the given statement is an "if and only if" statement, there are two statements to be proved:

> If $(A, +, \cdot, {}', 0, 1)$ is a Boolean algebra, then $1 \in A$ and $xy' \in A$
>   for all $x, y \in A$.   (1)
>
> If $1 \in A$ and $xy' \in A$ for all $x, y \in A$, then
> $(A, +, \cdot, {}', 0, 1)$ is a Boolean algebra.   (2)

To prove (1), we can use the laws as specified by the definition of "Boolean algebra" (Definition 9.3.1) and the laws derived in Theorem 9.3.6 that elements of a Boolean algebra must obey. To prove that $(A, +, \cdot, {}', 0, 1)$ is a Boolean algebra, we will verify that the laws specified by Definition 9.3.1 are satisfied. Before reading on, you should review Definition 9.3.1 and Theorem 9.3.6.

Finding a Solution

First let's try to prove (1). We assume that $(A, +, \cdot, {}', 0, 1)$ is a Boolean algebra and prove that

- $1 \in A$

and

- $xy' \in A$ for all $x, y \in A$.

Definition 9.3.1 says that a Boolean algebra contains 1. Since $(A, +, \cdot, {}', 0, 1)$ is a Boolean algebra, $1 \in A$.

Now suppose that $x, y \in A$. Definition 9.3.1 says that $'$ is a unary operator on A. This means that $y' \in A$. Definition 9.3.1 also says that $\cdot$ is a binary operator on A. This means that $xy' \in A$. This completes the proof of (1).

Now let's try to prove (2). This time we assume that $1 \in A$ and $xy' \in A$ for all $x, y \in A$ and try to prove that $(A, +, \cdot, {}', 0, 1)$ is a Boolean algebra. According to Definition 9.3.1, we must prove that

> A contains distinct elements 0 and 1.   (3)
>
> $+$ and $\cdot$ are binary operators on A.   (4)
>
> $'$ is a unary operator on A.   (5)
>
> The associative laws hold.   (6)

The commutative laws hold.   (7)

The distributive laws hold.   (8)

The identity laws hold.   (9)

The complement laws hold.   (10)

A contains 1 by assumption. To prove (3), we must show that $0 \in A$. We have only two assumptions about A: $1 \in A$ and if $x, y \in A$, then $xy' \in A$. All we can do at this point is combine these assumptions; that is, take $x = y = 1$ and examine the conclusion: $11' \in A$. Now Theorem 9.3.6e [applied to the Boolean algebra $(S, +, \cdot, ', 0, 1)$] says that $1' = 0$. Substituting for $1'$, we know now that $10 \in A$. But Theorem 9.3.6b says that for any x, $x0 = 0$. Thus $10 = 0$ is in A. Success! A contains 1 and 0. 0 and 1 are distinct because they are elements of the Boolean algebra $(S, +, \cdot, ', 0, 1)$. Therefore (3) is proved.

To prove (4), we must show that $+$ and $\cdot$ are binary operators on A; that is, if $x, y \in A$, then $x + y$ and xy are in A. Consider proving that $\cdot$ is a binary operator on A. What we know is that if $x, y \in A$, then $xy' \in A$, which is close to what we want to prove. If we could somehow replace y' by y in the expression xy', we could conclude that $xy \in A$. What we would like to do is assume that $x, y \in A$, then deduce

$$x, y' \in A, \tag{11}$$

and then conclude that

$$xy = xy'' \in A.$$

To deduce (11), we need to show if $y \in A$, then $y' \in A$. But this is (5). Detour! Let's work on (5).

We will assume that $y \in A$ and try to prove that $y' \in A$. If we could get rid of that pesky x (in the hypothesis $x, y \in A$ implies $xy' \in A$), we would have exactly what we want. We can effectively eliminate x by taking $x = 1$ since $1y = y$. Formally, we argue as follows. Let y be in A. Since $1 \in A$, $y' = 1y' \in A$. ($y' = 1y'$ by Definition 9.3.1b and 9.3.1d.) We have proved (5).

Now back to (4). Let $x, y \in A$. By the just proved (5), $y' \in A$. By the given condition, $xy = xy'' \in A$. ($y = y''$ by Theorem 9.3.6d.) We have proved that $\cdot$ is a binary operator on A.

De Morgan's laws (Theorem 9.3.6f), in effect, allow us to interchange $+$ and $\cdot$, so we can use them to prove that if $x, y \in A$, then $x + y \in A$. Formally we argue as follows. Suppose that $x, y \in A$. By (5), x' and y' are both in A. Since we have already proved that $\cdot$ is a binary operator on A, $x'y' \in A$. By (5), $(x'y')' \in A$. By De Morgan's laws (Theorem 9.3.6f) and Theorem 9.3.6d, $x + y = x'' + y'' = (x'y')' \in A$. Therefore $+$ is a binary operator on A. We have proved (4).

The next statement to prove is (6), which is to verify the associative laws

$$(x + y) + z = x + (y + z), \qquad (xy)z = x(yz) \qquad \text{for all } x, y, z \in A.$$

Now $(S, +, \cdot, ', 0, 1)$ is a Boolean algebra and so the associative laws hold in S. Since A is a subset of S, the associative laws surely hold in A. Thus (6) holds. For the same reason, properties (7) through (10) also hold in A. Therefore $(A, +, \cdot, ', 0, 1)$ is a Boolean algebra.

Formal Solution

Suppose that $(A, +, \cdot, ', 0, 1)$ is a Boolean algebra. Then $1 \in A$. Suppose that $x, y \in A$. Then $y' \in A$. Thus $xy' \in A$.

Now suppose that $1 \in A$ and $xy' \in A$ for all $x, y \in A$. Taking $x = y = 1$, we obtain $0 = 11' \in A$. Taking $x = 1$, we obtain $y' = 1y' \in A$. Thus $'$ is a unary operator on A. Replacing y by y', we obtain $xy = xy'' \in A$. Thus $\cdot$ is a binary operator on A. Now $x + y = x'' + y'' = (x'y')' \in A$. Thus $+$ is a binary operator on A. Parts a–e of Definition 9.3.1 automatically hold in A since they hold in S. Therefore $(A, +, \cdot, ', 0, 1)$ is a Boolean algebra.

Summary of Problem-Solving Techniques

- When trying to construct a proof, write out carefully what is assumed and what is to be proved.

- When trying to construct a proof, look at closely related definitions and theorems.

- To prove that something is a Boolean algebra, go directly to the definition (Definition 9.3.1).

- Consider proving statements in an order different from that given. In this problem, it was easier to prove statement (5) before proving statement (4).

- Try various substitutions for the variables in a universally quantified statement. (After all, "universally quantified" means that the statement holds true *for all* values.) By taking $x = y = 1$ in the statement

$$xy' \in A \qquad \text{for all } x, y \in A,$$

we were able to prove that $0 \in A$.

9.4 BOOLEAN FUNCTIONS AND SYNTHESIS OF CIRCUITS

A circuit is constructed to carry out a specified task. If we want to construct a combinatorial circuit, the problem can be given in terms of inputs and outputs. For example, suppose that we want to construct a combinatorial circuit to compute the **exclusive-OR** of x_1 and x_2. We can state the problem by listing the inputs and outputs that define the exclusive-OR. This is equivalent to giving the desired logic table.

TABLE 9.4.1
The Exclusive-OR

x_1	x_2	$x_1 \oplus x_2$
1	1	0
1	0	1
0	1	1
0	0	0

DEFINITION 9.4.1

The *exclusive-OR* of x_1 and x_2 written $x_1 \oplus x_2$ is defined by Table 9.4.1.

A logic table, with one output, is a function, The domain is the set of inputs and the range is the set of outputs. For the exclusive-OR function given in Table 9.4.1, the domain is the set

$$\{(1, 1), (1, 0), (0, 1), (0, 0)\}$$

and the range is the set

$$Z_2 = \{0, 1\}.$$

If we could develop a formula for the exclusive-OR function of the form

$$x_1 \oplus x_2 = X(x_1, x_2),$$

where X is a Boolean expression, we could solve the problem of constructing the combinatorial circuit. We could merely construct the circuit corresponding to X.

Functions that can be represented by Boolean expressions are called **Boolean functions.**

DEFINITION 9.4.2

Let $X(x_1, \ldots, x_n)$ be a Boolean expression. A function f of the form

$$f(x_1, \ldots, x_n) = X(x_1, \ldots x_n)$$

is called a *Boolean function.*

EXAMPLE 9.4.3

The function $f: Z_2^3 \to Z_2$ defined by

$$f(x_1, x_2, x_3) = x_1 \wedge (\overline{x}_2 \vee x_3)$$

is a Boolean function. The inputs and outputs are given in the following table.

x_1	x_2	x_3	$f(x_1, x_2, x_3)$
1	1	1	1
1	1	0	0
1	0	1	1
1	0	0	1
0	1	1	0
0	1	0	0
0	0	1	0
0	0	0	0

□

In the next example we show how an arbitrary function $f: Z_2^n \to Z_2$ can be realized as a Boolean function.

EXAMPLE 9.4.4

Show that the function f given by the table is a Boolean function.

x_1	x_2	x_3	$f(x_1, x_2, x_3)$
1	1	1	1
1	1	0	0
1	0	1	0
1	0	0	1
0	1	1	0
0	1	0	1
0	0	1	0
0	0	0	0

Consider the first row of the table and the combination

$$x_1 \wedge x_2 \wedge x_3. \tag{9.4.1}$$

Notice that if $x_1 = x_2 = x_3 = 1$, as indicated in the first row of the table, then (9.4.1) is 1. The values of x_i given by any other row of the table give (9.4.1) the value 0. Similarly, for the fourth row of the table we may construct the combination

$$x_1 \wedge \overline{x}_2 \wedge \overline{x}_3. \tag{9.4.2}$$

Expression (9.4.2) has the value 1 for the values of x_i given by the fourth row of the table, whereas the values of x_i given by any other row of the table give (9.4.2) the value 0.

The procedure is clear. We consider a row R of the table where the output is 1. We then form the combination $x_1 \wedge x_2 \wedge x_3$ and place a bar over each x_i whose value is 0 in row R. The combination formed is 1 if and only if the x_i have the values given in row R. Thus, for row 6, we obtain the combination

$$\overline{x}_1 \wedge x_2 \wedge \overline{x}_3. \tag{9.4.3}$$

Next, we OR the terms (9.4.1)–(9.4.3) to obtain the Boolean expression

$$(x_1 \wedge x_2 \wedge x_3) \vee (x_1 \wedge \overline{x}_2 \wedge \overline{x}_3) \vee (\overline{x}_1 \wedge x_2 \wedge \overline{x}_3). \tag{9.4.4}$$

We claim that $f(x_1, x_2, x_3)$ and (9.4.4) are equal. To verify this, first suppose that $x_1, x_2,$ and x_3 have values given by a row of the table for which $f(x_1, x_2, x_3) = 1$. Then one of (9.4.1)–(9.4.3) is 1, so the value of (9.4.4) is 1. On the other hand, if x_1, x_2, x_3 have values given by a row of the table for which $f(x_1, x_2, x_3) = 0$, all of (9.4.1)–(9.4.3) are 0, so the value of (9.4.4) is 0. Thus f and the Boolean expression (9.4.4) agree on Z_2^3; therefore,

$$f(x_1, x_2, x_3) = (x_1 \wedge x_2 \wedge x_3) \vee (x_1 \wedge \overline{x}_2 \wedge \overline{x}_3) \vee (\overline{x}_1 \wedge x_2 \wedge \overline{x}_3),$$

as claimed.   □

After one more definition, we will show that the method of Example 9.4.4 can be used to represent any function $f: Z_2^n \to Z_2$.

DEFINITION 9.4.5

A *minterm* in the symbols $x_1, \ldots, x_n$ is a Boolean expression of the form

$$y_1 \wedge y_2 \wedge \cdots \wedge y_n,$$

where each y_i is either x_i or $\overline{x}_i$.

THEOREM 9.4.6

If $f: Z_2^n \to Z_2$, then f is a Boolean function. If f is not identically zero, let $A_1, \ldots, A_k$ denote the elements A_i of Z_2^n for which $f(A_i) = 1$. For each $A_i = (a_1, \ldots, a_n)$, set

$$m_i = y_1 \wedge \cdots y_n,$$

where

$$y_j = \begin{cases} x_j & \text{if } a_j = 1 \\ \overline{x}_j & \text{if } a_j = 0. \end{cases}$$

Then

$$f(x_1, \ldots, x_n) = m_1 \vee m_2 \vee \cdots \vee m_k. \tag{9.4.5}$$

Proof. If $f(x_1, \ldots, x_n) = 0$ for all x_i, then f is a Boolean function, since 0 is a Boolean expression.

Suppose that f is not identically zero. Let $m_i(a_1, \ldots, a_n)$ denote the value obtained from m_i by replacing each x_j with a_j. It follows from the definition of m_i that

$$m_i(A) = \begin{cases} 1 & \text{if } A = A_i \\ 0 & \text{if } A \neq A_i. \end{cases}$$

Let $A \in Z_2^n$. If $A = A_i$ for some $i \in \{1, \ldots, k\}$, then $f(A) = 1, m_i(A) = 1$, and

$$m_1(A) \vee \cdots \vee m_k(A) = 1.$$

On the other hand, if $A \neq A_i$ for any $i \in \{1, \ldots, k\}$, then $f(A) = 0, m_i(A) = 0$ for $i = 1, \ldots, k$, and

$$m_1(A) \vee \cdots \vee m_k(A) = 0.$$

Therefore, (9.4.5) holds. ∎

DEFINITION 9.4.7

The representation (9.4.5) of a Boolean function $f: Z_2^n \to Z_2$ is called the *disjunctive normal form* of the function f.

EXAMPLE 9.4.8

Design a combinatorial circuit that computes the exclusive-OR of x_1 and x_2.

The logic table for the exclusive-OR function $x_1 \oplus x_2$ is given in Table 9.4.1. The disjunctive normal form of this function is

$$x_1 \oplus x_2 = (x_1 \wedge \overline{x}_2) \vee (\overline{x}_1 \wedge x_2). \tag{9.4.6}$$

The combinatorial circuit corresponding to (9.4.6) is given in Figure 9.4.1.

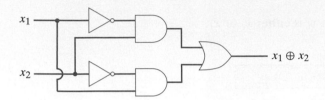

FIGURE 9.4.1   A combinatorial circuit for the exclusive-OR.   □

Suppose that a function is given by a Boolean expression such as

$$f(x_1, x_2, x_3) = (x_1 \vee x_2) \wedge x_3$$

and we wish to find the disjunctive normal form of f. We could write the logic table for f and then use Theorem 9.4.6. Alternatively, we can deal directly with the Boolean expression by using the definitions and results of Sections 9.2 and 9.3. We begin by distributing the term x_3 as follows:

$$(x_1 \vee x_2) \wedge x_3 = (x_1 \wedge x_3) \vee (x_2 \wedge x_3).$$

Although this represents the Boolean expression as a combination of terms of the form $y \wedge z$, it is not in disjunctive normal form, since each term does not contain all of the symbols x_1, x_2, and x_3. However, this is easily remedied, as follows:

$$
\begin{aligned}
(x_1 \wedge x_3) \vee (x_2 \wedge x_3) &= (x_1 \wedge x_3 \wedge 1) \vee (x_2 \wedge x_3 \wedge 1) \\
&= (x_1 \wedge x_3 \wedge (x_2 \vee \overline{x}_2)) \vee (x_2 \wedge x_3 \wedge (x_1 \vee \overline{x}_1)) \\
&= (x_1 \wedge x_2 \wedge x_3) \vee (x_1 \wedge \overline{x}_2 \wedge x_3) \\
&\quad \vee (x_1 \wedge x_2 \wedge x_3) \vee (\overline{x}_1 \wedge x_2 \wedge x_3) \\
&= (x_1 \wedge x_2 \wedge x_3) \vee (x_1 \wedge \overline{x}_2 \wedge x_3) \\
&\quad \vee (\overline{x}_1 \wedge x_2 \wedge x_3).
\end{aligned}
$$

This expression is the disjunctive normal form of f.

Theorem 9.4.6 has a dual. In this case the function f is expressed as

$$f(x_1, \ldots, x_n) = M_1 \wedge M_2 \wedge \cdots \wedge M_k. \tag{9.4.7}$$

Each M_i is of the form

$$y_1 \vee \cdots \vee y_n \tag{9.4.8}$$

where y_j is either x_j or $\overline{x}_j$. A term of the form (9.4.8) is called a **maxterm** and the representation of f (9.4.7) is called the **conjunctive normal form**. Exercises 24–28 explore maxterms and the conjunctive normal form in more detail.

ᔓ ᔓ ᔓ
Exercises

In Exercises 1–10, find the disjunctive normal form of each function and draw the combinatorial circuit corresponding to the disjunctive normal form.

1.

x	y	$f(x, y)$
1	1	1
1	0	0
0	1	1
0	0	1

2.

x	y	$f(x, y)$
1	1	0
1	0	1
0	1	0
0	0	1

3.

x	y	z	$f(x, y, z)$
1	1	1	1
1	1	0	1
1	0	1	0
1	0	0	1
0	1	1	0
0	1	0	0
0	0	1	1
0	0	0	1

4.

x	y	z	$f(x, y, z)$
1	1	1	1
1	1	0	1
1	0	1	0
1	0	0	1
0	1	1	1
0	1	0	1
0	0	1	0
0	0	0	0

5.

x	y	z	$f(x, y, z)$
1	1	1	1
1	1	0	1
1	0	1	1
1	0	0	0
0	1	1	0
0	1	0	1
0	0	1	1
0	0	0	1

6.

x	y	z	$f(x, y, z)$
1	1	1	0
1	1	0	1
1	0	1	1
1	0	0	1
0	1	1	1
0	1	0	1
0	0	1	1
0	0	0	0

7.

x	y	z	$f(x, y, z)$
1	1	1	1
1	1	0	0
1	0	1	0
1	0	0	1
0	1	1	0
0	1	0	0
0	0	1	0
0	0	0	1

8.

x	y	z	$f(x, y, z)$
1	1	1	0
1	1	0	0
1	0	1	0
1	0	0	1
0	1	1	1
0	1	0	1
0	0	1	1
0	0	0	0

9.

w	x	y	z	$f(w, x, y, z)$
1	1	1	1	1
1	1	1	0	0
1	1	0	1	1
1	1	0	0	0
1	0	1	1	0
1	0	1	0	0
1	0	0	1	0
1	0	0	0	1
0	1	1	1	1
0	1	1	0	0
0	1	0	1	0
0	1	0	0	0
0	0	1	1	1
0	0	1	0	0
0	0	0	1	0
0	0	0	0	0

10.

w	x	y	z	$f(w, x, y, z)$
1	1	1	1	0
1	1	1	0	0
1	1	0	1	1
1	1	0	0	1
1	0	1	1	1
1	0	1	0	1
1	0	0	1	0
1	0	0	0	1
0	1	1	1	0
0	1	1	0	1
0	1	0	1	1
0	1	0	0	1
0	0	1	1	0
0	0	1	0	1
0	0	0	1	0
0	0	0	0	1

In Exercises 11–20, find the disjunctive normal form of each function using algebraic techniques. (We abbreviate $a \wedge b$ as ab.)

11. $f(x, y) = x \vee xy$

12. $f(x, y) = (x \vee y)(\overline{x} \vee \overline{y})$

13. $f(x, y, z) = x \vee y(x \vee \overline{z})$

14. $f(x, y, z) = (yz \vee x\overline{z})(x\overline{y} \vee z)$

15. $f(x, y, z) = (\overline{x}y \vee \overline{xz})(x \vee yz)$

16. $f(x, y, z) = x \vee (\overline{y} \vee (x\overline{y} \vee x\overline{z}))$

17. $f(x, y, z) = (x \vee \overline{x}y \vee \overline{x}y\overline{z})(xy \vee \overline{xz})(y \vee xy\overline{z})$

18. $f(x, y, z) = (\overline{x}y \vee \overline{xz})(\overline{xyz \vee y\overline{z}})(x\overline{y}z \vee x\overline{y} \vee x\overline{yz} \vee \overline{xyz})$

19. $f(w, x, y, z) = wy \vee (w\overline{y} \vee z)(x \vee \overline{w}z)$

20. $f(w, x, y, z) = (\overline{w}x\overline{y}z \vee x\overline{y}\,\overline{z})(\overline{\overline{w}yz \vee xy\overline{z} \vee yxz})(\overline{w}z \vee xy \vee \overline{w}\,\overline{y}z \vee xy\overline{z} \vee \overline{x}yz)$

21. How many Boolean functions are there from Z_2^n into Z_2?

Let F denote the set of all functions from Z_2^n into Z_2. Define

$$
\begin{aligned}
(f \vee g)(x) &= f(x) \vee g(x) & x \in Z_2^n \\
(f \wedge g)(x) &= f(x) \wedge g(x) & x \in Z_2^n \\
\overline{f}(x) &= \overline{f(x)} & x \in Z_2^n \\
0(x) &= 0 & x \in Z_2^n \\
1(x) &= 1 & x \in Z_2^n.
\end{aligned}
$$

22. How many elements does F have?

23. Show that $(F, \vee, \wedge, {}^{-}, 0, 1)$ is a Boolean algebra.

24. By dualizing the procedure of Example 9.4.4, explain how to find the conjunctive normal form of a Boolean function from Z_2^n into Z_2.

25. Find the conjunctive normal form of each function in Exercises 1–10.

26. By using algebraic methods, find the conjunctive normal form of each function in Exercises 11–20.

27. Show that if $m_1 \vee \cdots \vee m_k$ is the disjunctive normal form of $f(x_1, \ldots, x_n)$, then $\overline{m}_1 \wedge \cdots \wedge \overline{m}_k$ is the conjunctive normal form of $\overline{f(x_1, \ldots, x_n)}$.

28. Using the method of Exercise 27, find the conjunctive normal form of $\overline{f}$ for each function f of Exercises 1–10.

29. Show that the disjunctive normal form (9.4.5) is unique; that is, show that if we have a Boolean function

$$f(x_1, \ldots, x_n) = m_1 \vee \cdots \vee m_k = m_1' \vee \cdots \vee m_j'$$

where each m_i, m_i' is a minterm, then $k = j$ and the subscripts on the m_i' may be permuted so that $m_i = m_i'$ for $i = 1, \ldots, k$.

9.5 APPLICATIONS

In the preceding section we showed how to design a combinatorial circuit using AND, OR, and NOT gates that would compute an arbitrary function from Z_2^n into Z_2, where $Z_2 = \{0, 1\}$. In this section we consider using other kinds of gates to implement a circuit. We also consider the problem of efficient design. We will conclude by looking at several useful circuits having multiple outputs. Throughout this section, we write ab for $a \wedge b$.

Before considering alternatives to AND, OR, and NOT gates, we must give a precise definition of gate.

DEFINITION 9.5.1

A *gate* is a function from Z_2^n into Z_2.

EXAMPLE 9.5.2

The AND gate is the function $\wedge$ from Z_2^2 into Z_2 defined as in Definition 9.1.1. The NOT gate is the function $^-$ from Z_2 into Z_2 defined as in Definition 9.1.3.　□

We are interested in gates that allow us to construct arbitrary combinatorial circuits.

DEFINITION 9.5.3

A set of gates $\{g_1, \ldots, g_k\}$ is said to be *functionally complete* if, given any positive integer n and a function f from Z_2^n into Z_2, it is possible to construct a combinatorial circuit that computes f using only the gates $g_1, \ldots g_k$.

EXAMPLE 9.5.4

Theorem 9.4.6 shows that the set of gates {AND, OR, NOT} is functionally complete. □

It is an interesting fact that we can eliminate either AND or OR from the set {AND, OR, NOT} and still obtain a functionally complete set of gates.

THEOREM 9.5.5

The sets of gates

$$\{\text{AND, NOT}\} \qquad \{\text{OR, NOT}\}$$

are functionally complete.

Proof. We will show that the set of gates {AND, NOT} is functionally complete and leave the problem of showing that the other set is functionally complete for the exercises (see Exercise 1).

We have

$$x \vee y = \overline{\overline{x}} \vee \overline{\overline{y}} \qquad \text{involution law}$$
$$= \overline{\overline{x}\,\overline{y}} \qquad \text{De Morgan's law.}$$

Therefore, an OR gate can be replaced by one AND gate and three NOT gates. (The combinatorial circuit is shown in Figure 9.5.1.)

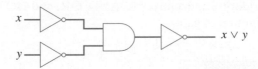

FIGURE 9.5.1   A combinatorial circuit using only AND and NOT gates that computes $x \vee y$.

Given any function $f: Z_2^n \to Z_2$, by Theorem 9.4.6 we can construct a combinatorial circuit C using AND, OR, and NOT gates that computes f. But Figure 9.5.1 shows that each OR gate can be replaced by AND and NOT gates. Therefore, the circuit C can be modified so that it consists only of AND and NOT gates. Thus the set of gates {AND, NOT} is functionally complete.   ∎

Although none of AND, OR, or NOT singly forms a functionally complete set (see Exercises 2–4), it is possible to define a new gate that by itself forms a functionally complete set.

DEFINITION 9.5.6

A *NAND gate* receives inputs x_1 and x_2, where x_1 and x_2 are bits, and produces output denoted $x_1 \uparrow x_2$, where

$$x_1 \uparrow x_2 = \begin{cases} 0 & \text{if } x_1 = 1 \text{ and } x_2 = 1 \\ 1 & \text{otherwise.} \end{cases}$$

A NAND gate is drawn as shown in Figure 9.5.2.

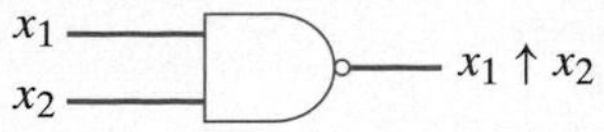

$$x_1 \uparrow x_2$$

FIGURE 9.5.2   NAND gate.

Many basic circuits used in digital computers today are built from NAND gates.

THEOREM 9.5.7

The set {NAND} *is a functionally complete set of gates.*

Proof. First we observe that

$$x \uparrow y = \overline{xy}.$$

Therefore,

$$\overline{x} = \overline{xx} = x \uparrow x \qquad (9.5.1)$$

$$x \vee y = \overline{\overline{x}\,\overline{y}} = \overline{x} \uparrow \overline{y} = (x \uparrow x) \uparrow (y \uparrow y). \qquad (9.5.2)$$

Equations (9.5.1) and (9.5.2) show that both OR and NOT can be written in terms of NAND. By Theorem 9.5.5, the set {OR, NOT} is functionally complete. It follows that the set {NAND} is also functionally complete.   ∎

EXAMPLE 9.5.8

Design combinatorial circuits using NAND gates to compute the functions $f_1(x) = \overline{x}$ and $f_2(x, y) = x \vee y$.

The combinatorial circuits, derived from equations (9.5.1) and (9.5.2), are shown in Figure 9.5.3.

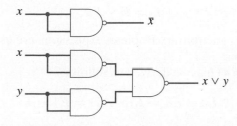

FIGURE 9.5.3   Combinatorial circuits using only NAND gates that compute $\overline{x}$ and $x \vee y$.   □

Consider the problem of designing a combinatorial circuit using AND, OR, and NOT gates to compute the function f.

x	y	z	$f(x, y, z)$
1	1	1	1
1	1	0	1
1	0	1	0
1	0	0	1
0	1	1	0
0	1	0	0
0	0	1	0
0	0	0	0

The disjunctive normal form of f is

$$f(x, y, z) = xyz \vee xy\overline{z} \vee x\overline{y}\,\overline{z}. \tag{9.5.3}$$

The combinatorial circuit corresponding to (9.5.3) is shown in Figure 9.5.4.

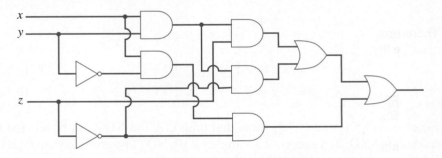

FIGURE 9.5.4   A combinatorial circuit that computes $f(x, y, z) = xyz \vee xy\overline{z} \vee x\overline{y}\,\overline{z}$.

The circuit in Figure 9.5.4 has nine gates. As we will show, it is possible to design a circuit having fewer gates. The problem of finding the best circuit is called the **minimization problem**. There are many definitions of "best."

To find a simpler combinatorial circuit equivalent to that in Figure 9.5.4, we attempt to simplify the Boolean expression (9.5.3) that represents it. The equations

$$Ea \vee E\overline{a} = E \tag{9.5.4}$$

$$E = E \vee Ea, \tag{9.5.5}$$

where E represents an arbitrary Boolean expression, are useful in simplifying Boolean expressions.

Equation (9.5.4) may be derived as follows:

$$Ea \vee E\overline{a} = E(a \vee \overline{a}) = E1 = E$$

using the properties of Boolean algebras. Equation (9.5.5) is essentially the absorption law (Theorem 9.3.6c).

Using (9.5.4) and (9.5.5), we may simplify (9.5.3) as follows:

$$xyz \vee xy\overline{z} \vee x\overline{y}\,\overline{z} = xy \vee x\overline{y}\,\overline{z} \qquad \text{by (9.5.4)}$$
$$= xy \vee xy\overline{z} \vee x\overline{y}\,\overline{z} \qquad \text{by (9.5.5)}$$
$$= xy \vee x\overline{z}. \qquad \text{by (9.5.4).}$$

A further simplification,

$$xy \vee x\overline{z} = x(y \vee \overline{z}), \qquad\qquad (9.5.6)$$

is possible using the distributive law (Definition 9.3.1c). The combinatorial circuit corresponding to (9.5.6), which requires only three gates, is shown in Figure 9.5.5.

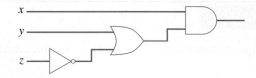

FIGURE 9.5.5   A three-gate combinatorial circuit equivalent to that of Figure 9.5.4.

EXAMPLE 9.5.9

The combinatorial circuit in Figure 9.4.1 uses five AND, OR, and NOT gates to compute the exclusive-OR $x \oplus y$ of x and y. Design a circuit that computes $x \oplus y$ using fewer AND, OR, and NOT gates.

Unfortunately, (9.5.4) and (9.5.5) do not help us simplify the disjunctive normal form $x\overline{y} \vee \overline{x}y$ of $x \oplus y$. Thus we must experiment with various Boolean rules until we produce an expression that requires fewer than five gates. One solution is provided by the expression

$$(x \vee y)\overline{x}\,\overline{y}$$

whose implementation requires only four gates. This combinatorial circuit is shown in Figure 9.5.6.

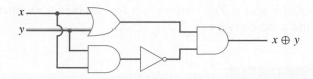

FIGURE 9.5.6   A four-gate combinatorial circuit that computes the exclusive-OR $x \oplus y$ of x and y.   □

The set of gates available determines the minimization problem. Since the state of technology determines the available gates, the minimization problem changes through time. In the 1950s, the typical problem was to minimize circuits consisting of AND, OR, and NOT gates. Solutions such as the Quine–McCluskey method and the method of Karnaugh maps were provided. The reader is referred to [Mendelson] for the details of these methods.

Advances in solid-state technology have made it possible to manufacture very small components, called **integrated circuits**, which are themselves entire circuits. Thus circuit design today consists of combining basic gates such as AND, OR, NOT, and NAND gates and integrated circuits to compute the desired functions. Boolean algebra remains an essential tool, as a glance at a book on logic design such as [McCalla] will show.

We conclude this section by considering several useful combinatorial circuits having multiple outputs. A circuit with n outputs can be characterized by n Boolean expressions, as the next example shows.

EXAMPLE 9.5.10

Write two Boolean expressions to describe the combinatorial circuit of Figure 9.5.7.

The output y_1 is described by the expression

$$y_1 = \overline{ab}$$

and y_2 is described by the expression

$$y_2 = bc \vee \overline{ab}.$$

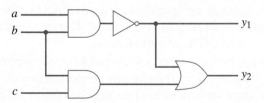

FIGURE 9.5.7   A combinatorial circuit with two outputs.   □

Our first circuit is called a **half adder**.

DEFINITION 9.5.11

A *half adder* accepts as input two bits x and y and produces as output the binary sum cs of x and y. The term cs is a two-bit binary number. We call s the sum bit and c the carry bit.

EXAMPLE 9.5.12   *Half-Adder Circuit*

Design a half-adder combinatorial circuit.

The table for the half-adder circuit is as follows:

x	y	c	s
1	1	1	0
1	0	0	1
0	1	0	1
0	0	0	0

This function has two outputs c and s. We observe that $c = xy$ and $s = x \oplus y$. Thus we obtain the half-adder circuit of Figure 9.5.8. We used the circuit of Figure 9.5.6 to realize the exclusive-OR.

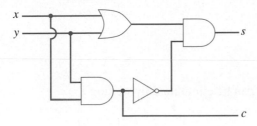

FIGURE 9.5.8   A half-adder circuit.   □

A **full adder** sums three bits and is useful for adding two bits and a third carry bit from a previous addition.

DEFINITION 9.5.13

A *full adder* accepts as input three bits x, y, and z and produces as output the binary sum cs of x, y, and z. The term cs is a two-bit binary number.

EXAMPLE 9.5.14   *Full-Adder Circuit*

Design a full-adder combinatorial circuit.
    The table for the full-adder circuit is as follows:

x	y	z	c	s
1	1	1	1	1
1	1	0	1	0
1	0	1	1	0
1	0	0	0	1
0	1	1	1	0
0	1	0	0	1
0	0	1	0	1
0	0	0	0	0

Checking the eight possibilities, we see that

$$s = x \oplus y \oplus z;$$

hence we can use two exclusive-OR circuits to compute s.

To compute c, we first find the disjunctive normal form

$$c = xyz \vee xy\overline{z} \vee x\overline{y}z \vee \overline{x}yz \tag{9.5.7}$$

of c. Next, we use (9.5.4) and (9.5.5) to simplify (9.5.7) as follows:

$$
\begin{aligned}
xyz \vee xy\overline{z} \vee x\overline{y}z \vee \overline{x}yz &= xy \vee x\overline{y}z \vee \overline{x}yz \\
&= xy \vee xyz \vee x\overline{y}z \vee \overline{x}yz \\
&= xy \vee xz \vee \overline{x}yz \\
&= xy \vee xz \vee xyz \vee \overline{x}yz \\
&= xy \vee xz \vee yz.
\end{aligned}
$$

Additional gates can be eliminated by writing

$$c = xy \vee z(x \vee y).$$

We obtain the full-adder circuit given in Figure 9.5.9.

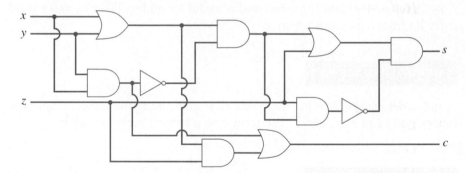

FIGURE 9.5.9   A full-adder circuit.   □

Our last example shows how we may use half-adder and full-adder circuits to construct a circuit to add binary numbers.

EXAMPLE 9.5.15   *A Circuit to Add Binary Numbers*

Using half-adder and full-adder circuits, design a combinatorial circuit that computes the sum of two three-bit numbers.

We will let $M = x_3 x_2 x_1$ and $N = y_3 y_2 y_1$ denote the numbers to be added and let $z_4 z_3 z_2 z_1$ denote the sum. The circuit that computes the sum of M and N is drawn in Figure 9.5.10. It is an implementation of the standard algorithm for adding numbers, since the "carry bit" is indeed *carried* into the next binary addition.   □

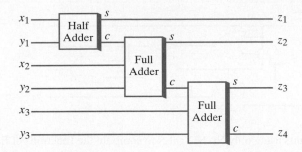

FIGURE 9.5.10   A combinatorial circuit that computes the sum of two three-bit numbers.

If we were using three-bit registers for addition, so that the sum of two three-bit numbers would have to be no more than three bits, we could use the z_4 bit in Example 9.5.15 as an overflow flag. If $z_4 = 1$, overflow occurred; if $z_4 = 0$, there was no overflow.

In the next chapter (Example 10.1.3), we will discuss a sequential circuit that makes use of a primitive internal memory to add binary numbers.

෨ ෨ ෨

Exercises

1. Show that the set of gates {OR, NOT} is functionally complete.

Show that each set of gates in Exercises 2–5 is not functionally complete.

2. {AND}   **3.** {OR}   **4.** {NOT}   **5.** {AND, OR}

6. Draw a circuit using only NAND gates that computes xy.

7. Write xy using only $\uparrow$.

8. Prove or disprove: $x \uparrow (y \uparrow z) = (x \uparrow y) \uparrow z$, for all $x, y, z \in Z_2$.

Write Boolean expressions to describe the multiple output circuits in Exercises 9–11.

9.

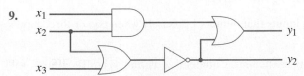

10.

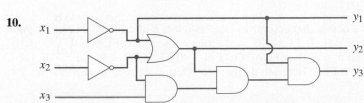

11.

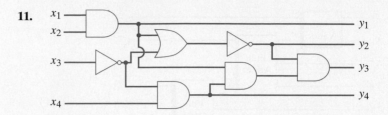

12. Design circuits using only NAND gates to compute the functions of Exercises 1–10, Section 9.4.

13. Can you reduce the number of NAND gates used in any of your circuits for Exercise 12?

14. Design circuits using as few AND, OR, and NOT gates as you can to compute the functions of Exercises 1–10, Section 9.4.

15. Design a half-adder circuit using only NAND gates.

★ **16.** Design a half-adder circuit using five NAND gates.

A *NOR gate* receives inputs x_1 and x_2, where x_1 and x_2 are bits, and produces output denoted $x_1 \downarrow x_2$, where

$$x_1 \downarrow x_2 = \begin{cases} 0 & \text{if } x_1 = 1 \text{ or } x_2 = 1 \\ 1 & \text{otherwise.} \end{cases}$$

17. Write xy, $x \vee y$, $\overline{x}$, and $x \uparrow y$ in terms of $\downarrow$.

18. Write $x \downarrow y$ in terms of $\uparrow$.

19. Write the logic table for the NOR function.

20. Show that the set of gates {NOR} is functionally complete.

21. Design circuits using only NOR gates to compute the functions of Exercises 1–10, Section 9.4.

22. Can you reduce the number of NOR gates used in any of your circuits for Exercise 21?

23. Design a half-adder circuit using only NOR gates.

★ **24.** Design a half-adder circuit using five NOR gates.

25. Design a circuit with three inputs that outputs 1 precisely when two or three inputs have value 1.

26. Design a circuit that multiplies the binary numbers $x_2 x_1$ and $y_2 y_1$. The output will be of the form $z_4 z_3 z_2 z_1$.

27. A **2's module** is a circuit that accepts as input two bits b and FLAGIN and outputs bits c and FLAGOUT. If FLAGIN $= 1$, then $c = \overline{b}$ and FLAGOUT $= 1$. If FLAGIN $= 0$ and $b = 1$, then FLAGOUT $= 1$. If FLAGIN $= 0$ and $b = 0$, then FLAGOUT $= 0$. If FLAGIN $= 0$, then $c = b$. Design a circuit to implement a 2's module.

The *2's complement* of a binary number can be computed by using the following algorithm.

ALGORITHM 9.5.16   *Finding the 2's Complement*

This algorithm computes the 2's complement $C_N C_{N-1} \cdots C_2 C_1$ of the binary number $M = B_N B_{N-1} \cdots B_2 B_1$. The number M is scanned from right to left and the bits are copied until 1 is found. Thereafter, if $B_i = 0$, we set $C_i = 1$ and if $B_i = 0$, we set $C_i = 0$. The flag F indicates whether a 1 has been found ($F = $ **true**) or not ($F = $ **false**).

Input: $B_N B_{N-1} \cdots B_1$
Output: $C_N C_{N-1} \cdots C_1$

```
procedure twos_complement(B)
  F := false
  i := 1
  while F and i ≤ n do
    begin
    Cᵢ := Bᵢ
    if Bᵢ = 1 then
      F := true
    i := i + 1
    end
  while i ≤ n do
    begin
    Cᵢ := Bᵢ ⊕ 1
    i := i + 1
    end
  return(C)
end twos_complement
```

Find the 2's complement of the numbers in Exercises 28–30 using Algorithm 9.5.16.

28. 101100 **29.** 11011 **30.** 011010110

31. Using 2's modules, design a circuit that computes the 2's complement $y_3 y_2 y_1$ of the three-bit binary number $x_3 x_2 x_1$.

★ **32.** Let $*$ be a binary operator on a set S containing 0 and 1. Write a set of axioms for $*$, modeled after rules that NAND satisfies, so that if we define

$$\overline{x} = x * x$$
$$x \vee y = (x * x) * (y * y)$$
$$x \wedge y = (x * y) * (x * y),$$

then $(S, \vee, \wedge, ^-, 0, 1)$ is a Boolean algebra.

★ **33.** Let $*$ be a binary operator on a set S containing 0 and 1. Write a set of axioms for $*$, modeled after rules that NOR satisfies, and definitions for $^-$, $\vee$, and $\wedge$ so that $(S, \vee, \wedge, ^-, 0, 1)$ is a Boolean algebra.

★ **34.** Show that $\{\rightarrow\}$ is functionally complete (see Definition 1.2.4).

★ **35.** Let $B(x, y)$ be a Boolean expression in the variables x and y that uses only the operator $\leftrightarrow$ (see Definition 1.2.8).

(a) Show that if B contains an even number of x's, the values of $B(\overline{x}, y)$ and $B(x, y)$ are the same for all x and y.

(b) Show that if B contains an odd number of x's, the values of $B(\overline{x}, y)$ and $\overline{B(x, y)}$ are the same for all x and y.

(c) Use parts (a) and (b) to show that $\{\leftrightarrow\}$ is *not* functionally complete.

This exercise was contributed by Paul Pluznikov.

↷ NOTES

General references on Boolean algebras are [Hohn; and Mendelson]. [Mendelson] contains over 150 references on Boolean algebras and combinatorial circuits. Books on logic design include [Kohavi; McCalla; and Ward].

[Hailperin] gives a technical discussion of Boole's mathematics. Additional references are also provided. Boole's book, *The Laws of Thought*, has been reprinted (see [Boole]).

Because of our interest in applications of Boolean algebra, most of our discussion was limited to the Boolean algebra $(Z_2, \vee, \wedge, ^-, 0, 1)$. However, versions of most of our results remain valid for arbitrary, finite Boolean algebras.

Boolean expressions in the symbols $x_1, \ldots, x_n$ over an arbitrary Boolean algebra $(S, \vee, \wedge, ^-, 0, 1)$ are defined recursively as

$$\{x \mid x \in S\}, x_1, \ldots, x_n \text{ are Boolean expressions.}$$

If X_1 and X_2 are Boolean expressions, so are

$$(X_1), \quad \overline{X}_1, \quad X_1 \vee X_2, \quad X_1 \wedge X_2.$$

A **Boolean function** over S is defined as a function from S^n to S of the form

$$f(x_1, \ldots, x_n) = X(x_1, \ldots, x_n),$$

where X is a Boolean expression in the symbols $x_1, \ldots, x_n$ over S. A disjunctive normal form can be defined for f. Another result is that if X and Y are Boolean expressions over S and

$$X(x_1, \ldots, x_n) = Y(x_1, \ldots, x_n)$$

for all $x_i \in S$, then Y is derivable from X using the definition (Definition 9.3.1) of a Boolean algebra. Other results are that any finite Boolean algebra has 2^n elements and that if two Boolean algebras both have 2^n elements they are essentially the same. It follows that any finite Boolean algebra is essentially Example 9.3.3, the Boolean algebra of subsets of a finite, universal set U. The proofs of these results can be found in [Mendelson].

↷ CHAPTER REVIEW

Section 9.1

Combinatorial circuit
Sequential circuit
AND gate

OR gate
NOT gate (inverter)
Logic table of a combinatorial circuit
Boolean expression
Literal

Section 9.2

Properties of $\wedge$, $\vee$, and $\overline{}$: associative laws; commutative laws; distributive laws; identity laws; complement laws (see Theorem 9.2.1)

Equal Boolean expressions

Equivalent combinatorial expressions

Combinatorial expressions are equivalent if and only if the Boolean expressions that represent them are equal.

Section 9.3

Boolean algebra

x': Complement of x

Properties of Boolean algebras: Idempotent laws; bound laws; absorption laws; involution law; 0 and 1 laws; De Morgan's laws

Dual of statement involving Boolean expressions

The dual of a theorem about Boolean algebras is also a theorem.

Section 9.4

Exclusive-OR

Boolean function

Minterm: $y_1 \wedge y_2 \wedge \cdots \wedge y_n$, where each y_i is x_i or $\overline{x}_i$

Disjunctive normal form

How to write a Boolean function in disjunctive normal form (Theorem 9.4.6)

Maxterm: $y_1 \vee y_2 \vee \cdots \vee y_n$, where each y_i is x_i or $\overline{x}_i$

Conjunctive normal form

Section 9.5

Gate

Functionally complete set of gates

The sets of gates {AND, NOT} and {OR, NOT} are functionally complete.

NAND gate

The set {NAND} is a functionally complete set of gates.

Minimization problem

Integrated circuit

Half-adder circuit

Full-adder circuit

✍ CHAPTER SELF-TEST

Section 9.1

1. Write a Boolean expression that represents the combinatorial circuit and write the logic table.

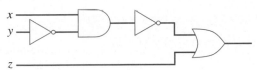

2. Find the value of the Boolean expression

$$(x_1 \wedge x_2) \vee (\overline{x}_2 \wedge x_3)$$

   if $x_1 = x_2 = 0$ and $x_3 = 1$.

3. Find a combinatorial circuit corresponding to the Boolean expression of Exercise 2.

4. Show that the circuit is not a combinatorial circuit.

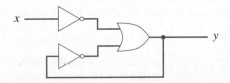

Section 9.2

Are the combinatorial circuits in Exercises 5 and 6 equivalent? Explain.

5.

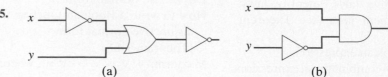

(a)   (b)

6.

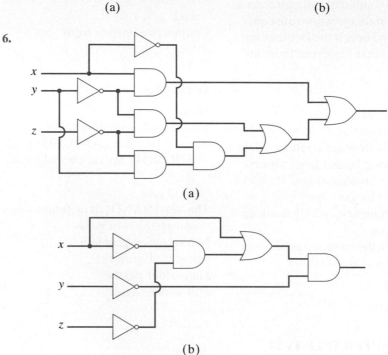

(a)

(b)

Prove or disprove the equations in Exercises 7 and 8.

7. $(x \wedge y) \vee (\overline{x} \wedge z) \vee (\overline{x} \wedge y \wedge \overline{z}) = y \vee (\overline{x} \wedge z)$

8. $(x \wedge y \wedge z) \vee (\overline{x \vee z}) = (x \wedge z) \vee (\overline{x} \wedge \overline{z})$

Section 9.3

9. If U is a universal set and $S = \mathcal{P}(U)$, the power set of U, then

$$(S, \cup, \cap, {}^{-}, \emptyset, U)$$

is a Boolean algebra. State the bound and absorption laws for this Boolean algebra.

10. Prove that in any Boolean algebra, $(x(x + y \cdot 0))' = x'$, for all x and y.

11. Write the dual of the statement of Exercise 10 and prove it.

12. Let U be the set of positive integers. Let S be the collection of finite subsets of U. Why does $(S, \cup, \cap, {}^{-}, \emptyset, U)$ fail to be a Boolean algebra?

Section 9.4

In Exercises 13–16, find the disjunctive normal form of a Boolean expression having a logic table the same as the given table and draw the combinatorial circuit corresponding to the disjunctive normal form.

13.

x_1	x_2	x_3	y
1	1	1	0
1	1	0	0
1	0	1	0
1	0	0	1
0	1	1	0
0	1	0	0
0	0	1	0
0	0	0	0

14.

x_1	x_2	x_3	y
1	1	1	0
1	1	0	1
1	0	1	0
1	0	0	1
0	1	1	0
0	1	0	0
0	0	1	0
0	0	0	0

15.

x_1	x_2	x_3	y
1	1	1	1
1	1	0	0
1	0	1	0
1	0	0	1
0	1	1	0
0	1	0	0
0	0	1	0
0	0	0	1

16.

x_1	x_2	x_3	y
1	1	1	0
1	1	0	1
1	0	1	0
1	0	0	1
0	1	1	1
0	1	0	0
0	0	1	1
0	0	0	0

Section 9.5

17. Write the logic table for the circuit

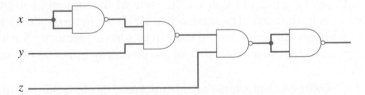

18. Find a Boolean expression in disjunctive normal form for the circuit of part (a) of Exercise 6. Use algebraic methods to simplify the disjunctive normal form. Draw the circuit corresponding to the simplified expression.

19. Design a circuit using only NAND gates to compute $x \oplus y$.

20. Design a full-adder circuit that uses two half adders and one OR gate.

AUTOMATA, GRAMMARS, AND LANGUAGES

*I*t's a state of mind.
You can't move away from
a state of mind.

—from The Late George Apley

In Chapter 9 we discussed combinatorial circuits in which the output depended only on the input. These circuits have no memory. In this chapter we begin by discussing circuits in which the output depends not only on the input but also on the state of the system at the time the input is introduced. The state of the system is determined by previous processing. In this sense, these circuits have memory. Such circuits are called *sequential circuits* and are obviously important in computer design.

Finite-state machines are abstract models of machines with a primitive internal memory. A finite-state automaton is a special kind of finite-state machine that is closely linked to a particular type of language. In the latter part of this chapter, we will discuss finite-state machines, finite-state automata, and languages in some detail.

10.1 SEQUENTIAL CIRCUITS AND FINITE-STATE MACHINES

Operations within a digital computer are carried out at discrete intervals of time. Output depends on the state of the system as well as on the input. We will assume that the state of the system changes only at time $t = 0, 1, \ldots$. A simple way to introduce sequencing in circuits is to introduce a **unit time delay**.

DEFINITION 10.1.1

A *unit time delay* accepts as input a bit x_t at time t and outputs x_{t-1}, the bit received as input at time $t - 1$. The unit time delay is drawn as shown in Figure 10.1.1.

As an example of the use of the unit time delay, we discuss the **serial adder**.

FIGURE 10.1.1
Unit time delay.

DEFINITION 10.1.2

A *serial adder* accepts as input two binary numbers

$$x = 0x_N x_{N-1} \cdots x_0 \qquad \text{and} \qquad y = 0y_N y_{N-1} \cdots y_0$$

and outputs the sum $z_{N+1} z_N \cdots z_0$ of x and y. The numbers x and y are input sequentially in pairs, $x_0, y_0; \ldots; x_N, y_N; 0, 0$. The sum is output $z_0, z_1, \ldots, z_{N+1}$.

EXAMPLE 10.1.3   *Serial-Adder Circuit*

A circuit, using a unit time delay, that implements a serial adder is shown in Figure 10.1.2.

Let us show how the serial adder computes the sum of

$$x = 010 \qquad \text{and} \qquad y = 011.$$

We begin by setting $x_0 = 0$ and $y_0 = 1$. (We assume that at this instant $i = 0$. This can be arranged by first setting $x = y = 0$.) The state of the system is shown in Figure 10.1.3a. Next, we set $x_1 = y_1 = 1$. The unit time delay sends $i = 0$ as the third bit to the full adder. The state of the system is shown in Figure 10.1.3b. Finally, we set $x_2 = y_2 = 0$. This time the unit time delay sends $i = 1$ as the third bit to the full adder. The state of the system is shown in Figure 10.1.3c. We obtain the sum $z = 101$.

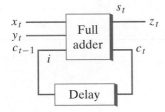

FIGURE 10.1.2
A serial-adder circuit.

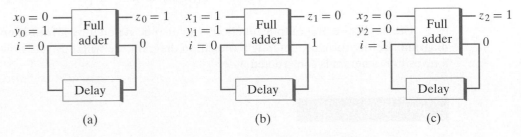

(a) (b) (c)

FIGURE 10.1.3   Computing $010 + 011$ with the serial-adder circuit.   □

A **finite-state machine** is an abstract model of a machine with a primitive internal memory.

DEFINITION 10.1.4

A *finite-state machine M* consists of

(a) A finite set $\mathcal{I}$ of *input symbols*.
(b) A finite set $\mathcal{O}$ of *output symbols*.
(c) A finite set $\mathcal{S}$ of *states*.
(d) A *next-state function* f from $\mathcal{S} \times \mathcal{I}$ into $\mathcal{S}$.
(e) An *output function* g from $\mathcal{S} \times \mathcal{I}$ into $\mathcal{O}$.
(f) An *initial state* $\sigma \in \mathcal{S}$.

We write $M = (\mathcal{I}, \mathcal{O}, \mathcal{S}, f, g, \sigma)$.

EXAMPLE 10.1.5

Let $\mathcal{I} = \{a, b\}$, $\mathcal{O} = \{0, 1\}$, and $\mathcal{S} = \{\sigma_0, \sigma_1\}$. Define the pair of functions $f: \mathcal{S} \times \mathcal{I} \to \mathcal{S}$ and $g: \mathcal{S} \times \mathcal{I} \to \mathcal{O}$ by the rules given in Table 10.1.1.

TABLE 10.1.1

	f		g	
$\mathcal{I}$ a b a b $\mathcal{S}$				
σ_0	σ_0	σ_1	0	1
σ_1	σ_1	σ_1	1	0

Then $M = (\mathcal{I}, \mathcal{O}, \mathcal{S}, f, g, \sigma_0)$ is a finite-state machine.
Table 10.1.1 is interpreted to mean

$$f(\sigma_0, a) = \sigma_0 \qquad g(\sigma_0, a) = 0,$$
$$f(\sigma_0, b) = \sigma_1 \qquad g(\sigma_0, b) = 1,$$
$$f(\sigma_1, a) = \sigma_1 \qquad g(\sigma_1, a) = 1,$$
$$f(\sigma_1, b) = \sigma_1 \qquad g(\sigma_1, b) = 0.$$

$\square$

The next-state and output functions can also be defined by a **transition diagram**. Before formally defining a transition diagram, we will illustrate how a transition diagram is constructed.

EXAMPLE 10.1.6

Draw the transition diagram for the finite-state machine of Example 10.1.5.

The transition diagram is a digraph. The vertices are the states (see Figure 10.1.4). The initial state is indicated by an arrow as shown. If we are in state σ and inputting i causes output o and moves us to state σ', we draw a directed edge from vertex σ to vertex σ' and label it i/o. For example, if we are in state σ_0, and we input a, Table 10.1.1 tells us that we output 0 and remain in state σ_0. Thus we draw a directed loop on vertex σ_0 and label it $a/0$ (see Figure 10.1.4). On the other hand, if we are in state σ_0 and we input b, we output 1 and move to state σ_1. Thus we draw a directed edge from σ_0 to σ_1 and label it $b/1$. By considering all such possibilities, we obtain the transition diagram of Figure 10.1.4.

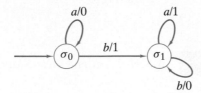

FIGURE 10.1.4 A transition diagram. □

DEFINITION 10.1.7

Let $M = (\mathcal{I}, \mathcal{O}, \mathcal{S}, f, g, \sigma)$ be a finite-state machine. The *transition diagram* of M is a digraph G whose vertices are the members of $\mathcal{S}$. An arrow designates the initial state σ. A directed edge (σ_1, σ_2) exists in G if there exists an input i with $f(\sigma_1, i) = \sigma_2$. In this case, if $g(\sigma_1, i) = o$, the edge (σ_1, σ_2) is labeled i/o.

We can regard the finite-state machine $M = (\mathcal{I}, \mathcal{O}, \mathcal{S}, f, g, \sigma)$ as a simple computer. We begin in state σ, input a string over $\mathcal{I}$, and produce a string of output.

DEFINITION 10.1.8

Let $M = (\mathcal{I}, \mathcal{O}, \mathcal{S}, f, g, \sigma)$ be a finite-state machine. An *input string* for M is a string over $\mathcal{I}$. The string

$$y_1 \cdots y_n$$

is the *output string* for M corresponding to the input string

$$\alpha = x_1 \cdots x_n$$

if there exist states $\sigma_0, \ldots, \sigma_n \in \mathcal{S}$ with

$$\sigma_0 = \sigma$$
$$\sigma_i = f(\sigma_{i-1}, x_i) \qquad \text{for } i = 1, \ldots, n;$$
$$y_i = g(\sigma_{i-1}, x_i) \qquad \text{for } i = 1, \ldots, n.$$

EXAMPLE 10.1.9

Find the output string corresponding to the input string

$$aababba \qquad\qquad (10.1.1)$$

for the finite-state machine of Example 10.1.5.

Initially, we are in state σ_0. The first symbol input is a. We locate the outgoing edge in the transition diagram of M (Figure 10.1.4) from σ_0 labeled a/x, which tells us that if a is input, x is output. In our case, 0 is output. The edge points to the next state, σ_0. Next, a is input again. As before, we output 0 and remain in state σ_0. Next, b is input. In this case, we output 1 and change to state σ_1. Continuing in this way, we find that the output string is

$$0011001. \tag{10.1.2}$$

☐

EXAMPLE 10.1.10   *A Serial Adder Finite-State Machine*

Design a finite-state machine that performs serial addition.

We will represent the finite-state machine by its transition diagram.

Since the serial adder accepts pairs of bits, the input set will be

$$\{00, 01, 10, 11\}.$$

The output set is

$$\{0, 1\}.$$

Given an input xy, we take one of two actions: either we add x and y, or we add x, y, and 1, depending on whether the carry bit was 0 or 1. Thus there are two states, which we will call C (carry) and NC (no carry). The initial state is NC. At this point, we can draw the vertices and designate the initial state in our transition diagram (see Figure 10.1.5).

Next, we consider the possible inputs at each vertex. For example, if 00 is input to NC, we should output 0 and remain in state NC. Thus NC has a loop labeled 00/0. As another example, if 11 is input to C, we compute $1 + 1 + 1 = 11$. In this case we output 1 and remain in state C. Thus C has a loop labeled 11/1. As a final example, if we are in state NC and 11 is input, we should output 0 and move to state C. By considering all possibilities, we arrive at the transition diagram of Figure 10.1.6.

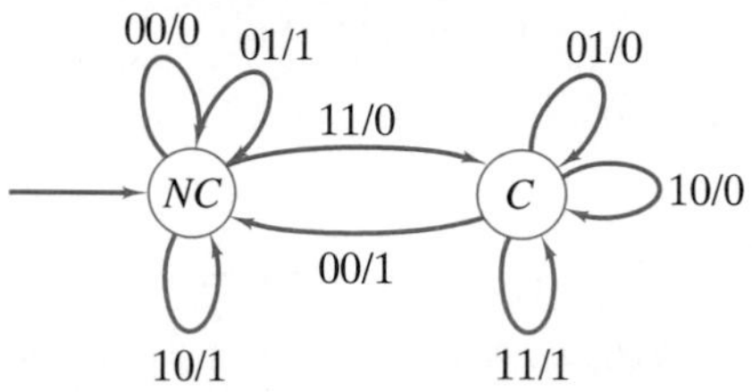

FIGURE 10.1.5

Two states for the serial adder finite-state machine.

FIGURE 10.1.6   A finite-state machine that performs serial addition.   ☐

EXAMPLE 10.1.11   *The SR Flip-Flop*

A **flip-flop** is a basic component of digital circuits since it serves as a one-bit memory cell. The **SR flip-flop** (or **set-reset flip-flop**) can be defined by the table

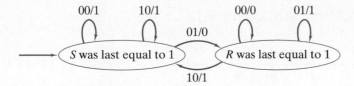

FIGURE 10.1.7   The *SR* flip-flop as a finite-state machine.

S	*R*	*Q*
1	1	Not allowed
1	0	1
0	1	0
0	0	$\begin{cases} 1 & \text{if } S \text{ was last equal to 1} \\ 0 & \text{if } R \text{ was last equal to 1} \end{cases}$

The *SR* flip-flop "remembers" whether *S* or *R* was last equal to 1. (If $Q = 1$, *S* was last equal to 1; if $Q = 0$, *R* was last equal to 0.) We can model the *SR* flip-flop as a finite-state machine by defining two states: "*S* was last equal to 1" and "*R* was last equal to 1" (see Figure 10.1.7). We define the input to be the new values of *S* and *R*; the notation *SR* means that $S = s$ and $R = r$. We define *Q* to be the output. We have arbitrarily designated the initial state as "*S* was last equal to 1." A sequential circuit implementation of the *SR* flip-flop is shown in Figure 10.1.8.

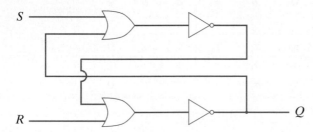

FIGURE 10.1.8   A sequential circuit implementation of the *SR* flip-flop.   □

Exercises

In Exercises 1–5, draw the transition diagram of the finite-state machine $(\mathcal{I}, \mathcal{O}, \mathcal{S}, f, g, \sigma_0)$.

1. $\mathcal{I} = \{a, b\}, \mathcal{O} = \{0, 1\}, \mathcal{S} = \{\sigma_0, \sigma_1\}$

$\overset{\mathcal{I}}{\underset{\mathcal{S}}{\diagdown}}$	*f* a	*f* b	*g* a	*g* b
σ_0	σ_1	σ_1	1	1
σ_1	σ_0	σ_1	0	1

2. $\mathcal{I} = \{a, b\}, \mathcal{O} = \{0, 1\}, \mathcal{S} = \{\sigma_0, \sigma_1\}$

$\mathcal{S}$ \ $\mathcal{I}$	f : a	b	g : a	b
σ_0	σ_1	σ_0	0	0
σ_1	σ_0	σ_0	1	1

3. $\mathcal{I} = \{a, b\}, \mathcal{O} = \{0, 1\}, \mathcal{S} = \{\sigma_0, \sigma_1, \sigma_2\}$

$\mathcal{S}$ \ $\mathcal{I}$	f : a	b	g : a	b
σ_0	σ_1	σ_1	0	1
σ_1	σ_2	σ_1	1	1
σ_2	σ_0	σ_0	0	0

4. $\mathcal{I} = \{a, b, c\}, \mathcal{O} = \{0, 1\}, \mathcal{S} = \{\sigma_0, \sigma_1, \sigma_2\}$

$\mathcal{S}$ \ $\mathcal{I}$	f : a	b	c	g : a	b	c
σ_0	σ_0	σ_1	σ_2	0	1	0
σ_1	σ_1	σ_1	σ_0	1	1	1
σ_2	σ_2	σ_1	σ_0	1	0	0

5. $\mathcal{I} = \{a, b, c\}, \mathcal{O} = \{0, 1, 2\}, \mathcal{S} = \{\sigma_0, \sigma_1, \sigma_2, \sigma_3\}$

$\mathcal{S}$ \ $\mathcal{I}$	f : a	b	c	g : a	b	c
σ_0	σ_1	σ_0	σ_2	1	1	2
σ_1	σ_0	σ_2	σ_2	2	0	0
σ_2	σ_3	σ_3	σ_0	1	0	1
σ_3	σ_1	σ_1	σ_0	2	0	2

In Exercises 6–10, find the sets $\mathcal{I}$, $\mathcal{O}$, and $\mathcal{S}$, the initial state, and the table defining the next-state and output functions for each finite-state machine.

6.

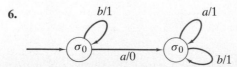

7.

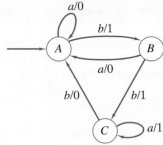

8.

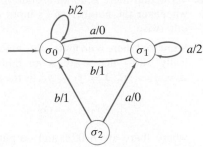

9.

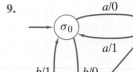

10.

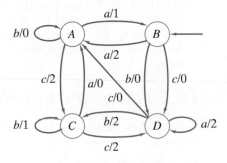

In Exercises 11–20, find the output string for the given input string and finite-state machine.

11. *abba*; Exercise 1

12. *abba*; Exercise 2

13. *aabbaba*; Exercise 3

14. *aabbcc*; Exercise 4

15. *aabaab*; Exercise 5

16. *aaa*; Exercise 6

17. *aabbabaab*; Exercise 7

18. *baaba*; Exercise 8

19. *bbababbabaaa*; Exercise 9

20. *cacbccbaabac*; Exercise 10

In Exercises 21–26, design a finite-state machine having the given properties. The input is always a bit string.

21. Outputs 1 if an even number of 1's have been input; otherwise, outputs 0

22. Outputs 1 if k 1's have been input, where k is a multiple of 3; otherwise, outputs 0

23. Outputs 1 if two or more 1's are input; otherwise, outputs 0

24. Outputs 1 whenever it sees 101; otherwise, outputs 0

25. Outputs 1 when it sees 101 and thereafter; otherwise, outputs 0

26. Outputs 1 when it sees the first 0 and until it sees another 0; thereafter, outputs 0; in all other cases, outputs 0

27. Let $\alpha = x_1 \cdots x_n$ be a bit string. Let $\beta = y_1 \cdots y_n$ where

$$y_i = \begin{cases} a & \text{if } x_i = 0 \\ b & \text{if } x_i = 1 \end{cases}$$

for $i = 1, \ldots, n$. Let $\gamma = y_n \cdots y_1$.

Show that if γ is input to the finite-state machine of Figure 10.1.4, the output is the 2's complement of α (see Algorithm 9.5.16 for a description of 2's complement).

★ **28.** Show that there is no finite-state machine that receives a bit string and outputs 1 whenever the number of 1's input equals the number of 0's input and outputs 0 otherwise.

★ **29.** Show that there is no finite-state machine that performs serial multiplication. Specifically, show that there is no finite-state machine that inputs binary numbers $X = x_1 \cdots x_n$, $Y = y_1 \cdots y_n$, as the sequence of two-bit numbers

$$x_n y_n, \quad x_{n-1} y_{n-1}, \quad \ldots, \quad x_1 y_1, \quad 00, \quad \ldots, \quad 00,$$

where there are n 00's, and outputs $z_{2n}, \ldots, z_1$, where $Z = z_1 \cdots z_{2n} = XY$. *Example:* If there is such a machine, to multiply 101×1001 we would input 11,00,10, 01,00,00,00,00. The first pair 11 is the pair of rightmost bits ($10\underline{1}$, $100\underline{1}$); the second pair 00 is the next pair of bits ($1\underline{0}1$, $10\underline{0}1$); and so on. We pad the input string with four pairs of 00's—the length of the longest number 1001 to be multiplied. Since $101 \times 1001 = 101101$, it is alleged that we obtain the output shown in the adjacent table.

Input	Output
11	1
00	0
10	1
01	1
00	0
00	1
00	0
00	0

10.2 *FINITE-STATE AUTOMATA*

A **finite-state automaton** is a special kind of finite-state machine. Finite-state automata are of special interest because of their relationship to languages, as we shall see in Section 10.5.

DEFINITION 10.2.1

A *finite-state automaton* $A = (\mathcal{I}, \mathcal{O}, \mathcal{S}, f, g, \sigma)$ is a finite-state machine in which the set of output symbols is $\{0, 1\}$ and where the current state determines the last output. Those states for which the last output was 1 are called *accepting states*.

EXAMPLE 10.2.2

Draw the transition diagram of the finite-state machine A defined by the table. The initial state is σ_0. Show that A is a finite-state automaton and determine the set of accepting states.

$\mathcal{S}$ \ $\mathcal{I}$	f : a	f : b	g : a	g : b
σ_0	σ_1	σ_0	1	0
σ_1	σ_2	σ_0	1	0
σ_2	σ_2	σ_0	1	0

The transition diagram is shown in Figure 10.2.1. If we are in state σ_0, the last output was 0. If we are in either state σ_1 or σ_2, the last output was 1; thus A is a finite-state automaton. The accepting states are σ_1 and σ_2.

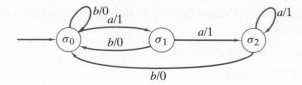

FIGURE 10.2.1   The transition diagram for Example 10.2.2.   □

Example 10.2.2 shows that the finite-state machine defined by a transition diagram will be a finite-state automaton if the set of output symbols is {0, 1} and if, for each state σ, all incoming edges to σ have the same output label.

The transition diagram of a finite-state automaton is usually drawn with the accepting states in double circles and the output symbols omitted. When the transition diagram of Figure 10.2.1 is redrawn in this way, we obtain the transition diagram of Figure 10.2.2.

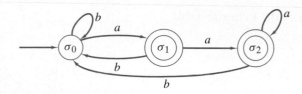

FIGURE 10.2.2   The transition diagram of Figure 10.2.1 redrawn with accepting states in double circles and output symbols omitted.

EXAMPLE 10.2.3

Draw the transition diagram of the finite-state automaton of Figure 10.2.3 as a transition diagram of a finite-state machine.

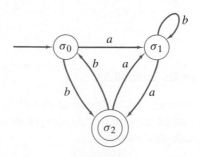

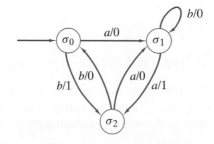

FIGURE 10.2.3   A finite-state automaton.

FIGURE 10.2.4   The finite-state automaton of Figure 10.2.3 redrawn as a transition diagram of a finite-state machine.

Since σ_2 is an accepting state, we label all its incoming edges with output 1 (see Figure 10.2.4). The states σ_0 and σ_1 are not accepting, so we label all their incoming edges with output 0. We obtain the transition diagram of Figure 10.2.4.   □

As an alternative to Definition 10.2.1, we can regard a finite-state automaton A as consisting of

1. A finite set $\mathcal{I}$ of *input symbols*.
2. A finite set $\mathcal{S}$ of *states*.
3. A *next-state function* f from $\mathcal{S} \times \mathcal{I}$ into $\mathcal{S}$.
4. A subset $\mathcal{A}$ of $\mathcal{S}$ of *accepting states*.
5. An *initial state* $\sigma \in \mathcal{S}$.

If we use this characterization, we write $A = (\mathcal{I}, \mathcal{S}, f, \mathcal{A}, \sigma)$.

EXAMPLE 10.2.4

The transition diagram of the finite-state automaton $A = (\mathcal{I}, \mathcal{S}, f, \mathcal{A}, \sigma)$ where

$$\mathcal{I} = \{a, b\}; \quad \mathcal{S} = \{\sigma_0, \sigma_1, \sigma_2\}; \quad \mathcal{A} = \{\sigma_2\}; \quad \sigma = \sigma_0;$$

and f is given by the following table

	f	
$\mathcal{S}$ \ $\mathcal{I}$	a	b
σ_0	σ_0	σ_1
σ_1	σ_0	σ_2
σ_2	σ_0	σ_2

is shown in Figure 10.2.5.

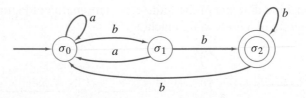

FIGURE 10.2.5   The transition diagram for Example 10.2.4.   □

If a string is input to a finite-state automaton, we will end at either an accepting or a nonaccepting state. The status of this final state determines whether the string is **accepted** by the finite-state automaton.

DEFINITION 10.2.5

Let $A = (\mathcal{I}, \mathcal{S}, f, \mathcal{A}, \sigma)$ be a finite-state automaton. Let $\alpha = x_1 \cdots x_n$ be a string over $\mathcal{I}$. If there exist states $\sigma_0, \ldots, \sigma_n$ satisfying

(a) $\sigma_0 = \sigma$;
(b) $f(\sigma_{i-1}, x_i) = \sigma_i$ for $i = 1, \ldots, n$;
(c) $\sigma_n \in \mathcal{A}$;

we say that α is *accepted by* A. The null string is accepted if and only if $\sigma \in \mathcal{A}$. We let $\mathrm{Ac}(A)$ denote the set of strings accepted by A and we say that A *accepts* $\mathrm{Ac}(A)$.

Let $\alpha = x_1 \cdots x_n$ be a string over $\mathcal{I}$. Define states $\sigma_0, \ldots, \sigma_n$ by conditions (a) and (b) above. We call the (directed) path $(\sigma_0, \ldots, \sigma_n)$ the path *representing* α in A.

It follows from Definition 10.2.5 that if the path P represents the string α in a finite-state automaton A, then A accepts α if and only if P ends at an accepting state.

EXAMPLE 10.2.6

Is the string *abaa* accepted by the finite-state automaton of Figure 10.2.2?

We begin at state σ_0. When a is input, we move to state σ_1. When b is input, we move to state σ_0. When a is input, we move to state σ_1. Finally, when the last symbol a is input, we move to state σ_2. The path $(\sigma_0, \sigma_1, \sigma_0, \sigma_1, \sigma_2)$ represents the string *abaa*. Since the final state σ_2 is an accepting state, the string *abaa* is accepted by the finite-state automaton of Figure 10.2.2.   □

EXAMPLE 10.2.7

Is the string $\alpha = abbabba$ accepted by the finite-state automaton of Figure 10.2.3?

The path representing α terminates at σ_1. Since σ_1 is not an accepting state, the string α is not accepted by the finite-state automaton of Figure 10.2.3.   □

We next give two examples illustrating design problems.

EXAMPLE 10.2.8

Design a finite-state automaton that accepts precisely those strings over $\{a, b\}$ that contain no a's.

The idea is to use two states:

    A: An a was found.
    NA: No a's were found.

The state NA is the initial state and the only accepting state. It is now a simple matter to draw the edges (see Figure 10.2.6). Notice that the finite-state automaton correctly accepts the null string.   □

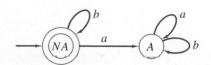

FIGURE 10.2.6
A finite-state automaton that accepts precisely those strings over $\{a, b\}$ that contain no a's.

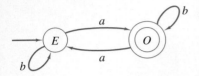

FIGURE 10.2.7
A finite-state automaton that accepts precisely those strings over $\{a, b\}$ that contain an odd number of a's.

EXAMPLE 10.2.9

Design a finite-state automaton that accepts precisely those strings over $\{a, b\}$ that contain an odd number of a's.

This time the two states are

E: An even number of a's was found.
O: An odd number of a's was found.

The initial state is E and the accepting state is O. We obtain the transition diagram shown in Figure 10.2.7. $\square$

A finite-state automaton is essentially an algorithm to decide whether or not a given string is accepted. As an example, we convert the transition diagram of Figure 10.2.7 to an algorithm.

ALGORITHM 10.2.10

This algorithm determines whether a string over $\{a, b\}$ is accepted by the finite-state automaton whose transition diagram is given in Figure 10.2.7.

Input: n, the length of the string ($n = 0$ designates the null string)
$s_1 s_2 \cdots s_n$, the string

Output: "Accept" if the string is accepted
"Reject" if the string is not accepted

```
procedure fsa(s, n)
  state := 'E'
  for i := 1 to n do
    begin
    if state = 'E' and si = 'a' then
      state := 'O'
    if state = 'O' and si = 'a' then
      state := 'E'
    end
  if state = 'O' then
    return("Accept")
  else
    return("Reject")
end fsa
```

If two finite-state automata accept precisely the same strings, we say that the automata are **equivalent**.

DEFINITION 10.2.11

The finite-state automata A and A' are *equivalent* if $\mathrm{Ac}(A) = \mathrm{Ac}(A')$.

EXAMPLE 10.2.12

It can be verified that the finite-state automata of Figures 10.2.6 and 10.2.8 are equivalent (see Exercise 33). $\square$

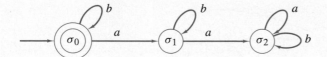

Figure 10.2.8   A finite-state automaton equivalent to that in Figure 10.2.6.

If we define a relation R on a set of finite-state automata by the rule ARA' if A and A' are equivalent (in the sense of Definition 10.2.11), R is an equivalence relation. Each equivalence class consists of a set of mutually equivalent finite-state automata.

Exercises

In Exercises 1–3, show that each finite-state machine is a finite-state automaton and redraw the transition diagram as the diagram of a finite-state automaton.

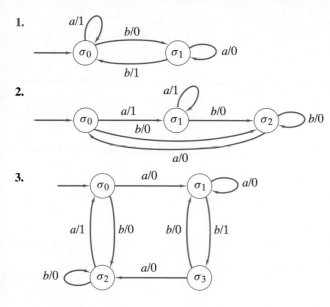

In Exercises 4–6, redraw the transition diagram of the finite-state automaton as the transition diagram of a finite-state machine.

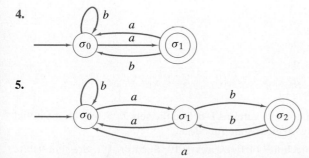

6.

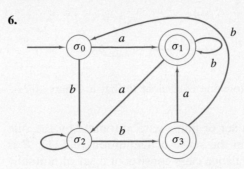

In Exercises 7–9, draw the transition diagram of the finite-state automaton $(\mathcal{I}, \mathcal{S}, f, \mathcal{A}, \sigma_0)$.

7. $\mathcal{I} = \{a, b\}, \mathcal{S} = \{\sigma_0, \sigma_1, \sigma_2\}, \mathcal{A} = \{\sigma_0\}$

		f
$\mathcal{I}$ $\diagdown$ $\mathcal{S}$	a	b
σ_0	σ_1	σ_0
σ_1	σ_2	σ_0
σ_2	σ_0	σ_2

8. $\mathcal{I} = \{a, b\}, \mathcal{S} = \{\sigma_0, \sigma_1, \sigma_2\}, \mathcal{A} = \{\sigma_0, \sigma_2\}$

		f
$\mathcal{I}$ $\diagdown$ $\mathcal{S}$	a	b
σ_0	σ_1	σ_1
σ_1	σ_0	σ_2
σ_2	σ_0	σ_1

9. $\mathcal{I} = \{a, b, c\}, \mathcal{S} = \{\sigma_0, \sigma_1, \sigma_2, \sigma_3\}, \mathcal{A} = \{\sigma_0, \sigma_2\}$

		f	
$\mathcal{I}$ $\diagdown$ $\mathcal{S}$	a	b	c
σ_0	σ_1	σ_0	σ_2
σ_1	σ_0	σ_3	σ_0
σ_2	σ_3	σ_2	σ_0
σ_3	σ_1	σ_0	σ_1

10. For each finite-state automaton in Exercises 1–6, find the sets $\mathcal{I}, \mathcal{S}$, and $\mathcal{A}$, the initial state, and the table defining the next-state function.

11. Which of the finite-state machines of Exercises 1–10, Section 10.1, are finite-state automata?

12. What must the table of a finite-state machine M look like in order for M to be a finite-state automaton?

In Exercises 13–17, determine whether the given string is accepted by the given finite-state automaton.

13. *abbaa*; Figure 10.2.2

14. *abbaa*; Figure 10.2.3

15. *aabaabb*; Figure 10.2.5

16. *aaabbbaab*; Exercise 5

17. *aaababbab*; Exercise 6

18. Show that a string α over $\{a, b\}$ is accepted by the finite-state automaton of Figure 10.2.2 if and only if α ends with a.

19. Show that a string α over $\{a, b\}$ is accepted by the finite-state automaton of Figure 10.2.5 if and only if α ends with bb.

★ **20.** Characterize the strings accepted by the finite-state automata of Exercises 1–9.

In Exercises 21–31, draw the transition diagram of a finite-state automaton that accepts the given set of strings over $\{a, b\}$.

21. Even number of a's

22. Exactly one b

23. At least one b

24. Exactly two a's

25. At least two a's

26. Contains m a's, where m is a multiple of 3

27. Starts with *baa*

★ **28.** Contains *abba*

29. Every b is followed by a

★ **30.** Ends with *aba*

★ **31.** Starts with *ab* and ends with *baa*

32. Write algorithms, similar to Algorithm 10.2.10, that decide whether or not a given string is accepted by the finite-state automata of Exercises 1–9.

33. Give a formal argument to show that the finite-state automata of Figures 10.2.6 and 10.2.8 are equivalent.

34. Let L be a finite set of strings over $\{a, b\}$. Show that there is a finite-state automaton that accepts L.

35. Let L be the set of strings accepted by the finite-state automaton of Exercise 6. Let S denote the set of all strings over $\{a, b\}$. Design a finite-state automaton that accepts $S - L$.

36. Let L_i be the set of strings accepted by the finite-state automaton $A_i = (\mathcal{I}, \mathcal{S}_i, f_i, \mathcal{A}_i, \sigma_i)$, $i = 1, 2$. Let
$$A = (\mathcal{I}, \mathcal{S}_1 \times \mathcal{S}_2, f, \mathcal{A}, \sigma)$$
where
$$f\left((S_1, S_2), x\right) = \left(f_1(S_1, x), f_2(S_2, x)\right),$$
$$\mathcal{A} = \{(A_1, A_2) \mid A_1 \in \mathcal{A}_1 \text{ and } A_2 \in \mathcal{A}_2\},$$
$$\sigma = (\sigma_1, \sigma_2).$$
Show that $\mathrm{Ac}(A) = L_1 \cap L_2$.

37. Let L_i be the set of strings accepted by the finite-state automaton $A_i = (\mathcal{I}, \mathcal{S}_i, f_i, \mathcal{A}_i, \sigma_i)$, $i = 1, 2$. Let
$$A = (\mathcal{I}, \mathcal{S}_1 \times \mathcal{S}_2, f, \mathcal{A}, \sigma)$$
where
$$f\left((S_1, S_2), x\right) = \left(f_1(S_1, x), f_2(S_2, x)\right),$$
$$\mathcal{A} = \{(A_1, A_2) \mid A_1 \in \mathcal{A}_1 \text{ or } A_2 \in \mathcal{A}_2\},$$
$$\sigma = (\sigma_1, \sigma_2).$$
Show that $\mathrm{Ac}(A) = L_1 \cup L_2$.

In Exercises 38–42, let $L_i = \text{Ac}(A_i)$, $i = 1, 2$. Draw the transition diagrams of the finite-state automata that accept $L_1 \cap L_2$ and $L_1 \cup L_2$.

38. A_1 given by Exercise 4; A_2 given by Exercise 5
39. A_1 given by Exercise 4; A_2 given by Exercise 6
40. A_1 given by Exercise 5; A_2 given by Exercise 6
41. A_1 given by Exercise 6; A_2 given by Exercise 6
42. A_1 given by Figure 10.5.7, Section 10.5; A_2 given by Exercise 6

10.3 LANGUAGES AND GRAMMARS

According to *Webster's New Collegiate Dictionary*, language is a "body of words and methods of combining words used and understood by a considerable community." Such languages are often called **natural languages** to distinguish them from **formal languages**, which are used to model natural languages and to communicate with computers. The rules of a natural language are very complex and difficult to characterize completely. On the other hand, it is possible to specify completely the rules by which certain formal languages are constructed. We begin with the definition of a formal language.

DEFINITION 10.3.1

Let A be a finite set. A (*formal*) *language* L over A is a subset of A^*, the set of all strings over A.

EXAMPLE 10.3.2

Let $A = \{a, b\}$. The set L of all strings over A containing an odd number of a's is a language over A. As we saw in Example 10.2.9, L is precisely the set of strings over A accepted by the finite-state automaton of Figure 10.2.7.   □

One way to define a language is to give a list of rules that the language is assumed to obey.

DEFINITION 10.3.3

A *phrase-structure grammar* (or, simply, *grammar*) G consists of
   (a) A finite set N of *nonterminal symbols*.
   (b) A finite set T of *terminal symbols* where $N \cap T = \emptyset$.
   (c) A finite subset P of $[(N \cup T)^* - T^*] \times (N \cup T)^*$, called the set of *productions*.
   (d) A *starting symbol* $\sigma \in N$.
We write $G = (N, T, P, \sigma)$.
   A production $(A, B) \in P$ is usually written

$$A \to B.$$

Definition 10.3.3c states that in the production $A \to B$, $A \in (N \cup T)^* - T^*$ and $B \in (N \cup T)^*$; thus A must include at least one nonterminal symbol, whereas B can consist of any combination of nonterminal and terminal symbols.

EXAMPLE 10.3.4

Let

$$N = \{\sigma, S\}$$

$$T = \{a, b\}$$

$$P = \{\sigma \to b\sigma, \sigma \to aS, S \to bS, S \to b\}.$$

Then $G = (N, T, P, \sigma)$ is a grammar.   □

Given a grammar G, we can construct a language $L(G)$ from G by using the productions to derive the strings that make up $L(G)$. The idea is to start with the starting symbol and then repeatedly use productions until a string of terminal symbols is obtained. The language $L(G)$ is the set of all such strings obtained. Definition 10.3.5 gives the formal details.

DEFINITION 10.3.5

Let $G = (N, T, P, \sigma)$ be a grammar.

If $\alpha \to \beta$ is a production and $x\alpha y \in (N \cup T)^*$, we say that $x\beta y$ is *directly derivable* from $x\alpha y$ and write

$$x\alpha y \Rightarrow x\beta y.$$

If $\alpha_i \in (N \cup T)^*$ for $i = 1, \ldots, n$, and α_{i+1} is directly derivable from α_i for $i = 1, \ldots, n - 1$, we say that α_n is *derivable from* α_1 and write

$$\alpha_1 \Rightarrow \alpha_n.$$

We call

$$\alpha_1 \Rightarrow \alpha_2 \Rightarrow \cdots \Rightarrow \alpha_n$$

the *derivation of α_n (from α_1).* By convention, any element of $(N \cup T)^*$ is derivable from itself.

The *language generated by G,* written $L(G)$, consists of all strings over T derivable from σ.

EXAMPLE 10.3.6

Let G be the grammar of Example 10.3.4.

The string $abSbb$ is directly derivable from $aSbb$, written

$$aSbb \Rightarrow abSbb,$$

by using the production $S \to bS$.

The string $bbab$ is derivable from σ, written

$$\alpha \Rightarrow bbab.$$

The derivation is

$$\sigma \Rightarrow b\sigma \Rightarrow bb\sigma \Rightarrow bbaS \Rightarrow bbab.$$

The only derivations from σ are

$$\sigma \Rightarrow b\sigma$$

$$\vdots$$

$$\Rightarrow b^n \sigma \qquad n \geq 0$$

$$\Rightarrow b^n a S$$

$$\vdots$$

$$\Rightarrow b^n a b^{m-1} S$$

$$\Rightarrow b^n a b^m \qquad n \geq 0, \quad m \geq 1.$$

Thus $L(G)$ consists of the strings over $\{a, b\}$ containing precisely one a that end with b. $\qquad \square$

An alternative way to state the productions of a grammar is by using **Backus normal form** (or **Backus-Naur form** or **BNF**). In BNF the nonterminal symbols typically begin with "$<$" and end with "$>$." The production $S \to T$ is written $S ::= T$. Productions of the form

$$S ::= T_1, \quad S ::= T_2, \quad \ldots, \quad S ::= T_n$$

may be combined as

$$S ::= T_1 \mid T_2 \mid \cdots \mid T_n.$$

The bar "$\mid$" is read "or."

EXAMPLE 10.3.7   *A Grammar for Integers*

An integer is defined as a string consisting of an optional sign ($+$ or $-$) followed by a string of digits (0 through 9). The following grammar generates all integers.

$$<\text{digit}> ::= 0 \mid 1 \mid 2 \mid 3 \mid 4 \mid 5 \mid 6 \mid 7 \mid 8 \mid 9$$

$$<\text{integer}> ::= <\text{signed integer}> \mid <\text{unsigned integer}>$$

$$<\text{signed integer}> ::= + <\text{unsigned integer}> \mid - <\text{unsigned integer}>$$

$$<\text{unsigned integer}> ::= <\text{digit}> \mid <\text{digit}> <\text{unsigned integer}>$$

The starting symbol is $<\text{integer}>$.

For example, the derivation of the integer -901 is

$$<\text{integer}> \Rightarrow <\text{signed integer}>$$

$$\Rightarrow - <\text{unsigned integer}>$$

$$\Rightarrow - <\text{digit}> <\text{unsigned integer}>$$

$$\Rightarrow - <\text{digit}> <\text{digit}> <\text{unsigned integer}>$$

$$\Rightarrow - <\text{digit}> <\text{digit}> <\text{digit}>$$

$$\Rightarrow -9 <\text{digit}> <\text{digit}>$$

$$\Rightarrow -90 <\text{digit}>$$

$$\Rightarrow -901.$$

In the notation of Definition 10.3.3, this language consists of

1. The set $N = \{<\text{digit}>, <\text{integer}>, <\text{signed integer}>,$ $<\text{unsigned integer}>\}$ of nonterminal symbols.
2. The set $T = \{0, 1, 2, 3, 4, 5, 6, 7, 8, 9, +, -\}$ of terminal symbols.
3. The productions

$$<\text{digit}> \rightarrow 0, \ldots, <\text{digit}> \rightarrow 9,$$

$$<\text{integer}> \rightarrow <\text{signed integer}>,$$

$$<\text{integer}> \rightarrow <\text{unsigned integer}>,$$

$$<\text{signed integer}> \rightarrow + <\text{unsigned integer}>,$$

$$<\text{signed integer}> \rightarrow - <\text{unsigned integer}>,$$

$$<\text{unsigned integer}> \rightarrow <\text{digit}>,$$

$$<\text{unsigned integer}> \rightarrow <\text{digit}><\text{unsigned integer}> .$$

4. The starting symbol $<\text{integer}>$. $\qquad\qquad\square$

Computer languages, such as FORTRAN, Pascal, and C, are typically specified in BNF. Example 10.3.7 shows how an integer constant in a computer language might be specified in BNF.

Grammars are classified according to the types of productions that define the grammars.

DEFINITION 10.3.8

Let G be a grammar and let λ denote the null string.
- (a) If every production is of the form

$$\alpha A\beta \rightarrow \alpha\delta\beta, \qquad \text{where } \alpha, \beta \in (N \cup T)^*, \quad A \in N,$$
$$\delta \in (N \cup T)^* - \{\lambda\}, \qquad\qquad (10.3.1)$$

  we call G a *context-sensitive* (or *type* 1) *grammar*.
- (b) If every production is of the form

$$A \rightarrow \delta, \qquad \text{where } A \in N, \quad \delta \in (N \cup T)^*, \qquad (10.3.2)$$

  we call G a *context-free* (or *type* 2) *grammar*.
- (c) If every production is of the form

$$A \rightarrow a \text{ or } A \rightarrow aB \text{ or } A \rightarrow \lambda, \qquad \text{where } A, B \in N, \quad a \in T,$$

  we call G a *regular* (or *type* 3) *grammar*.

According to (10.3.1), in a context-sensitive grammar, we may replace A by δ if A is in the context of α and β. In a context-free grammar, (10.3.2) states that we may replace A by δ anytime. A regular grammar has especially simple substitution rules: We replace a nonterminal symbol by a terminal symbol, by a terminal symbol followed by a nonterminal symbol, or by the null string.

Notice that a regular grammar is a context-free grammar and that a context-free grammar with no productions of the form $A \rightarrow \lambda$ is a context-sensitive grammar.

Some definitions allow a to be replaced by a string of terminals in Definition 10.3.8c; however, it can be shown (see Exercise 32) that the two definitions produce the same languages.

EXAMPLE 10.3.9

The grammar G defined by

$$T = \{a, b, c\}, \qquad N = \{\sigma, A, B, C, D, E\},$$

with productions

$$\sigma \to aAB, \quad \sigma \to aB, \quad A \to aAC, \quad A \to aC, \quad B \to Dc,$$
$$D \to b, \quad CD \to CE, \quad CE \to DE, \quad DE \to DC, \quad Cc \to Dcc,$$

and starting symbol σ is context-sensitive. For example, the production $CE \to DE$ says that we can replace C by D if C is followed by E and the production $Cc \to Dcc$ says that we can replace C by Dc if C is followed by c.

We can derive DC from CD since

$$CD \Rightarrow CE \Rightarrow DE \Rightarrow DC.$$

The string $a^3 b^3 c^3$ is in $L(G)$, since we have

$$\sigma \Rightarrow aAB \Rightarrow aaACB \Rightarrow aaaCCDc \Rightarrow aaaDCCc \Rightarrow aaaDCDcc$$
$$\Rightarrow aaaDDCcc \Rightarrow aaaDDDccc \Rightarrow aaabbbccc.$$

It can be shown (see Exercise 33) that

$$L(G) = \{a^n b^n c^n \mid n = 1, 2, \ldots\}. \qquad \square$$

It is natural to allow the language $L(G)$ to inherit a property of a grammar G. The next definition makes this concept precise.

DEFINITION 10.3.10

A language L is *context-sensitive* (respectively, *context-free*, *regular*) if there is a context-sensitive (respectively, context-free, regular) grammar G with $L = L(G)$.

EXAMPLE 10.3.11

According to Example 10.3.9, the language

$$L = \{a^n b^n c^n \mid n = 1, 2, \ldots\}$$

is context-sensitive. It can be shown (see [Hopcroft, p. 127]) that there is no context-free grammar G with $L = L(G)$; hence L is not a context-free language. $\qquad \square$

EXAMPLE 10.3.12

The grammar G defined by

$$T = \{a, b\}, \qquad N = \{\sigma\},$$

with productions

$$\sigma \rightarrow a\sigma b, \qquad \sigma \rightarrow ab$$

and starting symbol σ, is context-free. The only derivations of σ are

$$\sigma \Rightarrow a\sigma b$$

$$\vdots$$

$$\Rightarrow a^{n-1}\sigma b^{n-1}$$

$$\Rightarrow a^{n-1}abb^{n-1} = a^n b^n.$$

Thus $L(G)$ consists of the strings over $\{a, b\}$ of the form $a^n b^n$, $n = 1, 2 \ldots$. This language is context-free. In Section 10.5 (see Example 10.5.6), we will show that $L(G)$ is not regular. $\qquad \square$

It follows from Examples 10.3.11 and 10.3.12 that the set of context-free languages that do not contain the null string is a proper subset of the set of context-sensitive languages and that the set of regular languages is a proper subset of the set of context-free languages. It can also be shown that there are languages that are not context-sensitive.

EXAMPLE 10.3.13

The grammar G defined in Example 10.3.4 is regular. Thus the language

$$L(G) = \{b^n ab^m \mid n = 0, 1, \ldots; m = 1, 2, \ldots\}$$

it generates is regular. $\qquad \square$

EXAMPLE 10.3.14

The grammar of Example 10.3.7 is context-free, but not regular. However, if we change the productions to

$$< \text{integer} > ::= + < \text{unsigned integer} > | - < \text{unsigned integer} > |$$

$$0 < \text{digits} > | 1 < \text{digits} > | \cdots | 9 < \text{digits} >$$

$$< \text{unsigned integer} > ::= 0 < \text{digits} > | 1 < \text{digits} > | \cdots | 9 < \text{digits} >$$

$$< \text{digits} > ::= 0 < \text{digits} > | 1 < \text{digits} > | \cdots | 9 < \text{digits} > | \lambda,$$

the resulting grammar is regular. Since the language generated is unchanged, it follows that the set of strings representing integers is a regular language. $\qquad \square$

Example 10.3.14 motivates the following definition.

DEFINITION 10.3.15

Grammars G and G' are *equivalent* if $L(G) = L(G')$.

EXAMPLE 10.3.16

The grammars of Examples 10.3.7 and 10.3.14 are equivalent.   □

If we define a relation R on a set of grammars by the rule $G R G'$ if G and G' are equivalent (in the sense of Definition 10.3.15), R is an equivalence relation. Each equivalence class consists of a set of mutually equivalent grammars.

We close this section by briefly introducing another kind of grammar that can be used to generate fractal curves.

DEFINITION 10.3.17

A *context-free interactive Lindenmayer grammar* consists of

(a) A finite set N of *nonterminal symbols.*
(b) A finite set T of *terminal symbols* where $N \cap T = \emptyset$.
(c) A finite set P of *productions* $A \rightarrow B$, where $A \in N \cup T$ and $B \in (N \cup T)^*$.
(d) A *starting symbol* $\sigma \in N$.

The difference between a context-free interactive Lindenmayer grammar and a context-free grammar is that a context-free interactive Lindenmayer grammar allows productions of the form $A \rightarrow B$, where A is a terminal or a nonterminal. (In a context-free grammar, A must be a nonterminal.)

The rules for deriving strings in a context-free interactive Lindenmayer grammar are different from the rules for deriving strings in a phrase-structure grammar (see Definition 10.3.5). In a context-free interactive Lindenmayer grammar, to derive the string β from the string α, *all* symbols in α must be replaced *simultaneously*. The formal definition follows.

DEFINITION 10.3.18

Let $G = (N, T, P, \sigma)$ be a context-free interactive Lindenmayer grammar. If

$$\alpha = x_1 \cdots x_n$$

and there are productions

$$x_i \rightarrow \beta_i$$

in P, for $i = 1, \ldots n$, we write

$$\alpha \Rightarrow \beta_1 \cdots \beta_n$$

and say that $\beta_1 \cdots \beta_n$ is *directly derivable* from α. If α_{i+1} is directly derivable from α_i for $i = 1, \ldots, n-1$, we say that α_n is *derivable from* α_1 and write

$$\alpha_1 \Rightarrow \alpha_n.$$

We call

$$\alpha_1 \Rightarrow \alpha_2 \Rightarrow \cdots \Rightarrow \alpha_n$$

the *derivation of* α_n *(from* α_1*)*. The *language generated by* G, written $L(G)$, consists of all strings over T derivable from σ.

EXAMPLE 10.3.19   *The von Koch Snowflake*

Let

$$N = \{D\}$$
$$T = \{d, +, -\}$$
$$P = \{D \rightarrow D - D + + D - D, D \rightarrow d, + \rightarrow +, - \rightarrow -\}.$$

We regard $G(N, T, P, D)$ as a context-free Lindenmayer grammar. As an example of a derivation from D, we have

$$D \Rightarrow D - D + + D - D \Rightarrow d - d + + d - d.$$

Thus $d - d + + d - d \in L(G)$.

We now impose a meaning on the strings in $L(G)$. We interpret the symbol d as a command to draw a straight line of a fixed length in the current direction; we interpret $+$ as a command to turn right by $60°$; and we interpret $-$ as a command to turn left by $60°$. If we begin at the left and the first move is horizontal to the right, when the string $d - d + + d - d$ is interpreted, we obtain the curve shown in Figure 10.3.1a.

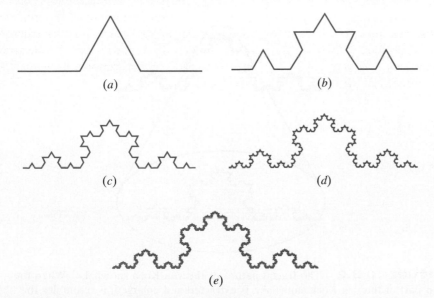

(a)

(b)

(c)

(d)

(e)

FIGURE 10.3.1   von Koch snowflakes.

The next-longest string in $L(G)$ is

$$d - d + + d - d - d - d + + d - d + + d - d + + d - d - d - d + + d - d$$

whose derivation is

$$
\begin{aligned}
D \Rightarrow\ & D - D + + D - D \\
\Rightarrow\ & D - D + + D - D - D - D + + D - D + + D \\
& - D + + D - D - D - D + + D - D \\
\Rightarrow\ & d - d + + d - d - d - d + + d - d + + d \\
& - d + + d - d - d - d + + d - d.
\end{aligned}
$$

No shorter string is possible because *all* symbols must be replaced simultaneously using productions (Definition 10.3.18). If we replace some D's by d and other D's by $D - D + + D - D$, we cannot derive any string from the resulting string, let alone a terminal string, since d does not occur on the left side of any production.

When the string

$$d - d + + d - d - d - d + + d - d + + d - d + + d - d - d - d + + d - d$$

is interpreted, we obtain the curve shown in Figure 10.3.1b.

The curves obtained by interpreting the next-longest strings in $L(G)$ are shown in Figure 10.3.1c–e. These curves are known as **von Koch snowflakes**.

□

Curves such as the von Koch snowflake are called **fractal curves** (see [Peitgen]). A characteristic of fractal curves is that a part of the whole resembles the whole. For example, as shown in Figure 10.3.2, when the part of the von Koch snowflake indicated is extracted and enlarged, it resembles the original.

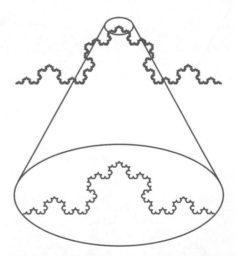

FIGURE 10.3.2   The fractal nature of the von Koch snowflake. When the top part of the von Koch snowflake is extracted and enlarged, it resembles the original.

Context-free and context-sensitive interactive Lindenmayer grammars were invented in 1968 by A. Lindenmayer (see [Lindenmayer]) to model the growth of plants. As Example 10.3.19 suggests, these grammars can be used in computer graphics to generate images (see [Prusinkiewicz 1986, 1988; Smith]). It can be shown (see [Wood, p. 503]) that the class of languages generated by context-sensitive Lindenmayer grammars is exactly the same as the class of languages generated by phrase-structure grammars.

༄ ༄ ༄

Exercises

In Exercises 1–6, determine whether the given grammar is context-sensitive, context-free, regular, or none of these. Give all characterizations that apply.

1. $T = \{a, b\}$, $N = \{\sigma, A\}$, with productions

$$\sigma \to b\sigma, \quad \sigma \to aA, \quad A \to a\sigma,$$
$$A \to bA, \quad A \to a, \quad \sigma \to b,$$

and starting symbol σ.

2. $T = \{a, b, c\}$, $N = \{\sigma, A, B\}$, with productions

$$\sigma \to AB, \quad AB \to BA, \quad A \to aA,$$
$$B \to Bb, \quad A \to a, \quad B \to b,$$

and starting symbol σ.

3. $T = \{a, b\}$, $N = \{\sigma, A, B\}$, with productions

$$\sigma \to A, \quad \sigma \to AAB, \quad Aa \to ABa, \quad A \to aa,$$
$$Bb \to ABb, \quad AB \to ABB, \quad B \to b,$$

and starting symbol σ.

4. $T = \{a, b, c\}$, $N = \{\sigma, A, B\}$, with productions

$$\sigma \to BAB, \quad \sigma \to ABA, \quad A \to AB, \quad B \to BA,$$
$$A \to aA, \quad A \to ab, \quad B \to b,$$

and starting symbol σ.

5. $<S> ::= b <S> \mid a <A> \mid a$

$<A> ::= a <S> \mid b <B>$

$<B> ::= b <A> \mid a <S> \mid b$

with starting symbol $<S>$.

6. $T = \{a, b\}$, $N = \{\sigma, A, B\}$, with productions

$$\sigma \to AA\sigma, \quad AA \to B, \quad B \to bB, \quad A \to a,$$

and starting symbol σ.

In Exercises 7–11, show that the given string α is in $L(G)$ for the given grammar G by giving a derivation of α.

7. *bbabbab*, Exercise 1

8. *abab*, Exercise 2

9. *aabbaab*, Exercise 3

10. *abbbaabab*, Exercise 4

11. *abaabbabba*, Exercise 5

12. Write the grammars of Examples 10.3.4 and 10.3.9 and Exercises 1–4 and 6 in BNF.

★ **13.** Let G be the grammar of Exercise 1. Show that $\alpha \in L(G)$ if and only if α is nonnull and contains an even number of a's.

★ **14.** Let G be the grammar of Exercise 5. Characterize $L(G)$.

In Exercises 15–24, write a grammar that generates the strings having the given property.

15. Strings over $\{a, b\}$ starting with a

16. Strings over $\{a, b\}$ ending with ba

17. Strings over $\{a, b\}$ containing ba

★ **18.** Strings over $\{a, b\}$ not ending with ab

19. Integers with no leading 0's

20. Floating-point numbers (numbers such as .294, 89., 67.284)

21. Exponential numbers (numbers including floating-point numbers and numbers such as 6.9E3, 8E12, 9.6E–4, 9E–10)

22. Boolean expressions in $X_1, \ldots, X_n$

23. All strings over $\{a, b\}$

24. Strings $x_1 \cdots x_n$ over $\{a, b\}$ with $x_1 \cdots x_n = x_n \cdots x_1$

Each grammar in Exercises 25–31 is proposed as generating the set L of strings over $\{a, b\}$ that contain equal numbers of a's and b's. If the grammar generates L, prove that it does so. If the grammar does not generate L, give a counterexample and prove that your counterexample is correct. In each grammar, S is the starting symbol.

25. $S \to aSb \mid bSa \mid \lambda$

26. $S \to aSb \mid bSa \mid SS \mid \lambda$

27. $S \to aB \mid bA \mid \lambda, B \to b \mid bA, A \to a \mid aB$

28. $S \to abS \mid baS \mid aSb \mid bSa \mid \lambda$

29. $S \to aSb \mid bSa \mid abS \mid baS \mid Sab \mid Sba \mid \lambda$

30. $S \to aB \mid bA, A \to a \mid SA, B \to b \mid SB$

31. $S \to aSbS \mid bSaS \mid \lambda$

★ **32.** Let G be a grammar and let λ denote the null string. Show that if every production is of the form

$$A \to \alpha \text{ or } A \to \alpha B \text{ or } A \to \lambda, \qquad \text{where } A, B \in N, \quad \alpha \in T^* - \{\lambda\},$$

there is a regular grammar G' with $L(G) = L(G')$.

★ **33.** Let G be the grammar of Example 10.3.9. Show that

$$L(G) = \{a^n b^n c^n \mid n = 1, 2, \ldots\}.$$

34. Show that the language

$$\{a^n b^n c^k \mid n, k \in \{1, 2, \ldots\}\}$$

is a context-free language.

35. Let

$$N = \{S, D\}$$

$$T = \{d, +, -\}$$

$$P = \{S \rightarrow D + D + D + D,$$

$$D \rightarrow D + D - D - DD + D + D - D \mid d, + \rightarrow +, - \rightarrow -\}.$$

Regard $G = (N, T, P, S)$ as a context-free Lindenmayer grammar. Interpret the symbol d as a command to draw a straight line of a fixed length in the current direction; interpret $+$ as a command to turn right by $90°$; and interpret $-$ as a command to turn left by $90°$. Generate the two smallest strings in $L(G)$ and draw the corresponding curves. These curves are known as *quadratic Koch islands*.

★ **36.** The following figure,

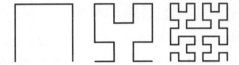

shows the first three stages of the *Hilbert curve*. Define a context-free Lindenmayer grammar that generates strings that when appropriately interpreted generate the Hilbert curve.

10.4 *NONDETERMINISTIC FINITE-STATE AUTOMATA*

In this section and the next, we show that regular grammars and finite-state automata are essentially the same in that either is a specification of a regular language. We begin with an example that illustrates how we can convert a finite-state automaton to a regular grammar.

EXAMPLE 10.4.1

Write the regular grammar given by the finite-state automaton of Figure 10.2.7.

The terminal symbols are the input symbols $\{a, b\}$. The states E and O become the nonterminal symbols. The initial state E becomes the starting symbol. The productions correspond to the directed edges. If there is an edge labeled x from S to S', we write the production

$$S \rightarrow xS'.$$

In our case, we obtain the productions

$$E \rightarrow bE, \qquad E \rightarrow aO, \qquad O \rightarrow aE, \qquad O \rightarrow bO. \qquad (10.4.1)$$

In addition, if S is an accepting state, we include the production

$$S \rightarrow \lambda.$$

In our case, we obtain the additional production

$$O \rightarrow \lambda. \qquad (10.4.2)$$

Then the grammar $G = (N, T, P, E)$, with $N = \{O, E\}$, $T = \{a, b\}$, and P consisting of the productions (10.4.1) and (10.4.2), generates the language $L(G)$, which is the same as the set of strings accepted by the finite-state automaton of Figure 10.2.7.   □

> ### THEOREM 10.4.2

Let A be a finite-state automaton given as a transition diagram. Let σ be the initial state. Let T be the set of input symbols and let N be the set of states. Define productions

$$S \to x S'$$

if there is an edge labeled x from S to S', and

$$S \to \lambda$$

if S is an accepting state. Let G be the regular grammar

$$G = (N, T, P, \sigma).$$

Then the set of strings accepted by A is equal to $L(G)$.

Proof. First we show that $\mathrm{Ac}(A) \subseteq L(G)$. Let $\alpha \in \mathrm{Ac}(A)$. If α is the null string, then σ is an accepting state. In this case, G contains the production

$$\sigma \to \lambda.$$

The derivation

$$\sigma \Rightarrow \lambda. \tag{10.4.3}$$

shows that $\alpha \in L(G)$.

Now suppose $\alpha \in \mathrm{Ac}(A)$ and α is not the null string. Then $\alpha = x_1 \cdots x_n$ for some $x_i \in T$. Since α is accepted by A, there is a path $(\sigma, S_1, \ldots, S_n)$, where S_n is an accepting state, with edges successively labeled $x_1, \ldots, x_n$. It follows that G contains the productions

$$\sigma \to x_1 S_1$$

$$S_{i-1} \to x_i S_i \qquad \text{for } i = 2, \ldots, n.$$

Since S_n is an accepting state, G also contains the production

$$S_n \to \lambda.$$

The derivation

$$\begin{aligned}
\sigma &\Rightarrow x_1 S_1 \\
&\Rightarrow x_1 x_2 S_2 \\
&\;\;\vdots \\
&\Rightarrow x_1 \cdots x_n S_n \\
&\Rightarrow x_1 \cdots x_n
\end{aligned} \tag{10.4.4}$$

shows that $\alpha \in L(G)$.

We complete the proof by showing that $L(G) \subseteq \text{Ac}(A)$. Suppose that $\alpha \in L(G)$. If α is the null string, α must result from the derivation (10.4.3) since a derivation that starts with any other production would yield a nonnull string. Thus the production $\sigma \to \lambda$ is in the grammar. Therefore, σ is an accepting state in A. It follows that $\alpha \in \text{Ac}(A)$.

Now suppose $\alpha \in L(G)$ and α is not the null string. Then $\alpha = x_1 \cdots x_n$ for some $x_i \in T$. It follows that there is a derivation of the form (10.4.4). If, in the transition diagram, we begin at σ and trace the path $(\sigma, S_1, \ldots, S_n)$, we can generate the string α. The last production used in (10.4.4) is $S_n \to \lambda$; thus the last state reached is an accepting state. Therefore, α is accepted by A, so $L(G) \subseteq \text{Ac}(A)$. The proof is complete. ∎

Next, we consider the reverse situation. Given a regular grammar G, we want to construct a finite-state automaton A so that $L(G)$ is precisely the set of strings accepted by A. It might seem, at first glance, that we can simply reverse the procedure of Theorem 10.4.2. However, the next example shows that the situation is a bit more complex.

EXAMPLE 10.4.3

Consider the regular grammar defined by

$$T = \{a, b\}, \qquad N = \{\sigma, C\}$$

with productions

$$\sigma \to b\sigma, \qquad \sigma \to aC, \qquad C \to bC, \qquad C \to b$$

and starting symbol σ.

The nonterminals become states with σ as the initial state. For each production of the form

$$S \to xS'$$

we draw an edge from state S to state S' and label it x. The productions

$$\sigma \to b\sigma, \qquad \sigma \to aC, \qquad C \to bC$$

give the graph shown in Figure 10.4.1. The production $C \to b$ is equivalent to the two productions

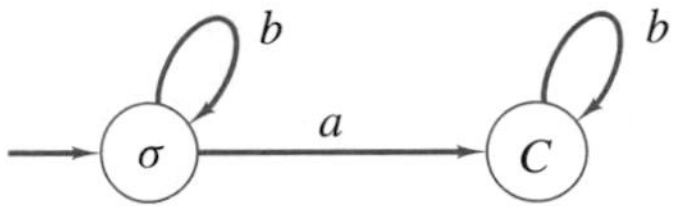

$$C \to bF, \qquad F \to \lambda,$$

where F is an additional nonterminal symbol. The productions

$$\sigma \to b\sigma, \qquad \sigma \to aC, \qquad C \to bC, \qquad C \to bF$$

FIGURE 10.4.1
The graph corresponding to the productions $\sigma \to b\sigma$, $\sigma \to aC$, $C \to bC$.

give the graph shown in Figure 10.4.2. The production

$$F \to \lambda$$

tells us that F should be an accepting state (see Figure 10.4.2). ☐

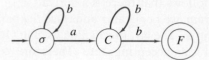

FIGURE 10.4.2   The nondeterministic finite-state automaton corresponding to the grammar $\sigma \to b\sigma, \sigma \to aC, C \to bC, C \to b$.

Unfortunately, the graph of Figure 10.4.2 is not a finite-state automaton. There are several problems. Vertex C has no outgoing edge labeled a and vertex F has no outgoing edges at all. Also, vertex C has two outgoing edges labeled b. A diagram like that of Figure 10.4.2 defines another kind of automaton called a **nondeterministic finite-state automaton**. The reason for the word "nondeterministic" is that when we are in a state where there are multiple outgoing edges all having the same label x, if x is input the situation is nondeterministic—we have a choice of next states. For example, if in Figure 10.4.2 we are in state C and b is input, we have a choice of next states—we can either remain in state C or go to state F.

DEFINITION 10.4.4

A *nondeterministic finite-state automaton A* consists of

(a) A finite set $\mathcal{I}$ of *input symbols*.
(b) A finite set $\mathcal{S}$ of *states*.
(c) A *next-state function* f from $\mathcal{S} \times \mathcal{I}$ into $\mathcal{P}(\mathcal{S})$.
(d) A subset $\mathcal{A}$ of $\mathcal{S}$ of *accepting states*.
(e) An *initial state* $\sigma \in \mathcal{S}$.

We write $A = (\mathcal{I}, \mathcal{S}, f, \mathcal{A}, \sigma)$.

The only difference between a nondeterministic finite-state automaton and a finite-state automaton is that in a finite-state automaton the next-state function takes us to a uniquely defined state, whereas in a nondeterministic finite-state automaton the next-state function takes us to a set of states.

EXAMPLE 10.4.5

For the nondeterministic finite-state automaton of Figure 10.4.2, we have

$$\mathcal{I} = \{a, b\}, \qquad \mathcal{S} = \{\sigma, C, F\}, \qquad \mathcal{A} = \{F\}.$$

The initial state is σ and the next-state function f is given by

S \\ $\mathcal{I}$	a	b
	f	
σ	$\{C\}$	$\{\sigma\}$
C	$\emptyset$	$\{C, F\}$
F	$\emptyset$	$\emptyset$

$\square$

We draw the transition diagram of a nondeterministic finite-state automaton similarly to that of a finite-state automaton. We draw an edge from state S to each state in the set $f(S, x)$ and label each x.

EXAMPLE 10.4.6

The transition diagram of the nondeterministic finite-state automaton

$$\mathcal{I} = \{a, b\}, \qquad \mathcal{S} = \{\sigma, C, D\}, \qquad \mathcal{A} = \{C, D\}$$

with initial state α and next-state function

S \\ $\mathcal{I}$	a	b
	f	
σ	$\{\sigma, C\}$	$\{D\}$
C	$\emptyset$	$\{C\}$
D	$\{C, D\}$	$\emptyset$

is shown in Figure 10.4.3.

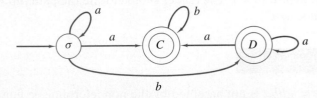

FIGURE 10.4.3   The transition diagram of the nondeterministic finite-state automaton of Example 10.4.6. $\square$

A string α is accepted by a nondeterministic finite-state automaton A if there is some path representing α in the transition diagram of A beginning at the initial state and ending in an accepting state. The formal definition follows.

DEFINITION 10.4.7

Let $A = (\mathcal{I}, \mathcal{S}, f, \mathcal{A}, \sigma)$ be a nondeterministic finite-state automaton. The null string is *accepted* by A if and only if $\sigma \in \mathcal{A}$. If $\alpha = x_1 \cdots x_n$ is a nonnull string over $\mathcal{I}$ and there exist states $\sigma_0, \ldots, \sigma_n$ satisfying

    (a) $\sigma_0 = \sigma$;

    (b) $\sigma_i \in f(\sigma_{i-1}, x_i)$ for $i = 1, \ldots, n$;

    (c) $\sigma_n \in \mathcal{A}$;

we say that α is *accepted* by A. We let $\mathrm{Ac}(A)$ denote the set of strings accepted by A and we say that A *accepts* $\mathrm{Ac}(A)$.

If A and A' are nondeterministic finite-state automata and $\mathrm{Ac}(A) = \mathrm{Ac}(A')$, we say that A and A' are *equivalent*.

If $\alpha = x_1 \cdots x_n$ is a string over $\mathcal{I}$ and there exist states $\sigma_0, \ldots, \sigma_n$ satisfying conditions (a) and (b), we call the path $(\sigma_0, \ldots, \sigma_n)$ a *path representing* α in A.

EXAMPLE 10.4.8

The string

$$\alpha = bbabb$$

is accepted by the nondeterministic finite-state automaton of Figure 10.4.2, since the path $(\sigma, \sigma, \sigma, C, C, F)$, which ends at an accepting state, represents α. Notice that the path $P = (\sigma, \sigma, \sigma, C, C, C)$ also represents α, but that P does not end at an accepting state. Nevertheless, the string α is accepted because there is at least one path representing α that ends at an accepting state. A strings β will fail to be accepted if no path represents β or every path representing β ends at a nonaccepting state. $\square$

EXAMPLE 10.4.9

The string $\alpha = aabaabbb$ is accepted by the nondeterministic finite-state automaton of Figure 10.4.3. The reader should locate the path representing α, which ends at state C. $\square$

EXAMPLE 10.4.10

The string $\alpha = abba$ is not accepted by the nondeterministic finite-state automaton of Figure 10.4.3. Starting at σ, when we input a, there are two choices: Go to C or remain at σ. If we go to C, when we input two b's, our moves are determined and we remain at C. But now when we input the final a, there is no edge along which to move. On the other hand, suppose that when we input the first a, we remain at σ. Then, when we input b, we move to D. But now when we input the next b, there is no edge along which to move. Since there is no path representing α in Figure 10.4.3, the string α is not accepted by the nondeterministic finite-state automaton of Figure 10.4.3. $\square$

We formulate the construction of Example 10.4.3 as a theorem.

THEOREM 10.4.11

Let $G = (N, T, P, \sigma)$ be a regular grammar. Let

$$\mathcal{I} = T$$

$$\mathcal{S} = N \cup \{F\}, \text{ where } F \notin N \cup T$$

$$f(S, x) = \{S' \mid S \rightarrow xS' \in P\} \cup \{F \mid S \rightarrow x \in P\}$$

$$\mathcal{A} = \{F\} \cup \{S \mid S \rightarrow \lambda \in P\}.$$

Then the nondeterministic finite-state automaton $A = (\mathcal{I}, \mathcal{S}, f, \mathcal{A}, \sigma)$ accepts precisely the strings $L(G)$.

Proof. The proof is essentially the same as the proof of Theorem 10.4.2 and is therefore omitted. ∎

It may seem that a nondeterministic finite-state automaton is a more general concept than a finite-state automaton; however, in the next section we will show that given a nondeterministic finite-state automaton A, we can construct a finite-state automaton that is equivalent to A.

ॐ ॐ ॐ

Exercises

In Exercises 1–5, draw the transition diagram of the nondeterministic finite-state automaton $(\mathcal{I}, \mathcal{S}, f, \mathcal{A}, \sigma_0)$.

1. $\mathcal{I} = \{a, b\}, \mathcal{S} = \{\sigma_0, \sigma_1, \sigma_2\}, \mathcal{A} = \{\sigma_0\}$

$\mathcal{S}$ \ $\mathcal{I}$	a	b
σ_0	$\emptyset$	$\{\sigma_1, \sigma_2\}$
σ_1	$\{\sigma_2\}$	$\{\sigma_0, \sigma_1\}$
σ_2	$\{\sigma_0\}$	$\emptyset$

2. $\mathcal{I} = \{a, b\}, \mathcal{S} = \{\sigma_0, \sigma_1, \sigma_2\}, \mathcal{A} = \{\sigma_0, \sigma_1\}$

$\mathcal{S}$ \ $\mathcal{I}$	a	b
σ_0	$\{\sigma_1\}$	$\{\sigma_0, \sigma_2\}$
σ_1	$\emptyset$	$\{\sigma_2\}$
σ_2	$\{\sigma_1\}$	$\emptyset$

3. $\mathcal{I} = \{a, b\}, \mathcal{S} = \{\sigma_0, \sigma_1, \sigma_2, \sigma_3\}, \mathcal{A} = \{\sigma_1\}$

$\mathcal{S}$ \\ $\mathcal{I}$	a	b
σ_0	$\emptyset$	$\{\sigma_3\}$
σ_1	$\{\sigma_1, \sigma_2\}$	$\{\sigma_3\}$
σ_2	$\emptyset$	$\{\sigma_0, \sigma_1, \sigma_3\}$
σ_3	$\emptyset$	$\emptyset$

4. $\mathcal{I} = \{a, b, c\}, \mathcal{S} = \{\sigma_0, \sigma_1, \sigma_2\}, \mathcal{A} = \{\sigma_0\}$

$\mathcal{S}$ \\ $\mathcal{I}$	a	b	c
σ_0	$\{\sigma_1\}$	$\emptyset$	$\emptyset$
σ_1	$\{\sigma_0\}$	$\{\sigma_2\}$	$\{\sigma_0, \sigma_2\}$
σ_2	$\{\sigma_0, \sigma_1, \sigma_2\}$	$\{\sigma_0\}$	$\{\sigma_0\}$

5. $\mathcal{I} = \{a, b, c\}, \mathcal{S} = \{\sigma_0, \sigma_1, \sigma_2, \sigma_3\}, \mathcal{A} = \{\sigma_0, \sigma_3\}$

$\mathcal{S}$ \\ $\mathcal{I}$	a	b	c
σ_0	$\{\sigma_1\}$	$\{\sigma_0, \sigma_1, \sigma_3\}$	$\emptyset$
σ_1	$\{\sigma_2, \sigma_3\}$	$\emptyset$	$\emptyset$
σ_2	$\emptyset$	$\{\sigma_0, \sigma_3\}$	$\{\sigma_1, \sigma_2\}$
σ_3	$\emptyset$	$\emptyset$	$\{\sigma_0\}$

For each nondeterministic finite-state automaton in Exercises 6–10, find the sets $\mathcal{I}$, $\mathcal{S}$, and $\mathcal{A}$, the initial state, and the table defining the next-state function.

6. 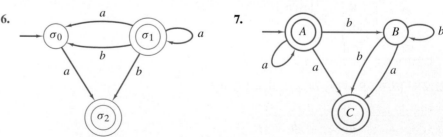   **7.**

8.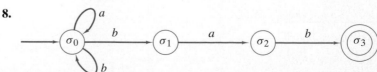

9.

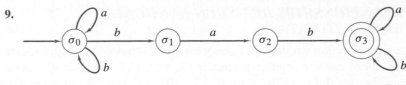

10.

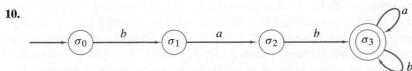

11. Write the regular grammars given by the finite-state automata of Exercises 4–9, Section 10.2.

12. Represent the grammars of Exercises 1 and 5, Section 10.3, and Example 10.3.14 by nondeterministic finite-state automata.

13. Is the string *bbabbb* accepted by the nondeterministic finite-state automaton of Figure 10.4.2? Prove your answer.

14. Is the string *bbabab* accepted by the nondeterministic finite-state automaton of Figure 10.4.2? Prove your answer.

15. Show that a string α over $\{a, b\}$ is accepted by the nondeterministic finite-state automaton of Figure 10.4.2 if and only if α contains exactly one *a* and ends with *b*.

16. Is the string *aaabba* accepted by the nondeterministic finite-state automaton of Figure 10.4.3? Prove your answer.

17. Is the string *aaaab* accepted by the nondeterministic finite-state automaton of Figure 10.4.3? Prove your answer.

18. Characterize the strings accepted by the nondeterministic finite-state automaton of Figure 10.4.3.

19. Show that the strings accepted by the nondeterministic finite-state automaton of Exercise 8 are precisely those strings over $\{a, b\}$ that end *bab*.

★ 20. Characterize the strings accepted by the nondeterministic finite-state automata of Exercises 1–7, 9, and 10.

Design nondeterministic finite-state automata that accept the strings over $\{a, b\}$ having the properties specified in Exercises 21–29.

21. Starting either *abb* or *ba*

22. Ending either *abb* or *ba*

23. Containing either *abb* or *ba*

★ 24. Containing *bab* and *bb*

25. Having each *b* preceded and followed by an *a*

26. Starting with *abb* and ending with *ab*

★ 27. Starting with *ab* but not ending with *ab*

28. Not containing *ba* or *bbb*

★ 29. Not containing *abba* or *bbb*

30. Write regular grammars that generate the strings of Exercises 21–29.

10.5 *RELATIONSHIPS BETWEEN LANGUAGES AND AUTOMATA*

In the preceding section we showed (Theorem 10.4.2) that if A is a finite-state automaton, there exists a regular grammar G, with $L(G) = \text{Ac}(A)$. As a partial converse, we showed (Theorem 10.4.11) that if G is a regular grammar, there exists a nondeterministic finite-state automaton A with $L(G) = \text{Ac}(A)$. In this section we show (Theorem 10.5.4) that if G is a regular grammar, there exists a finite-state automaton A with $L(G) = \text{Ac}(A)$. This result will be deduced from Theorem 10.4.11 by showing that any nondeterministic finite-state automaton can be converted to an equivalent finite-state automaton (Theorem 10.5.3). We will first illustrate the method by an example.

EXAMPLE 10.5.1

Find a finite-state automaton equivalent to the nondeterministic finite-state automaton of Figure 10.4.2.

The set of input symbols is unchanged. The states consist of all subsets

$$\emptyset, \quad \{\sigma\}, \quad \{C\}, \quad \{F\}, \quad \{\sigma, C\}, \quad \{\sigma, F\}, \quad \{C, F\}, \quad \{\sigma, C, F\}$$

of the original set $\mathcal{S} = \{\sigma, C, F\}$ of states. The initial state is $\{\sigma\}$. The accepting states are all subsets

$$\{F\}, \quad \{\sigma, F\}, \quad \{C, F\}, \quad \{\sigma, C, F\}$$

of $\mathcal{S}$ that contain an accepting state of the original nondeterministic finite-state automaton.

An edge is drawn from X to Y and labeled x if $X = \emptyset = Y$ or if

$$\bigcup_{S \in X} f(S, x) = Y.$$

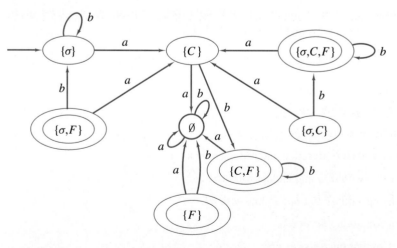

FIGURE 10.5.1   A finite-state automaton equivalent to the nondeterministic finite-state automaton of Figure 10.4.2.

We obtain the finite-state automaton of Figure 10.5.1. The states

$$\{\sigma, F\}, \quad \{\sigma, C\}, \quad \{\sigma, C, F\}, \quad \{F\},$$

which can never be reached, can be deleted. Thus we obtain the simplified, equivalent finite-state automaton of Figure 10.5.2.

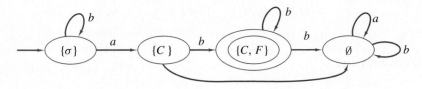

FIGURE 10.5.2   A simplified version of Figure 10.5.1 (with unreachable states deleted).   □

EXAMPLE 10.5.2

The finite-state automaton equivalent to the nondeterministic finite-state automaton of Example 10.4.6 is shown in Figure 10.5.3.

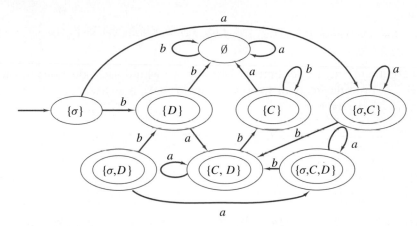

FIGURE 10.5.3   A finite-state automaton equivalent to the nondeterministic finite-state automaton of Figure 10.4.6.   □

We now formally justify the method of Examples 10.5.1 and 10.5.2.

THEOREM 10.5.3

Let $A = (\mathcal{I}, \mathcal{S}, f, \mathcal{A}, \sigma)$ be a nondeterministic finite-state automaton. Let
  (a) $\mathcal{S}' = \mathcal{P}(\mathcal{S})$.
  (b) $\mathcal{S}' = \mathcal{I}$.
  (c) $\sigma' = \{\sigma\}$.
  (d) $\mathcal{A}' = \{X \subseteq \mathcal{S} \mid X \cap \mathcal{A} \neq \emptyset\}$.
  (e) $f'(X, x) = \begin{cases} \emptyset & \text{if } X = \emptyset \\ \displaystyle\bigcup_{S \in X} f(S, x) & \text{if } X \neq \emptyset. \end{cases}$

Then the finite-state automaton $A' = (\mathcal{I}', \mathcal{S}', f', \mathcal{A}', \sigma')$ is equivalent to A.

Proof. Suppose that the string $\alpha = x_1 \cdots x_n$ is accepted by A. Then there exist states $\sigma_0, \ldots, \sigma_n \in \mathcal{S}$ with

$$\sigma_0 = \sigma;$$
$$\sigma_i \in f(\sigma_{i-1}, x_i) \qquad \text{for } i = 1, \ldots, n;$$
$$\sigma_n \in \mathcal{A}.$$

Set $Y_0 = \{\sigma_0\}$ and

$$Y_i = f'(Y_{i-1}, x_i) \qquad \text{for } i = 1, \ldots, n.$$

Since

$$Y_1 = f'(Y_0, x_1) = f'(\{\sigma_0\}, x_1) = f(\sigma_0, x_1),$$

it follows that $\sigma_1 \in Y_1$. Now

$$\sigma_2 \in f(\sigma_1, x_2) \subseteq \bigcup_{S \in Y_1} f(S, x_2) = f'(Y_1, x_2) = Y_2.$$

Again,

$$\sigma_3 \in f(\sigma_2, x_3) \subseteq \bigcup_{S \in Y_2} f(S, x_3) = f'(Y_2, x_3) = Y_3.$$

The argument may be continued (formally, we would use induction) to show that $\sigma_n \in Y_n$. Since σ_n is an accepting state in A, Y_n is an accepting state in A'. Thus, in A', we have

$$f'(\sigma', x_1) = f'(Y_0, x_1) = Y_1$$
$$f'(Y_1, x_2) = Y_2$$
$$\vdots$$
$$f'(Y_{n-1}, x_n) = Y_n.$$

Therefore, α is accepted by A'.

Now suppose that the string $\alpha = x_1 \cdots x_n$ is accepted by A'. Then there exist subsets $Y_0, \ldots, Y_n$ of $\mathcal{S}$ such that

$$Y_0 = \sigma' = \{\sigma\};$$
$$f'(Y_{i-1}, x_i) = Y_i \qquad \text{for } i = 1, \ldots, n;$$

there exists a state $\sigma_n \in Y_n \cap \mathcal{A}$.

Since

$$\sigma_n \in Y_n = f'(Y_{n-1}, x_n) = \bigcup_{S \in Y_{n-1}} f(S, x_n),$$

there exists $\sigma_{n-1} \in Y_{n-1}$ with $\sigma_n \in f(\sigma_{n-1}, x_n)$. Similarly, since

$$\sigma_{n-1} \in Y_{n-1} = f'(Y_{n-2}, x_{n-1}) = \bigcup_{S \in Y_{n-2}} f(S, x_{n-1}),$$

there exists $\sigma_{n-2} \in Y_{n-2}$ with $\sigma_{n-1} \in f(\sigma_{n-2}, x_{n-1})$. Continuing, we obtain

$$\sigma_i \in Y_i \qquad \text{for } i = 0, \ldots, n,$$

with

$$\sigma_i \in f(\sigma_{i-1}, x_i) \qquad \text{for } i = 1, \ldots, n.$$

In particular,

$$\sigma_0 \in Y_0 = \{\sigma\}.$$

Thus $\sigma_0 = \sigma$, the initial state in A. Since σ_n is an accepting state in A, the string α is accepted by A. ■

The next theorem summarizes these results and those of the preceding section.

THEOREM 10.5.4

A language L is regular if and only if there exists a finite-state automaton that accepts precisely the strings in L.

Proof. This theorem restates Theorems 10.4.2, 10.4.11, and 10.5.3. ■

EXAMPLE 10.5.5

Find a finite-state automaton A that accepts precisely the strings generated by the regular grammar G having productions

$$\sigma \to b\sigma, \quad \sigma \to aC, \quad C \to bC, \quad C \to b.$$

The starting symbol is σ, the set of terminal symbols is $\{a, b\}$, and the set of nonterminal symbols is $\{\sigma, C\}$.

The nondeterministic finite-state automaton A' that accepts $L(G)$ is shown in Figure 10.4.2. A finite-state automaton equivalent to A' is shown in Figure 10.5.1 and an equivalent simplified finite-state automaton A is shown in Figure 10.5.2. The finite-state automaton A accepts precisely the strings generated by G. □

We close this section by giving some applications of the methods and theory we have developed.

EXAMPLE 10.5.6   *A Language That Is Not Regular*

Show that the language

$$L = \{a^n b^n \mid n = 1, 2, \ldots\}$$

is not regular.

If L is regular, there exists a finite-state automaton A such that $\text{Ac}(A) = L$. Suppose that A has k states. The string $\alpha = a^k b^k$ is accepted by A. Consider the path P, which represents α. Since there are k states, some state σ is revisited on the part of the path representing a^k. Thus there is a cycle C, all of whose edges are labeled a, that contains σ. We change the path P to obtain a path P' as follows. When we arrive at σ in P, we follow C. After returning to σ on C, we continue on P to the end. If the length of C is j, the path P' represents the string $\alpha' = a^{j+k}b^k$. Since P and P' end at the same state σ' and σ' is an accepting state, α' is accepted by A. This is a contradiction, since α' is not of the form $a^n b^n$. Therefore, L is not regular.   □

EXAMPLE 10.5.7

Let L be the set of strings accepted by the finite-state automaton A of Figure 10.5.4. Construct a finite-state automaton that accepts the strings

$$L^R = \{x_n \cdots x_1 \mid x_1 \cdots x_n \in L\}.$$

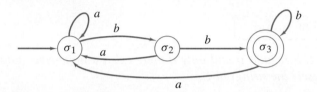

FIGURE 10.5.4   The finite-state automaton for Example 10.5.7 that accepts L.

We want to convert A to a finite-state automaton that accepts L^R. The string $\alpha = x_1 \cdots x_n$ is accepted by A if there is a path P in A representing α that starts at σ_1 and ends at σ_3. If we start at σ_3 and trace P in reverse, we end at σ_1 and process the edges in the order $x_n, \ldots, x_1$. Thus we need only reverse all arrows in Figure 10.5.4 and make σ_3 the starting state and σ_1 the accepting state (see Figure 10.5.5). The result is a nondeterministic finite-state automaton that accepts L^R.

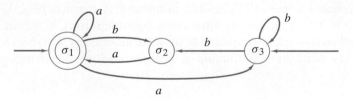

FIGURE 10.5.5   A nondeterministic finite-state automaton that accepts L^R.

After finding an equivalent finite-state automaton and eliminating the unreachable states, we obtain the equivalent finite-state automaton of Figure 10.5.6.

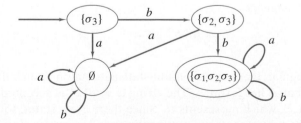

FIGURE 10.5.6   A finite-state automaton that accepts L^R.   □

EXAMPLE 10.5.8

Let L be the set of strings accepted by the finite-state automaton A of Figure 10.5.7. Construct a nondeterministic finite-state automaton that accepts the strings

$$L^R = \{x_n \cdots x_1 \mid x_1 \cdots x_n \in L\}.$$

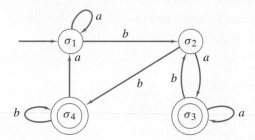

FIGURE 10.5.7   The finite-state automaton for Example 10.5.8 that accepts L.

If A had only one accepting state, we could use the procedure of Example 10.5.7 to construct the desired nondeterministic finite-state automaton. Thus we first construct a nondeterministic finite-state automaton equivalent to A with one accepting state. To do this we introduce an additional state σ_5. Then we arrange for paths terminating at σ_3 or σ_4 to optionally terminate at σ_5 (see Figure 10.5.8). The desired nondeterministic finite-state automaton is obtained from Figure 10.5.8 by the method of Example 10.5.7, (see Figure 10.5.9). Of course, if desired, we could construct an equivalent finite-state automaton. ☐

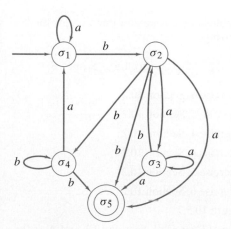

FIGURE 10.5.8   A nondeterministic finite-state automaton with one accepting state equivalent to the finite-state automaton of Figure 10.5.7.

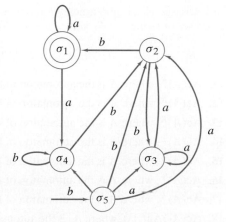

FIGURE 10.5.9   A nondeterministic finite-state automaton that accepts L^R.

꙳ ꙳ ꙳

Exercises

1. Find the finite-state automata equivalent to the nondeterministic finite-state automata of Exercises 1–10, Section 10.4.

In Exercises 2–6, find finite-state automata that accept the strings generated by the regular grammars.

2. Grammar of Exercise 1, Section 10.3

3. Grammar of Exercise 5, Section 10.3

4. $<S> ::= a<A> | a<B>$
   $<A> ::= a<B> | b<S> | b$
   $<B> ::= b<S> | b$
   with starting symbol $<S>$

5. $<S> ::= a<S> | a<A> | b<C> | a$
   $<A> ::= b<A> | a<C>$
   $<B> ::= a<S> | a$
   $<C> ::= a<B> | a<C>$
   with starting symbol $<S>$

6. $<S> ::= a<A> | a<B>$
   $<A> ::= b<S> | b$
   $<B> ::= a<B> | a<C>$
   $<C> ::= a<S> | b<A> | a<C> | a$
   with starting symbol $<S>$

7. Find finite-state automata that accept the strings of Exercises 21–29, Section 10.4.

8. By eliminating unreachable states from the finite-state automaton of Figure 10.5.3, find a simpler, equivalent finite-state automaton.

9. Show that the nondeterministic finite-state automaton of Figure 10.5.5 accepts a string α over $\{a, b\}$ if and only if α begins bb.

★ 10. Characterize the strings accepted by the nondeterministic finite-state automata of Figures 10.5.7 and 10.5.9.

In Exercises 11–21, find a nondeterministic finite-state automaton that accepts the given set of strings. If S_1 and S_2 are sets of strings, we let

$$S_1^+ = \{u_1 u_2 \cdots u_n \mid u_i \in S_1, n \in \{1, 2, \ldots\}\}; \quad S_1 S_2 = \{uv \mid u \in S_1, v \in S_2\}.$$

11. $\text{Ac}(A)^R$, where A is the automaton of Exercise 4, Section 10.2

12. $\text{Ac}(A)^R$, where A is the automaton of Exercise 5, Section 10.2

13. $\text{Ac}(A)^R$, where A is the automaton of Exercise 4, Section 10.2

14. $\text{Ac}(A)^+$, where A is the automaton of Exercise 4, Section 10.2

15. $\text{Ac}(A)^+$, where A is the automaton of Exercise 5, Section 10.2

16. $\text{Ac}(A)^+$, where A is the automaton of Exercise 6, Section 10.2

17. $\text{Ac}(A)^+$, where A is the automaton of Figure 10.5.7

18. $\text{Ac}(A_1)\text{Ac}(A_2)$, where A_1 is the automaton of Exercise 4, Section 10.2, and A_2 is the automaton of Exercise 5, Section 10.2

19. $\text{Ac}(A_1)\text{Ac}(A_2)$, where A_1 is the automaton of Exercise 5, Section 10.2, and A_2 is the automaton of Exercise 6, Section 10.2

20. $\text{Ac}(A_1)\text{Ac}(A_1)$, where A_1 is the automaton of Exercise 6, Section 10.2

21. $\text{Ac}(A_1)\text{Ac}(A_2)$, where A_1 is the automaton of Figure 10.5.7 and A_2 is the automaton of Exercise 5, Section 10.2

22. Find a regular grammar that generates the language L^R, where L is the language generated by the grammar of Exercise 5, Section 10.3.

23. Find a regular grammar that generates the language L^+, where L is the language generated by the grammar of Exercise 5, Section 10.3.

24. Let L_1 (respectively, L_2) be the language generated by the grammar of Exercise 5, Section 10.3 (respectively, Example 10.5.5). Find a regular grammar that generates the language $L_1 L_2$.

★ **25.** Show that the set

$$L = \{x_1 \cdots x_n \mid x_1 \cdots x_n = x_n \cdots x_1\}$$

of strings over $\{a, b\}$ is not a regular language.

26. Show that if L_1 and L_2 are regular languages over $\mathcal{I}$ and S is the set of all strings over $\mathcal{I}$, then each of $S - L_1$, $L_1 \cup L_2$, $L_1 \cap L_2$, L_1^+, and $L_1 L_2$ is a regular language.

★ **27.** Show, by example, that there are context-free languages L_1 and L_2 such that $L_1 \cap L_2$ is not context-free.

★ **28.** Prove or disprove: If L is a regular language, so is

$$\{u^n \mid u \in L, n \in \{1, 2, \ldots\}\}.$$

✑ NOTES

General references on automata, grammars, and languages are [Carroll; Cohen; Davis; Hopcroft; Kelley; McNaughton; Sudkamp; and Wood].

The systematic development of fractal geometry was begun by Benoit B. Mandelbrot (see [Mandelbrot, 1977, 1982]).

A finite-state machine has a primitive internal memory in the sense that it remembers which state it is in. By permitting an external memory on which the machine can read and write data, we can define more powerful machines. Other enhancements are achieved by allowing the machine to scan the input string in either direction and by allowing the machine to alter the input string. It is then possible to characterize the classes of machines that accept context-free languages, context-sensitive languages, and languages generated by phrase-structure grammars.

Turing machines form a particularly important class of machines. Like a finite-state machine, a Turing machine is always in a particular state. The input string to a Turing machine is assumed to reside on a tape that is infinite in both directions. A Turing machine scans one character at a time and after scanning a character, the machine either halts or does some, none, or all of: alter the character, move one position left or right, change states. In particular, the input string can be changed. A Turing machine T accepts a string α if, when α is input to T, T halts in an accepting state. It can be shown that a language L is generated by a phrase-structure grammar if and only if there is a Turing machine that accepts L.

The real importance of Turing machines results from the widely held belief that any function that can be computed by some, perhaps hypothetical, digital computer can be computed by some Turing machine. This last assertion is known as **Turing's hypothesis** or **Church's thesis**. Church's thesis implies that a Turing machine is the correct abstract model of a digital computer. These ideas also yield the following formal definition of algorithm. An **algorithm** is a Turing machine that, given an input string, eventually stops.

590 CHAPTER 10 / AUTOMATA, GRAMMARS, AND LANGUAGES

⟆ CHAPTER REVIEW

Section 10.1

Unit time delay
Serial adder
Finite-state machine
Input symbol
Output symbol
State
Next-state function
Output function
Initial state
Transition diagram
Input and output strings for a finite-state machine
SR flip-flop

Section 10.2

Finite-state automaton
Accepting state
String accepted by a finite-state automaton
Equivalent finite-state automata

Section 10.3

Natural language
Formal language
Phrase-structure grammar
Nonterminal symbol
Terminal symbol
Production
Starting symbol
Directly derivable string
Derivable string
Derivation
Language generated by a grammar
Backus normal form (= Backus-Naur form = BNF)
Context-sensitive grammar (= type 1 grammar)
Context-free grammar (= type 2 grammar)

Regular grammar (= type 3 grammar)
Context-sensitive language
Context-free language
Regular language
Context-free interactive Lindenmayer grammar
von Koch snowflake
Fractal curves

Section 10.4

Given a finite-state automaton A, how to construct a regular grammar G, such that the set of strings accepted by A is equal to the language generated by G (see Theorem 10.4.2)
Nondeterministic finite-state automaton
String accepted by a nondeterministic finite-state automaton
Equivalent nondeterministic finite-state automata
Given a regular grammar G, how to construct a nondeterministic finite-state automaton A such that the language generated by G is equal to the set of strings accepted by A (see Theorem 10.4.11)

Section 10.5

Given a nondeterministic finite-state automaton, how to construct an equivalent deterministic finite-state automaton (see Theorem 10.5.3)
A language L is regular if and only if there exists a finite-state automaton that accepts the strings in L.

⟆ CHAPTER SELF-TEST

Section 10.1

1. Draw the transition diagram of the finite-state machine $(\mathcal{I}, \mathcal{O}, \mathcal{S}, f, g, \sigma_0)$, where $\mathcal{I} = \{a, b\}$, $\mathcal{O} = \{0, 1\}$, and $\mathcal{S} = \{\sigma_0, \sigma_1\}$.

$\mathcal{I}$	f		g	
	a	b	a	b
$\mathcal{S}$				
σ_0	σ_1	σ_0	0	1
σ_1	σ_0	σ_1	1	0

2. Find the sets $\mathcal{I}$, $\mathcal{O}$, and $\mathcal{S}$, the initial state, and the table defining the next-state and output functions for the finite-state machine.

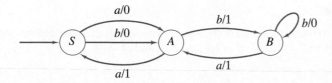

3. For the finite-state machine of Exercise 1, find the output string for the input string *bbaa*.

4. Design a finite-state machine, whose input is a bit string that outputs 0 when it sees 001 and thereafter; otherwise, it outputs 1.

Section 10.2

5. Draw the transition diagram of the finite-state automaton $(\mathcal{I}, \mathcal{S}, f, \mathcal{A}, S)$, where $\mathcal{I} = \{0, 1\}$, $\mathcal{S} = \{S, A, B\}$, and $\mathcal{A} = \{A\}$.

$\mathcal{I}$	f	
	0	**1**
$\mathcal{S}$		
S	A	S
A	S	B
B	A	S

6. Is the string 11010 accepted by the finite-state automaton of Exercise 5?

7. Draw the transition diagram of a finite-state automaton that accepts the set of strings over $\{0, 1\}$ that contain an even number of 0's and an odd number of 1's.

8. Characterize the set of strings accepted by the finite-state automaton

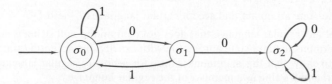

Section 10.3

9. Is the grammar

$$S \to aSb, \quad S \to Ab, \quad A \to aA, \quad A \to b, \quad A \to \lambda$$

context-sensitive, context-free, regular, or none of these? Give all characterizations that apply.

10. Show that the string $\alpha = aaaabbbb$ is in the language generated by the grammar of Exercise 9 by giving a derivation of α.

11. Characterize the language generated by the grammar of Exercise 9.

12. Write a grammar that generates all nonnull strings over $\{0, 1\}$ having an equal number of 0's and 1's.

Section 10.4

13. Draw the transition diagram of the nondeterministic finite-state automaton $(\mathcal{I}, \mathcal{S}, f, \mathcal{A}, \sigma_0)$, where $\mathcal{I} = \{a, b\}$, $\mathcal{S} = \{\sigma_0, \sigma_1, \sigma_2\}$, and $\mathcal{A} = \{\sigma_2\}$.

$\mathcal{I}$ $\diagdown$ $\mathcal{S}$	a	b
σ_0	$\{\sigma_0\}$	$\{\sigma_2\}$
σ_1	$\{\sigma_0, \sigma_1\}$	$\emptyset$
σ_2	$\{\sigma_2\}$	$\{\sigma_0, \sigma_1\}$

14. Find the sets $\mathcal{I}$, $\mathcal{S}$, and $\mathcal{A}$, the initial state, and the table defining the next-state function for the nondeterministic finite-state automaton

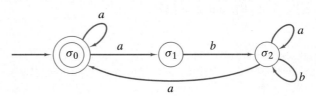

15. Is the string $aabaaba$ accepted by the nondeterministic finite-state automaton of Exercise 14? Explain.

16. Design a nondeterministic finite-state automaton that accepts all strings over $\{0, 1\}$ that begin 01 and contain 110.

Section 10.5

17. Find a finite-state automaton equivalent to the nondeterministic finite-state automaton of Exercise 13.

18. Find a finite-state automaton equivalent to the nondeterministic finite-state automaton of Exercise 14.

19. Explain how to construct a nondeterministic finite-state automaton that accepts the language

$$L_1 L_2 = \{\alpha\beta \mid \alpha \in L_1, \beta \in L_2\},$$

given finite-state automata that accept regular languages L_1 and L_2.

20. Prove that any regular language that does not contain the null string is accepted by a nondeterministic finite-state automaton with exactly one accepting state. Give an example to show that this statement is false for arbitrary regular languages (i.e., if we allow the null string as a member of the regular language).

11

COMPUTATIONAL GEOMETRY

But I proved beyond a shadow of a doubt and with geometric logic that a key did exist.

—from The Caine Mutiny

Computational geometry is concerned with the design and analysis of algorithms to solve geometry problems. Efficient geometric algorithms are useful in fields such as computer graphics, statistics, image processing, and very-large-scale-integration (VLSI) design. In this chapter we present an introduction to this fascinating subject.

The closest-pair problem furnishes an example of a problem from computational geometry: Given n points in the plane, find a closest pair. In addition to the closest-pair problem, we consider the problem of finding the convex hull.

11.1 THE CLOSEST-PAIR PROBLEM

The **closest-pair problem** is easily stated: Given n points in the plane, find a closest pair (see Figure 11.1.1). (We say *a* closest pair since it is possible that several pairs achieve the same minimum distance.) Our distance measure is ordinary Euclidean distance.

One way to solve this problem is to list the distance between each pair and choose the minimum in this list of distances. Since there are $C(n, 2) = n(n - 1)/2 = \Theta(n^2)$ pairs, this "list all" algorithm's time

---

† This section can be omitted without loss of continuity.

is $\Theta(n^2)$. We can do better; we will give a divide-and-conquer closest-pair algorithm whose worst-case time is $\Theta(n \lg n)$. We first discuss the algorithm and then give a more precise description using pseudocode.

Our algorithm begins by finding a vertical line l that divides the points into two nearly equal parts (see Figure 11.1.1). [If n is even, we divide the points into parts each having $n/2$ points. If n is odd, we divide the points into parts, one having $(n+1)/2$ points and the other having $(n-1)/2$ points.]

We then recursively solve the problem for each of the parts. We let δ_L be the distance between a closest pair in the left part; we let δ_R be the distance between a closest pair in the right part; and we let

$$\delta = \min\{\delta_L, \delta_R\}.$$

Unfortunately, δ may not be the distance between a closest pair from the original set of points because a pair of points, one from the left part and the other from the right part, might be closer than δ (see Figure 11.1.1). Thus we must consider distances between points on opposite sides of the line l.

We first note that if the distance between a pair of points is less than δ, the points must lie in the vertical strip of width 2δ centered at l (see Figure 11.1.1). (Any point not in this strip is at least δ away from every point on the other side of l.) Thus we can restrict our search for a pair closer than δ to points in this strip.

If there are n points in the strip and we check *all* pairs in the strip, the worst-case time to process the points in the strip is $\Theta(n^2)$. In this case the worst-case time of our algorithm will be $\Omega(n^2)$, which is at least as bad as exhaustive search; thus we must avoid checking all pairs in the strip.

We order the points in the strip in increasing order of their y-coordinates. We then examine the points in this order. When we examine a point p in the strip, any point q following p whose distance to p is less than δ must lie strictly within or on the base of the rectangle of height δ whose base contains p and whose vertical sides are at a distance δ from l (see Figure 11.1.2). (We need not compute the distance between p and points below p. These distances would already have been considered since we are examining the points in increasing order of their y-coordinates.) We will show that this rectangle contains at most eight points, including p itself, so if we compute the distances between p and the next seven points in the strip, we can be sure that we will compute the distance between p and all the points in the rectangle. Of course, if fewer than seven points follow p in the list, we compute the distances between p and all of the remaining points. By restricting the search in the strip in this way, the time spent processing the points in the strip is $O(n)$. (Since there are at most n points in the strip, the time spent processing the points in the strip is at most $7n$.)

We show that the rectangle of Figure 11.1.2 contains at most eight points. Figure 11.1.3 shows the rectangle of Figure 11.1.2 divided into eight equal squares. Notice that the length of a diagonal of a square is

$$\left(\left(\frac{\delta}{2}\right)^2 + \left(\frac{\delta}{2}\right)^2\right)^{1/2} = \frac{\delta}{\sqrt{2}} < \delta;$$

thus each square contains at most one point. Therefore, the $2\delta \times \delta$ rectangle contains at most eight points.

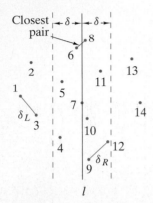

FIGURE 11.1.1

n points in the plane. The problem is to find a closest pair. For this set, the closest pair is 6 and 8. Line l divides the points into two approximately equal parts. The closest pair in the left half is 1 and 3, which is δ_L apart. The closest pair in the right half is 9 and 12, which is δ_R apart. Any pair (e.g., 6 and 8) closer together than $\delta = \min\{\delta_L, \delta_R\}$ must lie in the vertical strip of width 2δ centered at l.

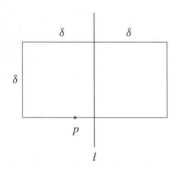

FIGURE 11.1.2

Any point q following p whose distance to p is less that δ must lie within the rectangle.

EXAMPLE 11.1.1

We show how the closest-pair algorithm finds a closest pair for the input of Figure 11.1.1.

The algorithm begins by finding a vertical line l that divides the points into two equal parts,

$$S_1 = \{1, 2, 3, 4, 5, 6, 7\}, \qquad S_2 = \{8, 9, 10, 11, 12, 13, 14\}.$$

For this input there are many possible choices for the dividing line. The particular line chosen here happens to go through point 7.

Next we recursively solve the problem for S_1 and S_2. The closest pair of points in S_2 is 1 and 3. We let δ_L denote the distance between points 1 and 3. The closest pair of points in S_2 is 9 and 12. We let δ_R denote the distance between points 9 and 12. We let

$$\delta = \min\{\delta_L, \delta_R\} = \delta_L.$$

We next order the points in the vertical strip of width 2δ centered at l in increasing order of their y-coordinates:

$$9, \quad 12, \quad 4, \quad 10, \quad 7, \quad 5, \quad 11, \quad 6, \quad 8.$$

We then examine the points in this order. We compute the distances between each point and the following seven points, or between each point and the remaining points if fewer than seven points follow it.

We first compute the distances from 9 to each of 12, 4, 10, 7, 5, 11, and 6. Since each of these distances exceeds δ, at this point we have not found a closer pair.

We next compute the distances from 12 to each of 4, 10, 7, 5, 11, 6, and 8. Since each of these distances exceeds δ, at this point we still have not found a closer pair.

We next compute the distances from 4 to each of 10, 7, 5, 11, 6, and 8. Since each of these distances exceeds δ, at this point we still have not found a closer pair.

We next compute the distances from 10 to each of 7, 5, 11, 6, and 8. Since the distance between points 10 and 7 is less than δ, we have discovered a closer pair. We update δ to the distance between points 10 and 7.

We next compute the distances from 7 to each of 5, 11, 6, and 8. Since each of these distances exceeds δ, we have not found a closer pair.

We next compute the distances from 5 to each of 11, 6, and 8. Since each of these distances exceeds δ, we have not found a closer pair.

We next compute the distances from 11 to each of 6 and 8. Since each of these distances exceeds δ, we have not found a closer pair.

We next compute the distance from 6 to 8. Since the distance between points 6 and 8 is less than δ, we have discovered a closer pair. We update δ to the distance between points 6 and 8. Since there are no more points in the strip to consider, the algorithm terminates. The closest pair is 6 and 8 and the distance between them is δ.   □

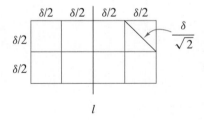

FIGURE 11.1.3
The large rectangle contains at most eight points because each square contains at most one point.

Before we give a formal statement of the closest-pair algorithm, there are several technical points to resolve.

In order to terminate the recursion, we check the number of points in the input and if there are three or fewer points, we find a closest pair directly. Dividing the input and using recursion only if there are four or more points ensures that each of the two parts contains at least one pair of points and, therefore, that there is a closest pair in each part.

Before invoking the recursive procedure, we sort the entire set of points by x-coordinate. This makes it easy to divide the points into two nearly equal parts.

We use mergesort (see Section 5.3) to sort by y-coordinate. However, instead of sorting each time we examine points in the vertical strip, we assume as in mergesort that each half is sorted by y-coordinate. Then we simply merge the two halves to sort all of the points by y-coordinate.

We can now formally state the closest-pair algorithm. To simplify the description, our version outputs the distance between a closest pair but not a closest pair. We leave this enhancement as an exercise (Exercise 5).

| ALGORITHM 11.1.2 | *Finding the Distance Between a Closest Pair of Points* |

Input: $p_1, \ldots, p_n$ ($n \geq 2$ points in the plane)

Output: δ, the distance between a closest pair of points

```
procedure closest_pair(p, n)
  sort p₁, . . . , pₙ by x-coordinate
  return(rec_cl_pair(p, 1, n))
end closest_pair

procedure rec_cl_pair(p, i, j)
  // The input is the sequence pᵢ, . . . , pⱼ of points in the plane
  // sorted by x-coordinate.

  // At termination of rec_cl_pair, the sequence is sorted by
  // y-coordinate.

  // rec_cl_pair returns the distance between a closest pair
  // in the input.

  // Denote the x-coordinate of point p by p.x.

  // trivial case (3 or fewer points)
  if j − i < 3 then
    begin
    sort pᵢ, . . . , pⱼ by y-coordinate
    directly find the distance δ between a closest pair
    return(δ)
    end
  // divide
  k := ⌊(i + j)/2⌋
  l := pₖ.x
  δ_L := rec_cl_pair(p, i, k)
  δ_R := rec_cl_pair(p, k + 1, j)
  δ := min{δ_L, δ_R}
```

```
// p_i, ..., p_k are now sorted by y-coordinate
// p_{k+1}, ..., p_j are now sorted by y-coordinate
merge p_i, ..., p_k and p_{k+1}, ..., p_j by y-coordinate
// assume that the result of the merge is stored back in
// p_i, ..., p_j

// now p_i, ..., p_j is sorted by y-coordinate

// store points in the vertical strip in v
t := 0
for k := i to j do
  if p_k.x > l − δ and p_k.x < l + δ then
    begin
    t := t + 1
    v_t := p_k
    end
// points in strip are v_1, ..., v_t
// look for closer pairs in strip
// compare each to next seven points
for k := 1 to t − 1 do
  for s := k + 1 to min{t, k + 7} do
    δ := min{δ, dist(v_k, v_s)}
return(δ)
end rec_cl_pair
```

We show that the worst-case time of the closest-pair algorithm is $\Theta(n \lg n)$. The procedure *closest_pair* begins by sorting the points by x-coordinate. If we use an optimal sort (e.g., mergesort), the worst-case sorting time will be $\Theta(n \lg n)$. Next *closest_pair* invokes *rec_cl_pair*. We let a_n denote the worst-case time of *rec_cl_pair* for input of size n. If $n > 3$, *rec_cl_pair* first invokes itself with input size $\lfloor n/2 \rfloor$ and $\lfloor (n + 1)/2 \rfloor$. Each of merge, extracting the points in the strip, and checking the distances in the strip takes time $O(n)$. Thus we obtain the recurrence

$$a_n \leq a_{\lfloor n/2 \rfloor} + a_{\lfloor (n+1)/2 \rfloor} + cn, \qquad n > 3.$$

This is the same recurrence that mergesort satisfies, so we conclude that *rec_cl_pair* has the same worst-case time $O(n \lg n)$ as mergesort. Since the worst-case time of the sorting of the points by x-coordinate is $\Theta(n \lg n)$ and the worst-case time of *rec_cl_pair* is $O(n \lg n)$, the worst-case time of *closest_pair* is $\Theta(n \lg n)$. In Section 11.2 we will show that any algorithm that finds a closest pair of points in the plane has worst-case time $\Omega(n \lg n)$; thus our algorithm is asymptotically optimal.

It can be shown (Exercise 10) that there are at most six points in the rectangle of Figure 11.1.2 when the base is included and the other sides are excluded. This result is the best possible since it is possible to place six points in the rectangle (Exercise 8). By considering the possible locations of the points in the rectangle, D. Lerner and R. Johnsonbaugh have shown that it suffices to compare each point in the strip with the next three points (rather than the next seven). This result is the best possible since checking the next two points does not lead to a correct algorithm (Exercise 7).

৵ৎ ৵ৎ ৵ৎ

Exercises

1. Describe how the closest-pair algorithm finds the closest pair of points if the input is (8, 4), (3, 11), (12, 10), (5, 4), (1, 2), (17, 10), (8, 7), (8, 9), (11, 3), (1, 5), (11, 7), (5, 9), (1, 9), (7, 6), (3, 7), (14, 7).

2. What can you conclude about input to the closest-pair algorithm when the output is zero for the distance between a closest pair?

3. Give an example of input for which the closest-pair algorithm puts some points on the dividing line l into the left half and other points on l into the right half.

4. Explain why in some cases, when dividing a set of points by a vertical line into two nearly equal parts, it is necessary for the line to contain some of the points.

5. Write a closest-pair algorithm that finds one closest pair as well as the distance between the pair of points.

6. Write an algorithm that finds the distance between a closest pair of points on a (straight) line.

7. Give an example of input for which comparing each point in the strip with the next two points gives incorrect output.

8. Give an example to show that it is possible to place six points in the rectangle of Figure 11.1.2 when the base is included and the other sides are excluded.

9. When we compute the distances between a point p in the strip and points following it, can we stop computing distances from p if we find a point q such that the distance between p and q exceeds δ? Explain.

★ 10. Show that there are at most six points in the rectangle of Figure 11.1.2 when the base is included and the other sides are excluded.

11. Write a $\Theta(n \lg n)$ algorithm that finds the distance δ between a closest pair and if $\delta > 0$ also finds *all* pairs δ apart.

12. Write a $\Theta(n \lg n)$ algorithm that finds the distance δ between a closest pair and *all* pairs less than 2δ apart.

†*11.2* A LOWER BOUND FOR THE CLOSEST-PAIR PROBLEM

In Section 11.1 we gave a $\Theta(n \lg n)$ algorithm that finds the distance between a closest pair among n elements in the plane. In this section we show that this result is optimal, that is, that any algorithm that finds the distance between a closest pair among n elements in the plane has worst-case time $\Omega(n \lg n)$.

We begin by showing that a related problem, that of determining whether n elements are distinct, has the worst-case time lower bound $\Omega(n \lg n)$. Since a closest-pair algorithm can be used to determine whether n elements are distinct (n elements are distinct if and only if the closest-pair distance is not zero), its worst-case time must be at least as large as $\Omega(n \lg n)$, the lower bound for the distinct elements problem.

---

† This section can be omitted without loss of continuity.

The worst-case time of any algorithm that solves the problem of determining whether n real numbers are distinct is $\Omega(n \lg n)$.

Proof. We prove the theorem by showing that any algorithm that solves the problem of determining whether n real numbers are distinct must, in effect, sort the numbers. Since sorting has worst-case time $\Omega(n \lg n)$ (Theorem 7.7.3), any algorithm that solves the problem of determining whether n real numbers are distinct must also have worst-case time $\Omega(n \lg n)$.

Suppose that an algorithm that solves the problem of determining whether n real numbers are distinct receives the input

$$x_1, \quad \ldots, \quad x_n$$

where the x_i are distinct. The output will be "Distinct." Suppose that the sorted order is

$$x_{k_1}, \quad x_{k_2}, \quad \ldots, \quad x_{k_n}. \tag{11.2.1}$$

We claim that the algorithm must compare x_{k_i} and $x_{k_{i+1}}$ for $i = 1, \ldots, n - 1$, so that the algorithm must "know" the sorted order of the input. We establish this claim by arguing by contradiction.

Suppose that the algorithm does not compare x_{k_j} and $x_{k_{j+1}}$ for some j. We alter the original input $x_1, \ldots, x_n$ by changing the value of x_{k_j} to $x_{k_{j+1}}$, but leaving all other x_i the same. We then rerun the algorithm. The result of each comparison will be the same as in the original run since the only comparison whose outcome would change involves x_{k_j} and $x_{k_{j+1}}$ and the algorithm does not compare this pair. Thus the output is again "Distinct." This is a contradiction since the input now has duplicates. Thus any algorithm that solves the problem of determining whether n real numbers are distinct must compare x_{k_i} and $x_{k_{i+1}}$ for $i = 1, \ldots, n - 1$.

To complete the proof, we show how we can convert an algorithm that solves the problem of determining whether n real numbers are distinct into a sorting algorithm. Since sorting has worst-case lower bound $\Omega(n \lg n)$, this will complete the proof of the theorem.

Let A be an algorithm that solves the problem of determining whether n real numbers are distinct. We modify A in the following way. We first construct vertices $1, \ldots, n$. Each time algorithm A compares x_j and x_k, we place a directed edge from j to k if $x_j < x_k$. For distinct input, if the sorted order is (11.2.1), we showed that A must compare x_{k_i} and $x_{k_{i+1}}$ for $i = 1, \ldots, n-1$. Thus there is a path from x_{k_1} to x_{k_n} that gives the sorted order. We can determine this path in the following way. First locate the unique vertex with no incoming edges. This is vertex k_1 that corresponds to x_{k_1}, the smallest element in the list. Remove all outgoing edges from k_1. Repeat this process; that is, locate the unique vertex with no incoming edges. This is vertex k_2 which corresponds to x_{k_2}, the second smallest element in the list. Continuing in this way, we produce the sorted order. Since sorting requires at least $Cn \lg n$ comparisons (Theorem 7.7.3), we conclude that our modified algorithm performs at least $Cn \lg n$ comparisons. Since the modifications to algorithm A do not involve comparisons of elements, this modified algorithm has exactly the same number of comparisons as does A. Thus algorithm A requires at least $Cn \lg n$ comparisons. The proof is complete. ∎

COROLLARY 11.2.2

The worst-case time of any algorithm that finds the distance between a closest pair among n elements in the plane is $\Omega(n \lg n)$.

Proof. Let t_n denote the worst-case time for an algorithm CP that returns the distance between a closest pair of n points in the plane. Consider the following algorithm that solves the problem of determining whether there are any duplicates among n numbers:

```
procedure dup(x, n)
   // Input is x₁, ..., xₙ.
   // Make the input into points in the plane.
   for i := 1 to n do
     aᵢ := (xᵢ, 0)
   if CP(a, n) = 0 then
     return("Duplicates")
   else
     return("No duplicates")
end dup
```

The worst-case time t_n' for *dup* is the time for the for loop plus the worst-case time for CP; that is,

$$t_n' = n + t_n.$$

By Theorem 11.2.1,

$$Cn \lg n \le t_n'.$$

Combining these last two statements, we obtain

$$\Omega(n \lg n) = Cn \lg n - n \le t_n' - n = t_n. \quad ∎$$

૩ ૩ ૩

Exercises

1. Write an algorithm that solves the problem of determining whether n real numbers are distinct. Make your algorithm as efficient as you can.

2. The *nearest-neighbor problem* is: Given n points S in the plane, one of which is designated p, and the sorted distances from p to q for all $q \ne p$, find a point s in $S, s \ne p$, nearest p. Show that the worst-case time of any algorithm that solves the nearest-neighbor problem is $\Omega(\lg n)$.

3. The *all-nearest-neighbors problem* is: Given n points S in the plane, and the distances from p to q for all $q \ne p$, for each point p in S, find a point q in $S, q \ne p$, nearest p. Show that the worst-case time of any algorithm that solves the all-nearest-neighbors problem is $\Omega(n \lg n)$.

4. Suppose that we are given a directed graph with vertices $1, \ldots, n$, which contains edges (p_i, p_{i+1}), $i = 1, \ldots, n-1$, for some permutation $p_1, \ldots, p_n$ of $1, \ldots, n$. Suppose also that if (p_i, p_j) is an edge, then $i < j$. (This situation is analogous to that in the proof of Theorem 11.2.1.) Give an algorithm whose output is $p_1, \ldots, p_n$. [There is an algorithm whose worst-case time is $\Theta(e + n)$, where e is the number of edges in the graph.]

11.3 AN ALGORITHM TO COMPUTE THE CONVEX HULL

A fundamental problem in computational geometry is that of computing the "bounding" points, formally the **convex hull**, of a finite set of points in the plane. (See Figure 11.3.1, where the points that form the convex hull are circled.) The convex hull has applications in many areas, including statistics, computer graphics, and image processing. For example, in statistics the points of a data set that determine its convex hull may be outliers that are not truly representative of the data, and so might be discarded. In this section we present Graham's Algorithm to compute the convex hull. Throughout this section, by "set of points" we mean "set of *distinct* points." We begin with the definitions.

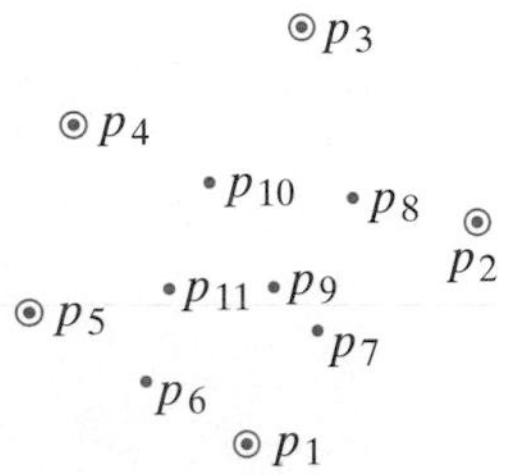

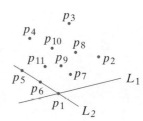

FIGURE 11.3.1
The convex hull of the points $p_1, \ldots, p_{11}$ is p_1, p_2, p_3, p_4, p_5.

DEFINITION 11.3.1

Given a finite set of points S in the plane, a point p in S is a *hull point* if there exists a (straight) line L through p such that all points in S except p lie on one side of L (and except for p, none lie on L).

EXAMPLE 11.3.2

In Figure 11.3.2, p_1 is a hull point since we can find a line L_1 through p_1 such that all other points lie strictly on one side of L_1. The point p_8 is not a hull point since every line L through p_8 has points on both sides of L. The point p_6 is also not a hull point. As shown in Figure 11.3.2, it is possible to place a line L_2 through p_6 such that one side of L_2 contains no points, but L_2 does not meet the conditions of Definition 11.3.1 since it contains points other than p_6. $\square$

The convex hull of a finite set of points S in the plane consists of the hull points listed in order as one travels around the boundary of S. In Figure 11.3.2, the convex hull is the sequence of points p_1, p_2, p_3, p_4, p_5. The next definition makes precise this notion of ordering the hull points.

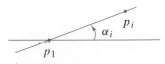

FIGURE 11.3.2
Hull points are p_1, p_2, p_3, p_4, p_5. All other points are not hull points.

DEFINITION 11.3.3

The *convex hull* of a finite set of points S in the plane is the sequence $p_1, p_2, \ldots, p_n$ of hull points of S listed in the following order. The point p_1 is the point with minimum y-coordinate. If several points have the same minimum y-coordinate, p_1 is the one with minimum x-coordinate. (Note that p_1 is a hull point.) For $i \geq 2$, let α_i be the angle from the horizontal to the line segment p_1, p_i (see Figure 11.3.3). The points $p_2, p_3, \ldots, p_n$ are ordered so that $\alpha_2, \alpha_3, \ldots, \alpha_n$ is an increasing sequence.

FIGURE 11.3.3
α_i is the angle from the horizontal to line segment p_1, p_i.

EXAMPLE 11.3.4

In Figure 11.3.4 the point p_1 has minimum y-coordinate, so that it is the first point listed in the convex hull. As shown, the angles from the horizontal to the line segments p_1, p_2; p_1, p_3; p_1, p_4; p_1, p_5 increase. Thus the convex hull of the set of points in Figure 11.3.4 is p_1, p_2, p_3, p_4, p_5.   □

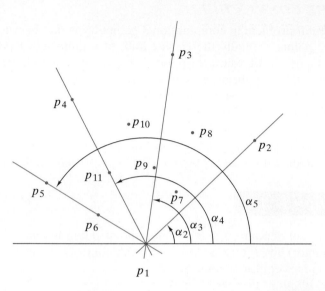

FIGURE 11.3.4   The convex hull is p_1, p_2, p_3, p_4, p_5 because these points are hull points and the corresponding angles $\alpha_2, \alpha_3, \alpha_4, \alpha_5$ increase, where α_i is the angle from the horizontal to the line segment p_1, p_i.

Definition 11.3.3 suggests an algorithm to compute the convex hull of a finite set of points S in the plane. First find the point p_1 with minimum y-coordinate. If several points have the same minimum y-coordinate, choose the one with minimum x-coordinate. Next sort *all* of the points p in S according to the angle from the horizontal to the line segment p_1, p. Finally, examine the points in order and discard those not on the convex hull. The result will be the convex hull. This is the strategy used by Graham's Algorithm. In order to turn this idea into an algorithm, two main issues must be addressed. We must describe a way to compare angles, and we must develop a method for testing whether points are on the convex hull. We deal first with the problem of comparing angles.

Suppose that we visit the distinct points p_1, p_0, p_2 in the plane in this order. If, after leaving p_0 we move to the left, we say that the points p_1, p_0, p_2 *make a left turn* (see Figure 11.3.5). More precisely, the points p_1, p_0, p_2 make a left turn if the angle from the line segment p_0, p_2 to the line segment p_0, p_1 measured counterclockwise is less than 180 degrees. Similarly, if after leaving p_0 we move to the right, we say that the points p_1, p_0, p_2 *make a right turn* (see Figure 11.3.5); that is, the points p_1, p_0, p_2 make a right turn if the angle from the line segment p_0, p_2 to the line segment p_0, p_1 measured counterclockwise is greater than 180 degrees.

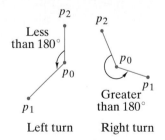

FIGURE 11.3.5
The first turn is a left turn because the angle from p_0, p_2 to p_0, p_1 is less that 180 degrees. The second turn is a right turn because the angle from p_0, p_2 to p_0, p_1 is greater than 180 degrees.

We can use analytic geometry methods to derive a test for whether the points p_1, p_0, p_2 make a left or right turn. Let (x_i, y_i) be the coordinates of point p_i, $i = 0, 1, 2$ (see Figure 11.3.6). First assume that $x_1 < x_0$. The equation of the line L through p_0 and p_1 is

$$y = y_0 + \frac{y_1 - y_0}{x_1 - x_0}(x - x_0).$$

Now p_1, p_0, p_2 make a left turn precisely when p_2 is above L or when

$$y_2 > y' = y_0 + \frac{y_1 - y_0}{x_1 - x_0}(x_2 - x_0).$$

We may rewrite this last inequality as

$$y_2 - y_0 > \frac{y_1 - y_0}{x_1 - x_0}(x_2 - x_0).$$

Multiplying by $x_1 - x_0$, which is negative, and moving all terms to one side of the inequality gives

$$(y_2 - y_0)(x_1 - x_0) - (y_1 - y_0)(x_2 - x_0) < 0.$$

If we define the **cross product** of the points p_0, p_1, p_2 to be

$$\text{cross}(p_0, p_1, p_2) = (y_2 - y_0)(x_1 - x_0) - (y_1 - y_0)(x_2 - x_0),$$

we have proved

If the points p_1, p_0, p_2 make a left turn, then $\text{cross}(p_0, p_1, p_2) < 0$.

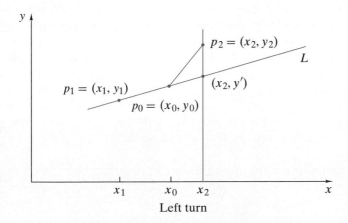

Left turn

FIGURE 11.3.6   Testing for whether the points p_1, p_0, p_2 make a left or right turn. The figure assumes that $x_1 < x_0$. L is the line through p_0 and p_1. As shown, in this case a left turn occurs precisely when p_2 is above L.

Similarly, we can show that

If the points p_1, p_0, p_2 make a right turn, then $\text{cross}(p_0, p_1, p_2) > 0$,

and

If the points p_1, p_0, p_2 are collinear, then $\text{cross}(p_0, p_1, p_2) = 0$.

We can also prove the converses of these statements. For example, to prove that, if $\text{cross}(p_0, p_1, p_2) < 0$, then the points p_1, p_0, p_2 make a left turn, we can argue as follows. Assume that $\text{cross}(p_0, p_1, p_2) < 0$. Since the points do not make a right turn [if they did, $\text{cross}(p_0, p_1, p_2)$ would be positive] and are not collinear [if they were, $\text{cross}(p_0, p_1, p_2)$ would be zero], they must make a left turn. Thus

The points p_1, p_0, p_2 make a left turn if and only if $\text{cross}(p_0, p_1, p_2) < 0$.

The points p_1, p_0, p_2 make a right turn if and only if $\text{cross}(p_0, p_1, p_2) > 0$.

The points p_1, p_0, p_2 are collinear if and only if $\text{cross}(p_0, p_1, p_2) = 0$.

$$(11.3.1)$$

We proved (11.3.1) assuming that $x_1 < x_0$. If $x_1 = x_0$ or $x_1 > x_0$, the conclusion (11.3.1) remains valid (see Exercises 2 and 3). We summarize these conclusions as a theorem.

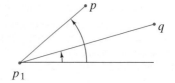

FIGURE 11.3.7
The case $p > q$ occurs when p, p_1, q make a left turn.

THEOREM 11.3.5

If p_0, p_1, p_2 are distinct points in the plane,

   (a) p_1, p_0, p_2 *make a left turn if and only if* $\text{cross}(p_0, p_1, p_2) < 0$.

   (b) p_1, p_0, p_2 *make a right turn if and only if* $\text{cross}(p_0, p_1, p_2) > 0$.

   (c) p_1, p_0, p_2 *are collinear if and only if* $\text{cross}(p_0, p_1, p_2) = 0$.

Proof. The proof precedes the statement of the theorem. ∎

Graham's Algorithm begins by finding the point p_1 with minimum y-coordinate. If several points have the same minimum y-coordinate, the algorithm chooses the one with minimum x-coordinate. Next the algorithm sorts *all* of the points p in S according to the angle from the horizontal to the line segment p_1, p. Finally, the algorithm examines the points in order and discards those not on the convex hull.

We may use the cross product to compare points $p \neq q$ in the sort. To compare the points p and q, we compute $\text{cross}(p_1, p, q)$. If $\text{cross}(p_1, p, q) < 0$, then p, p_1, q make a left turn. With regard to the angles with the horizontal, this means that $p > q$ (see Figure 11.3.7). If $\text{cross}(p_1, p, q) > 0$, then $p < q$. If $\text{cross}(p_1, p, q) = 0$, then p, p_1, q are collinear. In the latter case, we define $p > q$ if p is farther from p_1 than q; and we define $p < q$ if q is farther from p_1 than p.

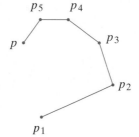

FIGURE 11.3.8
A situation in the convex hull algorithm when point p is examined. Before p is examined, the convex hull of the points so far examined is p_1, p_2, p_3, p_4, p_5. Since p_4, p_5, p make a left turn, p_5 is still on the convex hull of the points so far examined and, so, is retained. The current convex hull is $p_1, p_2, p_3, p_4, p_5, p$. The algorithm continues by examining the point following p.

We can also use the cross product to determine whether a point is not on the convex hull and so should be discarded. In examining the points in order, we will retain points that would be on the convex hull if there were no more points to examine. We then examine the next point p. For example, in Figure 11.3.8 suppose that we have retained $p_1, \ldots, p_5$ and the next point we examine is p. Since p_4, p_5, p make a left turn, we retain p_5. We then continue by examining the point following p.

As another example, suppose that we have retained $p_1, \ldots, p_5$ (see Figure 11.3.9). This time, since p_4, p_5, p make a right turn, we discard p_5. We then back up to examine p_3, p_4, p. Since these points also make a right turn, we discard p_4. We then back up to examine p_2, p_3, p. Since these points make a left turn, we retain p_3. We continue by examining the point following p. The pseudocode for Graham's Algorithm is given as Algorithm 11.3.6.

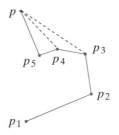

| ALGORITHM 11.3.6 | *Graham's Algorithm to Compute the Convex Hull* |

This algorithm computes the convex hull of the points $p_1, \ldots, p_n$ in the plane. The x- and y-coordinates of the point p are denoted $p.x$ and $p.y$, respectively.

Input: $p_1, \ldots, p_n$ and n

Output: $p_1, \ldots, p_k$ (the convex hull of $p_1, \ldots, p_n$) and k

FIGURE 11.3.9
A situation in the convex hull algorithm when point p is examined. Before p is examined, the convex hull of the points so far examined is p_1, p_2, p_3, p_4, p_5. Since p_4, p_5, p make a right turn, p_5 is discarded. This leaves the points p_3, p_4, p, which also make a right turn; thus, p_4 is also discarded. This leaves the points p_2, p_3, p, which make a left turn; thus p_3 is retained. The current convex hull is p_1, p_2, p_3, p. The algorithm continues by examining the point following p.

```
procedure graham_scan(p, n, k)
  // trivial case
  if n = 1 then
    begin
    k := 1
    return
    end
  // find the point with minimum y-coordinate
  min := 1
  for i := 2 to n do
    if p_i.y < p_min.y then
      min := 1
  // Among all such points, find the one with minimum
  // x-coordinate
  for i := 1 to n do
    if p_i.y = p_min.y and p_i.x < p_min.x then
      min := i
  swap(p_1, p_min)
  // sort on angle from horizontal to p_1, p_i
  sort p_2, ..., p_n
  // p_0 is an extra point added to prevent the algorithm from
  // backing up forever
  p_0 := p_n
  // discard points not on the convex hull
  k := 2
  for i := 3 to n do
    begin
    while p_{k-1}, p_k, p_i do not make a left turn do
      // discard p_k
      k := k - 1
    k := k + 1
    swap(p_i, p_k)
    end
end graham_scan
```

EXAMPLE 11.3.7

Figure 11.3.10 shows Graham's Algorithm in action.

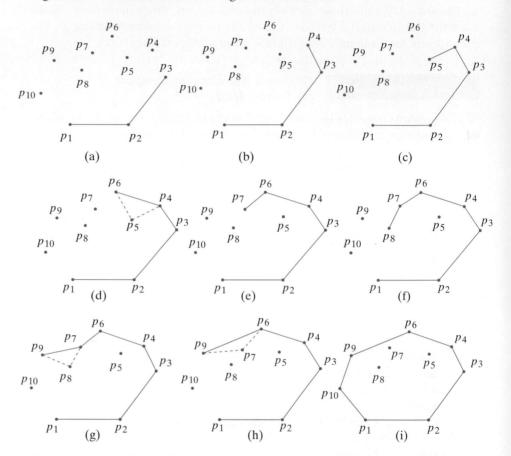

FIGURE 11.3.10 Graham's algorithm to compute the convex hull. Point p_1 has minimum y-coordinate and, among all such points, has minimum x-coordinate. The points sorted on the angle from the horizontal to p_1, p_i are $p_2, p_3, \ldots, p_{10}$. In (a), the algorithm begins by examining p_1, p_2, p_3. p_1, p_2, p_3 make a left turn, so p_2 is retained. Next in (b), p_2, p_3, p_4 are examined. p_2, p_3, p_4 make a left turn, so p_3 is retained. Next in (c), p_3, p_4, p_5 are examined. p_3, p_4, p_5 make a left turn, so p_4 is retained. Next in (d), p_4, p_5, p_6 are examined. p_4, p_5, p_6 make a right turn, so p_5 is discarded. The algorithm backs up to p_3, p_4, p_6. p_3, p_4, p_6 make a left turn, so p_4 is retained. Next in (e), p_4, p_6, p_7 are examined. p_4, p_6, p_7 make a left turn, so p_6 is retained. Next in (f), p_6, p_7, p_8 are examined. p_6, p_7, p_8 make a left turn, so p_7 is retained. Next in (g), p_7, p_8, p_9 are examined. p_7, p_8, p_9 make a right turn, so p_8 is discarded. In (h), the algorithm backs up to p_6, p_7, p_9. p_6, p_7, p_9 also make a right turn, so p_7 is discarded. The algorithm backs up to p_4, p_6, p_9. p_4, p_6, p_9 make a left turn, so p_6 is retained. Finally, in (i), the algorithm concludes by examining p_6, p_9, p_{10}. p_6, p_9, p_{10} make a left turn, so p_9 is retained. The convex hull is $p_1, p_2, p_3, p_4, p_6, p_9, p_{10}$. □

Graham's Algorithm begins with two for loops that each take time $\Theta(n)$. If we use an optimal sort such as mergesort, the worst-case sorting time will be $\Theta(n \lg n)$. The last while loop's *total* execution time is $O(n)$ since a point can be discarded at most once and there are n points. The for loop itself executes

$\Theta(n)$ times; thus the total time for the last for loop and the while loop is $\Theta(n)$. We see that the sorting time dominates so that the worst-case time for Graham's Algorithm is $\Theta(n \lg n)$. The time after the points are sorted is $\Theta(n)$, so that if the points come presorted, Graham's Algorithm can compute the convex hull in linear time.

Our last theorem shows that any algorithm that computes the convex hull of n points in the plane has worst-case time $\Omega(n \lg n)$, so that Graham's Algorithm is optimal.

THEOREM 11.3.8

Any algorithm that computes the convex hull of n points in the plane has worst-case time $\Omega(n \lg n)$.

Proof. Let A be an algorithm that computes the convex hull of a finite set of points in the plane. We show that in the worst case this algorithm spends as much time as a sorting algorithm and therefore has the same lower bound $\Omega(n \lg n)$ as does the sorting problem.

Consider an arbitrary sequence of real numbers

$$y_1, \quad y_2, \quad \ldots, \quad y_n,$$

where each y_i is between 0 and 1. We use the convex hull algorithm, algorithm A, to construct an algorithm, algorithm B, that sorts this sequence. First, algorithm B projects these real numbers over to the unit circle (see Figure 11.3.11). Next algorithm B calls algorithm A to find the convex hull $h_1, h_2, \ldots, h_n$ of the points on the circle. Finally, algorithm B outputs the y-coordinates of $h_1, h_2, \ldots, h_n$ in this order. Notice that the output of algorithm B is the sorted order of the input sequence

$$y_1, \quad y_2, \quad \ldots, \quad y_n.$$

Since algorithm B is a sorting algorithm, by Theorem 7.7.3 its worst-case time t_n satisfies

$$t_n \geq C n \lg n.$$

On the other hand, algorithm B consists of two $\Theta(n)$ loops (one to project the points to the unit circle, the other to output the y-coordinates of the convex hull) and the call to algorithm A that takes time s_n, say, in the worst case. Thus

$$t_n = 2n + s_n.$$

Therefore,

$$s_n = t_n - 2n \geq C n \lg n - 2n = \Omega(n \lg n)$$

and the proof is complete. ∎

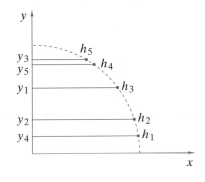

FIGURE 11.3.11
Sorting by using the convex hull algorithm. The points y_1, y_2, y_3, y_4, y_5 are first projected onto the unit circle. The resulting points on the unit circle are denoted as h_i. Next the convex hull algorithm is used to find the convex hull h_1, h_2, h_3, h_4, h_5. The y-coordinates of the convex hull (in the order h_1, h_2, h_3, h_4, h_5) are y_4, y_2, y_1, y_5, y_3, the sorted order of the y's.

✍ ✍ ✍

Exercises

1. Let S be a finite set of points in the plane. Let p_1 be the point with minimum y-coordinate. If several points have the same minimum y-coordinate, choose the one with minimum x-coordinate. Prove that p_1 is on the convex hull of S.

2. Prove Theorem 11.3.5 when $x_1 = x_0$.

3. Prove Theorem 11.3.5 when $x_1 > x_0$.

4. Use Graham's Algorithm to find the convex hull of the points $(10, 1), (7, 7), (3, 13),$ $(6, 10), (16, 4), (10, 5), (7, 13), (13, 8), (4, 4), (2, 2), (1, 8), (10, 13), (7, 1), (4, 8),$ $(12, 3), (16, 10), (14, 5), (10, 9)$.

5. Use Graham's Algorithm to find the convex hull of the points $(7, 8), (9, 8), (3, 11),$ $(5, 1), (7, 11), (9, 5), (9, 1), (6, 7), (4, 5), (2, 1), (10, 17), (7, 3), (7, 14), (4, 8), (11, 3),$ $(10, 12)$.

6. Suppose that Graham's Algorithm has been used to find the convex hull of a set of n points S. Show that if a point is added to S to obtain S', the convex hull of S' can be found in time $\Theta(n)$.

Exercises 7–10 concern *Jarvis's march*, another algorithm that computes the convex hull of a finite set of points in the plane. It begins, as does Graham's Algorithm, by finding the point with minimum y-coordinate. If several points have the same minimum y-coordinate, the one chosen is that with minimum x-coordinate. Next, Jarvis's march finds the point p_2 such that the angle from the horizontal to the segment p_1, p_2 is a minimum. (In case of ties, the point farthest from p_1 is selected.) After finding $p_1, \ldots, p_i$, Jarvis's march finds the point p_{i+1} such that p_{i-1}, p_i, p_{i+1} make the smallest left turn. (In case of ties, the point farthest from p_i is chosen.)

7. Show that Jarvis's march does find the convex hull.

8. Write pseudocode for Jarvis's march.

9. Find the worst-case time of Jarvis's march.

10. Are there sets on which Jarvis's march is faster than Graham's Algorithm? Explain.

✍ NOTES

[Preparata, 1985 and Edelsbrunner] are books on computational geometry.

The closest-pair algorithm in Section 11.1 is due to M. I. Shamos and appears in [Preparata, 1985]. [Preparata, 1985] also gives a $\Theta(n \lg n)$ algorithm to find the closest pair in an arbitrary number of dimensions.

Graham's Algorithm (see [Graham, 1972]) that appeared in 1972 was one of the first $\Theta(n \lg n)$ planar convex hull algorithms. Jarvis's march appears in [Jarvis]. Computing the convex hull of a set of points in a space of more than two dimensions is significantly trickier than computing the convex hull in the plane. The first optimal three-dimensional algorithm was given in 1977 by [Preparata, 1977]. In 1981, [Seidel] gave an n-dimensional convex hull algorithm that is optimal for even n. Finding more efficient convex hull algorithms in high-dimensional spaces continues to be a lively area of research.

✌ CHAPTER REVIEW

Section 11.1

Computational geometry
Closest-pair problem
Closest-pair algorithm

Section 11.2

The worst-case time of any algorithm
that solves the problem of determining whether n real numbers
are distinct is $\Omega(n \lg n)$ (Theorem
11.2.1).

The worst-case time of any algorithm
that finds the distance between a
closest pair among n elements in

the plane is $\Omega(n \lg n)$ (Corollary
11.2.2).

Section 11.3

Hull point
Convex hull
Cross product
Graham's Algorithm to compute the
convex hull
Any algorithm that computes the
convex hull of n points in the plane
has worst-case time $\Omega(n \lg n)$ (Theorem 11.3.8).

✌ CHAPTER SELF-TEST

Section 11.1

1. Describe how the closest-pair algorithm finds the closest pair of points if the input is that of Exercise 4, Section 11.3.

2. In order to terminate the recursion in the closest-pair algorithm, if there are three or fewer points in the input we find the closest pair directly. Why can't we replace "three" by "two"?

3. Show that there are at most four points in the lower half of the rectangle of Figure 11.1.3.

4. What would the worst-case time of the closest-pair algorithm (Algorithm 11.1.2) be if instead of merging $p_i, \ldots, p_k$ and $p_{k+1}, \ldots, p_j$, we used mergesort to sort $p_i, \ldots, p_j$?

Section 11.2

5. Show that the worst-case time of any algorithm that finds a closest pair among n elements in d-dimensional space is $\Omega(n \lg n)$.

6. Show that the worst-case time of any algorithm that finds all closest pairs among n elements in the plane is $\Omega(n \lg n)$.

7. State and prove a sharp lower bound for the worst-case time of any algorithm that solves the problem of determining whether n real numbers are all equal.

8. Show that the worst-case time of any algorithm that finds all pairs at a distance 2δ or less apart, where δ is the distance between a closest pair, is $\Omega(n \lg n)$.

Section 11.3

9. Let S be a finite set of points in the plane. Let p be the point in S with maximum x-coordinate. If several points have the same maximum x-coordinate, choose the one with maximum y-coordinate. Prove that p is on the convex hull of S.

10. Let S be a finite set of distinct points in the plane. Assume that S contains at least two points. Let p and q be points in S at a maximum distance apart. Prove that p and q are on the convex hull of S.

11. Use Graham's Algorithm to find the convex hull of the set of points of Exercise 1, Section 11.1.

12. Suppose that Graham's Algorithm has been used to find the convex hull of a set of $n \geq 2$ points. Show that if a point other than p_1 of Algorithm 11.3.6 is deleted from S to obtain S', the convex hull of S' can be found in time $\Theta(n)$.

Appendix: Matrices

It is a common practice to organize data into rows and columns. In mathematics, such an array of data is called a **matrix**. In this appendix we summarize some definitions and elementary properties of matrices. We begin with the definition of matrix.

DEFINITION A.1

A *matrix*

$$A = \begin{pmatrix} a_{11} & a_{12} & \dots & a_{1n} \\ a_{21} & a_{22} & \dots & a_{2n} \\ \vdots & \vdots & & \vdots \\ a_{m1} & a_{m2} & \dots & a_{mn} \end{pmatrix} \qquad \text{(Appendix.1)}$$

is a rectangular array of data.

If A has m rows and n columns, we say that the *size* of A is m by n (written $m \times n$).

We will often abbreviate equation (A.1) to $A = (a_{ij})$. In this equation, a_{ij} denotes the element of A appearing in the ith row and jth column.

EXAMPLE A.2

The matrix

$$A = \begin{pmatrix} 2 & 1 & 0 \\ -1 & 6 & 14 \end{pmatrix}$$

has two rows and three columns, so its size is 2×3. If we write $A = (a_{ij})$, we would have, for example,

$$a_{11} = 2, \qquad a_{21} = -1, \qquad a_{13} = 0. \qquad \square$$

DEFINITION A.3

Two matrices A and B are *equal*, written $A = B$, if they are the same size and their corresponding entries are equal.

EXAMPLE A.4

Determine w, x, y, and z so that

$$\begin{pmatrix} x + y & y \\ w + z & w - z \end{pmatrix} = \begin{pmatrix} 5 & 2 \\ 4 & 6 \end{pmatrix}.$$

According to Definition A.3, since the matrices are the same size, they will be equal provided that

$$x + y = 5 \qquad y = 2$$
$$w + z = 4 \qquad w - z = 6.$$

Solving these equations, we obtain

$$w = 5, \qquad x = 3, \qquad y = 2, \qquad z = -1. \qquad \square$$

We describe next some operations that can be performed on matrices. The **sum** of two matrices is obtained by adding the corresponding entries. The **scalar product** is obtained by multiplying each entry in the matrix by a fixed number.

DEFINITION A.5

Let $A = (a_{ij})$ and $B = (b_{ij})$ be two $m \times n$ matrices. The *sum* of A and B is defined as

$$A + B = (a_{ij} + b_{ij}).$$

The *scalar product* of a number c and a matrix $A = (a_{ij})$ is defined as

$$cA = (ca_{ij}).$$

If A and B are matrices, we define $-A = (-1)A$ and $A - B = A + (-B)$.

EXAMPLE A.6

If

$$A = \begin{pmatrix} 4 & 2 \\ -1 & 0 \\ 6 & -2 \end{pmatrix}, \qquad B = \begin{pmatrix} 1 & -3 \\ 4 & 4 \\ -1 & -3 \end{pmatrix},$$

then

$$A + B = \begin{pmatrix} 5 & -1 \\ 3 & 4 \\ 5 & -5 \end{pmatrix}, \qquad 2A = \begin{pmatrix} 8 & 4 \\ -2 & 0 \\ 12 & -4 \end{pmatrix}, \qquad -B = \begin{pmatrix} -1 & 3 \\ -4 & -4 \\ 1 & 3 \end{pmatrix}. \quad \square$$

Multiplication of matrices is another important matrix operation.

DEFINITION A.7

Let $A = (a_{ij})$ be an $m \times n$ matrix and let $B = (b_{jk})$ be an $n \times l$ matrix. The *matrix product* of A and B is defined as the $m \times l$ *matrix*

$$AB = (c_{ik}),$$

where

$$c_{ik} = \sum_{j=1}^{n} a_{ij} b_{jk}.$$

To multiply the matrix A by the matrix B, Definition A.7 requires that the number of columns of A be equal to the number of rows of B.

EXAMPLE A.8

Let

$$A = \begin{pmatrix} 1 & 6 \\ 4 & 2 \\ 3 & 1 \end{pmatrix}, \qquad B = \begin{pmatrix} 1 & 2 & -1 \\ 4 & 7 & 0 \end{pmatrix}.$$

The matrix product AB is defined since the number of columns of A is the same as the number of rows of B; both are equal to 2. Entry c_{ik} in the product AB is obtained by using the ith row of A and the kth column of B. For example, the entry c_{31} will be computed using the third row

$$(3 \quad 1)$$

of A and the first column

$$\begin{pmatrix} 1 \\ 4 \end{pmatrix}$$

of B. We then multiply, consecutively, each element in the third row of A by each element in the first column of B and then sum to obtain

$$3 \cdot 1 + 1 \cdot 4 = 7.$$

Since the number of columns of A is the same as the number of rows of B, the elements pair up correctly. Proceeding in this way, we obtain the product

$$AB = \begin{pmatrix} 25 & 44 & -1 \\ 12 & 22 & -4 \\ 7 & 13 & -3 \end{pmatrix}. \qquad \square$$

EXAMPLE A.9

The matrix product

$$\begin{pmatrix} a & b \\ c & d \end{pmatrix} \begin{pmatrix} x \\ y \end{pmatrix} \qquad \text{is} \qquad \begin{pmatrix} ax + by \\ cx + dy \end{pmatrix}. \qquad \square$$

DEFINITION A.10

Let A be an $n \times n$ matrix. If m is a positive integer, the mth *power* of A is defined as the matrix product

$$A^m = \underbrace{A \cdots A}_{m \ A's}.$$

EXAMPLE A.11

If

$$A = \begin{pmatrix} 1 & -3 \\ -2 & 4 \end{pmatrix},$$

then

$$A^2 = AA = \begin{pmatrix} 1 & -3 \\ -2 & 4 \end{pmatrix}\begin{pmatrix} 1 & -3 \\ -2 & 4 \end{pmatrix} = \begin{pmatrix} 7 & -15 \\ -10 & 22 \end{pmatrix},$$

$$A^4 = AAAA = A^2 A^2 = \begin{pmatrix} 7 & -15 \\ -10 & 22 \end{pmatrix}\begin{pmatrix} 7 & -15 \\ -10 & 22 \end{pmatrix} = \begin{pmatrix} 199 & -435 \\ -290 & 634 \end{pmatrix}.$$

$\square$

Exercises

1. Compute the sum

$$\begin{pmatrix} 2 & 4 & 1 \\ 6 & 9 & 3 \\ 1 & -1 & 6 \end{pmatrix} + \begin{pmatrix} a & b & c \\ d & e & f \\ g & h & i \end{pmatrix}.$$

In Exercises 2–8, let

$$A = \begin{pmatrix} 1 & 6 & 9 \\ 0 & 4 & -2 \end{pmatrix}, \qquad B = \begin{pmatrix} 4 & 1 & -2 \\ -7 & 6 & 1 \end{pmatrix}$$

and compute each expression.

2. $A + B$ **3.** $B + A$ **4.** $-A$ **5.** $3A$

6. $-2B$ **7.** $2B + A$ **8.** $B - 6A$

In Exercises 9–13, compute the products.

9. $\begin{pmatrix} 1 & 2 & 3 \\ -1 & 2 & 3 \\ 0 & 1 & 4 \end{pmatrix}\begin{pmatrix} 2 & 8 \\ -1 & 1 \\ 6 & 0 \end{pmatrix}$

10. $\begin{pmatrix} 1 & 6 \\ -8 & 2 \\ 4 & 1 \end{pmatrix}\begin{pmatrix} 4 & 1 \\ 7 & -6 \end{pmatrix}$

11. A^2, where $A = \begin{pmatrix} 1 & -2 \\ 6 & 2 \end{pmatrix}$

12.

$$(2 \quad -4 \quad 6 \quad 1 \quad 3)\begin{pmatrix} 1 \\ 3 \\ -2 \\ 6 \\ 4 \end{pmatrix}$$

13. $\begin{pmatrix} 2 & 4 & 1 \\ 6 & 9 & 3 \\ 1 & -1 & 6 \end{pmatrix}\begin{pmatrix} a & b \\ c & d \\ e & f \end{pmatrix}$

14. (a) Give the size of each matrix.

$$A = \begin{pmatrix} 1 & 4 & 6 \\ 0 & 1 & 7 \end{pmatrix}, \qquad B = \begin{pmatrix} 1 & 4 & 7 \\ 8 & 2 & 1 \\ 0 & 1 & 6 \end{pmatrix}, \qquad C = \begin{pmatrix} 4 & 2 \\ 0 & 0 \\ 2 & 9 \end{pmatrix}$$

(b) Using the matrices of part (a), decide which of the products

$$A^2, \quad AB, \quad BA, \quad AC, \quad CA, \quad AB^2, \quad BC, \quad CB, \quad C^2$$

are defined and then compute these products.

15. Determine x, y, and z so that the equation

$$\begin{pmatrix} x+y & 3x+y \\ x+z & x+y-2z \end{pmatrix} = \begin{pmatrix} -1 & 1 \\ 9 & -17 \end{pmatrix}$$

holds.

16. Determine w, x, y, and z so that the equation

$$\begin{pmatrix} 2 & 1 & -1 & 7 \\ 6 & 8 & 0 & 3 \end{pmatrix} \begin{pmatrix} x & 2x \\ y & -y+z \\ x+w & w-2y+x \\ z & z \end{pmatrix} = -\begin{pmatrix} 45 & 46 \\ 3 & 87 \end{pmatrix}$$

holds.

17. Define the $n \times n$ matrix $I_n = (a_{ij})$ by

$$a_{ij} = \begin{cases} 1 & \text{if } i = j \\ 0 & \text{if } i \neq j \end{cases}.$$

The matrix I_n is called the $n \times n$ **identity matrix.**
  Show that if A is an $n \times n$ matrix (such a matrix is called a **square matrix**), then

$$AI_n = A = I_n A.$$

An $n \times n$ matrix A is said to be *invertible* if there exists an $n \times n$ matrix B satisfying

$$AB = I_n = BA.$$

(The matrix I_n is defined in Exercise 17.)

18. Show that the matrix

$$\begin{pmatrix} 2 & 1 \\ 1 & 1 \end{pmatrix}$$

is invertible.

★ 19. Show that the matrix

$$\begin{pmatrix} a & b \\ c & d \end{pmatrix}$$

is invertible if and only if $ad - bc \neq 0$.

20. Suppose that we want to solve the system

$$AX = C \qquad \text{where} \qquad A = \begin{pmatrix} a_{11} & a_{12} \\ a_{21} & a_{22} \end{pmatrix}, \qquad X = \begin{pmatrix} x \\ y \end{pmatrix}, \qquad C = \begin{pmatrix} c_1 \\ c_2 \end{pmatrix}$$

for x and y.
  Show that if A is invertible, the system has a solution.

21. The **transpose** of a matrix $A = (a_{ij})$ is the matrix $A^T = (a'_{ji})$, where $a'_{ji} = a_{ij}$.
  Example:

$$\begin{pmatrix} 1 & 3 \\ 4 & 6 \end{pmatrix}^T = \begin{pmatrix} 1 & 4 \\ 3 & 6 \end{pmatrix}.$$

If A and B are $m \times k$ and $k \times n$ matrices, respectively, show that

$$(AB)^T = B^T A^T.$$

REFERENCES

AHO, A., J. HOPCROFT, and J. ULLMAN, *The Design and Analysis of Computer Algorithms,* Addison-Wesley, Reading, Mass., 1974.

AINSLIE, T., *Ainslie's Complete Hoyle,* Simon and Schuster, New York, 1975.

AKL, S. G., *The Design and Analysis of Parallel Algorithms,* Prentice Hall, Upper Saddle River, N.J., 1989.

APPEL, K., and W. HAKEN, "Every planar map is four-colorable," *Illinois J. Math.,* 21 (1977), 429–567.

BAASE, S., *Computer Algorithms: Introduction to Design and Analysis,* 2nd ed., Addison-Wesley, Reading, Mass., 1988.

BABAI, L., and T. KUCERA, "Canonical labelling of graphs in linear average time," *Proc. 20th Symposium on the Foundations of Computer Science,* 1979, 39–46.

BARKER, S. F., *The Elements of Logic,* 2nd ed., McGraw-Hill, New York, 1984.

BELL, R. C., *Board and Table Games from Many Civilizations,* rev. ed., Dover, New York, 1979.

BENTLEY, J., *Programming Pearls,* Addison-Wesley, Reading, Mass., 1986.

BERGE, C., *Graphs and Hypergraphs,* North-Holland, Amsterdam, 1979.

BERLEKAMP, E. R., J. H. CONWAY, and R. K. GUY, *Winning Ways,* Vols. 1 and 2, Academic Press, New York, 1982.

BONDY, J. A., and U. S. R. MURTY, *Graph Theory with Applications,* American Elsevier, New York, 1976.

BOOLE, G., *The Laws of Thought,* reprinted by Dover, New York, 1951.

BRASSARD, G., and P. BRATLEY, *Fundamentals of Algorithms*, Prentice Hall, Upper Saddle River, N.J., 1996.

BRUALDI, R. A., *Introductory Combinatorics*, 2nd ed., Prentice Hall, Upper Saddle River, N.J., 1992.

CARMONY, L., "Odd pie fights," *Math. Teacher,* 72 (1979), 61–64.

CARROLL, J., and D. LONG, *Theory of Finite Automata*, Prentice Hall, Upper Saddle River, N.J., 1989.

CHARTRAND, G., and L. LESNIAK, *Graphs and Diagraphs*, 2nd ed., Wadsworth, Belmont, Calif., 1986.

CHU, I. P., and R. JOHNSONBAUGH, "Tiling deficient boards with trominoes," *Math. Mag.,* 59 (1986), 34–40.

CODD, E. F., "A relational model of data for large shared databanks," *Comm. ACM,* 13 (1970), 377–387.

COHEN, D. I. A., *Introduction to Computer Theory,* Wiley, New York, 1986.

COPI, I. M., *Introduction to Logic,* 7th ed., Macmillan, New York, 1986.

CORMEN, T. H., C. E. LEISERSON, and R. L. RIVEST, *Introduction to Algorithms*, MIT Press, Cambridge, Mass., 1990.

CULL, P., and E. F. ECKLUND, JR., "Towers of Hanoi and analysis of algorithms," *Amer. Math. Monthly*, 92 (1985), 407–420.

DATE, C. J., *An Introduction to Database Systems,* 4th ed., Vol. 1, Addison-Wesley, Reading, Mass., 1986.

DAVIS, M. D., R. SIGAL, and E. J. WEYUKER, *Computability, Complexity, and Languages,* 2nd ed., Academic Press, San Diego, 1994.

DEO, N., *Graph Theory and Applications to Engineering and Computer Science,* Prentice Hall, Upper Saddle River, N.J., 1974.

DIJKSTRA, E. W., "A note on two problems in connexion with graphs," *Numer. Math.,* 1 (1959), 260–271.

DIJKSTRA, E. W., "Cooperating sequential processes," in *Programming Languages,* F. Genuys, ed., Academic Press, New York, 1968.

DOSSEY, J. A., A. D. OTTO, L. E. SPENCE, and C. VANDEN EYNDEN, *Discrete Mathematics*, Scott, Foresman, Glenview, Ill., 1987.

EDELSBRUNNER, H., *Algorithms in Combinatorial Geometry,* Springer-Verlag, New York, 1987.

EDGAR, W. J., *The Elements of Logic*, SRA, Chicago, 1989.

ENGLISH, E., and S. HAMILTON, "Network security under siege, the timing attack," *Computer*, (March 1996), 95–97.

ERBAS, C., and M. M. TANIK, "Generating solutions to the n-queens problem using 2-circulants," *Math. Mag.,* 68 (1995), 343–356.

EVEN, S., *Algorithmic Combinatorics*, Macmillan, New York, 1973.

EVEN, S., *Graph Algorithms*, Computer Science Press, Rockville, Md., 1979.

EZEKIEL, M., "The cobweb theorem," *Quart. J. Econom.,* 52 (1938), 225–280.

FORD, L. R., JR., and D. R. FULKERSON, *Flows in Networks,* Princeton University Press, Princeton, N.J., 1962.

FOWLER, P. A., "The Königsberg bridges—250 years later," *Amer. Math Monthly,* 95 (1988), 42–43.

FREY, P., "Machine-problem solving—Part 3: The alpha-beta procedure," *BYTE,* 5 (November 1980), 244–264.

FUKUNAGA, K., *Introduction to Statistical Pattern Recognition,* 2nd ed., Academic Press, New York, 1990.

GALLIER, J. H., *Logic for Computer Science,* Harper & Row, New York, 1986.

GARDNER, M., *Mathematical Puzzles and Diversions*, Simon and Schuster, 1959.

GARDNER, M., "A new kind of cipher that would take millions of years to break," *Sci. Amer.* (February 1977), 120–124.

GARDNER, M., *Mathematical Circus*, Alfred A. Knopf, New York, 1979.

GENESERETH, M. R., and N. J. NILSSON, *Logical Foundations of Artificial Intelligence,* Morgan Kaufmann, Los Altos, Calif., 1987.

GIBBONS, A., *Algorithmic Graph Theory,* Cambridge University Press, Cambridge, 1985.

GOLDBERG, S., *Introduction to Difference Equations,* Wiley, New York, 1958.

GOLOMB, S. W., "Checker boards and polyominoes," *Amer. Math. Monthly,* 61 (1954), 675–682.

GOLOMB, S., and L. BAUMERT, "Backtrack programming," *J. ACM,* 12 (1965), 516–524.

GOSE, E., R. JOHNSONBAUGH, and S. JOST, *Pattern Recognition and Image Analysis,* Prentice Hall, Upper Saddle River, N.J., 1996.

GRAHAM, R. L., "An efficient algorithm for determining the convex hull of a finite planar set," *Info. Proc. Lett.,* 1 (1972), 132–133.

GRAHAM, R. L., D. E. KNUTH, and O. PATASHNIK, *Concrete Mathematics: A Foundation for Computer Science,* Addison-Wesley, Reading, Mass., 1988.

GRIES, D., *The Science of Programming,* Springer-Verlag, New York, 1981.

HAILPERIN, T., "Boole's algebra isn't Boolean algebra," *Math. Mag.,* 54 (1981), 137–184.

HALMOS, P. R., *Naive Set Theory,* Springer-Verlag, New York, 1974.

HARARY, F., *Graph Theory,* Addison-Wesley, Reading, Mass., 1969.

HELL, P., "Absolute retracts in graphs," in *Graphs and Combinatorics,* R. A. Bari and F. Harary, eds., Lecture Notes in Mathematics, Vol. 406, Springer-Verlag, New York, 1974.

HILLIER, F. S., and G. J. LIEBERMAN, *Introduction to Operations Research,* Holden-Day, San Francisco, 1974.

HINZ, A. M., "The Tower of Hanoi," *Enseignement Math.,* 35 (1989), 289–321.

HOHN, F., *Applied Boolean Algebra,* 2nd ed., Macmillan, New York, 1966.

HOPCROFT, J. E., and J. D. ULLMAN, *Introduction to Automata Theory, Languages, and Computation,* Addison-Wesley, Reading, Mass., 1979.

HU, T. C., *Combinatorial Algorithms,* Addison-Wesley, Reading, Mass., 1982.

JACOBS, H. R., *Geometry,* W.H. Freeman, San Fransisco, 1974.

JARVIS, R. A., "On the identification of the convex hull of a finite set of points in the plane," *Info. Proc. Lett.,* 2 (1973), 18–21.

JONES, R. H., and N. C. STEELE, *Mathematics in Communication Theory,* Ellis Horwood, Chichester, England, 1989.

KELLEY, D., *Automata and Formal Languages,* Prentice Hall, Upper Saddle River, N.J., 1995.

KLEINROCK, L., *Queueing Systems,* Vol. 2: *Computer Applications,* Wiley, New York, 1976.

KLINE, M., *Mathematical Thought from Ancient to Modern Times,* Oxford University Press, New York, 1972.

KNUTH, D. E., *The Art of Computer Programming,* Vol. 1: *Fundamental Algorithms,* 2nd ed., Addison Wesley, Reading, Mass., 1973.

KNUTH, D. E., *The Art of Computer Programming,* Vol. 3: *Sorting and Searching,* Addison-Wesley, Reading, Mass., 1973.

KNUTH, D. E., "Algorithms," *Sci. Amer.* (April 1977), 63–80.

KNUTH, D. E., *The Art of Computer Programming,* Vol. 2: *Seminumeric Algorithms,* 2nd ed., Addison-Wesley, Reading, Mass., 1981.

KNUTH, D. E., "Algorithmic thinking and mathematical thinking," *Amer. Math. Monthly,* 92 (1985), 170–181.

KÖBLER, J., U. SCHÖNING, and J. TORÁN, *The Graph Isomorphism Problem: Its Structural Complexity,* Birkhäuser Verlag, Basel, Switzerland, 1993.

KOHAVI, Z., *Switching and Finite Automata Theory,* 2nd ed., McGraw-Hill, New York, 1978.

KÖNIG, D., *Theorie der endlichen und unendlichen Graphen,* Akademische Verlagsgesellschaft, Leipzig, 1936. (Reprinted in 1950 by Chelsea, New York.) (English translation: *Theory of Finite and Infinite Graphs,* Birkhäuser Boston, Cambridge, Mass., 1990.)

KROENKE, D., *Database Processing,* 2nd ed., Science Research Associates, Palo Alto, Calif., 1983.

KRUSE, R. L., *Data Structures and Program Design,* 2nd ed., Prentice Hall, Upper Saddle River, N.J., 1987.

KUROSAKA, R. T., "A ternary state of affairs," *BYTE,* 12 (February 1987), 319–328.

LEIGHTON, F. T., *Introduction to Parallel Algorithms and Architectures,* Morgan Kaufmann, San Mateo, Calif., 1992.

LESTER, B. P., *The Art of Parallel Programming,* Prentice Hall, Upper Saddle River, N.J., 1993.

LEWIS, T. G., and H. EL-REWINI, *Introduction to Parallel Computing,* Prentice Hall, Upper Saddle River, N.J., 1992.

LINDENMAYER, A., "Mathematical models for cellular interaction in development," parts I and II, *J. Theoret. Biol.,* 18 (1968), 280–315.

LIPSCHUTZ, S., *Theory and Problems of Set Theory and Related Topics,* Schaum, New York, 1964.

LIU, C. L., *Introduction to Combinatorial Mathematics,* McGraw-Hill, New York, 1968.

LIU, C. L., *Elements of Discrete Mathematics,* 2nd ed., McGraw-Hill, New York, 1985.

MANBER, U., *Introduction to Algorithms,* Addison-Wesley, Reading, Mass., 1989.

MANDELBROT, B. B., *Fractals: Form, Chance, and Dimension,* W.H. Freeman, San Francisco, 1977.

MANDELBROT, B. B., *The Fractal Geometry of Nature,* W.H. Freeman, San Francisco, 1982.

MARTIN, G. E., *Polyominoes: A Guide to Puzzles and Problems in Tiling,* Mathematical Association of America, Washington, D.C., 1991.

MCCALLA, T. R., *Digital Logic and Computer Design,* Merrill, New York, 1992.

MCNAUGHTON, R., *Elementary Computability, Formal Languages, and Automata,* Prentice Hall, Upper Saddle River, N.J., 1982.

MENDELSON, E., *Boolean Algebra and Switching Circuits,* Schaum, New York, 1970.

NADLER, M., and E. P. SMITH, *Pattern Recognition Engineering,* Wiley, New York, 1993.

NEWMAN, J. R., "Leonhard Euler and the Koenigsberg bridges," *Sci. Amer.* (July 1953), 66–70.

NIEVERGELT, J., J. C. FARRAR, and E. M. REINGOLD, *Computer Approaches to Mathematical Problems,* Prentice Hall, Upper Saddle River, N.J., 1974.

NILSSON, N. J., *Problem-Solving Methods in Artificial Intelligence,* McGraw-Hill, New York, 1971.

NIVEN, I., *Mathematics of Choice,* Mathematical Association of America, Washington, D.C., 1965.

NYHOFF, L., and S. LEESTMA, *Data Structures and Program Design in Pascal,* 2nd ed., Macmillan, New York, 1992.

ORE, O., *Graphs and Their Uses,* Mathematical Association of America, Washington, D.C., 1963.

PEARL, J., "The solution for the branching factor of the alpha-beta pruning algorithm and its optimality," *Comm. ACM*, 25 (1982), 559–564.

PEITGEN, H., and D. SAUPE, eds., *The Science of Fractal Images,* Springer-Verlag, New York, 1988.

PETRI, C., "Kommunikation mit Automaten," Ph.D. dissertation, University of Bonn, Bonn, West Germany, 1962 (in German). Translated by C. F. Green, Jr., "Communication with automata," Supplement to Technical Report RADC-TR–65–377, Vol. I, Rome Air Development Center, Griffiss Air Force Base, New York, 1966.

PFLEEGER, C. P., *Security in Computing,* Prentice Hall, Upper Saddle River, N.J., 1989.

PREPARATA, F. P., and S. J. HONG, "Convex hulls of finite sets of points in two and three dimensions," *Comm. ACM,* 20 (1977), 87–93.

PREPARATA, F. P., and M. I. SHAMOS, *Computational Geometry*, Springer-Verlag, New York, 1985.

Problem 1186, *Math. Mag.,* 58 (1985), 112–114.

PRODINGER, H., and R. TICHY, "Fibonacci numbers of graphs," *Fibonacci Quarterly,* 20 (1982), 16–21.

PRUSINKIEWICZ, P., "Graphical applications of L-systems," *Proc. of Graphics Interface 1986—Vision Interface*, (1986), 247–253.

PRUSINKIEWICZ, P., and J. HANAN, "Applications of L-systems to computer imagery," in *Graph Grammars and Their Application to Computer Science; Third International Workshop,* H. Ehrig, M. Nagl, A. Rosenfeld, and G. Rozenberg, eds., Springer-Verlag, New York, 1988.

QUINN, M. J., *Designing Efficient Algorithms for Parallel Computers,* McGraw-Hill, New York, 1987.

READ, R. C., and D. G. CORNEIL, "The graph isomorphism disease," *J. Graph Theory,* 1 (1977), 339–363.

REINGOLD, E., J. NIEVERGELT, and N. DEO, *Combinatorial Algorithms,* Prentice Hall, Upper Saddle River, N.J., 1977.

RIORDAN, J., *An Introduction to Combinatorial Analysis,* Wiley, New York, 1958.

RITTER, G. L., S. R. LOWRY, H. B. WOODRUFF, and T. L. ISENHOUR, "An aid to the superstitious," *Math. Teacher,* May 1977, 456–457.

ROBERTS, F. S., *Applied Combinatorics,* Prentice Hall, Upper Saddle River, N.J., 1984.

ROBINSON, J. A., "A machine oriented logic based on the resolution principle," *J. ACM,* 12 (1965), 23–41.

ROSS, K. A., and C. R. B. WRIGHT, *Discrete Mathematics,* 3rd ed., Prentice Hall, Upper Saddle River, N.J., 1992.

SAAD, Y., and M. H. SCHULTZ, "Topological properties of hypercubes," *IEEE Trans. Computers,* 37 (1988), 867–872.

SCHWENK, A. S., "Which rectangular chessboards have a knight's tour?" *Math. Mag.,* 64 (1991), 325–332.

SEIDEL, R., "A convex hull algorithm optimal for points in even dimensions," M.S. thesis, Tech. Rep. 81–14, Dept. of Comp. Sci., Univ. of British Columbia, Vancouver, Canada, 1981.

SHANNON, C. E., "A symbolic analysis of relay and switching circuits," *Trans. Amer. Inst. Electr. Engrs.*, 47 (1938), 713–723.

SLAGLE, J. R., *Artificial Intelligence: The Heuristic Programming Approach,* McGraw-Hill, New York, 1971.

SMITH, A. R., "Plants, fractals, and formal languages," *Computer Graphics,* 18 (1984), 1–10.

SOLOW, D., *How to Read and Do Proofs,* 2nd ed., Wiley, New York, 1990.

STANDISH, T. A., *Data Structures, Algorithms, and Software Principles in C,* Addison-Wesley, Reading, Mass., 1995.

STOLL, R. R., *Set Theory and Logic,* W.H. Freeman, San Francisco, 1963.

SUDKAMP, T. A., *Languages and Machines,* Addison-Wesley, Reading, Mass., 1988.

TARJAN, R. E., *Data Structures and Network Algorithms*, Society for Industrial and Applied Mathematics, Philadelphia, 1983.

TAUBES, G., "Small army of code-breakers conquers a 129-digit giant," *Science*, 264 (1994), 776–777.

TUCKER, A., *Applied Combinatorics*, 2nd ed., Wiley, New York, 1985.

ULLMAN, J. D., *Principles of Database Systems,* Computer Science Press, Rockville, Md., 1980.

VILENKIN, N. Y., *Combinatorics*, Academic Press, New York, 1971.

WAGON, S., "Fourteen proofs of a result about tiling a rectangle," *Amer. Math. Monthly*, 94 (1987), 601–617. (Reprinted in R. K. Guy and R. E. Woodrow, eds., *The Lighter Side of Mathematics,* Mathematical Association of America, Washington, D.C., 1994, 113–128.)

WARD, S. A., and R. H. HALSTEAD, JR., *Computation Structures,* MIT Press, Cambridge, Mass., 1990.

WILSON, R. J., *Introduction to Graph Theory,* 2nd ed., Academic Press, Orlando, Fla., 1979.

WOOD, D., *Theory of Computation,* Harper & Row, New York, 1987.

WOS, L., R. OVERBEEK, E. LUSK, and J. BOYLE, *Automated Reasoning,* Prentice Hall, Upper Saddle River, N.J., 1984.

HINTS AND SOLUTIONS TO SELECTED EXERCISES

Section 1.1

1. Is a proposition. Negation: $2 + 5 \neq 19$

4. Is a proposition. Negation: Audrey Meadows was not the original "Alice" in "The Honeymooners."

7. Is a proposition. Some even integer greater than 4 is not the sum of two primes.

9. True

12. True

15.

p	q	$p \wedge \overline{q}$
T	T	F
T	F	T
F	T	F
F	F	F

18.

p	q	$(p \wedge q) \wedge \overline{p}$
T	T	F
T	F	F
F	T	F
F	F	F

21.

p	q	$(p \vee q) \wedge (\overline{p} \vee q) \wedge (p \vee \overline{q}) \wedge (\overline{p} \vee \overline{q})$
T	T	F
T	F	F
F	T	F
F	F	F

23. $p \wedge q$; false

26. Today is Monday or it is raining.

29. (Today is Monday and it is raining) and it is not the case that (it is hot or today is Monday).

32. No. A judge ruled that the ordinance was "vague." Presumably, the intended wording was: "It shall be unlawful for any person to keep more than three [3] dogs *or* three [3] cats upon his property within the city."

Section 1.2

1. If Joey studies hard, then he will pass the discrete mathematics exam.

4. If Katrina passes discrete mathematics, then she will take the algorithms course.

7. If the program is readable, then it is well structured.

8. [For Exercise 1] If Joey passes the discrete mathematics exam, then he studied hard.

10. True   **13.** False   **16.** False

18. $p \rightarrow q$ **21.** $q \leftrightarrow (p \wedge \overline{r})$

22. If today is Monday, then it is raining.

25. It is not the case that today is Monday or it is raining if and only if it is hot.

28. Let $p: 4 < 6$ and $q: 9 > 12$.
Given statement: $p \rightarrow q$: false.
Converse: $q \rightarrow p$; if $9 > 12$, then $4 < 6$; true.
Contrapositive: $\overline{q} \rightarrow \overline{p}$; if $9 \leq 12$, then $4 \geq 6$; false.

31. Let $p: |4| < 3$ and $q: -3 < 4 < 3$.
Given statement: $q \rightarrow p$; true.
Converse: $p \rightarrow q$; if $|4| < 3$, then $-3 < 4 < 3$; true.
Contrapositive: $\overline{p} \rightarrow \overline{q}$, if $|4| \geq 3$, then $-3 \geq 4$ or $4 \geq 3$; true.

32. $P \not\equiv Q$ **35.** $P \not\equiv Q$ **38.** $P \not\equiv Q$ **41.** $P \not\equiv Q$

42.

p	q	$p\,impl\,q$	$q\,impl\,p$
T	T	T	T
T	F	F	F
F	T	F	F
F	F	T	T

Since $p\,impl\,q$ is true precisely when $q\,impl\,p$ is true, $p\,impl\,q \equiv q\,impl\,p$.

45.

p	q	$p \rightarrow q$	$\overline{p} \vee q$
T	T	T	T
T	F	F	F
F	T	T	T
F	F	T	T

Since $p \rightarrow q$ is true precisely when $\overline{p} \vee q$ is true, $p \rightarrow q \equiv \overline{p} \vee q$.

Section 1.3

1. Is a propositional function. The domain of discourse could be taken to be all integers.

4. Is a propositional function. The domain of discourse is the set of all movies.

7. 11 divides 77. True

10. For every positive integer n, n divides 77. False

12. Everyone is taller than everyone else. False

15. Someone is taller than someone else. True

17. $\exists x \forall y L(x, y)$. Probably true

20. $\forall x \exists y L(x, y)$. Unfortunately, probably false

22. False. A counterexample is $x = \frac{1}{2}$.

25. True. For example if $x = 0$, the conditional proposition, if $x > 1$, then $x^2 > x$, is true because the hypothesis is false.

28. False. A counterexample is $x = 2$, $y = 0$.

31. True. Take $x = y = 0$.

34. False. A counterexample is $x = y = 2$.

37. True. Take $x = 1$, $y = \sqrt{8}$.

40. True. Take $x = 0$. Then for all y, $x^2 + y^2 \geq 0$.

43. True. For any x, if we set $y = x - 1$, the conditional proposition, if $x < y$, then $x^2 < y^2$, is true because the hypothesis is false.

47. The correct interpretation is: Some man does not cheat on his wife. This proposition is true. This was Abby's intended meaning. She received some letters in which the writers interpreted the proposition as: No man cheats on his wife. Under this interpretation, the proposition is false.

Section 1.4

1. If three points are not collinear, then there is exactly one plane that contains them.

4. If x is a nonnegative real number and n is a positive integer, $x^{1/n}$ is the nonnegative number y satisfying $y^n = x$.

7. $x \cdot 0 + 0 = x \cdot 0$

because $b + 0 = b$ for all real numbers b

$$= x \cdot (0 + 0)$$

because $b + 0 = b$ for all real numbers b

$$= x \cdot 0 + x \cdot 0$$

because $a(b + c) = ab + ac$ for all real numbers a, b, c

Taking $a = b = x \cdot 0$ and $c = 0$, the preceding equation becomes $a + b = a + c$; therefore, $x \cdot 0 = b = c = 0$.

10. Valid $\quad p \rightarrow q$
$$\dfrac{p}{\therefore q}$$

13. Invalid $\quad (p \vee r) \rightarrow q$
$$\dfrac{q}{\therefore \overline{p} \rightarrow r}$$

15. Valid. If 64K is better than no memory at all, then we will buy a new computer. If 64K is better than no memory at all, then we will buy more memory. Therefore, if 64K is better than no memory at all, then we will buy a new computer and we will buy more memory.

18. Invalid. If we will not buy a new computer, then 64K is not better than no memory at all. We will buy a new computer. Therefore, 64K is better than no memory at all.

20. Invalid   **23.** Invalid

26. An analysis of the argument must take into account the fact that "nothing" is being used in two very different ways.

Section 1.5

1.

p	q	r	$p \vee q$	$\overline{p} \vee r$	$q \vee r$
T	T	T	T	T	T
T	T	F	T	F	T
T	F	T	T	T	T
T	F	F	T	F	F
F	T	T	T	T	T
F	T	F	T	T	T
F	F	T	F	T	T
F	F	F	F	T	F

2.

1. $\overline{p} \vee q \vee r$
2. $\overline{q}$
3. $\overline{r}$
4. $\overline{p} \vee r$ From 1 and 2
5. $\overline{p}$ From 3 and 4

5. First we note that $p \to q$ is logically equivalent to $\overline{p} \vee q$. We now argue as follows:

1. $\overline{p} \vee q$
2. $p \vee q$
3. q From 1 and 2

7. (For Exercise 2).

1. $\overline{p} \vee q \vee r$ Hypothesis
2. $\overline{q}$ Hypothesis
3. $\overline{r}$ Hypothesis
4. p Negation of conclusion
5. $\overline{p} \vee r$ From 1 and 2
6. $\overline{p}$ From 3 and 5

Now 4 and 6 combine to give a contradiction.

Section 1.6

1. BASIS STEP. $1 = 1^2$

INDUCTIVE STEP. Assume true for n.

$$1 + \cdots + (2n - 1) + (2n + 1) = n^2 + 2n + 1 = (n + 1)^2$$

4. BASIS STEP. $1^2 = (1 \cdot 2 \cdot 3)/6$

INDUCTIVE STEP. Assume true for n.

$$1^2 + \cdots + n^2 + (n + 1)^2 = \frac{n(n + 1)(2n + 1)}{6} + (n + 1)^2$$
$$= \frac{(n + 1)(n + 2)(2n + 3)}{6}$$

7. BASIS STEP. $1/(1 \cdot 3) = \dfrac{1}{3}$

INDUCTIVE STEP. Assume true for n.

$$\frac{1}{1 \cdot 3} + \cdots + \frac{1}{(2n - 1)(2n + 1)} + \frac{1}{(2n + 1)(2n + 3)}$$
$$= \frac{n}{2n + 1} + \frac{1}{(2n + 1)(2n + 3)}$$
$$= \frac{n + 1}{2n + 3}$$

10. BASIS STEP. $\cos x = \dfrac{\cos[(x/2) \cdot 2] \sin(x/2)}{\sin(x/2)}$

INDUCTIVE STEP. Assume true for n. Then

$$\cos x + \cdots + \cos nx + \cos(n + 1)x$$
$$= \frac{\cos[(x/2)(n + 1)] \sin(nx/2)}{\sin(x/2)} + \cos(n + 1)x. \quad (*)$$

We must show that the right-hand side of $(*)$ is equal to

$$\frac{\cos[(x/2)(n + 2)] \sin[(n + 1)x/2]}{\sin(x/2)}.$$

This is the same as showing that [after multiplying by the term $\sin(x/2)$]

$$\cos\left[\frac{x}{2}(n + 1)\right] \sin\frac{nx}{2} + \cos(n + 1)x \sin\frac{x}{2}$$
$$= \cos\left[\frac{x}{2}(n + 2)\right] \sin\left[\frac{(n + 1)x)}{2}\right].$$

If we let $\alpha = (x/2)(n + 1)$ and $\beta = x/2$, we must show that

$$\cos\alpha \sin(\alpha - \beta) + \cos 2\alpha \sin\beta = \cos(\alpha + \beta)\sin\alpha.$$

This last equation can be verified by reducing each side to terms involving α and β.

12. BASIS STEP. $\dfrac{1}{2} \leq \dfrac{1}{2}$

INDUCTIVE STEP. Assume true for n.

$$\frac{1 \cdot 3 \cdot 5 \cdots (2n - 1)(2n + 1)}{2 \cdot 4 \cdot 6 \cdots (2n)(2n + 2)} \geq \frac{1}{2n} \cdot \frac{2n + 1}{2n + 2}$$
$$= \frac{2n + 1}{2n} \cdot \frac{1}{2n + 2}$$
$$\geq \frac{1}{2n + 2}$$

15. BASIS STEP ($n = 4$). $2^4 = 16 \geq 16 = 4^2$

INDUCTIVE STEP. Assume true for n.

$$(n + 1)^2 = n^2 + 2n + 1 \leq 2^n + 2n + 1$$
$$\leq 2^n + 2^n \quad \text{by Exercise 14}$$
$$= 2^{n+1}$$

18. BASIS STEP. $7^1 - 1 = 6$ is divisible by 6.

INDUCTIVE STEP. Suppose that 6 divides $7^n - 1$. Now

$$7^{n+1} - 1 = 7 \cdot 7^n - 1 = 7^n - 1 + 6 \cdot 7^n.$$

Since 6 divides both $7^n - 1$ and $6 \cdot 7^n$, it divides their sum, which is $7^{n+1} - 1$.

21. BASIS STEP. $3^1 + 7^1 - 2 = 8$ is divisible by 8.

INDUCTIVE STEP. Suppose that 8 divides $3^n + 7^n - 2$. Now

$$3^{n+1} + 7^{n+1} - 2 = 3(3^n + 7^n - 2) + 4(7^n + 1).$$

By the inductive assumption, 8 divides $3^n + 7^n - 2$. We can use mathematical induction to show that 2 divides $7^n + 1$ for all $n \geq 1$ (the argument is similar to that given in the hint for Exercise 18). It then follows that 8 divides $4(7^n + 1)$. Since 8 divides both $3(3^n + 7^n - 2)$ and $4(7^n + 1)$, it divides their sum, which is $3^{n+1} + 7^{n+1} - 2$.

23. At the Inductive Step when the $(n + 1)$st line is added, because of the assumptions, the line will intersect each of the other n lines. Now, imagine traveling along the $(n + 1)$st line. Each time we pass through one of the original regions, it is divided into two regions.

26. $\dfrac{5}{6} = \dfrac{1}{2} + \dfrac{1}{3} = \dfrac{1}{2} + \dfrac{1}{4} + \dfrac{1}{12}$

30. We may assume that $p/q > 1$. Choose the largest integer n satisfying

$$\frac{1}{1} + \frac{1}{2} + \cdots + \frac{1}{n} \leq \frac{p}{q}.$$

If we obtain an equality, p/q is in Egyptian form, so suppose that

$$\frac{1}{1} + \frac{1}{2} + \cdots + \frac{1}{n} < \frac{p}{q}. \qquad (*)$$

Set

$$D = \frac{p}{q} - \left(\frac{1}{1} + \cdots + \frac{1}{n}\right).$$

Clearly, $D > 0$. Since n is the largest integer satisfying $(*)$,

$$\frac{1}{1} + \frac{1}{2} + \cdots + \frac{1}{n} + \frac{1}{n+1} \geq \frac{p}{q}.$$

Thus

$$D = \frac{p}{q} - \left(\frac{1}{1} + \cdots + \frac{1}{n}\right)$$

$$\leq \left(\frac{1}{1} + \cdots + \frac{1}{n} + \frac{1}{n+1}\right) - \left(\frac{1}{1} + \cdots + \frac{1}{n}\right)$$

$$= \frac{1}{n+1}.$$

In particular, $D < 1$. By Exercise 28, D may be written in Egyptian form:

$$D = \frac{1}{n_1} + \cdots + \frac{1}{n_k},$$

where the n_i are distinct. Since

$$\frac{1}{n_i} \leq D \leq \frac{1}{1 + n},$$

$n < n + 1 \leq n_i$ for $i = 1, \ldots, k$. It follows that

$$1, \quad 2, \quad \ldots, \quad n, \quad n_1, \quad \ldots, \quad n_k$$

are distinct. Thus

$$\frac{p}{q} = D + \frac{1}{1} + \cdots + \frac{1}{n} = \frac{1}{n_1} + \cdots + \frac{1}{n_k} + \frac{1}{1} + \cdots + \frac{1}{n}$$

is represented in Egyptian form.

33. A tromino can cover the square to the left of the missing square (shown shaded) as shown

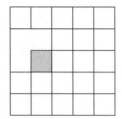

or in a symmetric fashion by reversing "up" and "down." In the first case, it is impossible to cover the two squares in the top row at the extreme left. In the second case, it is impossible to cover the two squares in the bottom row at the extreme left. Therefore, it is impossible to tile the board with trominoes.

36. **BASIS STEP** ($n = 7$ or 11). Exercise 35 gives a solution if $n = 7$.

If $n = 11$, we first rotate the board so that the missing square is located in the 7×7 board shown in the following figure. Exercise 35 shows how to tile this subboard. Exercise 34 shows that the two 6×4 boards can be tiled. It is straightforward to verify that the 5×5 board with a corner square missing can be tiled. We have shown that any deficient 11×11 board can be tiled with trominoes.

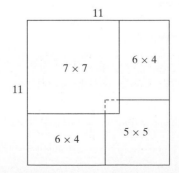

INDUCTIVE STEP. Suppose that n is odd, $n > 11$, 3 divides $n^2 - 1$, and that deficient $k \times k$ boards, where k is odd, $n > k > 5$, and 3 divides $k^2 - 1$, can be tiled. The following figure shows how to tile a deficient $n \times n$ board. We first rotate the board so that the missing square is located in the $(n - 6) \times (n - 6)$ subboard. Now $n - 6$ is odd, $n - 6 > 5$, and 3 divides $(n - 6)^2 - 1$; so, by the inductive assumption, this deficient $(n - 6) \times (n - 6)$ subboard can be tiled. Since n is odd, $n - 7$ is even; thus, by Exercise 34, the two $6 \times (n - 7)$ subboards can be tiled. By Exercise 35, the deficient 7×7 subboard can be tiled. Therefore, we can tile the $n \times n$ board.

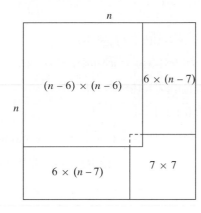

39. (a) $S_1 = 0 \neq 2$;

$$2 + \cdots + 2n + 2(n + 1) = S_n + 2n + 2$$
$$= (n + 2)(n - 1) + 2n + 2$$
$$= (n + 3)n$$
$$= S_{n+1}.$$

(b) We must have $S'_n = S'_{n-1} + 2n$; thus

$$S'_n = S'_{n-1} + 2n = [S'_{n-2} + 2(n - 1)] + 2n$$
$$= S'_{n-2} + 2n + 2(n - 1)$$
$$= S'_{n-3} + 2n + 2(n - 1) + 2(n - 2) = \cdots$$
$$= S'_1 + 2[n + (n - 1) + \cdots + 2]$$
$$= C' + 2\left[\frac{n(n + 1)}{2} - 1\right] = n^2 + n + C.$$

44. Suppose, by way of contradiction, that some statement $S(n)$ is false. Let X be the set of positive integers n for which $S(n)$ is false. Now apply the Well-Ordering Theorem.

Chapter 1 Self-Test

1. False

2.

p	q	r	$(p \wedge q) \vee (p \vee \bar{r})$
T	T	T	T
T	T	F	T
T	F	T	T
T	F	F	T
F	T	T	T
F	T	F	T
F	F	T	T
F	F	F	T

3. I take hotel management and either I do not take recreation supervision or I take popular culture.

4. $p \vee (q \wedge \bar{r})$

5. If Leah gets an A in discrete mathematics, then Leah studies hard.

6. Converse: If Leah studies hard, then Leah gets an A in discrete mathematics. Contrapositive: If Leah does not study hard, then Leah does not get an A in discrete mathematics.

7. True   **8.** $(\bar{r} \vee q) \rightarrow \bar{q}$

9. The statement is not a proposition. The truth value cannot be determined without knowing what "the team" refers to.

10. The statement is a propositional function. When we substitute a particular team for the variable "team," the statement becomes a proposition.

11. For all positive integers n, n and $n + 2$ are prime. The proposition is false. A counterexample is $n = 7$.

12. For some positive integer n, n and $n + 2$ are prime. The proposition is true. For example, if $n = 5$, n and $n + 2$ are prime.

13. Suppose that if four teams play seven games, no pair of teams plays at least two times; or, equivalently, if four teams play seven games, each pair of teams plays at most one time. If the teams are A, B, C, and D and each pair of teams plays at most one time, the most games that can be played are:

A and B; A and C; A and D; B and C;

B and D; C and D.

Thus at most six games can be played. This is a contradiction. Therefore, if four teams play seven games, some pair of teams plays at least two times.

14. Axioms are statements that are assumed true. Definitions are used to create new concepts in terms of existing ones.

15. In a direct proof, the negated conclusion is not assumed, whereas in a proof by contradiction, the negated conclusion is assumed.

16. The argument is invalid. If p and r are true and q is false, the hypotheses are true, but the conclusion is false.

17.
$$(p \vee q) \rightarrow r \equiv \overline{p \vee q} \vee r$$
$$\equiv \overline{p}\,\overline{q} \vee r$$
$$\equiv (\overline{p} \vee r)(\overline{q} \vee r)$$

18.
$$(p \vee \overline{q}) \rightarrow \overline{r}s \equiv \overline{p \vee \overline{q}} \vee \overline{r}s$$
$$\equiv \overline{p}q \vee \overline{r}s$$
$$\equiv (\overline{p} \vee \overline{r})(\overline{p} \vee s)(q \vee \overline{r})(q \vee s)$$

19.
1. $\overline{p} \vee q$
2. $\overline{q} \vee \overline{r}$
3. $p \vee \overline{r}$
4. $\overline{p} \vee \overline{r}$ From 1 and 2
5. $\overline{r}$ From 3 and 4

20.
1. $\overline{p} \vee q$
2. $\overline{q} \vee \overline{r}$
3. $p \vee \overline{r}$
4. r Negation of conclusion
5. $\overline{p} \vee \overline{r}$ From 1 and 2
6. $\overline{r}$ From 3 and 5

Now 4 and 6 give a contradiction.

In Exercises 21–24, only the Inductive Step is given.

21. $2 + 4 + \cdots + 2n + 2(n+1) = n(n+1) + 2(n+1)$
$$= (n+1)(n+2)$$

22. $2^2 + 4^2 + \cdots + (2n)^2 + [2(n+1)]^2$
$$= \frac{2n(n+1)(2n+1)}{3} + [2(n+1)]^2$$
$$= \frac{2(n+1)(n+2)[2(n+1)+1]}{3}$$

23. $\dfrac{1}{2!} + \dfrac{2}{3!} + \cdots + \dfrac{n}{(n+1)!}$
$$+ \frac{n+1}{(n+2)!} = 1 - \frac{1}{(n+1)!} + \frac{n+1}{(n+2)!}$$
$$= 1 - \frac{1}{(n+2)!}$$

24. $2^{n+2} = 2 \cdot 2^{n+1} < 2[1 + (n+1)2^n] = 2 + (n+1)2^{n+1}$
$$= 1 + [1 + (n+1)2^{n+1}]$$
$$< 1 + [2^{n+1} + (n+1)2^{n+1}]$$
$$= 1 + (n+2)2^{n+1}$$

Section 2.1

1. $\{1, 2, 3, 4, 5, 7, 10\}$ **4.** $\{2, 3, 5\}$

7. $\emptyset$ **10.** U

13. $\{6, 8\}$ **16.** $\{1, 2, 3, 4, 5, 7, 10\}$

17. $\{(1, a), (1, b), (1, c), (2, a), (2, b), (2, c)\}$

20. $\{(a, a), (a, b), (a, c), (b, a), (b, b), (b, c), (c, a),$
$(c, b), (c, c)\}$

21. $\{(1, a, \alpha), (1, a, \beta), (2, a, \alpha), (2, a, \beta)\}$

24. $\{(a, 1, a, \alpha), (a, 2, a, \alpha), (a, 1, a, \beta), (a, 2, a, \beta)\}$

25. $\{\{1\}\}$

28. $\{\{a, b, c, d\}\}, \{\{a, b, c\}, \{d\}\},$
$\{\{a, b, d\}, \{c\}\}, \{\{a, c, d\}, \{b\}\}, \{\{b, c, d\}, \{a\}\},$
$\{\{a, b\}, \{c\}, \{d\}\}, \{\{a, c\}, \{b\}, \{d\}\},$
$\{\{a, d\}, \{b\}, \{c\}\},$
$\{\{b, c\}, \{a\}, \{d\}\}, \{\{b, d\}, \{a\}, \{c\}\}, \{\{c, d\}, \{a\}, \{b\}\},$
$\{\{a, b\}, \{c, d\}\}, \{\{a, c\}, \{b, d\}\}, \{\{a, d\}, \{b, c\}\},$
$\{\{a\}, \{b\}, \{c\}, \{d\}\}$

29. True   **32.** True

33. Equal   **36.** Equal

38. $\emptyset, \{a\}, \{b\}, \{a, b\}$. All but $\{a, b\}$ are proper subsets.

41. $2^n - 1$ **43.** False. $X = \{1, 2\}, Y = \{2, 3\}$.

46. False. $X = \{1, 2, 3\}, Y = \{2\}, Z = \{3\}$.

49. False. $X = \{1\},$ **52.** True
$Y = \{1, 2\}$.

55. True   **57.** $A \subseteq B$

60. $B \subseteq A$ **61.** $\{1, 4, 5\}$

65. $|A| + |B|$ counts the elements in A and B but counts the elements in $A \cap B$ twice.

68. P is the set of primes.

Section 2.2

1. (a) c (b) c (c) $cddcdc$

4. (a) 12   (b) 23   (c) 7
(d) 46   (e) 1   (f) 3
(g) 3   (h) 21

8. (a) 15   (b) 155
(c) $2n + 3(n-1)n/2$ (d) Yes
(e) No

11. (a) $1, 3, 5, 7, 9, 11, 13$
(b) $1, 5, 9, 13, 17, 21, 25$
(c) $n_k = 2k - 1$ (d) $s_{n_k} = 4k - 3$

13. (a) 88   (b) 1140
(c) 48   (d) 3168

16. $b_1 = 2, b_2 = 3, b_3 = 5, b_4 = 8, b_5 = 12, b_6 = 257$

19. Let $s_0 = 0$. Then

$$\sum_{k=1}^{n} a_k b_k = \sum_{k=1}^{n} (s_k - s_{k-1}) b_k$$

$$= \sum_{k=1}^{n} s_k b_k - \sum_{k=1}^{n} s_{k-1} b_k$$

$$= \sum_{k=1}^{n} s_k b_k - \sum_{k=1}^{n} s_k b_{k+1} + s_n b_{n+1}$$

$$= \sum_{k=1}^{n} s_k (b_k - b_{k+1}) + s_n b_{n+1}.$$

22. 00, 01, 10, 11

25. 000, 010, 001, 011, 100, 110, 101, 111, 00, 01, 11, 10,
0, 1, λ

28. BASIS STEP ($n = 1$).   In this case, {1} is the only
nonempty subset of {1}, so the sum is

$$\frac{1}{1} = 1 = n.$$

INDUCTIVE STEP.   Assume that the statement is true
for n. We divide the subsets of $\{1, \ldots, n, n+1\}$ into two
classes:

C_1 = class of nonempty, subsets that do not
contain $n + 1$.

C_2 = class of subsets that contain $n + 1$.

By the inductive assumption,

$$\sum_{C_1} \frac{1}{n_1 \cdots n_k} = n.$$

Since a set in C_2 consists of $n + 1$ together with a subset
(empty or nonempty) of $\{1, \ldots, n\}$,

$$\sum_{C_2} \frac{1}{(n+1)n_1 \cdots n_k} = \frac{1}{n+1} + \frac{1}{n+1} \sum_{C_1} \frac{1}{n_1 \cdots n_k}.$$

[The term $1/(n + 1)$ results from the subset $\{n + 1\}$.] By
the inductive assumption,

$$\frac{1}{n+1} + \frac{1}{n+1} \sum_{C_1} \frac{1}{n_1 \cdots n_k}$$

$$= \frac{1}{n+1} + \frac{1}{n+1} \cdot n = 1.$$

Therefore

$$\sum_{C_2} \frac{1}{(n+1)n_1 \cdots n_k} = 1.$$

Finally,

$$\sum_{C_1 \cup C_2} \frac{1}{n_1 \cdots n_k} = \sum_{C_1} \frac{1}{n_1 \cdots n_k} + \sum_{C_2} \frac{1}{(n+1)n_1 \cdots n_k}$$

$$= n + 1.$$

Section 2.3

1. 9

7. 100010

13. 11000

19. 58

25. [For Exercise 1] 9

31. 3DBF9

4. 32

10. 110010000

16. 1001000

22. 2563

28. FE

33. 2010 cannot represent a number in binary because 2 is an
illegal symbol in binary. 2010 could represent a number in
either decimal or hexadecimal.

35. 51

41. (For Exercise 7) 42

38. 4570

44. (For Exercise 35) 33

47. 9450 cannot represent a number in binary because 9, 4,
and 5 are illegal symbols in binary. 9450 cannot represent a
number in octal because 9 is an illegal symbol in octal. 9450
does represent a number in either decimal or hexadecimal.

Section 2.4

1. {(8840, Hammer,)(9921, Pliers), (452, Paint),
(2207, Carpet)}

4. $\{(a, a), (b, b)\}$

5.

a	6
b	2
a	1
a	1

8.

Maine	Augusta
Maryland	Annapolis
Massachusetts	Boston
Michigan	Lansing
Minnesota	St. Paul
Mississippi	Jackson
Missouri	Jefferson City
Montana	Helena

9.

12.

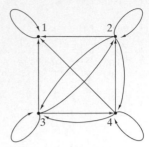

13. $\{(a, b), (a, c), (b, a), (b, d), (c, c), (c, d)\}$

16. $\{(b, c), (c, b), (d, d)\}$

17. [For Exercise 1] domain = $\{8840, 9921, 452, 2207\}$, range = {Hammer, Pliers, Paint, Carpet}

19. $\{(1, 1), (1, 4), (2, 2), (2, 5), (3, 3), (4, 1),$
$(4, 4), (5, 2), (5, 5)\}$

22. $\{1, 2, 3, 4, 5\}$

25. $R = R^{-1} = \{(1, 1), (1, 2), (1, 3), (1, 4),$

$(1, 5), (2, 1), (2, 2), (2, 3), (2, 4),$

$(3, 1), (3, 2), (3, 3), (4, 1), (4, 2), (5, 1)\}$

domain R = range R = domain R^{-1}

= range $R^{-1} = \{1, 2, 3, 4, 5\}$

28. Antisymmetric   **29.** Antisymmetric

32. Reflexive, symmetric, antisymmetric, transitive, partial order

34. Reflexive, antisymmetric, transitive, partial order

37. $R_1 \circ R_2 = \{(1, 1), (1, 2), (2, 1), (2, 2), (3, 1),$

$(3, 2), (4, 2)\}$

$R_2 \circ R_1 = \{(1, 1), (1, 2), (3, 4), (4, 1), (4, 2)\}$

38. $\{(1, 1), (2, 2), (3, 3), (4, 4), (1, 2), (2, 3), (2, 1), (3, 2)\}$

41. $\{(1, 1), (2, 1), (2, 2)\}$

43. False. Let $R = \{(1, 2)\}$, $S = \{(2, 3)\}$.

46. True   **49.** True   **52.** True

55. False. Let $R = \{(1, 2)\}$, $S = \{(2, 1)\}$.

58. True

Section 2.5

1. Equivalence relation: $[1] = [3] = \{1, 3\}$, $[2] = \{2\}$, $[4] = \{4\}$, $[5] = \{5\}$

4. Equivalence relation: $[1] = [3] = [5] = \{1, 3, 5\}$, $[2] = \{2\}$, $[4] = \{4\}$

7. Not an equivalence relation (neither transitive nor reflexive)

9. $\{(1, 1), (1, 2), (2, 1), (2, 2), (3, 3), (3, 4),$
$(4, 3), (4, 4)\}$, $[1] = [2] = \{1, 2\}$, $[3] = [4] = \{3, 4\}$

12. $\{(1, 1), (1, 2), (1, 3), (2, 1), (2, 2), (2, 3), (3, 1), (3, 2),$
$(3, 3), (4, 4)\}$, $[1] = [2] = [3] = \{1, 2, 3\}$, $[4] = \{4\}$

16. $\{1\}, \{1, 3\}, \{1, 4\}, \{1, 3, 4\}$

18.   (a) {San Francisco, San Diego, Los Angeles},
{Pittsburgh, Philadelphia}, {Chicago}

21. $R = \{(x, x) \mid x \in X\}$

24.   (b) $(1, 1), (1, 2), (1, 3), (1, 4), (1, 5), (1, 6), (1, 7),$
$(1, 8), (1, 9), (1, 10), (2, 1), (3, 1), (4, 1), (5, 1),$
$(6, 1), (7, 1), (8, 1), (9, 1), (10, 1)$

27.   (a) We show, symmetry only. Let $(x, y) \in R_1 \cap R_2$. Then $(x, y) \in R_1$ and $(x, y) \in R_2$. Since R_1 and R_2 are symmetric, $(y, x) \in R_1$ and $(y, x) \in R_2$. Thus $(y, x) \in R_1 \cap R_2$ and, therefore, $R_1 \cap R_2$ is symmetric.

(b) A is an equivalence class of $R_1 \cap R_2$ if and only if there are equivalence classes A_1 of R, and A_2 of R_2 such that $A = A_1 \cap A_2$.

30. (b) Torus

31. $\rho(R_1) = \{(1, 1), (2, 2), (3, 3), (4, 4), (1, 2), (3, 4), (4, 2)\}$

$\sigma(R_1) = \{(1, 1), (2, 1), (1, 2), (3, 4), (4, 3), (4, 2), (2, 4)\}$

$\tau(R_1) = \{(1, 1), (1, 2), (3, 4), (4, 2), (3, 2)\}$

$\tau(\sigma(\rho(R_1))) = \{(x, y) \mid x, y \in \{1, 2, 3, 4\}\}$

34. Let $(x, y), (x, z) \in \tau(R)$. Then $(x, y) \in R^m$ and $(y, z) \in R^n$. Thus $(x, z) \in R^{m+n}$. Therefore, $(x, z) \in \tau(R)$ and $\tau(R)$ is transitive.

37. Suppose that R is transitive. If $(x, y) \in \tau(R) = \cup\{R^n\}$, then there exist $x = x_0, \ldots, x_n = y \in X$ such that $(x_{i-1}, x_i) \in R$ for $i = 1, \ldots, n$. Since R is transitive, it follows that $(x, y) \in R$. Thus $R \supseteq \tau(R)$. Since we always have $R \subseteq \tau(R)$, it follows that $R = \tau(R)$.

Suppose that $\tau(R) = R$. By Exercise 34, $\tau(R)$ is transitive. Therefore, R is transitive.

38. True

41. False. Let $R_1 = \{(1, 2), (2, 3)\}$, $R_2 = \{(1, 3), (3, 4)\}$.

44. True

Section 2.6

1.

	α	β	Σ	δ
1	0	0	0	1
2	1	0	1	0
3	0	1	1	0

4.

	1	2	3	4	5
1	0	1	0	0	0
2	0	0	1	0	0
3	0	0	0	1	0
4	0	0	0	0	1
5	0	0	0	0	0

7. $R = \{(a, w), (a, y), (c, y), (d, w), (d, x),$
    $(d, y), (d, z)\}$

10. The test is, whenever the ijth entry is 1, $i \neq j$, then the jith entry is *not* 1.

13. [For Exercise 7] d

$$
\begin{array}{c}
w \\ x \\ y \\ z
\end{array}
\begin{pmatrix}
1 & 0 & 0 & 1 \\
0 & 0 & 0 & 1 \\
1 & 0 & 1 & 1 \\
0 & 0 & 0 & 1
\end{pmatrix}
$$

14. (a) $A_1 = \begin{pmatrix} 1 & 1 \\ 1 & 0 \\ 1 & 0 \end{pmatrix}$

   (b) $A_2 = \begin{pmatrix} 0 & 1 & 0 \\ 1 & 1 & 1 \end{pmatrix}$

   (c) $A_1 A_2 = \begin{pmatrix} 1 & 2 & 1 \\ 0 & 1 & 0 \\ 0 & 1 & 0 \end{pmatrix}$

   (d) We change each nonzero entry in part (c) to 1 to obtain

$$
\begin{pmatrix}
1 & 1 & 1 \\
0 & 1 & 0 \\
0 & 1 & 0
\end{pmatrix}
$$

   (e) $\{(1, b), (1, a), (1, c), (2, b), (3, b)\}$

17. Each column that contains 1 in row x corresponds to an element of the equivalence class containing x.

Section 2.7

1. $\{(1089, \text{Suzuki}, \text{Zamora}), (5620, \text{Kaminski}, \text{Jones}), (9354, \text{Jones}, \text{Yu}), (9551, \text{Ryan}, \text{Washington}), (3600, \text{Beaulieu}, \text{Yu}), (0285, \text{Schmidt}, \text{Jones}), (6684, \text{Manacotti}, \text{Jones})\}$

5. EMPLOYEE [Name]
Suzuki, Kaminski, Jones, Ryan, Beaulieu,
Schmidt, Manacotti

8. BUYER [Name]
United Supplies, ABC Unlimited, JCN Electronics,
Danny's, Underhanded Sales, DePaul University

11. TEMP := BUYER [Part No = 20A8]
TEMP [Name]
Underhanded Sales, Danny's, ABC Unlimited

14. TEMP1 := BUYER [Name = Danny's]
TEMP2 := TEMP1 [Part No = Part No] SUPPLIER
TEMP2 [Dept]
04, 96

17. TEMP1 := BUYER [Name = JCN Electronics]
TEMP2 := TEMP1 [Part No = Part No] SUPPLIER
TEMP3 := TEMP2 [Dept = Dept] DEPARTMENT
TEMP4 := TEMP3 [Manager = Manager] EMPLOYEE
TEMP4 [Name]
Kaminski, Schmidt, Manacotti

22. Let R_1 and R_2 be two n-ary relations. Suppose that the set of elements in the ith column of R_1 and the set of elements in the ith column of R_2 come from a common domain for $i = 1, \ldots, n$. The *union* of R_1 and R_2 is the n-ary relation $R_1 \cup R_2$.
TEMP1 := DEPARTMENT [Dept = 23]
TEMP2 := DEPARTMENT [Dept = 96]
TEMP3 := TEMP1 *union* TEMP2
TEMP4 := TEMP3 [Manager = Manager] EMPLOYEE
TEMP4 [Name]
Kaminski, Schmidt, Manacotti, Suzuki

Section 2.8

1. It is a function from X to Y; domain $= X$, range $= \{a, b, c\}$; it is neither one-to-one nor onto.

4. It is not a function (from X to Y).

6. Example 2.8.15

9. $f \circ g = \{(1, x), (2, z), (3, x)\}$

12. (a)
   $f \circ f = \{(a, a), (b, b), (c, a)\}$

   $f \circ f \circ f = \{(a, b), (b, a), (c, b)\}$

   (b) $f^9 = f, f^{623} = f$

15. The greatest common divisor of m and n must be 1.

In the solutions to Exercises 18 and 21, $a : b$ means store item a in cell b.

18. $53 : 9, 13 : 2, 281 : 6, 743 : 7, 377 : 3, 20 : 10, 10 : 0,$
$796 : 4$

21. $714 : 0, 631 : 6, 26 : 5, 373 : 1, 775 : 8, 906 : 13,$
$509 : 2, 2032 : 7, 42 : 4, 4 : 3, 136 : 9, 1028 : 10$

24. During a search, if we stop the search at an empty cell, we may not find the item even if it is present. The cell may be empty because an item was deleted. One solution is to mark deleted cells and consider them nonempty during a search.

25. False. Take $g = \{(a, x), (b, x)\}$ and $f = \{(x, 1)\}$.

28. False. Take $f = \{(a, z), (b, z)\}$ and $g = \{(1, a)\}$.

31. $g(S) = \{a\}, g(T) = \{a, c\}, g^{-1}(U) = \{1\},$
$g^{-1}(V) = \{1, 2, 3\}$

34. If $x \in X$, then $x \in f^{-1}(f(\{x\}))$.
Thus $\cup\{S \mid S \in \mathcal{S}\} = X$. Suppose that

$$a \in f^{-1}(\{y\}) \cap f^{-1}(\{z\})$$

for some $y, z \in Y$. Then $f(a) = y$ and $f(a) = z$. Thus $y = z$. Therefore, $\mathcal{S}$ is a partition of X. The equivalence relation that generates this partition is given in Exercise 33.

37. [The case: If g is one-to-one, then $f \circ g$ is one-to-one implies that f is one-to-one.]

Suppose that f is not one-to-one. Then there exist distinct $x_1, x_2 \in X$ with $f(x_1) = f(x_2)$. Let $A = \{1, 2\}$. Let $g = \{(1, x_1), (2, x_2)\}$. Now g is one-to-one, but $f \circ g$ is not one-to-one, which is a contradiction.

39. If $x \in X \cap Y$, $C_{X \cap Y}(x) = 1 = 1 \cdot 1 = C_X(x)C_Y(x)$. If $x \notin X \cap Y$, then $C_{X \cap Y}(x) = 0$. Since either $x \notin X$ or $x \notin Y$, either $C_X(x) = 0$ or $C_Y(x) = 0$. Thus $C_X(x)C_Y(x) = 0 = C_{X \cap Y}(x)$.

42. If $x \in X - Y$, then

$$C_{X-Y}(x) = 1 = 1 \cdot [1 - 0] = C_X(x)[1 - C_Y(x)].$$

If $x \notin X - Y$, then either $x \notin X$ or $x \in Y$. In case $x \notin X$,

$$C_{X-Y}(x) = 0 = 0 \cdot [1 - C_Y(x)] = C_X(x)[1 - C_Y(x)].$$

In case $x \in Y$,

$$C_{X-Y}(x) = 0 = C_X(x)[1 - 1] = C_X(x)[1 - C_Y(x)].$$

Thus the equation holds for all $x \in U$.

45. $C_{X \triangle Y}(x) = C_X(x) + C_Y(x) - 2C_X(x)C_Y(x)$

48. A set is equivalent to itself by the identity function.

If X is equivalent to Y, there is a one-to-one, onto function f from X to Y. Now f^{-1} is a one-to-one, onto function from Y to X.

If X is equivalent to Y, there is a one-to-one, onto function f from X to Y. If Y is equivalent to Z, there is a one-to-one, onto function g from Y to Z. Now $g \circ f$ is a one-to-one, onto function from X to Z.

51. Assume that X is equivalent $\mathcal{P}(X)$. Then there is a one-to-one, onto function f from X to $\mathcal{P}(X)$. Let

$$Y = \{x \in X \mid x \notin f(x)\}.$$

Then $f(y) = Y$ for some $y \in X$. Consider the possibilities $y \in Y$ and $y \notin Y$.

53. f is a commutative, binary operator.

56. f is not a binary operator since $f(x, 0)$ is not defined.

58. $g(x) = -x$

61. Each row must contain exactly one 1 for the relation to be a function.

64. True

67. If n is an odd integer, $n = 2k - 1$ for some integer k. Now

$$\frac{n^2}{4} = \frac{(2k-1)^2}{4} = \frac{4k^2 - 4k + 1}{4} = k^2 - k + \frac{1}{4}.$$

Since $k^2 - k$ is an integer,

$$\left\lfloor \frac{n^2}{4} \right\rfloor = k^2 - k.$$

The result now follows because

$$\frac{n-1}{2}\frac{n+1}{2} = \frac{(2k-1)-1}{2}\frac{2k-1+1}{2}$$
$$= \frac{(2k-1)^2 - 1}{4}$$
$$= \frac{4k^2 - k}{4} = k^2 - k.$$

69. April, July

Chapter 2 Self-Test

1. $\emptyset$

2. $256, 255$

3. $A \subseteq B$

4. Yes

5. (a) 14
(b) 18
(c) 192
(d) $a_{n_k} = 4k$

6. $\displaystyle\sum_{k=-1}^{n-2} (n - k - 2)r^{k+2}$

7. (a) $b_5 = 35, b_{10} = 120$
(b) $(n + 1)^2 - 1$
(c) Yes
(d) No

8. (a) $ccddccccdd$
(b) $cccddccddc$
(c) 5
(d) 20

9. 150

10. 11010110, 1AE

11. 1000010

12. 3129

13. Reflexive, symmetric, transitive

14. Symmetric

15. $R = \{(1, 1), (2, 2), (3, 3), (4, 4), (1, 2),$
$(2, 1), (2, 3)\}$

16. All counterexample relations are on $\{1, 2, 3\}$.
(a) False. $R = \{(1, 1)\}$
(b) True
(c) True
(d) False. $R = \{(1, 1)\}$.

17. Yes. It is reflexive, symmetric, and transitive.

18. $[3] = \{3, 4\}$. There are two equivalence classes.

19. $\{(a, a), (b, b), (b, d), (b, e), (d, b), (d, d),$
$(d, e), (e, b), (e, d), (e, e), (c, c)\}$

20. (a) R is reflexive because any eight-bit string has the same number of zeros as itself.

R is symmetric because, if s_1 and s_2 have the same number of zeros, then s_2 and s_1 have the same number of zeros.

To see that R is transitive, suppose that s_1 and s_2 have the same number of zeros and that s_2 and s_3 have the same number of zeros Then s_1 and s_3 have the same number of zeros. Therefore, R is an equivalence relation.

(b) There are nine equivalence classes.

(c) 11111111, 01111111, 00111111, 00011111, 00001111, 00000111, 00000011, 00000001, 00000000

21. $\begin{pmatrix} 1 & 0 \\ 1 & 1 \\ 0 & 1 \end{pmatrix}$

22. $\begin{pmatrix} 1 & 1 & 0 \\ 1 & 0 & 1 \end{pmatrix}$

23. $\begin{pmatrix} 1 & 1 & 0 \\ 2 & 1 & 1 \\ 1 & 0 & 1 \end{pmatrix}$

24. $\begin{pmatrix} 1 & 1 & 0 \\ 1 & 1 & 1 \\ 1 & 0 & 1 \end{pmatrix}$

25. ASSIGNMENT [Team]
Blue Sox, Mutts, Jackalopes

26. PLAYER [Name, Age]
Johnsonbaugh, 22; Glover, 24; Battey, 18; Cage, 30; Homer, 37; Score 22; Johnsonbaugh, 30; Singleton, 31

27. TEMP1 := PLAYER [Position]
TEMP2 := TEMP1 [ID Number = PID] ASSIGNMENT
TEMP2 [Team]
Mutts, Jackalopes

28. TEMP1 := PLAYER [Age ≥ 30]
TEMP2 := TEMP1 [ID Number = PID] ASSIGNMENT
TEMP2 [Team]
Blue Sox, Mutts

29. f is not one-to-one. f is onto.

30. $x = y = 2.3$

31. Define f from $X = \{1, 2\}$ to $\{3\}$ by $f(1) = f(2) = 3$. Define g from $\{1\}$ to X by $g(1) = 1$.

32. ($a : b$ means store item a in cell b.) 1 : 1, 784 : 4, 18 : 5, 329 : 6, 43 : 7, 281 : 8, 620 : 9, 1141 : 10, 31 : 11, 684 : 12

Section 3.1

1. If $a < b$ then $x := a$, else $x := b$.
If $c < x$ then $x := c$,

Section 3.2

1. Input: The sequence $s_1, s_2, \ldots, s_n$ and the length n of s

Output: *small*, the smallest element in this sequence

procedure *find_small*(s, n)
  $small := s_1$
  $i := 2$
  while $i \le n$ **do**
    begin
    if $s_i < small$ **then**
      $small := s_i$
    $i := i + 1$
    end
  return$(small)$
end *find_small*

4. Input: The sequence $s_1, s_2, \ldots, s_n$ and the length n of s

Output: *small*, the smallest element in this sequence
         large, the largest element in this sequence

procedure *small_large* $(s, n, small, large)$
  $small := s_1$
  $large := s_1$
  $i := 2$
  while $i \le n$ **do**
    begin
    if $s_i < small$ **then**
      $small := s_i$
    if $s_i > large$ **then**
      $large := s_i$
    $i := i + 1$
    end
end *small_large*

7. Input: The sequence $s_1, s_2, \ldots, s_n$, the length n of s, and *key*, the value to find

Output: The index of the first occurrence of the value *key* in the sequence

procedure *find*(s, n, key)
  $i := 1$
  while $i \le n$ **do**
    begin
    if $s_i = key$ **then**
      return(i)
    $i := i + 1$
    end
  return(0)
end *find*

10. Input: The sequence $s_1, s_2, \ldots, s_n$ and the length n of s

   Output: The index of the first item that is greater than its predecessor. If none, the output is 0.

```
procedure check_order(s, n)
  i := 2
  while i ≤ n do
    begin
    if s_i > s_{i-1} then
      return(i)
    i := i + 1
    end
  return(0)
end check_order
```

12. In this algorithm, we let 0^k denote k zeros.

   Input: $s_n s_{n-1} \cdots s_1, t_m t_{m-1} \cdots t_1$, two decimal numbers; and n and m

   Output: The product of s and t

```
procedure product(s, t, n, m)
  for i := 1 to m do
    begin
    carry := 0
    for j := 1 to n do
      begin
      xy := decimal representation
        of carry + t_i s_j
      u_{ij} := y
      carry := x
      end
    u_{i,n+i} := carry
    end
  prod := 0
  for i := 1 to m do
    prod := prod + u_{i,n+1} \cdots u_{i1} 0^{i-1}
  return(prod)
end product
```

15. Input: A, the $m \times n$ matrix of the relation R; and m and n

   Output: **true**, if R is a function
   false, if R is not a function

```
procedure is_function(A, m, n)
  for i := 1 to m do
    begin
    sum := 0
    for j := 1 to n do
      sum := sum + A_{ij}
    if sum ≠ 1 then
      return(false)
    end
  return(true)
end is_function
```

Section 3.3

1. $45 = 6 \cdot 7 + 3$ **4.** $221 = 7 \cdot 13 + 0$

7. Divide 90 by 60 to obtain
$90 = 60 \cdot 1 + 30.$
Thus $\gcd(90, 60) = \gcd(60, 30)$.
Divide 60 by 30 to obtain $60 = 30 \cdot 2 + 0$.
Thus $\gcd(60, 30) = \gcd(30, 0) = 30$.
Therefore, $\gcd(60, 90) = 30$.

10. 15   **13.** 1   **16.** 1

17. Since $c \mid m, m = cq_1$ for some integer q_1.
Since $c \mid n, n = cq_2$ for some integer q_2. Now

$$m - n = cq_1 - cq_2 = c(q_1 - q_2).$$

Therefore, $c \mid (m - n)$.

20. Let m be a common divisor of a and b. By Theorem 3.3.4a, m divides $a + b$. Thus m is a common divisor of a and $a + b$.

   Let m be a common divisor of a and $a + b$. By Theorem 3.3.4b, m divides $(a + b) - a = b$. Thus m is a common divisor of a and b.

   Since the set of common divisors of a and $a + b$ is equal to the set of common divisors of a and b, $\gcd(a, b) = \gcd(a, a + b)$.

23. If p divides a, we are done; so suppose that p does not divide a. We must show that p divides b. Since p is prime, $\gcd(p, a) = 1$. By Exercise 21, there are integers s and t such that

$$1 = sp + ta.$$

Multiplying both sides of this equation by b, we obtain

$$b = spb + tab.$$

By Theorem 3.3.4c, p divides spb and p divides tab. By Theorem 3.3.4a, p divides $spb + tab = b$.

26. Input: a and b (nonnegative integers, not both zero)

   Output: Greatest common divisor of a and b.

```
procedure sub_gcd(a, b)
  while true do
    begin
    if a < b then
      swap(a, b)
    if b = 0 then
      return(a)
    a := a - b
    end
end sub_gcd
```

Section 3.4

1. (a)   At line 2, since $4 \neq 0$, we proceed to line 4. The algorithm is invoked with input 3.

(b)   At line 2, since $3 \neq 0$, we proceed to line 4. The algorithm is invoked with input 2.

(c)   At line 2, since $2 \neq 0$, we proceed to line 4. The algorithm is invoked with input 1.

(d)   At line 2, since $1 \neq 0$, we proceed to line 4. The algorithm is invoked with input 0.

(e) At lines 2 and 3, since $0 = 0$, we return 1.

Execution resumes in part (d) at line 4 after computing $0! (= 1)$. We return $0! \cdot 1 = 1$.

Execution resumes in part (c) at line 4 after computing $1! (= 1)$. We return $1! \cdot 2 = 2$.

Execution resumes in part (b) at line 4 after computing $2! (= 2)$. We return $2! \cdot 3 = 6$.

Execution resumes in part (a) at line 4 after computing $3! (= 6)$. We return $3! \cdot 4 = 24$.

4. At line 1, since $5 < 0$ is false, we move to line 3. At line 3, since $b = 0$, we return 5.

7. (a)   Input: n

Output: $2 + 4 + \cdots + 2n$

```
1. procedure sum(n)
2.   if n = 1 then
3.     return(2)
4.   return(sum(n − 1) + 2n)
   end(sum)
```

(b) **BASIS STEP** $(n = 1)$ If n is equal to 1, we correctly return 2.

INDUCTIVE STEP.   Assume that the algorithm correctly computes the sum when the input is $n - 1$. Now suppose that the input to this algorithm is $n > 1$. At line 2, since $n \neq 1$, we proceed to line 4, where we invoke this algorithm with input $n - 1$. By the inductive assumption, the value returned, $sum(n - 1)$, is equal to

$$2 + \cdots + 2(n - 1).$$

At line 4, we then return

$$sum(n - 1) + 2n = 2 + \cdots 2(n - 1) + 2n,$$

which is the correct value.

10.   Input: n

Output: $n!$

```
procedure factorial(n)
  fact := 1
  for i := 2 to n do
    fact = i · fact
  return(fact)
end factorial
```

13. After one month, there is still just one pair because a pair does not become productive until after one month. Therefore, $a_1 = 1$. After two months, the pair alive in the beginning becomes productive and adds one additional pair. Therefore, $a_2 = 2$. The increase in pairs of rabbits $a_n - a_{n-1}$ from month $n - 1$ to month n is due to each pair alive in month $n - 2$ producing an additional pair. That is, $a_n - a_{n-1} = a_{n-2}$. Since $\{a_n\}$ satisfies the same recurrence relation as $\{f_n\}$ and the same initial conditions, $a_n = f_n$, $n \geq 1$.

16. **BASIS STEP** $(n = 2)$

$$f_2^2 = 4 = 1 \cdot 3 + 1 = f_1 f_3 + (-1)^2$$

INDUCTIVE STEP

$$\begin{aligned}
f_n f_{n+2} + (-1)^{n+1} &= f_n(f_{n+1} + f_n) + (-1)^{n+1} \\
&= f_n f_{n+1} + f_n^2 + (-1)^{n+1} \\
&= f_n f_{n+1} + f_{n-1} f_{n+1} \\
&\quad + (-1)^n + (-1)^{n+1} \\
&= f_{n+1}(f_n + f_{n-1}) = f_{n+1}^2
\end{aligned}$$

19. **BASIS STEPS** $(n = 1, 2)$. $f_1 (= 1)$ is odd and 2 is not divisible by 3. $f_2 (= 2)$ is even and 3 is divisible by 3. Therefore, the statement is true for $n = 1, 2$.

INDUCTIVE STEP.   Assume that the statement is true for all $k < n$. We must prove that the statement is true for n. We can assume that $n > 2$. We consider two cases: $n + 1$ is divisible by 3 and $n + 1$ is not divisible by 3.

If $n + 1$ is divisible by 3, then neither n nor $n - 1$ is divisible by 3. By the inductive assumption, one of f_{n-1} and f_{n-2} are odd. Since $f_n = f_{n-1} + f_{n-2}$, f_n is even. Therefore, if $n + 1$ is divisible by 3, f_n is even.

Now suppose that $n + 1$ is not divisible by 3; then exactly one of n or $n - 1$ is divisible by 3. By the inductive assumption, one of f_{n-1} and f_{n-2} is odd and the other is even. Since $f_n = f_{n-1} + f_{n-2}$, f_n is odd. Therefore, if $n + 1$ is not divisible by 3, f_n is odd.

We have shown that f_n is even if and only if $n + 1$ is divisible by 3, so the Inductive Step is complete.

22. We prove the first formula only. The second is proved in a similar manner.

BASIS STEP

$$f_1 = 1 = 2 - 1 = f_2 - 1$$

INDUCTIVE STEP

$$\begin{aligned}
\sum_{k=1}^{n+1} f_{2k-1} &= \sum_{k=1}^{n} f_{2k-1} + f_{2n+1} \\
&= (f_{2n} - 1) + f_{2n+1} = f_{2n+2} - 1
\end{aligned}$$

26. BASIS STEP

$$\frac{dx}{dx} = 1 = 1x^{1-1}$$

INDUCTIVE STEP

$$\frac{dx^{n+1}}{dx} = \frac{d(x \cdot x^n)}{dx} = x\frac{dx^n}{dx} + x^n\frac{dx}{dx}$$

$$= xnx^{n-1} + x^n \cdot 1 = (n+1)x^n$$

Section 3.5

1. $\Theta(n)$ **4.** $\Theta(n^2)$

7. $\Theta(n^2)$ **10.** $\Theta(n)$

13. $\Theta(n^2)$ **16.** $\Theta(n)$

19. $\Theta(n^2)$ **22.** $\Theta(n^3)$

25. $\Theta(\lg n)$

30. $n! = n(n-1)\cdots 2 \cdot 1$

$$\leq n \cdot n \cdots n = n^n$$

33. True

36. False. A counterexample is $f(n) = n$ and $g(n) = 2n$.

39. True

42. False. A counterexample is $f(n) = 1$ and $g(n) = 1/n$.

43. $f_n = \begin{cases} 1 & \text{if } n \text{ is even} \\ 0 & \text{if } n \text{ is odd} \end{cases}$

46. The given inequality is equivalent, in turn, to

$$(n/2)[(\lg n) - 1] \geq (n \lg n)/4$$

$$2[(\lg n) - 1] \geq \lg n$$

$$\lg n \geq 2.$$

The last inequality is an equality if $n = 4$. Since $\lg n$ is an increasing function, the last inequality holds for $n \geq 4$. Thus the given inequality holds for $n \geq 4$.

49.   (a) The sum of the areas of the rectangles below the curve is equal to

$$\frac{1}{2} + \frac{1}{3} + \cdots + \frac{1}{n}.$$

This area is less than the area under the curve, which is equal to

$$\int_1^n \frac{1}{x}dx = \log_e n.$$

The given inequality now follows immediately.

(b) The sum of the areas of the rectangles whose bases are on the x-axis and whose tops are above the curve is equal to

$$1 + \frac{1}{2} + \cdots + \frac{1}{n-1}.$$

Since this area is greater than the area under the curve, the given inequality follows immediately.

(c) Part (a) shows that

$$1 + \frac{1}{2} + \cdots + \frac{1}{n} = O(\log_e n).$$

Since $\log_e n = \Theta(\lg n)$ (see Exercise 29),

$$1 + \frac{1}{2} + \cdots + \frac{1}{n} = O(\lg n).$$

Similarly, we can conclude from part (b) that

$$1 + \frac{1}{2} + \cdots + \frac{1}{n} = \Omega(\lg n).$$

Therefore

$$1 + \frac{1}{2} + \cdots + \frac{1}{n} = \Theta(\lg n).$$

52. (a) True

(b) False. A counterexample is

$$f(n) = 1, g(n) = 2 + (-1)^n.$$

(c) False. A counterexample is

$$f(n) = n, g(n) = n^2.$$

(d) True

(d) False. A counterexample is

$$f(n) = 1, g(n) = 2 + (-1)^n.$$

55. Multiply both sides of the inequality in Exercise 54 by $\lg e$ and use the change of base formula for logarithms.

Section 3.6

1.

a \ b	0	1	2	3	4	5	6	7	8	9	10	11	12	13
0	—	0	0	0	0	0	0	0	0	0	0	0	0	0
1	0	1	1	1	1	1	1	1	1	1	1	1	1	1
2	0	1	1	2	1	2	1	2	1	2	1	2	1	2
3	0	1	2	1	2	3	1	2	3	1	2	3	1	2
4	0	1	1	2	1	2	2	3	1	2	2	3	1	2
5	0	1	2	3	2	1	2	3	4	3	1	2	3	4
6	0	1	1	1	2	2	1	2	2	2	3	3	1	2
7	0	1	2	2	3	3	2	1	2	3	3	4	4	3
8	0	1	1	3	1	4	2	2	1	2	2	4	2	5

a	b	n (= Number of Divisions)
1	0	0
2	1	1
3	2	2
5	3	3
8	5	4
13	8	5

4. Using induction on n, we prove that when the pair f_{n+1}, f_n is input to the Euclidean algorithm, exactly n divisions are required.

BASIS STEP $(n = 1)$. Table 3.6.2 shows that when the pair f_2, f_1 is input to the Euclidian algorithm, one division is required.

INDUCTIVE STEP. Assume that when the pair f_{n+1}, f_n is input to the Euclidean algorithm, n divisions are required. We must show that when the pair f_{n+2}, f_{n+1} is input to the Euclidean algorithm, $n + 1$ divisions are required.

At line 7, we divide f_{n+2} by f_{n+1} to obtain

$$f_{n+2} = f_{n+1} + f_n.$$

The algorithm then repeats using the values of f_{n+1} and f_n. By the inductive assumption, exactly n additional divisions are required. Thus a total of $n + 1$ divisions are required.

Section 3.7

1. FKKGEJAIMWQ

4. $a = c^s \bmod z = 411^{569} \bmod 713 = 500$

5. $z = pq = 17 \cdot 23 = 391$

8. $c = a^n \bmod z = 101^{31} \bmod 391 = 186$

11. This algorithm computes $a^n \bmod z$.

   Input: a, n, z

  Output: $a^n \bmod z$

```
procedure exp_mod(a, n, z)
exp := 1
x := a mod z
while n > 0 do
  begin
  if n is odd then
    exp := (exp · x) mod z
  x := (x · x) mod z
  n := n/2
  end
return(exp)
end exp_mod
```

14. Using the Euclidean algorithm, we obtain

$$660 = 22 \cdot 29 + 22, \ 29 = 22 \cdot 1 + 7, \ 22 = 3 \cdot 7 + 1.$$

Therefore

$$1 = 22 - 3 \cdot 7 = 22 - 3(29 - 22) = 4 \cdot 22 - 3 \cdot 29$$
$$= 4(660 - 22 \cdot 29) - 3 \cdot 29 = 4 \cdot 660 - 91 \cdot 29.$$

Thus $s = 91 \cdot 659 \bmod 660 = 569$.

Chapter 3 Self-Test

1. At line 1, we set x to 12. At line 2, since $b > x$ $(3 > 12)$ is false, we move to line 3. At line 3, since $c > x$ $(0 > 12)$ is false, we terminate the algorithm. The value of x is 12, the maximum of the given values.

2. 1. $x := a, y := b, z := c$
   2. If $y < x$, swap (x, y)
   3. If $z < x$, swap (x, z)
   4. If $y > z$, swap (y, z)

3. 1. If $a = b$, output "No" and stop.
   2. If $a = c$, output "No" and stop.
   3. If $b = c$, output "No" and stop.
   4. Output "Yes."

4. If the set S is an infinite set, the algorithm will not terminate, so it lacks the finiteness and output properties. Line 1 is not precisely stated since *how* to list the subsets of S and their sums is not specified; thus the algorithm lacks the precision property. The order of the subsets listed in line 1 depends on the method used to generate them, so the algorithm lacks the uniqueness property. Since line 2 depends on the order of the subsets generated in line 1, the uniqueness property is lacking here as well.

5. At line 2, we set *large* to 7. At line 3, we set i to 2. At line 6, since $s_2 > large$ $(9 > 7)$ is true, we set *large* to 9. At line 6, since $s_3 > large$ $(17 > 9)$ is true, we set *large* to 17. At line 6, since $s_4 > large$ $(7 > 17)$ is false, we return the value 17.

6.   Input: The $n \times n$ matrix A of a relation R and n

Output: **true**, if R is symmetric
  false, if R is not symmetric

```
procedure is_symmetric(A, n)
  for i := 1 to n − 1 do
    for j := i + 1 to n do
      if A_{ij} ≠ A_{ji} then
      return(false)
  return(true)
end is_symmetric
```

7.   Input: $n \times n$ matrix A and n

Output: A^T

```
procedure transpose(A, n)
  for i := 1 to n − 1 do
    for j := i + 1 to n do
      swap(A_{ij}, A_{ji})
end transpose
```

8.   Input: $s_1, \ldots, s_n$ and n

Output: All repeated values

```
procedure repeat(s, n)
  i := 1
  j := 2
  while j ≤ n do
    begin
    if s_i = s_j then
      begin
      print s_i
      while j ≤ n and s_i = s_j do
        j := j + 1
      end
    i := j
    j := j + 1
    end
end repeat
```

9. $333 = 24 \cdot 13 + 21$ **10.**   12

11. 2   **12.** $\gcd(b, r)$

13. Since $n \ne 2$, we proceed immediately to line 3, where we divide the board into four 4×4 boards. At line 4, we rotate the board so that the missing square is in the upper left quadrant and call the algorithm to tile the upper left subboard. The following figure shows the status of the execution at this point:

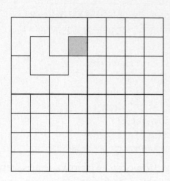

Next, at line 5, we place one tromino in the center. We then proceed to lines 5–9, where we call the algorithm to tile each of the remaining subboards. We obtain the tiling:

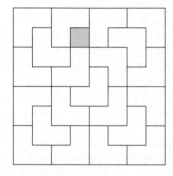

14. $t_4 = 3, t_5 = 5$

15.   Input: n, an integer greater than or equal to 1

Output: t_n

```
procedure tribonacci(n)
1.  if n = 1 or n = 2 or n = 3 then
2.    return(1)
3.  return(tribonacci(n − 1)
      + tribonacci(n − 2)
      + tribonacci(n − 3))
  end tribonacci
```

16. BASIS STEP ($n = 1, 2, 3$).   If $n = 1, 2, 3$, at lines 1 and 2 we output the correct value, 1. Therefore, the algorithm is correct in these cases.

INDUCTIVE STEP.   Assume that $n > 3$ and that the algorithm correctly computes t_k, if $k < n$. Since $n > 3$, we proceed to line 3. We then call this algorithm to compute t_{n-1}, t_{n-2}, and t_{n-3}. By the inductive assumption, the values computed are correct. The algorithm then computes $t_{n-1} + t_{n-2} + t_{n-3}$. But the recurrence relation shows that this value is equal to t_n. Therefore, the algorithm computes the correct value for t_n.

17. $\Theta(n^3)$ **18.** $\Theta(n^4)$ **19.** $\Theta(n^2)$

20. Input: $n \times n$ matrices A and B and n

Output: **true**, if $A = B$; **false**, if $A \neq B$

```
procedure equal_matrix(A, B, n)
  for i := 1 to n do
    for j := 1 to n do
      if Aᵢⱼ ≠ Bᵢⱼ then
        return(false)
  return(true)
end equal_matrix
```

The worst-case time is $\Theta(n^2)$.

21. We show that 14 divisions are required by the Euclidean algorithm in the worst case for numbers in the range 0 to 1000. By Theorem 3.6.1, if the pair $a, b, a > b$, requires 15 divisions, then $a \geq f_{16} = 1597$. Therefore, at most 14 divisions are required. By Exercise 4, Section 3.6, when the pair $f_{15} = 987$, $f_{14} = 610$ is input to the Euclidean algorithm, exactly 14 divisions are required. Therefore, exactly 14 divisions are required by the Euclidean algorithm in the worst case for numbers in the range 0 to 1000.

22. Two divisions are required by the Euclidean algorithm (Algorithm 3.3.8) to compute gcd(2, 76652913). At line 4, we swap a and b so that $a = 76652913$ and $b = 2$. Since $b \neq 0$, we proceed to line 7. The first division occurs when we divide 76652913 by 2 to obtain

$$76652913 = 2 \cdot 38326456 + 1.$$

At lines 8 and 9 we set a to 2 and b to 1. We then return to line 5.

Since $b \neq 0$, we proceed to line 7. The second division occurs when we divide 2 by 1 to obtain

$$2 = 1 \cdot 2 + 0.$$

At lines 8 and 9 we set a to 1 and b to 0. We then return to line 5.

This time $b = 0$ and the algorithm terminates. Two divisions occurred.

23. 323 (see Exercise 4, Section 3.6)

24. Since

$$\log_{3/2} \frac{2(100,000,000)}{3} = \log_{3/2} 100^4 + \log_{3/2} \frac{2}{3}$$

$$= 4 \log_{3/2} 100 - 1$$

$$= 4(11.357747) - 1$$

$$= 44.430988,$$

an upper bound for the number of divisions required by the Euclidean algorithm for integers in the range 0 to 100,000,000 is 44.

25. $z = pq = 13 \cdot 7 = 221$, $\phi = (p-1)(q-1) = 12 \cdot 16 = 192$

26. $ns \bmod \phi = 19 \cdot 91 \bmod 192 = 1729 \bmod 192 = 1$

27. $c = a^n \bmod z = 144^{19} \bmod 221 = 53$

28. $a = c^s \bmod z = 28^{91} \bmod 221 = 63$

Section 4.1

1. $2 \cdot 4$ **4.** $8 \cdot 4 \cdot 5$ **7.** 6^2

10. $6 + 2$ **13.** 11 **16.** $10 \cdot 5$

19. $26^3 10^2$, $26 \cdot 25 \cdot 24 \cdot 10 \cdot 9$

22. $3 \cdot 2^6$ **25.** $2^8 - 1$ **27.** $5 \cdot 4 \cdot 3$

30. $3 \cdot 4 \cdot 3$ **33.** 5^3 **36.** $4 \cdot 3$

39. $5^3 - 4^3$ **41.** $200 - 5 + 1$ **44.** 40

47. One one-digit number contains 7. The distinct two-digit numbers that contain 7 are $17, 27, \ldots, 97$ and $70, 71, \ldots, 76, 78, 79$. There are 18 of these. The distinct three-digit numbers that contain 7 are 107 and $1xy$, where xy is one of the two-digit numbers listed above. The answer is $1 + 18 + 19$.

50. $5 + (8 + 7 + \cdots + 1) + (7 + 6 + \cdots + 1)$

53. $10!$ **56.** $(3!)(5!)(2!)(3!)$ **60.** 2^{10}

63. We count the number of antisymmetric relations on $\{1, 2, \ldots, n\}$ by computing the number of ways to construct the matrix of an antisymmetric relation.

Each element of the diagonal can be either 0 or 1. Thus there are 2^n ways to assign values to the diagonal.

For i and j satisfying $1 \leq i < j \leq n$, we can assign the entries in row i, column j and row j, column i in three ways:

Row i, Column j	Row j, Column i
0	0
1	0
0	1

Since there are $(n^2 - n)/2$ values of i and j satisfying $1 \leq i < j \leq n$, we can assign the off-diagonal values in $3^{(n^2-n)/2}$ ways. Therefore, there are

$$2^n \cdot 3^{(n^2-n)/2}$$

antisymmetric relations on an n-element set.

65. $2^5 + 2^7 - 2^4$ (According to Exercise 64, the total number of possibilities = number of strings that begin 100 + number of strings that have the fourth bit 1 − number of strings that begin 100 and have the fourth bit 1.)

68. $5 \cdot 4 + 3 \cdot 5 \cdot 4 - 2 \cdot 4$ (According to Exercise 64, the total number of possibilities = number in which Connie is chairperson + number in which Alice is an officer − number in which Connie is chairperson and Alice is an officer.)

Section 4.2

1. $4! = 24$

4. $abc, acb, bac, bca, cab, cba, abd, adb, bad,$
$bda, dab, dba, acd, adc, cad, cda, dac, dca,$
$bcd, bdc, cbd, cdb, dbc, dcb$

7. $P(11, 3) = 11 \cdot 10 \cdot 9$ **10.** $3!$

13. $4!$ contain the substring AE and $4!$ contain the substring EA; therefore, the total number is $2 \cdot 4!$.

16. We first count the number N of strings that contain either the substring AB or the substring BE. The answer to the exercise will be: Total number of strings − N or $5! - N$.

According to Exercise 64, Section 4.1, the number of strings that contain AB or BE = number of strings that contain AB + number of strings that contain BE − number of strings that contain AB and BE. A string contains AB and BE if and only if it contains ABE and the number of such strings $3!$. The number of strings that contain AB = number of strings that contain BE = $4!$. Thus the number of strings that contain AB or BE is $4! + 4! - 3!$. The solution to the exercise is

$$5! - (2 \cdot 4! - 3!).$$

19. $8! P(9, 5) = 8!(9 \cdot 8 \cdot 7 \cdot 6 \cdot 5)$

21. $10!$

24. Fix a seat for a Jovian. There are $7!$ arrangements for the remaining Jovians. For each of these arrangements, we can place the Martians in five of the eight in-between positions, which can be done in $P(8, 5)$ ways. Thus there are $7! P(8, 5)$ such arrangements.

25. $C(4, 3) = 4$

28. $C(11, 3)$

31. $C(13, 5)$

34. A committee that has at most one man has exactly one man or no men. There are $C(6, 1)C(7, 3)$ committees with exactly one man. There are $C(7, 4)$ committees with no men. Thus the answer is $C(6, 1)C(7, 3) + C(7, 4)$.

37. $C(10, 4)C(12, 3)C(4, 2)$

40. First, we count the number of eight-bit strings with no two 0's in a row. We divide this problem into counting the number of such strings with exactly eight 1's, with exactly seven 1's, and so on.

There is one eight-bit string with no two 0's in a row that has exactly eight 1's. Suppose that an eight-bit string with no two 0's in a row has exactly seven 1's. The 0 can go in any one of eight positions; thus there are eight such strings. Suppose that an eight-bit string with no two 0's in a row has exactly six 1's. The two 0's must go in two of the blanks shown:

$$\_1\_1\_1\_1\_1\_1\_.$$

Thus the two 0's can be placed in $C(7, 2)$ ways. Thus there are $C(7, 2)$ such strings. Similarly, there are $C(6, 3)$ eight-bit strings with no two 0's in a row that have exactly five 1's and there are $C(5, 4)$ eight-bit strings with no two 0's in a row that have exactly four 1's in a row. If a string has less than four 1's, it will have two 0's in a row. Therefore, the number of eight-bit strings with no two 0's in a row is

$$1 + 8 + C(7, 2) + C(6, 3) + C(5, 4).$$

Since there are 2^8 eight-bit strings, there are

$$2^8 - [1 + 8 + C(7, 2) + C(6, 3) + C(5, 4)]$$

eight-bit strings that contain at least two 0's in a row.

41. $1 \cdot 48$ (The four aces can be chosen in one way and the fifth card can be chosen in 48 ways.)

44. First, we count the number of hands containing cards in spades and hearts. Since there are 26 spades and hearts, there are $C(26, 5)$ ways to select five cards from among these 26. However, $C(13, 5)$ contain only spades and $C(13, 5)$ contain only hearts. Therefore, there are

$$C(26, 5) - 2C(13, 5)$$

ways to select five cards containing cards in spades and hearts.

Since there are $C(4, 2)$ ways to select two suits, the number of hands containing cards of exactly two suits is

$$C(4, 2)[C(26, 5) - 2C(13, 5)].$$

47. There are nine consecutive patterns: A2345, 23456, 34567, 45678, 56789, 6789T, 789TJ, 89TJQ, 9TJQK. Corresponding to the four possible suits, there are four ways for each pattern to occur. Thus there are $9 \cdot 4$ hands that are consecutive and of the same suit.

50. $C(52, 13)$

53. $1 \cdot C(48, 9)$ (Select the aces, then select the nine remaining cards.)

56. There are $C(13, 4)C(13, 4)C(13, 4)C(13, 1)$ hands that contain four spades, four hearts, four diamonds, and one club. Since there are four ways to select the three suits to have four cards each, there are $4C(13, 4)^3 C(13, 1)$ hands that contain four cards of three suits and one card of the fourth suit.

58. 2^{10} **61.** 2^9

63. $C(50, 4)$

66. $C(50, 4) - C(46, 4)$ (Total number − number with no defectives)

70. Order the $2n$ items. The first item can be paired in $2n - 1$ ways. The next (not yet selected) item can be paired in $2n - 3$ ways, and so on.

73. The solution counts *ordered* hands.

78. Use Theorems 2.5.1 and 2.5.9.

Section 4.3

1. 1357   **4.** 12435

7. [For Exercise 1] At lines 9–14, we find the rightmost s_m not at its maximum value. In this case, $m = 4$. At line 16, we increment s_m. This makes the last digit 7. Since m is the rightmost position, at lines 18 and 19, we do nothing. The next combination is 1357.

9. 123, 124, 125, 126, 134, 135, 136, 145, 146, 156, 234, 235, 236, 245, 246, 256, 345, 346, 356, 456

12. 12, 21

14.   Input: r, n

   Output: A list of all r-combinations of $\{1, 2, \ldots, n\}$ in increasing lexicographic order

```
procedure r_comb1(r, n)
  s₀ := -1    // assume that s₀ exists
  for i := 1 to r do
    sᵢ := i
  output s
  while true do
    begin
    m := r
    max_val := n
    while sₘ = max_val do
      begin
      m := m - 1
      max_val := max_val - 1
      end
    if m = 0 then
      return
    sₘ := sₘ + 1
    for j := m + 1 to r do
      sⱼ := sⱼ₋₁ + 1
    output s
    end
end r_comb1
```

17.   Input: $r, s_k, s_{k+1}, \ldots, s_n$, a string α, k, and n

   Output: A list of all r-combinations of $\{s_k, s_{k+1}, \ldots, s_n\}$ each prefixed by α. [To list all r-combinations of $\{s_1, s_2, \ldots, s_n\}$, invoke this procedure as $r\_comb2(r, s, 1, n, \lambda)$, where λ is the null string.]

```
procedure r_comb2(r, s, k, n, α)
  if r = 0 then
    begin
    output α
    return
    end
  if r = n then
    begin
    output α, sₖ, sₖ₊₁, ..., sₙ
    return
    end
  // output r-combinations containing sₖ
  β := α followed by sₖ
  r_comb2(r - 1, s, k + 1, n - 1, β)
  // output r-combinations not containing sₖ
  r_comb2(r, s, k + 1, n - 1, α)
end r_comb2
```

Section 4.4

1. $5!$ **4.** $10!/(5! \cdot 3! \cdot 2!)$

5. $C(10 + 3 - 1, 10)$

8. $C(9 + 2 - 1, 9)$

11. Four, since the possibilities are $(0, 0)$, $(2, 1)$, $(4, 2)$, and $(6, 3)$ where the pair (r, g) designates r red and g green balls.

12. $C(15 + 3 - 1, 15)$

15. $C(13 + 2 - 1, 13)$

18. $C(12 + 4 - 1, 12)$
$- [C(7 + 4 - 1, 7) + C(6 + 4 - 1, 6)$
$+ C(3 + 4 - 1, 3) + C(2 + 4 - 1, 2)$
$- C(1 + 4 - 1, 1)]$

21. $52!/(13!)^4$ **24.** $C(20, 5)$ **27.** $C(20, 5)^2$

30. $C(15 + 6 - 1, 15)$ **33.** $C(10 + 12 - 1, 10)$

36. Apply the result of Example 4.4.9 to the inner $k - 1$ nested loops of that example. Next, write out the number of iterations for $i_1 = 1$; then $i_1 = 2$; and so on. By Example 4.4.9, this sum is equal to $C(k + n - 1, k)$.

Section 4.5

1. $x^4 + 4x^3y + 6x^2y^2 + 4xy^3 + y^4$

3. $C(11, 7)x^4y^7$ **6.** 5,987,520

9. $C(7, 3) + C(5, 2)$, since
$$(a + \sqrt{ax} + x)^2(a + x)^5$$
$$= [(a + x) + \sqrt{ax}]^2(a + x)^5$$
$$= (a + x)^7 + 2\sqrt{ax}(a + x)^6 + ax(a + x)^5.$$

10. $C(10 + 3 - 1, 10)$ **13.** 1 8 28 56 70 56 28 8 1

16. [Inductive Step only] Assume that the theorem is true for n.

$$(a+b)^{n+1} = (a+b)(a+b)^n$$

$$= (a+b) \sum_{k=0}^{n} C(n,k)a^{n-k}b^k$$

$$= \sum_{k=0}^{n} C(n,k)a^{n+1-k}b^k$$

$$+ \sum_{k=0}^{n} C(n,k)a^{n-k}b^{k+1}$$

$$= \sum_{k=0}^{n} C(n,k)a^{n+1-k}b^k$$

$$+ \sum_{k=1}^{n+1} C(n,k-1)a^{n+1-k}b^k$$

$$= C(n,0)a^{n+1}b^0 + \sum_{k=1}^{n} C(n,k)a^{n+1-k}b^k$$

$$+ C(n,n)a^0b^{n+1}$$

$$+ \sum_{k=1}^{n} C(n,k-1)a^{n+1-k}b^k$$

$$= C(n+1,0)a^{n+1}b^0$$

$$+ \sum_{k=1}^{n}[C(n,k)+C(n,k-1)]a^{n+1-k}b^k$$

$$+ C(n+1,n+1)a^0b^{n+1}$$

$$= C(n+1,0)a^{n+1}b^0$$

$$+ \sum_{k=1}^{n} C(n+1,k)a^{n+1-k}b^k$$

$$+ C(n+1,n+1)a^0b^{n+1}$$

$$= \sum_{k=0}^{n+1} C(n+1,k)a^{n+1-k}b^k$$

19. The number of solutions in nonnegative integers of

$$x_1 + x_2 + \cdots + x_{k+2} = n - k$$

is $C(k+2+n-k-1, n-k) = C(n+1, k+1)$. The number of solutions is also the number of solutions $C(k+1+n-k-1, n-k) = C(n,k)$ with $x_{k+2} = 0$ plus the number of solutions

$$C(k+1+n-k-1-1, n-k-1) = C(n-1, k)$$

with $x_{k+2} = 1$ plus $\cdots$ plus the number of solutions $C(k+1+0-1, 0) = C(k,k)$ with $x_{k+2} = n - k$. The result now follows.

22. Take $a = 1$ and $b = 2$ in the Binomial Theorem.

25. $x^3 + 3x^2y + 3x^2z + 3xy^2 + 6xyz + 3xz^2 + y^3 + 3y^2z + 3yz^2 + z^3$

28. Set $a = 1$ and $b = x$ and replace n by $n-1$ in the Binomial Theorem to obtain

$$(1+x)^{n-1} = \sum_{k=0}^{n-1} C(n-1, k)x^k.$$

Now multiply by n to obtain

$$n(1+x)^{n-1} = n\sum_{k=0}^{n-1} C(n-1,k)x^k$$

$$= n\sum_{k=1}^{n} C(n-1, k-1)x^{k-1}$$

$$= \sum_{k=1}^{n} \frac{n(n-1)!}{(n-k)!\,(k-1)!}x^{k-1}$$

$$= \sum_{k=1}^{n} \frac{n!}{(n-k)!\,k!}kx^{k-1}$$

$$= \sum_{k=1}^{n} C(n,k)kx^{k-1}.$$

31. The solution is by induction on k. We omit the Basis Step. Assume that the statement is true for k. After k iterations, we obtain the sequence defined by

$$a'_j = \sum_{i=0}^{k-1} a_{i+j}\frac{B_i}{2^n}.$$

Let $B'_0, \ldots, B'_k$ denote the row after $B_0, \ldots, B_{k-1}$ in Pascal's triangle. Smoothing a' by c to obtain a'' yields

$$a''j = \frac{1}{2}\left(a'_j + a'_{j+1}\right)$$

$$= \frac{1}{2^{n+1}}\left(\sum_{i=0}^{k-1} a_{i+j}B_i + \sum_{i=0}^{k-2} a_{i+j+1}B_i\right)$$

$$= \frac{1}{2^{n+1}}\left(a_jB_0 + \sum_{i=1}^{k-1} a_{i+j}B_i + \sum_{i=0}^{k-2} a_{i+j+1}B_i + a_{k+j}B_{k-1}\right)$$

$$= \frac{1}{2^{n+1}}\left(a_jB_0 + \sum_{i=1}^{k-1} a_{i+j}B_i + \sum_{i=1}^{k-1} a_{i+j}B_{i-1} + a_{k+j}B_{k-1}\right)$$

$$= \frac{1}{2^{n+1}}\left(a_jB'_0 + \sum_{i=1}^{k-1} a_{i+j}B'_i + a_{k+j}B'_k\right)$$

$$= \frac{1}{2^{n+1}}\sum_{i=0}^{k} a_{i+j}B'_i,$$

and the Inductive Step is complete.

Section 4.6

1. There are 12 possible names for the 13 persons. We can consider the assignment of names to people to be that of assigning pigeonholes to the pigeons. By the Pigeonhole Principle, some name is assigned to at least two persons.

4. Yes. Connect processors 1 and 2, 2 and 3, 2 and 4, 3 and 4. Processor 5 is not connected to any processors. Now only processors 3 and 4 are directly connected to the same number of processors.

7. Let a_i denote the position of the ith unavailable item. Consider

$$a_1, \ldots, a_{30}; \quad a_1+3, \ldots, a_{30}+3; \quad a_1+6, \ldots, a_{30}+6.$$

These 90 numbers range in value from 1 to 86. By the second form of the Pigeonhole Principle, two of these numbers are the same. If $a_i = a_j + 3$, two are three apart. If $a_i = a_j + 6$, two are six apart. If $a_i +3 = a_j +6$, two are three apart.

11. $n + 1$

12. Suppose that $k \le m/2$. Clearly, $k \ge 1$. Since $m \le 2n+1$,

$$k \le \frac{m}{2} \le n + \frac{1}{2} < n + 1.$$

Suppose that $k > m/2$. Then

$$m - k < m - \frac{m}{2} = \frac{m}{2} < n + 1.$$

Because m is the largest element in X, $k < m$. Thus $k + 1 \le m$ and so $1 \le m - k$. Therefore, the range of a is contained in $\{1, \ldots, n\}$.

13. The second form of the Pigeonhole Principle applies.

14. Suppose that $a_i = a_j$. Then either $i \le m/2$ and $j > m/2$ or $j \le m/2$ and $i > m/2$. We may assume that $i \le m/2$ and $j > m/2$. Now

$$i + j = a_i + m - a_j = m.$$

24. When we divide a by b, the possible remainders are $0, 1, \ldots, b - 1$. Consider what happens after b divisions.

28. We suppose that the board has three rows and seven columns. We call two squares in one column that are the same color a *colorful pair*. By the Pigeonhole Principle, each column contains at least one colorful pair. Thus the board contains seven colorful pairs, one in each column. Again by the Pigeonhole Principle, at least four of these seven colorful pairs are the same color, say red. Since there are three pairs of rows and four red colorful pairs, a third application of the Pigeonhole Principle shows that at least two columns contain red colorful pairs in the same rows. These colorful pairs determine a rectangle whose four corner squares are red.

Chapter 4 Self-Test

1. 2^4

2. $6 \cdot 9 \cdot 7 + 6 \cdot 9 \cdot 4 + 6 \cdot 7 \cdot 4 + 9 \cdot 7 \cdot 4$

3. $2^n - 2$

4. $6 \cdot 5 \cdot 4 \cdot 3 + 6 \cdot 5 \cdot 4 \cdot 3 \cdot 2$

5. $6!/(3!\,3!) = 20$

6. We construct the strings by a three-step process. First, we choose positions for A, C, and E [$C(6, 3)$ ways]. Next, we place A, C, and E in these positions. We can place C one way (last), and we can place A and E two ways (AE or EA). Finally, we place the remaining three letters (3! ways). Therefore, the total number of strings is $C(6, 3) \cdot 2 \cdot 3!$.

7. Two suits can be chosen in $C(4, 2)$ ways. We can choose three cards of one suit in $C(13, 3)$ ways and we can choose three cards of the other suit in $C(13, 3)$ ways. Therefore, the total number of hands is $C(4, 2)C(13, 3)^2$.

8. We must select either three or four defective disks. Thus the total number of selections is $C(5, 3)C(95, 1)+C(5, 4)$.

9. 12567   10. 234567   11. 6427153

12. 631245   13. $8!/(3!2!)$

14. We count the number of strings in which no I appears before any L and then subtract from the total number of strings.

We construct strings in which no I appears before any L by a two-step process. First, we choose positions for N, O, and S; then we place the I's and L's. We can choose positions for N, O, and S in $8 \cdot 7 \cdot 6$ ways. The I's and L's can then be placed in only one way because the L's must come first. Thus there are $8 \cdot 7 \cdot 6$ strings in which no I appears before any L.

Exercise 13 shows that there are $8!/(3!\,2!)$ strings formed by ordering the letters *ILLINOIS*. Therefore there are

$$\frac{8!}{3!\,2!} - 8 \cdot 7 \cdot 6$$

strings formed by ordering the letters *ILLINOIS* in which some I appears before some L.

15. $12!(3!)^4$

16. $C(11 + 4 - 1, 4 - 1)$

17. $(s - r)^4 = C(4, 0)s^4$
$$+ C(4, 1)s^3(-r) + C(4, 2)s^2(-r)^2$$
$$+ C(4, 3)s(-r)^3 + C(4, 4)(-r)^4$$
$$= s^4 - 4s^3r + 6s^2r^2 - 4sr^3 + r^4$$

18. $2^3 \cdot 8!/(3!\,1!\,4!)$

19. If we set $a = 2$ and $b = -1$ in the Binomial Theorem, we obtain

$$1 = 1^n = [2 + (-1)]^n = \sum_{k=0}^{n} C(n, k)2^{n-k}(-1)^k.$$

20. $C(n, 1) = n$

21. Let the 15 individual socks be the pigeons and let the 14 types of pairs be the pigeonholes. Assign each sock (pigeon) to its type (pigeonhole). By the Pigeonhole Principle, some pigeonhole will contain at least two pigeons (the matched socks).

22. There are $3 \cdot 2 \cdot 3 = 18$ possible names for the 19 persons. We can consider the assignment of names to people to be that of assigning pigeonholes to the pigeons. By the Pigeonhole Principle, some name is assigned to at least two persons.

23. Let a_i denote the position of the ith available item. The 220 numbers

$$a_1, \ldots, a_{110}; \quad a_1 + 19, \ldots, a_{110} + 19$$

range from 1 to 219. By the Pigeonhole Principle, two are the same.

24. Each point has an x-coordinate that is either even or odd and a y-coordinate that is either even or odd. Since there are four possibilities and there are five points, by the Pigeonhole Principle at least two points, $p_i = (x_i, y_i)$ and $p_j = (x_j, y_j)$ have

- Both x_i and x_j even or both x_i and x_j odd.

and

- Both y_i and y_j even or both y_i and y_j odd.

Therefore $x_i + x_j$ is even and $y_i + y_j$ is even. In particular, $(x_i + x_j)/2$ and $(y_i + y_j)/2$ are integers. Thus the midpoint of the pair p_i and p_j has integer coordinates.

Section 5.1

1. $a_n = a_{n-1} + 4; a_1 = 3$

4. $A_n = (1.14)A_{n-1}$ **5.** $A_0 = 2000$

6. $A_1 = 2280, A_2 = 2599.20, A_3 = 2963.088$

7. $A_n = (1.14)^n 2000$

8. We must have $A_n = 4000$ or $(1.14)^n 2000 = 4000$ or $(1.14)^n = 2$. Taking the logarithm of both sides, we must have $n \log 1.14 = \log 2$. Thus

$$n = \frac{\log 2}{\log 1.14} = 5.29.$$

18. We count the number of n-bit strings not containing pattern 000.

- Begin with 1. In this case, if the remaining $(n - 1)$-bit string does not contain 000, neither will the n-bit string. There are S_{n-1} such $(n - 1)$-bit strings.
- Begin with 0. There are two cases to consider.

  1. Begin with 01. In this case, if the remaining $(n - 2)$-bit string does not contain 000, neither will the n-bit string. There are S_{n-2} such $(n - 2)$-bit strings.

  2. Begin with 00. Then the third bit must be a 1 and if the remaining $(n - 3)$-bit string does not contain 000, neither will the n-bit string. There are S_{n-3} such $(n - 3)$-bit strings.

Since the cases are mutually exclusive and cover all n-bit strings $(n > 3)$ not containing 000, we have $S_n = S_{n-1} + S_{n-2} + S_{n-3}$ for $n > 3$. $S_1 = 2$ (there are two 1-bit strings), $S_2 = 4$ (there are four 2-bit strings), and $S_3 = 7$ (there are eight 3-bit strings but one of them is 000).

19. There are S_{n-1} n-bit strings that begin 1 and do not contain the pattern 00 and there are S_{n-2} n-bit strings that begin 0 (since the second bit must be 1) and do not contain the pattern 00. Thus $S_n = S_{n-1} + S_{n-2}$. Initial conditions are $S_1 = 2, S_2 = 3$.

22. $S_1 = 2, S_2 = 4, S_3 = 7, S_4 = 12$

25. $C_3 = 5, C_4 = 14, C_5 = 42$

29. Let P_n denote the number of ways to divide a convex $(n + 2)$-sided polygon, $n \geq 1$, into triangles by drawing $n - 1$ lines through the corners that do not intersect in the interior of the polygon. We note that $P_1 = 1$.

Suppose that $n > 1$ and consider a convex $(n + 2)$-sided polygon (see the following figure).

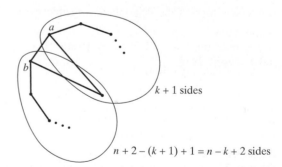

We choose one edge ab and construct a partition of the polygon by a two-step procedure. First we select a triangle to which side ab belongs. This triangle divides the original polygon into two polygons: one having $k + 1$ sides, for some k satisfying $1 \leq k \leq n$, and the other having $n - k + 2$ sides (see the preceding figure). By definition, the $(k + 1)$-sided polygon can be partitioned in P_{k-1} ways and the $(n - k + 2)$-sided polygon can be partitioned in P_{n-k} ways. (For the degenerate cases $k = 1$ and $k = n$, we set

$P_0 = 1$.) Therefore, the total number of ways to partition the $(n + 2)$-sided polygon is

$$P_n = \sum_{k=1}^{n} P_{k-1} P_{n-k}.$$

Since the sequence $P_1, P_2, \ldots$ satisfies the same recurrence relation as the Catalan sequence $C_1, C_2, \ldots$ and $P_0 = P_1 = 1 = C_0 = C_1$, it follows that $P_n = C_n$ for all $n \geq 1$.

33. [For $n = 3$]

Step 1—move disk 3 from peg 1 to peg 3.

Step 2—move disk 2 from peg 1 to peg 2.

Step 3—move disk 3 from peg 3 to peg 2.

Step 4—move disk 1 from peg 1 to peg 3.

Step 5—move disk 3 from peg 2 to peg 1.

Step 6—move disk 2 from peg 2 to peg 3.

Step 7—move disk 3 from peg 1 to peg 3.

35. Let α and β be the angles shown in Figure 5.1.6. The geometry of the situation shows that the price tends to stabilize if and only if $\alpha + \beta > 180°$. This last condition holds if and only if $-\tan \beta < \tan \alpha$. Since $b = -\tan \beta$ and $k = \tan \alpha$, we conclude that the price stabilizes if and only if $b < k$.

37. $A(2, 2) = 7$, $A(2, 3) = 9$

40. $A(3, n) = 2^{n+3} - 3$

43. If $m = 0$,

$$A(m, n + 1) = A(0, n + 1)$$
$$= n + 2 > n + 1$$
$$= A(0, n) = A(m, n).$$

The last inequality follows from Exercise 41.

44. Use Exercises 38 and 39.

47. We prove the statement by using induction on x. The inductive step will itself require induction on y.

Exercise 44 shows that the equation is true for $x = 0, 1, 2$ and for all y.

BASIS STEP ($x = 2$).   See Exercise 44.

INDUCTIVE STEP (CASE x IMPLIES CASE $x+1$).
Assume that $x \geq 2$ and

$$A(x, y) = AO(x, 2, y + 3) - 3 \quad \text{for all } y \geq 0.$$

We must prove that

$$A(x + 1, y) = AO(x + 1, 2, y + 3) - 3 \quad \text{for all } y \geq 0.$$

We establish this last equation by induction on y.

BASIS STEP ($y = 0$).   We must prove that

$$A(x + 1, 0) = AO(x + 1, 2, 3) - 3.$$

Now

$AO(x + 1, 2, 3) - 3$
$= AO(x, 2, AO(x + 1, 2, 2)) - 3$ by definition
$= AO(x, 2, 4) - 3$ by Exercise 46
$= A(x, 1)$ by the inductive
                        assumption on x
$= A(x + 1, 0)$ by (5.1.11).

INDUCTIVE STEP (CASE y IMPLIES CASE $y+1$).
Assume that

$$A(x + 1, y) = AO(x + 1, 2, y + 3) - 3.$$

We must prove that

$$A(x + 1, y + 1) = AO(x + 1, 2, y + 4) - 3.$$

Now

$AO(x + 1, 2, y + 4) - 3$
$= AO(x, 2, AO(x + 1, 2, y + 3)) - 3$ by definition
$= AO(x, 2, A(x + 1, y) + 3) - 3$ by the inductive
                            assumption on y
$= A(x, A(x + 1, y))$ by the inductive
                            assumption on x
$= A(x + 1, y + 1)$ by (5.1.12).

50. Suppose that we have n dollars. If we buy orange juice the first day, we have $n - 1$ dollars left, which may be spent in R_{n-1} ways. Similarly, if the first day we buy milk or beer, there are R_{n-2} ways to spend the remaining dollars. Since these cases are disjoint, $R_n = R_{n-1} + 2R_{n-2}$.

53. $S_3 = \frac{1}{2}$, $S_4 = \frac{3}{4}$

55. A function f from $X = \{1, \ldots, n\}$ into X will be denoted $(i_1, i_2, \ldots, i_n)$, which means that $f(k) = i_k$. The problem then is to count the number of ways to select $i_1, \ldots, i_n$ so that if i occurs, so do $1, 2, \ldots, i - 1$.

We shall count the number of such functions having exactly j 1's. Such functions can be constructed in two steps: Pick the positions for the j 1's; then place the other numbers. There are $C(n, j)$ ways to place the 1's. The remaining numbers must be selected so that if i appears, so do $1, \ldots, i - 1$. There are F_{n-j} ways to select the remaining numbers, since the remaining numbers must be selected from $\{2, \ldots, n\}$. Thus there are $C(n, j)F_{n-j}$

functions of the desired type having exactly j 1's. Therefore, the total number of functions from X into X having the property that if i is in the range of f, then so are $1, \ldots, i-1$, is

$$\sum_{j=1}^{n} C(n, j) F_{n-j} = \sum_{j=1}^{n} C(n, n-j) F_{n-j}$$

$$= \sum_{j=0}^{n-1} C(n, j) F_j.$$

58. Use equation (4.5.4) to write
$S(k, n) = \sum_{i=1}^{n} S(k-1, i)$.

61. We use the terminology of Exercise 74, Section 4.2. Choose one of $n+1$ people, say P. There are $s_{n, j-1}$ ways for P to sit alone. (Seat the other n people at the other $k-1$ tables.) Next we count the number of arrangements in which P is not alone. Seat everyone but P at k tables. This can be done in $s_{n,k}$ ways. Now P can be seated to the right of someone in n ways. Thus there are $ns_{n,k}$ arrangements in which P is not alone. The recurrence relation now follows.

64. Let A_n denote the amount at the end of n years and let i be the interest rate expressed as a decimal. The discussion following Example 5.1.3 shows that

$$A_n = (1 + i)^n A_0.$$

The value of n required to double the amount satisfies

$$2A_0 = (1 + i)^n A_0 \quad \text{or} \quad 2 = (1 + i)^n.$$

If we take the natural logarithm (logarithm to the base e) of both sides of this equation, we obtain

$$\ln 2 = n \ln(1 + i).$$

Thus

$$n = \frac{\ln 2}{\ln(1 + i)}.$$

Since $\ln 2 = 0.6931472\ldots$ and $\ln(1 + i)$ is approximately equal to i for small values of i, n is approximately equal to $0.69\ldots/i$, which, in turn, is approximately equal to $70/r$.

66. 1, 3, 2; 2, 3, 1; $E_3 = 2$

69. We count the number of rise/fall permutations of $1, \ldots, n$ by considering how many have n in the second, fourth, ..., positions.

Suppose that n is in the second position. Since any of the remaining numbers is less than n, any of them may be placed in the first position. Thus we may select the number to be placed in the first position in $C(n-1, 1)$ ways and, after selecting it, we may arrange it in $E_1 = 1$ way. The last $n-2$ positions can be filled in E_{n-2} ways since any rise/fall permutation of the remaining $n-2$ numbers gives a rise/fall permutation of $1, \ldots, n$. Thus the number

of rise/fall permutations of $1, \ldots, n$ with n in the second position is $C(n-1, 1)E_1 E_{n-2}$.

Suppose that n is in the fourth position. We may select numbers to be placed in the first three positions in $C(n-1, 3)$ ways. After selecting the three items, we may arrange them in E_3 ways. The last $n-4$ numbers can be arranged in E_{n-4} ways. Thus the number of rise/fall permutations of $1, \ldots, n$ with n in the fourth position is $C(n-1, 3)E_3 E_{n-4}$.

In general, the number of rise/fall permutations of $1, \ldots, n$ with n in the $(2j)$th position is

$$C(n-1, 2j-1)E_{2j-1} E_{n-2j}.$$

Summing over all j gives the desired recurrence relation.

Section 5.2

1. Yes; order 1 **4.** No

7. No **10.** Yes; order 3

11. $a_n = 2(-3)^n$ **14.** $a_n = 2^{n+1} - 4^n$

17. $a_n = (2^{2-n} + 3^n)/5$ **20.** $a_n = 2(-4)^n + 3n(-4)^n$

23. $R_n = [(-1)^n + 2^{n+1}]/3$

26. Let d_n denote the deer population at time n. The initial condition is $d_0 = 0$. The recurrence relation is

$$d_n = 100n + 1.2d_{n-1}, \quad n > 0.$$

$$d_n = 100n + 1.2d_{n-1} = 100n + 1.2[100(n-1) + 1.2d_{n-2}]$$

$$= 100n + 1.2 \cdot 100(n-1) + 1.2^2 d_{n-2}$$

$$= 100n + 1.2 \cdot 100(n-1)$$
$$\quad + 1.2^2[100(n-2) + 1.2d_{n-3}]$$

$$= 100n + 1.2 \cdot 100(n-1)$$
$$\quad + 1.2^2 \cdot 100(n-2) + 1.2^3 d_{n-3}$$

$$\vdots$$

$$= \sum_{i=0}^{n-1} 1.2^i \cdot 100(n-i) + 1.2^n d_0$$

$$= \sum_{i=0}^{n-1} 1.2^i \cdot 100(n-i)$$

$$= 100n \sum_{i=0}^{n-1} 1.2^i - 1.2 \cdot 100 \sum_{i=1}^{n-1} i \cdot 1.2^{i-1}$$

$$= \frac{100n(1.2^n - 1)}{1.2 - 1}$$

$$\quad - 120\frac{(n-1)1.2^n - n1.2^{n-1} + 1}{(1.2 - 1)^2}, \quad n > 0.$$

29. Set $b_n = a_n/n!$ to obtain $b_n = -2b_{n-1} + 3b_{n-2}$. Solving gives $a_n = n! \, b_n = (n!/4)[5 - (-3)^n]$.

32. We establish the inequality by using induction on n.

The Base Cases $n = 1$ and $n = 2$ are left to the reader. Now assume that the inequality is true for values less than $n + 1$. Then

$$f_{n+1} = f_n + f_{n-1}$$

$$\geq \left(\frac{1-\sqrt{5}}{2}\right)^{n-1} + \left(\frac{1+\sqrt{5}}{2}\right)^{n-2}$$

$$= \left(\frac{1+\sqrt{5}}{2}\right)^{n-2}\left(\frac{1+\sqrt{5}}{2} + 1\right)$$

$$= \left(\frac{1+\sqrt{5}}{2}\right)^{n-2}\left(\frac{1+\sqrt{5}}{2}\right)^{2}$$

$$= \left(\frac{1+\sqrt{5}}{2}\right)^{n},$$

and the Inductive Step is complete.

34. $a_n = b2^n + d4^n + 1$

37. $a_n = b/2^n + d3^n - (4/3)2^n$

40. The argument is identical to that given in Theorem 5.2.11.

43. Recursively invoking this algorithm to move the $n - k_n$ disks at the top of peg 1 to peg 2 takes $T(n - k_n)$ moves. Moving the k_n disks on peg 1 to peg 4 requires $2^{k_n} - 1$ moves (see Example 5.2.4). Recursively invoking this algorithm to move the $n - k_n$ disks on peg 2 to peg 4 again takes $T(n - k_n)$ moves. The recurrence relation now follows.

46. From the inequality

$$\frac{k_n(k_n + 1)}{2} \leq n,$$

we can deduce $k_n \leq \sqrt{2n}$. Since

$$n - k_n \leq \frac{k_n(k_n + 1)}{2},$$

it follows that $r_n \leq k_n$. Therefore,

$$T(n) = (k_n + r_n - 1)2^{k_n} + 1$$
$$< 2k_n 2^{k_n} + 1$$
$$\leq 2\sqrt{2n}\,2^{\sqrt{2n}} + 1$$
$$= O(4^{\sqrt{n}}).$$

Section 5.3

1. At line 2, since $i > j \, (1 > 5)$ is false, we proceed to line 4, where we set k to 3. At line 5, since *key* ('G') is not equal to s_3('J'), we proceed to line 7. At line 7, *key* $< s_k$('G' $<$ 'J') is true, so at line 8 we set j to 2. We then invoke this algorithm with $i = 1$, $j = 2$ to search for *key* in

$$s_1 = \text{'}C\text{'}, \qquad s_2 = \text{'}G\text{'}.$$

At line 2, since $i > j \, (1 > 2)$ is false, we proceed to line 4, where we set k to 1. At line 5, since *key* ('G') is not equal to s_1('C'), we proceed to line 7. At line 7, *key* $< s_k$('G' $<$ 'C') is false, so at line 10 we set i to 2. We then invoke this algorithm with $i = j = 2$ to search for *key* in

$$s_2 = \text{'}G\text{'}.$$

At line 2, since $i > j \, (2 > 2)$ is false, we proceed to line 4, where we set k to 2. At line 5, since *key* ('G') is equal to s_2('G'), we return 2, the index of *key* in the sequence s.

4. At line 2, since $i > j \, (1 > 5)$ is false, we proceed to line 4, where we set k to 3. At line 5, since *key* ('Z') is not equal to s_3('J'), we proceed to line 7. At line 7, *key* $< s_k$('Z' $<$ 'J') is false, so at line 10 we set i to 4. We then invoke this algorithm with $i = 4$, $j = 5$ to search for *key* in

$$s_4 = \text{'}M\text{'}, \qquad s_5 = \text{'}X\text{'}.$$

At line 2, since $i > j \, (4 > 5)$ is false, we proceed to line 4, where we set k to 4. At line 5, since *key* ('Z') is not equal to s_4('M') we proceed to line 7. At line 7, *key* $< s_k$('Z' $<$ 'M') is false, so at line 10 we set i to 5. We then invoke this algorithm with $i = j = 5$ to search for *key* in

$$s_5 = \text{'}X\text{'}.$$

At line 2, since $i > j \, (5 > 5)$ is false, we proceed to line 4, where we set k to 5. At line 5, since *key* ('Z') is not equal to s_5('X'), we proceed to line 7. At line 7, *key* $< s_k$('Z' $<$ 'X') is false, so at line 10 we set i to 6. We then invoke this algorithm with $i = 6$, $j = 5$.

At line 2, since $i > j \, (6 > 5)$ is true, we return 0 to indicate that we failed to find *key*.

7. Algorithm B is superior if $2 \leq n \leq 15$. (For $n = 1$ and $n = 16$, the algorithms require equal numbers of comparisons.)

10. Suppose that the sequences are $a_1, \ldots, a_n$ and $b_1, \ldots, b_n$.

(a) $a_1 < b_1 < a_2 < b_2 <$ (b) $a_n < b_1$
  $\cdots$

13. 11

17. Algorithm 5.3.11 computes a^n by using the formula $a^n = a^m a^{n-m}$.

18. $b_n = b_{\lfloor n/2 \rfloor} + b_{\lfloor (n+1)/2 \rfloor} + 1, b_1 = 0$

19. $b_2 = 1, b_3 = 2, b_4 = 3$

20. $b_n = n - 1$

21. We prove the formula by using mathematical induction. The Basis Step, $n = 1$, has already been established.

Assume that $b_k = k - 1$ for all $k < n$. We show that $b_n = n - 1$. Now

$$b_n = b_{\lfloor n/2 \rfloor} + b_{\lfloor (n+1)/2 \rfloor} + 1$$

$$= \left\lfloor \frac{n}{2} \right\rfloor - 1 + \left\lfloor \frac{n+1}{2} \right\rfloor - 1 + 1$$

by the inductive assumption

$$= \left\lfloor \frac{n}{2} \right\rfloor + \left\lfloor \frac{n+1}{2} \right\rfloor - 1 = n - 1.$$

34. If $n = 1$, then $i = j$ and we return before reaching line 7c, 11, or 15. Therefore, $b_1 = 0$. If $n = 2$, then $j = i + 1$. There is one comparison at line 7c and we return before reaching line 11 or 15. Therefore, $b_2 = 1$.

35. $b_3 = 3, b_4 = 4$

36. When $n > 2$, $b_{\lfloor (n+1)/2 \rfloor}$ comparisons are required for the first recursive call and $b_{\lfloor n/2 \rfloor}$ comparisons are required for the second recursive call. Two additional comparisons are required at lines 11 and 15. The recurrence relation now follows.

37. Suppose that $n = 2^k$. Then (5.3.12) becomes

$$b_{2^k} = 2b_{2^{k-1}} + 2.$$

Now

$$b_{2^k} = 2b_{2^{k-1}} + 2$$

$$= 2[2b_{2^{k-2}} + 2] + 2$$

$$= 2^2 b_{2^{k-2}} + 2^2 + 2 = \cdots$$

$$= 2^{k-1} b_{2^1} + 2^{k-1} + 2^{k-2} + \cdots + 2$$

$$= 2^{k-1} + 2^{k-1} + \cdots + 2$$

$$= 2^{k-1} + 2^k - 2$$

$$= n - 2 + \frac{n}{2} = \frac{3n}{2} - 2.$$

38. We use the following fact, which can be verified by considering the cases x even and x odd:

$$\left\lceil \frac{3x}{2} - 2 \right\rceil + \left\lceil \frac{3(x+1)}{2} - 2 \right\rceil = 3x - 2 \text{ for } x = 1, 2, \ldots.$$

Let a_n denote the number of comparisons required by the algorithm in the worst case. The cases $n = 1$ and $n = 2$ may be directly verified. (The case $n = 2$ is the Basis Step.)

INDUCTIVE STEP. Assume that $a_k \leq \lceil (3k/2) - 2 \rceil$ for $2 \leq k < n$. We must show that the inequality holds for $k = n$.

If n is odd, the algorithm partitions the array into subclasses of sizes $(n - 1)/2$ and $(n + 1)/2$. Now

$$a_n = a_{(n-1)/2} + a_{(n+1)/2} + 2$$

$$\leq \left\lceil \frac{(3/2)(n-1)}{2} - 2 \right\rceil +$$

$$\left\lceil \frac{(3/2)(n+1)}{2} - 2 \right\rceil + 2$$

$$= \frac{3(n-1)}{2} - 2 + 2 = \frac{3n}{2} - \frac{3}{2}$$

$$= \left\lceil \frac{3n}{2} - 2 \right\rceil.$$

The case n even is treated similarly.

47. If $n = 2^k$,

$$a_{2^k} = 3a_{2^{k-1}} + 2^k,$$

so

$$a_n = a_{2^k} = 3a_{2^{k-1}} + 2^k$$

$$= 3[3a_{2^{k-2}} + 2^{k-1}] + 2^k$$

$$= 3^2 a_{2^{k-2}} + 3 \cdot 2^{k-1} + 2^k$$

$$\vdots$$

$$= 3^k a_{2^0} + 3^{k-1} \cdot 2^1 + 3^{k-2} \cdot 2^2 + \cdots$$

$$+ 3 \cdot 2^{k-1} + 2^k$$

$$= 3^k + 2(3^k - 2^k)$$

$$= 3 \cdot 3^k - 2 \cdot 2^k$$

$$= 3 \cdot 3^{\lg n} - 2n.$$

Line $(*)$ results from the equation

$$(a - b)\left(a^{k-1}b^0 + a^{k-2}b^1 + \cdots + a^1 b^{k-2} + a^0 b^{k-1}\right)$$

$$= a^k - b^k$$

with $a = 3$ and $b = 2$.

49. $b_n = b_{\lfloor (1+n)/2 \rfloor} + b_{\lfloor n/2 \rfloor} + 3$

52. $b_n = 4n - 3$

55. We will show that $b_n \leq b_{n+1}, n = 1, 2, \ldots$. We have the recurrence relation

$$b_n = b_{\lfloor (1+n)/2 \rfloor} + b_{\lfloor n/2 \rfloor} + c_{\lfloor (1+n)/2 \rfloor, \lfloor n/2 \rfloor}.$$

BASIS STEP. $b_2 = 2b_1 + c_{1,1} \geq 2b_1 \geq b_1$

INDUCTIVE STEP. Assume that the statement holds for $k < n$. In case n is even, we have $b_n = 2b_{n/2} + c_{n/2, n/2}$; so

$$b_{n+1} = b_{(n+2)/2} + b_{n/2} + c_{(n+2)/2, n/2}$$

$$\geq b_{n/2} + b_{n/2} + c_{n/2, n/2} = b_n.$$

The case n odd is similar.

57. procedure $ex57(s, i, j)$

    if $i = j$ **then**

      return

    $m := \lfloor (i + j)/2 \rfloor$

    call $ex57(s, i, m)$

    call $ex57(s, m + 1, j)$

    call $combine(s, i, m, j)$

  end $ex57$

60. We prove the inequality by using mathematical induction.

BASIS STEP. $a_1 = 0 \leq 0 = b_1$

INDUCTIVE STEP. Assume that $a_k \leq b_k$ for $k < n$. Then

$$a_n \leq a_{\lfloor n/2 \rfloor} + a_{\lfloor (n+1)/2 \rfloor} + 2 \lg n$$

$$\leq b_{\lfloor n/2 \rfloor} + b_{\lfloor (n+1)/2 \rfloor} + 2 \lg n = b_n.$$

63. Let $c = a_1$. If n is a power of m, say $n = m^k$, then

$$a_n = a_{m^k} = a_{m^{k-1}} + d$$

$$= [a_{m^{k-2}} + d] + d$$

$$= a_{m^{k-2}} + 2d$$

$$\vdots$$

$$= a_{m^0} + kd = c + kd.$$

An arbitrary value of n falls between two powers of m, say

$$m^{k-1} < n \leq m^k.$$

This last inequality implies that

$$k - 1 < \log_m n \leq k.$$

Since the sequence a is increasing,

$$a_{m^{k-1}} \leq a_n \leq a_{m^k}.$$

Now

$$\Omega(\log_m n) = c + (-1 + \log_m n) \leq c + (k - 1)d$$

$$= a_{m^{k-1}} \leq a_n$$

and

$$a_n \leq a_{m^k} = c + kd$$

$$\leq c + (1 + \log_m n)d = O(\log_m n).$$

Thus $a_n = \Theta(\log_m n)$. By Exercise 29, Section 3.5, $a_n = \Theta(\lg n)$.

Chapter 5 Self-Test

1. (a) $3, 5, 8, 12$ (b) $a_1 = 3$ (c) $a_n = a_{n-1} + n$

2. $A_n = (1.17)A_{n-1}$, $A_0 = 4000$

3. Let X be an n-element set and choose $x \in X$. Let k be a fixed integer, $0 \leq k \leq n - 1$. We can select a k-element subset Y of $X - \{x\}$ in $C(n - 1, k)$ ways. Having done this, we can partition Y in P_k ways. This partition together with $X - Y$ partitions X. Since all partitions of X can be generated in this way, we obtain the desired recurrence relation.

4. If the first domino is placed as shown, there are a_{n-1} ways to cover the $2 \times (n - 1)$ board that remains.

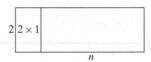

If the first two dominoes are placed as shown, there are a_{n-2} ways to cover the $2 \times (n-2)$ board that remains.

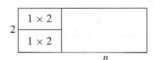

It follows that $a_n = a_{n-1} + a_{n-2}$.

By inspection, $a_1 = 1$ and $a_2 = 2$. Since $\{a_n\}$ satisfies the same recurrence relation as the Fibonacci sequence and $a_1 = f_1$ and $a_2 = f_2$, it follows that $a_i = f_i$ for $i = 1, 2, \ldots$.

5. Yes

6. $a_n = 2(-2)^n - 4n(-2)^n$

7. $a_n = 3 \cdot 5^n + (-2)^n$

8. Consider a string of length n that contains an even number of 1's that begins with 0. The string that follows the 0 may be any string of length $n - 1$ that contains an even number of 1's, and there are c_{n-1} such strings. A string of length n that contains an even number of 1's that begins with 2 can be followed by any string of length $n - 1$ that contains an even number of 1's, and there are c_{n-1} such strings. A string of length n that contains an even number of 1's that begins with 1 can be followed by any string of length $n - 1$ that contains an odd number of 1's. Since there are 3^{n-1} strings altogether of length $n - 1$ and c_{n-1} of these contain an even number of 1's, there are $3^{n-1} - c_{n-1}$ strings of length $n - 1$ that contain an odd number of 1's. It follows that

$$c_n = 2c_{n-1} + 3^{n-1} - c_{n-1} = c_{n-1} + 3^{n-1}.$$

An initial condition is $c_1 = 2$, since there are two strings (0 and 2) that contain an even number (namely, zero) of 1's.

We may solve the recurrence relation by iteration:

$$c_n = c_{n-1} + 3^{n-1} = c_{n-2} + 3^{n-2} + 3^{n-1}$$

$$\vdots$$

$$= c_1 + 3^1 + 3^2 + \cdots + 3^{n-1}$$

$$= 2 + \frac{3^n - 3}{3 - 1} = \frac{3^n + 1}{2}.$$

9. $b_n = b_{n-1} + 1, b_0 = 0$

10. $b_1 = 1, b_2 = 2, b_3 = 3$

11. $b_n = n$

12. $n(n+1)/2 = O(n^2)$. The given algorithm is faster than the straightforward technique and is, therefore, preferred.

Section 6.1

1. Since an odd number of edges touch some vertices (c and d), there is no path from a to a that passes through each edge exactly one time.

4. $(a, c, e, b, c, d, e, f, d, b, a)$

7. $V = \{v_1, v_2, v_3, v_4\}$. $E = \{e_1, e_2, e_3, e_4, e_5, e_6\}$. e_1 and e_6 are parallel edges. e_5 is a loop. There are no isolated vertices. G is not a simple graph. e_1 is incident on v_1 and v_2.

10.

K_3

K_5

13. Bipartite. $V_1 = \{v_1, v_2, v_5\}, V_2 = \{v_3, v_4\}$.

16. Not bipartite

19. Bipartite. $V_1 = \{v_1\}, V_2 = \{v_2, v_3\}$.

20.

$K_{2,3}$

$K_{3,3}$

22. (b, c, a, d, e)

25. Two classes

30.

00   01

10   11

33. n

36.

39.

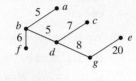

Section 6.2

1. Cycle, simple cycle

4. Cycle, simple cycle

7. Simple path

10.

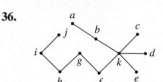

13.

16. Suppose that there is such a graph with vertices a, b, c, d, e, f. Suppose that the degrees of a and b are 5. Since the graph is simple, the degrees of $c, d, e,$ and f are each at least 2; thus there is no such graph.

19. $(a, a), (b, c, g, b), (b, c, d, f, g, b),$
$(b, c, d, e, f, g, b), (c, g, f, d, c),$
$(c, g, f, e, d, c), (d, f, e, d)$

22. Every vertex has degree 4.

24. v_1 v_2 v_3 v_1 v_2 v_1 v_2 v_1 v_2 v_1 v_2

G_1 G_2 G_3 G_4 v_3 v_3 v_3 v_3

v_1 v_2 G_5 G_6 G_7 G_8

G_9

27. There are 17 subgraphs.

28. No Euler cycle

31. No Euler cycle

34. For

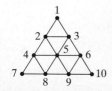

an Euler cycle is $(10, 9, 6, 5, 9, 8, 5, 4, 8, 7, 4, 2, 5, 3, 2, 1, 3, 6, 10)$. The method generalizes.

37. $m = n = 2$ or $m = n = 1$

39. d and e are the only vertices of odd degree.

42. The argument is similar to that of the proof of Theorem 6.2.23.

45. True. In the path, for all repeated a,

$$(\dots, a, \dots, b, a, \dots)$$

eliminate $a, \dots, b$.

47. Suppose that $e = (v, w)$ is in a cycle. Then there is a path P from v to w not including e. Let x and y be vertices in $G - \{e\}$. Since G is connected, there is a path P' in G from v to w. Replace any occurrence of e in P' by P. The resulting path from v to w lies in $G - \{e\}$. Therefore, $G - \{e\}$ is connected.

50. The union of all connected subgraphs containing G' is a component.

53. Let G be a simple, disconnected graph with n vertices having the maximum number of edges. Show that G has two components. If one component has i vertices, show that the components are K_i and K_{n-i}. Use Exercise 11, Section 6.1, to find a formula for the number of edges in G as a function of i. Show that the maximum occurs when $i = 1$.

55.

58. Modify the proofs of Theorems 6.2.17 and 6.2.18.

61. Use Exercises 58 and 60.

64. We first count the number of paths

$$(v_0, v_1, \dots, v_k)$$

of length $k \geq 1$. The first vertex v_0 may be chosen in n ways. Each subsequent vertex may be chosen in $n-1$ ways (since it must be different from its predecessor). Thus the number of paths of length k is $n(n-1)^k$.

The number of paths of length k, $1 \leq k \leq n$, is

$$\sum_{k=1}^{n} n(n-1)^k = n(n-1)\frac{(n-1)^k - 1}{(n-1) - 1}$$
$$= \frac{n(n-1)[(n-1)^k - 1]}{n-2}.$$

68. If v is a vertex in V, the path consisting of v and no edges is a path from v to v; thus vRv for every vertex v in V. Therefore, R is reflexive.

Suppose that vRw. Then there is a path $(v_0, \dots, v_n)$, where $v_0 = v$ and $v_n = w$. Now $(v_n, \dots, v_0)$ is a path from w to v, and thus wRv. Therefore, R is symmetric.

Suppose that vRw and wRx. Then there is a path P_1 from v to w and a path P_2 from w to x. Now P_1 followed by P_2 is a path from v to x, and thus vRx. Therefore, R is transitive.

Since R is reflexive, symmetric, and transitive on V, R is an equivalence relation on V.

70. 2

73. Let s_n denote the number of paths of length n from v_1 to v_1. We show that the sequences $s_1, s_2, \dots$ and $f_1, f_2, \dots$ satisfy the same recurrence relation and the same initial conditions, from which it follows that $s_n = f_n$ for $n \geq 1$.

If $n = 1$, there is one path of length 1 from v_1 to v_1, namely the loop on v_1; thus, $s_1 = f_1$.

If $n = 2$, there are two paths of length 2 from v_1 to v_1: (v_1, v_1, v_1) and (v_1, v_2, v_1); thus, $s_2 = f_2$.

Assume that $n > 2$. Consider a path of length n from v_1 to v_1. The path must begin with the loop (v_1, v_1) or the edge (v_1, v_2).

If the path begins with the loop, the remainder of the path must be a path of length $n-1$ from v_1 to v_1. Since there are s_{n-1} such paths, there are s_{n-1} paths of length n from v_1 to v_1 that begin $(v_1, v_1, \dots)$.

If the path begins with the edge (v_1, v_2), the next edge in the path must be (v_2, v_1). The remainder of the path must be a path of length $n - 2$ from v_1 to v_1. Since there are s_{n-2} such paths, there are s_{n-2} paths of length n from v_1 to v_1 that begin $(v_1, v_2, v_1, \dots)$.

Since any path of length $n > 2$ from v_1 to v_1 begins with the loop (v_1, v_1) or the edge (v_1, v_2), it follows that

$$s_n = s_{n-1} + s_{n-2}.$$

Because the sequences $s_1, s_2, \dots$ and $f_1, f_2, \dots$ satisfy the same recurrence relation and the same initial conditions, it follows that $s_n = f_n$ for $n \geq 1$.

75. Suppose that every vertex has an out edge. Choose a vertex v_0. Follow an edge out of v_0 to a vertex v_1. (By assumption such an edge exists.) Continue to follow an edge out of v_i to a vertex v_{i+1}. Since there are a finite number of vertices, we will eventually return to a previously visited vertex. At this point, we will have discovered a cycle, which is a contradiction. Therefore a dag has at least one vertex with no out edges.

Section 6.3

1. $(d, a, e, b, c, h, g, f, j, i, d)$

3. We would have to eliminate two edges each at b, d, i, and k, leaving $19 - 8 = 11$ edges. A Hamiltonian cycle would have 12 edges.

6. $(a, b, c, j, i, m, k, d, e, f, l, g, h, a)$

9.

12. If n is even and $m > 1$ or if m is even and $n > 1$, there is a Hamiltonian cycle. The sketch shows the solution in case n is even.

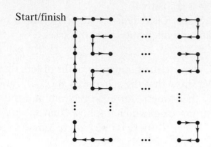

Start/finish

If $n = 1$ or if $m = 1$, there is no cycle and, in particular, there is no Hamiltonian cycle. Suppose that n and m are both odd and that the graph has a Hamiltonian cycle. Since there are nm vertices, this cycle has nm edges; therefore, the Hamiltonian cycle contains an odd number of edges. However, we note that in a Hamiltonian cycle, there must be as many "up" edges as "down" edges and as many "left" edges as "right" edges. Thus a Hamiltonian cycle must have an even number of edges. This contradiction shows that if n and m are both odd, the graph does not have a Hamiltonian cycle.

15. When $m = n$ and $n > 1$

18. Any cycle C in the n-cube has even length since the vertices in C alternate between an even and an odd number of 1's.

Suppose that the n-cube has a simple cycle of length m. We just observed that m is even. Now $m > 0$ by definition. Since the n-cube is a simple graph, $m \neq 2$. Therefore, $m \geq 4$.

Now suppose that $m \geq 4$ and m is even. Let G be the first $m/2$ members of the Gray code G_{n-1}. Then $0G$, $1G^R$ describes a simple cycle of length m in the n-cube.

21.

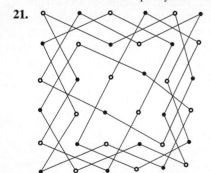

Section 6.4

1. $7; (a, b, c, f)$ **4.** $7; (b, c, f, j)$

6. An algorithm can be modeled after Example 6.4.2.

9. Modify Algorithm 6.4.1 so that it begins by assigning the weight ∞ to each nonexistent edge. The algorithm then continues as written. At termination, $L(z)$ will be equal to ∞ if there is no pith from a to z.

Section 6.5

1.

	a	b	c	d	e
a	0	1	1	1	1
b	1	0	1	0	0
c	1	1	0	1	1
d	1	0	1	0	1
e	1	0	1	1	0

4.

	v_1	v_2	v_3	v_4	v_5	v_6
v_1	0	1	1	0	0	0
v_2	1	0	1	0	0	0
v_3	1	1	0	0	0	0
v_4	0	0	0	0	0	0
v_5	0	0	0	0	0	1
v_6	0	0	0	0	1	0

7.

	x_1	x_2	x_3	x_4	x_5	x_6	x_7	x_8
a	1	0	1	0	1	1	0	0
b	1	1	0	0	0	0	0	0
c	0	1	0	1	1	0	1	0
d	0	0	0	1	0	1	0	1
e	0	0	1	0	0	0	1	1

10.

	e_1	e_2	e_3	e_4	e_5	e_6	e_7	e_8
1	1	0	0	0	0	0	0	0
2	1	1	0	1	1	1	0	0
3	0	1	1	0	0	0	0	0
4	0	0	1	1	0	0	0	0
5	0	0	0	0	1	0	1	0
6	0	0	0	0	0	1	1	1
7	0	0	0	0	0	0	0	1

13.

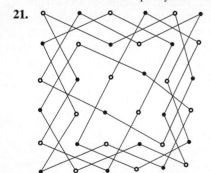

16.

19. [For K_5]

$$\begin{pmatrix} 4 & 3 & 3 & 3 & 3 \\ 3 & 4 & 3 & 3 & 3 \\ 3 & 3 & 4 & 3 & 3 \\ 3 & 3 & 3 & 4 & 3 \\ 3 & 3 & 3 & 3 & 4 \end{pmatrix}$$

22. The graph is not connected.

25.

28. G is not connected.

29. Because of the symmetry of the graph, if v and w are vertices in K_5, there are the same number of paths of length n from v to v as there are from w to w. Thus all the diagonal elements of A^n are equal. Similarly, all the off-diagonal elements of A^n are equal.

32. If $n \geq 2$,

$$d_n = 4a_{n-1} \qquad \text{by Exercise 30}$$
$$= 4\left(\frac{1}{5}\right)[4^{n-1} + (-1)^n] \qquad \text{by Exercise 31.}$$

The formula can be directly verified for $n = 1$.

Section 6.6

1. The graphs are not isomorphic, since they do not have the same number of vertices.

4. The graphs are not isomorphic, since G_2 has a vertex (c') of degree 4, but G_1 has no vertex of degree 4.

7. The graphs are not isomorphic, since G_1 has a vertex of degree 2, but G_2 does not.

10. The graphs are isomorphic: $f(a) = e'$, $f(b) = c'$, $f(c) = a'$, $f(d) = g'$, $f(e) = d'$, $f(f) = f'$, $f(g) = b'$, $f(h) = h'$. Define

$$g((x, y)) = (f(x), f(y)).$$

In Exercises 12–18, we use the notation of Definition 6.6.1.

12. If $(v_0, v_1, \ldots, v_k)$ is a simple cycle of length k in G_1, then $(f(v_0), f(v_1), \ldots, f(v_k))$ is a simple cycle of length k in G_2. [The vertices $f(v_i)$, $i = 1, \ldots k - 1$, are distinct, since f is one-to-one.]

15. In the hint to Exercise 12, we showed that if $C = (v_0, v_1, \ldots, v_k)$ is a simple cycle of length k in G_1, then $(f(v_0), f(v_1), \ldots, f(v_k))$, which here we denote $f(C)$, is a simple cycle of length k in G_1. Let $C_1, C_2, \ldots, C_n$ denote the n simple cycles of length k in G_1. Then $f(C_1), f(C_2), \ldots, f(C_n)$ are n simple cycles of length k in G_2. Moreover, since f is one-to-one, $f(C_1), f(C_2), \ldots, f(C_n)$ are distinct.

18. The property is an invariant. If $(v_0, v_1, \ldots, v_n)$ is an Euler cycle in G_1, then, since g is onto, $(f(v_0), f(v_1), \ldots, f(v_n))$ is an Euler cycle in G_2.

21.

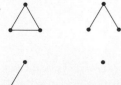

24.

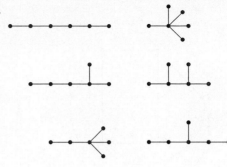

26.

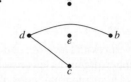

29.

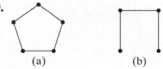

(a)   (b)

32. Define $g((v, w)) = (f(v), f(w))$.

33. $f(a) = a'$, $f(b) = b'$, $f(c) = c'$, $f(d) = b'$

36. $f(a) = a'$, $f(b) = b'$, $f(c) = c'$, $f(d) = a'$

Section 6.7

1.

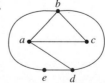

4.

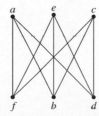

is $K_{3,3}$

6. Planar

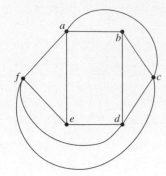

9. $2e = 2 + 2 + 2 + 3 + 3 + 3 + 4 + 4 + 5$,
so $e = 14$. $f = e - v + 2 = 14 - 9 + 2 = 7$

12. A graph with five or fewer vertices and a vertex of degree 2 is homeomorphic to a graph with four or fewer vertices. Such a graph cannot contain a homeomorphic copy of $K_{3,3}$ or K_5.

15. If K_5 is planar, $e \le 3v - 6$ becomes $10 \le 3 \cdot 5 - 6 = 9$.

18.

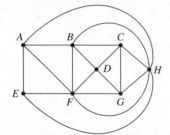

22.

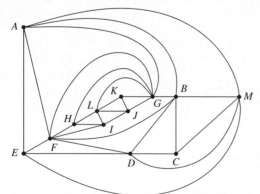

25.

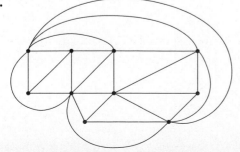

28. It contains

31. Assume that G does not have a vertex of degree 5. Show that $2e \ge 6v$. Now use Exercise 13 to deduce a contradiction.

Section 6.8

1.

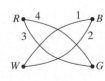

G_1 G_2

4.

G_1 G_2

7.   (a)

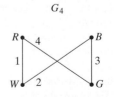

G_1 G_2

G_3 G_4

G_5 G_6

G_7

(b) Solutions are: G_1, G_5; G_1, G_7; G_2, G_4; G_2, G_6; G_3, G_6; and G_3, G_7.

13. One edge can be chosen in $C(2+4-1, 2) = 10$ ways. The three edges labeled 1 can be chosen in $C(3+10-1, 3) = 220$ ways. Thus the total number of graphs is 220^4.

15.

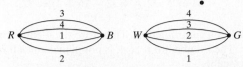

19. According to Exercise 14, not counting loops every vertex must have degree at least 4. In Figure 6.8.5, not counting loops vertex W has degree 3 and, therefore, Figure 6.8.5 does not have a solution to the modified version of Instant Insanity. Figure 6.8.3 gives a solution to regular Instant Insanity for Figure 6.8.5.

Chapter 6 Self-Test

1. $V = \{v_1, v_2, v_3, v_4\}$. $E = \{e_1, e_2, e_3\}$. e_1 and e_2 are parallel edges. There are no loops. v_1 is an isolated vertex. G is not a simple graph. e_3 is incident on v_2 and v_4. v_2 is incident on e_1, e_2, and e_3.

2. There are vertices (a and e) of odd degree.

3.

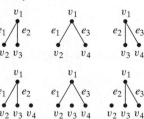

4. If we let V_1 denote the set of vertices containing an even number of 1's and V_2 the set of vertices containing an odd number of 1's, each edge is incident on one vertex in V_1 and one vertex in V_2. Therefore, the n-cube is bipartite.

5. It is a cycle.

6.

7.

8. No. There are vertices of odd degree.

9. $(v_1, v_2, v_3, v_4, v_5, v_7, v_6, v_1)$

10. (000, 001, 011, 010, 110, 111, 101, 100, 000)

11. A Hamiltonian cycle would have seven edges. Suppose that the graph has a Hamiltonian cycle. We would have to eliminate three edges at vertex b and one edge at vertex f. This leaves $10 - 4 = 6$ edges, not enough for a Hamiltonian

cycle. Therefore, the graph does not have a Hamiltonian cycle.

12. In a minimum-weight Hamiltonian cycle, every vertex must have degree 2. Therefore, edges $(a, b), (a, j), (j, i), (i, h), (g, f), (f, e)$, and (e, d) must be included. We cannot include edge (b, h) or we will complete a cycle. This implies that we must include edges (h, g) and (b, c). Since vertex g now has degree 2, we cannot include edges (c, g) or (g, d). Thus we must include (c, d). This is a Hamiltonian cycle and the argument shows that it is unique. Therefore, it is minimal.

13. 9   **14.** 11   **15.** (a, e, f, i, g, z) **16.** 12

17.

	v_1	v_2	v_3	v_4	v_5	v_6	v_7
v_1	0	1	0	0	0	1	0
v_2	1	0	1	1	0	1	1
v_3	0	1	0	1	0	0	0
v_4	0	1	1	0	1	0	0
v_5	0	0	0	1	0	1	1
v_6	1	1	0	0	1	0	1
v_7	0	1	0	0	1	1	0

18.

	e_1	e_2	e_3	e_4	e_5	e_6	e_7	e_8	e_9	e_{10}	e_{11}
v_1	1	0	0	0	0	0	1	0	0	0	0
v_2	1	1	0	1	1	1	0	0	0	0	0
v_3	0	1	1	0	0	0	0	0	0	0	0
v_4	0	0	1	1	0	0	0	0	0	1	0
v_5	0	0	0	0	0	0	0	1	1	1	0
v_6	0	0	0	0	0	1	1	0	1	0	1
v_7	0	0	0	0	1	0	0	1	0	0	1

19. The number of paths of length 3 from v_2 to v_3

20. No. Each edge is incident on at least one vertex.

21. The graphs are isomorphic. The orderings v_1, v_2, v_3, v_4, v_5 and w_3, w_1, w_4, w_2, w_5 produce equal adjacency matrices.

22. The graphs are isomorphic. The orderings $v_1, v_2, v_3, v_4, v_5, v_6$ and $w_3, w_6, w_2, w_5, w_1, w_4$ produce equal adjacency matrices.

23.

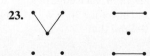

24.

25. The graph is planar;

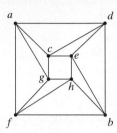

26. The graph is not planar; the following subgraph is homeomorphic to K_5:

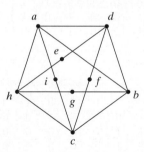

27. A simple, planar, connected graph with e edges and v vertices satisfies $e \leq 3v - 6$ (see Exercise 13, Section 6.7). If $e = 31$ and $v = 12$, the inequality is not satisfied, so such a graph cannot be planar.

28. For $n = 1, 2, 3$, it is possible to draw the n-cube in the plane without having any of its edges cross:

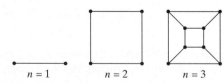

   We argue by contradiction to show that the 4-cube is not planar. Suppose that the 4-cube is planar. Since every cycle has at least four edges, each face is bounded by at least four edges. Thus the number of edges that bound faces is at least $4f$. In a planar graph, each edge belongs to at most two bounding cycles. Therefore $2e \geq 4f$. Using Euler's formula for graphs, we find that

$$2e \geq 4(e - v + 2).$$

For the 4-cube, we have $e = 32$ and $v = 16$, so Euler's formula becomes

$$64 = 2 \cdot 32 \geq 4(32 - 16 + 2) = 72,$$

which is a contradiction. Therefore the 4-cube is not planar. The n-cube, for $n > 4$, is not planar since it contains the 4-cube.

29.

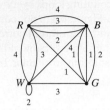

30. See the hints for Exercises 31 and 32.

31.

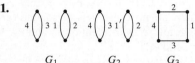

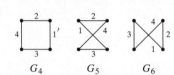

We denote the two edges incident on B and G labeled 1 in the graph of Exercise 29 is 1 and $1'$ here.

32. The puzzle of Exercise 29 has four solutions. Using the notation of Exercise 31, the solutions are G_1, G_5; G_2, G_5; G_3, G_6; and G_4, G_6.

Section 7.1

1. a-1; b-1; c-1; d-1; e-2; f-3; g-3; h-4; i-2; j-3; k-0

4. Height $= 4$

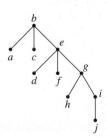

7. PEN

10. SALAD

11. 0111100010

14. 0110000100100001111

17.

20. Another tree is shown in the hint for Exercise 17.

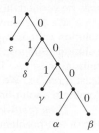

24. Let T be a tree. Root T at some arbitrary vertex. Let V be the set of vertices on even levels and let W be the set of vertices on odd levels. Since each edge is incident on a vertex in V and a vertex in W, T is a bipartite graph.

27. e, g

30. The radius is the eccentricity of a center. It is not necessarily true that $2r = d$ (see Figure 7.1.5).

Section 7.2

1. Kronos

4. Apollo, Athena, Hermes, Heracles

7. $b; d$

10. $e, f, g, j; j$

13. a, b, c, d, e

17. They are siblings.

22.

25.

27. A single vertex is a "cycle" of length 0.

30. Each component of a forest is connected and acyclic and, therefore, a tree.

33. Suppose that G is connected. Add parallel edges until the resulting graph G^* has $n - 1$ edges. Since G^* is connected and has $n - 1$ edges, by Theorem 7.2.3, G^* is acyclic. But adding an edge in parallel introduces a cycle. Contradiction.

36.

Section 7.3

1.

4. The path (h, f, e, g, b, d, c, a)

7.

10. The two-queens problem clearly has no solution. For the three-queens problem, by symmetry, the only possible first column positions are upper left and second from top. If the first move is first column, upper left, the second move must be to the bottom of the second column. Now no move is possible for the third column. If the first move is first column, second from top, there is no move possible in column two. Therefore, there is no solution to the three-queens problem.

13. False. Consider K_4.

16. First, show that the graph T constructed is a tree. Now use induction on the level of T to show that T contains all the vertices of G.

19. Suppose that x is incident on vertices a and b. Removing x from T produces a disconnected graph with two components U and V. Vertices a and b belong to different components—say, $a \in U$ and $b \in V$. There is a path P from a to b in T'. As we move along P, at some point we encounter an edge $y = (v, w)$ with $v \in U$, $w \in V$. Since adding y to $T - \{x\}$ produces a connected graph, $(T - \{x\}) \cup \{y\}$ is a spanning tree. Clearly, $(T' - \{y\}) \cup \{x\}$ is a spanning tree.

22. Suppose that T has n vertices. If an edge is added to T, the resulting graph T' is connected. If T' were acyclic, T' would be a tree with n edges and n vertices. Thus, T' contains a cycle. If T' contains two or more cycles, we would be able to produce a connected graph T'' by deleting two or more edges from T'. But now T'' would be a tree with n vertices and fewer than $n - 1$ edges—an impossibility.

23.

	e_1	e_2	e_6	e_5	e_3	e_4	e_7	e_8
$(abca)$	1	0	0	0	1	1	0	0
$(acda)$	0	1	0	0	1	0	0	1
$(acdb)$	0	0	1	0	0	1	0	1
$(bcdeb)$	0	0	0	1	0	1	1	1

26.

Input: A graph $G = (V, E)$ with n vertices

Output: **true**, if G is connected
         false, if G is not connected

```
procedure is_connected(V, E)
    T := bfs(V, E)
    // T = (V', E') is the
    // spanning tree returned by bfs
    if |V'| = n then
        return(true)
    else
        return(false)
end is_connected
```

Section 7.4

1.   **4.**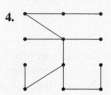

10. If v is the first vertex examined by Prim's Algorithm, the edge will be in the minimal spanning tree constructed by the algorithm.

13. Suppose that G has two minimal spanning trees T_1 and T_2. Then, there exists an edge x in T_1 that is not in T_2. By Exercise 19, Section 7.3, there exists an edge y in T_2 that is not in T_1 such that $T_3 = (T_1 - \{x\}) \cup \{y\}$ and $T_4 = (T_2 - \{y\}) \cup \{x\}$ are spanning trees. Since x and y have different weights, either T_3 or T_4 has weight less than T_1. This is a contradiction.

14. False

16. False. Consider K_5 with the weight of every edge equal to 1.

20. Input: The edges E of an n-vertex connected, weighted graph. If e is an edge, $w(e)$ is equal to the weight of e; if e is not an edge, $w(e)$ is equal to ∞ (a value greater than any actual weight).

Output: A minimal spanning tree.

```
procedure kruskal(E, w, n)
  V' := ∅
  E' := ∅
  T' := (V', E')
  while |E'| < n − 1 do
    begin
      among all edges that if added to T'
        would not complete a cycle, choose
        e = (v_i, v_j) of minimum weight
      E' := E' ∪ {e}
      V' := V' ∪ {v_i, v_j}
      T' := (V', E')
    end
  return(T')
end kruskal
```

23. Terminate Kruskal's Algorithm after k iterations. This groups the data into $n - k$ classes.

27. We show that $a_1 = 7$ and $a_2 = 3$ provides a solution. We use induction on n to show that the greedy solution gives an optimal solution for $n \geq 1$. The cases $n = 1, 2, \ldots, 8$ may be verified directly.

We first show that if $n \geq 9$, there is an optimal solution containing at least one 7. Let S' be an optimal solution. Suppose that S' contains no 7's. Since S' contains at most two 1's (since S' is optimal), S' contains at least three 3's. We replace three 3's by one 7 and two 1's to obtain a solution S. Since $|S| = |S'|$, S is optimal.

If we remove a 7 from S, we obtain a solution S^* to the $(n-7)$-problem. If S^* were not optimal, S could not be optimal. Thus S^* is optimal. By the inductive assumption, the greedy solution GS^* to the $(n-7)$-problem is optimal, so $|S^*| = |GS^*|$. Notice that 7 together with GS^* is the greedy solution GS to the n-problem. Since $|GS| = |S|$, GS is optimal.

30. We show that if the greedy algorithm is optimal for $n = 1, 2, \ldots, k$, where

$$k = \sum_{i=1}^{m-1} \left(\frac{a_{i+1}}{\gcd(a_{i+1}, a_i)} - 1 \right) a_i,$$

then the greedy algorithm is optimal for all n for denominations

$$1 = a_1 < a_2 < \cdots < a_m.$$

The converse is straightforward.

We use induction on n to prove that the greedy algorithm is optimal for n. The Basis Steps are $n = 1, \ldots, k$, which are satisfied because of the hypothesis.

Assume that $n > k$. Let S be an optimal solution for n, and let G_n be the greedy solution for n. Also, let $|S|$ and $|G_n|$ denote the number of stamps in these solutions. We claim that S contains at least one stamp of denomination a_m. Suppose, by way of contradiction, that S contains no stamp of denomination a_m. Now S contains at most

$$\frac{a_{i+1}}{\gcd(a_{i+1}, a_i)} - 1$$

stamps of denomination a_i for $i = 1, \ldots, m - 1$. Otherwise, we could replace

$$\frac{a_{i+1}}{\gcd(a_{i+1}, a_i)}$$

stamps of denomination a_i by

$$\frac{a_i}{\gcd(a_{i+1}, a_i)}$$

stamps of denomination a_{i+1} and thereby reduce the number of stamps in S. Since S is an optimal solution, this is impossible. Therefore S contains at most

$$\frac{a_{i+1}}{\gcd(a_{i+1}, a_i)} - 1$$

stamps of denomination a_i for $i = 1, \ldots, m-1$. It follows that we can make at most k cents postage. Since $n > k$, this is a contradiction. Therefore S contains at least one stamp of denomination a_m.

Let S' be S with one stamp of denomination a_m removed. Then S' is optimal for $n - a_m$ cents postage. By the inductive assumption, the greedy algorithm is optimal for $n - a_m$ cents postage. Therefore $|G_{n-a_m}| = |S'|$. It follows that $|G_n| = |S|$, and the proof is complete.

Section 7.5

1.

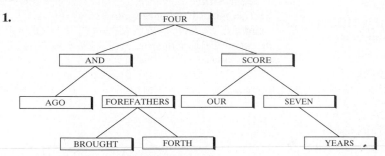

4. False. Consider

5.

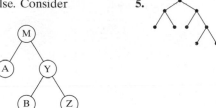

8. $mi + 1$, $(m-1)i + 1$ **11.** $t - 1$

14. Balanced   **17.** Balanced

18. A tree of height 0 has one vertex, so $N_0 = 1$. In a balanced binary tree of height 1, the root must have at least one child. If the root has exactly one child, the number of vertices will be minimized. Therefore, $N_1 = 2$. In a balanced binary tree of height 2, there must be a path from the root to a terminal vertex of length 2. This accounts for three vertices. But for the tree to be balanced, the root must have two children. Therefore, $N_2 = 4$.

21. Suppose that there are n vertices in a balanced binary tree of height h. Then

$$n \ge N_h = f_{h+2} - 1 > \left(\frac{3}{2}\right)^{h+2} - 1,$$

for $h \ge 3$. The equality comes from Exercise 20 and the last inequality comes from Exercise 20, Section 3.4. Therefore,

$$n + 1 > \left(\frac{3}{2}\right)^{h+2}.$$

Taking the logarithm to the base $\frac{3}{2}$ of each side, we obtain

$$\log_{3/2}(n+1) > h + 2.$$

Therefore,

$$h < [\log_{3/2}(n+1)] - 2 = O(\lg n).$$

Section 7.6

1.
preorder	inorder	postorder
$ABDCE$	$BDAEC$	$DBECA$

4.
preorder	inorder	postorder
$ABCDE$	$EDCBA$	$EDCBA$

6.

prefix: $* + AB - CD$

postfix: $AB + CD - *$

9.

prefix: $- * + * + ABCDE + * + ABCD$

postfix: $AB + C * D + E * AB + C * D + -$

11.

prefix: $- + ABC$

usual infix: $A + B - C$

parened infix: $((A + B) - C)$

14.

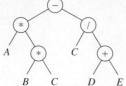

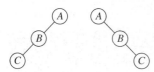

prefix: $- * A * BC / C + DE$
usual infix: $A * B * C - C / (D + E)$
parened infix: $((A * (B * C)) - (C / (D + E)))$

16. -4 **19.** 0

22.

25.

28.   Input: PT, the root of a binary tree

   Output: PT, the root of the modified binary tree

procedure $swap\_children(PT)$
  if PT is empty **then**
   return
  swap the left and right children of PT
  $l :=$ left child of PT
  $swap\_children(l)$
  $r :=$ right child of PT
  $swap\_children(r)$
end $swap\_children$

30. Define an *initial segment* of a string to be the first $i \geq 1$ characters for some i. Define $r(x) = 1$, for $x = A$, $B, \ldots, Z$; and $r(x) = -1$, for $x = +, -, *, /$. If $x_1 \cdots x_n$ is a string over $\{A, \ldots, Z, +, -, *, /\}$, define

$$r(x_1 \cdots x_n) = r(x_1) + \cdots + r(x_n).$$

Then a string s is a postfix string if and only if $r(s) = 1$ and $r(s') \geq 1$, for all initial segments s' of s.

33.   Input: PT, the root of a nonempty tree

   Output: Each node of the tree has a field *in_cover* that is set to **true** if that node is in the vertex cover, or to **false** if that node is not in the vertex cover.

procedure $tree\_cover(PT)$
  $flag :=$ **false**
  $ptr :=$ first child of PT
  while ptr is **not** empty **do**
   begin
   $tree\_cover(ptr)$
   if *in_cover* of $ptr =$ **false then**
    $flag :=$ **true**
   $ptr :=$ next sibling of ptr
   end
  in_cover of $PT := flag$
end $tree\_cover$

Section 7.7

1.

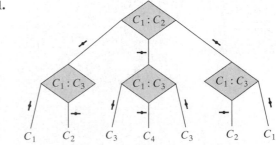

4. See following page.

7. There are 28 possible outcomes to the fourteen-coins puzzle. A tree of height 3 has at most 27 terminal vertices; thus at least four weighings are required in the worst case. In fact, there is an algorithm that uses four weighings in the worst case: We begin by weighing four coins against four coins. If the coins do not balance, we proceed as in the solution given for Exercise 4 (for the twelve-coins puzzle). In this case, at most three weighings are required. If the coins do balance, we disregard these coins; our problem then is to find the bad coin from among the remaining six coins. The six-coins puzzle can be solved in at most three weighings in the worst case, which, together with the initial weighing, requires four weighings in the worst case.

10. The decision tree analysis shows that at least $\lceil \lg 5! \rceil = 7$ comparisons are required to sort five items in the worst case. The following algorithm sorts five items using at most seven comparisons in the worst case.

Given the sequence $a_1, \ldots, a_5$, we first sort a_1, a_2 (one comparison) and then a_3, a_4 (one comparison). (We assume now that $a_1 < a_2$ and $a_3 < a_4$.) We then compare a_2 and a_4. Let us assume that $a_2 < a_4$. (The case $a_2 > a_4$ is symmetric and for this reason that part of the algorithm is omitted.) At this point we know that

$$a_1 < a_2 < a_4 \qquad \text{and} \qquad a_3 < a_4$$

Exercise 4, Section 7.7:

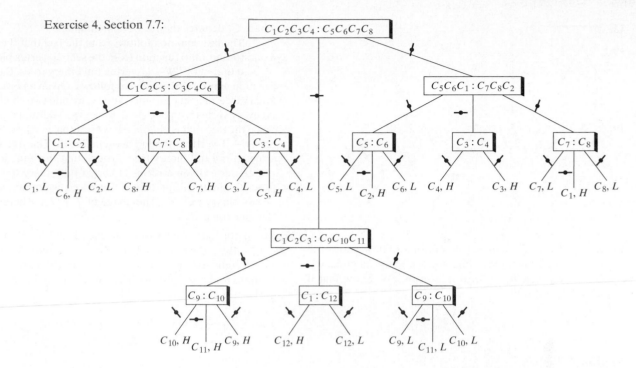

Next we determine where a_5 belongs among a_1, a_2, and a_4 by first comparing a_5 with a_2. If $a_5 < a_2$, we next compare a_5 with a_1; but if $a_5 > a_2$, we next compare a_5 with a_4. In either case, two additional comparisons are required. At this point, a_1, a_2, a_4, a_5 is sorted. Finally, we insert a_3 in its proper place. If we first compare a_3 with the second smallest item among a_1, a_2, a_4, a_5, only one additional comparison will be required for a total of seven comparisons. To justify this last statement, we note that the following arrangements are possible after we insert a_5 in its correct position:

$$a_5 < a_1 < a_2 < a_4$$
$$a_1 < a_5 < a_2 < a_4$$
$$a_1 < a_2 < a_5 < a_4$$
$$a_1 < a_2 < a_4 < a_5.$$

If a_3 is less than the second item, only one additional comparison is needed (with the first item) to locate the correct position for a_3. If a_3 is greater than the second item, at most one additional comparison is needed to locate the correct position for a_3. In the first three cases, we need only compare a_3 with either a_2 or a_5 to find the correct position for a_3 since we already know that $a_3 < a_4$. In the fourth case, if a_3 is greater than a_2, we know that it goes between a_2 and a_4.

12. We can consider the numbers as contestants and the internal vertices as winners where the larger value wins.

15. Suppose we have an algorithm that finds the largest value

among $x_1, \ldots, x_n$. Let $x_1, \ldots, x_n$ be the vertices of a graph. An edge exists between x_i and x_j if the algorithm compares x_i and x_j. The graph must be connected. The least number of edges necessary to connect n vertices is $n - 1$.

18. By Exercise 14, Tournament Sort requires $2^k - 1$ comparisons to find the largest element. By Exercise 16, Tournament Sort requires k comparisons to find the second largest element. Similarly Tournament Sort requires at most k comparisons to find the third largest, at most k comparisons to find the fourth largest, and so on. Thus the total number of comparisons is at most

$$[2^k - 1] + (2^k - 1)k \le 2^k + k2^k$$
$$\le k2^k + k2^k$$
$$= 2 \cdot 2^k k = 2n \lg n.$$

Section 7.8

1. Isomorphic. $f(v_1) = w_1$, $f(v_2) = w_5$, $f(v_3) = w_3$, $f(v_4) = w_4$, $f(v_5) = w_2$, $f(v_6) = w_6$.

4. Not isomorphic. T_2 has a simple path of length 2 from a vertex of degree 1 to a vertex of degree 1, but T_1 does not.

7. Isomorphic as rooted trees. $f(v_1) = w_1$, $f(v_2) = w_4$, $f(v_3) = w_3$, $f(v_4) = w_2$, $f(v_5) = w_6$, $f(v_6) = w_5$, $f(v_7) = w_7$, $f(v_8) = w_8$. Also isomorphic as free trees.

10. Not isomorphic as binary trees. The root of T_1 has a left child but the root of T_2 does not. Isomorphic as rooted trees and as free trees.

13.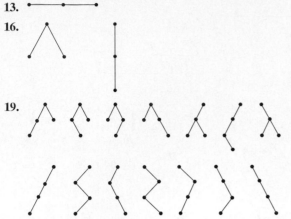

16.

19.

22. Let b_n denote the number of nonisomorphic n-vertex full binary trees. Since every full binary tree has an odd number of vertices, $b_n = 0$, if n is even. We show that if $n = 2i + 1$ is odd,

$$b_n = C_i,$$

where C_i denotes the ith Catalan number.

The last equation follows from the fact that there is a one-to-one, onto function from the set of i-vertex binary trees to the set of $(2i + 1)$-vertex full binary trees. Such a function may be constructed as follows. Given an i-vertex binary tree, at every terminal vertex, we add two children. At every vertex with one child, we add an additional child. Since the tree that is obtained has i internal vertices, there are $2i + 1$ vertices total (Theorem 7.5.4). The tree constructed is a full binary tree. Notice that this function is one-to-one. Given a $(2i + 1)$-vertex full binary tree T', if we eliminate all the terminal vertices, we obtain an i-vertex binary tree T. The image of T is T'. Therefore, the function is onto.

25. There are four comparisons at lines 1 and 3. By Exercise 24, the call *bin_tree_isom*(lc_r_1, lc_r_2) requires $6(k - 1) + 2$ comparisons. The call *bin_tree_isom*(rc_r_1, rc_r_2) requires four comparisons. Thus the total number of comparisons is

$$4 + 6(k - 1) + 2 + 4 = 6k + 4.$$

Section 7.9

1.

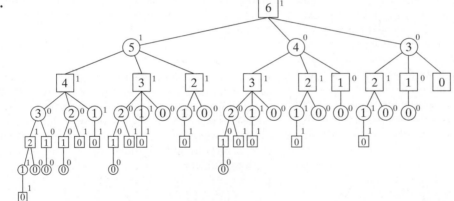

The first player always wins. The winning strategy is to first take one token; then, whatever the second player does, leave one token.

4. The second player always wins. If two piles remain, leave piles with equal numbers of tokens. If one pile remains, take it.

7. Suppose that the first player can win in nim. The first player can always win in nim′ by adopting the following strategy: Play nim′ exactly like nim unless the move would leave an odd number of singleton piles and no other pile. In this case, leave an even number of piles.

Suppose that the first player can always win in nim′. The first player can always win in nim by adopting the following strategy: Play nim exactly like nim′ unless the move would leave an even number of singleton piles and no other pile. In this case, leave an odd number of piles.

9.

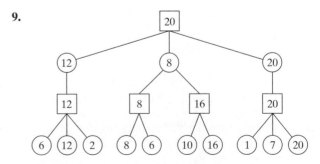

12. The value of the root is 3.

14. [For Exercise 11]

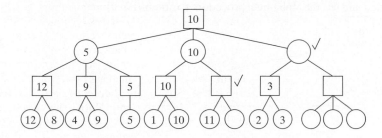

15. $3 - 2 = 1$ **18.** $4 - 1 = 3$

19.

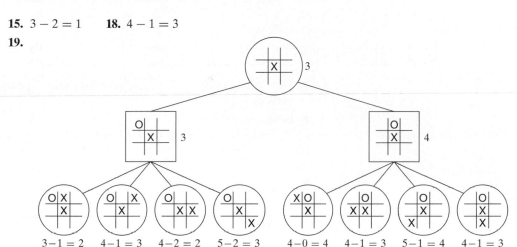

O will move to a corner.

22.   Input: The root PT of a game tree, the type PT_type
of PT (*box or circle*), the level PT_level of PT,
the maximum level n to which the search is to
be conducted, an evaluation function E, and
a number ab_val (which is either the alpha- or
beta-value of the parent of PT). (The initial
call sets ab_val to ∞ if PT is a box vertex, or
to $-\infty$ if PT is a circle vertex.)

Output: The game tree with PT evaluated

```
procedure alpha_beta_prune(PT,
  PT_type, PT_level, n, E, ab_val)
  if PT_level = n then
    begin
    contents(PT) := E(PT)
    return
    end
  if PT_type = box then
    begin
    contents(PT) := -∞
    for each child C of PT do
      alpha_beta_prune (C, circle,
        PT_level + 1, n, E, contents(PT))
      c_val := contents(C)
```

```
      if c_val ≥ ab_val then
        begin
        contents(PT) := ab_val
        return
        end
      if c_val > contents(PT) then
        contents(PT) := c_val
    end
  else
    begin
    contents(PT) := ∞
    for each child C of PT do
      alpha_beta_prune(C, box,
        PT_level + 1, n, E, contents(PT))
      c_val := contents(C)
      if c_val ≤ ab_val then
        begin
        contents(PT) := ab_val
        return
        end
      if c_val < contents(PT) then
        contents(PT) := c_val
    end
  end alpha_beta_prune
```

23. We first obtain the values 6, 6, 7 for the children of the root. Then we order the children of the root with the rightmost child first and use the alpha-beta procedure to obtain

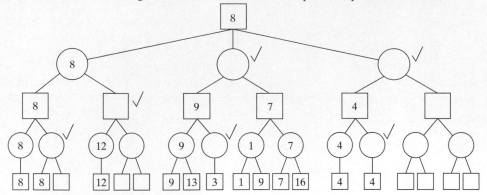

Chapter 7 Self-Test

1.

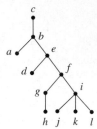

2. *a*-2, *b*-1, *c*-0, *d*-3, *e*-2, *f*-3, *g*-4, *h*-5, *i*-4, *j*-5, *k*-5, *l*-5

3. 5

4.

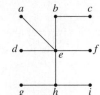

5. (a) *b*   (b) *a, c*

(c) *d, a, c, h, j, k, l*   (d)

6. True. See Theorem 7.2.3.

7. True. A tree of height 6 or more must have seven or more vertices.

8. False.

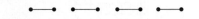

9.

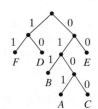

10.

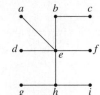

11.

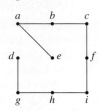

12.

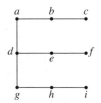

13.

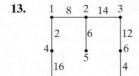

14. $(1, 4), (1, 2), (2, 5), (2, 3), (3, 6), (6, 9), (4, 7), (7, 8)$

15. $(6, 9), (3, 6), (2, 3), (2, 5), (1, 2), (1, 4), (4, 7), (7, 8)$

16. Consider a "shortest-path algorithm" in which at each step we select an available edge having minimum weight incident on the most recently added vertex (see the discussion preceding Theorem 7.4.5).

17. /\   **18.** 16

19.

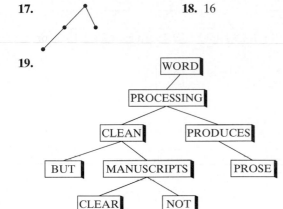

20. We first compare MORE with the word WORD in the root. Since MORE is less than WORD, we go to the left child. Next, we compare MORE with PROCESS-ING. Since MORE is less than PROCESSING, we go to the left child. Since MORE is greater than CLEAN, we go to the right child. Since MORE is greater than MANUSCRIPTS, we go to the right child. Since MORE is less than NOT, we go to the left child. Since MORE is less than NECESSARILY, we attempt to go to the left child. Since there is no left child, we conclude that MORE is not in the tree.

21. *ABFGCDE*   **22.** *BGFAEDC*   **23.** *GFBEDCA*

24.

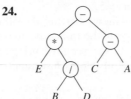

postfix: $EBD/ \quad * \quad CA \quad - \quad -$
parened infix: $((E * (B/D)) - (C - A))$

25. An algorithm that requires at most two weighings can be represented by a decision tree of height at most 2. However, such a tree has at most nine terminal vertices. Since there are 12 possible outcomes, there is no such algorithm. Therefore, at least three weighings are required in the worst case to identify the bad coin and determine whether it is heavy or light.

26.

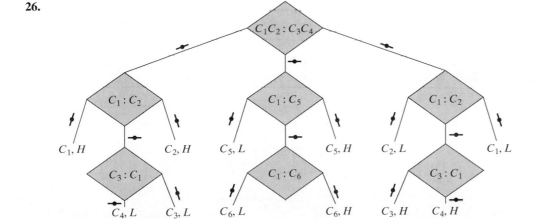

27. According to Theorem 7.7.3, any sorting algorithm requires at least $Cn \lg n$ comparisons in the worst case. Since Professor Sabic's algorithm uses at most $100n$ comparisons, we must have $Cn \lg n \le 100n$ for all $n \ge 1$. If we cancel n, we obtain $C \lg n \le 100$ for all $n \ge 1$, which is false. Therefore, the professor does not have a sorting algorithm that uses at most $100n$ comparisons in the worst case for all $n \ge 1$.

28. In the worst case, three comparisons are required to sort three items using an optimal sort (see Example 7.7.2).

If $n = 4$, binary insertion sort sorts three items (three comparisons—worst case) and then inserts the fourth item in the sorted three-item list (two comparisons—worst case) for a total of five comparisons in the worst case.

If $n = 5$, binary insertion sort sorts four items (five comparisons—worst case) and then inserts the fifth item in the sorted four-item list (three comparisons—worst case) for a total of eight comparisons in the worst case.

If $n = 6$, binary insertion sort sorts five items (eight comparisons—worst case) and then inserts the sixth item in the sorted five-item list (three comparisons—worst case) for a total of eleven comparisons in the worst case.

The decision tree analysis shows that any algorithm requires at least five comparisons in the worst case to sort four items. Thus binary insertion sort is optimal if $n = 4$.

The decision tree analysis shows that any algorithm requires at least seven comparisons in the worst case to sort five items. It is possible, in fact, to sort five items using seven comparisons in the worst case. Thus binary insertion sort is not optimal if $n = 5$.

The decision tree analysis shows that any algorithm requires at least ten comparisons in the worst case to sort six items. It is possible, in fact, to sort six items using ten comparisons in the worst case. Thus binary insertion sort is not optimal if $n = 6$.

29. True. If f is an isomorphism of T_1 and T_2 as rooted trees, f is also an isomorphism of T_1 and T_2 as free trees.

30. False.

31. Isomorphic. $f(v_1) = w_6$, $f(v_2) = w_2$, $f(v_3) = w_5$, $f(v_4) = w_7$, $f(v_5) = w_4$, $f(v_6) = w_1$, $f(v_7) = w_3$, $f(v_8) = w_8$.

32. Not isomorphic. T_1 has a vertex (v_3) on level 1 of degree 3, but T_2 does not.

33. $3 - 1 = 2$

34. Let each row, column, or diagonal that contains one X and two blanks count 1. Let each row, column, or diagonal

that contains two X's and one blank count 5. Let each row, column, or diagonal that contains three X's count 100. Let each row, column, or diagonal that contains one 0 and two blanks count -1. Let each row, column, or diagonal that contains two 0's and one blank count -5. Let each row, column, or diagonal that contains three 0's count -100. Sum the values obtained.

35.

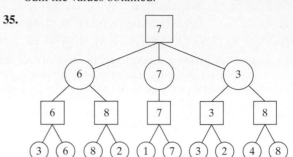

36.

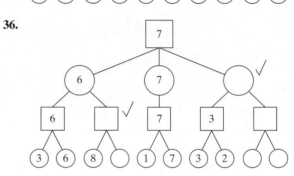

Section 8.1

1. (b, c) is 6, 3; (a, d) is 4, 2; (c, e) is 6, 1; (c, z) is 5, 2. The value of the flow is 5.

4. Add edges (a, w_1), (a, w_2), (a, w_3), (A, z), (B, z), and (C, z) each having capacity ∞.

7.

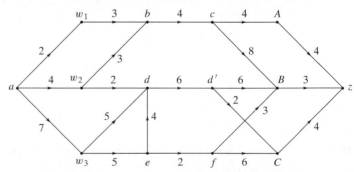

Section 8.2

1. 1

4. $(a, w_1)-6$, $(a, w_2)-0$, $(a, w_3)-3$, $(w_1, b)-6$, $(w_2, b)-0$, $(w_3, d) - 3$, $(d, c) - 3$, $(b, c) - 2$, $(b, A) - 4$, $(c, A) - 2$, $(c, B) - 3$, $(A, z) - 6$, $(B, z) - 3$

10.

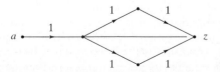

7.

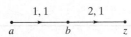

Figure for problem 7: a network flow diagram with vertices a, w_1, w_2, w_3, b, c, d, e, f, A, B, C, z, with edge labels: w_1 to b: $3, 2$; b to c: $4, 4$; c to A: $4, 4$; a to w_1: $2, 2$; w_1 to w_2: $3, 2$; a to w_2: $4, 4$; w_2 to d: $2, 2$; d to B: $6, 6$; c to B: $8, 0$; A to z: $\infty, 4$; B to z: $\infty, 8$; a to w_3: $7, 7$; w_3 to w_2: $5, 5$; w_3 to e: $5, 2$; e to d: $4, 0$; d to f: $2, 1$; e to f: $2, 2$; f to C: $6, 0$; d to C: $3, 2$; C to z: $\infty, 1$.

10. $(a, A - 7{:}00) - 3000$, $(a, A - 7{:}15) - 3000$, $(a, A - 7{:}30) - 2000$, $(A - 7{:}00, B - 7{:}30) - 1000$, $(A - 7{:}00, C - 7{:}15) - 2000$, $(A - 7{:}15, B - 7{:}45) - 1000$, $(A - 7{:}15, C - 7{:}30) - 2000$, $(A - 7{:}30, C - 7{:}45) - 2000$, $(B - 7{:}30, D - 7{:}45) - 1000$, $(C - 7{:}15, D - 7{:}30) - 2000$, $(B - 7{:}45, D - 8{:}00) - 1000$, $(C - 7{:}30, D - 7{:}45) - 2000$, $(C - 7{:}45, D - 8{:}00) - 2000$, $(D - 7{:}45, z) - 3000$, $(D - 7{:}30, z) - 2000$, $(D - 8{:}00, z) - 3000$. All other edges have flow equal to 0.

13.

Figure for problem 13: network with vertices a, b, c, d, e, z, with edge labels: a to b: $3, 2$; b to c: $2, 2$; c to z: $4, 4$; a to d: $5, 4$; d to c: $2, 2$; d to e: $2, 2$; e to z: $4, 2$.

16. The maximum flow is 9.

19. Suppose that the sum of the capacities of the edges incident on a is U. Each iteration of Algorithm 8.2.5 increases the flow by 1. Since the flow cannot exceed U, eventually the algorithm must terminate.

Section 8.3

1. 8; minimal

4. $P = \{a, b, d\}$

7. $P = \{a, d\}$

10. $P = \{a, w_1, w_2, w_3, b, d, e\}$

13. $P = \{a, w_1, w_2, w_3, b, c, d, d', e, f, A, B, C\}$

16. $P = \{a, b, c, f, g, h, j, k, l, m\}$

17.

Figure for problem 17: $a \xrightarrow{1, 1} b \xrightarrow{2, 1} z$

with $C_{ab} = 1$, $C_{bz} = 2$, $m_{ab} = 1$, $m_{bz} = 2$.

20. Alter Algorithm 8.2.4.

23. False. Consider the flow

$a \xrightarrow{1, 1} b \xrightarrow{2, 1} z$

and the cut $P = \{a, b\}$.

Section 8.4

1. $P = \{a, A, B, D, J_2, J_5\}$

4. All unlabeled edges are $1, 0$. There is no complete matching.

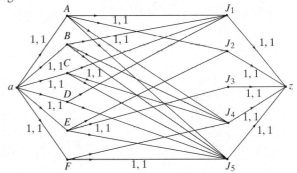

Figure for problem 4: bipartite network from a through vertices A, B, C, D, E, F and J_1, J_2, J_3, J_4, J_5 to z, with edges labeled $1, 1$.

8. Each row and column has at most one label.

12. If $\delta(G) = 0$, then $|S| - |R(S)| \leq 0$, for all $S \subseteq V$. By Theorem 8.4.7, G has a complete matching.

If G has a complete matching, then $|S| - |R(S)| \leq 0$, for all $S \subseteq V$, so $\delta(G) \leq 0$. If $S = \emptyset$, $|S| - |R(S)| = 0$, so $\delta(G) = 0$.

Section 8.5

1.

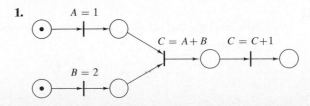

Figure for problem 1: a Petri net with places and transitions, labeled $A = 1$, $B = 2$, $C = A + B$, $C = C + 1$.

4. Computer operating systems, communication protocols, office information systems, industrial process control, database systems (especially, distributed database systems)

7. Let M_1 denote the marking that results from M by firing t_1. The only transition enabled in M_1 is t_2. Let M_2 denote the marking that results from M_1 by firing t_2. The only transition enabled in M_2 is t_3. If t_3 is fired, we obtain the marking M. It follows that M is live and bounded.

8. (a) t_1

(b)

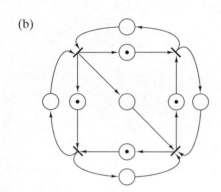

(c) Yes (d) Yes (e) Yes

(f)

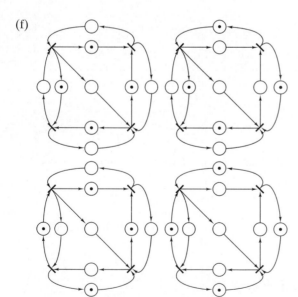

(g) Any marking that places two tokens in at least one place

11. (a) t_1

(b)

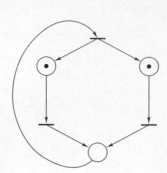

(c) Yes (d) No (e) No

(f) Designate the input place to t_i as p_i. The markings reachable from M are of two types:

    1. p_3 has at least one token.

    2. p_3 has no tokens and p_1 and p_2 each have at least one token.

(g)

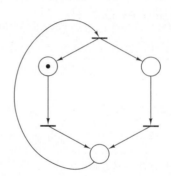

13. Figure 8.5.5

17. 8 and 9

20. If a vertex in a simple directed cycle C is fired, one token is removed from one edge in C, and one token is added to one edge in C; therefore, the token count of C does not change.

23.

Chapter 8 Self-Test

1. In each edge, the flow is less than or equal to the capacity and, except for the source and sink, the flow into each vertex v is equal to the flow out of v.

2. 3   **3.** 3   **4.** 3   **5.** (a, b, e, f, g, z)

6. Change the flows to $F_{a,b} = 2$, $F_{e,b} = 1$, $F_{e,f} = 1$, $F_{f,g} = 1$, $F_{g,z} = 1$.

7. $F_{a,b} = 3$, $F_{b,c} = 3$, $F_{c,d} = 4$, $F_{d,z} = 4$, $F_{a,e} = 2$, $F_{e,f} = 2$, $F_{f,c} = 2$, $F_{f,g} = 1$, $F_{g,z} = 1$, and all other edge flows zero.

8. $F_{a,b} = 0$, $F_{b,c} = 5$, $F_{c,d} = 5$, $F_{d,z} = 8$, $F_{e,b} = 3$, $F_{g,d} = 3$, $F_{a,e} = 8$, $F_{e,f} = 3$, $F_{f,g} = 3$, $F_{a,h} = 4$,

$F_{e,i} = 2$, $F_{j,z} = 6$, $F_{h,i} = 4$, $F_{i,j} = 6$, and all other edge flows zero.

9. a—True, b—False, c—False, d—True

10. 6

11. No. The capacity of $(P, \overline{P})$ is 6, but the capacity of $(P', \overline{P'})$, $P' = \{a, b, c, e, f\}$, is 5.

12. $P = \{a, b, c, e, f, g, h, i\}$

13.

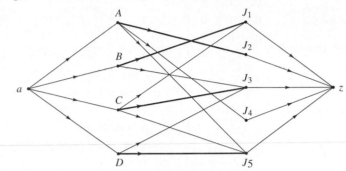

14. See the solution to Exercise 13.

15. $A - J_2$, $B - J_1$, $C - J_3$, $D - J_5$ is a complete matching.

16. $P = \{a\}$　　　　**17.** t_1 and t_2

18.

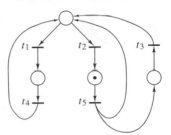

19. No

20. Yes

Section 9.1

1. $\overline{x_1 \wedge x_2}$

x_1	x_2	$\overline{x_1 \wedge x_2}$
1	1	0
1	0	1
0	1	1
0	0	1

4.

x_1	x_2	x_3	$((x_1 \wedge x_2) \vee \overline{(x_1 \wedge x_3)}) \wedge \overline{x_3}$
1	1	1	0
1	1	0	1
1	0	1	0
1	0	0	1
0	1	1	0
0	1	0	1
0	0	1	0
0	0	0	1

7. If $x = 1$, the output y is undetermined: Suppose that $x = 1$ and $y = 0$. Then the input to the AND gate is $1, 0$. Thus the output of the AND gate is 0. Since this is then NOTed, $y = 1$. Contradiction. Similarly, if $x = 1$ and $y = 1$, we obtain a contradiction.

10. 0

13. 1

16. Is a Boolean expression. x_1, x_2, and x_3 are Boolean expressions by (9.1.2). $x_2 \vee x_3$ is a Boolean expression by (9.1.3c). $\overline{(x_2 \vee x_3)}$ is a Boolean expression by (9.1.3a). $x_1 \wedge \overline{(x_2 \vee x_3)}$ is a Boolean expression by (9.1.3d).

19. Not a Boolean expression

22.

25. $(A \wedge B) \vee (C \wedge \overline{A})$

A	B	C	$(A \wedge B) \vee (C \wedge \overline{A})$
1	1	1	1
1	1	0	1
1	0	1	0
1	0	0	0
0	1	1	1
0	1	0	0
0	0	1	1
0	0	0	0

28. $(A \wedge (C \vee (D \wedge C))) \vee (B \wedge (\overline{D} \vee (C \wedge A) \vee \overline{C}))$

30.

A	B	$(A \vee \overline{B}) \wedge A$
1	1	1
1	0	1
0	1	0
0	0	0

33.

Section 9.2

1.

x_1	x_2	$\overline{x_1 \wedge x_2}$	$\overline{x}_1 \vee \overline{x}_2$
1	1	0	0
1	0	1	1
0	1	1	1
0	0	1	1

4.

x_1	x_2	x_3	$\overline{x}_1 \vee (\overline{x}_2 \vee x_3)$	$\overline{(x_1 \wedge x_2)} \vee x_3$
1	1	1	1	1
1	1	0	0	0
1	0	1	1	1
1	0	0	1	1
0	1	1	1	1
0	1	0	1	1
0	0	1	1	1
0	0	0	1	1

6.

x_1	$x_1 \vee x_1$
1	1
0	0

9.

x_1	x_2	x_3	$x_1 \wedge \overline{(x_2 \wedge x_3)}$	$(x_1 \wedge \overline{x}_2) \vee (x_1 \wedge \overline{x}_3)$
1	1	1	0	0
1	1	0	1	1
1	0	1	1	1
1	0	0	1	1
0	1	1	0	0
0	1	0	0	0
0	0	1	0	0
0	0	0	0	0

11.

x	$\overline{\overline{x}}$
1	1
0	0

14. False. Take $x_1 = 1$, $x_2 = 1$, $x_3 = 0$.

16.

a	b	c	$a \vee (b \wedge c)$	$(a \vee b) \wedge (a \vee c)$
1	1	1	1	1
1	1	0	1	1
1	0	1	1	1
1	0	0	1	1
0	1	1	1	1
0	1	0	0	0
0	0	1	0	0
0	0	0	0	0

18. The Boolean expressions that represent the circuits are $(A \wedge \overline{B}) \vee (A \wedge C)$ and $A \wedge (\overline{B} \vee C)$. The expressions are equal by Theorem 9.2.1c. Therefore, the switching circuits are equivalent.

21.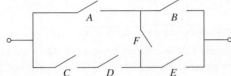

Section 9.3

2. One can show that the Associative and Distributive Laws hold for lcm and gcd directly. The Commutative Law clearly holds. To see that the Identity Laws hold, note that

$$\text{lcm}(x, 1) = x \quad \text{and} \quad \gcd(x, 6) = x.$$

Since

$$\text{lcm}(x, 6/x) = 6 \quad \text{and} \quad \gcd(x, 6/x) = 1,$$

the Complement Laws hold. Therefore, $(S, +, \cdot, ', 1, 6)$ is a Boolean algebra.

4. We show only

$$x \cdot (x + z) = (x \cdot y) + (x \cdot z) \quad \text{for all } x, y, z \in S_n.$$

Now

$$x \cdot (y + z) = \min\{x, \max\{y, z\}\},$$
$$(x \cdot y) + (x \cdot z) = \max\{\min\{x, y\}, \min\{x, z\}\}.$$

We assume that $y \le z$. (The argument is similar if $y > z$.) There are three cases to consider: $x < y$, $y \le x \le z$, and $z < x$.

If $x < y$, we obtain

$$x \cdot (y + z) = \min\{x, \max\{y, z\}\}$$
$$= \min\{x, z\} = x = \max\{x, x\}$$
$$= \max\{\min\{x, y\}, \min\{x, z\}\}$$
$$= (x \cdot y) + (x \cdot z).$$

If $y \le x \le z$, we obtain

$$x \cdot (y + z) = \min\{x, \max\{y, z\}\}$$
$$= \min\{x, z\} = x = \max\{y, x\}$$
$$= \max\{\min\{x, y\}, \min\{x, z\}\}$$
$$= (x \cdot y) + (x \cdot z).$$

If $z < x$, we obtain

$$x \cdot (y + z) = \min\{x, \max\{y, z\}\}$$
$$= \min\{x, z\} = z = \max\{y, z\}$$
$$= \max\{\min\{x, y\}, \min\{x, z\}\}$$
$$= (x \cdot y) + (x \cdot z).$$

7. If $X \cup Y = U$ and $X \cap Y = \emptyset$, then $Y = \overline{X}$.

8. $xy + x0 = x(x + y)y$

11. $x + y' = 1$ if and only if $x + y = x$.

14. $x(x + y0) = x$

15. [For Exercise 12]

$$0 = x + y = (x + x) + y$$
$$= x + (x + y) = x + 0 = x$$

Similarly, $y = 0$.

18. [For part (c)]

$$x(x + y) = (x + 0)(x + y)$$
$$= x + 0y = x + y0 = x + 0 = x$$

21. First, show that if $ba = ca$ and $ba' = ca'$, then $b = c$. Now take $a = x$, $b = x + (y + z)$, and $c = (x + y) + z$ and use this result.

23. If the prime p divides n, p^2 does not divide n.

Section 9.4

In these hints, $a \wedge b$ is written ab.

1. $xy \vee \overline{x}y \vee \overline{x}\,\overline{y}$

4. $xyz \vee xy\overline{z} \vee x\overline{y}\,\overline{z} \vee \overline{x}yz \vee \overline{x}y\overline{z}$

7. $xyz \vee x\overline{y}\,\overline{z} \vee \overline{x}\,\overline{y}\,\overline{z}$

10. $wx\overline{y}z \vee wx\overline{y}\,\overline{z} \vee w\overline{x}yz \vee w\overline{x}y\overline{z} \vee w\overline{x}\,\overline{y}\,\overline{z}$
$\vee \overline{w}xyz \vee \overline{w}x\overline{y}z \vee \overline{w}x\overline{y}\,\overline{z} \vee \overline{w}\,\overline{x}yz \vee \overline{w}\,\overline{x}\,\overline{y}\,\overline{z}$

11. $xy \vee x\overline{y}$ **14.** $xy\overline{z}$

17. $xyz \vee \overline{x}yz \vee xy\overline{z} \vee \overline{x}y\overline{z}$

20. 0

22. 2^{2^n}

25. [For Exercise 3]

$$(\overline{x} \vee y \vee \overline{z})(x \vee \overline{y} \vee \overline{z})(x \vee \overline{y} \vee z)$$

28. [For Exercise 3]

$$(\overline{x} \vee \overline{y} \vee \overline{z})(\overline{x} \vee \overline{y} \vee z)(\overline{x} \vee y \vee z)(x \vee y \vee \overline{z})(x \vee y \vee z)$$

Section 9.5

1. AND can be expressed in terms of OR and NOT: $xy = \overline{\overline{x} \vee \overline{y}}$.

2. A combinatorial circuit consisting only of AND gates would always output 0 when all inputs are 0.

5. We use induction on n to show that there is no n-gate combinatorial circuit consisting of only AND and OR gates that computes $f(x) = \overline{x}$.

   If $n = 0$, the input x equals the output x, and so it is impossible for a 0-gate circuit to compute f. The Basis Step is proved.

   Suppose that there is no n-gate combinatorial circuit consisting of only AND and OR gates that computes f. Consider an $(n + 1)$-gate combinatorial circuit consisting of only AND and OR gates. The input x first arrives at either an AND or an OR gate. Suppose that x first arrives at an AND gate. (The argument is similar if x first arrives at an OR gate and is omitted.) Because the circuit is a combinatorial circuit, the other input to the AND gate is either x itself, the constant 1, or the constant 0. If both inputs to the AND gate are x itself, then the output of the AND gate is equal to the input. In this case, the behavior of the circuit is unchanged if we remove the AND gate and connect x to what was the output line of the AND gate. But we now have an equivalent n-gate circuit, which, by the inductive hypothesis, cannot compute f. Thus the $(n + 1)$-gate circuit cannot compute f.

   If the other input to the AND gate is the constant 1, the output of the AND gate is again equal to the input and we can argue as in the previous case that the $(n + 1)$-gate circuit cannot compute f.

   If the other input to the AND gate is the constant 0, the AND gate always outputs 0 and, so, changing the value of x does not affect the output of the circuit. In this case, the circuit cannot compute f. The Inductive Step is complete. Therefore no n-gate combinatorial circuit consisting of only AND and OR gates can compute $f(x) = \overline{x}$. Thus {AND, OR} is not functionally complete.

6.

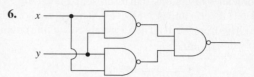

9. $y_1 = x_1 x_2 \vee \overline{(x_2 \vee x_3)}$; $y_2 = \overline{x_2 \vee x_3}$

12. [For Exercise 3] The dnf may be simplified to $xy \vee x\overline{z} \vee \overline{x}\,\overline{y}$ and then rewritten as $x(y \vee \overline{z}) \vee \overline{x}\,\overline{y} = (x\overline{\overline{y}z}) \vee \overline{x}\,\overline{y} = \overline{\overline{x\overline{\overline{y}z}}\,\overline{\overline{x}\,\overline{y}}}$, which gives the circuit

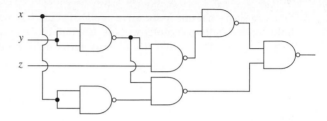

15.

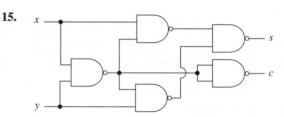

17. $xy = (x \downarrow x) \downarrow (y \downarrow y)$;
$x \vee y = (x \downarrow y) \downarrow (x \downarrow y)$; $\overline{x} = x \downarrow x$;
$x \uparrow y = [(x \downarrow x) \downarrow (y \downarrow y)] \downarrow [(x \downarrow x) \downarrow (y \downarrow y)]$

20. Since

$$\overline{x} = x \downarrow x, \quad x \vee y = (x \downarrow y) \downarrow (x \downarrow y),$$

and {NOT, OR} is functionally complete, {NOR} is functionally complete.

23.

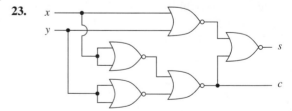

25. The logic table is

x	y	z	Output
1	1	1	1
1	1	0	1
1	0	1	1
1	0	0	0
0	1	1	1
0	1	0	0
0	0	1	0
0	0	0	0

27. The logic table is

b	FLAGIN	c	FLAGOUT
1	1	0	1
1	0	1	1
0	1	1	1
0	0	0	0

Thus $c = b \oplus$ FLAGIN and FLAGOUT $= b \vee$ FLAGIN. We obtain the circuit

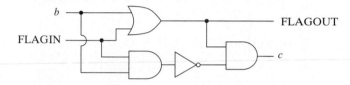

28. 010100

31.

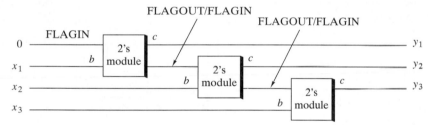

34. Writing the truth tables shows that

$$\overline{x} = x \to 0, \qquad x \vee y = (x \to 0) \to y.$$

Therefore a NOT gate can be replaced by one $\to$ gate, and an OR gate can be replaced by two $\to$ gates. Since the set {NOT, OR} is functionally complete, it follows that the set {$\to$} is functionally complete.

Chapter 9 Self-Test

1.

x	y	z	$\overline{(x \wedge \overline{y})} \vee z$
1	1	1	1
1	1	0	1
1	0	1	1
1	0	0	0
0	1	1	1
0	1	0	1
0	0	1	1
0	0	0	1

2. 1

3.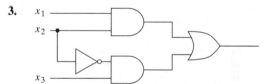

4. Suppose that x is 1. Then the upper input to the OR gate is 0. If y is 1, then the lower input to the OR gate is 0. Since both inputs to the OR gate are 0, the output y of the OR gate is 0, which is impossible. If y is 0, then the lower input to the OR gate is 1. Since an input to the OR gate is 1, the output y of the OR gate is 1, which is impossible. Therefore, if the input to the circuit is 1, the output is not uniquely determined. Thus the circuit is not a combinatorial circuit.

5. The circuits are equivalent. The logic table for either circuit is

x	y	Output
1	1	0
1	0	1
0	1	0
0	0	0

6. The circuits are not equivalent. If $x = 0$, $y = 1$, and $z = 0$, the output of circuit (a) is 1, but the output of circuit (b) is 0.

7. The equation is true. The logic table for either expression is

x	y	z	Value
1	1	1	1
1	1	0	1
1	0	1	0
1	0	0	0
0	1	1	1
0	1	0	1
0	0	1	1
0	0	0	0

8. The equation is false. If $x = 1$, $y = 0$, and $z = 1$, then

$$(x \wedge y \wedge z) \vee \overline{(x \vee z)} = 0,$$

but

$$(x \wedge z) \vee (\overline{x} \wedge \overline{z}) = 1.$$

9. Bound laws:

$$X \cup U = U \qquad X \cap \emptyset = \emptyset \qquad \text{for all } X \in S.$$

14. $x_1 x_2 \overline{x}_3 \vee x_1 \overline{x}_2 \overline{x}_3$

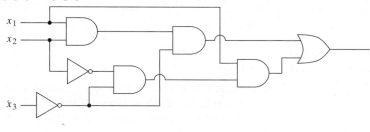

15. $x_1 x_2 x_3 \vee x_1 \overline{x}_2 \overline{x}_3 \vee \overline{x}_1 \overline{x}_2 \overline{x}_3$

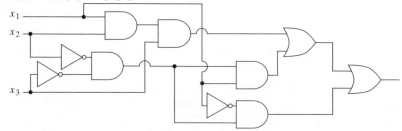

Absorption laws:

$$X \cup (X \cap Y) = X \qquad X \cap (X \cup Y) = X \text{ for all } X, Y \in S.$$

10.
$$\begin{aligned}
(x(x + y \cdot 0))' &= (x(x + 0))' && \text{(Bound law)} \\
&= (x \cdot x)' && \text{(Identity law)} \\
&= x' && \text{(Idempotent law)}
\end{aligned}$$

11. Dual: $(x + x(y + 1))' = x'$

$$\begin{aligned}
(x + x(y + 1))' &= (x + x \cdot 1)' && \text{(Bound law)} \\
&= (x + x)' && \text{(Identity law)} \\
&= x' && \text{(Idempotent law)}
\end{aligned}$$

12. $^-$ is not a unary operator on S. For example, $\overline{\{1, 2\}} \notin S$.

In Exercises 13–16, $a \wedge b$ is written ab.

13. $x_1 \overline{x}_2 \overline{x}_3$

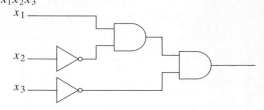

16. $x_1x_2\overline{x}_3 \vee x_1\overline{x}_2\overline{x}_3 \vee \overline{x}_1x_2x_3 \vee \overline{x}_1\overline{x}_2x_3$

17.

x	y	z	Output
1	1	1	1
1	1	0	0
1	0	1	1
1	0	0	0
0	1	1	0
0	1	0	0
0	0	1	1
0	0	0	0

18. Disjunctive normal form: $x\overline{y}z \vee x\overline{y}\,\overline{z} \vee \overline{x}y\overline{z} \vee \overline{x}\,\overline{y}\,\overline{z}$

$$(x\overline{y}z \vee x\overline{y}\,\overline{z}) \vee \overline{x}y\overline{z} \vee \overline{x}\,\overline{y}\,\overline{z} = x\overline{y} \vee (\overline{x}y\overline{z} \vee \overline{x}\,\overline{y}\,\overline{z})$$

$$= x\overline{y} \vee \overline{x}\,\overline{z}$$

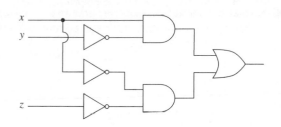

19.

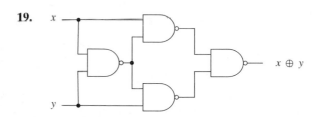

20.

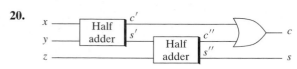

Section 10.1

1.

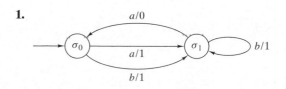

4.

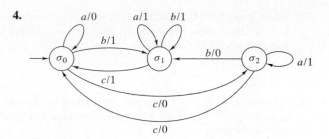

6. $\mathcal{I} = \{a, b\}$; $\mathcal{O} = \{0, 1\}$; $\mathcal{S} = \{\sigma_0, \sigma_1\}$; initial state $= \sigma_0$

$\mathcal{I}$ $\diagdown$ $\mathcal{S}$	a	b	a	b
σ_0	σ_1	σ_0	0	1
σ_1	σ_1	σ_1	1	1

9. $\mathcal{I} = \{a, b\}$; $\mathcal{O} = \{0, 1\}$; $\mathcal{S} = \{\sigma_0, \sigma_1, \sigma_2, \sigma_3\}$; initial state $= \sigma_0$

$\mathcal{I}$ $\diagdown$ $\mathcal{S}$	a	b	a	b
σ_0	σ_1	σ_2	0	0
σ_1	σ_0	σ_2	1	0
σ_2	σ_3	σ_0	0	1
σ_3	σ_1	σ_3	0	0

11. 1110

14. 001110

17. 001110001

20. 020022201020

21.

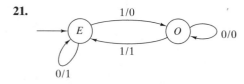

24.

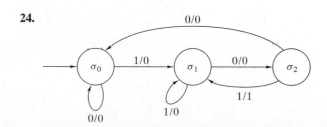

27. When γ is input, the machine outputs $x_n, x_{n-1}, \ldots$ until $x_i = 1$. Thereafter, it outputs $\overline{x}_i$. However, according to Algorithm 9.5.16, this is the 2's complement of α.

Section 10.2

1. All incoming edges to σ_0 output 1 and all incoming edges to σ_1 output 0; hence the finite-state machine is a finite-state automaton.

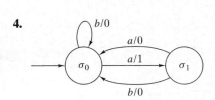

4.

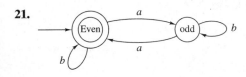

21.

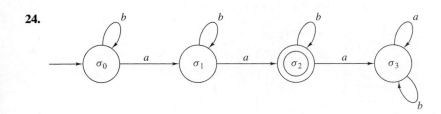

24.

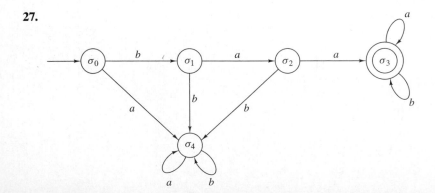

27.

7.

10. [For Exercise 1] $\mathcal{I} = \{a, b\}$; $\mathcal{S} = \{\sigma_0, \sigma_1\}$; $\mathcal{A} = \{\sigma_0\}$; initial state $= \sigma_0$

$\mathcal{I}$ $\diagdown$ $\mathcal{S}$	a	b
σ_0	σ_0	σ_1
σ_1	σ_1	σ_0

13. Accepted

16. Accepted

18. No matter which state we are in, after an a we move to an accepting state; however, after a b we move to a nonaccepting state.

30.

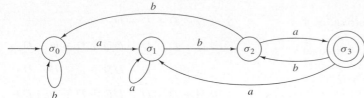

32. [For Exercise 1] This algorithm determines whether a string over $\{a, b\}$ is accepted by the finite-state automaton whose transition diagram is given in Exercise 1.

Input: n, the length of the string ($n = 0$ designates the null string)
$s_1 \cdots s_n$, the string

Output: "Accept" if the string is accepted
"Reject" if the string is not accepted

```
procedure ex32(s, n)
  state := 'σ₀'
  for i := 1 to n do
    begin
    if state = 'σ₀' and sᵢ = 'b' then
      state := 'σ₁'
    if state = 'σ₁' and sᵢ = 'b' then
      state := 'σ₀'
    end
  if state = 'σ₀' then
    return("Accept")
  else
    return("Reject")
end ex32
```

35. Make each accepting state nonaccepting and each nonaccepting state accepting.

38. Using the construction given in Exercises 36 and 37, we obtain the following finite-state automaton that accepts $L_1 \cap L_2$. (We designate the states in Exercise 5 with primes.)

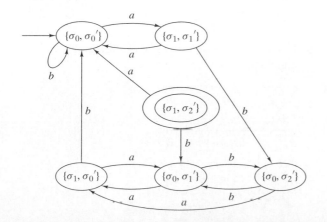

The finite-state automaton that accepts $L_1 \cup L_2$ is the same as the finite-state automaton that accents $L_1 \cap L_2$ except that the set of accepting states is

$$\{(\sigma_1, \sigma_0'), \quad (\sigma_1, \sigma_1'), \quad (\sigma_1, \sigma_2'), \quad (\sigma_0, \sigma_2')\}.$$

41. Use the construction of Exercises 36 and 37.

Section 10.3

1. Regular, context-free, context-sensitive

4. Context-free, context-sensitive

7. $\sigma \Rightarrow b\sigma \Rightarrow bb\sigma \Rightarrow bbaA \Rightarrow bbabA \Rightarrow bbabbA \Rightarrow bbabba\sigma \Rightarrow bbabbab$

10. $\sigma \Rightarrow ABA \Rightarrow ABBA \Rightarrow ABBAA$
$\Rightarrow ABBaAA \Rightarrow abBBaAA \Rightarrow abbBaAA$
$\Rightarrow abbbaAA \Rightarrow abbbaabA \Rightarrow abbbaabab$

12. [For Exercise 1]

$$<\sigma> ::= b<\sigma> \mid a<A> \mid b$$

$$<A> ::= a<\sigma> \mid b<A> \mid a$$

15. $S \to aA, A \to aA, A \to bA, A \to a,$
$A \to b, S \to a$

18. $S \to aA, S \to bS, S \to \lambda, A \to aA,$
$A \to bB, A \to \lambda, B \to aA, B \to bS$

21. $<\text{exp number}> ::= <\text{integer}> E <\text{integer}> \mid$

$<\text{float number}> \mid$

$<\text{float number}> E <\text{integer}>$

24. $S \to aSa, S \to bSb, S \to a, S \to b, S \to \lambda$

25. If a derivation begins $S \Rightarrow aSb$, the resulting string begins with a and ends with b. Similarly, if a derivation begins $S \Rightarrow bSa$, the resulting string begins with b and ends with a. Therefore, the grammar does not generate the string $abba$.

28. If a derivation begins $S \Rightarrow abS$, the resulting string begins ab. If a derivation begins $S \Rightarrow baS$, the resulting string begins ba. If a derivation begins $S \Rightarrow aSb$, the resulting string starts with a and ends with b. If a derivation begins $S \Rightarrow bSa$, the resulting string begins with b and ends with a. Therefore, the grammar does not generate the string $aabbabba$.

31. The grammar does generate L, the set of all strings over $\{a, b\}$ with equal numbers of a's and b's.

Any string generated by the grammar has equal numbers of a's and b's since whenever any of the productions are used in a derivation, equal numbers of a's and b's are added to the string.

To prove the converse, we consider an arbitrary string α in L, and we use induction on the length $|\alpha|$ of α to show that α is generated by the grammar. The Basis Step is $|\alpha| = 0$. In this case, α is the null string, and $S \Rightarrow \lambda$ is a derivation of α.

Let α be a nonnull string, and suppose that any string in L whose length is less than $|\alpha|$ is generated by the grammar. We first consider the case that α starts with a. Then α can be written $\alpha = a\alpha_1 b\alpha_2$, where α_1 and α_2 have equal numbers of a's and b's. By the inductive hypothesis, there are derivations $S \Rightarrow \alpha_1$ and $S \Rightarrow \alpha_2$ of α_1 and α_2. But now

$$S \Rightarrow aSbS \Rightarrow a\alpha_1 b\alpha_2$$

is a derivation of α. Similarly, if α starts with b, there is a derivation of α. The Inductive Step is finished and the proof is complete.

32. Replace each production

$$A \rightarrow x_1 \cdots x_n B,$$

where $n > 1$, $x_i \in T$, and $B \in N$, with the productions

$$A \rightarrow x_1 A_1$$
$$A_1 \rightarrow x_2 A_2$$
$$\vdots$$
$$A_{n-1} \rightarrow x_n B,$$

where $A_1, \ldots, A_{n-1}$ are additional nonterminal symbols.

35. $S \Rightarrow D + D + D + D \Rightarrow d + d + d + d$

$$S \Rightarrow D + D + D + D$$
$$\Rightarrow D + D - D - DD + D + D - D+$$
$$D + D - D - DD + D + D - D+$$
$$D + D - D - DD + D + D - D+$$
$$D + D - D - DD + D + D - D$$
$$\Rightarrow d + d - d - dd + d + d - d+$$
$$d + d - d - dd + d + d - d+$$
$$d + d - d - dd + d + d - d+$$
$$d + d - d - dd + d + d - d$$

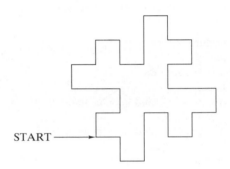

START ———→

Section 10.4

1.

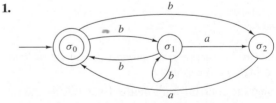

4.

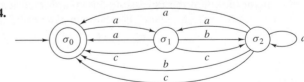

6. $\mathcal{I} = \{a, b\}$; $\mathcal{S} = \{\sigma_0, \sigma_1, \sigma_2\}$; $\mathcal{A} = \{\sigma_1, \sigma_2\}$; initial state $= \sigma_0$

$\mathcal{S}$ ＼ $\mathcal{I}$	a	b
σ_0	$\{\sigma_1, \sigma_2\}$	$\emptyset$
σ_1	$\{\sigma_1\}$	$\{\sigma_0, \sigma_2\}$
σ_2	$\emptyset$	$\emptyset$

9. $\mathcal{I} = \{a, b\}$; $\mathcal{S} = \{\sigma_0, \sigma_1, \sigma_2, \sigma_3\}$; $\mathcal{A} = \{\sigma_3\}$; initial state $= \sigma_0$

$\mathcal{I}$ $\mathcal{S}$	a	b
σ_0	$\{\sigma_0\}$	$\{\sigma_0, \sigma_1\}$
σ_1	$\{\sigma_2\}$	$\emptyset$
σ_2	$\emptyset$	$\{\sigma_3\}$
σ_3	$\{\sigma_3\}$	$\{\sigma_3\}$

11. [For Exercise 5] $N = \{\sigma_0, \sigma_1, \sigma_2\}$, $T = \{a, b\}$,

$$\sigma_0 \to a\sigma_1, \quad \sigma_0 \to b\sigma_0, \quad \sigma_1 \to a\sigma_0, \quad \sigma_1 \to b\sigma_2,$$

$$\sigma_2 \to b\sigma_1, \quad \sigma_2 \to a\sigma_0, \quad \sigma_2 \to \lambda$$

14. No. For the first three characters, *bba*, the moves are determined and we end at C. From C, no edge contains an a; therefore, *bbabab* is not accepted.

17. Yes. The path $(\sigma, \sigma, \sigma, \sigma, C, C)$, which represents the string *aaaab*, ends at C, which is an accepting state.

21.

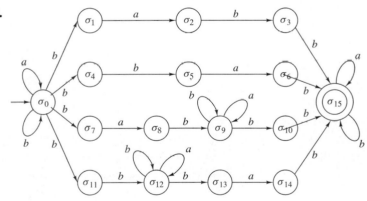

24.

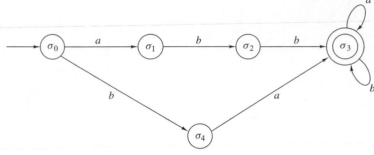

27.

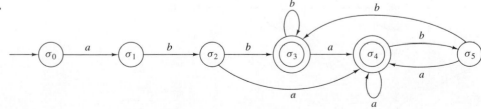

30. [For Exercise 21] $\sigma_0 \to a\sigma_1$, $\sigma_0 \to b\sigma_4$, $\sigma_1 \to b\sigma_2$, $\sigma_2 \to b\sigma_3$, $\sigma_3 \to a\sigma_3$, $\sigma_3 \to b\sigma_3$, $\sigma_4 \to a\sigma_3$, $\sigma_3 \to \lambda$

Section 10.5

1. [For Exercise 1]

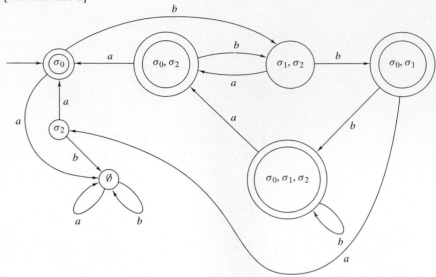

2.

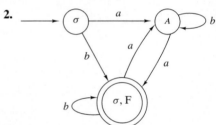

5.

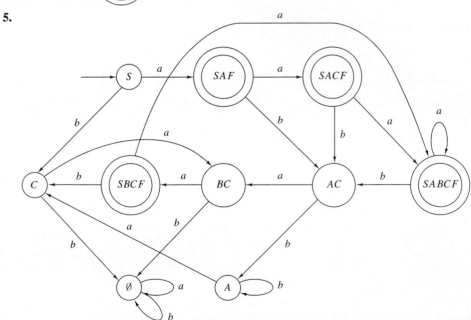

7. [For Exercise 21]

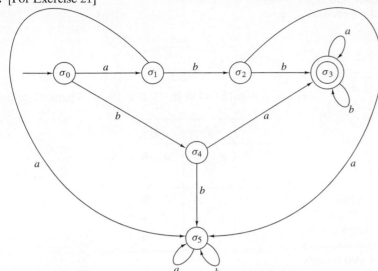

10. Figure 10.5.7 accepts the string ba^n, $n \geq 1$, and strings that end b^2 or aba^n, $n \geq 1$. Using Example 10.5.8, we see that Figure 10.5.9 accepts the string $a^n b$, $n \geq 1$, and strings that start b^2 or $a^n ba$, $n \geq 1$.

11.

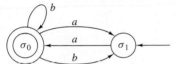

14.

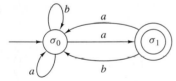

17.

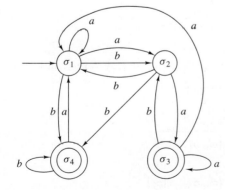

20.

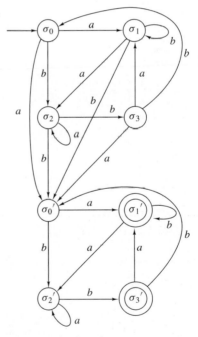

22. $\sigma_0 \rightarrow a\sigma_1$, $\sigma_0 \rightarrow b\sigma_2$, $\sigma_0 \rightarrow a$, $\sigma_1 \rightarrow a\sigma_0$, $\sigma_1 \rightarrow a\sigma_2$, $\sigma_1 \rightarrow b\sigma_1$, $\sigma_1 \rightarrow b$, $\sigma_2 \rightarrow b\sigma_0$

25. Suppose that L is regular. Then there exists a finite-state automaton A with $L = \text{Ac}(A)$. Suppose that A has k states. Consider the string $a^k bba^k$ and argue as in Example 10.5.6.

28. The statement is false. Consider the regular language $L = \{a^n b \mid n \geq 0\}$, which is accepted by the finite-state automaton

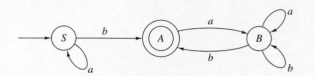

The language

$$L' = \{u^n \mid u \in L, n \in \{1, 2, \ldots\}\}$$

is not regular. Suppose that L' is regular. Then there is a finite-state automaton A that accepts L'. In particular, A accepts $a^n b$ for every n. It follows that for sufficiently large n, the path representing $a^n b$ contains a cycle of length k. Since A accepts $a^n b a^n b$, A also accepts $a^{n+k} b a^n b$, which is a contradiction.

Chapter 10 Self-Test

1.

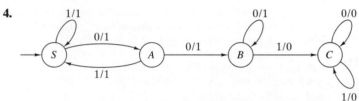

2. $\mathcal{I} = \{a, b\}; \mathcal{O} = \{0, 1\}; \mathcal{S} = \{S, A, B\}$; initial state $= S$

	f		g	
$\mathcal{I}$	a	b	a	b
$\mathcal{S}$				
S	A	A	0	0
A	S	B	1	1
B	A	B	1	0

3. 1101

4.

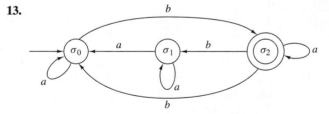

5.

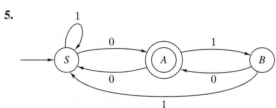

6. Yes

7.

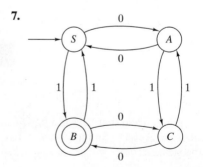

8. Every 0 is followed by a 1.   **9.** Context-free

10. $S \Rightarrow aSb \Rightarrow aaSbb \Rightarrow aaaSbbb \Rightarrow aaaAbbbb \Rightarrow aaaaAbbbb \Rightarrow aaaabbbb$

11. $a^i b^j, j \leq 2 + i, j \geq 1, i \geq 0$

12. $S \to ASB, S \to AB, AB \to BA, BA \to AB, A \to a, B \to b$

13.

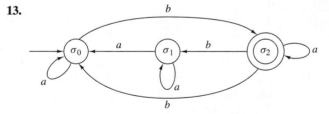

14. $\mathcal{I} = \{a, b\}; \mathcal{S} = \{\sigma_0, \sigma_1, \sigma_2\}; \mathcal{A} = \{\sigma_0\}$; initial state $= \sigma_0$

$\mathcal{I}$	a	b
$\mathcal{S}$		
σ_0	$\{\sigma_0, \sigma_1\}$	$\emptyset$
σ_1	$\emptyset$	$\{\sigma_2\}$
σ_2	$\{\sigma_0, \sigma_2\}$	$\{\sigma_2\}$

15. Yes, since the path

$$(\sigma_0, \sigma_0, \sigma_1, \sigma_2, \sigma_2, \sigma_2, \sigma_2, \sigma_0)$$

represents $aabaaba$ and σ_0 is an accepting state.

16.

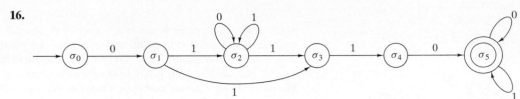

17.

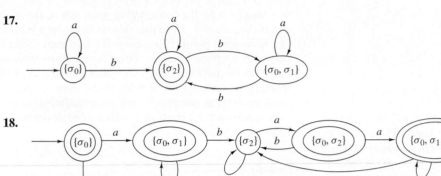

18.

19. Combine the nondeterministic finite-state automata that accept L_1 and L_2 in the following way. Let S be the start state of L_2. For each edge of the form (S_1, S_2) labeled a in L_1 where S_2 is an accepting state, add an edge (S_1, S) labeled a. The start state of the nondeterministic finite-state automaton is the start state of L_1. The accepting states of the nondeterministic finite-state automaton are the accepting states of L_2.

20. Let A' be a nondeterministic finite-state automaton that accepts a regular language that does not contain the null string. Add a state F. For each edge, (σ, σ') labeled a in A' where σ' is accepting, add the edge (σ, F) labeled a. Make F the only accepting state. The resulting nondeterministic finite-state automaton A has one accepting state. We claim that $\mathrm{Ac}(A) = \mathrm{Ac}(A')$.

We show that $\mathrm{Ac}(A) \subseteq \mathrm{Ac}(A')$. [The argument that $\mathrm{Ac}(A') \subseteq \mathrm{Ac}(A)$ is similar and omitted.] Suppose that $\alpha \in \mathrm{Ac}(A)$. There is a path

$$(\sigma_0, \sigma_1, \ldots, \sigma_{n-1}, \sigma_n)$$

that represents α in A, with σ_n an accepting state. Since $\alpha \neq \lambda$, there is a last symbol a in α. Thus the edge (σ_{n-1}, σ_n) is labeled a. Now the path

$$(\sigma_0, \sigma_1, \ldots, \sigma_{n-1}, F)$$

represents α in A' and terminates in an accepting state. Therefore, $\alpha \in \mathrm{Ac}(A')$.

To see that the statement is false for an arbitrary regular language, consider the regular language

$$L = \{\lambda\} \cup \{0^i \mid i \text{ is odd }\}$$

and a nondeterministic finite-state automaton A with start state S that accepts L. Since $\lambda \in L$, S is an accepting state. If S has a loop labeled 0, then A accepts all strings of 0's; therefore, there is no loop at S labeled 0. Thus there is an edge (S, S'), $S \neq S'$, labeled 0. Since $0 \in L$, S' is accepting. Therefore, A has at least two accepting states.

Section 11.1

1. The 16 points sorted by x-coordinate are: $(1,2)$, $(1,5)$, $(1,9)$, $(3,7)$, $(3,11)$, $(5,4)$, $(5,9)$, $(7,6)$, $(8,4)$, $(8,7)$, $(8,9)$, $(11,3)$, $(11,7)$, $(12,10)$, $(14,7)$, $(17,10)$, so the dividing point is $(7,6)$. We next find $\delta_L = \sqrt{8}$, the minimum distance among the left-side points $(1,2)$, $(1,5)$, $(1,9)$, $(3,7)$, $(3,11)$, $(5,4)$, $(5,9)$, $(7,6)$, and $\delta_R = 2$, the minimum distance among the right-side points $(8,4)$, $(8,7)$, $(8,9)$, $(11,3)$, $(11,7)$, $(12,10)$, $(14,7)$, $(17,10)$. Thus $\delta = \min\{\delta_L, \delta_R\} = 2$. The points, sorted by y-coordinate in the vertical strip, are $(8,4)$, $(7,6)$, $(8,7)$, $(8,9)$. In this case we compare each point in the strip to the all the following points. The distances from $(8,4)$ to $(7,6)$, $(8,7)$, $(8,9)$ are not less than 2, so δ is not updated at this point. The distance from $(7,6)$ to $(8,7)$ is $\sqrt{2}$, so δ is updated to $\sqrt{2}$. The distance from $(7,6)$ to $(8,9)$ and from $(8,7)$ to $(8,9)$ is greater than $\sqrt{2}$, so δ remains $\sqrt{2}$. Therefore the distance between the closest pair is $\sqrt{2}$.

4. Consider the extreme case when all of the points are on the vertical line.

7.

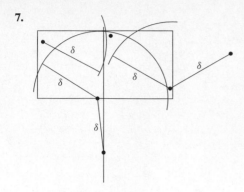

10. Let B be either of the left or right $\delta \times \delta$ squares that make up the $\delta \times 2\delta$ rectangle (see Figure 11.1.2). We argue by contradiction and assume that B contains four or more points. We partition B into four $\delta/2 \times \delta/2$ squares as shown in Figure 11.1.3. Then each of the four squares contains at most one point, and therefore exactly one point. Subsequently we refer to these four squares as the subsquares of B.

   The figure

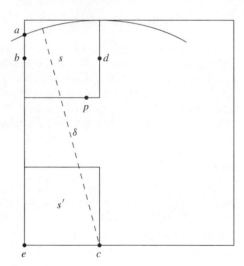

shows the following construction. We reduce the size of the subsquares, if possible, so that

   • Each subsquare contains one point.
   • The subsquares are the same size.
   • The subsquares are as small as possible.

Since at least one point is not in a corner of B, the subsquares do not collapse to points and so at least one point is on a side of a subsquare s interior to B. We choose such a point and call it p. We select a subsquare s' nearest p. We label the two corner points of s' on the side farthest from p, e and c. We draw a circle of radius δ with center at c and let a be the (noncorner) point where this circle meets the side of s. Note that this circle meets a side of s in a noncorner point. Choose a point b in s on the same side as a between a and e. Let d be the corresponding point on the opposite side of s. Now the length of the diameter of rectangle $R = bdce$ is less than δ; hence, R contains at most one point. This is contradiction since R contains p and the point in s'. Therefore B contains at most three points.

Section 11.2

1.   Input: $x_1, \ldots, x_n$ and n

   Output: "Yes" if $x_1, \ldots, x_n$ are distinct, and "No" otherwise

```
procedure check_distinct(x, n)
   sort x₁, ..., xₙ
   for i := 1 to n − 1 do
      if xᵢ = xᵢ₊₁ then
         return("No")
   return("Yes")
end check_distinct
```

4. Initialize a list L to empty. Find the vertex v with no incoming edges and add it to the end of L. Delete v and all incident edges. Repeat; that is, find the vertex v with no incoming edges and add it to the end of L. Continue until no vertices remain. Output L.

Section 11.3

1. Let L be the horizontal line through p_1. By the choice of p_1, no points of S lie below L. If p_1 is the only point of S on L, p_1 is a hull point. If other points of S lie on L, they all lie to the right of p_1 (by the choice of p_1). In this case, if we rotate L clockwise slightly about p_1, L will contain only p_1 and all other points of S will lie above L. Again we conclude that p_1 is a hull point.

4. The points [sorted with respect to $(7,1)$] are $(7,1)$, $(10,1)$, $(16,4)$, $(12,3)$, $(14,5)$, $(16,10)$, $(13,8)$, $(10,5)$, $(10,9)$, $(10,13)$, $(7,7)$, $(7,13)$, $(6,10)$, $(3,13)$, $(4,8)$, $(1,8)$, $(4,4)$, $(2,2)$. The following table shows each triple that is examined in the while loop, whether it makes a left turn, and the action taken with respect to the triple:

Triple	Left Turn?	Discard Middle Point?
$(7,1), (10,1), (16,4)$	Yes	No
$(10,1), (16,4), (12,3)$	Yes	No
$(16,4), (12,3), (14,5)$	No	Yes
$(10,1), (16,4), (14,5)$	Yes	No
$(16,4), (14,5), (16,10)$	No	Yes
$(10,1), (16,4), (16,10)$	Yes	No
$(16,4), (16,10), (13,8)$	Yes	No
$(16,10), (13,8), (10,5)$	Yes	No
$(13,8), (10,5), (10,9)$	No	Yes
$(16,10), (13,8), (10,9)$	No	Yes
$(16,4), (16,10), (10,9)$	Yes	No
$(16,10), (10,9), (10,13)$	No	Yes
$(16,4), (16,10), (10,13)$	Yes	No
$(16,10), (10,13), (7,7)$	Yes	No
$(10,13), (7,7), (7,13)$	No	Yes
$(16,10), (10,13), (7,13)$	Yes	No
$(10,13), (7,13), (6,10)$	Yes	No
$(7,13), (6,10), (3,13)$	No	Yes
$(10,13), (7,13), (3,13)$	No	Yes
$(16,10), (10,13), (3,13)$	Yes	No
$(10,13), (3,13), (4,8)$	Yes	No
$(3,13), (4,8), (1,8)$	No	Yes
$(10,13), (3,13), (1,8)$	Yes	No
$(3,13), (1,8), (4,4)$	Yes	No
$(1,8), (4,4), (2,2)$	No	Yes
$(3,13), (1,8), (2,2)$	Yes	No

The convex hull is $(7,1), (10,1), (16,4), (16,10), (10,13), (3,13), (1,8), (2,2)$.

7. After finding $p_1, \ldots, p_i$, Jarvis's march finds the point p_{i+1} such that p_{i-1}, p_i, p_{i+1} make the smallest left turn. It follows that if the line L through p_i, p_{i+1} is rotated clockwise slightly about p_i, L will contain only p_i, and all other points of S will lie on one side of L. Thus p_i is a hull point. By construction, Jarvis's march finds all hull points. Thus Jarvis's march does find the convex hull.

10. Yes. Jarvis's march is faster when "most" points are not on the convex hull.

Chapter 11 Self-Test

1. The 18 points sorted by x-coordinate are: $(1,8), (2,2),$

$(3,13), (4,4), (4,8), (6,10), (7,1), (7,7), (7,13), (10,1),$ $(10,5), (10,9), (10,13), (12,3), (13,8), (14,5), (16,4),$ $(16,10),$ so the dividing point is $(7,13)$. We next find $\delta_L = \sqrt{8}$, the minimum distance among the left-side points $(1,8), (2,2), (3,13), (4,4), (4,8), (6,10), (7,1), (7,7),$ $(7,13),$ and $\delta_R = \sqrt{5}$, the minimum distance among the right-side points $(10,1), (10,5), (10,9), (10,13), (12,3),$ $(13,8), (14,5), (16,4), (16,10)$. Thus $\delta = \min\{\delta_L, \delta_R\} = \sqrt{5}$. The points, sorted by y-coordinate in the vertical strip, are $(7,1), (7,7), (6,10), (7,13)$. In this case we compare each point in the strip to all the following points. Since no pair is closer than $\sqrt{5}$, the algorithm does not update δ. Therefore the distance between the closest pair is $\sqrt{5}$.

2. If we replace "three" by "two," when there are three points, the algorithm would be called recursively with inputs of sizes 1 and 2. But a set consisting of one point has no pair—let alone a closest pair.

3. Each $\delta/2 \times \delta/2$ box contains at most one point, so there are at most four points in the lower half of the rectangle.

4. $\Theta(n(\lg n)^2)$

5. Let t_n denote the worst-case time for an algorithm that finds a closest pair among n elements in d-dimensional space. Then t_n is also the worst-case asymptotic time for an algorithm CP that returns the distance between a closest pair of points in d-dimensional space, since we can add one line to the original algorithm that computes and returns the distance between a closest pair. Consider the following algorithm that solves the problem of determining whether there are any duplicates among n numbers:

```
procedure dup(x, n)
  // Input is x_1, ..., x_n.
  // Make the input into points in d − space.
  for i := 1 to n do
    a_i := (x_i, 0, ..., 0)
  if CP(a, n) = 0 then
    return("Duplicates")
  else
    return("No duplicates")
end dup
```

The worst-case time t'_n for dup is the time for the for loop plus the worst-case time for CP; that is,

$$t'_n = n + 6_n.$$

By Theorem 11.2.1,

$$Cn \lg n \leq t'_n.$$

Combining these last two statements, we obtain

$$\Omega(n \lg n) = Cn \lg n - n \leq t'_n - n = t_n.$$

6. We may add one line without changing the asymptotic time to an algorithm that finds all closest pairs so that it finds the distance between a closest pair. By Corollary 11.2.2, this requires time $\Omega(n \lg n)$.

7. The worst-case time of any algorithm that determines whether n real numbers are all equal is $\Omega(n)$ since any algorithm must examine each element at least once. This lower bound is sharp since the following algorithm solves this problem in time $\Theta(n)$:

> **procedure** *all_equal*(x, n)
> // determine whether $x_1, \ldots, x_n$ are equal
>   **for** $i := 1$ **to** $n - 1$ **do**
>   **if** $x_i \neq x_{i+1}$ **then**
>   **return**("Not all equal")
>   **return**("All equal")
> **end** *all_equal*

8. The statement follows from the fact that such an algorithm can be modified without changing its asymptotic worst-case time to determine whether the input contains duplicates and, by Theorem 11.2.1, any algorithm that determines whether duplicates exist has worst-case time $\Omega(n \lg n)$. Duplicates exist if and only if the distance between every output pair is zero; thus, we need only check one pair to determine whether there are duplicates or not.

9. Let L be the vertical line through p. By the choice of p, no points of S lie to the right of L. If p is the only point of S on L, p is a hull point. If other points of S lie on L, they all lie below p. In this case, if we rotate L clockwise slightly about p, L will contain only p and all other points of S will be to the left of L. Again we conclude that p is a hull point.

10. Let L be the line segment joining p and q. Let L' be the line through p perpendicular to L. There can be no other point r of S on L' or on the side of L' opposite q, for if there were such a point r, the distance from r to q would exceed the distance from p to q, which is impossible. Thus p is a hull point. Similarly, q is a hull point.

11. The points [sorted with respect to $(1, 2)$] are $(1, 2)$, $(11, 3)$, $(8, 4)$, $(14, 7)$, $(5, 4)$, $(11, 7)$, $(17, 10)$, $(7, 6)$, $(8, 7)$, $(12, 10)$, $(8, 9)$, $(5, 9)$, $(3, 7)$, $(3, 11)$, $(1, 5)$, $(1, 9)$. The following table shows each triple that is examined in the while loop, whether it makes a left turn, and the action taken with respect to the triple:

Triple	Left Turn?	Discard Middle Point?
$(1, 2), (11, 3), (8, 4)$	Yes	No
$(11, 3), (8, 4), (14, 7)$	No	Yes
$(1, 2), (11, 3), (14, 7)$	Yes	No
$(11, 3), (14, 7), (5, 4)$	Yes	No
$(14, 7), (5, 4), (11, 7)$	No	Yes
$(11, 3), (14, 7), (11, 7)$	Yes	No
$(14, 7), (11, 7), (17, 10)$	No	Yes
$(11, 3), (14, 7), (17, 10)$	No	Yes
$(1, 2), (11, 3), (17, 10)$	Yes	No
$(11, 3), (17, 10), (7, 6)$	Yes	No
$(17, 10), (7, 6), (8, 7)$	No	Yes
$(11, 3), (17, 10), (8, 7)$	Yes	No
$(17, 10), (8, 7), (12, 10)$	No	Yes
$(11, 3), (17, 10), (12, 10)$	Yes	No
$(17, 10), (12, 10), (8, 9)$	Yes	No
$(12, 10), (8, 9), (5, 9)$	No	Yes
$(17, 10), (12, 10), (5, 9)$	Yes	No
$(12, 10), (5, 9), (3, 7)$	Yes	No
$(5, 9), (3, 7), (3, 11)$	No	Yes
$(12, 10), (5, 9), (3, 11)$	No	Yes
$(17, 10), (12, 10), (3, 11)$	No	Yes
$(11, 3), (17, 10), (3, 11)$	Yes	No
$(17, 10), (3, 11), (1, 5)$	Yes	No
$(3, 11), (1, 5), (1, 9)$	No	Yes
$(17, 10), (3, 11), (1, 9)$	Yes	No

The convex hull is $(1, 2)$, $(11, 3)$, $(17, 10)$, $(3, 11)$, $(1, 9)$.

12. Run the part of Graham's Algorithm that follows the sort on the remaining points.

Section Appendix

1. $\begin{pmatrix} 2+a & 4+b & 1+c \\ 6+d & 9+e & 3+f \\ 1+g & -1+h & 6+i \end{pmatrix}$

2. $\begin{pmatrix} 5 & 7 & 7 \\ -7 & 10 & -1 \end{pmatrix}$

5. $\begin{pmatrix} 3 & 18 & 27 \\ 0 & 12 & -6 \end{pmatrix}$

8. $\begin{pmatrix} -2 & -35 & -56 \\ -7 & -18 & 13 \end{pmatrix}$

9. $\begin{pmatrix} 18 & 10 \\ 14 & -6 \\ 23 & 1 \end{pmatrix}$

12. (-4)

14. (a) $2 \times 3, 3 \times 3, 3 \times 2$

(b)

$$AB = \begin{pmatrix} 33 & 18 & 47 \\ 8 & 9 & 43 \end{pmatrix}$$

$$AC = \begin{pmatrix} 16 & 56 \\ 14 & 63 \end{pmatrix}$$

$$CA = \begin{pmatrix} 4 & 18 & 38 \\ 0 & 0 & 0 \\ 2 & 17 & 75 \end{pmatrix}$$

$$AB^2 = \begin{pmatrix} 177 & 215 & 531 \\ 80 & 93 & 323 \end{pmatrix}$$

$$BC = \begin{pmatrix} 18 & 65 \\ 34 & 25 \\ 12 & 54 \end{pmatrix}$$

17. Let $A = (b_{ij})$, $I_n = (a_{jk})$, $AI_n = (c_{ik})$. Then

$$c_{ik} = \sum_{j=1}^{n} b_{ij}a_{jk} = b_{ik}a_{kk} = b_{ik}.$$

Therefore, $AI_n = A$. Similarly, $I_nA = A$.

20. The solution is $X = A^{-1}C$.

INDEX